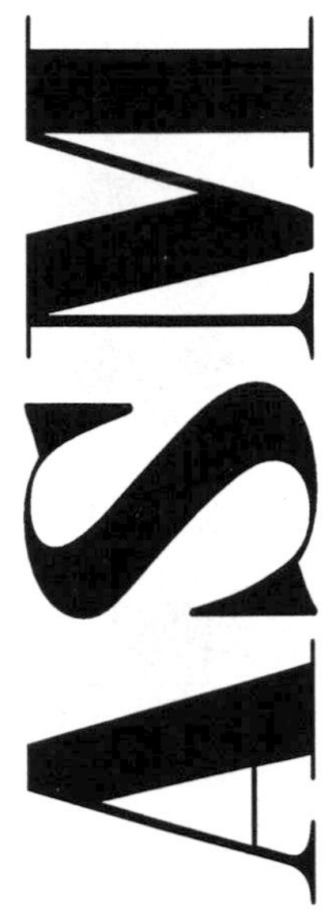

ENGINEERING MATERIALS Reference Book

Second Edition

Editor
Michael Bauccio

Acquisitions/Editorial
Veronica Flint

Technical Editor
Sunniva Collins

Production Project Manager
Suzanne E. Frueh

Production Coordinator
Randall L. Boring

First printing, May 1994

This book is a collective effort involving hundreds of technical specialists. It brings together a wealth of information from world-wide sources to help scientists, engineers, and technicians solve current and long-range problems.

Comments, criticisms, and suggestions are invited, and should be forwarded to ASM International.

Library of Congress Cataloging-in-Publication Data

ASM engineered materials reference book/edited by Michael L. Bauccio.—2nd ed. p. cm.
ISBN: 0-87170-502-8
SAN: 204-7586
1. Materials—Handbooks, manuals, etc.
I. Bauccio, Michael. TA403.4.A84 1994
620.1'1—dc20
94-1474 CIP

ASM International®
Materials Park, OH 44073-0002

Printed in the United States of America

Foreword

This two-volume handbook is a reference for engineering designers (materials engineers, metallurgists, etc.) involved in the complex and continuous process of materials selection.

Many types of materials are used for the construction of commercial and military equipment. The primary materials used are metallics (for example, steel, aluminum, titanium, and nickel), non-metallics (such as plastics and ceramics), and composites. This volume covers the properties and design applications for engineered materials—which include the non-metallics and composites. Each of these classes of materials has a wide range of properties, presenting the designer with the formidable challenge of first choosing the proper class of material for a specific structure, and then determining the most appropriate material within the class selected. In making these decisions, the designer must review mechanical properties, corrosion resistance properties, and economic data for each material under consideration for completion of each structural engineering project.

Because the materials selection process depends to a large degree on mechanical and corrosion resistance properties, these handbooks emphasize the presentation of property data. Pertinent information on materials properties will assist the designer in selecting the most appropriate material for a specific application.

Michael L. Bauccio
Editor

About the Editor

Michael L. Bauccio has over 14 years of engineering experience and is an engineer with The Boeing Company, Seattle WA. Before joining Boeing he was an engineering specialist with Bell Helicopter Textron, Inc., Fort Worth TX. Mr. Bauccio has a BS in biological sicences from Loyola University, New Orleans LA; an MS in pharmacology and toxicology from St. John's University, Jamaica NY; and an MS in materials science and chemical engineering from the Sever Institute of Technology at Washington University, St. Louis MO. Mr. Bauccio is the founder-president of the St. Louis Institute of Chemists, and past president of the Technical group committee on Corrosion for Aerospace Equipments, for the National Association of corrosion engineers (NACE). A contributing author to the Metals Handbook, Volume 13, Corrosion, Mr. Bauccio is a registered Engineer-in-Training (E.I.T.) in the State of Washington.

Table of Contents

Glossary of Terms

A

A. The symbol for a repeating unit in a polymer chain.

ABA copolymers. Block copolymers with three sequences, but only two domains.

A-basis. The "A" mechanical property value is the value above which at least 99% of the population of values is expected to fall, with a confidence of 95%. Also called A-allowable. See also *B-basis*, *S-basis*, and *typical-basis*.

abhesive. A material that resists adhesion. A film or coating applied to surfaces to prevent sticking, heat sealing, and so on, such as a parting agent or mold release agent.

ablation. A self-regulating heat and mass transfer process in which incident thermal energy is expended by sacrificial loss of material.

ablative plastic. A material that absorbs heat (with a low material loss and char rate) through a decomposition process (pyrolysis) that takes place at or near the surface exposed to the heat. This mechanism essentially provides thermal protection (insulation) of the subsurface materials and components by sacrificing the surface layer. Ablation is an exothermic process.

ABL bottle. An internal pressure test vessel about 460 mm (18 in.) in diameter and 610 mm (24 in.) long used to determine the quality and properties of the filament-wound material in the vessel.

ABS. See *acrylonitrile-butadiene-styrene resins*.

absolute humidity. The weight of water vapor present in a unit volume of air, such as grams per cubic foot, or grams per cubic meter. The amount of water vapor is also reported in terms of weight per unit weight of dry air, such as grams per pound of dry air, but this value differs from values calculated on a volume basis and should not be referred to as absolute humidity. It is designated as humidity ratio, specific humidity, or moisture content.

absolute viscosity. See *viscosity*.

absorption. (1) The taking up of a liquid or gas by capillary, osmotic, or solvent action. (2) The capacity of a solid to receive and retain a substance, usually a liquid or gas, with the formation of an apparently homogeneous mixture. (3) Transformation of radiant energy to a different form of energy by interaction with matter. (4) The process by which a liquid is drawn into and tends to fill permeable pores in a porous solid body; also, the increase in mass of a porous solid body resulting from the penetration of a liquid into its permeable pores. See also *adsorption*.

AC. See *acetal (AC) copolymers*, *acetal (AC) homopolymers*, and *acetal (AC) resins*.

accelerated-life test. A method designed to approximate, in a short time, the deteriorating effect obtained under normal long-term service conditions. See also *artificial aging*.

accelerator. A material that, when mixed with a catalyst or a resin, speeds up the chemical reaction between the catalyst and the resin (usually in the polymerizing of resins or vulcanization of rubbers). Also called promoter.

acceptance test. A test, or series of tests, conducted by the procuring agency, or an agent thereof, upon receipt, to determine whether an individual lot of materials conforms to the purchase order or contract or to determine the degree of uniformity of the material supplied by the vendor, or both. Compare to *preproduction test* and *qualification test*.

accumulator. An auxiliary cylinder and piston (plunger) mounted on injection molding or blowing machines and used to provide faster molding cycles. In blow molding, the accumulator cylinder is filled (during the time between parison deliveries, or "shots") with melted plastic coming from the main (primary) extruder. The plastic melt is stored, or "accumulated," in this auxiliary cylinder until the next shot or parison is required. At that time the piston in the accumulator cylinder forces the molten plastic into the dies that form the parison.

acetal (AC) copolymers. A family of highly crystalline thermoplastics prepared by copolymerizing trioxane with small amounts of a comonomer that randomly distributes carbon-carbon bonds in the polymer chain. These bonds, as well as hydroxyethyl terminal units, give the acetal copolymers a high degree of thermal stability and resistance to strong alkaline environments.

acetal (AC) homopolymers. Highly crystalline linear polymers formed by polymerizing formaldehyde and capping it with acetate end groups.

acetal (AC) resins. Thermoplastics (polyformaldehyde and polyoxymethylene resins) produced by the *addition polymerization* of aldehydes by means of the carbonyl function, yielding unbranched polyoxymethylene chains of great length. The acetal resins, among the strongest and stiffest of all thermoplastics, are also characterized by good fatigue life, resilience, low moisture sensitivity, high solvent and chemical resistance, and good electrical properties. They may be processed by conventional injection molding and extrusion techniques and fabricated by welding methods used for other plastics.

acid-acceptor. A compound that acts as a stabilizer by chemically combining with acid that may be initially present in minute quantities in a plastic, or that may be formed by the decomposition of the resin.

acid refractory. Siliceous ceramic materials of a high melting temperature, such as silica brick, used for metallurgical furnace linings. Compare with *basic refractories*.

acrylate resins. See *acrylic resins*.

acrylic plastic. A thermoplastic polymer made by the polymerization of esters of acrylic acid or its derivatives.

acrylic resins. Polymers of acrylic or methacrylic esters, sometimes modified with nonacrylic monomers such as the ABS group. The acrylates may be methyl, ethyl, butyl, or 2-ethylhexyl. Usual methacrylates are methyl, ethyl, butyl, laural, and stearyl. The resins may be in the form of molding powders or casting syrups, and are noted for their exceptional clarity and optical properties. Acrylics are widely used in lighting fixtures because they are either slow burning or self-extinguishing and do not produce harmful smoke or gases in the presence of flame.

acrylonitrile. A monomer with the structure (CH_2:CHCN). It is most useful in copolymers. Its copolymer with butadiene is nitrile rubber; acrylonitrile-butadiene copolymers with styrene (SAN) are tougher than polystyrene. Acrylonitrile is also used as a synthetic fiber and as a chemical intermediate.

acrylonitrile-butadiene-styrene (ABS) resins. A family of thermoplastics based on acrylonitrile, butadiene, and styrene, combined by a variety of methods involving polymerization, graft polymerization, physical mixing, and combinations thereof. The standard grades of ABS resins are rigid, hard, and tough, but not brittle, and possess good impact strength, heat resistance, low-temperature properties, chemical resistance, and electrical properties.

activation. The (usually) chemical process of making a surface more receptive to bonding with a coating or an encapsulating material.

addition polymerization. A chemical reaction in which simple molecules (monomers) are linked to each other to form long-chain molecules (polymers) by chain reaction.

additive. A substance added to another substance, usually to improve properties, such as plasticizers, initiators, light stabilizers, and flame retardants. See also *filler*.

adhere. To cause two surfaces to be held together by adhesion.

adherend. A body held to another body by an adhesive. See also *substrate*.

adherend preparation. See *surface preparation*.

adhesion. The state in which two surfaces are held together by interfacial forces, which may consist of valence forces, interlocking action, or both. See also *mechanical adhesion* and *specific adhesion*.

adhesion promoter. A coating applied to a substrate, before it is coated with an adhesive, to improve the adhesion of the substrate. Also called primer.

adhesion promotion. The chemical process of preparing a surface to provide for a uniform, well-bonded interface.

adhesive. A substance capable of holding materials together by surface attachment. Adhesive is a general term and includes, among others, cement, glue, mucilage, and paste. These terms are loosely used interchangeably. Various descriptive adjectives are applied to the term adhesive to indicate certain physical characteristics. See also *hot-melt adhesive*, *pressure-sensitive adhesive*, *structural adhesive*, *ultraviolet/electron beam cured adhesive*, and *water-based adhesive*.

adhesive, anaerobic. See *anaerobic adhesive*.

adhesive bond. Attractive forces, generally physical in character, between an adhesive and the base materials.

adhesive bonding. A materials joining process in which an adhesive, placed between the faying surfaces (adherends) solidifies to produce an adhesive bond.

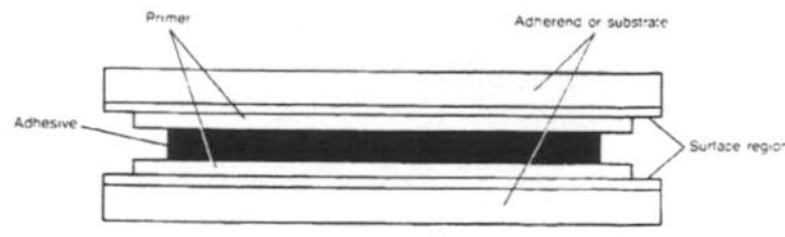

Idealized adhesively bonded assembly

Adhesive bonded structure

adhesive, cold-setting. See *cold-setting adhesive*.

adhesive, contact. See *contact adhesive*.

adhesive dispersion. A two-phase system in which one phase is suspended in a liquid. Compare with *emulsion*.

adhesive failure. Rupture of an adhesive bond such that the separation appears to be at the adhesive-adherend interface. Sometimes termed failure in adhesion. Compare with *cohesive failure*.

adhesive film. A synthetic resin adhesive, with or without a film carrier fabric, usually of the thermosetting type, in the form of a thin film of resin, used under heat and pressure as an interleaf in the production of bonded structures.

adhesive, gap-filling. See *gap-filling adhesive*.

adhesive, heat-activated. See *heat-activated adhesive*.

adhesive, heat-sealing. See *heat-sealing adhesive*.

adhesive, hot-melt. See *hot-melt adhesive*.

adhesive, hot-setting. See *hot-setting adhesive*.

adhesive, intermediate-temperature-setting. See *intermediate-temperature-setting adhesive*.

adhesive joint. Location at which two adherends are held together with a layer of adhesive. See also *bond*.

adhesive, pressure-sensitive. See *pressure-sensitive adhesive*.

adhesive strength. The strength of the bond between an adhesive and an adherend.

adhesive, structural. See *structural adhesive*.

adhesive system. An integrated engineering process that analyzes the total environment of a potential bonded assembly to select the most suitable adhesive, application method, and dispensing equipment.

admixture. (1) The addition and homogeneous dispersion of discrete components, before cure of a polymer. (2) A material other than water, aggregates, hydraulic cement, and fiber reinforcement used as an ingredient of concrete or mortar and added to the batch immediately before or during its mixing. (3) Material added to (cement) mortars as a water-repellent or coloring agent or to retard or hasten setting.

adsorption. The adhesion of the molecules of gases, dissolved substances, or liquids in more or less concentrated form, to the surfaces of solids or liquids with which they are in contact. The concentration of a substance at a surface or interface of another substance.

advanced ceramics. Ceramic materials that exhibit superior mechanical properties, corrosion/oxidation resistance, or electrical, optical, and/or magnetic properties. This term includes many monolithic ceramics as well as particulate-, whisker-, and fiber-reinforced glass, glass-ceramics, and ceramic-matrix composites. Also known as engineering, fine, or technical ceramics. Contrast with *traditional ceramics*.

advanced composites. Composite materials that are reinforced with continuous fibers having a modulus higher than that of fiberglass fibers. The term includes metal-matrix and ceramic-matrix composites, as well as carbon-carbon composites.

afterbake. See *postcure*.

aggregate. (1) A hard, coarse material usually of mineral origin used with an epoxy binder (or other resin) in plastic tools. Also used in flooring or as a surface medium. (2) A dense mass of particles held together by strong intermolecular or atomic cohesive forces. (3) Granular material, such as sand, gravel, crushed stone, or iron blast-furnace slag, used with a cementing medium to form hydraulic-cement concrete or mortar.

aging. (1) The effect on materials of exposure to an environment for a prolonged interval of time. (2) The process of exposing materials to an environment for a prolonged interval of time in order to predict in-service lifetime. (3) Generally, the degradation of properties or function with time. In capacitors, the loss of dielectric constant, K, by dielectric relaxation. Expressed as a percent change per decade of time.

air-assist forming. A method of thermoforming in which air flow or air pressure is employed to preform plastic sheet partially just before the final pull-down onto the mold using vacuum.

air-bubble void. Air entrapment within a molded item or between the plies of reinforcement or within a bond line or encapsulated area; localized, noninterconnected, and spherical in shape.

air gap. In extrusion coating, the distance from the die opening to the nip formed by the pressure roll and the chill roll.

air setting. The characteristic of some materials, such as refractory cements, core pastes, binders, and plastics, to take permanent set at normal air temperatures.

air-slip forming. A variation of vacuum snap-back *thermoforming* in which the male mold is enclosed in a box such that when the mold moves forward toward the hot plastic, air is trapped between the mold and the plastic sheet. As the mold advances, the plastic is kept away from it by this air cushion until the full travel of the mold is completed, at which point a vacuum is applied, destroying the cushion and forming the part against the plug.

air vent. A small outlet to prevent entrapment of gases in a molding or tooling fixture.

alcohols. Characterized by the hydroxyl (–OH) group they contain, alcohols are valuable starting points for the manufacture of synthetic resins, synthetic rubbers, and plasticizers.

aldehydes. Volatile liquids with sharp, penetrating odors that are slightly less soluble in water than are corresponding alcohols.

aliphatic hydrocarbons. Saturated hydrocarbons having an open-chain structure, for example, gasoline and propane.

alkyd. Resin used in coatings. Reaction products of polyhydric alcohols and polybasic acids.

alkyd plastic. Thermoset plastic based on resins composed principally of polymeric esters, in which the recurring ester groups are an integral part of the main polymer chain, and in which ester groups occur in most cross links that may be present between chains.

alkylation. (1) A chemical process in which an alkyl radical is introduced into an organic compound by substitution or addition. (2) A refinery process for chemically combining isoparaffin with olefin hydrocarbons.

allophanate. Reactive product of an isocyanate and the hydrogen atoms in a urethane.

alloprene. Chlorinated rubber.

allotropy. (1) A near synonym for *polymorphism.* Allotropy is generally restricted to describing polymorphic behavior in elements, terminal phases, and alloys whose behavior closely parallels that of the predominant constituent element. (2) The existence of a substance, especially an element, in two or more physical states (for example, crystals).

alloy. A blend of polymers or copolymers with other polymers or elastomers under selected conditions; for example, styrene-acrylonitrile. Also called polymer blend.

allyl plastic. A thermoset plastic based on resins made by *addition polymerization* of monomers containing allyl groups; for example, diallyl phthalate (DAP).

allyl resins. A family of thermoset resins made by *addition polymerization* of compounds containing the group CH_2:CH–CH_2, such as esters of allyl alcohol and dibasic acids. They are available as monomers, partially polymerized prepolymers, or molding compounds. Members of the family are diallyl phthalate (DAP), diallyl isophthalate (DAIP), diallyl maleate (DAM), and diallyl chlorendate (DAC).

alpha (α) cellulose. A very pure cellulose prepared by special chemical treatment.

alpha (α) loss peak. In dynamic mechanical or dielectric measurement, the first peak in the damping curve below the melt, in order of decreasing temperature or increasing frequency.

alternating copolymer. A copolymer in which each repeating unit is joined to another repeating unit in the polymer chain (–A–B–A–B–).

alternating stress amplitude. A test parameter of a dynamic fatigue test; one-half the algebraic difference between the maximum and minimum stress in one cycle.

alumina. A compound in the form of a white powder or colorless hexagonal crystals, the composition of which is Al_2O_3. Used in aluminum production, spark plugs, artificial gems, and manufacture of abrasives, refractories, electrical insulators, and structural ceramics. See also *corundum*.

aluminum nitride (AlN). A high-thermal-conductivity ceramic used as an electronic substrate. Also a key component in the production of *Sialons*.

aluminum oxide (Al_2O_3). See *alumina*.

ambient. The environment that surrounds and contacts a system or component.

amine adduct. Product of the reaction of an amine with a deficiency of a substance containing epoxy groups.

amino. Relating to or containing an $-NH_2$ or –NH group.

amino resins. Resins made by the polycondensation of a compound containing amino groups, such as urea or melamine, with an aldehyde, such as formaldehyde, or an aldehyde-yielding material. Melamine-formaldehyde and urea-formaldehyde resins are the most important family members. The resins can be dispersed in water to form colorless syrups. With appropriate catalysts, they can be cured at elevated temperatures.

amorphous plastic. A plastic that has no crystalline component, no known order or pattern of molecule distribution, and no sharp melting point.

amylaceous. Pertaining to, or of the nature of, starch; starchy.

anaerobic adhesive. An adhesive that cures only in the absence of air after being confined between assembled parts.

anchorage. Part of an insert that is molded inside the plastic and held fast by the shrinkage of the plastic.

anchor pattern. A pattern made by blast cleaning abrasives on an adherend surface in preparation for adhesive application prior to bonding. Pattern is examined in profile.

andalusite. A mineral of composition $Al_2O_3 \cdot SiO_2$ used in the production of aluminosilicate bricks for use in blast furnaces, steel ladles, and torpedo ladles.

anelastic deformation. Any portion of the total deformation of a body that occurs as a function of time when load is applied and which disappears completely after a period of time when the load is removed.

angle-ply laminate. A laminate having fibers of adjacent plies, oriented at alternating angles.

angle press. A hydraulic molding press equipped with horizontal and vertical rams, specially designed for the production of complex plastic moldings having deep undercuts.

angle wrap. Tape fabric that is wrapped on a starter dam mandrel at an angle to the centerline.

aniline. An important organic base ($C_6H_5NH_2$) made by reacting chlorobenzene with aqueous ammonia in the presence of a catalyst. It is used in the production of aniline formaldehyde resins and in the manufacture of certain rubber accelerators and antioxidants.

aniline-formaldehyde resins. Members of the aminoplastics family made by the condensation of formaldehyde and aniline in an acid solution. The resins are thermoplastic and are used to a limited extent in the production of molded and laminated insulating materials. Products made from these resins have high dielectric strength and good chemical resistance.

aniline point. As applied to a petroleum product, the lowest temperature at which the product is completely miscible with an equal volume of freshly distilled aniline. The aniline point is a guide to the oil composition.

anisotropic. Exhibiting different properties when tested along axes in different directions. In magnetics, capable of being magnetized more readily in one direction than in a transverse direction.

anisotropic conductive adhesive. An adhesive that can be made conductive in the vertical, or *z*, axis while remaining an insulator in the horizontal, or *x* and *y*, axes.

anisotropic laminate. One in which the properties are different in different directions along the laminate plane.

anisotropy. The characteristic of exhibiting different values of a property in different directions with respect to a fixed reference system in the material.

anisotropy of laminates. The difference of the properties along the directions parallel to the length or width of the lamination planes and perpendicular to the lamination.

anneal (glass). To prevent or remove objectionable stresses in glassware by controlled cooling from a suitable temperature.

annealing (glass). A controlled cooling process for glass designed to reduce thermal residual stress to a commercially acceptable level, and, in some cases, modify structure.

annealing (plastics). Heating to a temperature at which the molecules have significant mobility, permitting them to reorient to a configuration having less residual stress. In semicrystalline polymers, heating to a temperature at which retarded crystallization or recrystallization can occur.

annealing point (glass). That temperature at which internal stresses in a glass are substantially relieved in a matter of minutes.

annealing range (glass). The range of glass temperature in which stress in glass can be relieved at a commercially practical rate. For purposes of comparing glasses, the annealing range is assumed to correspond with the temperature between the annealing point and the strain point.

antiferromagnetic material. A material wherein interatomic forces hold the elementary atomic magnets (electron spins) of a solid in alignment, a state similar to that of a *ferromagnetic material* but with the difference that equals numbers of elementary magnets (spins) face in opposite directions and are antiparallel, causing the solid to be weakly magnetic, that is, paramagnetic, instead of ferromagnetic.

antioxidant. Any additive for the purpose of reducing the rate of oxidation and subsequent deterioration of a material.

antistatic agents. Agents that, when added to a plastic molding material or applied to the surface of a molded object, make it less conductive, thus hindering the fixation of dust or the buildup of electrical charge.

aramid. A manufactured organic fiber in which the fiber-forming substance is a long-chain synthetic aromatic polyamide in which at least 85% of the amide linkages are directly attached to two aromatic rings. Aramid fibers, most notably Kevlar fibers, were the first with a high enough tensile modulus and strength to be used as a reinforcement in advanced composites. See also *Kevlar*.

arc resistance. Ability to withstand exposure to an electric voltage. The total time in seconds that an intermittent arc may play across a plastic surface without rendering the surface conductive.

areal weight. The weight of a fiber reinforcement per unit area (width × length) of tape or fabric.

aromatic. Unsaturated hydrocarbon with one or more benzene ring structures in the molecule.

aromatic polyester. A polyester derived from monomers in which all the hydroxyl and carboxyl groups are directly linked to aromatic nuclei.

artificial aging. The exposure of a plastic to conditions that accelerate the effects of time. Such conditions include heating, exposure to cold, flexing, application of electric field, exposure to chemicals, ultraviolet light radiation, and so forth. Typically, the conditions chosen for such testing reflect the conditions under which the plastic article will be used. Usually, the length of time the article is exposed to these test conditions is relatively short. Properties such as dimensional stability, mechanical fatigue, chemical resistance, stress cracking resistance, dielectric strength, and so forth, are evaluated in such testing. See also *aging*.

artificial weathering. The exposure of plastics to cyclic laboratory conditions, consisting of high and low temperatures, high and low relative humidities, and ultraviolet radiant energy, with or without direct water spray and moving air (wind), in an attempt to produce changes in the properties of the plastics similar to those observed after long-term continuous exposure outdoors. The laboratory exposure conditions are usually more intensified than those encountered in actual outdoor exposure in an attempt to achieve an accelerated effect. Also called accelerated aging.

ash content. Proportion of the solid residue remaining after a reinforcing substance has been incinerated (charred or intensely heated).

aspect ratio. The ratio of length to diameter of a reinforcing fiber.

assembly adhesive. An adhesive that can be used for bonding parts together, such as in the manufacture of a boat, an airplane, furniture, and the like. The term assembly adhesive is commonly used in the wood industry to distinguish such adhesives (formerly called joint glues) from those used in making plywood (sometimes called veneer glues). It describes adhesives used in fabricating finished structures or goods, or subassemblies thereof, as differentiated from adhesives used in the production of sheet materials for sale as such, for example, plywood or laminates.

assembly time. The time interval between the spreading of the adhesive on the adherend and the application of pressure or heat, or both, to the assembly. For assemblies involving multiple layers or parts, the assembly time begins with the spreading of the adhesive on the first adherend. See also *closed assembly time* and *open assembly time*.

A-stage. An early stage in the reaction of certain thermosetting resins in which the material is fusible and still soluble in certain liquids. Synonym for resole. Compare with *B-stage* and *C-stage*.

atactic stereoisomerism. A chain of molecules in which the position of the side chains or side atoms is more or less random. See also *isotactic stereoisomerism* and *syndiotactic stereoisomerism*.

attenuation. (1) The fractional decrease of the intensity of an energy flux, including the reduction of intensity resulting from geometrical spreading, absorption, and scattering. (2) The diminution of vibrations or energy over time or distance. The process of making thin and slender, as applied to the formation of fiber from molten glass. (3) The exponential decrease with distance in the amplitude of an electrical signal traveling along a very long, uniform transmission line, due to conductor and dielectric losses.

autoclave. A closed vessel for conducting and completing either a chemical reaction under pressure and heat or other operation, such as cooling. Widely used for bonding and curing reinforced plastic laminates.

autoclave molding. A process in which, after lay-up, winding, or wrapping, an entire assembly is placed in a heated autoclave, usually at 340 to 1380 kPa (50 to 200 psi). Additional pressure permits higher density and improved removal of volatiles from the resin. Lay-up is usually vacuum bagged with a bleeder and release cloth.

automatic mold. A mold for injection or compression molding of plastics that repeatedly goes through the entire cycle, including ejection, without human assistance.

automatic press. A hydraulic press for compression molding or an injection machine that operates continuously, being controlled mechanically (toggle) or hydraulically, or by a combination of these methods.

average molecular weight. The molecular weight of the most typical chain in a given plastic; it is characteristic of neither the longest nor the shortest chain.

axial strain. The linear strain in a plane parallel to the longitudinal axis of the specimen.

axial winding. In filament-wound reinforced plastics, a winding with the filament parallel to, or at a small angle to, the axis (0° helix angle). See also *polar winding*.

B

B. The symbol for a repeating unit in a copolymer chain.

backing plate. In plastic injection molding equipment, a heavy steel plate that is used as a support for the cavity blocks, guide pins, and bushings. In blow molding equipment, it is the steel plate on which the cavities (that is, the bottle molds) are mounted.

back pressure. Resistance of a plastic, because of its viscosity, to continued flow when the mold is closing.

back-pressure relief port. An opening in an extrusion die for plastics that allows for the escape of excess material.

back taper. Reverse draft used in a mold to prevent the molded plastic article from drawing freely. See also *undercut*.

bagging. Applying an impermeable layer of film over an uncured part and sealing the edges so that a vacuum can be drawn.

bag molding. A method of molding or bonding plastics or composites involving the application of fluid pressure, usually by means of air, steam, water, or vacuum, to a flexible cover that, sometimes in conjunction with the rigid die, completely encloses the material to be bonded. Also called blanket molding. See also *vacuum bag molding*.

bag side. The side of a plastic or composite part that is cured against the vacuum bag.

Bakelite. A proprietary name for a phenolic thermosetting resin used as a plastic mounting material for metallographic samples.

balanced construction. In woven reinforcements, equal parts of warp and fill fibers. Construction in which reactions to tension and compression loads result in extension or compression deformations only and in which flexural loads produce pure bending of equal magnitude in axial and lateral directions.

balanced design. In filament-wound reinforced plastics, a winding pattern so designed that the stresses in all filaments are equal.

balanced-in-plane contour. In a filament-wound part, a head contour in which the filaments are oriented within a plane and the radii of curvature are adjusted to balance the stresses along the filaments with the pressure loading.

balanced laminate. A laminate in which all laminae at angles other than 0° and 90° occur only in ± pairs (not necessarily adjacent) and are symmetrical around the centerline). See also *symmetrical laminate*.

balanced twist. An arrangement of twists in a combination of two or more reinforcing strands that does not kink or twist when the yarn produced is held in the form of an open loop.

ball clay. A secondary clay, commonly characterized by the presence of organic matter, high plasticity, high dry strength, long vitrification range, and a light color when fired. Used extensively in traditional ceramics, such as whiteware, wall tile, and china.

banbury. An apparatus for compounding polymeric materials. It is composed of a pair of contrarotating rotors that masticate the materials to form a homogeneous blend. This internal-type mixer produces excellent mixing.

band density. In filament winding of composites, the quantity of fiberglass reinforcements per inch of band width, expressed as strands (or filaments) per inch.

band thickness. In filament winding of composites, the thickness of the reinforcement as it is applied to the mandrel.

band width. In filament winding of composites, the width of the reinforcement as it is applied to the mandrel.

Barcol hardness. A hardness value obtained by measuring the resistance to penetration of a sharp steel point under a spring load. The instrument, called the Barcol impressor, gives a direct reading on a 0 to 100 scale. The hardness value is often used as a measure of the degree of cure of a plastic.

bare glass. Glass in the form of yarns, rovings, and fabrics from which the sizing or finish has been removed. Also, such glass before the application of sizing or finish.

barium titanate ($BaTiO_3$). The basic raw material used to make high dielectric constant ceramic capacitors. Used also in high thermal conductivity, thick-film ceramic pastes.

barrier coat. An exterior coating applied to a composite filament-wound structure to provide protection. In fuel tanks, a coating applied to the inside of the tank to prevent fuel from permeating the side wall.

barrier film. The layer of film used during cure to permit removal of air and volatiles from a reinforced plastic or a composite lay-up while minimizing resin loss.

barrier plastics. A general term applied to a group of lightweight, transparent, impact-resistant plastics, usually rigid copolymers of high acrylonitrile content. Barrier plastics are generally characterized by gas, aroma, and flavor barrier characteristics approaching those of metal and glass.

basic refractories. Refractories whose major constituent is lime, magnesia, or both, and which may react chemically with acid refractories, acid slags, or acid fluxes at high temperatures. Basic refractories are used for furnace linings. Compare with *acid refractory*.

basket weave. In this type of woven reinforcement, two or more warp threads go over and under two or more filling threads in a repeat pattern. The basket weave is less stable than the *plain weave* but produces a flatter and stronger fabric. It is also a more pliable fabric than the plain weave and maintains a certain degree of porosity without too much sleaziness, although not as much as the plain weave.

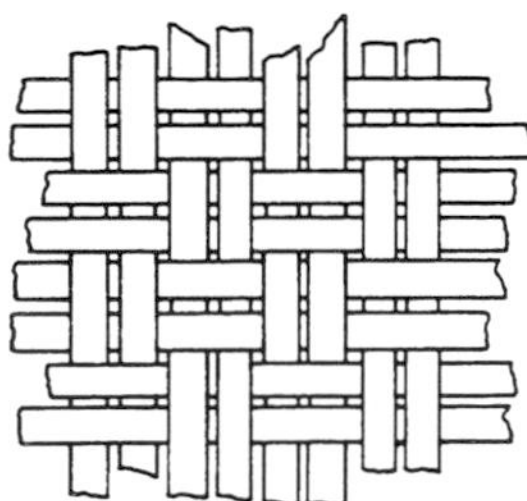

Basket weave construction

batch. A quantity of materials formed during the same process or in one continuous process and having identical characteristics throughout. See also *lot*.

batt. Felted fabrics. Structures built by the interlocking action of compressing fibers, without spinning, weaving, or knitting.

B-basis. The "B" mechanical property value is the value above which at least 90% of the population of values is expected to fall, with a confidence of 95%. See also *A-basis*, *S-basis*, and *typical basis*.

bearing strain. The ratio of the deformation of the bearing hole, in the direction of the applied force, to the pin diameter. Also, the stretch or deformation strain for a sample under bearing load.

bearing strength. The maximum bearing stress that can be sustained. Also, the bearing stress at that point on the stress-strain curve at which the tangent is equal to the bearing stress divided by *n*% of the bearing hole diameter.

bearing stress. The applied load in pounds divided by the bearing area. Maximum bearing stress is the maximum load in pounds sustained by the specimen during the test, divided by the original bearing area.

bearing test. A method of determining the response to stress (load) of sheet products that are subjected to riveting, bolting, or a similar fastening procedure. The purpose of the test is to determine the *bearing strength* of the material and to measure the *bearing stress* versus the deformation of the hole created by a pin or rod of circular cross section that pierces the sheet perpendicular to the surface.

bearing yield strength. The *bearing stress* at which a material exhibits a specified limiting deviation from the proportionality of *bearing stress* to *bearing strain*.

bending stress. A stress system that simultaneously imposes a compressive component at one surface, graduating to an imposed tensile component at the opposite surface of a glass section.

bending-twisting coupling. A property of certain classes of laminates that exhibit twisting curvatures when subjected to bending moments.

bend test. A test for ductility performed by bending or folding, usually by steadily applied forces, but in some instances by blows, to a specimen having a cross section substantially uniform over a length several times as great as the largest dimension of the cross section.

benzene ring. The six-carbon ring structure found in benzene, C_6H_6, and in organic compounds formed from benzene by replacement of one or more hydrogen atoms by other chemical atoms or radicals.

beryllia. A colorless to white powder of the composition BeO used in the manufacture of hot-pressed ceramic parts— most notably basic refractories and substrates (heat sinks) in electronics. Also known as beryllium oxide.

beryllium oxide (BeO). See *beryllia*.

beta (β) gage. A gage consisting of two facing elements, a β-ray-emitting source, and a β-ray detector. Also called beta-ray gage.

beta (β) loss peak. In dynamic mechanical or dielectric measurement, the second peak in the damping curve below the melt, in order of decreasing temperature or increasing frequency.

bias fabric. Fabric consisting of warp and fill fibers at an angle to the length of the fabric.

biaxial load. A loading condition in which a specimen is stressed in two directions in its plane.

biaxial winding. In filament winding, a type of winding in which the helical band is laid in sequence, side by side, with crossover of the fibers eliminated.

bidirectional laminate. A reinforced plastic laminate with the fibers oriented in two directions in its plane. A cross laminate. See also *unidirectional laminate*.

binder. (1) The resin or cementing constituent (of a plastic compound) that holds the other components together. The agent applied to fiber mat or preforms to bond the fibers before laminating or molding. (2) A component of an adhesive composition that is primarily responsible for the adhesive forces which hold two bodies together. See also *extenders* and *filler*.

biscuit. See *cull* and *preform*.

bismaleimide (BMI). A type of polyimide that cures by an addition rather than a condensation reaction, thus avoiding problems with volatiles formation, and which is produced by a vinyl-type polymerization of a prepolymer terminated with two maleimide groups. Intermediate in temperature capability between epoxy and polyimide.

bitumen. Asphaltlike polymer.

black marking. Black smudges on the surface of a pultruded plastic product that results from excessive pressures in the die when the pultrusion is rubbing against it or unchromed die surfaces, and that cannot be removed by cleaning or scrubbing or by wiping with solvent.

bladder. An elastomeric lining for the containment of hydroproof or hydroburst pressurization medium in filament-wound structures.

blanket. Fiber or fabric plies that have been laid up in a complete assembly and placed on or in the mold all at one time (flexible bag process). Also, the type of bag in which the edges are sealed against the mold.

blanking. The cutting of flat sheet stock to shape on a punch press by sharply striking the stock with a punch while it is supported on a mating die. Also called die cutting.

bleed. To give up color when in contact with water or a solvent. Undesirable movement of certain materials in a plastic, such as plasticizers in vinyl, to the surface of the finished article or into an adjacent material; also called migration.

bleeder cloth. A woven or nonwoven layer of material used in the manufacture of composite parts to allow the escape of excess gas and resin during cure. The bleeder cloth is removed after the curing process and is not part of the final composite.

bleeding. (1) The removal of excess resin from a laminate during cure. The diffusion of color from a plastic part into the surrounding surface or part. (2) Separation of oil (or other fluid) from a grease.

bleedout. The excess liquid resin that migrates to the surface of a winding. Primarily pertinent to filament winding.

bleed-out. The spread of adhesive away from the bond area.

blind hole. A hole that is not drilled entirely through.

blister. An elevation of the surface of an adherend, the shape of which somewhat resembles a blister on the human skin. Its boundaries may be indefinitely outlined, and it may have burst and become flattened. A blister may be caused by insufficient adhesive; inadequate curing time, temperature, or pressure; or trapped air, water, or solvent vapor.

block copolymer. An essentially linear copolymer consisting of a small number of repeated sequences of polymeric segments of different chemical structure.

blocked curing agent. A curing agent or hardener rendered unreactive, which can be reactivated as desired by physical or chemical means. Compare with *hardener*.

blocking. (1) An undesired adhesion between touching layers of a material, such as occurs under moderate pressure during storage or use. (2) The process of shaping a gather of glass in a cavity of wood or metal. (3) The process of stirring and fining glass by immersion of a wooden block or other source of bubbles. (4) The process of reprocessing to remove surface imperfections. (5) The mounting of optical glass blanks in a shell for grinding and polishing operations. (6) The process wherein a furnace is idled at reduced temperatures. (7) The process of setting refractory blocks in a furnace.

bloom. A noncontinuous surface coating on plastic products that comes from ingredients such as plasticizers, lubricants, antistatic agents, and so on, which are incorporated into the plastic resin, or that occurs by atmospheric contamination. Bloom is the result of ingredients in the plastic coming out of solution and migrating to the surface.

blowing agent. A compounding ingredient used to produce gas by chemical or thermal action, or both, in the manufacture of hollow or cellular plastic articles.

blow molding. A method of fabricating plastics in which a warm plastic parison (hollow tube) is placed between the two halves of a mold (cavity) and forced to assume the shape of that mold cavity by use of air pressure. The air pressure is introduced through the inside of the parison and forces the plastic against the surface of the mold, which defines the shape of the product.

blown-film extrusion. Technique for making film by extruding the plastic through a circular die, followed by expansion (by the pressure of internal air admitted through the center of the mandrel), cooling, and collapsing of the bubble.

blown tubing. A thermoplastic film that is produced by extruding a tube, applying a slight internal pressure to the tube to expand it while still molten, and subsequently cooling to set the tube. The tube is then flattened through guides and wound up flat on rolls. The size of blown tubing is determined by the flat width in inches as wound rather than by the diameter, as in the case of rigid types of tubing.

blow pin. Part of the tooling used to form hollow plastic objects or containers by the blow molding process. It is a tubular tool through which air pressure is introduced into the parison to create the air pressure necessary to form the parison into the shape of the mold. In some blow molding systems, it is a part of, or an extension of, the core pin.

blow pressure. The air pressure required to form the parison into the shape of the mold cavity, in a plastic blow molding operation.

blow rate. The rate of speed at which air enters, or the time required for air to enter, the parison during the *blow molding* cycle.

blow-up ratio. In *blow molding* of plastics, the ratio of the diameter of the product (usually its greatest diameter) to the diameter of the parison from which the product is formed. In blown-film extrusion, the ratio between the diameter of the final film tube and the diameter of the die orifice.

blueing. A mold blemish in the form of a blue oxide film on the polished surface of a mold due to abnormally high mold temperatures.

BMC. See *bulk molding compound.*

BMI. See *bismaleimide*.

bolster. Space or filler in a mold for making plastics.

bond. In an adhesive bonded or diffusion bonded joint, the line along which the faying surfaces are joined together.

bond angle. The angle formed by the bonds of one atom to other atoms; for example, 109.5° for C–C bonds.

bond face. The part or surface of an adherend that serves as a substrate for an adhesive.

bonding force. The force that holds two atoms together; it results from a decrease in energy as two atoms are brought closer to one another.

bond length. The average distance between the centers of two atoms; for example, 0.154 nm (1.54 Å) for C–C bonds.

bond line. The interface between adhesive and adherend in an adhesive bonded joint.

bond strength. (1) The unit load applied to tension, compression, flexure, peel, impact, cleavage, or shear required to break an adhesive assembly with failure occurring in or near the plane of the bond. The term adherence is frequently used in place of bond strength. (2) The force required to pull a coating free of a substrate.

borate glass. A glass in which the essential glass former is boron oxide instead of silica.

boron carbide. A black crystalline powder of high hardness, the composition of which is either B_6C or B_4C (the latter being a composite of B_4C and carbon in graphitic form). Applications include loose abrasives and hot pressed shot blast nozzles and other wear-resistant components.

boron fiber. A fiber produced by vapor deposition of elemental boron, usually onto a tungsten filament core, to impart strength and stiffness.

boron nitride (hexagonal). A white fluffy powder of composition BN with high chemical and thermal stability and high electrical resistance. Used as a lubricant for high-pressure bearings and in the hot pressed condition for mechanical and electrical parts. See also *cubic boron nitride*.

borosilicate glass. Any silicate glass having at least 5% of boron oxide (B_2O_3).

boss. (1) A relatively short protrusion or projection from the surface of a forging or casting, often cylindrical in shape. Usually intended for drilling and tapping for attaching parts. (2) Projection on a plastic part designed to add strength, to facilitate alignment during assembly, or to provide for a fastening.

bottom blow. A specific type of *blow molding* technique for plastics that forms hollow arteries by injecting the blowing air into the parison from the bottom of the mold (as opposed to introducing the blowing air at a container opening).

bottom plate. In making of plastic parts, the part of the mold that contains the heel radius and the push-up.

bow. A condition of longitudinal curvature in pultruded plastic parts.

braiding. Intertwining two or more systems of yarns in the bias direction to form an integrated structure.

branched polymer. In the molecular structure of polymers, a main chain with attached side chains, in contrast to a linear polymer. Two general types are recognized, short-chain, and long-chain branching.

branching. The presence of molecular branches in a polymer. The generation of branch crystals during the crystallization of a polymer.

breakdown voltage. The voltage required, under specific conditions, to cause the failure of an insulating material. See also *dielectric strength* and *arc resistance*.

breaker plate. In plastic forming, a perforated plate located at the rear end of an extruder or at the nozzle end of an injector cylinder. It often supports the screens that prevent foreign particles from entering the die, and is used to keep unplasticized material out of the nozzle and to improve distribution of color particles.

breaking extension. The elongation necessary to cause rupture of an adhesively bonded test specimen. The tensile strain at the moment of rupture.

breaking factor. The breaking load divided by the original width of an adhesively bonded test specimen, expressed in lb/in.

breaking length. A measure of the breaking strength of reinforcing yarn. The length of a specimen the weight of which is equal to the breaking load.

breakout. Fiber separation or break on surface plies at drilled or machined composite material edges.

breather. A loosely woven material that serves as a continuous vacuum path over a part but is not in contact with the resin.

breathing. The opening and closing of a mold to allow gas to escape early in the plastic molding cycle. Also called degassing; sometimes called bumping, in phenolic molding. When referring to plastic sheeting, the term breathing indicates permeability to air.

bridging. Condition in which reinforcing fibers for composites do not move into or conform to radii and corners during molding, resulting in voids and dimensional control problems.

broad goods. Fiber woven to form fabric up to 1270 mm (50 in.) wide for reinforcement of plastics. It may or may not be impregnated with resin and is usually furnished in rolls of 25 to 140 kg (50 to 300 lb).

B-stage. An intermediate stage in the reaction of certain thermosetting resins in which the material softens when heated and swells when in contact with certain liquids, but may not entirely fuse or dissolve. The resin in an uncured thermosetting adhesive, usually in this stage. Synonym for resitol. Compare with *A-stage* and *C-stage*.

B-stage resin. In a thermosetting reaction, a resin that is in an intermediate state of cure when it is sticky, or tacky, and capable of further flow. The cure is normally complete during the laminating cycle. See also *C-stage resin* and *prepreg*.

bubble. A spherical, internal void or globule of air or other gas trapped within a plastic. See *void*.

bubbler. In forming of plastics, a device inserted into a mold force, cavity, or core, that allows water to flow deep inside the hole into which it is inserted and to discharge through the open end of the hole. Uniform cooling of the mold and of isolated mold sections can be achieved in this manner.

bubbler mold cooling. In injection molding of plastics, a method of uniformly cooling a mold; a stream of cooling liquid flows continuously into a cooling cavity equipped with a coolant outlet normally positioned at the opposite end.

buckling. A mode of failure generally characterized by an unstable lateral material deflection due to compressive action on the structural element involved.

buckyball. An inorganic solid crystal structure which forms from carbon. Buckyballs have a spherical structure with their surfaces made up of from 28 to 450 carbon atoms in hexagonal and pentagonal arrays, with 60 carbon atoms being the most stable. This soccer ball shaped configuration is reminiscent of R. Buckminster Fuller's geodesic domes, hence the name Buckminster Fullerene or "buckyball."

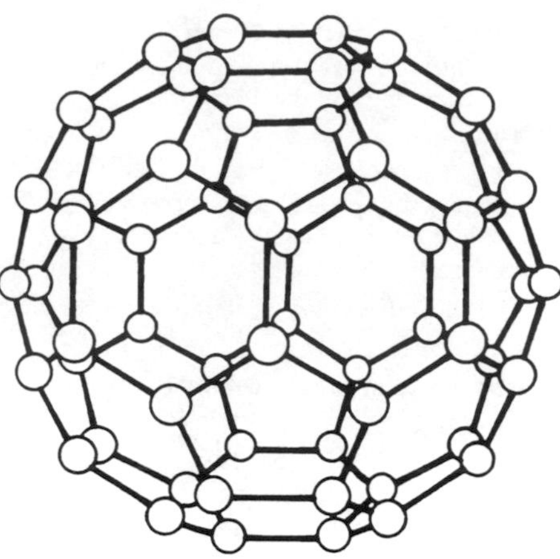

Buckyball crystal structure

built-up laminated wood. An assembly made by joining layers of lumber with mechanical fastenings so that the grain of all laminations is essentially parallel.

bulk adherend. With respect to interphase, the adherend, unaltered by the adhesive. Compare with *bulk adhesive*.

bulk adhesive. With respect to interphase, the adhesive, unaltered by the adherend. Compare with *bulk adherend*.

bulk density. (1) The weight of an object or material divided by its material volume less the volume of its open pores. (2) The density of a plastic molding material in loose form (granular, nodular, and so forth), expressed as a ratio of weight to volume.

bulk factor. The ratio of the volume of a raw plastic molding compound or powdered plastic to the volume of the finished, solid piece produced from it. The ratio of the density of the solid plastic object to the apparent, or bulk, density of the loose molding powder.

bulk molding compound (BMC). Thermosetting resin combined with strand reinforcement, fillers, and so on, into a viscous compound for compression or injection molding.

bundle. A general term for a collection of essentially parallel filaments or fibers used in composite materials.

burned. Showing evidence of thermal decomposition or charring through some discoloration, distortion, destruction, or conversion of the surface of the plastic, sometimes to a carbonaceous char.

burning rate. The tendency of plastic articles to burn at given temperatures.

butadiene. A gas (CH_2:CH·CH:CH_2), insoluble in water but soluble in alcohol and ether, obtained from the cracking of petroleum, from coal tar benzene, or from acetylene produced from coke and lime. It is widely used in the formation of copolymers with styrene, acrylonitrile, vinyl chloride, and other monomeric substances and imparts flexibility to the resultant moldings.

butadiene-styrene plastic. A synthetic resin derived from the copolymerization of butadiene gas and styrene liquids.

butt fusion. A method of joining pipe, sheet, or other similar forms of a thermoplastic resin in which the ends of the two pieces to be joined are heated to the molten state and then rapidly pressed together.

butt joint. A type of edge joint in which the edge faces of the two adherends are at right angles to the other faces of the adherends.

butylene plastics. Plastics based on resins made by the polymerization of butene or the copolymerization of butene with one or more unsaturated compounds, the butene being the greatest amount by weight.

C

calcined gypsum. A dry powder, primarily calcium sulfate hemihydrate, resulting from calcination of gypsum. Cementitious base for production of most gypsum plasters. Also called plaster of paris or stucco.

calender. The passing of plastic sheet material between sets of pressure rollers to produce a smooth finish and a desired thickness.

capacitance (*C*). That property of a system of conductors and dielectrics which permits the storage of electrically separated charges when potential differences exist between the conductors. It is the ratio of a quantity, Q, of electricity to a potential difference, V. A capacitance value is always positive. The units are farads when the charge is expressed in coulombs and the potential in volts: $C = Q/V$.

capacitor. A device that can store an electrical charge when voltage is applied. Its impedance is inversely proportional to the frequency of the voltage impressed; that is, it offers little resistance or impedance to high frequencies, but much to low frequencies.

caprolactam. A cyclic amide type of compound containing six carbon atoms. When the ring is opened, caprolactam is polymerizable into a nylon resin known as type 6 nylon or polycaprolactam.

carbanion ion. Negatively charged organic compound (ion).

carbon. The element that provides the backbone for all organic polymers. Graphite is a crystalline form of carbon. Diamond is the densest crystalline form of carbon.

carbon black. A black pigment produced by the incomplete burning of natural gas or oil. It is widely used as a filler, particularly in the rubber industry. Because it possesses useful ultraviolet protective properties, it is also much used in molding compounds intended for outside weathering applications.

carbon-carbon composites. *Advanced composites* that consist of continuous carbon or graphite fibers in a carbon or graphite matrix. These materials have many of the desirable high-temperature properties of conventional carbons and graphites, including high strength, high modulus, high fracture toughness, and low creep. In addition, the high thermal conductivity and low coefficient of thermal expansion, coupled with high strength, produced a material with low sensitivity to thermal shock. Carbon-carbon composites are used for a variety of military and aerospace applications.

carbon fiber. Fiber produced by the pyrolysis of organic precursor fibers, such as rayon, polyacrylonitrile (PAN), and pitch, in an inert environment. The term is often used interchangeably with the term graphite; however, carbon fibers and graphite fibers differ. The basic differences lie in the temperature at which the fibers are made and heat treated, and in the amount of elemental carbon produced. Carbon fibers typically are carbonized in the region of 1315 °C (2400 °F) and assay at 93 to 95% carbon, while graphite fibers are graphitized at 1900 to 2480 °C (3450 to 4500 °F) and assay at more than 99% elemental carbon. See also *pyrolysis*.

carbonium ion. Positively charged organic compound (ion).

carbonization. The conversion of an organic substance into elemental carbon in an inert atmosphere at temperatures ranging from 800 to 1600 °C (1470 to 2910 °F) and higher, but usually at about 1315 °C (2400 °F). Range is influenced by precursor, processing of the individual manufacturer, and properties desired.

casein. A protein material precipitated from skimmed milk by the action of either rennet or dilute acid. Rennet casein finds its main application in the manufacture of plastics. Acid casein is a raw material used in a number of industries, including the manufacture of adhesives.

casein adhesive. An aqueous colloidal dispersion of *casein* that may be prepared with or without heat; that may contain modifiers, inhibitors, and secondary binders to provide specific adhesive properties; and that includes a subclass, usually identified as casein glue, that is based on a dry blend of casein, lime, and sodium salts, mixed with water prepared without heat.

cast. To form a "plastic" object by pouring a fluid monomer-polymer solution into an open mold where it finishes polymerizing. Forming plastic film and sheet by pouring the liquid resin onto a moving belt or by precipitation in a chemical bath.

cast film. A film made by depositing a layer of liquid plastic onto a surface and stabilizing this form by the evaporation of solvent, by fusing after deposition, or by allowing a melt to cool. Cast films are usually made from solutions or dispersions.

catalyst. (1) A substance capable of changing the rate of a reaction without itself undergoing any net change. (2) A substance that markedly speeds up the *cure* of a plastic compound when added in minor quantity, compared to the amounts of primary reactants.

catenary. A measure of the difference in length of reinforcing strands in a specified length of *roving* caused by unequal tension. The tendency of some strands in a taut, horizontal roving to sag more than the others.

caul. In adhesive bonding, a sheet of material employed singly or in pairs in the hot or cold pressing of assemblies being bonded. A caul is used to protect either the faces of the assembly or the press platens, or both, against marring and staining in order to prevent sticking, facilitate press loading, impart a desired surface texture or finish, and provide uniform pressure distribution. A caul may be made of any suitable material such as aluminum, stainless steel, hardboard, fiberboard, or plastic, the length and width dimensions generally being the same as those of the plates of the press where it is used.

caul plates. In fabrication of composites, smooth metal plates, free of surface defects, that are the same size and shape as a composite lay-up, and that contact the lay-up during the curing process in order to transmit normal pressure and temperature, and to provide a smooth surface on the finished laminate.

cavity. The space inside a mold into which a resin or molding compound is poured or injected. The female portion of a mold. The portion of the mold that encloses the molded article (often referred to as the die). Depending on the number of such depressions, molds are designated as single cavity or multiple cavity.

cavity retainer plates. In forming of plastics, plates in a mold that hold the cavities and forces. These plates are at the mold parting line and usually contain the guide pins and bushings. Also called force retainer plates.

cell. (1) In honeycomb core, a cell is a single honeycomb unit, usually in a hexagonal shape. See also the figure accompanying *honeycomb*. (2) A single cavity formed by gaseous displacement in a plastic material. See also *cellular plastic*.

cell size. The diameter of an inscribed circle within a cell of *honeycomb* core.

cellular adhesive. Synonym for foamed adhesive.

cellular plastic. A plastic with greatly decreased density because of the presence of numerous cells or bubbles dispersed throughout its mass. See also *cell* (2), *foamed plastics*, and *syntactic cellular plastics*.

cellulose acetate. An acetic acid ester of cellulose. It is obtained by the action, under rigidly controlled conditions, of acetic acid and acetic anhydride on purified cellulose usually obtained from cotton linters. All three available hydroxyl groups in each glucose unit of the cellulose can be acetylated, but in the material normally used for plastics, it is usual to acetylate fully and then lower the acetyl value (expressed as acetic acid) to 52 to 56% by partial hydrolysis. When compounded with suitable plasticizers, it gives a tough thermoplastic material.

cellulose acetate butyrate. An ester of cellulose made by the action of a mixture of acetic and butyric acids and their anhydrides on purified cellulose. It is used in the manufacture of plastics that are similar in general properties to cellulose acetate but are tougher and have better moisture resistance and dimensional stability.

cellulose ester. A derivative of cellulose in which the free hydroxyl groups attached to the cellulose chain have been replaced wholly in part by acetic groups; for example, nitrate, acetate, or stearate groups. Esterification is effected by the use of a mixture of an acid with its anhydride in the presence of a catalyst, such as sulfuric acid. Mixed esters of cellulose, such as cellulose acetate butyrate, are prepared by the use of mixed acids and mixed anhydrides. Esters and mixed esters, a wide range of which are known, differ in their compatibility with plasticizers, in molding properties, and in physical characteristics. These esters and mixed esters are used in the manufacture of thermoplastic molding compositions.

cellulose nitrate. A nitric acid ester of cellulose manufactured by the action of a mixture of sulfuric acid and nitric acid on cellulose, such as purified cotton linters. The type of cellulose nitrate used for celluloid manufacture usually contains 10.8 to 11.1% nitrogen. The latter figure is the nitrogen content of the dinitrate. Also called nitrocellulose.

cellulose propionate. An ester of cellulose made by the action of propionic acid and its anhydride on purified cellulose. It is used as the basis of a thermoplastic molding material.

cellulosic plastics. Plastics based on cellulose compounds, such as esters (cellulose acetate) and ethers (ethyl cellulose).

center-gated mold. An injection or transfer mold in which the cavity is filled with plastic molding material, through a sprue or gate, directly into the center of the part.

centrifugal casting. A method of forming thermoplastic resins in which the granular resin is placed in a rotatable container, heated to a molten condition by the transfer of heat through the walls of the container, and rotated such that the centrifugal force induced will force the molten resin to conform to the configuration of the interior surface of the container. Used to fabricate large-diameter pipes and similar cylindrical items.

ceramic (adjective). (1) Of or pertaining to ceramics, that is, inorganic nonmetallic as opposed to organic or metallic. (2) Pertaining to products manufactured from inorganic nonmetallic substances which are subjected to a high temperature during manufacture or use. (3) Pertaining to the manufacture or use of such articles or materials, such as ceramic process or ceramic science.

ceramic(s) (noun). Any of a class of inorganic nonmetallic products which are subjected to a high temperature during manufacture or use (high temperature usually means a temperature above a barely visible red, approximately 540 °C, or 1000 °F). Typically, but not exclusively, a ceramic is a metallic oxide, boride, carbide, or nitride, or a mixture of compound of such materials; that is, it includes anions that play important roles in atomic structures and properties. See also *advanced ceramics*, *electronic ceramics*, *refractories*, *structural ceramics*, and *traditional ceramics*.

ceramic color glaze. An opaque colored glass of satin or gloss finish obtained by spraying the clay body with a compound of metallic oxides, chemicals, and clays. It is fired at high temperatures, fusing the glaze to the body, making them inseparable.

ceramic glass decorations. Ceramic glass enamels fused to glassware at temperatures above 245 °C (800 °F) to produce a decoration.

ceramic glass enamels. Predominantly colored, silicate glass fluxes used to decorate glassware. Also referred to as ceramic enamels or glass enamels.

ceramic-matrix composites. *Advanced composites* that consist of ceramic fibers or whiskers in a ceramic matrix. Typical reinforcing materials include carbon fibers and silicon carbide fibers or whiskers. Matrix materials commonly used are glass, glass-ceramics, alumina, and silicon nitride. Ceramic-matrix composites are resistant to both oxidation and wear.

ceramic process. The production of articles or coatings from essentially inorganic, nonmetallic materials, the article or coating being made permanent and suitable for utilitarian and decorative purposes by the action of heat at temperatures sufficient to cause sintering, solid-state reactions, bonding, or conversion partially or wholly to the glassy state.

ceramic whiteware. A fired ware consisting of glazed or unglazed ceramic body which is commonly white and of fine texture. This term designates such product classifications as tile, china, porcelain, semivitreous ware, and earthenware.

C-glass. A glass with a soda-lime-borosilicate composition that is used for its chemical stability in corrosive environments.

chain length. In plastics, the length of the stretched linear macromolecule, most often expressed by the number of identical links.

chain transfer agent. In plastics, a molecule from which an atom, such as hydrogen, may be readily abstracted by a free radical.

chalcogenide. A binary or ternary compound containing a chalcogen (sulfur, selenium, or tellurium) and a more electropositive element. Ternary molybdenum chalcogenides, M_x - Mo_6X_8, where M is a cation and X is a chalcogen, are superconducting materials.

chalking. (1) Dry, chalklike appearance of deposit on the surface of a plastic. (2) The development of loose removable powder at the surface of an organic coating usually caused by weathering.

charge. The weight of plastic material used to load a mold at one time or during one cycle.

Charpy test. An impact test in which a V-notched, keyhole-notched, or U-notched specimen, supported at both ends, is struck behind the notch by a striker mounted at the lower end of a bar that can swing as a pendulum. The energy that is absorbed in fracture is calculated from the height to which the striker would have risen had there been no specimen and the height to which it actually rises after fracture of the specimen. Contrast with *Izod test*.

charring. The heating of a reinforced plastic or composite in air to reduce the polymer matrix to ash, allowing the fiber content to be determined by weight.

chase. In plastic part making, an enclosure of any shape, used to shrink-fit parts of a mold cavity in place, prevent spreading or distortion in hobbing, or enclose an assembly of two or more parts in a split cavity block.

chelate. (1) Five- or six-membered ring formation based on intramolecular attraction of H, O, or N atoms. (2) A molecular structure in which a heterocyclic ring can be formed by the unshared electrons of neighboring atoms. (3) A *coordination compound* in which a heterocyclic ring is formed by a metal bound to two atoms of the associated *ligand*. See also *complexation*.

chemical blowing agent. In processing of plastics, an agent that readily decomposes to produce a gas.

chemically strengthened. Glass that has been ion-exchanged to produce a compressive stress layer at the treated surface.

chemical vapor deposited (CVD) carbon. Carbon deposited on a substrate by pyrolysis of a hydrocarbon, such as methane.

chill mark. A wrinkled surface condition on glassware resulting from uneven cooling in the forming process.

chill roll. A cored roll, usually temperature controlled by circulating water, that cools the web before winding. For chill roll plastic (cast) film, the surface of the roll is highly polished. In extrusion coating, either a polished or matte surface may be used, depending on the surface desired on the finished coating.

chill roll extrusion. The extruded plastic film is cooled while being drawn around two or more highly polished chill rolls cored for water cooling for exact temperature control. Also called cast film extrusion.

chin. (1) Area along an edge or corner where the ceramic or glass material has broken off. (2) An imperfection due to breakage of a small fragment out of an otherwise regular ceramic or glass surface.

china. A glazed or unglazed vitreous ceramic whiteware used for nontechnical purposes. This term designates such products as dinnerware, sanitary ware, and artware when they are vitreous.

chips. Minor damage to a pultruded surface of a composite material that removes material but does not cause a crack or craze.

chlorinated hydrocarbon. An organic compound having chlorine atoms in its chemical structure. Trichloroethylene, methyl chloroform, and methylene chloride are chlorinated hydrocarbon solvents; polyvinyl chloride is a plastic.

chlorofluorohydrocarbon plastics. Plastics based on polymers made with monomers composed of chlorine, fluorine, hydrogen, and carbon only.

chord modulus. The slope of the chord drawn between any two points on a *stress-strain curve*. See also *modulus of elasticity*.

chromia. Formula Cr_2O_3, a compound having many properties and derivatives similar to those of *alumina*. Useful either pure or impure (e.g., as chrome ore) in both basic and high-alumina refractories.

CIL flow test. A method of determining the rheology or flow properties of thermoplastic resins. In this test, the amount of the molten resin that is forced through a specified size orifice per unit of time when a specified variable force is applied gives a relative indication of the flow properties of various resins.

circuit. (1) In filament winding of composites, one complete traverse of a winding band from one arbitrary point along the winding path to another point on a plane through the starting point and perpendicular to the axis. (2) The interconnection of a number of components in one or more closed paths to perform a desired electrical or electronic function.

circuit board. In electronics, a sheet of insulating material laminated to foil that is etched to produce a circuit pattern on one or both sides. Also called printed circuit board or printed wiring board.

circumferential ("circ") winding. In filament-wound reinforced plastics, a winding with the filaments essentially perpendicular to the axis (90° or level winding).

CIS stereoisomer. In engineering plastics, a stereoisomer in which side chains or side atoms are arranged on the same side of a double bond present in a chain of atoms.

clamping pressure. In injection molding and transfer molding of plastics, the pressure that is applied to the mold to keep it closed in opposition to the fluid pressure of the compressed molding material, within the mold cavity (cavities) and the runner system. In blow molding, the pressure exerted on the two mold halves (by the locking mechanism of the blowing table) to keep the mold closed during formation of the container. Normally, this pressure or force is expressed in tons.

clay. A natural mineral aggregate, consisting essentially of hydrous aluminum silicates. It is plastic when sufficiently wetted, rigid when dried en masse, and vitreous when fired to a sufficiently high temperature.

cleavage. (1) Fracture of a crystal by crack propagation across a crystallographic plane of low index. (2) The tendency to cleave or split along definite crystallographic planes. (3) Breakage of covalent bonds.

cleavage crack. Damage produced by the translation of a hard, sharp object across a glass surface. This fracture system typically includes a plastically deformed groove on the damaged surface, together with median and lateral cracks emanating from this groove.

cleavage strength. In testing of adhesive bonded assemblies, the tensile load in terms of kgf/mm (lbf/in.) of width required to cause the separation of a test specimen 25 mm (1 in.) in length.

closed assembly time. The time interval between completion of assembly of the parts for adhesive bonding and the application of pressure or heat, or both, to the assembly.

closed-cell cellular plastics. Cellular plastics in which almost all the cells are non-interconnecting.

closure. In fabricating of reinforced plastics, the complete coverage of a mandrel with one layer (two plies) of fiber. When the last tape circuit that completes mandrel coverage lays down adjacent to the first without gaps or overlaps, the wind pattern is said to have closed.

cloth. See *woven fabric* and *nonwoven fabric*.

coagulation. Precipitation of a polymer dispersed in a latex.

co-curing. The act of curing a composite laminate and simultaneously bonding it to some other prepared surface, or curing together an inner and outer tube of similar or dissimilar fiber-resin combinations after each has been wound or wrapped separately. See also *secondary bonding*.

coefficient of expansion. A measure of the change in length or volume of an object, specifically, a change measured by the increase in length or volume of an object per unit length or volume.

coefficient of friction. The dimensionless ratio of the friction force (F) between two bodies to the normal force (N) pressing these bodies together: μ (or f) = (F/N).

coefficient of thermal expansion. (1) Change in unit of length (or volume) accompanying a unit change of temperature, at a specified temperature. (2) The linear or volume expansion of a given material per degree rise of temperature, expressed at an arbitrary base temperature or as a more complicated equation applicable to a wide range of temperatures.

cohesion. (1) The state in which the particles of a single substance are held together by primary or secondary valence forces. As used in the adhesive field, the state in which the particles of the adhesive (or adherend) are held together. (2) Force of attraction between the molecules (or atoms) within a single phase. Contrast with *adhesion*.

cohesive blocking. The blocking of two similar, potentially adhesive faces.

cohesive failure. Failure of an adhesive joint occurring primarily in an adhesive layer.

cohesive strength. (1) The hypothetical stress causing tensile fracture without plastic deformation. (2) The stress corresponding to the forces between atoms. (3) Intrinsic strength of an adhesive.

coke. Carbonaceous residue resulting from the pyrolysis of pitch.

cold drawing. Technique for using standard metalworking equipment and systems for forming thermoplastic sheet at room temperature.

cold flow. The distortion that takes place in polymeric materials under continuous load at temperatures within the working range of the material without a phase or chemical change.

cold molding. A procedure in which a plastic is shaped at room temperature and subsequently cured by baking.

cold parison blow molding. A plastic forming technique in which parisons are extruded or injection molded separately and then stored for subsequent transportation to the blow molding machine for blowing. See also *blow molding*.

cold pressing. A bonding operation in which a plastic assembly is subjected to pressure without the application of heat.

cold-press molding. A plastic molding process in which inexpensive plastic male and female molds are used with room temperature curing resins to produce accurate parts. Limited runs are possible.

cold-runner molding. In plastic part making, a mold in which the sprue-and-runner system (the manifold section) is insulated from the rest of the mold and temperature-controlled to keep the plastic in the manifold fluid. This mold design eliminates scrap loss from sprues and runners.

cold-setting adhesive. An adhesive that sets at temperatures below 20 °C (68 °F). See also *hot-setting adhesive*, *intermediate-temperature-setting adhesive*, and *room-temperature-setting adhesive*.

cold slug. The first plastic material to enter an injection mold; so called because in passing through a sprue orifice it is cooled below the effective molding temperature.

cold-slug well. In plastic part making, the space provided directly opposite the sprue opening in an injection mold to trap the cold slug.

cold stretch. A pulling operation with little or no heat, usually on extruded filaments, to increase tensile properties of composite materials.

collapse. Inadvertent densification of cellular plastic material during manufacture resulting from the breakdown of cell structure.

colligative properties. Properties of plastics based on the number of molecules present. Most important are certain solution properties extensively used in molecular weight characterization.

collimated. Rendered parallel.

collimated roving. *Roving* for reinforced plastics that has been made using a special process (usually parallel wound), such that the strands are more parallel than in standard roving.

colloidal. A state of suspension in a liquid medium in which extremely small particles are suspended and dispersed but not dissolved.

color concentrate. A measured amount of dye or pigment incorporated into a predetermined amount of plastic. The pigmented or colored plastic is then mixed into larger quantities of plastic material to be used for molding. This mixture is added to the bulk of plastic in measured quantity in order to produce a precise, predetermined color of finished articles to be molded.

combination mold. See *family mold*.

combing. Lining up of reinforcing fibers.

compaction. (1) The act of forcing particulate or granular material together (consolidation) under pressure or impact to yield a relatively dense mass or formed object. Usually followed by drying, curing, or firing in refractory or other ceramic or powder metallurgy processing. (2) In ceramics or powder metallurgy, the preparation of a compact or object produced by the compression of a powder, generally while confined in a die, with or without the inclusion of lubricants, binders, etc., and with or without the concurrent applications of heat. (3) In reinforced plastics and composites, the application of a temporary vacuum bag and vacuum to remove trapped air and compact the lay-up.

compatibility. The ability of two or more substances combined with one another to form a homogeneous composition having useful plastic properties; for example, the suitability of a sizing or finish for use with certain general resin types. Nonreactivity or negligible reactivity between materials in contact.

complexation. The formation of complex chemical species by the coordination of groups of atoms termed ligands to a central ion, commonly a metal ion. Generally, the ligand coordinates by providing a pair of electrons that forms an ionic or covalent bond to the central ion. See also *chelate*, *coordination compound*, and *ligand*.

complex modulus. The ratio of stress to strain in which each is a vector that may be represented by a complex number. May be measured in tension or flexure, compression, or shear.

complex shear modulus. The vectorial sum of the shear modulus and the loss modulus.

complex Young's modulus. The vectorial sum of Young's modulus and the loss modulus.

compliance. Tensile compliance is the reciprocal of Young's modulus. Shear compliance is the reciprocal of shear modulus. The term is also used in the evaluation of stiffness and deflection.

composite material. A combination of two or more materials (reinforcing elements, fillers, and composite matrix binder), differing in form or composition on a macroscale. The constituents retain their identities, that is, they do not dissolve or merge completely into one another although they act in concert. Normally, the components can be physically identified and exhibit an interface between one another. See also *carbon-carbon composites*, *ceramic-matrix composites*, *metal-matrix composites*, and *resin-matrix composites*.

compound. (1) In chemistry, a substance of relatively fixed composition and properties, whose ultimate structural unit (molecule or repeat unit) is comprised of atoms of two or more elements. The number of atoms of each kind in this ultimate unit is determined by natural laws and is part of the identification of the compound. (2) In reinforced plastics and composites, the intimate admixture of a polymer with other ingredients, such as fillers, softeners, plasticizers, reinforcements, catalysts, pigments, or dyes. A thermoset compound usually contains all the ingredients necessary for the finished product, while a thermoplastic compound may require subsequent addition of pigments, blowing agents, and so forth.

compression molding. A technique of thermoset molding in which the plastic molding compound (generally preheated) is placed in the heated open mold cavity, the mold is closed under pressure (usually in a hydraulic press), causing the material to flow and completely fill the cavity, and then pressure is held until the material has cured.

compression ratio. In an extruder screw, the ratio of the volume available in the first flight at the hopper to the volume at the last flight, at the end of the screw.

compressive modulus. The ratio of compressive stress to compressive strain below the proportional limit. Theoretically equal to Young's modulus determined from tensile experiments.

compressive strength. The maximum compressive stress that a material is capable of developing, based on original area of cross section. If a material fails in compression by a shattering fracture, the compressive strength has a very definite value. If a material does not fail in compression by a shattering fracture, the value obtained for compressive strength is an arbitrary value depending upon the degree of distortion that is regarded as indicating complete failure of the material.

compressive stress. A stress that causes an elastic body to deform (shorten) in the direction of the applied load. Contrast with *tensile stress*.

condensation. A chemical reaction in which two or more molecules combine, with the resulting separation of water or some other simple substance; the process is called polycondensation if a polymer is formed. See also *polymerization*.

condensation polymerization. In plastics technology, a stepwise chemical reaction in which two or more molecules combine, often but not necessarily accompanied by the separation of water or some other simple substance. If a polymer is formed, the process is called polycondensation. See also *polymerization*.

condensation ratio. A resin formed by polycondensation, for example, the alkyd, phenolaldehyde, and urea-formaldehyde resins.

conditioning. Subjecting a material to a prescribed environmental and/or stress history before testing.

conductivity, electrical. The reciprocal of volume resistivity. The electrical or thermal conductance of a unit cube of any material (conductivity per unit volume).

conductivity, thermal. The time rate of heat flow through unit thickness of an infinite slab of a homogeneous material in a direction perpendicular to the surface, induced by unit temperature difference. Recommended SI units: W/m · K.

configurations. In plastics technology, related chemical structures produced by the cleavage and reforming of covalent bonds.

conformations. Different shapes of polymer molecules resulting from rotation about single covalent bonds in the polymer chain.

consistency. In adhesives, that property of a liquid adhesive by virtue of which it tends to resist deformation. Consistency is not a fundamental property but is composed of viscosity, plasticity, and other phenomena. See also *viscosity* and *viscosity coefficient*.

consolidation. In metal-matrix or thermoplastic composites, a processing step in which fiber and matrix are compressed by one of several methods to reduce voids and achieve desired density.

constituent. (1) One of the ingredients that make up a chemical system. (2) In composites, the principal constituents are the fibers and the matrix.

contact adhesive. An adhesive that is apparently dry to the touch and that will adhere to itself simultaneously upon contact. An adhesive that, when applied to both adherends and allowed to dry, develops a bond when the adherends are brought together without sustained pressure.

contact bond adhesive. Synonym for contact adhesive.

contact molding. A process for molding reinforced plastics in which reinforcement and resin are placed on a mold. Cure is either at room temperature using a catalyst-promoter system or by heating in an oven, without additional pressure. Also referred to as hand lay-up.

contact pressure resins. Liquid resins that thicken or polymerize upon heating, and, when used for bonding laminates, require little or no pressure.

contact stress. The tensile stress component imposed at a glass surface immediately surrounding the contact area between the glass surface and an object generating a locally applied force.

contaminant. An impurity or foreign substance present in a material or environment that affects one or more properties of a material.

continuous filament yarn. Yarn formed by twisting two or more continuous filaments into a single, continuous strand.

continuous furnace. A glass-making furnace in which the level of glass remains substantially constant because the feeding of batch continuously replaces the glass withdrawn.

cooling channels. In plastic part making, channels or passageways located within the body of a mold through which a cooling medium can be circulated to control temperature on the mold surface. May also be used for heating a mold by circulating steam, hot oil, or other heated fluid through channels, as in the molding of thermosetting and some thermoplastic materials.

cooling fixture. In plastic part making, block of metal or wood shaped to hold a molded part to maintain the proper shape or dimensional accuracy of the part after it is removed from the mold until it is cooled enough to retain its shape without further appreciable distortion. Also called a shrink fixture.

coordination catalysis. Ziegler-type of catalysis for processing plastics. See also *Ziegler-Natta catalysts*.

coordination compound. A compound with a central atom or ion bound to a group of ions or molecules surrounding it. Also called coordination complex. See also *chelate*, *complexation*, and *ligand*.

copolymer. A long-chain molecule formed by the reaction of two or more dissimilar monomers. See also *polymer*.

copolymerization. See *polymerization*.

core. (1) In plastic part making, a channel in a mold for circulation of heat transfer media. (2) The part of a complex mold that molds undercut parts. Cores are usually withdrawn before the main sections of the mold are opened.

core crush. A collapse, distortion, or compression of the core of a *sandwich construction*.

core depression. A localized indentation or gouge in the core of a *sandwich construction*.

cored mold. In plastic part making, a mold incorporating passages for electrical heating elements, steam, or water.

core pin. In plastic part making, a pin used to mold a hole.

core pin plate. In plastic part making, a plate holding core pins.

coring. In plastic part making, the removal of excess material from the cross section of a molded part to attain a more uniform wall thickness.

corrosion resistance. The ability of a material to withstand contact with ambient natural factors or those of a particular, artificially created atmosphere, without degradation or change in properties. For metals, this could be pitting or rusting; for organic materials, it could be crazing.

corundum. (1) A naturally occurring alumina usually of relatively high purity but containing associated minerals, such as diaspore and various silicates. Commonly coarsely crystalline, but sometimes microcrystalline. (2) Native alumina occurring as rhombohedral crystals and also in masses and variously colored grains. Corundum and its artificial counterparts are abrasives especially suited to the grinding of metals.

cottoning. The formation of weblike filaments of adhesive between the applicator and substrate surface.

count. For fabric used in composite fabrication, the number of warp and filling yarns per inch in woven cloth. For yarn, size based on relation of length and weight.

coupling agent. In fabricating composites, any chemical designed to react with both the reinforcement and matrix phases of a composite material to form or promote a stronger bond at the interface.

crack. (1) A fracture type discontinuity characterized by a sharp tip and high ratio of length and width to opening displacement. (2) A line of fracture without complete separation.

crack growth. Rate of propagation of a crack through a material due to a static or dynamic applied load.

cratering. Depressions on coated plastic surfaces caused by excess lubricant. Cratering results when paint is too thin and later ruptures, leaving pinholes and other voids. Use of less thinner in the coating can reduce or eliminate cratering, as can less lubricant on the part.

crazing. Region of ultrafine cracks, which may extend in a network on or under the surface of a resin or plastic material. May appear as a white band. Often found in a filament-wound pressure vessel or bottle. In many plastics, craze growth precedes crack growth because crazes are load bearing.

creel. In composites fabrication, a spool, along with its supporting structure, that holds the required number of roving balls or supply packages in a desired position for unwinding for the next processing step, that is, weaving, braiding, or filament winding.

creep. Time-dependent strain occurring under stress. The creep strain occurring at a diminishing rate is called primary creep; that occurring at a minimum and almost constant rate, secondary creep; and that occurring at an accelerating rate, tertiary creep.

creep rate. The slope of the creep-time curve at a given time. Deflection with time under a given static load.

creep recovery. The time-dependent decrease in strain in a solid, following the removal of force.

creep-rupture strength. The stress that causes fracture in a creep test at a given time, in a specified constant environment. This is sometimes referred to as the stress-rupture strength. In glass technology, this is termed the static fatigue strength.

crescent crack. Damage having the appearance of a crescent, produced in a glass surface by the frictive translation of a hard, blunt object across the glass surface. The crescent shape is concave toward the direction of translation on the damaged surface.

crimp. The waviness of a fiber or fabric used in composites fabrication, which determines the capacity of fibers to cohere under light pressure. Measured by the number of crimps or waves per unit length.

critical damping. In dynamic mechanical measurement of plastics, that damping required for the borderline condition between oscillatory and nonoscillatory behavior.

critical length. In composites fabrication, the minimum fiber length required for shear loading to its ultimate strength by the matrix.

critical longitudinal stress. The longitudinal stress necessary to cause internal slippage and separation of a spun yarn in a fiber-reinforced plastic. The stress necessary to overcome the interfiber friction developed as a result of twist.

critical strain. In mechanical testing, the strain at the *yield point*.

cross laminate. In composites fabrication, a laminate in which some of the layers of material are oriented approximately at right angles to the remaining layers with respect to the grain, or strongest direction in tension. A bidirectional laminate. See also *parallel laminate*.

cross link. Intermolecular bonds produced between long-chain molecules in a material to increase molecular size and weight by chemical or electron bombardment, resulting in a change in physical properties in the material, usually improved properties.

cross linking. With thermosetting and certain thermoplastic polymers, the setting up of chemical links between the molecular chains. When extensive, as in most thermosetting resins, cross linking makes an infusible supermolecule of all the chains. In rubbers, the cross linking is just enough to join all molecules into a network.

cross-linking, degree of. The fraction of cross-linked polymeric units in the entire system.

cross-ply laminate. A *laminate* with plies usually oriented at 0° and 90° only.

crosswise direction. In testing of plastics, crosswise refers to the cutting of specimens and to the application of load. For rods and tubes, crosswise is any direction perpendicular to the long axis. For other shapes or materials that are stronger in one direction than in another, crosswise is the direction that is weaker. For materials that are equally strong in both directions, crosswise is an arbitrarily designated direction at right angles to the lengthwise direction.

crowfoot satin. In this type of composite fabric weave, there is a three-by-one interlacing; that is, a filling thread floats over three warp threads and then under one. This type of fabric looks different on one side than the other. Fabrics with this weave are more pliable than either the *plain* or *basket weave* and, consequently, are easier to form around curves.

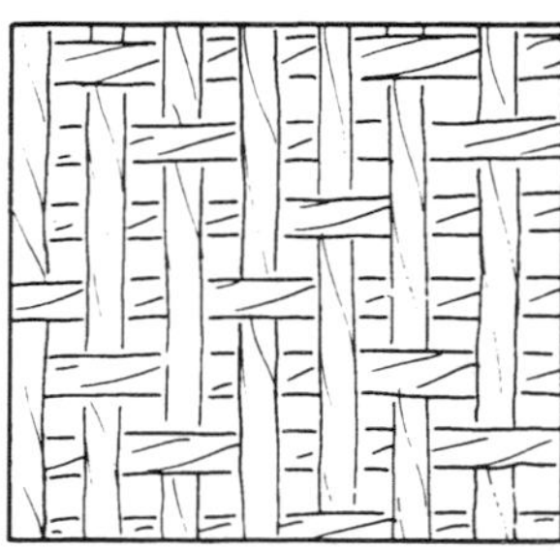

Crowfoot satin

crystalline plastic. A material having an internal structure in which the atoms are arranged in an orderly three-dimensional configuration. More accurately referred to as a semicrystalline plastic because only a portion of the molecules are in crystalline form.

crystallinity. A regular arrangement of the atoms of a solid in space. In most polymers, including cellulose, this state is usually imperfectly achieved. The crystalline regions (ordered regions) are submicroscopic volumes in which there is some degree of regularity in the arrangement of the component molecules. In these regions there is sufficient geometric order to obtain definite x-ray diffraction patterns.

C-stage. In processing of plastics, the final stage in the reaction of certain thermosetting resins in which the material is practically insoluble and infusible. Also called resite. The resin in a fully cured thermoset molding is in this stage. See also *A-stage* and *B-stage*.

cubic boron nitride (CBN). An extremely hard ceramic material synthesized by high-pressure sintering of hexagonal *boron nitride*. CBN is used in the machining and grinding of ferrous materials such as tool steels, cast irons, hardfacing alloys, and surface-hardened steels.

cull. Plastic material in a transfer chamber after the mold has been filled. Unless there is a slight excess in the charge, the operator cannot be sure the cavity is filled. Charge is generally regulated to control thickness of the cull.

cure. (1) To change the physical properties of a material (usually from a liquid to a solid) by chemical reaction or by the action of heat and catalysts, alone or in combination, with or without pressure. (2) To irreversibly change, usually at elevated temperatures, the properties of a thermosetting resin by chemical reaction, that is, by condensation, ring closure, or addition. Cure may be accomplished by the addition of curing (cross linking) agents, with or without heat and pressure.

cure cycle. The time/temperature/pressure cycle used to cure a thermosetting resin system or prepreg.

cure monitoring, electrical. Use of electrical techniques to detect changes in the electrical properties and/or mobility of the resin molecules during cure. A measuring of resin cure.

cure stress. A residual internal stress produced during the curing cycle of a composite structure. Normally, these stresses originate when different components of a wet lay-up have different thermal coefficients of expansion.

curing agent. A catalytic or reactive agent that causes cross linking of a plastic. Also called a hardener.

curing temperature. The temperature to which an adhesive or an assembly is subjected to cure the adhesive. The temperature attained by the adhesive in the process of curing (adhesive curing temperature) may differ from the temperature of the atmosphere surrounding the assembly (assembly curing temperature). See also *drying temperature* and *setting temperature*.

curing time. In plastic part making, the time between the instant of cessation of relative movement between the moving parts of a mold and the instant that pressure is released. Also called molding time.

curtain coating. A method of coating that may be employed with low-viscosity resins or solutions, suspensions, or emulsions of resins in which the substrate to be coated is passed through and perpendicular to a freely falling liquid curtain (or waterfall). The flow rate of the falling liquid and the linear speed of the substrate passing through the curtain are coordinated in accordance with the thickness of coating desired.

cut layers. With laminated plastics, a condition of the surface of machined or ground rods and tubes and of sanded sheets in which cut edges of the surface layer or lower laminations are revealed.

cut-off. The line where the two halves of a mold come together. Also called flash groove and pinch-off.

CVD carbon. See *chemical vapor deposited (CVD) carbon*.

cyanate resins. Thermosetting resins that are derived from bisphenols or polyphenols, and are available as monomers, oligomers, blends, and solutions. Also known as cyanate esters, cyanic esters, and triazine esters.

cyanoacrylate. A thermoplastic monomer adhesive characterized by excellent polymerizing and bonding strength.

cycle. One complete operation of a plastic molding press from closing time, for example, in one cycle to closing time in the next cycle.

cyclic hydrocarbons. Cyclic or ring compounds; benzene (C_6H_6) is a classic example.

D

dam. In composites fabrication, a boundary support or ridge used to prevent excessive edge bleeding or resin runout of a laminate and to prevent crowning of the bag during cure.

damage tolerance. (1) A design measure of crack growth rate. Cracks in damage-tolerant designed structures are not permitted to grow to critical size during expected service life. (2) The ability of a part component, such as an aerospace engine, to resist failure due to the presence of flaws, cracks, or other damage for a specified period of usage. The damage tolerance approach is used extensively in the aerospace industry.

damping. The loss in energy, as dissipated heat, that results when a material or material system is subjected to an oscillatory load or displacement.

daylight. The distance, in the open position, between the moving and the fixed tables or the platens of a hydraulic press. In the case of a multiplaten press, daylight is the distance between adjacent platens. Daylight provides space for removal of the molded/formed part from the mold/die.

dead-burned. Term applied to ceramic materials that have been fired to a temperature sufficiently high to render them relatively resistant to moisture and contraction.

debond. In composites, a deliberate separation of a bonded joint or interface, usually for repair or rework purposes. Also, an unbonded or nonadhered region; a separation at the fiber-matrix interface due to strain incompatibility. In the United Kingdom, the term often refers to accidental damage. See also *disbond* and *delamination*.

debossed. An indented or depressed design or lettering that is molded into a plastic article so that it is below the main outside surface of that article.

debulking. Compacting of a thick composite laminate under moderate heat and pressure and/or vacuum to remove most of the air, to ensure seating on the tool, and to prevent wrinkles.

deckle rod. In forming plastics, a small rod, or similar device, inserted at each end of the extrusion coating die that is used to adjust the length of the die opening.

deep-draw mold. A mold having a core that is appreciably longer than the wall thickness.

deflashing. A finishing technique used to remove the flash (excess, unwanted material) on a plastic molding.

deflection temperature under load (DTUL). In testing of plastics, the temperature at which a simple cantilever beam deflects a given amount under load. Formerly called heat distortion temperature.

deformation point. The temperature observed during the measurement of expansivity of glass by the interferometer method at which viscous flow exactly counteracts thermal expansion. The deformation point generally corresponds to a viscosity in the range of 1010 to 1011 Pa · s.

deformation under load. The dimensional change of a material under load for a specified time following the instantaneous elastic deformation caused by the initial application of the load. See also *cold flow* and *creep*.

degassing. Opening and closing of a plastics mold to allow gases to escape during molding.

degree of polymerization. Number of structural units, or mers, in the average polymer molecule in a sample measure of molecular weight.

degree of saturation. The ratio of the weight of water vapor associated with a pound of dry air to the weight of water vapor associated with a pound of dry air saturated at the same temperature.

delamination. The separation of layers in a laminate because of failure of the adhesive, either in the adhesive itself or at the interface between the adhesive and the adherend.

deliquescence. The absorption of atmospheric water vapor by a crystalline solid until the crystal eventually dissolves into a saturated solution.

denier. A yarn and filament numbering system in which the yarn number is numerically equal to the weight in grams of 9000 meters (1 denier = 1.111×10^{-7} kg/m). Used for continuous filaments. The lower the denier, the finer the yarn.

densification process. Consolidation of a loose or bulky material.

deposition. The process of applying a material to a base by means of vacuum, electrical, chemical, screening, or vapor methods, often with the assistance of a temperature and pressure container.

depth dose. The variation of absorbed dose with distance from the incident surface of a material exposed to radiation. Depth-dose profiles give information about the distribution of absorbed energy in a specific material.

desiccant. A substance that can be used to dry materials because of its affinity for water.

desizing. The process of eliminating sizing, which is generally starch, from *gray* (also *greige*) *goods* before applying special finishes or bleaches (for yarn such as glass or cotton). Also, removing lubricant *size* following the weaving of a cloth.

desorption. A process in which an absorbed material is released from another material. See also *absorption, adsorption,* and *chemisorption*.

destaticization. A treatment of plastic materials that minimizes the effects of static electricity on the surface either by treating the surface with specific materials or by incorporating materials in the molding compound. Minimization of surface static electricity prevents dust and dirt from being attracted to and/or clinging to the surface of the article.

devitrification. (1) Crystallization in glass. (2) The formation of crystals (seeds) in a glass melt, usually occurring when the melt is too cold. These crystals can appear as defects in glass fibers.

dewpoint temperature. The temperature at which condensation of water vapor in a space begins for a given state of humidity and pressure as the vapor temperature is reduced; the temperature corresponding to saturation (100% relative humidity) for a given absolute humidity at constant pressure.

D-glass. A high boron content glass made especially for laminates requiring a precisely controlled dielectric constant.

diadic polyamide. Polyamide produced by condensation of diamine and a dicarboxylic acid.

diamond. A highly transparent mineral composed entirely of carbon (allotropic form of carbon) having a cubic structure. The hardest material known, it is used as a gemstone and as an abrasive in cutting and grinding applications. Natural diamonds are produced deep within the earth's crust at extremely high pressures and temperatures. Synthetic diamonds are synthesized by subjecting carbon, in the form of graphite, to high temperatures and pressures using large special-purpose presses.

diamond film. A carbon-composed film, usually deposited by chemical vapor deposition or related process, that has the following three characteristics: (1) a crystalline morphology that can be visually discerned by scanning electron or optical microscopy, (2) a single-phase crystalline structure identifiable by x-ray and/or electron diffraction, and (3) a Raman spectrum typical for crystalline diamond. See also *diamondlike film*.

diamondlike film. A hard, noncrystalline carbon film, usually grown by chemical vapor deposition or related techniques, that contains predominantly sp^2 carbon-carbon bonds. See also *diamond film*.

diaphragm gate. A gate used in molding annular or tubular plastic articles. The gate forms a solid web across the opening of the part. Also called disc gate.

die adapter. In forming plastics, the part of an extrusion die that holds the die block.

die cutting. Cutting shapes from metal sheet stock by sharply striking it with a shaped knife-edge known as a steel rule die. Clicking and dinking are other names for die cutting of this kind.

dielectric. A nonconductor of electricity. The ability of a material to resist the flow of an electrical current.

dielectric curing. The curing of a synthetic thermosetting resin by the passage of an electric charge (produced from a high-frequency generator) through the resin.

dielectric heating. The heating of plastic materials by dielectric loss in a high-frequency electrostatic field.

dielectric loss. A loss of energy evidenced by the rise in heat of a dielectric placed in an alternating electric field. It is usually observed as a frequency-dependent conductivity.

dielectric monitoring. Monitoring the cure of thermosets by tracking the changes in their electrical properties during material processing.

dielectric strength. (1) The maximum voltage that a dielectric can withstand, under specified conditions, without resulting in a voltage breakdown (usually expressed as volts per unit dimension). (2) A measure of the ability of a dielectric (insulator) to withstand a potential difference across it without electric discharge.

dielectrometry. Use of electrical techniques to measure the changes in loss factor (dissipation) and in capacitance during cure of the resin in a laminate. Also called dielectric spectroscopy.

die-parting line. A lengthwise flash or depression on the surface of a pultruded plastic part. The line occurs where separate pieces of the die join together to form the cavity.

die swell ratio. In forming plastics, the ratio of the outer parison diameter (or parison thickness) to the outer diameter of the die (or die gap). Die swell ratio is influenced by head construction, land length, extrusion speed, and temperature. See also *parison* and *parison swell*.

differential scanning calorimetry (DSC). Measurement of energy absorbed (endotherm) or produced (exotherm). May be applied to melting, crystallization, resin curing, loss of solvents, and other processes involving an energy change. May also be applied to processes involving a change in heat capacity, such as the glass transition.

differential thermal analysis (DTA). An experimental analysis technique in which a specimen and a control are heated simultaneously and the difference in their temperatures is monitored. The difference in temperatures provides information on relative heat capacities, presence of solvents, changes in structure (that is, phase changes, such as melting of one component in a resin system), and chemical reactions. See also *differential scanning calorimetry*.

diffusion. (1) Spreading of a constituent in a gas, liquid, or solid, tending to make the composition of all parts uniform. (2) The spontaneous movement of atoms or molecules to new sites within a material. (3) The movement of a material, such as a gas or liquid, in the body of a plastic. If the gas or liquid is absorbed on one side of a piece of plastic and given off on the other side, the phenomenon is called permeability. Diffusion and permeability are not due to holes or pores in the plastic but are caused and controlled by chemical mechanisms.

dilatant. A reversible increase in viscosity with increasing shear stress. Compare with *pseudoplastic behavior*, *rheopectic material*, and *thixotropy*.

diluent. In an *organosol*, a liquid component that has little or no solvating action on the resin. Its purpose is to modify the action of the dispersant. The term diluent is commonly used in place of the term plasticizer.

dimensional stability. (1) A measure of dimensional change caused by such factors as temperature, humidity, chemical treatment, age, or stress (usually expressed as Δ units per unit). (2) Ability of a plastic part to retain the precise shape in which it was molded, cast, or otherwise fabricated.

dimer. A substance (comprising molecules) formed from two molecules of a *monomer*.

dimerization. The formation of a *dimer*.

dip casting. In forming plastics, the process of submerging a hot mold into a resin. After cooling, the product is removed from the mold.

dip coating. Applying a plastic coating by dipping the article to be coated into a tank of melted resin or plastisol, then chilling the adhering metal.

diphenyl oxide resins. Thermosetting resins based on diphenyl oxide and possessing excellent handling properties and heat resistance.

disbond. In adhesive bonded structures, an area within the bonded interface between two adherends in which an adhesion failure or separation has occurred. Also, colloquially, an area of separation between two laminae in the finished laminate (in this case, the term delamination is normally preferred). See also *debond*.

dished. Showing a symmetrical distortion of a flat or curved section of a plastic object so that, as normally viewed, it appears concave, or more concave than intended.

displacement angle. In filament winding, the advancement distance of the winding ribbon on the equator after one complete circuit.

disproportionation. In the processing of plastics, termination by chain transfer between macroradicals to produce a saturated and an unsaturated polymer molecule.

distortion. In fabric, the displacement of fill fiber from the 90° angle (right angle) relative to the warp fiber. In a laminate, the displacement of the fibers (especially at radii), relative to their idealized location, due to motion during lay-up and cure.

doctor blade or bar. In forming plastics, a straight piece of material used to spread resin, as in application of a thin film of resin for use in hot-melt prepregging or for use as an adhesive film. Also called paste metering blade.

doctor roll. In applying adhesives, a roller mechanism that is revolving at a different surface speed, or in an opposite direction, resulting in a wiping action for regulating the adhesive supplied to the spreader roll.

doily. In filament winding of composites, the planar reinforcement applied to a local area between windings to provide extra strength in an area where a cutout is to be made, for example, port openings. Usually placed at the knuckle joints of cylinder to dome.

domain. A morphological term used in noncrystalline systems, such as block copolymers, in which the chemically different sections of the chain separate, generating two or more amorphous phases.

dome. In filament winding, the portion of a cylindrical container that forms the spherical or elliptical shell ends of the container.

domed. Showing a symmetrical distortion of a flat or curved section of a plastic object so that, as normally viewed, it appears convex, or more convex than intended.

dopant. (1) An impurity introduced under highly controlled conditions in very small but accurately known quantities into a semiconductor material, such as silicon. Dopants modify the electrical characteristics of the silicon by creating *p* or *n* regions and hence *pn* junctions. (2) A material added to a polymer to change a physical property.

dosimeter. A device for measuring radiation-induced signals that can be related to absorbed dose (or energy deposited) by radiation in materials and is calibrated in terms of the appropriate quantities and units. Also called dose meter.

doubler. In filament winding of composites, a local area with extra reinforcement, wound integrally with the part, or wound separately and fastened to the part. See also *tabs*.

double-shot molding. In forming plastics, a means of producing two-color parts and/or two different thermoplastic materials by successive molding operations.

double spread. The application of adhesive to both adherends of a joint.

draft. An angle or taper on the vertical surfaces of a mold that facilitates removal of the parts from a mold or die cavity.

draft angle. The angle of a taper on a mandrel or mold that facilitates removal of the finished plastic part.

drape. In fabricating composites, the ability of a fabric or prepreg to conform to a contoured surface.

drape forming. Method of forming thermoplastic sheet in which the sheet is clamped into a movable frame, heated, and draped over high points of a male mold. Vacuum is then pulled to complete the forming operation. Also known as basic male mold forming.

draw-down ratio. In forming plastics, the ratio of the thickness of the die opening to the final thickness of the product.

drawing. The process of stretching a thermoplastic to reduce its cross-sectional area, thus creating a more orderly arrangement of polymer chains with respect to each other.

drawn fiber. Fiber for reinforced plastics with a certain amount of orientation imparted by the drawing process by which it is formed.

dry. To change the physical state of an adhesive on an adherend by the loss of solvent constituents by evaporation or absorption, or both. See also *cure* and *set*.

dry blend. Refers to a plastic molding compound containing all necessary ingredients mixed in a way that produces a dry, free flowing, particulate material. This term is commonly used in connection with polyvinyl chloride molding compounds.

dry bond adhesive. See *contact adhesive*.

dry coloring. Method commonly used by fabricators for coloring plastics by tumble blending uncolored particles of the plastic material with selected dyes and pigments.

dry fiber. In composites fabrication, a condition in which fibers are not fully encapsulated by resin during *pultrusion*.

drying. Removal, by evaporation, of uncombined water or other volatile substance from a ceramic raw material or product, usually expedited by low-temperature heating.

drying temperature. The temperature to which an adhesive on an adherend, an adhesive in an assembly, or the assembly itself is subjected to dry the adhesive. The temperature attained by the adhesive in the process of drying (adhesive drying temperature) may differ from the temperature of the atmosphere surrounding the assembly (assembly drying temperature). See also *curing temperature* and *setting temperature*.

drying time. The period of time during which an adhesive on an adherend or an assembly is allowed to dry with or without the application of heat or pressure, or both. See also *curing time*, *joint-conditioning time*, and *setting time*.

dry laminate. A laminate containing insufficient resin for complete bonding of the reinforcement. See also *resin-starved area*.

dry lay-up. Construction of a laminate by the layering of preimpregnated reinforcement (partly cured resin) in a female mold or on a male mold, usually followed by bag molding or autoclave molding. See also *vacuum bag molding*.

dry strength. The strength of an adhesive joint determined immediately after drying under specified conditions or after a period of conditioning in the standard laboratory atmosphere. See also *wet strength*.

dry tack. The property of certain adhesives, particularly nonvulcanizing rubber adhesives, to adhere on contact to themselves at a stage in the evaporation of volatile constituents, even though they seem dry to the touch. Synonym for aggressive tack.

dry winding. In composites fabrication, filament winding using preimpregnated roving, as differentiated from wet winding, in which unimpregnated roving is pulled through a resin bath just before being wound onto a mandrel. See also *wet winding*.

DSC. See *differential scanning calorimetry*.

DTA. See *differential thermal analysis*.

DTUL. See *deflection temperature under load*.

ductility. The ability of a material to deform plastically without fracturing.

dullness. A lack of pultruded surface gloss or shine in plastic parts. Can be caused by insufficient cure locally or in large areas, resulting in the dull band created on a pultruded part within the die when the pultrusion process is interrupted briefly.

dwarf width. A condition in which the crosswise (of the direction of pultrusion) dimension of a flat surface of a part is less than that which the die would normally yield for a particular plastic or composite. The condition is usually caused by a partial blockage of the pultrusion die cavity caused by buildup, or particles of the composite adhering to the cavity surface. This condition is commonly called a lost edge, when the flat surface has a free edge that is altered by the buildup.

dwell. In forming plastics and composites, a pause in the application of pressure or temperature to a mold, made just before it is completely closed, to allow the escape of gas from the molding material. In filament winding, the time that the traverse mechanism is stationary while the mandrel continues to rotate to the appropriate point for the traverse to begin a new pass. In a standard autoclave cure cycle, an intermediate step in which the resin matrix is held at a temperature below the cure temperature for a specified period of time sufficient to produce a desired degree of staging. Used primarily to control resin flow.

dynamic mechanical measurement. A technique in which either the modulus and/or damping of a substance under oscillatory load or displacement is measured as a function of temperature, frequency, or time, or a combination thereof.

dynamic modulus. The ratio of stress to strain under cyclic conditions (calculated from data obtained from either free or forced vibration tests, in shear, compression, or tension).

E

earthenware. A glazed or unglazed nonvitreous clay-based ceramic *whiteware*.

edge distance ratio. In a *bearing test*, the distance from the center of the bearing hole to the edge of a specimen in the direction of the principal stress, divided by the diameter of the hole.

edge joint. A joint made by bonding the edge faces of two adherends.

E-glass. A family of glasses with a calcium aluminoborosilicate composition and a maximum alkali content of 2.0%. A general-purpose fiber that is most often used in reinforced plastics, and is suitable for electrical laminates because of its high resistivity. Also called electric glass.

eight-harness satin. A type of fabric weave. The fabric has a seven-by-one weave pattern in which a filling thread floats over seven warp threads and then under one. Like the crowfoot weave, it looks different on one side than on the other. This weave is more pliable than any of the others and is especially adaptable to forming around compound curves, such as on radomes.

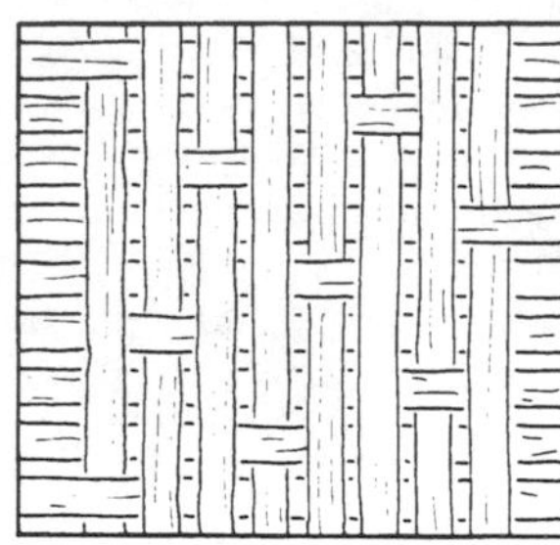

Eight-harness satin

ejection mark. A surface mark on a plastic part caused by the ejector pin when it pushes the part out of the molded cavity.

elastic deformation. A change in dimensions directly proportional to and in phase with an increase or decrease in applied force.

elasticity. The property of a material by virtue of which deformation caused by stress disappears upon removal of the stress. A perfectly elastic body completely recovers its original shape and dimensions after release of stress.

elastic limit. The maximum stress which a material is capable of sustaining without any permanent strain (deformation) remaining upon complete release of the stress. A material is said to have passed its elastic limit when the load is sufficient to initiate plastic, or nonrecoverable, deformation. See also *proportional limit.*

elastic recovery. (1) The fraction of a given deformation that behaves elastically. A perfectly elastic material has an elastic recovery of 1; a perfectly plastic material has an elastic recovery of 0. (2) In hardness testing, the shortening of the original dimensions of the indentation upon release of the applied load.

elastic true strain (ε_e). Elastic component of the *true strain.*

elastomer. (1) A material that substantially recovers its original shape and size at room temperature after removal of a deforming force. A material that shows reversible elasticity up to very high strain levels. (2) Any elastic, rubberlike substance, such as natural or synthetic rubber.

elastomeric tooling. A tooling system that uses the thermal expansion of rubber materials to form reinforced plastic or composite parts during cure.

electrical dissipation factor. The ratio of the power loss in a dielectric material to the total power transmitted through it; thus, the imperfection of the dielectric. Equal to the tangent of the loss angle or to the ratio of the dielectric loss to the dielectric constant.

electrical porcelain. Vitrified *whiteware* having an electrical insulating function.

electrical resistivity. The electrical resistance offered by a material to the flow of current, times the cross-sectional area of current flow and per unit length of current path; the reciprocal of the conductivity. Also called resistivity or specific resistance.

electroformed molds. A mold for forming plastics made by electroplating metal on the reverse pattern of the cavity. Molten steel may then be sprayed on the back of the mold to increase its strength.

electromagnetic interference. Interference related to accumulated electrostatic charge in a nonconductor.

electronic ceramics. *Advanced ceramics* widely used in the electronics industry. Examples include: alumina (Al_2O_3), alumina-titania (Al_2O_3-TiO_2), beryllia (BeO), and aluminum nitride (AlN) for substrates; barium titanate ($BaTiO_3$) for capacitors; lead zirconate titanate (PZT), lead magnesium niobate (PMN), and lead magnesium titanium niobate (PMTN) for actuators and transducers; zirconia (ZrO_2)-based ceramics for oxygen sensors; zinc oxide (ZnO)-based materials for varistors; nickel oxide (NiO) and iron oxide (Fe_2O_3) for temperature sensors; and a variety of glassy materials for a host of devices and packaging. Semiconducting ceramics include materials, such as silicon carbide (SiC), that have properties similar to silicon but that retain their semiconductive properties at elevated properties.

electronic treating. A method of oxidizing a polyolefin part or film to render it printable by passing it between electrodes and subjecting it to a high voltage corona discharge.

elongation. (1) A term used in mechanical testing to describe the amount of extension of a test piece when stressed. (2) In tensile testing, the increase in the gage length, measured after fracture of the specimen within the gage length, usually expressed as a percentage of the original gage length. See also *elongation, percent.*

elongation at break. Elongation recorded at the moment of rupture of a specimen, often expressed as a percentage of the original length.

elongation, percent. The extension of a uniform section of a specimen expressed as a percentage of the original gage length:

$$\text{Elongation, \%} = \frac{(L_x - L_o)}{L_o} \cdot 100$$

where L_o is the original gage length and L_x is the final gage length.

embossing. (1) Technique used to create depressions of a specific pattern in plastic film and sheeting. Such embossing in the form of surface patterns can be achieved on molded parts by the treatment of the mold surface with photoengraving or another process. (2) Raising a design in relief against a surface.

emulsion. (1) A two-phase liquid system in which small droplets of one liquid (the internal phase) are immiscible in, and are dispersed uniformly throughout, a second continuous liquid phase (the external phase). The internal phase is sometimes described as the dispense phase. (2) A stable dispersion of one liquid in another, generally by means of an emulsifying agent that has affinity for both the continuous and discontinuous phases. The emulsifying agent, discontinuous phase, and continuous phase can together produce another phase that serves as an enveloping (encapsulating) protective phase around the discontinuous phase.

emulsion polymerization. Polymerization of monomers dispersed in an aqueous emulsion.

encapsulated adhesive. An adhesive in which the particles or droplets of one of the reactive components are enclosed in a protective film (microcapsules) to prevent cure until the film is destroyed by suitable means.

encapsulation. The enclosure of an item in plastic. Sometimes used specifically in reference to the enclosure of capacitors or circuit board modules.

end. In composites fabrication, a strand of *roving* consisting of a given number of filaments gathered together. The group of filaments is considered an end, or strand, before twisting, and a yarn after twist has been applied. An individual warp, yarn, thread, fiber, or roving.

endurance limit. The maximum stress that a material can withstand for an infinitely large number of fatigue cycles. See also *fatigue limit* and *fatigue strength*.

endurance ratio. The ratio of the *endurance limit* for completely reversed flexural stress to the tensile strength of a given material.

engineering adhesive. A bonding agent intended to join metal, plastics, wood, glass, ceramics, or rubber. The term differentiates such bonding agents from glues used to join paper and other nondurables.

engineering ceramics. Same as *advanced ceramics*.

engineering plastics. A general term covering all plastics, with or without fillers or reinforcements, that have mechanical, chemical, and thermal properties suitable for use as construction materials, machine components, and chemical processing equipment components. Included are acrylonitrile-butadiene-styrene, acetal, acrylic, fluorocarbon, nylon, phenoxy, polybutylene, polyaryl ether, polycarbonate, polyether (chlorinated), polyether sulfone, polyphenylene oxide, polysulfone, polyimide, rigid polyvinyl chloride, polyphenylene sulfide, thermoplastic urethane elastomers, and many other reinforced plastics.

engineering strain (*e*). A term sometimes used for average linear strain or conventional strain in order to differentiate it from *true strain*. In tension testing it is calculated by dividing the change in the gage length by the original gage length.

engineering stress (*s*). A term sometimes used for conventional stress in order to differentiate it from *true stress*. In tension testing, it is calculated by dividing the breaking load applied to the specimen by the original cross-sectional area of the specimen.

engobe. A slip coating applied to a ceramic body for imparting color, opacity, or other characteristics, and subsequently covered with a *glaze*.

environment. The aggregate of all conditions (such as contamination, temperature, humidity, radiation, magnetic and electric fields, shock, and vibration) that externally influence the performance of a material or component.

environmental stress cracking. The susceptibility of a plastic resin to cracking or crazing when in the presence of such chemicals as surface-active agents, or other environments.

epichlorohydrin. The basic epoxidizing resin intermediate in the production of epoxy resins. It contains an epoxy group and is highly reactive with polyhydric phenols such as bisphenol A.

epoxide. Compound containing the oxirane structure, a three-member ring containing two carbon atoms and one oxygen atom. The most important members are ethylene oxide and propylene oxide.

epoxy plastic. A thermoset polymer containing one or more epoxide groups and curable by reaction with amines, alcohols, phenols, carboxylic acids, acid anhydrides, and mercaptans. An important matrix resin in reinforced composites and in structural adhesives.

epoxy resin. A viscous liquid or the brittle, solid-containing epoxide groups that can be cross linked into final form by means of a chemical reaction with a variety of setting agents used with or without heat.

epoxy smear. See *resin smear*.

equilibrium centrifugation. In resinography, a method for determining the distribution of a molecular weight by spinning a solution of the specimen at a speed such that the molecules of the specimen are not removed from the solvent but are held at a point where the (centrifugal) force tending to remove them is balanced by the dispersive forces caused by thermal agitation.

ester. The reaction product of an alcohol and an acid.

ethylene plastics. Plastics based on polymers of ethylene or copolymers of ethylene with other monomers, the ethylene being in greatest amount by mass.

even tension. In composite reinforcing fibers, the process whereby each end of roving is kept in the same degree of tension as the other ends making up that ball of roving. See also *catenary*.

exotherm. The temperature/time curve of a chemical reaction or a phase change giving off heat, particularly the polymerization of casting resins. The amount of heat given off. The term has not been standardized with respect to sample size, ambient temperature, degree of mixing, and so forth.

expandable plastic. A plastic that can be made cellular by thermal, chemical, or mechanical means. Foam plastics such as expandable polystyrene and foam polyurethane are examples.

exposed underlayer. In composites fabrication, the underlying layer of mat or roving not covered by surface mat in a pultrusion. This condition can be caused by reinforcement shifting, too narrow a surface mat, too wide an underlying mat, uneven slitting of the surface mat, necking down of the surface mat, or excessive tension in pulling the surface mat off the spindle.

extend. The addition of fillers or low-cost materials to plastic resins as an economy measure. To add inert materials to improve void-filling characteristics and reduce crazing.

extenders. (1) Low-cost materials used to dilute or extend high-cost resins without significant lessening of properties. (2) Substances, generally having some adhesive action, added to an adhesive to reduce the amount of primary binder required per unit area. See also *binder*, *diluent*, *filler*, and *thinner*.

extensibility. The ability of a material to extend or elongate upon application of sufficient force, expressed as a percent of the original length.

extensional-bending composite. A property of certain classes of laminates that exhibit bending curvatures when subjected to extensional loading.

extensometer. An instrument for measuring changes in length over a given gage length caused by application or removal of a force. Commonly used in tension testing.

extruder. A machine that accepts solid particles, such as pellets or powder, or liquid (molten) feed, conveys it through a surrounding barrel by means of a rotating screw, and pumps it, under pressure, through an orifice.

extrusion (ceramics). (1) The process of forcing a mixture of plastic binder and ceramic powder through the opening(s) of a die at relatively high pressure. The material may thus be compacted and emerges in elongated cylindrical or ribbon (or wire, etc.) form having the cross section of the die opening. Ordinarily followed by drying, curing, activating, or firing. (2) The process of forming clay products by forcing the plastic material through a die.

extrusion (plastics). Compacting a plastic material into a powder or granules in a uniform melt and forcing it through an orifice in a more or less continuous fashion to yield a desired shape. While held in the desired shape, the melt must be cooled to a solid state.

extrusion blow molding. The most common blow molding process, in which a parison is extruded from a plastic melt and is then entrapped between the halves of a mold. The parison is expanded, under air pressure, against the mold cavity to form the part, and is then cooled, removed, and trimmed. See also *blow molding*.

extrusion coating. Using a resin to coat a substrate by extruding a thin film of molten resin and pressing it onto or into the substrate, or both, without the use of an adhesive.

F

fabric fill face. That side of a woven fabric used in reinforced plastics on which the greatest number of yarns are perpendicular to the *selvage*.

fabric, nonwoven. See *nonwoven fabric*.

fabric prepreg batch. In composites processing, a *prepreg* containing fabric from one fabric batch and impregnated with one batch of resin in one continuous operation.

fabric warp face. That side of a woven fabric used in reinforced plastics on which the greatest number of yarns are parallel to the *selvage*.

fabric, woven. See *woven fabric*.

fadeometer. An apparatus for determining the resistance of resins and other materials to fading.

fairing. A secondary structure in airframes and ship hulls, the major function of which is to streamline the airflow or flow of fluid by producing a smooth outline and reducing drag.

family mold. A multicavity mold used for forming of plastic in which each of the cavities form one of the component parts of the assembled finished object. The term is often applied to a mold in which parts from different customers are grouped together for economy. Sometimes called combination mold.

fat. An organic ester, the product of a reaction between a fatty acid and glycerol. Fat can be of animal or vegetable materials or can be made synthetically.

fatigue. The phenomenon leading to fracture under repeated or fluctuating stresses having a maximum value less than the ultimate tensile strength of the material. *Fatigue failure* generally occurs at loads which applied statically would produce little perceptible effect. Fatigue fractures are progressive, beginning as minute cracks that grow under the action of the fluctuating stress.

fatigue ductility (D_f). The ability of a material to deform plastically before fracturing, determined from a constant-strain amplitude, low-cycle fatigue test. Usually expressed in percent in direct analogy with elongation and reduction in area ductility measures.

fatigue ductility exponent (c). The slope of a log-log plot of the plastic strain range and the fatigue life.

fatigue failure. Failure that occurs when a specimen undergoing *fatigue* completely fractures into two parts or has softened or been otherwise significantly reduced in stiffness by thermal heating or cracking.

fatigue limit. The maximum stress that presumably leads to fatigue fracture in a specified number of stress cycles. The value of the *maximum stress* and the *stress ratio* also should be stated. See also *endurance limit*.

fatigue strength. The maximum cyclical stress a material can withstand for a given number of cycles before failure occurs.

faying surface. The surfaces of materials in contact with each other and joined or about to be joined.

feathering. The tapering of an *adherend* on one side to form a wedge section, as used in an adhesively bonded scarf joint.

feldspar. A group of alumina silicate minerals consisting chiefly of microcline ($K_2O \cdot Al_2O_3 \cdot 6SiO_2$), albite ($Na_2O \cdot Al_2O_3 \cdot 6SiO_2$), and/or anorthite ($CaO \cdot Al_2O_3 \cdot 2SiO_2$) used in the production of glass and ceramic whiteware.

felt. A fibrous material used in reinforced plastics made up of interlocked fibers by mechanical or chemical action, moisture, or heat. Made from fibers such as asbestos, cotton, glass, and so forth. See also *batt*.

FEP. See *fluorinated ethylene propylene*.

ferrimagnetic material. (1) A material that macroscopically has properties similar to those of a *ferromagnetic material* but that microscopically also resembles an antiferromagnetic material in that some of the elementary magnetic moments are aligned antiparallel. If the moments are of different magnitudes, the material may still have a large resultant magnetization. (2) A material in which unequal magnetic moments are lined up antiparallel to each other. Permeabilities are of the same order of magnitude as those of ferromagnetic materials, but are lower than they would be if all atomic moments were parallel and in the same direction. Under ordinary conditions the magnetic characteristics of ferrimagnetic materials are quite similar to those of ferromagnetic material.

ferrites. A term referring to magnetic oxides in general, and especially to material having the formula $MOFe_2O_3$, where M is a divalent metal ion or a combination of such ions. Certain ferrites, magnetically "soft" in character, are useful for core applications at radio and higher frequencies because of their advantageous magnetic properties and high volume resistivity. Other ferrites, magnetically "hard" in character, have desirable permanent magnet properties.

ferromagnetic material. A material that in general exhibits the phenomena of hysteresis and saturation, and whose permeability is dependent on the magnetizing force. Microscopically, the elementary magnets are aligned parallel in volumes called domains. The unmagnetized condition of a ferromagnetic material results from the overall neutralization of the magnetization of the domains to produce zero external magnetization.

fiber. A general term used to refer to filamentary materials. Often, fiber is used synonymously with filament. It is a general term for a filament with a finite length that is at least 100 times its diameter, which is typically 0.10 to 0.13 mm (0.004 to 0.005 in.). In most cases it is prepared by drawing from a molten bath, spinning, or depositing on a substrate. Fibers can be continuous or specific short lengths (discontinuous), normally no less than 3.2 mm (1/8 in.).

fiber bridging. Reinforcing fiber material that bridges an inside radius of a pultruded product. This condition is caused by shrinkage stresses around such a radius during cure.

fiber content. The amount of fiber present in reinforced plastics and composites, usually expressed as a percentage volume fraction or weight fraction.

fiber count. The number of fibers per unit width of ply present in a specified section of a reinforced plastic or composite.

fiber diameter. The measurement (expressed in micrometers or microinches) of the diameter of individual filaments used in reinforced plastics or composites.

fiber direction. The orientation or alignment of the longitudinal axis of a fiber with respect to a stated reference axis.

fiber exposure. A condition in which reinforcing fibers within the base material are exposed in machined, abraded, or chemically attacked areas.

fiberglass. An individual filament made by drawing molten glass. A continuous filament is a glass fiber of great or indefinite length. A staple fiber is a glass fiber of relatively short length, generally less than 430 mm (17 in.), the length depending on the forming or spinning process used. The four main glasses used for fiberglass are high-alkali glass (A-glass), electrical grade glass (E-glass), a modified E-glass that is chemically resistant (ECR-glass), and high-strength glass (S-glass).

fiberglass reinforcement. Major material used to reinforce plastic. Available as mat, roving, fabric, and so forth, it is incorporated into both thermosets and thermoplastics.

fiber pattern. Visible fibers on the surface of laminates or molding. The thread size and weave of glass cloth.

fiber-reinforced composite. A material consisting of two or more discrete physical phases, in which a fibrous phase is dispersed in a continuous matrix phase. The fibrous phase may be macro-, micro-, or submicroscopic, but it must retain its physical identity so that it could conceivably be removed from the matrix intact.

fiber-reinforced plastic (FRP). A general term for a plastic that is reinforced with cloth, mat, strands, or any other fiber form. See also *resin-matrix composites*.

fiber show. Strands or bundles of fibers that are not covered by plastic because they are at or above the surface of a reinforced plastic or composite.

fiber wash. Splaying out of woven or nonwoven fibers from the general reinforcement direction. Fibers are carried along with bleeding resin during cure.

fibrillation. Production of fiber from film.

filament. The smallest unit of fibrous material. The basic units formed during drawing and spinning, which are gathered into strands of fiber for use as reinforcements. Filaments usually are of extreme length and very small diameter, usually less than 25 μm (1 mil). Normally, filaments are not used individually. Some textile filaments can function as a reinforcing yarn when they are of sufficient strength and flexibility.

filament winding. A process for fabricating a reinforced plastic or composite structure in which continuous reinforcements (filament, wire, yarn, tape, and the like), either previously impregnated with a matrix material or impregnated during the winding, are placed over a rotating and removable form or mandrel in a prescribed way to meet certain stress conditions. Generally the shape is a surface of revolution and may or may not include end closures. When the required number of layers is applied, the wound form is cured and the mandrel is removed. See also *helical winding* and *polar winding*.

fill. Reinforcing yarn oriented at right angles to the warp in a woven fabric.

fill-and-wipe. Technique used with plastic parts that are molded with depressed designs; after application of paint, the surplus is wiped off, leaving paint only in the depressed areas. Sometimes called wipe-ins.

filler. (1) A relatively inert substance added to a plastic to alter its physical, mechanical, thermal, electrical, or other properties, or to lower cost or density. Sometimes the term is used specifically to mean particulate additives. See also *inert filler* and *reinforced plastics*. (2) A relatively nonadhesive substance added to an adhesive to improve its working properties, permanence, strength, or other qualities. See also *binder*.

filler sheet. A sheet of deformable or resilient material that, when placed between the assembly to be adhesively bonded and pressure applicator, or when distributed within a stack of assemblies, aids in providing uniform application of pressure over the area to be bonded.

fillet. A rounded filling or adhesive that fills the corner or angle where two adherends are joined.

filling yarn. The transverse threads or fibers in a woven fabric used in reinforced plastics or composites. Those fibers running perpendicular to the warp. Also called weft.

film adhesive. A synthetic resin adhesive, usually of thermosetting type, in the form of a thin, dry film of resin with or without a paper or glass carrier.

fin. Excess material left on a molded plastic object at those places where the molds or dies mated. Also, the web of material remaining in holes or opening in a molded part, which must be removed in finishing.

fine ceramics. Same as *advanced ceramics*.

fines. Very small particles (usually under 200 mesh) accompanying larger grains, usually of molding powder.

finish. (1) A mixture of materials for treating glass or other fibers. It contains a coupling agent to improve the bond of resin to the fiber and usually includes a lubricant to prevent abrasion, as well as a binder to promote strand integrity. With graphite or other filaments, it may perform any or all of the above functions. (2) To complete the secondary work on a molded plastic part so that it is ready for use. Operations such as filling, deflashing, buffing, drilling, tapping, and degating are commonly called finishing operations.

firebrick. A refractory brick, often made from *fireclay*, that is able to withstand high temperature (1500 to 1600 °C, or 2700 to 2900 °F) and is used to line furnaces, ladles, or other molten metal containment components.

fireclay. A mineral aggregate that has as its essential constituent the hydrous silicates of aluminum with or without free silica. It is used in commercial refractory products.

firing. The controlled heat treatment of ceramic ware in a kiln or furnace, during the process of manufacture, to develop the desired properties.

firing range. (1) The range of fired temperature within which a ceramic composition develops properties which render it commercially useful. (2) The time-temperature interval in which a porcelain enamel or ceramic coating is satisfactorily matured.

first-degree blocking. An adherence between adhesively bonded surfaces under test of such degree that when the upper specimen is lifted, the lower specimen will cling thereto, but may be parted with no evidence of damage to either surface.

first-order transition. A change of state associated with crystallization, melting, or a change in crystal structure of a polymer.

fisheye. A small, globular mass that has not completely blended into the surrounding pultruded material. This condition is particularly evident in a transparent or translucent material.

flake. A term used to denote the dry, unplasticized base of cellulosic plastics.

flame resistance. Ability of a material to extinguish flame once the source of heat is removed. See also *self-extinguishing resin*.

flame retardants. Certain chemicals that are used to reduce or eliminate the tendency of a resin to burn.

flame spraying. Method of applying a plastic coating in which finely powdered fragments of the plastic, together with suitable fluxes, are projected through a cone of flame onto a surface.

flame treating. A method of rendering inert thermoplastic objects receptive to inks, lacquers, paints, adhesives, and so forth, in which the object is bathed in an open flame to promote oxidation of the surface of the article.

flammability. Measure of the extent to which a material will support combustion.

flash. The portion of the charge that flows from or is extruded from the mold cavity during the molding. Extra plastic attached to a molding along the parting line, which must be removed before the part is considered finished.

flash mold. A mold in which the mold faces are perpendicular to the clamping action of the press, so that the greater the clamping force, the tighter the mold seam.

flat glass. A general term covering sheet glass, plate glass, and various forms of rolled glass.

flats. A longitudinal, flat area on a normally convex surface of a pultruded plastic, caused by shifting of the reinforcement, lack of sufficient reinforcement, or local fouling of the die surface.

flexibilizer. An additive that makes a finished plastic more flexible or tough. See also *plasticizer*.

flexible molds. Molds made of rubber or elastomeric plastics, used for casting plastics. They can be stretched to remove cured pieces having undercuts.

flexural modulus. The ratio, within the elastic limit, of the applied stress on a reinforced plastic test specimen in flexure to the corresponding strain in the outermost fibers of the specimen.

flexural strength. (1) A property of solid material that indicates its ability to withstand a flexural or transverse load. (2) The maximum stress that can be borne by the surface fibers in a beam in bending. The flexural strength is the unit resistance to the maximum load before failure by bending, usually expressed in force per unit area.

flexure stress. The tensile component of the bending stress produced on the surface of a glass section opposite to that experiencing a locally impinging force.

floating chase. In forming of plastics, a mold member, free to move vertically, that fits over a lower plug or cavity, and into which an upper plug telescopes.

flock. A material obtained by reducing textile fibers to fragments as by cutting, tearing, or grinding, to give various degrees of comminution. Flock can either be fibers in entangled, small masses or beads, usually of irregular broken fibers, or comminuted (powdered) fibers.

flocking. A method of coating by spraying finely dispersed textile powders or fibers.

flow. (1) The movement of resin under pressure, allowing it to fill all parts of a mold. (2) Movement of an adhesive during the bonding process, before the adhesive is set.

flow marks. Wavy surface appearance of an object molded from thermoplastic resins, caused by improper flow of the resin into the mold.

flow molding. The technique of producing leatherlike materials by placing a die-cut plastic blank (solid or expanded vinyl or vinyl-coated substrate) in a mold cavity (usually silicone rubber molds) and applying power via a high-frequency radio frequency generator to melt the plastic such that it flows into the mold to the desired shape and with the desired texture.

fluidized-bed coating. A method of applying a coating of a thermoplastic resin to a heated article that is immersed in a dense-phase fluidized bed of powdered resin and thereafter heated in an oven to provide a smooth, pinhole-free coating.

fluorinated ethylene propylene (FEP). A member of the fluorocarbon family of plastics that is a copolymer of tetrafluoroethylene and hexafluoropropylene, possessing most of the properties of polytetrafluoroethylene, and having a melt viscosity low enough to permit conventional thermoplastic processing. Available in pellet form for molding and extrusion, and as dispersions for spray or dip coating processes.

fluorocarbon plastics. Plastics based on polymers made with monomers composed of fluorine and carbon only.

fluorocarbons. The family of plastics including polytetrafluoroethylene, polychlorotrifluoroethylene, polyvinylidene, and fluorinated ethylene propylene. They are characterized by good thermal and chemical resistance, nonadhesiveness, low dissipation factor, and low dielectric constant. They are available in a variety of forms, such as molding materials, extrusion materials, dispersions, film, or tape, depending on the particular fluorocarbon.

fluorohydrocarbon plastics. Plastics based on polymers made with monomers composed of fluorine, hydrogen, and carbon only.

fluoroplastics. Plastics based on polymers with monomers containing one or more atoms of fluorine, or copolymers of such monomers with other monomers, with the fluorine-containing monomer(s) being in greatest amount by mass.

fluted core. An integrally woven reinforcement material consisting of ribs between two skins in a unitized *sandwich construction*.

flux. A plastics composition additive incorporated during processing to improve flow. For example, coumarone-indene resins are used as a flux during the milling of vinyl polymers. Also a term indicating a state of fluidity.

foamed plastics. Resins in sponge form, flexible or rigid, with cells closed or interconnected and density over a range from that of the solid parent resin to 0.030 g/cm^3. Compressive strength of rigid foams is fair, making them useful as core materials for *sandwich constructions*. Also, chemical cellular plastics, the structures of which are produced by gases generated from the chemical interaction of their constituents. See also *expandable plastics*.

foaming agent. Chemicals added to plastics and rubbers that generate inert gases, such as nitrogen, upon heating, causing the resin to form a cellular structure.

foil decorating. Molding paper, textile, or plastic foils printed with compatible inks directly into a plastic part so that the foil is visible below the surface of the part as integral decoration.

folded chain. The conformation of a flexible polymer when present in a crystal. The molecule exits and reenters the same crystal, frequently generating folds.

force plug. The male half of the mold for making plastics that enters the cavity, exerting pressure on the resin and causing it to flow. Also called punch. Sometimes called a core, plunger, or ram.

form-and-spray. Technique for thermoforming plastic sheet into an end-product and then backing up the sheet with *spray-up* reinforced plastics.

forming. A process in which the shape of plastic pieces such as sheets, rods, or tubes is changed to a desired configuration. See also *thermoforming*. The use of the term forming in plastics technology does not include such operations as molding, casting, or extrusion, in which shapes or pieces are made from molding materials or liquids.

forsterite. A magnesium silicate mineral of composition $2MgO \cdot SiO_2$, which is usually produced synthetically as a ceramic raw material, but it may also be a reaction-produced phase in fired ceramics used in refractory brick.

four-harness satin. A fabric weave, also called *crowfoot satin* because the weaving pattern, when laid out on cloth design paper, resembles the imprint of crow's foot. In this type of weave there is a three-by-one interlacing. That is, a filling thread floats over the three warp threads and then under one. The two sides of the fabric have different appearances. Fabrics with this weave are more pliable than either the plain or basket weaves and, consequently, are easier to form around curves. See also the figure accompanying *crowfoot satin*.

FP fiber. A polycrystalline all-alumina fiber (>99% Al_2O_3). A ceramic fiber useful for high-temperature (1370 to 1650 °C, or 2500 °F) composites.

fracture. The separation of a body. Defined both as rupture of the surface without complete separation of the laminate and as complete separation of a body because of external or internal forces. Fractures in continuous fiber reinforced composites can be divided into three basic fracture types: intralaminar, interlaminar, and translaminar. Translaminar fractures are those oriented transverse to the laminated plane in which conditions of fiber fracture are generated. Interlaminar fracture, on the other hand, describes failures oriented between plies, whereas intralaminar fractures are those located internally within a *ply*. See also accompanying figures.

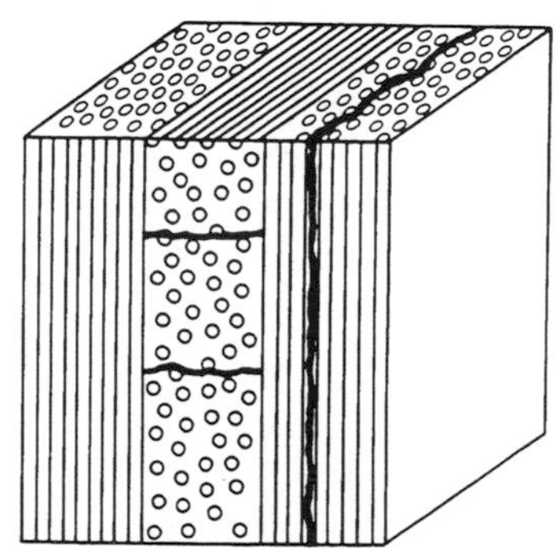

Intralaminar fracture

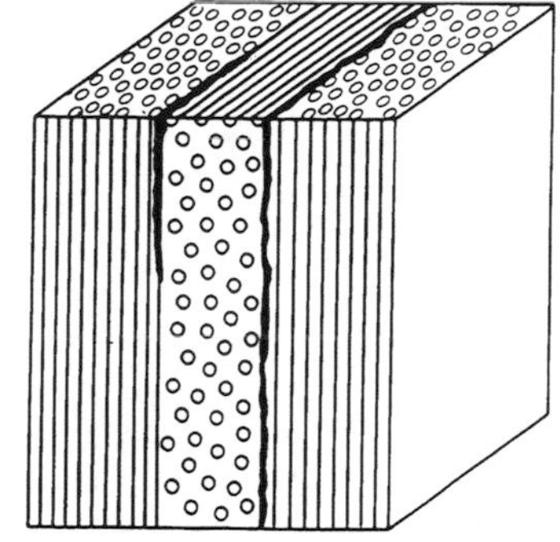

Interlaminar fracture

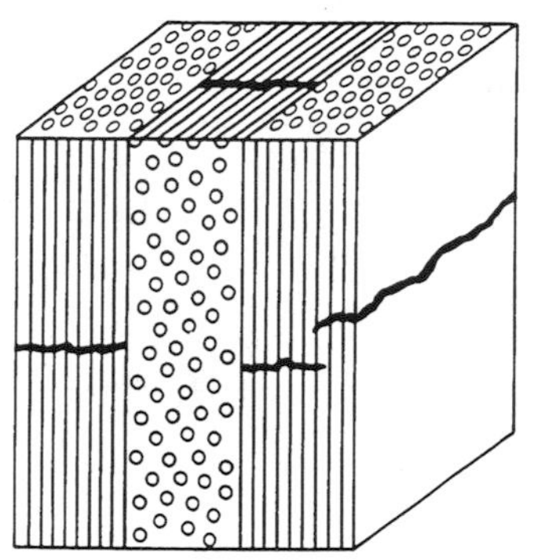

Translaminar fracture

fracture ductility. The true plastic strain of fracture.

fracture mechanics. A quantitative analysis for evaluating structural behavior in terms of applied stress, crack length, and specimen or machine component geometry. See also *linear elastic fracture mechanics*.

fracture strength. The normal stress at the beginning of fracture. Calculated from the load at the beginning of fracture during a tension test and the original cross-sectional area of the specimen.

fracture stress. The true, normal stress on the minimum cross-sectional area at the beginning of fracture. The term usually applies to tension tests of unnotched specimens.

fracture toughness. A generic term for measures of resistance to extension of a crack. The term is sometimes restricted to results of *fracture mechanics* tests, which are directly applicable in fracture control. However, the term commonly includes results from simple tests of notched or precracked specimens not based on fracture mechanics analysis. Results from tests of the latter type are often useful for fracture control, based on either service experience or empirical correlations with fracture mechanics tests. See also *stress-intensity factor*.

free bend. The bend obtained by applying forces to the ends of a specimen without the application of force at the point of maximum bending.

free radical. Any molecule or atom that possesses one unpaired electron. In chemical notation, a free radical is symbolized by a single dot (to denote the odd electron) to the right of the chemical symbol.

free-radical polymerization. A type of *polymerization* in which the propagating species is a long-chain free radical initiated by the introduction of free radicals from thermal or photochemical decomposition of an initiator molecule.

free rotation. The rotation of atoms, particularly carbon atoms, about a single bond. Because the energy requirement is only a few kcal, the rotation is said to be free if sufficient thermal energy is available.

free vibration. A technique for performing dynamic mechanical measurements in which the sample is deformed, released, and allowed to oscillate freely at the natural resonant frequency of the system. Elastic modulus is calculated from the measured resonant frequency, and damping is calculated from the rate at which the amplitude of the oscillation decays.

friction. The resisting force tangential to the common boundary between two bodies when, under the action of an external force, one body moves or tends to move relative to the surface of the other.

friction coefficient. See *coefficient of friction*.

friction welding. A method of welding thermoplastic materials in which the heat necessary to soften the components is provided by friction.

frit. A glass produced by *fritting*, which contains fluxing material and is employed as a constituent in a glaze, body, or other ceramic composition.

fritting. The rapid chilling of the molten glassy material to produce frit.

frost line. In the extrusion of polyethylene lay-flat film, a ring-shape zone located at the point where the film reaches its final diameter. This zone is characterized by a frosty appearance on the film caused when the film temperature falls below the softening range of the resin.

frothing. A technique for applying urethane foam in which blowing agents or tiny air bubbles are introduced under pressure into the liquid mixture of foam ingredients.

FRP. See *fiber-reinforced plastic*.

full-contour length. The length of a fully extended polymer chain.

functional group. A chemical radical or structure that has characteristic properties; examples are hydroxyl and carboxyl groups.

functionality. The average number of reaction sites on an individual polymer chain.

furan resins. Dark-colored thermosetting resins available primarily as liquids ranging from low-viscosity polymers to thick, heavy syrups, which cure to highly cross-linked, brittle substances. Made primarily by polycondensation of furfuryl alcohol in the presence of strong acids, sometimes in combination with formaldehyde or furfuryldehyde.

furfural resins. A dark-colored synthetic resin of the thermosetting variety obtained by the condensation of furfural with phenol or its homologs. It is used in the manufacture of molding materials, adhesives, and impregnating varnishes. Properties include high resistance to acids and alkalies.

fused silica. A glass made either by flame hydrolysis of silicon tetrachloride or by melting silica, usually in the form of granular quartz.

fusion. In vinyl dispersions, the heating of a dispersion to produce a homogeneous mixture.

fuzz. Accumulation of short, broken filaments after passing glass strands, yarns, or rovings over a contact point. Often, weighted and used as an inverse measure of abrasion resistance.

G

gage length. The original length of that portion of the specimen over which strain, change of length and other characteristics are measured.

gamma transition. See *glass transition*.

gap. (1) In filament winding, the space between successive windings in which windings are usually intended to lay next to each other. Separations between fibers within a filament winding band. (2) The distance between adjacent plies in a layup of unidirectional tape materials.

gap-filling adhesive. An adhesive subject to reduce shrinkage upon setting, used as a sealant.

garnet. A generic name for a related group of mineral silicates which have the general chemical formula $A_3B_2(SiO_4)_3$, where A is Fe^{2+}, Mn^{2+}, Mg, or Ca, and B is Al, Fe^{3+}, Cr^{3+}, or Ti^{3+}. Garnet is used for coating abrasive paper or cloth, for bearing pivots in watches, for electronics, and the finer specimens for gemstones. The hardness of garnet varies from Mohs 6 to 8 (1360 Knoop), the latter being used for abrasive applications.

gate. In injection and transfer molding of plastics, the orifice through which the melt enters the mold cavity. The gate can have a variety of configurations, depending on product design.

gate mark. A surface discontinuity on a molded plastic part caused by the gate through which material enters the cavity.

gel. (1) A colloidal state comprised of interdispersed solid and liquid, in which the solid particles are themselves interconnected or interlaced in three dimensions. (2) A two-phase colloidal system consisting of a solid and a liquid in more solid form than a *sol*. (3) The initial jellylike solid phase that develops during the formation of a resin from a liquid. With respect to vinyl plastisols, a state between liquid and solid that occurs in the initial states of heating, or upon prolonged storage. In general, gels have very low strengths and do not flow like a liquid. They are soft, flexible, and may rupture under their own weight unless supported externally. In a cross-linked thermoplastic, gel is the fraction of polymeric material present in the network.

gelation. The point in a resin cure when the resin viscosity has increased to a point such that it barely moves when probed with a sharp instrument.

gelation time. (1) That interval of time, in connection with the use of synthetic thermosetting resins, extending from the introduction of a catalyst into a liquid adhesive system until the start of *gel* formation. (2) The time under application of load for a resin to reach a solid state.

gel coat. A quick-setting resin applied to the surface of a mold and gelled before *layup*. The gel coat becomes an integral part of the finished laminate, and is usually used to improve surface and bonding.

gelling agent. See *thickener*.

gel point. (1) The point at which a thermosetting system attains an infinite value of its average molecular weight. (2) The viscosity at which a liquid begins to exhibit pseudoelastic properties. This stage may be conveniently observed from the inflection point on a viscosity-time plot.

gel time. The period of time from the initial mixing of the reactants of a liquid material composition to the point in time when *gelation* occurs, as defined by a specific test method.

geodesic. The shortest distance between two points on a surface.

geodesic isotensoid. Constant stress level in any given *filament* at all points in its path.

geodesic-isotensoid contour. In filament-wound reinforced plastic pressure vessels, a dome contour in which the filaments are placed on geodesic paths so that the filaments exhibit uniform tensions throughout their length under pressure loading.

glass. An amorphous solid made by fusing silica (SiO_2) with a basic oxide. Its characteristic properties are its transparency, its hardness and rigidity at ordinary temperatures, its capacity for plastic working, and its resistance to weathering and to most chemicals.

glass-ceramics. A family of fine-grained crystalline materials made by a process of controlled crystallization from special glass compositions containing nucleating agents. By definition, glass-ceramics are ≥50% crystalline by volume and generally are >90% crystalline. Depending on heat treatment and composition, crystal sizes range from <0.1 m to 10 μm. The range of glass-ceramic compositions is extremely broad, requiring only the ability to form a glass and control its crystallization. Glass-ceramics are used in cookware, stovetops, and heat exchangers.

glass cloth. Woven glass fiber material. See also *scrim*.

glass fiber. A fiber spun from an inorganic product of fusion that has cooled to a rigid condition without crystallizing. See also *fiberglass*.

glass filament. A form of glass that has been drawn to a small diameter and extreme length. It is standard practice in the fiberglass industry to refer to a specific filament diameter by a specific alphabet designation. Fine fibers, which are used in textile applications, range from D (~6 μm) through G (~10 μm). Conventional plastics reinforcement, however, uses filament diameters that range from G to T (~24 μm).

glass filament bushing. The unit through which molten glass is drawn in making glass filaments.

glass finish. A material applied to the surface of a glass reinforcement to improve the bond between the glass and the plastic resin matrix.

glass flake. Thin, irregularly shaped flakes of glass used as a reinforcement in composites.

glass former. An oxide that forms a glass easily. Also, one that contributes to the network of silica glass when added to it.

glass, percent by volume. The product of the specific gravity of a laminate and the percent glass by weight, divided by the specific gravity of the glass.

glass stress. In a filament-wound part, usually a pressure vessel, the stress calculated using the load and the cross-sectional area of the reinforcement only.

glass transition. The reversible change in an amorphous polymer or in amorphous regions of a partially crystalline polymer from, or to, a viscous or rubbery condition to, or from, a hard and relatively brittle one.

glass transition temperature (T_g). The temperature at which an amorphous polymer (or the amorphous regions in a partially crystalline polymer) changes from a hard and relatively brittle condition to a viscous or rubbery condition. In this temperature region, many physical properties, such as hardness, brittleness, thermal expansion, and specific heat, undergo significant, rapid changes.

glassy. A state or matter that is amorphous or disordered like a liquid in structure, hence capable of continuous composition variation and lacking a true melting point, but softening gradually with increasing temperature. Glasses of commerce are mainly complex silicates in chemical combination with numerous other oxidic substances; made by melting the source materials together, forming in various ways while fluid, and allowing to cool.

glaze. A ceramic coating matured to the glassy state on a formed ceramic article, or the material or mixture from which the coating is made.

glue. Originally, a hard gelatin obtained from hides, tendons, cartilage, bones, and so on, of animals and also an adhesive prepared from this substance by heating with water. Through general use, the term is now synonymous with the terms bond and adhesive. See also *adhesive, gum, mucilage, paste, resin*, and *sizing*.

glue-laminated wood. An assembly made by bonding layers of veneer or lumber with an adhesive so that the grain of all laminations is essentially parallel.

glue line. Synonym for *bond line*.

glue line thickness. Thickness of layer of cured adhesive.

gob. (1) A portion of hot glass delivered by a feeder. (2) A portion of hot glass gathered on a punty or pipe.

gob process. A process whereby glass is delivered to a forming unit in *gob* form.

graft copolymers. A chain of one type of polymer to which side chains of a different type are attached or grafted.

granular structure. Nonuniform appearance of finished plastic material due to retention of, or incomplete fusion of, particles of composition, either within the mass or on the surface.

graphite. A crystalline allotropic form of carbon. (2) Uncombined carbon in cast irons.

graphite fiber. A fiber made from a pitch or polyacrylonitrile (PAN) precursor by an oxidation, carbonization, and graphitization process (which provides a graphitic structure). See also *carbon fiber*.

graphitization. The process of pyrolyzation in an inert atmosphere at temperatures in excess of 1925 °C (3500 °F), usually as high as 2480 °C (4500 °F), and sometimes as high as 5400 °C (9750 °F), converting carbon to its crystalline allotropic form. Temperature depends on precursor and properties desired.

green strength. The mechanical strength of material, that, while cure is not complete, allows removal from the mold and handling without tearing or permanent distortion.

greenware. A term for formed ceramic articles in the unfired condition.

greige, gray goods. Any fabric before finishing, as well as any yarn or fiber before bleaching or dyeing; therefore, fabric with no finish or size.

gum. Any of a class of colloidal substances exuded by or prepared from plants, sticky when moist, composed of complex carbohydrates and organic acids, which are soluble or swell in water. The term gum is sometimes used loosely to denote various materials that exhibit gummy characteristics under certain conditions, for example, gum balata, gum benzoin, and gum asphaltum. Gums are included by some in the category of natural resins. See also *adhesive, glue*, and *resin*.

gypsum. The mineral consisting primarily of fully hydrated calcium sulfate, $CaSO_4 \cdot 2H_2O$ or calcium sulfate dihydrate. Most gypsum is ground for wallboard and plasters or used as a flux and fining agent in glass manufacture.

H

hackle. (1) A line on a glass crack surface, running parallel to the local direction of cracking, separating parallel but noncoplanar portions of the crack surface. (2) A finely structured fracture surface marking that gives a matte or roughened appearance to the surface, having varying degrees of coarseness. Finely structured hackle is variously known as fine hackle, frosted area, gray area, matte, mist, and stippled area. Coarsely structured hackle is also known as *striation*.

hackle marks. Fine ridges on the fracture surface of the glass, parallel to the direction of propagation of the fracture.

hairline craze. Multiple fine surface separation cracks in composites that exceed 6 mm (1/4 in.) in length and do not penetrate in depth the equivalent of a full ply of reinforcement. See also *crazing*.

halocarbon plastics. Plastics based on resins made by the polymerization of monomers composed only of carbon and a halogen or halogens.

hand lay-up. In composites processing, the process of manually placing (and working) successive plies of reinforcing material or resin-impregnated reinforcement in position on a mold.

handling strength. A low level of strength initially obtained by an adhesive that allows specimens to be handled, moved, or unclamped without causing disruption of the curing process or affecting bond strength.

hardener. A substance or mixture added to a plastic composition to promote or control the curing action by taking part in it.

hardness. A measure of the resistance of a material to surface indentation or abrasion; may be thought of as a function of the stress required to produce some specified type of surface deformation. There is no absolute scale for hardness; therefore, to express hardness quantitatively, each type of test has its own scale of arbitrarily defined hardness. Indentation hardness can be measured by *Brinell, Rockwell, Vickers, Knoop*, and *Scleroscope hardness tests*.

harness satin. A fabric weaving pattern producing a satin appearance. See also *eight-harness satin* and *four-harness satin*.

HDPE. See *high-density polyethylene*.

head-to-head. On a polymer chain, a type of configuration in which the functional groups are on adjacent carbon atoms.

head-to-tail. On a polymer chain, a type of configuration in which the functional groups on adjacent polymers are as far apart as possible.

heat-activated adhesive. A dry adhesive that is rendered tacky or fluid by application of heat, or heat and pressure, to the assembly.

heat buildup. In processing of plastics, the rise in temperature in a part resulting from the dissipation of applied strain energy as heat or from applied mold cure heat.

heat cleaned. A condition in which glass or other fibers are exposed to elevated temperatures to remove preliminary sizings or binders not compatible with the resin system to be applied.

heat-deflection temperature. The temperature at which a standard plastic test bar deflects a specified amount under a stated load. Now called deflection temperature under load (DTUL).

heat distortion point. The temperature at which a standard plastic test bar deflects a specified amount under a stated load. Now called deflection temperature.

heat-fail temperature. The temperature at which delamination of an adhesively bonded structure occurs under static loading in shear.

heat forming. See *thermoforming*.

heat mark. Extremely shallow depression or groove in the surface of a plastic visible because of a sharply defined rim or a roughened surface. See also *sinkmark*.

heat sealing. A method of joining plastic films by simultaneous application of heat and pressure to areas in contact.

heat-sealing adhesive. A thermoplastic film adhesive that is melted between the adherend surfaces by heat application to one or both of the surfaces.

heat sink. A material that absorbs or transfers heat away from a critical element or part.

helical winding. In filament-wound items, a winding in which a filament band advances along a helical path, not necessarily at a constant angle, except in the case of a cylinder.

hermetic. Sealed so that the object is gas-tight. The test for hermeticity is to fill the object with a test gas, often helium, and observe leak rates when the object is placed in a vacuum. Plastic encapsulation is not hermetic because it allows permeation by gases.

heterogeneous. Of a body of material or matter, comprised of more than one phase (solid, liquid, and gas) separated by boundaries; similarly of a solid, comprised of more than one chemical, crystalline, and/or glassy species, separated by boundaries.

heterogeneous nucleation. In the crystallization of polymers, the growth of crystals on vessel surfaces, dust, or added nucleating agents.

hexa. An abbreviated form of hexamethylenetetramene, a source of reactive methylene for curing *novolacs*.

high-cycle fatigue. *Fatigue* that occurs at relatively large numbers of cycles. The arbitrary, but commonly accepted, dividing line between high-cycle fatigue and *low-cycle fatigue* is considered to be about 10^4 to 10^5 cycles. In practice, this distinction is made by determining whether the dominant component of the *strain* imposed during cyclic loading is elastic (high cycle) or plastic (low cycle), which in turn depends on the properties of the metal and on the magnitude of the nominal *stress*.

high-density polyethylene (HDPE). This term generally includes polyethylene ranging in density from about 0.94 to 0.96 and over. While molecules in low density polyethylene are branched and linked randomly, those in the higher density polyethylenes are linked in longer chains with fewer side branches, resulting in more rigid material with greater strength, hardness, and chemical resistance, and higher softening temperature.

high-frequency heating. The heating of materials by dielectric loss in a high-frequency electrostatic field. The material is exposed between electrodes and is heated quickly and uniformly by absorption of energy from the electrical field.

high-impact polystyrene (HIPS). A thermoplastic resin produced from a styrene monomer and elastomers. It has good dimensional stability and low-temperature impact strength, high rigidity, and ease of processing.

high polymer. A macromolecular substance that, as indicated by the polymer by which it is identified, consists of molecules that are multiples of the low molecular unit and have a molecular weight of at least 20,000.

high-pressure laminates. Laminates molded and cured at pressures not lower than 6.9 MPa (1.0 ksi), and more commonly in the range of 8.3 to 13.8 MPa (1.2 to 2.0 ksi).

high-pressure molding. A plastic molding or laminating process in which the pressure used is greater than 1400 kPa (200 psi), but commonly 7000 kPa (1000 psi).

high-pressure spot. See *resin-starved area*.

hob. A master model used to sink the shape of a mold into a soft steel block.

homogeneous. A body of material or matter, alike throughout; hence, comprised of only one chemical composition and phase, without internal boundaries.

homogeneous nucleation. In the crystallization of polymers, the primary nucleated species generated by the polymer molecules.

homologous. Belonging to or consisting of a series of organic compounds differentiated by the number of methylene groups (CH_2).

homopolymer. A polymer resulting from polymerization of a single monomer.

honeycomb. Manufactured product of resin-impregnated sheet material (paper, fiberglass, and so on) or metal (aluminum, titanium, and corrosion-resistant alloys) foil, formed into hexagonal-shaped cells. Used as a core material in composite sandwich constructions. See also *sandwich construction*.

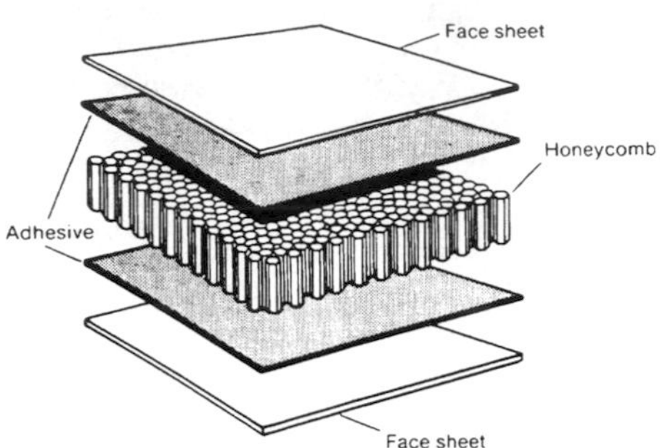

Honeycomb assembly

hoop stress. The circumferential stress in a material of cylindrical form subjected to internal or external pressure.

hopper dryer. A combination feeding and drying device for extrusion and injection molding of thermoplastics. Hot air flows upward through the hopper containing the feed pellets.

hopper loader. A curved pipe through which molding plastic powders are pneumatically conveyed from shipping drums to machine hoppers.

hot-gas welding. A technique for joining thermoplastic materials (usually sheet) in which the materials are softened by a jet of hot air from a welding torch and joined together at the softened points. Generally, a thin rod of the same material is used to fill and consolidate the gap.

hot heated manifold mold. A thermoplastic injection mold in which the portion of the mold (the manifold) that contains the runner system has its own heating elements, which keep the molding material in a plastic state ready for injection into the cavities, from which the manifold is insulated. See also *thermoplastic injection molding*.

hot-melt adhesive. An adhesive that is applied in a molten state and forms a bond after cooling to a solid state. A bonding agent that achieves a solid state and resultant strength by cooling, in contrast to other adhesives, which achieve the solid state through evaporation of solvents or chemical cure. Also, a thermoplastic resin that functions as an adhesive when melted between substrates and cooled.

hot-runner mold. A thermoplastic injection mold in which the runners are insulated from the chilled cavities and remain hot so that the center of the runner never cools during the molding cycle. Contrary to usual practice, the runners are not ejected with the molded pieces. Also called insulated-runner mold. See also *thermoplastic injection molding*.

hot-setting adhesive. An adhesive that requires a temperature at or above 100 °C (212 °F) to set.

hot stamping. Engraving operation for plastics in which a design is stamped with heated metal dies onto the face of the plastics.

hot working. Controlled mechanical operations for shaping a product at temperatures above the recrystallization temperature.

humidity ratio. In a mixture of water vapor and air, the mass of water vapor per unit mass of dry air.

hybrid. A composite laminate consisting of laminae of two or more composite material systems. A combination of two or more different fibers, such as carbon and glass or carbon and aramid, in a structure. Tapes, fabrics, and other forms may be combined; usually only the fibers differ. See also *interply hybrid* and *intraply hybrid*.

hydraulic press. A press in which the molding force is created by the pressure exerted by a fluid.

hydrocarbon plastics. Plastics based on resins made by the polymerization of monomers composed of carbon and hydrogen only.

hydrolysis. (1) Decomposition or alteration of a chemical substance by water. (2) In aqueous solutions of electrolytes, the reactions of cations with water to produce a weak base or of anions to produce a weak acid.

hydrolytic stability. The ability of an organic or polymeric material to withstand an irreversible change of state when exposed to elevated temperature and humidity.

hydromechanical press. A press in which forces are created partly by a mechanical system and partly by a hydraulic system.

hydrophilic. Having an affinity for water; easily wetted by water. Contrast with *hydrophobic*.

hydrophobic. Lacking an affinity for, repelling, or failing to absorb water; poorly wetted by water. Contrast with *hydrophilic*.

hydroxyl group. A chemical group consisting of one hydrogen atom and one oxygen atom.

hygroscopic. (1) Capable of attracting, absorbing, and retaining atmospheric moisture. (2) Possessing a marked ability to accelerate the condensation of water vapor; applied to condensation nuclei composed of salts that yield aqueous solutions of a very low equilibrium vapor pressure compared with that of pure water at the same temperature. (3) Pertaining to a substance whose physical characteristics are appreciably altered by effects of water vapor. (4) Pertaining to water absorbed by dry soil minerals from the atmosphere; the amounts depend on the physicochemical character of the surfaces, and increase with rising relative humidity.

hygrothermal effect. Change in properties of a material (particularly plastics) due to moisture absorption and temperature change.

hysteresis (magnetic). The lag of the magnetization of a substance behind any cyclic variation of the applied magnetizing field.

hysteresis (mechanical). The phenomenon of permanently absorbed or lost energy that occurs during any cycle of loading or unloading when a material is subjected to repeated loading.

hysteresis loop (magnetic). A closed curve that characterizes the magnetization/demagnetization characteristics as a function of the applied magnetic field for magnetic materials.

hysteresis loop (mechanical). In dynamic mechanical measurement, the closed curve representing successive stress-strain status of a material during a deformation cycle.

I

ignition loss. (1) The difference in weight before and after burning. (2) The burning off of *binder* or *size*.

immiscible. (1) Of two phases, the inability to dissolve in one another to form a single solution; mutually insoluble. (2) With respect to two or more fluids, not mutually soluble; incapable of attaining homogeneity.

impact strength. A measure of the resiliency or toughness of a solid. The maximum force or energy of a blow (given by a fixed procedure) which can be withstood without fracture, as opposed to fracture strength under a steady applied force.

impact test. A test for determining the energy absorbed in fracturing a test piece at high velocity, as distinct from static test. The test may be carried out in tension, bending, or torsion, and the test bar may be notched or unnotched. See also *Charpy test* and *Izod test*.

impact value. The energy absorbed by a specimen of standard design when sheared by a single blow from a testing machine hammer. Expressed in J/m^2 or ft · $lbf/in.^2$.

impregnate. In reinforcing plastics, to saturate the reinforcement with a resin.

impregnated fabric. A fabric impregnated with a synthetic resin. See also *prepreg*.

inclusion. A physical and mechanical discontinuity occurring within a material or part, usually consisting of solid, encapsulated foreign material. Inclusions are often capable of transmitting some structural stresses and energy fields, but to a noticeably different degree than from the parent material.

indentation hardness. (1) The resistance of a material to indentation. This is the usual type of hardness test, in which a pointed or rounded indenter is pressed into a surface under a substantially static load. (2) Resistance of a solid surface to the penetration of a second, usually harder, body under prescribed conditions. Numerical values used to express indentation hardness are not absolute physical quantities, but depend on the hardness scale used to express hardness. See also *Brinell hardness test*, *Knoop hardness test*, *Rockwell hardness test*, and *Vickers hardness test*.

induction bonding. The use of high-frequency (5 to 7 MHz) electromagnetic fields to heat a bonding agent placed between the plastic parts to be joined. The bonding agent consists of microsized ferromagnetic particles dispersed in a thermoplastic matrix, preferably the parent material of the parts to be bonded. When this binder is exposed to the high-frequency source, the ferromagnetic particles respond and melt the surrounding plastic matrix, which in turn melts the interface surfaces of the parts to be joined.

inert filler. A material added to a plastic to alter the end-item properties through physical rather than chemical means.

infrared (IR). Pertaining to that part of the electromagnetic spectrum between the visible light range and the radar range. Radiant heat is in this range, and infrared heaters are frequently used in the thermoforming and curing of plastics and composites. Infrared analysis is used for identification of polymer constituents.

inhibitor. A substance that retards some specific chemical reaction. Inhibitors are used in certain types of monomers and resins to prolong storage life.

initial modulus. The slope of the initial straight portion of a stress-strain or load-elongation curve. See also *Young's modulus*.

initial recovery. The decrease in strain in a solid during the removal of force before any creep recovery takes place, usually determined at constant temperature. Sometimes referred to as instantaneous recovery.

initial strain. The strain produced in a specimen by given loading conditions before creep occurs.

initial stress. The stress produced by strain in a specimen before stress relaxation occurs. Also called instantaneous stress.

initiator. Sources of free radicals, often peroxides or azo compounds. They are used in free-radical polymerizations, for curing thermosetting resins, and as cross-linking agents for elastomers and cross-linked polyethylene.

injection blow molding. A *blow molding* process in which an injection molded preform is used instead of an extruded parison.

injection molding. Method of forming a plastic to the desired shape by forcing the heat-softened plastic into a relatively cool cavity under pressure. See also *thermoplastic injection molding* and *thermoset injection molding*.

inorganic. Being or composed of matter other than hydrocarbons and their derivatives, or matter that is not of plant or animal origin. Contrast with *organic*.

inorganic pigments. Natural or synthetic metallic oxides, sulfides, and other salts that impart heat and light stability, weathering resistance, color, and migration resistance to plastics.

insert. An integral part of a plastic molding consisting of metal or other material that may be molded or pressed into position after the molding is completed.

insulator. A material of such low electrical conductivity that the flow of current through it can usually be neglected. Similarly, a material of low thermal conductivity, such as that used to insulate structures.

integral composite structure. Composite structure in which several structural elements, which would conventionally be assembled together by bonding or mechanical fasteners after separate fabrication, are instead laid up and cured as a single, complex, continuous structure. The term is sometimes applied more loosely to any composite structure not assembled by mechanical fasteners. All or some parts of the assembly may be cocured.

integral skin foam. Urethane foam with a cellular core structure and a relatively smooth skin.

interface. (1) The boundary between any two phases. Among the three phases (gas, liquid, and solid), there are five types of interfaces: gas-liquid, gas-solid, liquid-liquid, liquid-solid, and solid-solid. (2) The boundary or surface between two different, physically distinguishable media. With fibers, the contact area between the fibers and the sizing or finish. In a laminate, the contact area between the reinforcement and the laminating resin.

interference fit. (1) A joint or mating of two parts in which the male part has an external dimension larger than the internal dimension of the mating female part. Distension of the female by the male creates a stress, which supplies the bonding force for the joint. (2) Any of various classes of fit between mating parts where there is nominally a negative or zero allowance between the parts, and where there is either part interference or no gap when the mating parts are made to the respective extremes of individual tolerances that ensure the tightest fit between the parts.

interlaminar. Between two adjacent laminae, for example, an object (such as a flaw), an event (such as a fracture), or a potential field (such as shear stress). Compare with *intralaminar*.

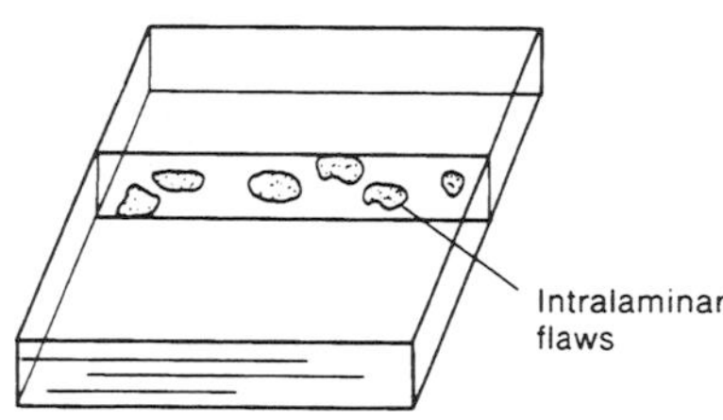

interlaminar shear. Shearing force tending to produce a relative displacement between two laminae in a laminate along the plane of their interface.

intermediate temperature setting adhesive. An adhesive that sets in the temperature range from 30 to 100 °C (87 to 211 °F).

internal shrinkage cracks. Longitudinal cracks in a composite pultrusion that are found within sections of roving reinforcement. This condition is caused by shrinkage strains during cure that show up in the roving portion of the pultrusion, where transverse strength is low.

interphase. The boundary region between a bulk resin or polymer and an adherend in which the polymer has a high degree of orientation to the adherend on a molecular basis. It plays a major role in the load transfer process between the bulk of the adhesive and the adherend or the fiber and the laminate matrix resin.

interply hybrid. A reinforced plastic laminate in which adjacent laminae are composed of different materials.

intralaminar. Within a single lamina, for example, an object (such as a void), and event (such as a fracture), or a potential field (such as a temperature gradient). Compare with *interlaminar*.

intraply hybrid. A reinforced plastic laminate in which more than one material is used within a specific layer.

intrinsic viscosity. For a polymer, the limiting value of infinite dilution of the ratio of the specific viscosity of the polymer solution to its concentration in moles per liter.

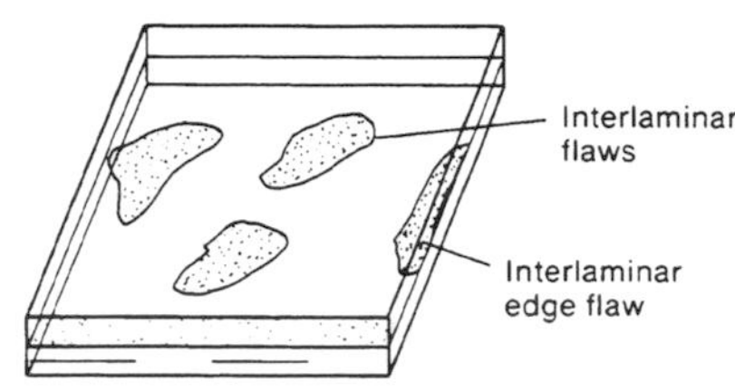

introfaction. The change in fluidity and wetting properties of a polymeric impregnating material, produced by the addition of an *introfier*.

introfier. A chemical that converts a colloidal solution into a molecular one. See also *introfaction*.

ion-exchange resins. Cross-linked polymers that form salts with ions from aqueous solutions.

ionomer resins. A polymer that has ethylene as its major component, but that contains both covalent and ionic bonds. The polymer exhibits very strong interchain ionic forces. The anions hang from the hydrocarbon chain, and the cations are metallic, for example, sodium, potassium, or magnesium. These resins have many of the same features as polyethylene, plus high transparency, tenacity, resilience, and increased resistance to oils, greases, and solvents. Fabrication is accomplished as it is with polyethylene.

iridescence. Loss of brilliance in metallized plastics and development of multicolor reflectance. Iridescence is caused by the cold flow of plastic or of coating and by excess heat during vacuum metallizing.

irradiation. (1) The exposure of a material or object to x-rays, gamma rays, ultraviolet rays, or other ionizing radiation. (2) The bombardment of plastics with a variety of subatomic particles, usually alpha-, beta-, or gamma-rays. Used to initiate the polymerization and copolymerization of plastics and in some cases to bring about changes in the physical properties of a plastic.

irreversible. Not capable of redissolving or remelting. Chemical reactions that proceed in a single direction and are not capable of reversal (as applied to thermosetting resins).

isocyanate plastics. Plastics based on resins made by the condensation of organic isocyanates with other compounds. Generally reacted with polyols on a polyester or polyether backbone molecule, with the reactants being joined by the formation of the urethane linkage. See also *polyurethane* and *urethane plastics*.

isomer. A compound, radical, ion, or nuclide that contains the same number of atoms of the same elements but differs in structural arrangement and properties. See also *stereoisomer*.

isotactic stereoisomerism. A type of polymeric molecular structure containing a sequence of regularly spaced asymmetric groups arranged in like configuration in a polymer chain. Isotactic (and syndiotactic) polymers are crystallizable.

isotropic. Having uniform properties in all directions. The measured properties of an isotropic material are independent of the axis of testing.

isotropy. The condition of having the same values of properties in all directions.

Izod test. A type of impact test in which a V-notched specimen, mounted vertically, is subjected to a sudden blow delivered by the weight at the end of a pendulum arm. The energy required to break off the free end is a measure of the impact strength or toughness of the material. Contrast with *Charpy test*.

J

jet molding. A processing technique for plastics characterized by the fact that most of the heat is applied to the material as it passes through the nozzle or jet, rather than in a heating cylinder, as is done in conventional processes.

jetting. The turbulent flow of resin from an undersized gate or thin section into a thicker mold cavity, as opposed to the laminar flow of material progressing radially from a gate to the extremities of the cavity.

jig. A mechanism for holding a part and guiding the tool during machining or assembly operation.

joint-aging time. See *joint-conditioning time*.

joint-conditioning time. In adhesive bonding, the time interval between the removal of the joint from the conditions of heat or pressure, or both, used to accomplish bonding and the attainment of approximately maximum bond strength. See also *curing time, drying time*, and *setting time*.

jute. A bast fiber obtained from the stems of several species of the plant *Corchorus* found mainly in India and Pakistan. Used as a filler for plastic molding materials, and as a reinforcement for polyester resins in the fabrication of reinforced plastics.

K

Kevlar. An organic polymer composed of aromatic polyamides (*aramids*) having a para-type orientation (parallel chain with bonds extending from each aromatic nucleus). Often used as a reinforcing fiber.

kiln. A large furnace used for baking, drying, or burning firebrick or refractories, or for calcining ores or other substances.

knife coating. A method of coating a substrate (usually paper or fabric) in which the substrate, in the form of a continuous moving web, is coated with a material, the thickness of which is controlled by an adjustable knife or bar set at a suitable angle to the substrate.

knitted fabrics. Fabrics produced by interlooping chains of yarn.

Knoop hardness number (HK). A number related to the applied load and to the projected area of the permanent impression made by a rhombic-based pyramidal diamond indenter having included edge angles of 172° 30′ and 130° 0′ computed from the equation:

$$HK = \frac{P}{0.07028\, d^2}$$

where P is applied load, kgf; and d is the long diagonal of the impression, mm. In reporting Knoop hardness numbers, the test load is stated.

Knoop hardness test. An indentation hardness test using calibrated machines to force a rhombic-based pyramidal diamond indenter having specified edge angles, under specified conditions, into the surface of the material under test and to measure the long diagonal after removal of the load.

knuckle area. In reinforced plastics, the area of transition between sections of different geometry in a filament-wound part, for example, where the skirt joins the cylinder of the pressure vessel. Also called Y-joint.

L

lack of resin fill-out. In reinforced plastics, a condition in which an area contains reinforcement not wetted with a sufficient quantity of resin. This condition usually occurs at the edge of a pultrusion.

lacquer. A coating formulation based on thermoplastic film-forming material dissolved in organic solvent. The coating dries primarily by evaporation of the solvent. Typical lacquers include those based on lac, nitrocellulose, other cellulose derivatives, vinyl resins, acrylic resins, and so forth.

ladder polymer. A polymer with two polymer chains cross linked at intervals.

lamella. The basic morphological unit of a crystalline polymer, usually ribbonlike or platelike in shape. Generally, about 10 nm thick, 1 μm long, and 0.1 μm wide, if ribbonlike.

lamellar thickness. A characteristic morphological parameter in plastics, usually estimated from x-ray studies or electron microscopy, that is usually 10 to 50 nm. The average thickness of lamellae in a specimen. See also *lamella*.

lamina. A single ply or layer in a laminate, which is made up of a series of layers.

laminate. To unite laminae with a bonding material, usually with pressure and heat (normally refers to flat sheets, but also includes rods and tubes); a product made by such bonding. Two or more layers of material bonded together. The term can apply to preformed layers joined by adhesives or by heat and pressure. The term also applies to composites of plastics films with other films, or with foil or paper, even though they have been made by spread coating or by extrusion coating. A reinforced laminate usually refers to superimposed layers of resin-impregnated or resin-coated fabrics or fibrous reinforcements that have been bonded, especially by heat and pressure. Products produced with little or no pressure, such as hand lay-ups, filament-wound structures, and spray-ups, are sometimes called contact-pressure laminates. A single resin-impregnated sheet of paper, fabric, or glass mat is not considered a laminate. Such a single-sheet construction may be called a lamina. See also *bidirectional laminate*,

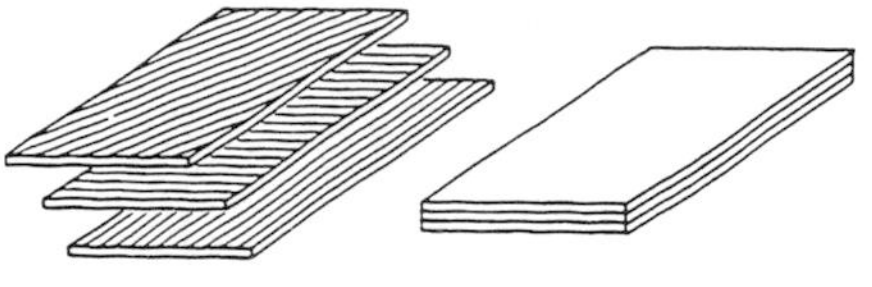

Lamination Laminate

cross laminate, *parallel laminate*, and *unidirectional laminate*.

laminate coordinates. A reference coordinate system used to describe the properties of a laminate, generally in the direction of principal axes, when they exist.

laminated glass. Two or more glasses that are fused together to produce a material with properties that would have been either difficult or impossible to obtain in a single glass. There are many multistep lamination processes. Usually a glass is formed during one process and is subsequently laminated with another glass during a separate process that involves melting the two glasses in separate furnaces and then passing them through a cladding delivery system.

laminate orientation. The geometric configuration of a cross-plied composite laminate with regard to the angles of cross-plying, the number of laminae at each angle, and the exact sequence of laminae lay-up.

laminate ply. One fabric-resin or fiber-resin layer (of a product) that is bonded to adjacent layers in the curing process.

laminate void. The absence of resin in an area that normally contains resin.

lampworking. Forming glass articles from tubing and cane by heating in a gas flame.

land. (1) The horizontal bearing surface of a semipositive or flash mold by which excess material escapes. (2) The bearing surface along the top of the flights of a screw in a screw extruder. (3) The surface of an extrusion die parallel to the direction of melt flow. (4) The land region of a nozzle used in injection molding.

Lanxide process. A composite material formation process which involves pressureless metal infiltration into a ceramic preform. The molten metal is exposed to an oxidizing atmosphere resulting in a matrix material composed of a mixture of the oxidation reaction product and unreacted metal.

lap. (1) In filament winding, the amount of overlay between successive windings, usually intended to minimize gapping. (2) In adhesive bonding, the distance one adherend covers another adherend.

lap joint. A joint made between two overlapping members.

latent curing agent. A curing agent for plastics that produces long-term stability at room temperature but rapid cure at elevated temperatures.

lateral crack. A crack produced beneath and generally paralleling a glass surface during the unloading phase of mechanical contact with a hard, sharp object. See also *cleavage crack*.

lattice pattern. A pattern of a filament winding with a fixed arrangement of open voids.

lay. The length of twist produced by stranding filaments, such as fibers, wires, or roving. The angle that such filaments make with the axis of the strand during a stranding operation. The length of twist of a filament is usually measured as the distance parallel to the axis of the strand between successive turns of the filament.

lay-up. Reinforcing material that is placed in position in the mold. The process of placing the reinforcing material in position in the mold. The resin-impregnated reinforcement. A description of the component materials, geometry, and so forth, of a laminate.

least count. In mechanical testing, the smallest value that can be read from an instrument having a graduated scale.

leno weave. A locking-type weave in which two or more warp threads cross over each other and interlace with one or more filling threads. It is used primarily to prevent the shifting of fibers in open-weave fabrics.

Leno weave

let-go. An area in laminated glass over which an initial adhesion between interlayer and glass has been lost.

level winding. See *circumferential winding*.

ligand. The molecule, ion, or group bound to the central atom in a *chelate* or a *coordination compound*.

light metal. One of the low-density metals, such as aluminum (~2.7 g/cm^3), magnesium (~1.7 g/cm^3), titanium (~4.4 g/cm^3), beryllium (~1.8 g/cm^3), or their alloys.

lime glass. A glass containing a substantial proportion of lime, usually associated with soda and silica.

limited-coordination specification (or standard). A specification (or standard) that has not been fully coordinated and accepted by all interested parties. Limited-coordination specifications and standards are issued to cover the need for requirements unique to one particular department. This applies primarily to military agency documents.

linear elastic fracture mechanics. A method of fracture analysis that can determine the stress (or load) required to induce fracture instability in a structure containing a cracklike flaw of known size and shape. See also *fracture mechanics* and *stress-intensity factor*.

linear expansion. The increase of a given dimension measured by the expansion of a specimen or component subject to a temperature gradient. See also *coefficient of thermal expansion*.

linear (tensile or compressive) strain. The change per unit length due to force in an original linear dimension. An increase in length is considered positive.

liner. In a filament-wound pressure vessel, the continuous, usually flexible coating on the inside surface of the vessel, used to protect the laminate from chemical attack or to prevent leakage under stress.

linters. Short fibers that adhere to the cottonseed after ginning. Used in rayon manufacture as fillers for plastics and as a base for the manufacture of cellulosic plastics.

liquid crystal polymer (LCP). A thermoplastic polymer that contains primarily benzene rings in its backbone, is melt processible, and develops high orientation during molding with resultant improvement in tensile strength and high-temperature capability. First commercial availability was as an aromatic polyamide. Available with or without fiber reinforcement.

liquid-phase sintering. Sintering of a compact or loose powder aggregate under conditions where a liquid phase is present during part of the sintering cycle.

liquid resin. An organic, polymeric liquid that becomes a solid when converted to its final state for use.

liquidus. The maximum temperature at which equilibrium exists between the molten glass and its primary crystalline phase.

load. In the case of testing machines, a force applied to a testpiece that is measured in units such as pound-force, newton, or kilogram-force.

locating ring. In injection molding machine, a ring that serves to align the nozzle of an injection cylinder with the entrance of the sprue bushing and the mold to the machine platen.

long-chain branching. A form of molecular branching found in addition polymers as a result of an internal transfer reaction. It primarily influences the melt flow properties.

longos. Low-angle helical or longitudinal filament windings.

long period. A morphological parameter for plastics obtained from small-angle x-ray scattering. It is usually equated to the sum of the *lamellar thickness* and the amorphous thickness.

loop tenacity. The tenacity or strength value obtained by pulling two loops, such as two links in a chain, against each other to demonstrate the susceptibility of a fibrous material to cutting or crushing; loop strength.

loss factor. The product of the dissipation factor and the dielectric constant of a dielectric material. See also *tan delta*.

loss modulus. A quantitative measure of energy dissipation in polymers, defined as the ratio of stress 90° out of phase with oscillating strain to the magnitude of strain. The loss modulus may be measured in tension or flexure, compression, or shear. See also *complex modulus*.

loss on ignition (LOI). (1) The fractional or percentage weight loss of a material on heating in air from an initial defined state (usually, dried) to a specified temperature, such as 1000 °C (1830 °F), and holding there for a specified period, such as 1 hour. Fixed procedures are designed, usually, such that LOI represents the loss of combined H_2O, CO_2, certain other volatile inorganics, and combustible organic matter. (2) Weight loss, usually expressed as a percent of the total, after burning off an organic sizing from glass fibers, or an organic resin from a glass fiber laminate.

lot. (1) A specific amount of material produced at one time using one process and constant conditions of manufacture, and offered for sale as a unit quantity. (2) A quantity of material that is thought to be uniform in one or more stated properties such as isotopic, chemical, or physical characteristics. (3) A quantity of bulk material of similar composition whose properties are under study. (4) A definite quantity of a product or material accumulated under conditions that are considered uniform for *sampling* purposes. Compare with *batch*.

low-cycle fatigue. *Fatigue* that occurs at relatively small numbers of cycles ($<10^4$ cycles). Low-cycle fatigue may be accompanied by some plastic, or permanent, deformation. Compare with *high-cycle fatigue*.

low-pressure laminates. In general, composite laminates molded and cured in the range of pressures from 2760 kPa (400 psi) down to and including pressure obtained by the mere contact of the plies.

low-pressure molding. The distribution of relatively uniform low pressure (1400 kPa, or 200 psi, or less) over a resin-bearing fibrous assembly of cellulose, glass, asbestos, or other material, with or without application of heat from an external source, to form a structure possessing specific physical properties.

low-profile resins. Special polyester resin systems for reinforced plastics that are combinations of thermoset and thermoplastic resins. Although the terms low-profile and low-shrink are sometimes used interchangeably, there is a difference. Low-shrink resins contain up to 30 wt% thermoplastic polymer, while low-profile resins contain from 30 to 50 wt%. Low shrink offers minimum surface waviness in the molded part (as low as 25 μm per 25 mm, or 1 mil per in., mold shrinkage); low profile offers no surface waviness (from 12.7 to 0 μm per 25 mm, or 0.5 to 0 mil per in., mold shrinkage).

low-shrink resins. See *low-profile resins*.

lyotropic liquid crystal. A type of liquid crystalline polymer that can be processed only from solution.

M

macerate. To chop or shred, as fabric, for use as a filler for a molding resin.

machine shot capacity. The maximum weight of thermoplastic resin that can be displaced or injected by the injection (molding) ram in a single stroke.

machining damage. Atypical or excessively large surface microcracks or damage on ceramics or glasses resulting from the machining process; for example, striations, scratches, impact cracks. Small surface and subsurface damage is intrinsic to the machining damage.

macro. In reinforced plastics, the gross properties of a composite as a structural element without consideration of the individual properties or the identity of the constituents.

macrograph. A graphic representation of the surface of a prepared specimen at a magnification not exceeding 25×. When photographed, the reproduction is known as a photomacrograph.

macroscopic. Visible at magnifications at or below 25×.

macroscopic stress. Residual stress in a material in a distance comparable to the gage length of strain measurement devices (as opposed to stresses within very small, specific regions, such as individual grains). Compare with *microscopic stress*.

macrostrain. The mean strain over any finite gage length of measurement large in comparison with interatomic distances. Macrostrain can be measured by several methods, including electrical-resistance strain gages and mechanical or optical extensometers. Elastic macrostrain can be measured by x-ray diffraction.

magnesia. Magnesium oxide (MgO), used principally in *basic refractories*.

magnetic ceramics. Inorganic nonmetallic materials having properties associated with the phenomena of magnetism, that is, these materials can produce or conduct magnetic lines of force capable of interacting with electric fields or other magnetic fields. Magnetic ceramics form the basis for numerous devices which rely on soft or hard (permanent) magnets. The soft magnets include materials such as the *ferrites* and *garnets*, while the hard magnets include magnetoplumbites and γ-Fe_2O_3. The applications are diverse, from items such as microwave components to recording tape. They are particularly useful for high-frequency devices and can be found in numerous television and radio applications. In thin films applied to nonmagnetic substrates, these types of ceramics form the basis for magnetic bubble memories for computers.

mandrel. (1) In blow molding of thermoplastics, part of the tooling that forms the inside of the container neck through which air is forced to form the hot parison to the shape of the mold. (2) In extrusion of thermoplastics, the solid, cylindrical part of the die that forms tubing or pipe. (3) In filament winding of reinforced plastic, the form (usually cylindrical) around which the filaments are wound.

man-made (synthetic) diamond. A manufactured diamond, darker, blockier, and considered to be more friable than most natural diamonds.

mat. A fibrous glass material used as a plastic reinforcement and consisting of randomly oriented chopped filaments, short fibers (with or without a carrier fabric), or swirled filaments loosely held together with a binder. Available in blankets of various widths, weights, and lengths. Also, a sheet formed by filament winding a single-hoop ply of fiber on a mandrel, cutting across its width and laying out a flat sheet.

matched metal die molding. A reinforced plastics manufacturing process in which matching male and female metal molds are used (for example, in compression molding) to form the part, with time, pressure, and heat.

materials characterization. The use of various analytical methods (spectroscopy, microscopy, chromatography, etc.) to describe those features of composition (both bulk and surface) and structure (including defects) of a material that are significant for a particular preparation, study of properties, or use. Test methods that yield information primarily related to materials properties, such as thermal, electrical, and mechanical properties, are excluded from this definition.

matrix. (1) The essentially homogeneous plastic resin in which the fiber reinforcement is embedded. Both thermoplastic and thermoset resins may be used. (2) The part of an adhesive that surrounds or engulfs embedded filler or reinforcing particles and filaments.

matrix metal. The continuous phase of a polyphase alloy, mechanical mixture, or *metal-matrix composite*; the physically continuous metallic constituent in which separate particles of another constituent are embedded.

maximum elongation. The elongation at the time of fracture, including both elastic and plastic deformation of the tensile specimen. Applicable to rubber, plastic, and some metallic materials. Maximum elongation is also called ultimate elongation or break elongation.

mechanical adhesion. Adhesion between surfaces in which the adhesive holds the parts together by interlocking action.

mechanical hysteresis. Energy absorbed in a complete cycle of loading and unloading within the elastic limit and represented by the closed loop of the stress-strain curves for loading and unloading. Sometimes referred to as elastic, but more properly, mechanical.

mechanically formed plastic. A cellular plastic having a structure produced by physically incorporated gases.

mechanical properties. The properties of a material that reveal its elastic and inelastic behavior when force is applied, thereby indicating its suitability for mechanical applications; for example, modulus of elasticity, tensile strength, elongation, hardness, and fatigue limit. Compare with *physical properties*.

melamine plastics. Thermosetting plastics made from melamine and formaldehyde resins.

melt index. The amount, in grams, of a thermoplastic resin that can be forced through a 2.0955 mm (0.0825 in.) orifice when subjected to 20.7 N (2160 gf) in 10 min at 190 °C (375 °F).

melting point (plastics). The term that refers to the first-order transition in crystalline polymers. The fixed point between the solid and liquid phases of a material when approached from the solid phase under a pressure of 101.325 kPa (1 atm).

melting temperature (glass). The range of furnace temperatures within which melting takes place at a commercially desirable rate and within which the resulting glass generally has a viscosity of $10^{0.5}$ to $10^{1.5}$ Pa · s ($10^{1.5}$ to $10^{2.5}$ P). To compare the melting temperatures of glasses, it is assumed that a glass at its melting temperature has a viscosity of 10 Pa · s (10^2 P).

melt strength. The strength of a plastic while in the molten state.

mer. The repeating structural unit of any polymer.

mesophase. An intermediate phase in the formation of carbon from a pitch precursor. This is a liquid crystal phase in the form of microspheres, which, upon prolonged heating above 400 °C (750 °F), coalesce, solidify, and form regions of extended order. Heating to above 2000 °C (3630 °F) leads to the formation of graphite structure.

metallic fiber. Manufactured fiber composed of metal, plastic-coated metal, metal-coated plastic, or a core completely covered by metal.

metallic whisker. A fiber composed of a single crystal of metal. See also *whisker*.

metallizing. (1) Forming a metallic coating by atomized spraying with molten metal or by *vacuum deposition*. Also called spray metallizing. (2) Applying an electrically conductive metallic layer to the surface of a nonconductor.

metal-matrix composites. *Advanced composites* that consist of a nonmetallic reinforcement incorporated into a metallic matrix. Reinforcements may constitute from 10 to 60 vol% of the composite. Continuous fiber or filament reinforcements include graphite, silicon carbide, boron, alumina, and refractory metals. Matrix materials include aluminum (the most common), titanium, magnesium, and ordered intermetallic compounds (NiAl and Ti_3Al).

methyl methacrylate. A colorless, volatile liquid derived from acetone, cyanohydrin, methanol, and dilute sulfuric acid and used in the production of *acrylic resins*.

M-glass. A high beryllia (BeO_2) content glass designed especially for high modulus of elasticity.

micelle. A submicroscopic unit of structure built up from ions or polymeric molecules.

micro. In relation to reinforced plastics and composites, the properties of the constituents only, that is, matrix, reinforcements, and interface, and their effects on the properties of the composite.

microcracking. Cracks formed in composites when thermal stresses locally exceed the strength of the matrix. Since most microcracks do not penetrate the reinforcing fibers, microcracks in a cross-plied laminate or in a laminate made from cloth prepreg are usually limited to the thickness of a single ply.

microdynamometer. An instrument for measuring mechanical force and observing the change in microscopic appearance of a small specimen.

micrograph. A graphic reproduction of the surface of a specimen at a magnification greater than 25×. If produced by photographic means it is called a photomicrograph (not a microphotograph).

microhardness. The hardness of a material as determined by forcing an indenter such as a Vickers or Knoop indenter into the surface of a material under very light load; usually, the indentations are so small that they must be measured with a microscope. Capable of determining hardnesses of different microconstituents within a structure, or of measuring steep hardness gradients.

microscopic stress. Residual stress in a material within a distance comparable to the grain size. See also *macroscopic stress*.

microstrain. The strain over a gage length comparable to interatomic distances. These are the strains being averaged by the *macrostrain* measurement. Microstrain is not measurable by existing techniques. Variance of the microstrain distribution can, however, be measured by x-ray diffraction.

microstress. Same as *microscopic stress*.

microstructure. The structure of an object, organism, or material as revealed by a microscope at magnifications greater than 25×.

milled fiber. Continuous glass strands hammer milled into very short glass fibers. Useful as inexpensive filler or anticrazing reinforcing fillers for adhesives and engineering plastics.

mock leno weave. An open fabric weave for composites that resembles a leno and is accomplished by a system of interlacings that draws a group of threads together and leaves a space between that group and the next. The warp threads do not actually cross each other as in a real leno and, therefore, no special attachments are required for the loom. This type of weave is generally used when a high thread count is required for strength and the fabric must remain porous.

mode. One of the three classes of crack (surface) displacements adjacent to the crack tip. These displacement modes are associated with stress-strain fields around the crack tip and are designated I, II, and III as shown in the accompanying figures.

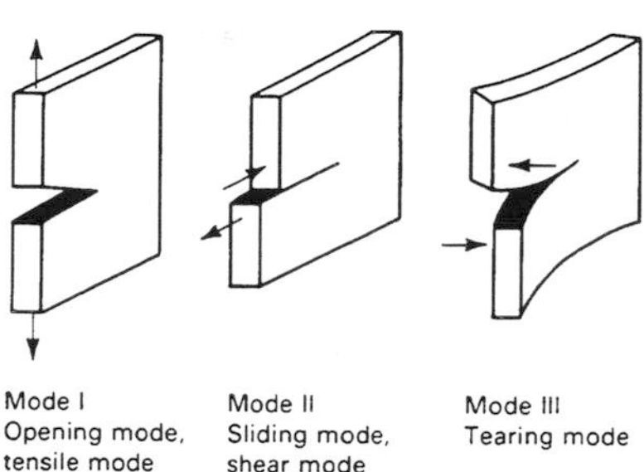

Mode I
Opening mode, tensile mode

Mode II
Sliding mode, shear mode

Mode III
Tearing mode

modified acrylic. A thermoplastic polymer that has been altered to eliminate mixing, curing ovens, and odor and that cures rapidly at room temperature.

modifier. Any chemically inert ingredient added to an adhesive formulation to change its properties. Compare with *filler*, *plasticizer*, and *extender*.

modulus of elasticity (*E*). (1) The measure of rigidity or stiffness of a material; the ratio of stress, below the proportional limit, to the corresponding strain. If a tensile stress of 13.8 MPa (2.0 ksi) results in an elongation of 1.0%, the modulus of elasticity is 13.8 MPa (2.0 ksi) divided by 0.01, or 1380 MPa (200 ksi). (2) In terms of the *stress-strain curve*, the modulus of elasticity is the slope of the stress-strain curve in the range of linear proportionality of stress to strain. Also known as *Young's modulus*. For materials that do not conform to Hooke's law throughout the elastic range, the slope of either the tangent to the stress-strain curve at the origin or at low stress, the secant drawn from the origin to any specified point on the stress-strain curve, or the chord connecting any two specific points on the stress-strain curve is usually taken to be the modulus of elasticity. In these cases, the modulus is referred to as the *tangent modulus*, *secant modulus*, or *chord modulus*, respectively.

modulus of resilience. The amount of energy stored in a material when loaded to its elastic limit. It is determined by measuring the area under the *stress-strain curve* up to the *elastic limit*. See also *resilience*.

modulus of rupture in bending (S_b). The value of maximum tensile or compressive stress (whichever causes failure) in the extreme fiber of a beam loaded to failure in bending computed from the flexure equation:

$$S_b = \frac{Mc}{I}$$

where M is maximum bending moment, computed from the maximum load and the original moment arm; c is the initial distance from the neutral axis to the extreme fiber where failure occurs; and I is the initial moment of inertia of the cross section about the neutral axis. See also *modulus of rupture*.

modulus of rupture in torsion (S_s). The value of maximum shear stress in the extreme fiber of a member of circular cross section loaded to failure in torsion computed from the equation:

$$S_s = \frac{Tr}{J}$$

where T is maximum twisting moment, r is original outer radius, and J is polar moment of inertia of the original cross section. See also *modulus of rupture*.

Mohs hardness. The hardness of a body according to a scale proposed by Mohs, based on ten minerals, each of which would scratch the one below it. These minerals, in decreasing order of hardness, are:

Mineral	Hardness
Diamond	10
Corundum	9
Topaz	8
Quartz	7
Orthoclase (feldspar)	6
Apatite	5
Fluorite	4
Calcite	3
Gypsum	2
Talc	1

moisture absorption. The pickup of water vapor from air by a polymeric material, in reference to vapor withdrawn from the air only, as distinguished from water absorption, which is the gain in weight due to the absorption of water by immersion.

moisture content. The amount of moisture in a polymeric material determined under prescribed conditions and expressed as a percent of the mass of the moist specimen, that is, the mass of the dry substance plus the moisture.

moisture equilibrium. The condition reached by a plastic sample when it no longer takes up moisture from, or gives up moisture to, the surrounding environment.

moisture regain. The moisture in a polymeric material determined under prescribed conditions and expressed as a percent of the weight of the moisture-free specimens. Moisture regain may result from either sorption or desorption, and differs from moisture content only in the basis used for calculation.

moisture vapor transmission (MVT). A rate at which water vapor passes through a polymeric material at a specified temperature and relative humidity.

mold. The cavity into which, or matrix on which, the plastic composition is placed and from which it takes form. To shape plastic parts or finished articles by heat and pressure. The assembly of all the parts that function collectively in the molding process.

molded edge. An edge on a plastic part that is not physically altered after molding for use in final form, particularly one that does not have fiber ends along its length.

molded net. Description of a molded plastic part that requires no additional processing to meet dimensional requirements.

molding. The forming of a polymer or composite into a solid mass of prescribed shape and size by the application of pressure and heat for a given time. The finished part.

molding compound. Plastic material in varying stages of pellet form or granulation (powder), consisting of resin, filler, pigments, reinforcements, plasticizers, and other ingredients, ready for use in the molding operation. Also called molding powder.

molding cycle. The period of time required for the complete sequence of operations on a molding press to produce one set of plastic moldings. The operations necessary to produce a set of moldings without reference to the total time.

molding pressure. The pressure applied to the ram of an injection machine, compression press, or transfer press to force the softened plastic to fill the mold cavities completely.

mold release agent. A lubricant, liquid, or powder (often silicon oils and waxes) used to prevent sticking of molded plastic articles in the cavity.

mold shrinkage. (1) The immediate shrinkage that a molded plastic part undergoes when it is removed from a mold and cooled to room temperature. (2) The difference in dimensions, expressed in inches per inch, between a molding and the mold cavity in which it was molded (at normal-temperature measurement). (3)

The incremental difference between the dimensions of the molding and the mold from which it was made, expressed as a percentage of the mold dimensions.

mold surface. The side of a laminate that faced the mold (tool) during cure in an autoclave or hydroclave.

molecular mass. The sum of the atomic mass of all atoms in a molecule. In *high polymers*, because the molecular masses of individual molecules vary widely, they must be expressed as averages. The average molecular mass of polymers may be expressed as number-average molecular mass or mass-average molecular map.

molecular weight. The sum of the atomic weights of all the atoms in a molecule. Atomic weights (and therefore molecular weights) are relative weights arbitrarily referred to an assigned atomic weight of exactly 12.0000 for the most abundant isotope of carbon, ^{12}C. See also *atomic weight*.

moly-manganese process. A common method of joining oxide ceramics (most commonly alumina, Al_2O_3) whereby a mixture of molybdenum and manganese powders (typically 10 at.% Mn) is applied on the ceramic surface, and then sintered to bond to the ceramic. The sintering temperature is usually above 1400 °C (2500 °F) in a hydrogen (H_2) atmosphere with a controlled dew point. Subsequent nickel plating in conjunction with this Mo-Mn metallized layer facilitates the application and adherence of brazing filler metals that would not bond to the original ceramic substrate.

monofilament. A single fiber or filament of indefinite length that is strong enough to function as a yarn in commercial textile operation.

monolayer. (1) The basic laminate unit from which cross-plied or other laminate types are constructed. (2) A "single" layer of atoms or molecules adsorbed on or applied to a surface.

monolithic. An object comprised entirely of one massive piece (although polycrystalline or even heterogeneous) as opposed to being built up of preformed units.

monolithic refractory. A refractory which may be installed *in situ* without joints to form an integral structure.

monomer. A single molecule that can react with like or unlike molecules to form a polymer. The smallest repeating structure of a polymer (mer). For addition polymers, this represents the original unpolymerized compound.

morphology. The overall physical form of the physical structure of a bulk polymer. Common units are lamellae, spherulites, and domains.

mounting resin. Thermosetting (e.g., Bakelite or diallyl phthalate) or thermoplastic (e.g., methyl methacrylate or polyvinyl chloride) resins used to mount metallographic specimens.

mucilage. An adhesive prepared from a gum and water, and also in a more general sense, a liquid adhesive that has a low order of bonding strength. See also *adhesive*, *glue*, *paste*, and *sizing*.

multicircuit welding. A filament winding that requires more than one circuit before the band repeats by laying adjacent to the first band.

multifilament yarn. A large number (500 to 2000) of fine, continuous filaments (consisting of 5 to 100 individual filaments), usually with some twist in the yarn to facilitate handling.

multiple-layer adhesive. A dry-film adhesive, usually supported, with a different adhesive composition on each side; designed to bond dissimilar materials such as the core to face bond of a sandwich composite. See also *honeycomb* and *sandwich construction*.

multiple-screw extruder. An extruder machine for processing thermoplastics that has two or four screws (conical or constant depth), in contrast to conventional single-screw extruders. Types include machines with intermeshing counter-rotating screws, intermeshing corotating screws, and nonintermeshing counter-rotating screws.

N

natural diamond. The densest form of crystallized carbon and the hardest substance known, natural diamond occurs most commonly as well-developed crystals in volcanic pipes or in alluvial deposits. Bort sometimes refers to all diamonds not suitable for gems, or it may refer to off-color flawed or impure diamonds not fit for use for gems or most other industrial applications, but suitable for the preparation of diamond grain and powder for use in lapping or the manufacture of most diamond grinding wheels. This type of bort is also called crushing bort or fragmented bort. See also *diamond* and *man-made diamond*.

NDE. See *nondestructive evaluation*.

NDI. See *nondestructive inspection*.

NDT. See *nondestructive testing*.

neat resin. Resin to which nothing (additives, reinforcements, and so on) has been added.

neck-in. In *extrusion coating* of plastics, the difference between the width of the extruded web as it leaves the die and the width of the coating on the substrate.

necking. (1) The reduction of the cross-sectional area of a material in a localized area by uniaxial tension or by stretching. (2) The reduction of the diameter of a portion of the length of a cylindrical shell or tube.

needle blow. A specific *blow molding* technique in which the blowing air is injected into the hollow article through a sharpened, hollow needle that pierces the parison.

needled mat. A *mat* for reinforcing plastics formed of strands cut to a short length, then felted together in a needle loom with or without a carrier.

nesting. In reinforced plastics, the placing of plies of fabric such that the yarns of one ply lie in the valleys between the yarns of the adjacent ply. Also called nested cloth.

netting analysis. The analysis of filament-wound structures that assumes that the stresses induced in the structure are carried entirely by the filaments, and the strength of the resin is neglected, and that assumes also that the filaments possess no bending or shearing stiffness and carry only the axial tensile loads.

nitrocellulose. See *cellulose nitrate*.

node. The connected portion of adjacent ribbons of honeycomb.

NOL ring. A parallel filament- or tape-wound hoop test specimen developed by the Naval Ordnance Laboratory (NOL) (now the Naval Surface Weapons Laboratory), for measuring various mechanical strength properties of a material, such as tension and compression, by testing the entire ring or segments of it. Also known as parallel fiber reinforced ring.

nominal stress. The stress at a point calculated on the net cross section without taking into consideration the effect on stress of geometric discontinuities, such as holes, grooves, fillets, and so forth. The calculation is made using simple elastic theory.

nominal value. A value assigned for the purpose of a convenient designation. A nominal value exists in name only. In dimensions, it is often an average number with a tolerance in order to fit together with adjacent parts.

nondestructive evaluation (NDE). Broadly considered synonymous with nondestructive inspection (NDI). More specifically, the quantitative analysis of NDI findings to determine whether the material will be acceptable for its function, despite the presence of discontinuities. With NDE, a discontinuity can be classified by its size, shape, type, and location, allowing the investigator to determine whether or not the flaw(s) is acceptable. Damage tolerant design approaches are based on the philosophy of ensuring safe operation in the presence of flaws.

nondestructive inspection (NDI). A process or procedure, such as ultrasonic or radiographic inspection, for determining the quality or characteristics of a material, part, or assembly, without permanently altering the subject or its properties. Used to find internal anomalies in a structure without degrading its properties or impairing its serviceability.

nondestructive testing (NDT). Broadly considered synonymous with *nondestructive inspection (NDI).*

nonhygroscopic. Lacking the property of absorbing and retaining an appreciable quantity of moisture (water vapor) from the air.

nonresonant forced and vibration technique. A technique for performing dynamic mechanical measurements in which a plastic sample is oscillated mechanically at a fixed frequency. Storage modulus and damping are calculated from the applied strain and the resultant stress and shift in phase angle.

nonrigid plastic. For purposes of general classification, a plastic that has a modulus of elasticity either in flexure or in tension of not over 70 MPa (10 ksi) at 23 °C (73 °F) and 50% relative humidity.

nonwoven fabric. A planar textile structure for reinforced plastics produced by loosely compressing together fibers, yarns, rovings, and so forth, with or without a scrim cloth carrier. Accomplished by mechanical, chemical, thermal, or solvent means, or combinations thereof.

normal stress. The stress component that is perpendicular to the plane on which the forces act. Normal stress may be either *tensile* or *compressive*.

notched specimen. A test specimen that has been deliberately cut or notched, usually in a V-shape, to induce and locate point of failure.

notch factor. Ratio of the resilience determined on a plain specimen to the resilience determined on a notched specimen.

notch sensitivity. The extent to which the sensitivity of a material to fracture is increased by the presence of a *stress concentration*, such as a notch, a sudden change in cross section, a crack, or a scratch. Low notch sensitivity is usually associated with ductile materials, and high notch sensitivity is usually associated with brittle materials.

notch strength. The maximum load on a notched tension-test specimen divided by the minimum cross-sectional area (the area at the root of the notch). Also called notch tensile strength.

novolac. A linear, thermoplastic, two-stage phenolic resin, which, in the presence of methylene or other cross-linking groups, reacts to form a thermoset phenolic. See also *phenolic resin*.

nozzle. The hollow-cored metal nose screwed into the injection end of either the heating cylinder of an injection machine or a transfer chamber when this is a separate structure. A nozzle is designed to form under pressure a seal between the heating cylinder or the transfer chamber and the mold. The shape of the front end of a nozzle may be either flat or spherical. See also *reciprocating-screw injection molding*.

nucleating agent. A foreign substance, often crystalline, usually added to a crystallizable polymer to increase its rate of solidification during processing.

nylon. The generic name, by common usage, for all synthetic *polyamides*.

nylon plastics. Plastics based on a resin composed principally of a long-chain synthetic polymeric amide that has recurring amide groups as an integral part of the main polymer chain. Numerical designations (nylon 6, nylon 6/6, and so on) refer to the monomeric amides from which they are made. Characterized by great toughness and elasticity. See also *polyamide*.

O

offset. The distance along the strain coordinate between the initial portion of a stress-strain curve and a parallel line that intersects the stress-strain curve at a value of stress (commonly 0.2%) that is used as a measure of the *yield strength*. Used for materials that have no obvious *yield point*. See also *stress-strain curve*.

offset yield strength. The stress at which the strain exceeds by a specific amount (the *offset*) an extension of the initial, approximately linear, proportional portion of the stress-strain curve. It is expressed in force per unit area.

olefin. A group of unsaturated hydrocarbons of the general formula C_nH_{2n} and named after the corresponding paraffins by the addition of -ene or sometimes -ylene to the root.

olefin plastics. Plastics based on polymers made by the polymerization of olefins with other monomers, the olefins being at least 50 mass %.

oligomer. A polymer consisting of only a few monomer units, for example, a dimer, trimer, tetramer, and so forth, or their mixtures.

one-component adhesive. An adhesive material incorporating a latent hardener or catalyst that is activated by heat.

one-shot molding. In the urethane foam field, a system in which the isocyanate, polyol, catalyst, and other additives are mixed together directly and a foam is produced immediately. Compare with *prepolymer molding*.

open assembly time. The time interval between the spreading of the adhesive on the adherend and the completion of the assembly of the parts for bonding.

open-cell cellular plastic. Foamed or cellular plastic with cells that are generally interconnected.

open-cell foam. Foamed or cellular polymeric material with cells that are generally interconnected. Closed cell refers to cells that are not interconnected.

optical glass. Glass of high quality having closely specified optical properties, used in the manufacture of plano or curved windows and refractive and reflective elements for precision instruments and devices. Most products are silicates designed for maximal transmittance in the visible spectrum. Some glasses are prepared from extremely pure raw materials under special conditions to yield high ultraviolet transmission. Among infrared transmitting glasses the chalcide glass As_2S_3, fused silica (waterfree, mostly prepared from natural quartz), and calcium-aluminate glasses are most common.

orange peel (ceramics). (1) A surface condition characterized by an irregular waviness of the porcelain enamel resembling an orange skin in texture; sometimes considered a defect. (2) A pitted texture of a fired glaze resembling the surface of rough orange peel.

orange peel (plastics). In injection molding of plastics, a part with an undesirable uneven surface somewhat resembling the skin of an orange.

organic. Being or composed of hydrocarbons or their derivatives, or matter of plant or animal origin. Contrast with *inorganic*.

organic acid. A chemical compound with one or more carboxyl radicals (COOH) in its structure; examples are butyric acid, $CH_3(CH_2)_2COOH$; maleic acid HOOCCH:CHCOOH; and benzoic acid, C_6H_5COOH.

organic fiber. A fiber derived or composed of matter originating in plant or animal life or composed of chemicals of hydrocarbon origin, either natural or synthetic.

organosol. A suspension of a finely divided resin in a volatile, organic liquid. The resin does not dissolve appreciably in the organic liquid at room temperature, but does at elevated temperatures. The liquid evaporates at an elevated temperature, and the residue upon cooling is a homogeneous plastic mass. Plasticizers may be dissolved in the volatile liquid.

orientation. The alignment of the crystalline structure in polymeric materials to produce a highly aligned structure. Orientation can be accomplished by cold drawing or stretching in fabrication. Orientation can generally be divided into two classes, uniaxial and biaxial.

oriented materials. Polymeric materials with molecules and/or macroconstituents that are aligned in a specific way. Oriented materials are anisotropic.

orthotropic. Having three mutually perpendicular planes of elastic symmetry.

outgassing. (1) The release of adsorbed or occluded gases or water vapor, usually by heating, as from a vacuum tube or other vacuum system. (2) Devolatilization of plastics or applied coatings during exposure to vacuum in vacuum metallizing. Resulting parts show voids or thin spots in plating with reduced and spotty brilliance. Additional drying prior to metallizing is helpful, but outgassing is inherent to plastic materials and coating ingredients, including plasticizers and volatile components.

out time. The time a *prepreg* is exposed to ambient temperature, namely, the total amount of time the prepreg is out of the freezer. The primary effects of out time are a decrease in the *drape* and *tack* of the prepreg and the absorption of moisture from the air.

ovaloid. A surface of revolution symmetrical about the polar axis that forms the end closure for a filament-wound cylinder.

oven dry. The condition of a plastic material that has been heated under prescribed conditions of temperature and humidity until there is no further significant changes in its mass.

ovenware. Glass and glass-ceramic materials able to withstand thermal downshock of 150 °C (270 °F) without breakage that are used for culinary oven use.

overglaze. A glass coating over another component or element, normally used to give physical or electrical protection.

overlay sheet. A nonwoven fibrous mat (of glass synthetic fiber, for example) used as the top layer in a cloth or mat *lay-up* to provide a smoother finish, minimize the appearance of the fibrous pattern, or permit machining or grinding to a precise dimension. Also called surfacing mat.

oxidation. In carbon/graphite fiber processing, the step of reacting the precursor polymer (rayon, polyacrylonitrile (PAN), or pitch) with oxygen, resulting in stabilization of the structure for the hot stretching operation.

P

PA. See *polyamide*.

package. Yarn, roving, and so forth for reinforcing plastics in the form of units capable of being unwound and suitable for handling, storing, shipping, and use.

PAEK. See *polyaryletherketone*.

PAI. See *polyamide-imide*.

PAN. See *polyacrylonitrile*.

parabolic reflector. A reflector for ultraviolet curing of adhesives that projects parallel light beams perpendicular to the assembly, but is not focused.

parallel laminate. (1) A composite laminate of woven fabric in which the plies are aligned in the same position as they were on the fabric roll. (2) A series of flat or curved cloth-resin layers stacked uniformly one on top of the other.

paramagnetic material. (1) A material whose specific permeability is greater than unity and is practically independent of the magnetizing force. (2) Material with a small positive susceptibility due to the interaction and independent alignment of permanent atomic and electronic magnetic moments with the applied field. Compare with *ferromagnetic material*.

paramagnetism. A property exhibited by substances that, when placed in a magnetic field, are magnetized parallel to the field to an extent proportional to the field (except at very low temperatures or in extremely large magnetic fields). Compare with *ferromagnetism*.

parison. The hollow plastic tube from which a plastic component is blow molded.

parison swell. In blow molding of plastics, the tendency of the *parison* to enlarge as it emerges from the die. It is expressed as the ratio of the cross-sectional area of the parison to the cross-sectional area of the die opening. See also *blow molding*.

partially stabilized zirconia. Zirconia (ZrO_2) which contains a mixture of cubic and tetragonal and/or monoclinic phases produced by the addition of small amounts of magnesium oxide (MgO), calcium oxide (CaO), or yttrium oxide (Y_2O_3).

particulate composite. Material consisting of one or more constituents suspended in a matrix of another material. These particles are either metallic or nonmetallic.

parting agent. See *mold release agent*.

parting line. (1) The intersection of the parting plane of a casting or plastic mold or the parting plane between forging dies with the mold or die cavity. (2) A raised line or projection on the surface of a casting, plastic part, or forging that corresponds to said intersection.

PAS. See *polyaryl sulfone*.

PBI. See *polybenzimidazole*.

PBT. See *polybutylene terephthalate*.

PC. See *polycarbonate*.

PEEK. See *polyether etherketone*.

peel ply. In composites fabrication, a layer of open-weave material, usually fiberglass or heat set nylon, applied directly to the surface of a prepreg lay-up. The peel ply is removed from the cured laminate immediately before bonding operations, leaving a clean, resin-rich surface that needs no further preparation for bonding, other than the application of a primer if one is required. See also *lay-up* and *prepreg*.

peel strength. The average load per unit width of bond line required to separate progressively one member from the other over the adhered surfaces at a separation angle of approximately 180° and a separation rate of 152 mm (6 in.) per minute. It is expressed in force per unit width of specimen (N/mm, or lbf/in.).

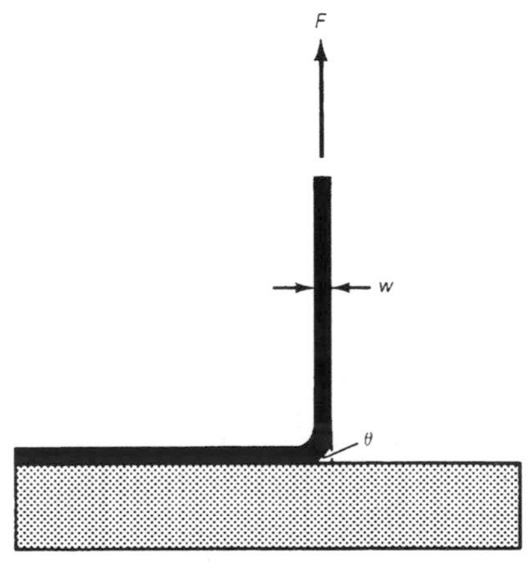

Peel test setup

peel test (adhesives). See *peel strength*.

PEI. See *polyether-imide*.

penetration. The entering of an adhesive into an adherend. This property of a system is measured by the depth of penetration of the adhesive into the adherend.

permanence. (1) The property of a plastic that describes its resistance to appreciable change in characteristics with time and environment. (2) The resistance of an adhesive bond to deteriorating influences.

permanent set. The deformation remaining after a specimen has been stressed a prescribed amount in tension, compression, or shear for a specified time period and released for a specified time period. For creep tests, the residual unrecoverable deformation after the load causing the creep has been removed for a substantial and specified period of time. Also, the increase in length, expressed as a percentage of the original length, by which an elastic material fails to return to its original length after being stressed for a standard period of time.

permeability. (1) The passage or diffusion (or rate of passage) of a gas, vapor, liquid, or solid through a material (often porous) without physically or chemically affecting it; the measure of fluid flow (gas or liquid) through a material. (2) A general term used to express various relationships between magnetic induction and magnetizing force. These relationships are either "absolute permeability," which is a change in magnetic induction divided by the corresponding change in magnetizing force, or "specific (relative) permeability," the ratio of the absolute permeability to the permeability of free space.

peroxy compounds. Polymeric compounds containing O–O linkage.

PESV. See *polyether sulfone*.

PET. See *polyether terephthalate*.

petrography. The study of nonmetallic matter under suitable microscopes to determine structural relationships and to identify the phases or minerals present. With transparent materials, the determination of the optical properties, such as the indices of refraction and the behavior in transmitted polarized light, are means of identification. With opaque materials, the color, hardness, reflectivity, shape, and etching behavior in polished sections are means of identification.

pH. The negative logarithm of the hydrogen-ion activity; it denotes the degree of acidity or basicity of a solution. At 25 °C (77 °F), 7.0 is the neutral value. Decreasing values below 7.0 indicates increasing acidity; increasing values above 7.0, increasing basicity. The pH values range from 0 to 14.

phenolic resin. A thermosetting resin produced by the condensation of an aromatic alcohol with an aldehyde, particularly of phenol with formaldehyde. Used in high-temperature applications with various fillers and reinforcements. The resin in a single-stage compound, because of its reactive groups, is capable of further polymerization by the application of heat. In a two-stage compound, the resin is essentially not reactive at normal stage temperatures, but contains a reactive additive that causes further polymerization with the application of heat.

phenoxy resin. A high-molecular-weight thermoplastic polyester resin (polyhydroxy ether) based on bisphenol-A and epichlorohydrin. The material is available in grades suitable for injection molding and extrusion, and for application as coatings and adhesives.

phenylsilane resins. Thermosetting copolymers of silicone and phenolic resins. Furnished in solution form.

phosphate glass. A glass in which the essential glass former is phosphorus pentoxide (P_2O_5) instead of silica. Also used in *glass-ceramics* and glass tableware.

photopolymer. A polymer that changes characteristics when exposed to light of a given frequency.

phthalate esters. The most widely used group of *plasticizers*, produced by the direct action of alcohol on phthalic anhydride. The phthalates are generally characterized by moderate cost, good stability, and good all-around properties.

physical blowing agent. A gas, such as a fluorocarbon, that is pumped into a mold during the forming of plastics.

physical catalyst. Radiant energy capable of promoting or modifying a chemical reaction.

physical properties. Properties of a material that are relatively insensitive to structure and can be measured without the application of force; for example, density, electrical conductivity, coefficient of thermal expansion, magnetic permeability, and lattice parameter. Does not include chemical reactivity. Compare with *mechanical properties*.

PI. See *polyimide*.

PIC. See *pressure-impregnation-carbonization*.

pick count. In reinforced plastics, the number of *tows*/mm (in.) of woven fabric in both the *warp* and *fill* directions.

piezoelectric. A material or crystal which becomes polarized and its surface becomes charged when a stress is applied to it. Conversely, if the material is subject to an electric field, it will expand in one direction and contract in another.

piezoelectric polymers. Polymers that spontaneously give an electric charge when mechanically stressed or that develop a mechanical response when an electric field is applied. Used as transducers or acoustic sensors.

pimple. An imperfection, such as a small protuberance of varied shape on the surface of a plastic product.

pinch-off. In blow molding of plastics, a raised edge around the cavity in the mold that seals off the part and separates the excess material as the mold closes around the *parison*.

pinholes. Small cavities that penetrate the surface of a cured composite or plastic part.

pit. A small, regular or irregular crater in the surface of a material created by exposure to the environment, for example, corrosion, wear, or thermal cycling.

pitch. A high molecular weight material that is a residue from the destructive distillation of coal and petroleum products. Pitches are used as base materials for the manufacture of certain high-modulus carbon fibers.

plain weave. A weaving pattern in which the warp and fill fibers alternate; that is, the repeat pattern is warp/fill/ warp/fill, and so on. The two sides, or faces, of a plain weave are identical. Properties are significantly reduced relative to a weaving pattern with fewer crossovers. See also *basket weave*, *crowfoot weave*, and *leno weave*.

Plain weave

planar. Lying essentially in a single plane.

planar helix winding. In *filament winding* of composites, a winding in which the filament path on each dome lies on a plane that intersects the dome, while a helical path over the cylindrical section is connected to the dome paths.

planar winding. In *filament winding* of composites, a winding in which the filament path lies on a plane that intersects the winding surface. See also *polar winding*.

plastic. A material that contains as an essential ingredient an organic polymer of large molecular weight; is solid in its finished state; and, at some stage in its manufacture or its processing into finished articles, can be shaped by flow. Although materials such as rubber, textiles, adhesives, and paint may in some cases meet this definition, they are not considered plastics. The terms plastic, resin, and polymer are somewhat synonymous, but the terms resins and polymers most often denote the basic material as polymerized, while the term plastic encompasses compounds containing plasticizers, stabilizers, fillers, and other additives.

plastic flow. (1) Deformation of plastics under the action of a sustained hot or cold force. (2) Flow of semisolids in the molding of plastics.

plastic foam. See preferred term *cellular plastic*.

plasticizer. (1) A material incorporated in a plastic to increase its workability, flexibility, or distensibility. Normally used in thermoplastics. (2) A material added to a plastic (or polymer) of lower molecular weight to reduce stiffness and brittleness, resulting in a lower glass transition temperature for the polymer. (3) A material added to an adhesive to cause a reduction in melt viscosity, lower the temperature of the second-order transition, or lower the elastic modulus of the solidified adhesive. See also *modifier*.

plastic memory. The tendency of a thermoplastic material that has been stretched while hot to return to its unstretched shape upon being reheated.

plastigel. A *plastisol* exhibiting gel-like flow properties. One having an effective yield value.

plastisols. Mixtures of vinyl resins and plasticizers that can be molded, cast, or converted to continuous films by the application of heat. If the mixtures contain volatile thinners as well, they are known as *organosols*.

plastometer. An instrument for determining the flow properties of a thermoplastic resin by forcing the molten resin through a die or orifice of specific size at a specified temperature and pressure.

plate glass. Flat glass formed by a rolling process, ground and polished on both sides, with surfaces essentially plane and parallel.

platen. The mounting plates of a plastic forming press, to which the entire mold assembly is bolted.

plied yarn. Yarn for reinforced plastics made by collecting two or more single yarns. Normally, the yarns are twisted together, though sometimes they are collected without twist.

plug-assist forming. A *thermoforming* process in which a plug or male mold is used to partially preform the plastic part before forming is completed using vacuum or pressure.

ply. A single layer in a *laminate*. In general, fabrics or felts consisting of one or more layers. Yarn resulting from a twisting operation (for example, three-ply yarn consists of three strands of yarn twisted together). In *filament winding*, a ply is a single pass (two plies forming one layer).

PMMA. See *polymethyl methacrylate*.

PMR polyimides. A novel class of high temperature resistant polymers. PMR represents *in situ* polymerization of monomer reactants. See also *polyimide (PI)*.

Poisson's ratio (ν). The absolute value of the ratio of transverse (lateral) strain to the corresponding axial strain resulting from uniformly distributed axial stress below the *proportional limit* of the material.

polar winding. In filament winding of composites, a winding in which the filament path passes tangent to the polar opening at one end of the chamber and tangent to the opposite side of the polar opening at the other end. A one-circuit pattern is inherent in the system.

polepiece. In reinforced plastics, the supporting part of the mandrel used in filament winding, usually on one of the axes of rotation.

polyacrylate. A thermoplastic resin made by the polymerization of an acrylic compound.

polyacrylonitrile (PAN). A base material or precursor used in the manufacture of certain carbon fibers. PAN-based carbon fibers have a ribbonlike structure and possess high strength (400 GPa, or 10^6 psi, tensile modulus).

polyamide (PA). A thermoplastic polymer in which the structural units are linked by amide or thio-amide groupings (repeated nitrogen and hydrogen groupings). Many polyamides are fiber forming.

polyamide-imide (PAI) resins. A family of polymers based on the combination of trimellitic anhydride with aromatic diamines. In the uncured form (orthoamic acid) the polymers are soluble in polar organic solvents. The imide linkage is formed by heating, producing an infusible resin with thermal stability up to 260 °C (500 °F). These resins, which also include graphite- (powder and fiber) and glass fiber-reinforced grades, are used in the automotive (friction and wear parts), aerospace (fasteners and housings), and electronic (connectors industries).

polyamide plastic. See *nylon plastics* and *polyamide (PA)*.

polyarylates (PAR). A family of aromatic polyesters derived from aromatic dicarboxylic acids and diphenols. These thermoplastics have an amorphous molecular structure and are tough, durable, and heat resistant. They have excellent dimensional and ultraviolet stability, electrical properties, flame retardance, and warp resistance.

polyaryletherketone (PAEK). A linear aromatic crystalline thermoplastic. A composite with a PAEK matrix may have a continuous-use temperature as high as 250 °C (480 °F). This definition is also germane to polyetheretherketone (PEEK), polyetherketone (PEK), and polyetherketoneketone (PEKK).

polyaryl sulfone (PAS). A thermoplastic resin consisting mainly of phenyl and biphenyl groups linked by thermally stable ether and sulfone groups. Its most outstanding property is resistance to high and low temperatures (from –240 to 260 °C, or –400 to 500 °F). It also has good impact resistance, resistance to chemical oils, and most solvents, and good electrical insulating properties. It can be processed by injection molding, extrusion, compression molding, ultrasonic welding, and machining.

polybenzimidazole (PBI). A thermoplastic resin that is strong and stable, has a high molecular weight, and contains recurring aromatic units. The resin is produced by the high-temperature, melt condensation reaction of aromatic bis-ortho-diamines and aromatic dicarboxylates (acids, esters, or amides). Although a wide variety of PBIs have been synthesized, the most common is poly (2,2′-(*m*-phenylene)-5-5′-bibenzimidazole. Parts made from PBI are used in petrochemical, geothermal, chemical process, power generation, aerospace, and transportation markets.

polybutylenes. A group of polymers consisting of isotactic, stereoregular, highly crystalline polymers based on butylene-1. Their properties are similar to those of *polypropylene* and linear polyethylene, with superior toughness, creep resistance, and flexibility.

polybutylene terephthalate (PBT). A member of the polyalkyleneterephthalate family that is produced by the transesterification of dimethyl terephthalate with butanediol by means of a catalyzed melt polycondensation. Properties include high strength, dimensional stability, low moisture absorption, good electrical characteristics, and resistance to heat and chemicals when suitably modified.

polycarbonate (PC). A thermoplastic polymer derived from the direct reaction between aromatic and aliphatic dihydroxy compounds with phosgene or by the ester exchange reaction with appropriate phosgene-derived precursors. Highest impact resistance of any transparent plastic.

polycondensation. See *condensation polymerization*.

polyester plastics. Plastics based on resins composed principally of polymeric esters, in which the recurring ester groups are an integral part of the main polymer chain, and in which ester groups occur in most cross links that may be present between chains. See also *thermoplastic polyesters*, *thermosetting polyesters*, and *unsaturated polyesters*.

polyetheretherketone (PEEK). A linear aromatic crystalline thermoplastic. A composite with a PEEK matrix may have a continuous-use temperature as high as 250 °C (480 °F). See also *polyaryletherketone (PAEK)*.

polyether-imide (PEI). An amorphous high-performance thermoplastic with a chemical structure that consists of repeating aromatic imide and ether units. PEIs are characterized by high strength and rigidity at room and elevated temperatures, long-term high heat resistance, highly stable dimensional and electrical properties, and broad chemical resistance. Their high glass-transition temperature (215 °C, or 419 °F) allows PEI to be used for short-term use at 200 °C (390 °F).

polyether sulfone (PES). A high-temperature engineering thermoplastic consisting of repeating phenyl groups linked by thermally stable ether and sulfone groups. The material has good transparency and flame resistance, and is one of the lowest smoke-emitting materials available. Both polymer and reinforced grades are available in granular form for extrusion or injection molding.

polyethylene (PE) plastics. Thermoplastic materials composed of ethylene. They are normally translucent, tough, waxy solids that are unaffected by water and by a large range of chemicals. In common usage, these plastics have no less than 85% ethylene and no less than 95% total olefins. See also *high-density polyethylene* and *ultrahigh molecular weight polyethylene*.

polyethylene terephthalate (PET). A saturated, thermoplastic polyester resin made by condensing ethylene glycol and terephthalic acid and used for fibers, films, and injection-molded parts. It is extremely hard, wear resistant, dimensionally stable, and resistant to chemicals, and it has good dielectric properties. Also known as polyethylene glycol terephthalate.

polyimide (PI). A polymer produced by reacting an aromatic dianhydride with an aromatic diamine. It is similar to a polyamide, differing only in the number of hydrogen molecules contained in the groupings. This polymer is suitable for use as a binder or adhesive, and its exceptional thermomechanical properties make it suitable for many high-temperature (≥230 °C, or 450 °F) applications. See also *thermoplastic polyimides (TPI)*.

polymer. A high molecular weight organic compound, natural or synthetic, with a structure that can be represented by a repeated small unit, the mer. Examples include polyethylene, rubber, and cellulose. Synthetic polymers are formed by addition or condensation polymerization of monomers. Some polymers are *elastomers*, some are *plastics*, and some are *fibers*. When two or more dissimilar monomers are involved, the product is called a copolymer. The chain lengths of commercial thermoplastics vary from ~1000 to >100,000 repeating units. Thermosetting polymers approach infinity after curing, but their resin precursors, often called prepolymers, may be relatively short—6 to 100 repeating units—before curing. The lengths of polymer chains, usually measured by molecular weight, have very significant effects on the performance properties of plastics and profound effects on processibility.

polymerization. A chemical reaction in which the molecules of a *monomer* are linked together to form large molecules with a molecular weight that is a multiple of the molecular weight of the original substance. When two or more monomers are involved, the process is called copolymerization.

polymer matrix. The resin portion of a reinforced or filled plastic. See also *resin-matrix composites*.

polymethyl methacrylate (PMMA). A thermoplastic polymer synthesized from methyl methacrylate. It is a transparent solid with exceptional optical properties. Available in the form of sheets, granules, solutions, and emulsions, it is used as facing material in certain composite constructions. See also *acrylic plastic*.

polyol. An alcohol having many hydroxyl groups. Also known as polyhydric alcohol or polyalcohol. In cellular plastics, usage, the term includes compounds containing alcoholic hydroxyl groups such as polyethers, glycols, polyesters, and castor oil used in urethane foams, and other polyurethanes. See also *alcohols*.

polyolefins. Plastics based on a polymer made with an *olefin* as essentially the sole *monomer*.

polyoxymethylene (POM). Acetal plastics based on polymers in which oxymethylene is essentially the sole repeating structural unit in the chains. See also *acetal (AC) resins*.

polyphenylene oxides (PPO). Thermoplastic, linear, noncrystalline polyethers obtained by the oxidative polycondensation of 2,6-dimethylphenol in the presence of a copper-amine complex catalyst. These resins have a useful temperature range from less than –170 to 190 °C (–275 to 375 °F) with intermittent use up to 205 °C (400 °F) possible, excellent electrical properties, unusual resistance to acids and bases, and processibility on conventional extrusion and injection molding equipment. Also known as polyphenylene ether (PPE).

polyphenylene sulfides (PPS). Crystalline polymers having a symmetrical, rigid backbone chain consisting of recurring para-substituted benzene rings and sulfur atoms. Known for excellent chemical resistance, thermal stability, and fire resistance. Its inertness to organic solvents, inorganic salts, and bases makes it corrosion resistant. Commercial engineering grades are always fiber reinforced.

polypropylenes (PP). Tough, lightweight thermoplastics made by the polymerization of high-purity propylene gas in the presence of an organometallic catalyst at relatively low pressures and temperatures. See also *reinforced polypropylene*.

polystyrenes (PS). Water-white homopolymer thermoplastics produced by the polymerization of styrene (vinyl benzene). Has outstanding electrical properties, good thermal and dimensional stability, and staining resistance. Because it is somewhat brittle, it is often copolymerized or blended with other materials to obtain desired properties. See also *high-impact polystyrene (HIPS)*.

polysulfide. A synthetic polymer containing sulfur and carbon linkages, produced from organic dihalides and sodium polysulfide. The material is elastomeric in nature, resistant to light, oil, and solvents, and impermeable to gases.

polysulfones (PSU). A family of sulfur-containing thermoplastics made by reacting bisphenol A and 4,4′-dichlorodiphenylsulfone with potassium hydroxide in dimethyl sulfoxide at 130 to 140 °C (265 to 285 °F). The structure of the polymer is benzene rings or phenylene units linked by three different chemical groups, a sulfone group, an ether linkage, and an isopropylidene group. Polysulfones are characterized by high strength, the highest service temperature of all melt-processible thermoplastics, low creep, good electrical characteristics, transparency, self-extinguishing properties, and resistance to greases, many solvents, and chemicals. They may be processed by extrusion, injection molding, or blow molding.

polyterephthalate. A thermoplastic polyester in which the terephthalate group is a repeating structural unit in the chain, the terephthalate being greater in amount than other dicarboxylates that may be present.

polytetrafluoroethylenes (PTFE). Members of the fluorocarbon family of plastics made by the polymerization of tetrafluoroethylene. PTFE is characterized by its extreme inertness to chemicals, very high thermal stability, and low frictional properties. Among the applications for these materials are bearings, fuel hoses, gaskets and tapes, and coatings for metal and fabric.

polyurethanes (PUR). A large family of polymers with widely varying properties and uses. All of these polymers are based on the reaction product of an organic isocyanate with compounds containing a hydroxyl group. The reaction product of an isocyanate with an alcohol is called a urethan, according to rules of chemical nomenclature, but the terms urethane and polyurethane are more widely used in the plastics industry. Polyurethanes may be thermosetting or thermoplastic, rigid or soft and flexible, cellular or solid. The properties of any of these types may be tailored within wide limits to suit the desired application.

polyvinyl acetals. Members of the family of vinyl plastics. Polyvinyl acetal is the general name for resins produced from a condensation of polyvinyl alcohol with an aldehyde. There are three main groups: polyvinyl acetal itself, *polyvinyl butyral*, and *polyvinyl formal*. Polyvinyl acetal resins are thermoplastics that can be processed by casting, extruding, molding, and coating, but their main uses are in adhesives, lacquers, coatings, and films.

polyvinyl acetate (PVAC). A thermoplastic material composed of polymers of vinyl acetate in the form of a colorless solid. It is obtainable in the form of granules, solutions, latices, and pastes and is used extensively in adhesives, for paper and fabric coatings, and in bases for inks and lacquers.

polyvinyl acetate emulsion adhesive. A latex adhesive in which the polymeric portion comprises polyvinyl acetate, copolymers based mainly on polyvinyl acetate, or a mixture of these and which may contain modifiers and secondary binders to provide specific properties.

polyvinyl alcohol (PVAL). A thermoplastic material composed of polymers of the hypothetical vinyl alcohol. Usually a colorless solid, insoluble in most organic solvents and oils, but soluble in water when the content of hydroxy groups in the polymer is sufficiently high. The product is normally granular. It is obtained by the partial hydrolysis or by the complete hydrolysis of polyvinyl esters, usually by the complete hydrolysis of polyvinyl acetate. It is mainly used for adhesives and coatings.

polyvinyl butyral (PVB). A thermoplastic material derived from a polyvinyl ester in which some or all of the acid groups have been replaced by hydroxyl groups, and some or all of these hydroxyl groups have been replaced by butyral groups by reaction with butyraldehyde. It is a colorless, flexible, tough solid used primarily in interlayers for laminated safety glass. See also *polyvinyl acetals*.

polyvinyl carbazole. A thermoplastic resin, brown in color, obtained by reacting acetylene with carbazole. The resin has excellent electrical properties and good heat and chemical resistance. It is used as an impregnant for paper capacitors.

polyvinyl chloride acetate. A thermoplastic material composed of copolymers of vinyl chloride and vinyl acetate. It is a colorless solid with good resistance to water as well as concentrated acids and alkalies. It is obtainable in the form of granules, solutions, and emulsions. Compounded with plasticizers, it yields a flexible material superior to rubber in aging properties. It is widely used for cable and wire coverings, in chemical plants, and in protective garments.

polyvinyl chloride (PVC). A thermoplastic material composed of polymers of vinyl chloride. It is a colorless solid with outstanding resistance to water, alcohols, and concentrated acids and alkalies. It is obtainable in the form of granules, solutions, latices, and pastes. Compounded with plasticizers, it yields a flexible material superior to rubber in aging properties. It is widely used for cable and wire coverings, in chemical plants, and in the manufacture of protective garments.

polyvinyl formal (PVF). One of the groups of polyvinyl acetal resins made by the condensation of formaldehyde in the presence of polyvinyl alcohol. It is used mainly in combination with cresylic phenolics, for wire coatings, and for impregnations, but can also be molded, extruded, or cast. It is resistant to greases and oils.

polyvinylidene chloride (PVDC). A thermoplastic material composed of polymers of vinylidene chloride (1,1-dichloroethylene). It is a white powder with a softening temperature at 185 to 200 °C (365 to 390 °F). The material is also supplied as a copolymer with acrylonitrile or vinyl chloride, giving products that range from the soft, flexible type to the rigid type. Also known as saran.

polyvinylidene fluoride. This recent member of the fluorocarbon family of plastics is a homopolymer of vinylidene fluoride. It is supplied as powders and pellets for molding and extrusion and in solution form for casting. The resin has good tensile and compressive strength and high impact strength.

POM. See *polyoxymethylene (POM)*.

porcelain. A glazed or unglazed vitreous ceramic whiteware used for technical purposes. This term designates such products as electrical, chemical, mechanical, structural, and thermal wares when they are vitreous. This term is frequently used as a synonym for *china*.

porcelain enamel. A substantially vitreous or glassy, inorganic coating (borosilicate glass) bonded to metal by fusion at a temperature above 425 °C (800 °F). Porcelain enamels are applied primarily to components made of sheet iron or steel, cast iron, aluminum, or aluminum-coated steels.

porosity. Fine holes or pores within a solid; the amount of these pores is expressed as a percentage of the total volume of the solid.

porous molds. Molds for forming plastics that are made up of bonded or fused aggregate (powdered metal, coarse pellets, and so forth) such that the resulting mass contains numerous open interstices of regular or irregular size allowing either air or liquids to pass through the mass of the mold.

porous region. A three-dimensional zone of porosity or microporosity of higher concentration than is normally found in the ceramic matrix.

porous seam. A two-dimensional area of porosity or microporosity of higher concentration than is normally found in the ceramic matrix.

positive mold. A mold for forming plastics designed to trap all the molding material when it closes.

postcure (adhesives). A treatment (normally involving heat) applied to an adhesive assembly following the initial cure, to modify specific properties. To expose an adhesive assembly to an additional cure, following the initial cure, for the purpose of modifying specific properties.

postcure (plastics). Additional elevated-temperature cure of a plastic, usually without pressure, to improve final properties and/or complete the cure, or decrease the percentage of volatiles in the compound. In certain resins, complete cure and ultimate mechanical properties are attached only by exposure of the cured resin to higher temperatures than those of curing. See also *cure*.

postforming. The forming, bending, or shaping of fully cured, C-staged thermoset laminates that have been heated to make them flexible. Upon cooling, the formed laminate retains the contours and shape of the mold over which it has been formed. See also *C-stage*.

pot. To embed a component or assembly in liquid resin, using a shell, can, or case that remains an integral part of the product after the resin is cured.

pottery. A generic term for all fired ceramic wares that contain clay, except for technical, structural, and refractory products. Specifically, however, pottery describes the low-temperature fired porous ware that is usually colored. The term is properly applied to the clay products of primitive peoples, or to decorative art products made of unrefined clays by using unsophisticated methods.

pot life. The length of time that a catalyzed thermosetting resin system retains a viscosity low enough to be used in processing. Also called working life.

powder molding. General term used to denote several techniques, such as injection molding, for producing objects of varying size and shape by melting plastic powder, usually against the inside of a mold. The molds are either stationary (for example, in variations of slush molding techniques) or rotating (for example, in variations of rotational molding).

PP. See *polypropylenes*.

PPE. Polyphenylene ether. See also *polyphenylene oxides*.

PPO. See *polyphenylene oxides*.

PPS. See *polyphenylene sulfides*.

preconditioning. Any preliminary exposure of a plastic to specified atmospheric conditions for the purpose of favorably approaching equilibrium with a prescribed atmosphere.

precure. The full or partial setting of a synthetic resin or adhesive in a joint before the clamping operation is complete or before pressure is applied. See also *cure*.

precursor. With respect to carbon or graphite fiber, the rayon, polyacrylonitrile (PAN), or pitch fibers from which carbon and graphite fibers are derived. See also *carbon fiber*, *pitch*, and *polyacrylonitrile (PAN)*.

prefit. A process for checking the fit of mating detail parts in an assembly prior to *adhesive bonding* to ensure proper bond lines. Mechanically fastened structures are sometimes prefitted to establish shimming requirements.

preform (ceramics). A porous ceramic mass in the shape of the desired final part that is infiltrated with metal to form ceramic-metal composite. See also *Lanxide process*.

preform (composites and plastics). A preshaped fibrous reinforcement formed by the distribution of chopped fibers or cloth by air, water flotation, or vacuum over the surface of a perforated screen to the approximate contour and thickness desired in the finished part. Also, a preshaped fibrous reinforcement of mat or cloth formed to the desired shape on a mandrel or mock-up before being placed in a mold press. (2) A compressed tablet or biscuit of plastic composition used for efficiency in handling and accuracy in weighing materials. (3) To make plastic molding powder into pellets or tablets.

preform binder. A resin applied to the chopped strands of a *preform*, usually during its formation, and cured so that the preform will retain its shape and can be handled.

pregel. An unintentional, extra layer of cured resin on part of the surface of a reinforced plastic. Not related to gel coat.

preheating. The heating of a polymeric compound before molding or casting to facilitate the operation or reduce the molding cycle.

preimpregnation. The practice of mixing resin and reinforcement and effecting partial cure before use or shipment to the user. See also *prepreg*.

premix. A plastic molding compound prepared prior to and apart from the molding operations and containing all components required for molding: resin, reinforcement, fillers, catalysts, release agents, and other ingredients.

premolding. In composites fabrication, the *lay-up* and partial *cure* at an intermediate cure temperature of a laminated or chopped-fiber detail part to stabilize its configuration for handling and assembly with other parts for final cure.

preplasticization. Technique of premelting injection molding plastic powders in a separate chamber, then transferring the melt to the injection chamber. See also *injection molding*.

preply. A composite material *lamina* in the raw-material stage, ready to be fabricated into a finished *laminate*. The lamina is usually combined with other raw laminae before fabrication. A preply includes a fiber system that is placed in position relative to all or part of the required matrix material to constitute the finished lamina. An organic matrix preply is called a *prepreg*.

prepolymer. A chemical intermediate with a molecular weight between that of the *monomer* or monomers and the final *polymer* or resin.

prepolymer molding. In the urethane foam field, a system in which a portion of the *polyol* is prereacted with the isocyanate to form a liquid *prepolymer* with a viscosity range suitable for pumping or metering. This component is supplied to end-users with a second premixed blend of additional polyol, catalyst, blowing agent, and so forth. When the two components are subsequently mixed, foaming occurs. See also *one-shot molding*.

prepreg. In composites fabrication, either ready-to-mold material in sheet form or ready-to-wind material in *roving* form, which may be cloth, mat, unidirectional fiber, or paper, impregnated with resin and stored for use. The resin is partially cured to a *B-stage* and supplied to the fabricator, who lays up the finished shape and completes the cure with heat and pressure. The two distinct types of prepreg available are commercial prepregs, in which the roving is coated with a hot melt or solvent system to produce a specific product to meet specific customer requirements; and wet prepreg, in which the basic resin is installed without solvents or preservatives but has limited room-temperature shelf-life.

preproduction test. A test or series of tests conducted by an adhesive manufacturer to determine conformity of an adhesive batch to established production standards, or by a fabricator to determine the quality of an adhesive before parts are produced, or by an adhesive specification custodian to determine conformance of an adhesive to the requirements of a specification not requiring qualification tests. Compare with *acceptance test* and *qualification test*.

press clave. In composites fabrication, a simulated autoclave made by using the platens of a press to seal the ends of an open chamber, providing both the force required to prevent loss of the pressurizing medium and the heat required to cure the laminate inside.

pressure bag molding. A process for molding reinforced plastics in which a tailored, flexible bag is placed over the contact lay-up on the mold, sealed, and clamped in place. Fluid pressure, usually provided by compressed air or water, is placed against the bag, and the part is cured. See also *vacuum bag molding*.

pressure break. In laminated plastics, a break in one or more outer sheets of the paper, fabric, or other base, which is visible through the surface layer of resin that covers it.

pressure bubble plug-assist forming. A *thermoforming* process in which a heated plastic sheet is positioned over the mold and air is blown up through the base plate channel causing the sheet to billow upward. The plug assist pushes the sheet into the mold cavity and vacuum is applied to transfer the sheet from the plug to the mold surface.

pressure forming. A thermoforming process in which pressure is used to push the sheet to be formed against the mold surface, as opposed to using a vacuum to suck the sheet flat against the mold.

pressure-impregnation-carbonization (PIC). A densification process for carbon-carbon composites involving pitch impregnation and carbonization under high temperature and isostatic pressure conditions. This process is carried out in hot isostatic pressing equipment.

pressure intensifier. In composites fabrication, a layer of flexible material (usually a high-temperature rubber) used to ensure the application of sufficient pressure to a location, such as a radius, in a *lay-up* being cured.

pressure-sensitive adhesive. A viscoelastic material that, in solvent-free form, remains permanently tacky. Such material will adhere instantaneously to most solid surfaces with the application of very light pressure.

primary nucleation. The mechanism by which crystallization is initiated in plastics, often by an added nucleation agent.

primer. A coating applied to a surface, prior to the application of an adhesive to improve the performance of the bond. The coating can be a low-viscosity fluid that is typically a 10% solution of the adhesive in an organic solvent, which can wet out the adherend surface to leave a coating over which the adhesive can readily flow. See also the figure accompanying *adhesive bonding*.

processing window. In forming of plastics, the range of processing conditions, such as stock (melt) temperature, pressure, shear rate, and so on, within which a particular grade of plastic can be fabricated with optimum or acceptable properties by a particular fabricating process, such as extrusion, injection molding, sheet molding, and so forth. The processing window for a particular plastic can vary significantly with design of the part and the mold, with the fabricating machinery used, and with the severity of the end-use stresses.

promoter. A chemical, itself a feeble catalyst, that greatly increases the activity of a given catalyst. Used in formulating plastics. See also *accelerator*.

proof. To test a component or system at its peak operating load or pressure.

proof load. A predetermined load, generally some multiple of the service load, to which a specimen or structure is submitted before acceptance for use.

proof pressure. The test pressure that pressurized components must sustain without detrimental deformation or damage. The proof pressure test is used to give evidence of satisfactory workmanship and material quality.

proof stress. (1) A specified stress to be applied to a member or structure to indicate its ability to withstand service loads. (2) The stress that will cause a specified small *permanent set* in a material.

propylene plastics. Plastics based on polymers of propylene or copolymers of propylene with other monomers, the propylene being the greatest amount by mass. See also *polypropylenes (PP)*.

prototype. A model suitable for use in the complete evaluation of form, design, performance, and material processing.

PS. See *polystyrenes*.

pseudoplastic behavior. A decrease in viscosity with increasing shear stress.

PSU. See *polysulfones*.

PTFE. See *polytetrafluoroethylenes*.

puckers. Local areas on prepreg material where the material has blistered and pulled away from the separator film or release paper. See also *prepreg*.

pulled surface. In laminated plastics, imperfections in the surface, ranging from a slight breaking or lifting in localized areas to pronounced separation of the surface from the body.

pulp molding. The process by which a resin-impregnated pulp material is preformed by application of a vacuum and subsequent molding or curing.

pultrusion. A continuous process for manufacturing composites that have a constant cross-sectional shape. The process consists of pulling a fiber-reinforcing material through a resin impregnation bath and then through a shaping die, where the resin is subsequently cured.

PUR. See *polyurethanes*.

purple plague. A failure mechanism in electronic components which involves the formation of brittle intermetallic compounds at aluminum wire/gold bonding pad intersections. Both moisture and temperature enhance the formation of such compounds. The term "purple plague" is used because one of the five compounds that can form appears purple to the eye when viewed through a microscope.

PVAC. See *polyvinyl acetate*.

PVAL. See *polyvinyl alcohol*.

PVB. See *polyvinyl butyral*.

PVC. See *polyvinyl chloride*.

PVDC. See *polyvinylidene chloride*.

PVDF. See *polyvinylidene fluoride*.

PVF. See *polyvinyl formal*.

pyrolysis. With respect to fibers, the thermal process by which organic precursor fiber materials, such as rayon, polyacrylonitrile (PAN), and pitch, are chemically changed into carbon fiber by the action of heat in an inert atmosphere. Pyrolysis temperatures can range from 800 to 2800 °C (1470 to 5070 °F), depending on the precursor. Higher processing graphitization temperatures of 1900 to 3000 °C (3450 to 5430 °F) generally lead to higher-modulus carbon fibers, usually referred to as graphite fibers. During the pyrolysis process, molecules containing oxygen, hydrogen, and nitrogen are driven from the precursor fiber, leaving continuous chains of carbon. See also *carbon fiber*.

Q

qualification test. A series of tests conducted by the procuring activity, or an agent thereof, to determine the conformance of materials, or materials system, to the requirements of a specification, normally resulting in a qualified products list under the specification. Generally, qualification under a specification requires a conformance to all tests in the specification, or it may be limited to conformance to a specific type or class, or both, under the specification. Compare with *acceptance test* and *preproduction test*.

qualified products list (QPL). A list of commercial plastic products that have been pretested and found to meet the requirements of a specification, especially a government specification.

quality. (1) The totality of features and characteristics of a product or service that bear on its ability to satisfy a given need (fitness-for-use concept of quality). (2) Degree of excellence of a product or service (comparative concept). Often determined subjectively by comparison against an ideal standard or against similar products or services available from other sources. (3) A quantitative evaluation of the features and characteristics of a product or service (quantitative concept).

quality factor. (1) In plastics, the ratio of elastic modulus to loss modulus, measured in tension, compression, flexure, or shear. This is a nondimensional term and is the reciprocal of tan delta. (2) The reciprocal of the electrical dissipation factor, which is the ratio of the power loss in a dielectric material to the total power transmitted through it; thus, the imperfection of the dielectric. Equal to the tangent of the loss angle or to the ratio of the dielectric loss to the dielectric constant.

quartz. One of several crystalline forms of silica (SiO_2); others include cristobalite and tridymite. All occur as minerals, also synthetic. Quartz is harder than most minerals, being 7 on the Mohs scale, and the crushed material is used as an abrasive. Fused quartz, or quartz glass is used for lightbulbs, optical glass, crucibles, and for tubes and rods in furnaces.

quartz glass. See *quartz*.

quasi-isotropic laminate. A laminate approximating isotropy by orientation of plies in several or more directions.

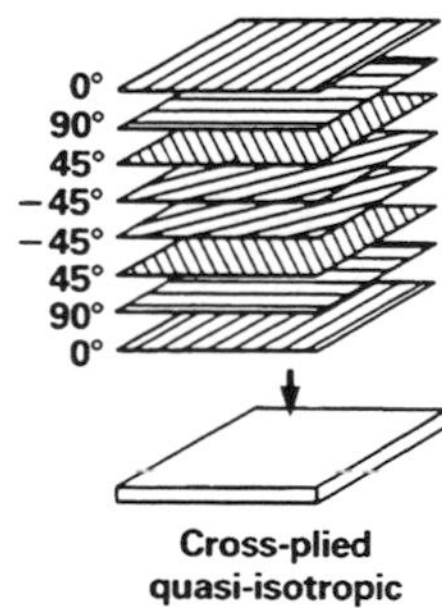

Cross-plied quasi-isotropic

quench. A process of shock-cooling thermoplastic materials from the molten state.

R

RA. See *reduction in area*.

radio frequency (RF) preheating. A method of preheating plastic molding materials to facilitate the molding operation and/or reduce molding cycle time. The frequencies most commonly used are those between 10 and 100 MHz.

radio frequency welding. A method of welding thermoplastics using a radio frequency field to apply the necessary heat. Also known as high-frequency welding.

radius of bend. The radius of the cylindrical surface of the pin or mandrel that comes in contact with the inside surface of the bend during bending. In the case of free or semiguided bends to 180° in which a shim or block is used, the radius of bend is one-half the thickness of the shim or block.

ram travel. In forming of plastics, the distance the injection ram moves to fill the mold in injection or transfer molding.

random pattern. (1) In *filament winding* of composites, a winding with no fixed pattern. If a large number of circuits is required for the pattern to repeat, a random pattern is approached. (2) A winding in which the filaments do not lie in an even pattern.

reaction injection molding (RIM). A process for molding polyurethane, epoxy, and other liquid chemical systems. Mixing two to four components in the proper chemical ratio is accomplished by a high-pressure impingement-type mixing head, from which the mixed material is delivered into the mold at low pressure, where it reacts (cures). See also *injection molding*.

reactive diluent. As used in epoxy formulations, a compound containing one or more epoxy groups that functions mainly to reduce the viscosity of the mixture.

reciprocating-screw injection molding. A combination injection and plasticating unit in which an extrusion device with a reciprocating screw is used to plasticate the material. The injection of material into a mold can take place by direct extrusion into the mold, by reciprocating the screw as an injection plunger, or by a combination of the two. When the screw serves as an injection plunger, this unit acts as a holding, measuring, and injection chamber. See also *injection molding*.

reduction in area (RA). The difference between the original cross-sectional area of a tensile specimen and the smallest area at or after fracture as specified for the material undergoing testing. Also known as reduction of area.

refractories. Nonmetallic (ceramic) materials that resist degradation by corrosive gases, liquids, or solids at elevated temperatures.

refractoriness. In *refractories*, the capability of maintaining a desired degree of chemical and physical identity at high temperatures and in the environment and conditions of use.

refractory. (1) A material (usually an inorganic, nonmetallic, ceramic material) of very high melting point with properties that make it suitable for such uses as furnace linings and kiln construction. (2) The quality of resisting heat.

regrind. Waste plastic material (such as sprues, runners, excess parison material, and reject parts from injection molding, blow molding, or extrusion operations) that has been reclaimed by shredding or granulating. Regrind is usually mixed with virgin compound, at a predetermined percentage, for remolding.

reinforced molding compound. A plastic reinforced with special fillers or fibers (glass, synthetic fibers, minerals, and so forth) to meet specific requirements.

reinforced plastics. Molded, formed, filament-wound, tape-wrapped, or shaped plastic parts consisting of resins to which reinforcing fibers, mats, fabrics, and so forth, have been added before the forming operation to provide some strength properties greatly superior to those of the base resin. See also *resin-matrix composites*.

reinforced polypropylene. Polypropylene that is reinforced with mineral fillers, such as talc, mica, and calcium carbonate, as well as glass and carbon fibers. The maximum concentration usually used is 50 wt%, although concentrates with higher levels of filler or reinforcement are available.

reinforced reaction injection molding (RRIM). A reaction injection molding with a reinforcement added. See also *reaction injection molding*.

reinforcement. A strong material bonded into a matrix to improve its mechanical properties. Reinforcements are usually long fibers, chopped fibers, whiskers, particulates, and so forth. The term is not synonymous with *filler*. The most commonly used reinforcement materials are glass (E-glass and S-glass), aramid, silicon carbide, boron, alumina, fused silica, alumina-boria-silica, and carbon/graphite.

relative humidity. (1) The ratio of the molecular fraction of water vapor present in the air to the molecular fraction of water vapor present in saturated air at the same temperature and barometric pressure. Approximately, it equals the ratio of the partial pressure or density of the water vapor in the air to the saturation pressure or density, respectively, at the same temperature. (2) The ratio, expressed as a percentage, of the amount of water vapor present in a given volume of air at a given temperature to the amount required to saturate the air at that temperature.

relative rigidity. In dynamic mechanical measurements of plastics, the ratio of modulus at any temperature, frequency, or time to the modulus at a reference temperature, frequency, or time.

relative viscosity. For a polymer in solution, the ratio of the absolute viscosities of the solution (of stated concentration) and of the pure solvent at the same temperature.

relaxation curve. A plot of either the remaining or relaxed stress as a function of time. See also *relaxation rate*.

relaxation rate. The absolute value of the slope of a relaxation curve at a given time.

relaxation time. The time required for a stress under a sustained constant strain to diminish by a stated fraction of its initial value.

relaxed stress. The initial stress minus the remaining stress at a given time during a stress-relaxation test.

release agent. In forming of plastics, a material that is applied as a thin film to the surface of a mold to keep the resin from bonding to the mold. Also called parting agent. See also *mold release agent*.

release film. In forming of plastics, an impermeable layer of film that does not bond to the resin being cured. See also *separator*.

reprocessed plastic. A thermoplastic prepared from melt-processed scrap or reject parts by a plastics processor, or from nonstandard or nonuniform virgin material. The term scrap does not necessarily connote feedstock that is less desirable or usable than virgin material. Reprocessed plastic may or may not be reformulated by the addition of fillers, plasticizers, stabilizers, or pigments.

resenes. The constituents of *rosin* that cannot be saponified with alcoholic alkali, but that contain carbon, hydrogen, and oxygen in the molecule. See also *saponification*.

residual strain. The strain associated with residual stress.

residual stress. The stress existing in a body at rest, in equilibrium, at uniform temperature, and not subjected to external forces. Often caused by the forming or thermal processing curing process.

resilience. (1) The amount of energy per unit volume released on unloading. (2) The capacity of a material, by virtue of high yield strength and low elastic modulus, to exhibit considerable elastic recovery on release of load.

resin. A solid, semisolid, or pseudosolid organic material that has an indefinite and often high molecular weight, exhibits a tendency to flow when subjected to stress, usually has a softening or melting range, and usually fractures conchoidally. (2) Liquid resin, that is, an organic polymeric liquid, which, when converted to its final state for use, becomes a resin. (3) An organic polymer that cross links to form a thermosetting plastic when mixed with a curing agent. (4) In reinforced plastics, the material used to bind together the reinforcement material; the matrix. See also *polymer*, *reinforced plastics*, and *resin-matrix composites*.

resin content. The amount of *resin* in a *laminate*, expressed as either a percentage of total weight or total volume.

resin-matrix composites. *Advanced composites* that contain a reinforcement (such as fibers or particles) supported by an organic (plastic) binder (matrix). Resin-matrix composite materials can be divided into two basic categories: continuous fiber reinforced and particulate/short fiber reinforced composites. Continuous fiber composites, which are the most common type of composite, are typically made up of 3 to 30 μm diam fibers that are oriented and surrounded in a supportive matrix material. Glass fiber reinforced organic-matrix composites are the most familiar and widely used, and have had extensive application in industrial, consumer, military, and aerospace markets. Carbon fiber reinforced resin-matrix composites are the most commonly applied (nonfiberglass) composites. A variety of thermosetting and thermoplastic polymers are used as matrix materials.

resinography. The science of morphology, structure, and related descriptive characteristics as correlated with composition or conditions and with properties or behavior of resins, polymers, plastics, and their products.

resinoid. Any of the class of thermosetting synthetic resins, either in their initial temporarily fusible state or in their final infusible state. See also *novolac* and *thermosetting*.

resinoid bond. An organic bond usually of the phenol formaldehyde resin type but sometimes consisting of other synthetic resins.

resin pocket. In plastics, an apparent accumulation of excess resin in a small, localized section visible on cut edges of molded surfaces, or internal to the structure and nonvisible. See also *resin-rich area*.

resin-rich area. A significant thickness of a nonreinforced resin layer of the same composition as that within the base material. In *reinforced plastics*, a localized area filled with resin and lacking reinforcing material. See also *resin pocket*.

resin-starved area. In *reinforced plastics*, a localized area of insufficient resin, usually identified by low glass, dry spots, or fiber showing on the surface.

resin system. A mixture of *resin* and ingredients such as catalyst, initiator, diluents, and so forth, required for the intended processing method and final product.

resin transfer molding (RTM). A process by which catalyzed resin is transferred or injected into an enclosed mold in which reinforcement has been placed. The fiberglass reinforcement is usually woven, nonwoven, or knitted fabric.

resistance. The opposition that a device or material offers to the flow of direct current, equal to the voltage drop across the element divided by the current through the element. Also called electrical resistance.

resistivity. See *electrical resistivity*.

resite. Synonym for *C-stage*.

resitol. Synonym for *B-stage*.

resole. Synonym for *A-stage*.

resole resin. Linear phenolic resin produced by alkaline condensate of phenol and formaldehyde.

resonant forced vibration technique. A technique for performing dynamic mechanical measurements on plastics, in which the sample is oscillated mechanically at the natural resonant frequency of the system. The amplitude of oscillation is maintained constant by the addition of makeup energy. Elastic modulus is calculated from the measured frequency. Damping is calculated from the additional energy required to maintain constant amplitude.

retarder. Synonym for *inhibitor*.

retrogradation. A change of starch pastes that are used in adhesive formulations from low to high consistency upon aging.

reverse helical winding. In *filament winding*, as the fiber delivery arm traverses one circuit, a continuous helix is laid down, reversing direction at the polar ends, in contrast to biaxial, compact, or sequential winding. The fibers cross each other at definite equators, the number depending on the helix. The minimum region of crossover is three.

reverse impact test. A test in which one side of a sheet of plastic is struck by a pendulum or falling object, and the reverse side is inspected for damage.

reworked plastic. A thermoplastic that is reground, pelletized, or solvated after having been previously processed by molding, extrusion, and so forth. In many specifications, the use of reworked material is limited to clean plastic that meets the requirements specified for virgin material and yields a product essentially equal in quality to one made from virgin material only.

rheology. The science of deformation and the flow of matter.

rheopectic material. A material that shows an increase in viscosity with time under a constant shear stress. After removal of the shear stress, the viscosity slowly returns to its original level. Compare with *thixotropy*.

rib. A reinforcing member designed into a plastic part to provide lateral, horizontal, hoop, or other structural support.

rigid plastics. For purposes of general classification, a plastic that has a modulus of elasticity either in flexure or in tension greater than 690 MPa (100 ksi) at 23 °C (73 °F) and 50% relative humidity.

rigid resin. A resin having a modulus high enough to be of practical importance, for example, ≥690 MPa (100 ksi).

RIM. See *reaction injection molding*.

rise time. In urethane foam molding, the time between the pouring of the urethane mix and the completion of foaming.

Rockwell hardness number. A number derived from the net increase in the depth of impression as the load on an indenter is increased from a fixed minor load to a major load and then returned to the minor load. Various scales of Rockwell hardness numbers have been developed based on the hardness of the materials to be evaluated. The scales are designated by alphabetic suffixes to the hardness designation. For example, 64 HRC represents the Rockwell hardness number of 64 on the Rockwell C scale.

Rockwell hardness test. An indentation hardness test using a calibrated machine that utilizes the depth of indentation, under constant load, as a measure of hardness. Either a 120° diamond cone with a slightly rounded point or a 1.6 or 3.2 mm (1/16 or 1/8 in.) diam steel ball is used as the indenter.

room temperature. A temperature in the range of ~20 to 30 °C (~68 to 85 °F). The term room temperature is usually applied to an atmosphere of unspecified relative humidity.

room-temperature setting adhesive. An adhesive that sets (to handling strength) within an hour at temperatures from 20 to 30 °C (68 to 86 °F) and later reaches full strength without heating. Compare with *cold-setting adhesive*, *hot-setting adhesive*, and *intermediate temperature setting adhesive*.

room-temperature vulcanizing (RTV). *Vulcanization* or curing at room temperature by chemical reaction, particularly of silicones and other rubbers.

root-mean-square end-to-end distance. A measure of the average size of a coiled polymer molecule, usually determined by light scattering.

rosin. A resin obtained as a residue in the distillation of crude turpentine from the sap of the pine tree (gum resin) or from an extract of the stumps and other parts of the tree (wood rosin); used in varnishes, lacquers, adhesives, and soldering fluxes. Compare with *resin*.

rotational casting. A method used to make hollow articles from thermoplastic materials. The material is charged into a hollow mold capable of being rotated in one or two planes. The hot mold fuses the material into a gel after the rotation has caused it to cover all surfaces. The mold is then chilled, and the product is stripped out.

rotational molding. The preferred term for a variation of the *rotational casting* process that uses dry, finely divided (35 mesh, or 500 μ) plastic powders, such as polyethylene, rather than fluid materials. After the powders are heated, they are fused against the mold walls forming a hollow item with uniform wall thickness.

roving. A number of yarns, strands, tows, or ends collected into a parallel bundle with little or no twist. Rovings are used to produce fiber-reinforced plastics. See also *woven roving*.

roving cloth. A textile fabric, coarse in nature and woven from roving, used in reinforced plastics.

row nucleation. The mechanism by which stress-induced crystallization is initiated in reinforcing fibers for composites, usually during fiber spinning or hot drawing.

RRIM. See *reinforced reaction injection molding*.

RTM. See *resin transfer molding*.

RTV. See *room-temperature vulcanizing*.

rubbers. Cross-linked polymers having glass transition temperatures below room temperature that exhibit highly elastic deformation and have high elongation.

runner system. All the sprues, runners, and gates through which plastic material flows from the nozzle of an injection machine or the pot of a transfer mold to the mold cavity. See also *resin transfer molding* and *transfer molding*.

rupture stress. The stress at failure. Also known as *fracture stress*.

rutile. A particular crystalline form of titanium dioxide (TiO_2) used as a pigment and as an additive or component in some ceramic, glass, and glaze manufacture.

S

safety glass. Glass so constructed, treated, or combined with other materials as to reduce, in comparison with ordinary sheet or plate glass, the likelihood of injury to persons by objects from exterior sources or by these safety glasses when they may be cracked or broken.

sag (plastics). (1) A local extension (often near the die face) of the parison during extrusion by gravitational forces. This causes necking down of the parison. (2) The flow of a molten sheet in a thermoforming operation.

sagging (adhesives). Run-off or flow-off of adhesive from an adherend surface due to application of excess or low-viscosity material.

sagging (ceramics and glasses). (1) A defect characterized by bending or slumping of a ceramic article fired at excessive temperature. (2) Process of forming glass by reheating until it conforms to the shape of the mold or form on which it rests. (3) A defect characterized by a wavy line or lines appearing on those surfaces of porcelain enamel that have been fired in a vertical position. (4) A defect characterized by irreversible downward bending in a ceramic article insufficiently supported during the firing cycle.

sample. (1) One or more units of a product (or a relatively small quantity of a bulk material) withdrawn from a *lot* or process stream and then tested or inspected to provide information about the properties, dimensions, or other quality characteristics of the lot or process stream. (2) A portion of a material intended to be representative of the whole.

sampling. The obtaining of a portion of a material that is adequate for making the required tests or analyses, and that is representative of that portion of the material from which it is taken.

sandwich construction. A panel composed of a lightweight core material, such as honeycomb, cellular or foamed plastic, and so forth, to which two relatively thin, dense, high-strength or high-stiffness faces, or skins, are adhered. See also the figure accompanying *honeycomb*.

sandwich testing. A method of heating a thermoplastic sheet, prior to forming, that consists of heating both sides of the sheet simultaneously.

satin finish. A type of finish having a satin or velvety appearance, specified for plastics or composites.

satin weave. See *four-harness satin*, *eight-harness satin*, and *harness satin*.

saturation pressure. The pressure, for a pure substance at any given temperature, at which vapor and liquid, or vapor and solid, coexist in stable equilibrium.

saw burn. Blackening or carbonization of a cut surface of a pultruded composite material or plastic section. Usually caused by cutting with a dull saw blade, cutting too slowly, or cutting a highly reinforced material with a diamond blade without water.

S-basis. The S-basis property allowable is the minimum value specified by the appropriate federal, military, Society of Automotive Engineers, American Society for Testing and Materials, or other recognized and approved specifications for the material.

scale (glass). A small particle of foreign material embedded in the surface of molded glass articles.

scale (plastics). A condition in which resin plates or particles are on the surface of a pultrusion. Scales can often be readily removed, sometimes leaving surface voids or depressions.

scarf joint. A joint made by cutting away similar angular segments on two adherends and bonding the adherends with the cut areas fitted together. See also *lap joint.*

schlieren. Regions of varying refraction in a transparent medium often caused by pressure or temperature differences and detectable especially by photographing the passage of a beam of light.

Scleroscope hardness test. A dynamic indentation hardness test using a calibrated instrument that drops a diamond-tipped hammer from a fixed height onto the surface of the material being tested. The height of rebound of the hammer is a measure of the hardness of the material.

scratch-resistant coatings. Coating applied to glass surfaces to reduce the effects of frictive damage. Examples are SnO_2 or TiO_2 coatings applied to glass containers.

screw plasticating injection molding. A technique in which the plastic is heated and converted from pellets to a viscous melt, and then forced into a mold by means of an extruder screw, which is an integral part of the molding machine. Machines are either single stage, in which plastication and injection are done in the same cylinder, or double stage, in which the material is plasticated in one cylinder and then fed to a second for injection into a mold. See also *injection molding.*

screws. See *extruder* (plastics).

scrim. A low-cost reinforcing fabric for composites made from continuous filament yarn in an open-mesh construction. Used in the processing of tape or other *B-stage* material to facilitate handling.

sealant. A material applied to a joint in paste or liquid form that hardens or cures in place, forming a seal against a gas or liquid entry.

secant modulus. The slope of the secant drawn from the origin to any specified point on the *stress-strain curve.* See also *modulus of elasticity.*

secondary bonding. The joining together, by the process of adhesive bonding, of two or more already cured composite parts, during which the only chemical or thermal reaction occurring is the curing of the adhesive itself.

secondary crystallization. In processing of plastics, the slow crystallization process that occurs after the main solidification process is complete. Often associated with impure molecules.

secondary nucleation. In processing of plastics, the mechanism by which crystals grow.

secondary-standard dosimetry system. A system that measures energy deposition indirectly. It requires conversion factors to account for such considerations as geometry, dose rate, relative stopping power, incident energy spectrum, or other effects, in order to interpret the response of the system. Thus, it requires calibration against a primary dosimetry system or by means of a standard radiation source. See also *dosimeter.*

secondary structure. In aircraft and aerospace applications, a structure that is not critical to flight safety.

segregation. A separation of components in a molded article usually denoted by wavy lines and color striations in thermoplastics. In thermoset plastics, usually a separation of resin and filler on the surface.

self-extinguishing resin. A resin formulation that will burn in the presence of a flame but will extinguish itself within a specified time after the flame is removed. The term is not universally accepted.

self-skinning foam. A urethane foam that produces a tough outer surface over a foam core upon curing.

self-vulcanizing. Pertaining to an adhesive that undergoes vulcanization without the application of heat. See also *vulcanization.*

selvage. The woven-edge portion of a fabric used in reinforced plastics that is parallel to the warp.

semiconductor. A solid crystalline material whose electrical conductivity is intermediate between that of a metal and an insulator, ranging from about 10^5 siemens to 10^{-7} siemens per meter, and is usually strongly temperature-dependent.

semicrystalline. In plastics, materials that exhibit localized crystallinity. See also *crystalline plastic.*

semipositive mold. A mold for forming plastics that combines the capabilities of a *flash mold* and *positive mold.* As the two halves of a semipositive mold begin to close, the mold acts much like a flash mold, because the excess material is allowed to escape around the loose-fitting plunger and cavity. As the plunger telescopes further into the cavity, the mold becomes a positive mold with very little clearance, and full pressure is exerted on the material, producing a part of maximum density. This type of mold uses to advantage the free flow of material in a flash mold and the capability of producing dense parts in the positive mold.

semirigid plastic. For purposes of general classification, a plastic that has a modulus of elasticity in flexure or in tension between 70 and 690 MPa (10 and 100 ksi) at 23 °C (70 °F) and 50% relative humidity.

separator. In processing of composites, a permeable layer that allows volatiles and air to escape from the laminate and excess resin to be bled from the laminate into the bleeder plies during cure. Porous Teflon-coated fiberglass is an example. Often placed between lay-up and bleeder to facilitate bleeder systems removal from laminate after cure. See also *bleeder* and *lay-up.*

set. To convert an adhesive into a fixed or hardened state by chemical or physical action, such as condensation, polymerization, oxidation, vulcanization, gelation, hydration, or evaporation of volatile constituents. See also *cure.*

setting temperature. The temperature to which an adhesive or an assembly is subjected to set the adhesive. The temperature attained by the adhesive in the process of setting (adhesive setting temperature) may differ from the temperature of the atmosphere surrounding the assembly (assembly setting temperature). See also *curing temperature* and *drying temperature.*

setting time. The period of time during which an adhesively bonded assembly is subjected to heat or pressure, or both, to set the adhesive. See also *drying time* and *joint-conditioning time.*

set up. To harden, as in the curing of a polymer resin.

S-glass. A magnesium aluminosilicate composition that is especially designed to provide very high tensile strength glass filaments. S-glass and S-2 glass fibers have the same glass composition but different finishes (coatings). S-glass is made to more demanding specifications, and S-2 is considered the commercial grade.

shape factor. For an elastomeric slab loaded in compression, the ratio of the loaded area to the force-free area.

shear. The type of force that causes or tends to cause two contiguous parts of the same body to slide relative to each other in a direction parallel to their plane of contact.

shear edge. The cutoff edge of a mold.

shear fracture. A mode of fracture in crystalline materials resulting from translation along slip planes that are preferentially oriented in the direction of the shearing stress.

shear modulus (*G*). The ratio of shear stress to the corresponding shear strain for shear stresses below the proportional limit of the material. Values of shear modulus are usually determined by torsion testing. Also known as modulus of rigidity.

shear rate. With regard to viscous fluids, the relative rate of flow or movement.

shear strain. The tangent of the angular change, caused by a force between two lines originally perpendicular to each other through a point in a body. Also called angular strain.

shear strength. The maximum shear stress that a material is capable of sustaining. Shear strength is calculated from the maximum load during a shear or torsion test and is based on the original cross-sectional area of the specimen.

shear stress. The stress developing in a polymer melt when the layers in a cross section are gliding along each other or along the wall of the channel (in laminar flow). Shear stress is equal to force divided by the area sheared, yielding pounds per square inch.

sheath. (1) The material, usually an extruded plastic or elastomer, applied outermost to a wire or cable. Also called a jacket. (2) A sheet metal or glass covering of a sintered billet to protect it from oxidation or other environmental contamination during hot working.

sheeting. A form of plastic in which the thickness is very small in proportion to length and width and in which the plastic is present as a continuous phase throughout, with or without filler.

sheet molding compound (SMC). A thermoset composite of fibers, usually an unsaturated polyester resin, and pigments, fillers, and other additives that have been compounded and processed into sheet form to facilitate handling in the subsequent compression molding operation. Generally cross linked with styrene.

shelf life. The length of time a material, substance, product, or reagent can be stored under specified environmental conditions and continue to meet all applicable specification requirements and/or remain suitable for its intended function.

shell tooling. A mold or bonding fixture for forming plastic parts that consists of a contoured surface shell supported by a substructure to provide dimensional stability.

Shore hardness. A measure of the resistance of material to indentation by a spring-loaded indenter during Scleroscope hardness testing. The higher the number, the greater the resistance. Normally used for rubber materials. See also *Scleroscope hardness test*.

short. An imperfection in a molded plastic part due to an incomplete fill. In reinforced plastics, this may be evident either from an absence of surface film in some areas or as lighter, unfused particles of material showing through a covering surface film, accompanied possibly by thin-skin blisters. In thermoplastics, also called short-shot.

short beam shear. A flexural test of a plastic specimen having a low test span-to-thickness ratio (for example, 4:1), such that failure is primarily in shear.

short-chain branching. The dominant form of molecular branching in addition polymers, usually formed by a "backbiting" transfer reaction and resulting, primarily, in *n*-butyl side chains, but also other short-pendant groups such as methyl and amyl. Such branching results in reduced levels of crystallinity.

shortness. A qualitative term that describes an adhesive that does not string cotton, or otherwise form filaments or threads during application.

short shot. Insufficient injection of material into the mold during forming of plastics or composites.

shot capacity. In forming of plastic parts, the maximum weight of material an injection machine can provide from one forward motion of the ram, screw, or plunger.

shrinkage (ceramics). The fractional reduction in dimensions or volume of a material or object when subjected to drying, calcining, or firing (sintering).

shrinkage (plastics). The relative change in dimension from the length measured on the mold when it is cold to the length of the molded plastic object 24 h after it has been taken out of the mold.

SI. See *silicones*.

Sialon. A generic term for a family of ceramic compositions produced by reacting silicon nitride (Si_3N_4) with aluminum oxide (Al_2O_3) and aluminum nitride (AlN).

side corings. In forming of plastics, projections that are used to core a hole in a direction other than the line of closing of a mold, and that must be withdrawn before the part is ejected from the mold. Also called side draw pins.

silica (SiO_2). The common oxide of silicon usually found naturally as quartz or in complex combination with other elements such as silicates. Various polymorphs and natural occurrences of silica include cristobalite, tridymite, cryptocrystalline chert, flint, chalcedony, and hydrated opal. Silica is the primary ingredient of sand, refractories, and glass.

silica gel. A precipitated colloidal mass or gel of indefinitely hydrated silica; also the dried or activated product of same. Useful as dessicant, scavenger, and catalyst substrate.

silicon carbide. Very hard, high-strength ceramic materials of the composition SiC used as an abrasive for grinding and cutting applications and as a refractory brick for furnace linings. Other applications include wear parts, heat exchangers, and radiant tubes for heat-treating furnaces.

silicones (SI). Plastics based on resins in which the main polymer chain consists of alternating silicon and oxygen atoms, with carbon-containing side groups. Derived from silica (sand) and methyl chlorides and furnished in different molecular weights, including liquids, solid resins, and elastomers.

silicon nitride. An *advanced ceramic* of the composition Si_3N_4 that combines high-temperature mechanical properties and resistance to oxidation and thermal shock. Used as cutting-tool materials in metal-machining applications and high-temperature engine components.

single-circuit winding. In forming of plastics and composites, a winding in which the filament path makes a complete traverse of the chamber, after which the following traverse lies immediately adjacent to the previous one. See also *filament winding*.

single-lap shear specimen. In adhesive testing, a specimen made by bonding the overlapped edges of two sheets or strips of material. In testing, a single-lap specimen is usually loaded in tension at the ends, thereby creating shear stresses at the joint interface.

sink mark. A shallow depression on the surface of an injection-molded plastic part due to the collapsing of the surface following local internal shrinkage after the gate seals.

sintering. The bonding of adjacent surfaces of particles in a mass of powder or a compact by heating. Sintering strengthens a powder mass and normally produces densification and, in powdered metals, recrystallization. See also *liquid phase sintering* and *solid-state sintering*.

size. In composites manufacturing, a treatment consisting of starch, gelatin, oil, wax, or other suitable ingredients applied to yarn or fibers at the time of formation to protect the surface and aid the process of handling and fabrication or to control the fiber characteristics. The treatment contains ingredients that provide surface lubricity and binding action, but unlike a finish, contains no coupling agent. Before final fabrication into a composite, the size is usually removed by heat cleaning, and a finish is applied.

sizing (adhesives). The process of applying a material on a surface in order to fill pores and thus reduce the absorption of the subsequently applied adhesive or coating or to otherwise modify the surface properties of the substrate to improve the adhesion, and also, the material used for this purpose. See also *primer*.

sizing (composites). See *size*.

sizing content. The percent of the total strand weight made up of sizing, usually determined by burning off or dissolving the organic sizing. Also known as loss on ignition. See also *size*.

skein. A continuous filament, strand, yarn, or roving for reinforced plastics, wound up to some measurable length and usually used to measure various physical properties.

skin. The relatively dense material that sometimes forms on the surface of a cellular plastic or sandwich construction.

slip (adhesives). In adhesively bonded components or specimens, the relative collinear displacement of the adherends on either side of the adhesive layer in the direction of the applied load.

slip (ceramics). (1) A slurry or suspension of fine clay or other ceramic powders in water, having the consistency of cream, that is used in slip casting or as a cement or glaze preparation. (2) A suspension of colloidal powder in an immiscible liquid (usually water).

slip angle. The angle at which a tensioned fiber will slide off a filament-wound dome. If the difference between the wind angle and the geodesic angle is less than the slip angle, the fiber will not slide off the dome. Slip angles for different fiber-resin systems vary and must be determined experimentally. See also *filament winding* and *wind angle*.

slip casting. The ceramic forming process consisting of filling or coating a porous mold with a *slip*, allowing to dry, and removing for subsequent firing.

slip coating. A ceramic material or mixture other than a glaze, applied to a ceramic body and fired to the maturity required to develop specified characteristics.

slip forming. A plastic sheet-forming technique in which some of the sheet material is allowed to slip through the mechanically operated clamping rings during a stretch-forming operation.

slippage. The movement of adherends with respect to each other during the adhesive bonding process.

sliver. A number of staple or continuous-filament fibers aligned in a continuous strand without twist. See also *continuous filament yarn*, *staple fiber*, and *strand*.

slot extrusion. A method of extruding plastic film or sheet in which the molten thermoplastic compound is forced through a slot die (T-shaped or coat hanger shape). Following extrusion, the film or sheet is cooled by passing it through a water bath or over water-cooled rolls.

sluffing. An occurrence during pultrusion of reinforced plastics in which scales peel off or become loose, either partially or entirely, from a pultrusion. Not to be confused with scraping, prying, or physically removing scale from a pultrusion. Sluffing is sometimes spelled sloughing. See also *pultrusion*.

slurry. (1) A thick mixture of liquid and solids, the solids being in suspension in the liquid. (2) Any pourable or pumpable suspension of a high content of insoluble particulate solids in a liquid medium, most often water.

slurry preforming. Method of preparing reinforced plastic preforms by wet processing techniques similar to those used in the pulp molding industry. For example, glass fibers suspended in water are passed through a screen that passes the water but retains the fibers in the form of a mat.

slush molding. Method for casting thermoplastics, in which the resin in liquid form is poured into a hot mold where a viscous skin forms. The excess slush is drained off, the mold is cooled, and the molding stripped out.

SMC. See *sheet molding compound*.

***S-N* curve.** A plot of stress (S) against the number of cycles to failure (N). The stress can be the maximum stress (S_{max}) or the alternating stress amplitude (S_a). The stress values are usually nominal stress; i.e., there is no adjustment for stress concentration. The diagram indicates the *S-N* relationship for a specified value of the mean stress (S_m) or the stress ratio (A or R) and a specified probability of survival. For N a log scale is almost always used. For S a linear scale is used most often, but a log scale is sometimes used. Also known as *S-N* diagram.

soda ash. Sodium carbonate (Na_2CO_3) obtained from trona, a hydrated sodium carbonate sodium bicarbonate ore ($Na_2CO_3 \cdot NaHCO_3 \cdot 2H_2O$). Soda ash is used in petroleum refining and for soaps and detergents. Its primary use, however, is in glass manufacture. Soda ash is the third major constituent of soda-lime-silica glasses and is the main source of Na_2O in any glass that contains soda. It acts as a flux, reducing the temperature required to melt the silica. See also *Solvay process*.

softening point. That temperature at which a glass fiber of uniform diameter elongates at a specific rate under its own weight when measured by standard ASTM test methods. The viscosity at the softening point depends on the density and surface tension. For example, for a glass of density 2.5 g/cm^2 and surface tension 300 dynes/cm, the softening point temperature corresponds to a viscosity of 106.6 Pa · s.

softening range. The range of temperatures within which a plastic changes from a rigid to a soft state. Actual values depend on the test method. Sometimes erroneously referred to as softening point.

sol. A colloidal suspension comprised of discrete or separate solid particles suspended in a liquid. Differs from a solution, though one merges into the other. Compare with *gel*.

sol-gel process. An important ceramic and glass-forming process in which a sol is converted to a gel by partial evaporation of the liquid phase and/or by neutralizing the electric charges on particles which cause them to repel each other. The gel is usually further processed (e.g., formed, dried, and fired).

solid-phase chemical dosimeter. An apparatus that measures radioactivity by using plastic, dyed plastic, or glass with an optical density, usually in the visible range, that changes when exposed to ionizing radiation. Examples currently in use include dyed polymethyl methacrylate (red perspex), undyed polyvinyl chloride, dyed polyamide (blue dye in a nylon matrix), and dyed polychlorostyrene (green dye in a chlorostyrene matrix). Solid-phase chemical dosimetry is generally considered to be a secondary-standard dosimetry system.

solid-phase forming. The use of metalworking technique to form thermoplastics in a solid phase. Procedure begins with a plastic blank that is heated and fabricated (that is, forged) by bulk deformation of the materials in constraining dies by the application of force. Also called solid-state stamping.

solid-state sintering. A sintering procedure for compacts or loose powder aggregates during which no component melts. Contrast with *liquid-phase sintering*.

solvation. The process of swelling, gelling, or dissolving a resin by a solvent or *plasticizer*.

Solvay process. A method for producing *soda ash* that involves the reaction of salt (NaCl) and limestone to form sodium carbonate (Na_2CO_3) with calcium chloride ($CaCl_2$) as a by-product.

solvent-activated adhesive. A dry-film adhesive that is rendered tacky by the application of a solvent just prior to use.

solvent adhesive. An adhesive having a volatile organic liquid as a vehicle. This term excludes water-base adhesives.

solvent molding. Process for forming thermoplastic articles by dipping a male mold in a solution or by dispersing the resin and drawing off the solvent, leaving a layer of plastic film adhering to the mold.

spalling. (1) The cracking or rupturing of a refractory unit, which usually results in the detachment of a portion of the unit. (2) A defect characterized by separation of the porcelain enamel from the aluminum base metal without apparent external cause. Spalling can result from the use of improper alloys or enamel formulations, incorrect pretreatment of the base metal, or faulty application and firing procedures.

specific adhesion. Adhesion between surfaces that are held together by valence forces of the same type as those that give rise to *cohesion*.

specific gravity (solids and liquids). The ratio of the density of a material to the density of some standard material, such as water, at a specified temperature. Also known as relative density.

specific humidity. In a mixture of water vapor and air, the mass of water vapor per unit mass of moist air.

specific properties. Material properties divided by material density.

specific viscosity. The relative viscosity of a solution of known concentration of a polymer minus one. It is usually determined for a low concentration of the polymer.

spherulite. A rounded aggregate of radiating lamellar crystals with appearance of a pom-pom in plastics. Spherulites contain amorphous material between the crystals and usually impinge on one another, forming polyhedrons. Spherulites are present in most crystalline plastics and may range in diameter from a few tenths of a micron to several millimeters.

spider. In a plastic molding press, that part of an ejector mechanism that operates the ejector pins. In extrusion, the membranes supporting a mandrel within the head/die assembly.

spinel. Any of a group of compounds of the same crystal type and general formula as magnesium aluminate, $MgAl_2O_4$ or $MgO{\cdot}Al_2O_3$. That compound itself, which is refractory and chemically near-neutral. Magnesium aluminate spinels are used as an addition to fired magnesia refractory bricks to improve thermal shock resistance. Lithium-based spinels are candidate materials for rechargeable lithium batteries.

spinneret. A type of extrusion die for plastics that consists of a metal plate with many tiny holes, through which a plastic melt is forced, to make fine fibers and filaments. Filaments may be hardened by cooling in air, water, and so forth, or by chemical action.

spiral-flow test. A method for determining the flow properties of a thermoplastic resin in which the resin flows along the path of a spiral cavity. The length of the material that flows into the cavity and its weight gives a relative indication of the flow properties of the resin.

spiral mold cooling. A method of cooling injection molds or similar molds for forming plastics in which the cooling medium flows through a spiral cavity in the body of the mold. In injection molds, the cooling medium is introduced at the center of the spiral, near the sprue section, because more heat is localized in this section.

splay. A fanlike surface defect near the gate on a plastic part.

split-ring mold. A mold for forming plastics in which a split-cavity block is assembled in a chase to permit the forming of undercuts in a molded plastic piece. These parts are ejected from the mold and then separated from the piece.

spodumene. A mineral of the composition $LiO_2 \cdot Al_2O_3{\cdot}4SiO_2$, which is the principal ore of lithium. It is also used as a melt accelerator in glass manufacture.

sprayed-metal molds. Molds for forming plastics made by spraying molten metal onto a master until a shell of predetermined thickness is achieved. The shell is then removed and backed with plaster, cement, casting resin, or other suitable material. Used primarily as a mold in the sheet-forming process.

spray lay-up. A wet lay-up for processing of reinforced plastic in which a stream of chopped fibers (usually glass) is fed into a stream of liquid resin in a mold. The direction of the fibers is random, as opposed to the mats or woven fabrics that can be used in hand lay-up. See also *hand lay-up* and *wet lay-up*.

spread. The quantity of adhesive per unit joint area applied to an adherend, usually expressed in pounds of adhesive per thousand square feet of joint area.

spreader. A streamlined metal block placed in the path of flow of the plastics material in the heating cylinder of extruders and injection molding machines to spread it into thin layers, thus forcing it into intimate contact with the heating areas.

spring constant. The force required to compress a spring or specimen 25 mm (1 in.) in a prescribed test procedure.

sprue. A single hole through which thermoset molding compounds are injected directly into the mold cavity.

spun roving. A heavy, low-cost glass or aramid fiber strand consisting of filaments that are continuous but doubled back on themselves. See also *roving*.

squeeze-out. Adhesive pressed out at the bond line due to pressure applied on the adherends.

stabilization. In carbon fiber forming, the process used to render the carbon fiber precursor infusible prior to carbonization.

stabilizers. Chemicals used in plastics formulation to help maintain physical and chemical properties during processing and service life. A specific type of stabilizer, known as an ultraviolet stabilizer, is designed to absorb ultraviolet rays and prevent them from attacking the plastic. Heat stabilizers are added to lessen the severity of thermal oxidation processes and their effect on properties.

stacking sequence. A description of a laminate that details the orientations of the plies and their sequence in the laminate. See also *laminate* and *quasi-isotropic laminate*.

staging. Heating a premixed resin system, such as in a *prepreg*, until the chemical reaction (curing) starts, but stopping the reaction before the gel point is reached. Staging is often used to reduce resin flow in subsequent press molding operations.

staple fibers. Fibers for reinforcing plastics that are of spinnable length manufactured directly or by cutting continuous filaments to short lengths (usually 13 to 50 mm, or 1/2 to 2 in., long, and 1 to 5 denier). See also *denier*.

star craze. Multiple fine surface separation cracks in pultruded reinforced plastics that appear to emanate from a central point and that exceed 6 mm (1/4 in.) in length, but do not penetrate the equivalent depth of a full ply of reinforcement. This condition is often caused by impact damage. See also *crazing*.

starved area. An area in a reinforced plastic part that has an insufficient amount of resin to wet out the reinforcement completely. This condition may be due to improper wetting, impregnation, or resin flow; excessive molding pressure; or incorrect bleeder cloth thickness.

starved joint. An adhesive bonded joint that has an insufficient amount of adhesive to produce a satisfactory bond. This condition may result from too thin a spread to fill the gap between the adherends, excessive penetration of the adhesive into the adherend, too short an assembly time, or the use of excessive pressure.

static modulus. The ratio of stress to strain under static conditions. It is calculated from static stress-strain tests, in shear, compression, or tension. Expressed in force per unit of area.

static stress. A stress in which the force is constant or slowly increasing with time, for example, test to failure without shock.

steam molding. A process used to mold plastic parts from preexpanded beads of polystyrene using steam as a source of heat to expand the blowing agent in the material. The steam in most cases is in direct, intimate contact with the beads. It may also be used indirectly, by heating mold surfaces that are in contact with the bead.

stereoisomer. An isomer in which atoms are linked in the same order but differ in their arrangement. See also *isomer* and *isotactic stereoisomerism*.

stereospecific plastics. Implies a specific or definite order of arrangement of molecules in space. This ordered regularity of the molecules in contrast to the branched or random arrangement found in other plastics permits close packing of the molecules and leads to high crystallinity (for example, in polypropylene).

stiffness. (1) The rate of stress with respect to strain; the greater the stress required to produce a given strain, the stiffer the material is said to be. (2) The ability of a material or shape to resist elastic deflection. For identical shapes, the stiffness is proportional to the modulus of elasticity. For a given material, the stiffness increases with increasing moment of inertia, which is computed from cross-sectional dimensions.

stoneware. A vitreous or semivitreous ceramic ware of fine texture, made primarily from either nonrefractory fireclay or some combination of clays, fluxes, and silica. Used for cookware, artware, and tableware.

stops. Metal pieces inserted between die halves used to control the thickness of a press-molded plastic part. Not a recommended practice, because the resin will receive less pressure, which can result in voids.

storage life. The period of time during which a liquid resin, packaged adhesive, or prepreg can be stored under specified temperature conditions and remain suitable for use. Also called shelf life.

storage modulus. A quantitative measure of elastic properties in polymers, defined as the ratio of the stress, in phase with the strain, to the magnitude of the strain. The storage modulus may be measured in tension or flexure, compression, or shear.

strain. The unit of change in the size or shape of a body due to force. Also known as nominal strain. The term is also used in a broader sense to denote a dimensionless number that characterizes the change in dimensions of an object during a deformation or flow process. See also *engineering strain*, *linear strain*, and *true strain*.

strain amplitude. The ratio of the maximum deformation, measured from the mean deformation to the free length of the unstrained test specimen.

strain gage. A device for measuring small amounts of strain in a stressed material based on the change in electrical resistance.

strain relaxation. Reduction in internal strain over time.

strand. Normally, an untwisted bundle or assembly of continuous filaments used as a unit to reinforce plastics, including slivers, tows, ends, yarn, and so forth. Sometimes a single fiber or filament is called a strand.

strand count. The number of *strands* in a plied *yarn* or *roving*.

strand integrity. The degree to which the individual filaments making up a *strand* or *end* are held together by the applied sizing. See also *size*.

strand tensile test. A tensile test of a single resin-impregnated *strand* of any fiber.

strength. The maximum nominal stress a material can sustain. Always qualified by the type of stress (tensile, compressive, or shear).

stress. The intensity of the internally distributed forces or components of forces that resist a change in the volume or shape of a material that is or has been subjected to external forces. Stress is expressed in force per unit area. Stress can be normal (tension or compression) or shear.

stress (glass). Any condition of tension or compression existing within the glass, particularly due to incomplete annealing, temperature gradient, or inhomogeneity.

stress concentration. On a macromechanical level, the magnification of the level of an applied stress in the region of a notch, void, hole, or inclusion.

stress-concentration factor. The ratio of the maximum stress in the region of a stress concentrator, such as a hole, to the stress in a similarly strained area without a stress concentrator.

stress crack. External or internal cracks in a plastic caused by tensile stresses less than that of its short-time mechanical strength, frequently accelerated by the environment to which the plastic is exposed. The stresses that cause cracking may be present internally or externally or may be combinations of these stresses. See also *crazing*.

stress-cracking failure. The failure of a plastic by cracking or crazing some time after it has been placed under load. Time-to-failure may range from minutes to years. Causes include molded-in stresses, postfabrication shrinkage or warpage, and hostile environment.

stress fracture. See *fracture stress*.

stress-induced crystallization. The production of crystals in a polymer by the action of stress, usually in the form of an elongation. It occurs in fiber-spinning and rubber elongation and is responsible for enhanced mechanical properties.

stress-intensity factor. A scaling factor, usually denoted by the symbol K, used in *linear-elastic fracture mechanics* to describe the intensification of applied stress at the tip of a crack of known size and shape. At the onset of rapid crack propagation in any structure containing a crack, the factor is called the critical stress-intensity factor, or the *fracture toughness*. Various subscripts are used to denote different loading conditions or fracture toughnesses:

K_c. Plane-stress fracture toughness. The value of stress intensity at which crack propagation becomes rapid in sections thinner than those in which plane-strain conditions prevail.

K_I. Stress-intensity factor for a loading condition that displaces the crack faces in a direction normal to the crack plane (also known as the opening mode of deformation).

K_{Ic}. Plane-strain fracture toughness. The minimum value of K_c for any given material and condition, which is attained when rapid crack propagation in the opening mode is governed by plane-strain conditions.

K_{Id}. Dynamic fracture toughness. The fracture toughness determined under dynamic loading conditions; it is used as an approximation of K_{Ic} for very tough materials.

K_{ISCC}. Threshold stress intensity factor for stress-corrosion cracking. The critical plane-strain stress intensity at the onset of stress-corrosion cracking under specified conditions.

K_Q. Provisional value for plane-strain fracture toughness.

K_{th}. Threshold stress intensity for stress-corrosion cracking. The critical stress intensity at the onset of stress-corrosion cracking under specified conditions.

ΔK. The range of the stress-intensity factor during a fatigue cycle. See also *fatigue crack growth rate*.

stress relaxation. The time-dependent decrease in stress in a solid under constant constraint at constant temperature.

stress-strain curve. A graph in which corresponding values of stress and strain from a tension, compression, or torsion test are plotted against each other. Values of stress are usually plotted vertically (ordinates or *y*-axis) and values of strain horizontally (abscissas or *x*-axis). Also known as deformation curve and stress-strain diagram. See also *engineering strain* and *engineering stress*.

striation. A fracture surface marking on glass consisting of a separation of the advancing crack front into separate fracture planes. Also known as coarse hackle, step fracture, or lance. Striations may also be called shark's teeth or whiskers.

stripper. A chemical solvent or acid that can remove an adhesive bond.

structural adhesive. Adhesive used for transferring required loads between adherends exposed to service environments typical for the structure involved.

structural bond. An adhesive bond that joins basic load-bearing parts of an assembly. The load may be either static or dynamic.

structural ceramics. *Advanced ceramics* used for load-bearing applications.

structural foams. Expanded plastic materials having integral solid skins and porous cores that exhibit outstanding rigidity. Structural foams involve a variety of thermoplastic resins as well as urethanes.

structural glass. (1) Flat glass, usually colored or opaque, and frequently ground and polished, used for structural purposes. (2) Glass block, usually hollow, used for structural purposes.

structural reaction injection molding (SRIM). A molding process that is similar in practice to resin transfer molding. SRIM derives its name from the RIM process from which the resin chemistry and injection techniques have been adapted. The term structural is added to indicate the reinforced nature of the composite components manufactured by this process. In the SRIM process a preformed reinforcement is placed in a closed mold, and a reactive resin mixture is mixed under high pressure in a specially designed mix head. Upon mixing, the reacting liquids flow at low pressure through a runner system and fill the mold cavity, impregnating the reinforcement material in the process. Once the mold cavity has filled, the resin quickly completes its reaction. A completed component can often be removed from the mold in as little as 1 min.

styrene-acrylonitrile (SAN). A copolymer of about 70% styrene and 30% acrylonitrile, with higher strength, rigidity, and chemical resistance than can be attained with polystyrene alone. These copolymers may be blended with butadiene as a terpolymer or grafted onto the butadiene to make acrylonitrile-butadiene-styrene resins. They are transparent and have high heat-deflection properties, excellent gloss, chemical resistance, hardness, rigidity, dimensional stability, and load-bearing capability.

styrene-maleic anhydride (SMA). Copolymers made by the copolymerization of styrene and maleic anhydride, with higher heat resistance than the parent styrenic and acrylonitrile-butadiene-styrene families. For structurally demanding applications, SMAs are reinforced with glass fibers. Most reinforced grades contain 20 wt% reinforcement.

styrene-rubber plastics. Plastics based on styrene polymers and rubbers, the styrene polymers being the greatest amount by mass.

substrate. (1) The material, workpiece, or substance on which the coating is deposited. (2) A material upon the surface of which an adhesive-containing substance is spread for any purpose, such as bonding or coating. A broader term than *adherend*. (3) In electronic devices, a body, board, or layer of material, on which some other active or useful material(s) or component(s) may be deposited or laid, for example, electronic circuitry laid on an alumina ceramic board. (4) In catalysts, the formed, porous, high-surface area carrier on which the catalytic agent is widely and thinly distributed for reasons of performance and economy.

superconductivity. A property of many metals, alloys, compounds, oxides, and organic materials at temperatures near absolute zero by virtue of which their electrical resistivity vanishes and they become strongly diamagnetic.

surface preparation. A physical or chemical treatment, or both, of an adherend to render it suitable for adhesive bonding. Also known as prebond treatment.

surface tension. (1) The force acting on the surface of a liquid, tending to minimize the area of the surface. (2) The force existing in a liquid-vapor phase interface that tends to diminish the area of the interface. This force acts at each point on the interface in the plane tangent to that point.

surface treatment. In composites fabrication, a material (size or finish) applied to fibrous material during the forming operation or in subsequent processes. For carbon fiber surface treatment, the process used to enhance bonding capability of fiber to resin. See also *size*.

surfacing mat. A very thin mat, usually 180 to 510 μm (7 to 10 mils) thick, of highly filamentized fiberglass, used primarily to produce a smooth surface on a reinforced plastic laminate, or for precise machining or grinding.

surfactant. (1) A chemical substance characterized by a strong tendency to form adsorbed interfacial films when in solution, emulsion, or suspension, thus producing effects such as low surface tension, penetration, boundary lubrication, wetting, and dispersing. (2) A compound that affects interfacial tensions between two liquids. It usually reduces surface tension.

symmetrical laminate. A *laminate* in which the stacking sequence of plies below its midplane is a mirror image of the stacking sequence above the midplane.

syndiotactic stereoisomerism. A polymer molecule in which side atoms or side groups alternate regularly on opposite sides of the chain.

syneresis. Spontaneous separation of a liquid from a gel due to contraction of the gel.

syntactic cellular plastics. Reinforced plastics made by mixing hollow microspheres of glass, epoxy, phenolic, and so forth, into fluid resins (with additives and curing agents) to form a moldable, curable, lightweight, fluid mass; as opposed to foamed plastic, in which the cells are formed by gas bubbles released in the liquid plastic by either chemical or mechanical action. Also known as syntactic foams.

T

tabs. Extra lengths of reinforced plastic or other material at the ends of a tensile specimen to promote failure away from the grips. Also called doublers.

tack. The property of an adhesive that enables it to form a bond of measurable strength immediately after adhesive and adherend are brought into contact under low pressure. See also *dry tack*, *tack range*, and *tacky-dry*.

tack range. The period of time in which an adhesive will remain in the *tacky-dry* condition after application to the adherend, under specified conditions of temperature and humidity.

tacky-dry. The condition of an adhesive when the volatile constituents have evaporated or been absorbed sufficiently to leave it in a desired tacky state.

talc. A whitish, greenish, or grayish hydrated magnesium silicate, $Mg_3Si_4O_{10}(OH)_2$, mineral which is extremely soft (hardness is 1 on the Mohs scale) and has a characteristic soapy or greasy feel. Used as an ingredient in *ceramic whiteware* and ceramic wall and floor tiles.

tan delta (tan δ). (1) The ratio of the loss modulus to the *storage modulus*, measured in compression, *K*; tension or flexure, *E*; or shear, *G*. See also *complex modulus* and *loss modulus*. (2) The ratio of the out-of-phase components of the dielectric constant (that is, the loss) to the in-phase component of the dielectric constant (that is, the permittivity). See also *loss factor*.

tangent modulus. The slope of the *stress-strain curve* at any specified stress or strain. See also *modulus of elasticity*.

tape. In composites manufacture, a unidirectional *prepreg* that consists of a thin sheet of fiber reinforced uncured resin usually wound on a cardboard core.

tape wrapped. In composites fabrication, wrapping of heated fabric tape onto a rotating mandrel, which is subsequently cooled to firm the surface for the next tape layer application.

technical ceramics. Same as *advanced ceramics*.

technical glass. A term that usually refers to glasses designed with some specific property essential for a mechanical, industrial, or scientific device.

telomer. A polymer composed of molecules having terminal groups incapable of reacting with additional monomers, under the conditions of the synthesis, to form larger polymer molecules of the same chemical type. See also *monomer* and *polymer*.

temper (ceramics). To moisten and mix clay, plaster, or mortar to proper consistency.

temper (glass). (1) The degree of residual stress in annealed glass measured polarimetrically or by polariscopic comparison with a reference standard. (2) Term sometimes employed in referring to *tempered glass*.

tempered glass. Glass that has been subjected to a thermal treatment characterized by rapid cooling to produce a compressively stressed surface layer. See also the figure accompanying the term *safety glass*.

tempering (glass). The process of rapidly cooling glass from near its softening point to induce compressive stresses on the surface balanced by interior tension, thereby imparting increased strength.

template (templet). (1) A guide or a pattern in manufacturing items. (2) A pattern used as a guide for cutting and laying plies of a *laminate*.

tensile strength. In tensile testing, the ratio of maximum load to original cross-sectional area. Also called *ultimate strength*. Compare with *yield strength*.

tensile stress. A stress that causes two parts of an elastic body, on either side of a typical stress plane, to pull apart. Contrast with *compressive stress*.

tensile testing. See *tension testing*.

tension. The force or load that produces elongation.

tension testing. A method of determining the behavior of materials subjected to uniaxial loading, which tends to stretch the material. A longitudinal specimen of known length and diameter is gripped at both ends and stretched at a slow, controlled rate until rupture occurs. Also known as tensile testing.

terpolymer. A polymeric system that contains three monomeric units.

terra cotta. A term applied to ornamental units of fired clay used as facing material for buildings.

tex. A unit for expressing linear density equal to the mass of weight in grams of 1000 meters of filament, fiber, yarn, or other textile strand.

textile fibers. Fibers or filaments used in reinforced plastics that can be processed into yarn or made into a fabric by interlacing in a variety of methods, including weaving, knitting, and braiding.

thermal aging. Exposure of a material or component to a given thermal condition or a programmed series of conditions for prescribed periods of time.

thermal conductivity. (1) Ability of a material to conduct heat. (2) The rate of heat flow, under steady conditions, through unit area, per unit temperature gradient in the direction perpendicular to the area. Usually expressed in English units as Btu per square feet per degrees Fahrenheit ($Btu/ft^2 \cdot °F$). It is given in SI units as watts per meter kelvin ($W/m \cdot K$).

thermal endurance. The time required at a selected temperature for a material or system of materials to deteriorate to some predetermined level of electrical, mechanical, or chemical performance under prescribed test conditions.

thermal expansion. The change in length of a material with change in temperature. See also *coefficient of thermal expansion*.

thermal expansion molding. In forming of plastics, a process in which elastomeric tooling details are constrained within a rigid frame to generate consolidation pressure by thermal expansion during the curing cycle of the autoclave molding process.

thermal stress cracking. The crazing and cracking of some thermoplastic resins from overexposure to elevated temperatures. See also *stress-cracking failure*.

thermoforming. The process of forming a thermoplastic sheet into a three-dimensional shape after heating it to the point at which it is soft and flowable, and then applying differential pressure to make the sheet conform to the shape of a mold or die positioned below the frame. When the thermoplastic material has been reinforced, it should be heated to the point that it is soft enough to be formed without cracking or breaking the reinforcing fibers. There are three basic mold types: female (concave), male (convex), and matched (a combination of the two). (1) In basic female forming, or straight vacuum forming, the heated sheet is positioned over the mold cavity and is pulled into the cavity by vacuum. (2) In basic male forming, also called drape forming, the heated plastic is drawn over the mold. As soon as a seal is created, the vacuum is activated, and the plastic is forced directly against the surface of the mold. See also *drape forming*. (3) In matched-mold thermoforming, the stamping force of matched male and female molds is used. The male mold pushes the heated sheet into the female cavity.

thermogravimetric analysis (TGA). The study of the change in mass of a material under various conditions of temperature and pressure.

thermoplastic. Capable of being repeatedly softened by an increase in temperature and hardened by a decrease in temperature. Those polymeric materials that, when heated, undergo a substantially physical rather than chemical change and that in the softened stage can be shaped into articles by molding or extrusion.

thermoplastic fluoropolymers. See *fluoroplastics*.

thermoplastic injection molding. A process in which melted plastic is injected into a mold cavity, where it cools and takes the shape of the cavity. Bosses, screw threads, ribs, and other details can be integrated, which allows the molding operation to be accomplished in one step. The finished part usually does not require additional work before assembling.

thermoplastic polyesters (TPES). A class of thermoplastic polymers in which the repeating units are joined by ester groups. The two important types are *polyethylene terephthalate (PET)*, which is widely used as film, fiber, and soda bottles; and *polybutylene terephthalate (PBT)*, which is primarily a molding compound.

thermoplastic polyimides (TPI). Fully imidized, linear polymers with exceptionally good thermomechanical performance characteristics. Aromatic TPIs are generally produced by the polycondensation reaction of aromatic dianhydrides with aromatic diamines or aromatic diisocyanates in a suitable reaction medium.

thermoplastic polyurethanes (TPUR). Linear (segmented) block copolymers. Such a copolymer consists of repeating groups of diisocyanate and short-chain diol, or chain extender, for a rigid block, and repeating groups of diisocyanate and long-chain diol, or polyol, for a flexible block. TPURs are commonly injection molded, blow molded, or extruded.

thermoset. A resin that is cured, set, or hardened, usually by heating, into a permanent shape. The polymerization reaction is an irreversible reaction known as cross linking. Once set, a thermosetting plastic cannot be remelted, although most soften with the application of heat.

thermoset injection molding. A process in which thermoset material that has been heated to a liquid state is caused to flow into a cavity or several cavities and held at an elevated temperature for a specific time. After cross linking is completed, the hardened part is removed from the open mold. The molds used are usually of through-hardened tool steel that has been chrome plated and highly polished.

thermosetting. Having the property of undergoing a chemical reaction by the action of heat, catalysts, ultraviolet light, and so on, leading to a relatively infusible state.

thermosetting polyesters. A class of resins produced by dissolving unsaturated, generally linear, alkyd resins in a vinyl-type active monomer such as styrene, methyl styrene, or diallyl phthalate. Cure is effected through vinyl polymerization using peroxide catalysts and promoters or heat to accelerate the reaction. One important commercial type is liquid resins that are cross linked with styrene and used either as impregnants for glass or carbon fiber reinforcements in laminates, filament-wound structures, and other built-up constructions, or as binders for chopped-fiber reinforcements in molding compounds, such as *sheet molding compound (SMC)* and *bulk molding compound (BMC)*. A second important type is liquid or solid resins that are cross linked with other esters in chopped-fiber and mineral-filled molding compounds, such as alkyd and diallyl phthalate. See also *unsaturated polyesters*.

thermotropic liquid crystal. A liquid crystalline polymer that can be processed using thermoforming techniques.

thinner. A volatile liquid added to an adhesive to modify the consistency or other properties. See also *diluent* and *extenders*.

thixotropic. Pertaining to the tendency of a fluid to decrease in viscosity as the time of exposure to a given shear rate increases. The shear stress versus shear strain rate curve of a thixotropic material should show a hysteresis loop. A purely pseudoplastic material will not give a hysteresis loop because this property is not time dependent.

thixotropy. A property of certain gels or adhesive systems to thin upon isothermal agitation (shearing) and to thicken upon subsequent rest.

thread. See *fiber*.

thread count. The number of yarns (threads) per inch in either the lengthwise (warp) or crosswise (fill, or weft) direction of woven fabrics used for reinforcing plastics.

tile. A ceramic surfacing unit, usually relatively thin in relation to facial area, made from clay or a mixture of clay and other ceramic materials, having either a glazed or unglazed face and fired above red heat in the course of manufacture to a temperature sufficiently high to produce specific physical properties and characteristics.

time profile. A plot of the modulus, damping, or both, of a material versus time.

titania. A white water-insoluble powder of composition TiO_2 that is produced commercially from the minerals ilmenite and rutile. Used in paints and cosmetics and as an ingredient in porcelain enamels, ceramic whiteware, and ophthalmic glasses. Pure TiO_2 is also used as thin- or thick-film semiconductors.

titanium carbide. Very hard, heat-resistant ceramic materials of the composition TiC used in cermets and tungsten carbide cutting tools. Chemical vapor deposited TiC coatings are also used to extend the life of cemented carbide cutting tools.

titanium nitride. A hard, high-melting-point ceramic (2950 °C, or 5342 °F) of the composition TiN that is used in cermets and as a coating material for cemented carbide cutting tools.

tolerance. The specified permissible deviation from a specified nominal dimension, or the permissible variation in size or other quality characteristic of a part.

tooling resin. Resins that have applications as tooling aids, coreboxes, prototypes, hammer forms, stretch forms, foundry patterns, and so forth. Epoxy and silicone are common examples.

torsion. (1) A twisting deformation of a solid or tubular body about an axis in which lines that were initially parallel to the axis become helices. (2) A twisting action resulting in shear stresses and strains.

torsional stress. The shear stress on a transverse cross section caused by a twisting action.

toughness. Ability of a material to absorb energy and deform plastically before fracturing. Toughness is proportional to the area under the *stress-strain curve* from the origin to the breaking point.

tow. An untwisted bundle of continuous filaments, usually referring to man-made fibers, particularly carbon and graphite, but also fiberglass and aramid. A tow designated as 140 K has 140,000 filaments.

T-peel strength. The average load per unit width of adhesive bond line required to produce progressive separation of two bonded, flexible adherends, under standard test conditions.

TPI. See *thermoplastic polyimides*.

TPUR. See *thermoplastic polyurethanes*.

tracer. In composites fabrication, fiber, tow, or yarn added to a prepreg for verifying fiber alignment and, in the case of woven materials, for distinguishing warp fibers from fill fibers.

traditional ceramics. Ceramic products that use clay or have a significant clay component in the batch.

transfer molding. A method of molding thermosetting materials in which the plastic is first softened by heat and pressure in a transfer chamber and then forced by high pressure through suitable sprues, runners, and gates into a closed mold for final shaping and curing. See also *resin transfer molding*.

transformation toughened zirconia. A generic term applied to stabilized zirconia systems in which the tetragonal symmetry is retained as the primary zirconia phase. The four most popular tetragonal phase stabilizers are ceria (CeO_2), calcia (CaO), magnesia (MgO), and yttria (Y_2O_3).

transition temperature. The temperature at which the properties of a material change. Depending on the material, the transition change may or may not be reversible.

trans stereoisomer. A stereoisomer in which atoms or groups of atoms are arranged on opposite sides of a chain of atoms.

transversely isotropic. (1) In reference to a material, exhibiting a special case of orthotropy in which properties are identical in two orthotropic dimensions but not the third. (2) Having identical properties in both transverse (short and long) but not in the longitudinal direction.

transverse strain. The linear strain in a plane perpendicular to the loading axis of a specimen.

tribology. (1) The science and technology of interacting surfaces in relative motion and of the practices related thereto. (2) The science concerned with the design, friction, lubrication, and wear of contacting surfaces that move relative to each other (as in bearings, cams, or gears, for example).

true strain. (1) The ratio of the change in dimension, resulting from a given load increment, to the magnitude of the dimension immediately prior to applying the load increment. (2) In a body subjected to axial force, the natural logarithm of the ratio of the gage length at the moment of observation to the original gage length. Also known as natural strain.

true stress. The value obtained by dividing the load applied to a member at a given instant by the cross-sectional area over which it acts.

tumbling. Finishing operation for small plastic articles by which gates, flash, and fins are removed and/or surfaces are polished by rotating them in a barrel with wooden pegs, sawdust, and polishing compounds.

tup impact test. A falling-weight (tup) impact test developed specifically for plastic pipe and fittings.

turns per inch (tpi). In composites, a measure of the amount of twist produced in a *yarn*, *tow*, or *roving* during its processing history. See also *twist*.

twill weave. A basic fabric weave for reinforced plastics characterized by a diagonal rib, or twill line. Each end floats over at least two consecutive picks, allowing a greater number of yarns per unit area than in a plain weave, while not losing a great deal of fabric stability. See also *plain weave*.

twin-sheet thermoforming. A technique for *thermoforming* hollow plastic objects by introducing high-pressure air between two sheets and blowing the sheets into the mold halves (vacuum is also applied).

twist. (1) In a yarn or other textile strand, the spiral turns about its axis per unit of length. Twist may be expressed as *turns per inch (tpi)*. Twist provides additional integrity to yarn before it is subjected to the weaving process, a typical twist consisting of up to one turn per inch. In many instances heavier yarns are needed for the weaving operation. This is normally accomplished by twisting together two or more single strands, followed by a plying operation. Plying essentially involves retwisting the twisted strands in the opposite direction from the original twist. The two types of twist normally used are known as S and Z, which indicate the direction in which the twisting is done. Usually, two or more strands twisted together with an S twist are plied with a Z twist in order to give a balanced yarn. Thus, the yarn properties, such as strength, bundle diameter, and yield, can be manipulated by the twisting and plying operations. (2) In pultruded parts, twist describes a condition of longitudinal, progressive rotation that can be easily detected for a noncircular cross section by placing the pultrusion on a plane surface, holding one end flat with the surface, and observing whether one edge or side of the other end does not lie parallel with that surface.

two-component adhesive. An adhesive supplied in two parts that are mixed before application. Such adhesives usually cure at room temperature.

typical-basis. An average property value. No statistical assurance is associated with this basis.

U

UHMWPE. See *ultrahigh molecular weight polyethylene*.

ultimate elongation. The elongation at rupture.

ultimate strength. The maximum stress (tensile, compressive, or shear) a material can sustain without fracture; determined by dividing maximum load by the original cross-sectional area of the specimen. Also known as nominal strength or maximum strength.

ultimate tensile strength. The ultimate or final (highest) stress sustained by a specimen in a tension test.

ultrahard tool materials. Very hard, wear-resistant materials—specifically, polycrystalline diamond and polycrystalline cubic boron nitride—that are fabricated into solid or layered cutting tool blanks for machining applications.

ultrahigh molecular weight polyethylene (UHMWPE). Those polyethylene resins having weight-average molecular weights ranging from 3×10^6 to 6×10^6. These materials have both the highest abrasion resistance and highest impact strength of any plastic. See also *high-density polyethylene*.

ultrasonic bonding. A method of joining plastics using vibratory mechanical pressure at ultrasonic frequencies. Electrical energy is changed to ultrasonic vibrations by means of either a magnetostrictive or piezoelectric transducer. The ultrasonic vibrations generate frictional heat, melting the plastics and allowing them to join.

ultraviolet (UV). Pertaining to the region of the electromagnetic spectrum from approximately 10 to 380 nm. The term ultraviolet without further qualification usually refers to the region from 200 to 380 nm.

ultraviolet degradation. The degradation caused by long-term exposure of a material to sunlight or other ultraviolet rays containing radiation.

ultraviolet/electron beam cured adhesives. Specially formulated adhesives that are cured by ultraviolet or electron beam energy sources.

ultraviolet stabilizer. Any chemical compound that, when admixed with a resin, selectively absorbs ultraviolet rays.

unbond. An area within an adhesively bonded interface between two adherends in which the intended bonding action failed to take place, or an area in which two layers of prepreg in a cured component do not adhere. Also used to denote specific areas deliberately prevented from bonding in order to simulate a defective bond, such as in the generation of quality standards specimens.

undercure. An undesirable condition of a molded plastic article resulting from the allowance of too little time and/or temperature or pressure for adequate hardening of the molding.

undercut. A protuberance or indentation that impedes the withdrawal of a molded part from a two-piece, rigid mold. Any such protuberance, or indentation, depending on the design of the mold.

uniaxial load. A condition in which a material or component is stressed in only one direction along its axis or centerline.

uniaxial strain. See *axial strain*.

uniaxial stress. A state of stress in which two of the three principal stresses are zero.

unidirectional laminate. A reinforced plastic laminate in which substantially all of the fibers are oriented in the same direction. See also *laminate* and compare with *quasi-isotropic laminate*.

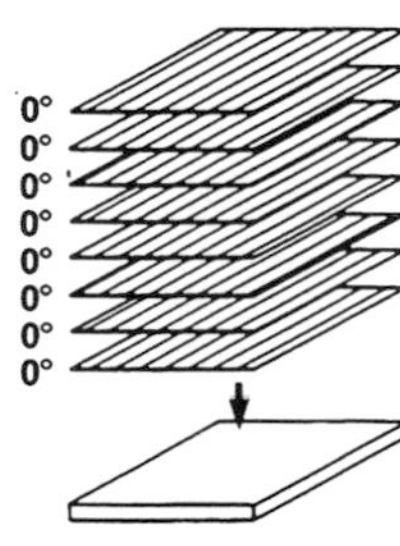

Unidirectional

uniform elongation. The elongation at maximum load and immediately preceding the onset of necking in a tensile test.

uniform strain. The strain occurring prior to the beginning of localization of strain (necking); the strain to maximum load in the tension test.

unimeric. Pertaining to a single molecule that is not monomeric, oligomeric, or polymeric, such as saturated hydrocarbons.

unsaturated compounds. Any chemical compound having more than one bond between two adjacent atoms, usually carbon atoms, and being capable of adding other atoms at that point to reduce it to a single bond; for example, *olefins*.

unsaturated polyesters. The family of polyesters characterized by vinyl unsaturation in the polyester backbone, which enables subsequent hardening or curing by copolymerization with a reactive monomer in which the polyester constituent has been dissolved.

unsymmetric laminate. A *laminate* having an arbitrary stacking sequence without midplane symmetry.

uranyl. The chemical name designating the UO_2^{2+} group and compounds containing this group.

urea-formaldehyde adhesive. (1) An aqueous colloidal dispersion of urea-formaldehyde polymer which may contain modifiers and secondary binders to provide specific adhesive properties. (2) A type of adhesive based on a dry urea-formaldehyde polymer and water. A curing agent is commonly used with this type of adhesive.

urethane hybrids. Urethane acrylic polymers that are formed by the reaction of two liquid components, an acrylesterol and a modified diphenyl-methane-4,4′-diisocyanate (MDI). The acrylesterol is a hybrid of a urethane (monoalcohol) and an acrylic (unsaturated monoalcohol). The liquid-modified MDI contains two or more isocyanate groups that can react with the hydroxyl portion of the acrylesterol molecule. Acrylamate resin systems are reinforced with glass (30 to 40%) and are used in automotive applications and recreational products. When reinforced with carbon mat or metallized glass cloth, these materials can be used in communication equipment, such as electromagnetic interference/radio frequency interference (EMI/ RFI) devices.

urethane plastics. Plastics based on resins made by the condensation of organic isocyanates with compounds or resins that contain hydroxyl groups. The resin is furnished as two component liquid monomers or prepolymers that are mixed in the field immediately before application. A great variety of materials are available, depending on the monomers used in the prepolymers and polyols and the type of diisocyanate employed. Extremely abrasion and impact resistant.

UV. See *ultraviolet*.

V

vacuum bag. A flexible bag in which pressure may be applied to an assembly (inside the bag) by means of evacuation of the bag. See also *vacuum bag molding*.

vacuum bag molding. A process for manufacturing reinforced plastics in which a sheet of flexible, transparent material plus a bleeder cloth and release film are placed over the lay-up on the mold and sealed at the edges. A vacuum is applied between the sheet and the lay-up. The entrapped air is mechanically worked out of the lay-up and removed by the vacuum, and the part is cured with temperature, pressure, and time. Also called *bag molding* or pressure bag molding. See also *lay-up*.

vacuum forming. A method of sheet forming in which the plastic sheet is clamped in a stationary frame, heated, and drawn down by a vacuum into a mold. In a broad sense, the term is sometimes used to refer to all sheet-forming techniques, including *drape forming* involving the use of vacuum and stationary molds.

vacuum hot pressing. A method of processing materials (especially metal and ceramic powders) at elevated temperatures, consolidation pressures, and low atmospheric pressures.

vacuum injection molding. A molding process for fabricating reinforced plastics which utilizes both a male and female mold in which reinforcements are placed, a vacuum is applied, and a room temperature curing liquid resin is introduced to saturate the reinforcement.

vacuum metallizing. A process in which surfaces are thinly coated by exposing them to a metal vapor under vacuum.

vacuum snapback thermoforming. A *thermoforming* process for production of plastic items with external deep draws, such as auto parts and luggage. First, the sheet is clamped over the female cavity. Air pressure is then introduced through the channel in the base plate, stretching the plastic. When the material has been sufficiently stretched, the pressure is turned off, and vacuum is turned on, pulling the plastic into the mold. There are many variations of this method, some of which employ plug assists. See also *plug-assist forming*.

vapor. The gaseous form of substances that are normally in the solid or liquid state, and that can be changed to these states either by increasing the pressure or decreasing the temperature.

varistor. A material, such as zinc oxide (ZnO), having an electrical resistance that is sensitive to changes in applied voltage.

veil. An ultrathin mat for reinforcing plastics similar to a *surfacing mat*, often composed of organic fibers as well as glass fibers. See also *mat*.

Vello process. A process for continuously drawing glass tubing (or cane) in which glass is fed downward to the draw through an annular orifice.

vent. A small hole or shallow channel in a mold that allows air or gas to exit as the plastic molding material enters.

vent cloth. A layer or layers of open-weave cloth used to provide a path for vacuum to "reach" the area over a *laminate* being cured, such that volatiles and air can be removed. Also causes the pressure differential that results in the application of pressure to the part being cured. Also called breather cloth.

venting. In autoclave curing of a composite part or assembly, turning off the vacuum source and venting the vacuum bag to the atmosphere. The pressure on the part is then the difference between pressure in the autoclave and atmospheric pressure. In injection molding, gases evolve from the melt and escape through vents machined in the barrel or mold.

vent mark. A small protrusion resulting from the entrance of metal into die vent holes.

vermiculite. A granular, clay mineral constituent that is used as a textural material in painting, as an aggregate in certain plaster formulations used in sculpture, or mixed with a resin to form a filler of relatively high compressive strength.

Vicat softening point. The temperature at which a flat-ended needle of 1 mm^2 (0.0015 $in.^2$) circular or square cross section will penetrate a thermoplastic specimen to a depth of 1 mm (0.040 in.) under a specified load, using a uniform rate of temperature rise.

Vickers hardness number (HV). A number related to the applied load and the surface area of the permanent impression made by a square-based pyramidal diamond indenter having included face angles of 136°, computed from:

$$\mathrm{HV} = 2P \sin\frac{\alpha/2}{d^2} = \frac{1.8544P}{d^2}$$

where P is applied load (kgf), d is mean diagonal of the impression (mm), and α is the face angle of the indenter (136°).

Vickers hardness test. A microindentation hardness test employing a 136° diamond pyramid indenter (Vickers) and variable loads, enabling the use of one hardness scale for all ranges of hardness—from very soft lead to tungsten carbide. Also known as diamond pyramid hardness test.

VI improver. An additive, usually a polymer, that reduces the variation of viscosity with temperature, thereby increasing the viscosity index of an oil.

vinyl acetate plastics. Plastics based on polymers of vinyl acetate or copolymers of vinyl acetate with other monomers, the vinyl acetate being the greatest amount by mass.

vinyl chloride plastics. Plastics based on polymers of vinyl chloride or copolymers of vinyl chloride with other monomers, the vinyl chloride being the greatest amount by mass.

vinyl esters. A class of thermosetting resins containing esters of acrylic and/or methacrylic acids, many of which have been made from epoxy resin. Cure is accomplished, as with unsaturated polyesters, by copolymerization with other vinyl monomers, such as styrene. Glass-reinforced vinyl esters are used in corrosion-resistant products, such as piping and storage tanks, used in the pulp and paper, chemical process, wastewater, and mining industries.

vinylidene chloride plastics. Plastics based on polymer resins made by the polymerization of vinylidene chloride or copolymerization of vinylidene chloride with other unsaturated compounds, the vinylidene chloride being the greatest amount by weight.

virgin filament. An individual *filament* that has not been in contact with any other fiber or any other hard material.

virgin material. A plastic material in the form of pellets, granules, powder, flock, or liquid that has not been subjected to use or processing other than that required for its initial manufacture.

viscoelasticity. A property involving a combination of elastic and viscous behavior that makes deformation dependent upon both temperature and strain rate. A material having this property is considered to combine the features of a perfectly elastic solid and a perfect fluid.

viscosity. The bulk property of a fluid, semifluid, or semisolid substance that causes it to resist flow. Viscosity is defined by the equation:

$$\eta = \frac{\tau}{(dv/ds)}$$

where τ is the shear stress, is the velocity, and s is the thickness of an element measured perpendicular to the direction of flow; (dv/ds) is known as the rate of shear. Newtonian viscosity is often called dynamic viscosity, or absolute viscosity. Kinematic viscosity, or static viscosity (ν), is the ratio of dynamic viscosity (η) to density (ρ) at a specified temperature and pressure ($\nu = \eta/\rho$). Recommended units of measure for dynamic viscosity are the pascal second (Pa · s) in SI units and poise (P) in English units. Recommended units of measure for kinematic viscosity are square meters per second (m^2/s) in SI units and the stoke, or centistoke (cSt) in English units.

viscosity coefficient. The shearing stress tangentially applied that will induce a velocity gradient in a material.

viscous. Possessing viscosity. This term is frequently used to imply high viscosity.

viscous deformation. Any portion of the total deformation of a body that occurs as a function of time when load is applied but that remains permanently when the load is removed. Generally referred to as an *elastic deformation*.

vitreous. Partially or completely comprised of a glass; often containing solid particles distributed therein.

vitrification. (1) The formation of a glassy or noncrystalline material. (2) That characteristic of a clay product resulting when the kiln temperature is sufficient to fuse grains and close the surface pores, forming an impervious mass. (3) The progressive reduction in porosity of a ceramic composition as a result of heat treatment, or the process involved.

vitrify. To render vitreous, generally by heating; usually, achieving enough glassy phase to render impermeable.

void. Air or gas that has been trapped and cured into a laminate. Porosity is an aggregation of microvoids. Voids are essentially incapable of transmitting structural stresses or nonradiative energy fields.

void content. Volume percentage of voids, usually less than 1% in a properly cured composite. The experimental determination is indirect, that is, it is calculated from the measured density of a cured laminate and the "theoretical" density of the starting material.

volatile content. The percent of *volatiles* that is driven off as a vapor from a plastic or an impregnated reinforcement.

volatiles. Materials, such as water and alcohol, in a sizing or resin formulation, that are capable of being driven off as a vapor at room temperature or at slightly elevated temperature. See also *size*.

volatilization. The conversion of a chemical substance from a liquid or solid state to a gaseous or vapor state by the application of heat, by reducing pressure, or by a combination of these processes. Also known as vaporization.

volume fraction. Fraction of a constituent material, such as fibers in a composite material, based on its volume.

volume resistance. The ratio of the direct voltage applied to two electrodes in contact with or embedded in a specimen to that portion of the current between them that is distributed through the volume of the specimen.

vulcanization. A chemical reaction in which the physical properties of a rubber are changed in the direction of decreased plastic flow, less surface tackiness, and increased tensile strength by reacting it with sulfur or other suitable agents. See also *self-vulcanizing*.

vulcanize. To subject to *vulcanization.*

W

wafer. A slice of a semiconductor crystalline ingot.

Wallner line. A fracture surface marking on ceramics and glasses, having a wavelike profile in the fracture surface. Such marks frequently appear as a series of curved lines, indicating the direction of propagation of the fracture from the concave to the convex side of a given Wallner line. Also known as ripple mark.

warp. The yarn running lengthwise in a woven fabric. A group of yarns in long lengths and approximately parallel. Also, in laminates, a change in dimension of a cured laminate from its original molded shape.

warpage. Dimensional distortion in a plastic object.

water-base adhesives. Solutions in which the polymer is totally soluble in water or alkaline water. Water-base adhesive solutions consist largely of natural adhesives. Materials that are soluble in water alone include animal glues, starch, dextrin, methylcellulose, and polyvinyl alcohol. Materials that are soluble in alkaline water include casein, rosin, carboxymethylcellulose, shellac, vinyl acetate, and acrylate copolymers containing carboxyl groups.

water break. The appearance of discontinous film of water on a surface signifying nonuniform wetting and usually associated with a surface contamination.

water break test. A test to determine if a surface is chemically clean by the use of a drop of water, preferably distilled water. If the surface is clean, the water will break and spread; a contaminated surface will cause the water to bead.

water-extended polyester. A casting formulation in which water is suspended in the polyester resin.

waterjet/abrasive waterjet machining. A hydrodynamic machining process that uses a high-velocity stream of water as a cutting tool. This process is limited to the cutting of nonmetallic materials when the jet stream consists solely of water. However, when fine abrasive particles are injected into the water stream, the process can be used to cut harder and denser materials. Abrasive waterjet machining has expanded the range of fluid jet machining to include the cutting of metals, glass, ceramics, and composite materials. Water pressures up to 410 MPa (60 ksi) are used. The coherent jet of water is propelled at speeds up to approximately 850 m/s (2800 ft/s).

weathering. Exposure of materials to the outdoor environment.

weave. The particular manner in which a fabric is formed by interlacing yarns. Usually assigned a style number. See also *basket weave*, *crowfoot satin*, *eight-harness satin*, *leno weave*, and *plain weave*.

web. A thin plastic sheet in process in a machine. The molten web is that which issues from the die. The substrate web is the substrate being coated.

webbing. Filaments or threads that sometimes form when adhesively bonded surfaces are separated.

weft. The transverse threads or fibers in a woven fabric. Those fibers running perpendicular to the warp. Also called *fill*, *filling yarn*, or woof.

weld line. The mark visible on a finished plastic part made by the meeting of two flow fronts of plastic material during molding. Also called weld mark, flow line, knit line, or stria.

wet installation. A bolted joint in which sealant is applied to the head and shank of the fastener such that after assembly a seal is provided between the fastener and the elements being joined. See also *sealant.*

wet lay-up. A method of making reinforced plastics by applying the resin system as a liquid when the reinforcement is put in place. Polyesters and vinyl esters are the most commonly used family of resins used for wet lay-up.

wet-out. In composites fabrication, the condition of an impregnated *roving* or *yarn* in which substantially all voids between the sized strands and filaments are filled with resin.

wet strength (adhesives). The strength of an adhesive joint determined immediately after removal from a liquid in which it has been immersed under specified conditions of time, temperature, and pressure. The term is commonly used alone to designate strength after immersion in water. In latex adhesives the term is used to describe the joint strength when the adherends are brought together with the adhesives still in the wet state. Compare with *dry strength.*

wet strength (composites). The strength of an organic matrix composite when the matrix resin is saturated with absorbed moisture, or is at a defined percentage of absorbed moisture less than saturation (saturation is an equilibrium condition in which the net rate of absorption under prescribed conditions falls essentially to zero).

wetting. (1) The spreading, and sometimes absorption, of a fluid on or into a surface. (2) A condition in which the interface tension between a liquid and a solid is such that the contact angle is 0° to 90°. (3) The phenomenon whereby a liquid filler metal or flux spreads and adheres in a thin continuous layer on a solid base metal. (4) The formation of a relatively uniform, smooth, unbroken, and adherent film of solder to a basis metal.

wetting agent. (1) A substance that reduces the surface tension of a liquid, thereby causing it to spread more readily on a solid surface. (2) A surface-active agent that produces *wetting* by decreasing the *cohesion* within the liquid.

wet winding. The process of *filament winding* of composites in which strands are impregnated with resin before or during winding onto the mandrel. See also *dry winding.*

whisker. (1) A short single crystal fiber or filament used as a reinforcement in a matrix. Whisker diameters range from 1 to 25 μm, with aspect ratios generally between 50 and 150. (2) Single-crystal growths resembling fine wire, which may extend to 0.64 mm (0.025 in.) high. They most frequently occur on printed circuit boards or electronic components that have been electroplated with tin. (3) Metallic filamentary growths, often microscopic, sometimes formed during electrodeposition and sometimes spontaneously during storage or service, after finishing.

whiskers (glass). See *striation.*

whiteware. A group of ceramic products characterized by a white or light colored body with a fine-grained structure which consist primarily of clay minerals, feldspars, and quartz. Most whiteware products are glazed—in whole or in part—and whiteware glazes may range from clear to completely opaque, white, or colored. Many whiteware products, such as tableware, are decorated with patterns or designs to enhance their beauty and appearance. Examples of whiteware products are sanitaryware, tableware, electrical porcelain, artware, stoneware, and tile.

width. In the case of a beam, the shorter dimension perpendicular to the direction in which the load is applied.

winding pattern. In filament winding of composites, the total number of individual circuits required for a winding path to begin repeating by laying down immediately adjacent to the initial circuit. A regularly recurring pattern of the filament path after a certain number of mandrel revolutions, leading eventually to the complete coverage of the mandrel. See also *filament winding*.

winding tension. In filament winding or tape wrapping of composites, the amount of tension on the reinforcement as it makes contact with the mandrel. See also *filament winding*.

window. A defect in thermoplastic film, sheet, or molding, caused by the incomplete plastication of a piece of material during processing. It appears as a globule in an otherwise blended mass. See also *fisheye*.

wire coating. The covering or coating of wire and cable in continuous length with insulating thermoplastics by extrusion.

wood veneer. A thin sheet of wood, generally within the thickness range from 0.3 to 6.3 mm (0.01 to 0.25 in.), to be used in a laminate.

woof. See *weft*.

working life. The period of time during which a liquid resin or adhesive, after mixing with catalyst, solvent, or other compounding ingredients, remains usable. See also *gelation time* and *pot life*.

woven fabric. A material (usually a planar structure) constructed by interlacing yarns, fibers, or filaments to form such fabric patterns as plain, harness satin, and leno weaves. See also *basket weave*, *crowfoot satin*, *eight-harness satin*, *leno weave*, and *plain weave*.

woven roving. A heavy glass fiber fabric made by weaving roving, or yarn bundles.

wrap-around bend. The bend obtained when a specimen is wrapped in a closed helix around a cylindrical mandrel.

wrinkle. A surface imperfection in laminated plastics that has the appearance of a crease or fold in one or more outer sheets of the paper, fabric, or other base. Also occurs in vacuum bag molding when the bag is improperly placed, causing a crease.

wrinkle depression. An undulation or series of undulations or waves on the surface of a pultruded composite part.

X

***x*-axis.** In reinforced plastic laminates, an axis in the plane of the laminate that is used as the 0° reference for designating the angle of a lamina. See also *laminate* and the figure accompanying *quasi-isotropic laminate*.

x-ray. A penetrating electromagnetic radiation, usually generated by accelerating electrons to high velocity and suddenly stopping them by collision with a solid body. Wavelengths of x-rays range from about 10^{-1} to 10^2 Å, the average wavelength used in research being about 1 Å. Also known as roentgen ray or x-radiation.

***xy*-plane.** In reinforced plastic laminates, the reference plane parallel to the plane of the laminate.

Y

yarn. An assemblage of twisted filaments, fibers, or strands, either natural or manufactured, to form a continuous length that is suitable for use in weaving into textile materials used to reinforce plastics.

yarn bundle. See *bundle*.

***y*-axis.** In composite laminates, the axis in the plane of the laminate that is perpendicular to the *x*-axis. See also *x-axis*.

yield. (1) Evidence of plastic deformation in structural materials. Also known as plastic flow or creep. See also *flow*. (2) The ratio of the number of acceptable items produced in a production run to the total number that were attempted to be produced. (3) Comparison of casting weight to the total weight of metal poured into the mold.

yield point. The first stress in a material, usually less than the maximum attainable stress, at which an increase in strain occurs without an increase in stress. Only certain materials—those which exhibit a localized, heterogeneous type of transition from elastic to plastic deformation—produce a yield point. If there is a decrease in stress after yielding, a distinction may be made between upper and lower yield points. The load at which a sudden drop in the flow curve occurs is called the upper yield point. The constant load shown on the flow curve is the lower yield point.

yield point elongation. In materials that exhibit a yield point, the difference between the elongation at the completion and at the start of discontinuous yield.

yield strength. The stress at which a material exhibits a specified deviation from proportionality of stress and strain. An offset of 0.2% is used for many materials, particularly metals. Compare with *tensile strength*.

yield stress. The stress level of highly ductile materials at which large strains take place without further increase in stress.

Young's modulus. A term used synonymously with modulus of elasticity. The ratio of tensile or compressive stresses to the resulting strain. See also *modulus of elasticity*.

yttria. A rare earth oxide of composition Y_2O_3 that is added in small amounts to ceramic compositions, most notably zirconia.

Z

***z*-axis.** In laminates, the reference axis normal to the *xy* plane of the laminate.

zero bleed. A laminate fabrication procedure that does not allow loss of resin during cure. Also describes prepreg made with the amount of resin desired in the final part, such that no resin has to be removed during cure. See also *laminate* and *prepreg*.

zero time. The time at which the given loading or constraint conditions are initially obtained in creep and stress-relaxation tests, respectively.

Ziegler-Natta catalysts. Initially, a catalyst consisting of an alkylaluminum compound with a compound of the titanium group of the periodic table, a typical combination being triethylaluminum and either titanium tetrachloride or titanium trichloride. Subsequently, an enormous variety of such mixtures is used in polymerization to provide stereospecificity (isotactic or syndiotactic). See also *isotactic stereoisomerism* and *syndiotactic stereoisomerism*.

zircon. The mineral zircon silicate ($ZrSiO_4$), a very high melting point acid refractory material used as a molding sand, and in ceramic products such as *refractories* and *whiteware*.

zirconia. A heavy white powder, the composition of which is ZrO_2, used in ceramic glazes, special glasses, refractory bricks, piezoelectric devices, and structural ceramics. See also *partially stabilized zirconia* and *transformation toughened zirconia*.

Selection of Engineered Structural Materials

Ceramics

Ceramic materials are solid compounds of both metallic and nonmetallic elements. Inorganic nonmetallic materials are the major constituents of ceramics, which include abrasives, porcelain, porcelain enamels, cements, glass (silicates), and carbon (graphite). The most significant characteristics of ceramic materials are high hardness, brittleness, greater resistance to high temperatures and other severe environments than metals or polymers, high compressive strength, high dielectric constant (which contributes to their good performance as electrical and thermal insulators), and low fracture toughness as compared to structural metallic materials.

Semiconductive ceramics have been developed for use in electronic components. The good electrical insulation properties of two semiconductive ceramics, silicon carbide (SiC) and zinc oxide (ZnO), make these materials useful varistors (Ref. 1). Varistors are variable (nonlinear) resistors that can be used as circuit protectors, shunts across contacts to prevent sparking, and control devices for current or voltage regulation (Ref. 2). The SiC-type semiconductors are high-temperature tolerant, as shown by the capability of these devices to operate reliably at 600 °C (1110 °F). These SiC semiconductors are useful in instrumentation and control electronics which are placed inside and on aircraft engines (Ref. 3).

High-alumina ceramics contain 80% (or more) Al_2O_3 and are used as bases for semiconductor transistors, integrated circuits, and other electronic components (Ref. 4, 5). Some of the specific types and applications of high Al_2O_3 ceramics are (Ref. 3):

- *85% Al_2O_3:* used in electronic equipment, shaft seal rings, valve trim, chokes, and nozzles
- *90% Al_2O_3, opaque:* used as a packaging material for light-sensitive electronics
- *94% Al_2O_3:* applied in transistor substrates, bases and headers; also, this ceramic is used for hybrid and integrated circuit packages (housings)
- *99.5% Al_2O_3, extra smooth*: used for electronic circuit substrates
- *99.9% Al_2O_3, translucent:* used for critical components in electronic parts

Alumina (Al_2O_3) is the oldest and most familiar of the engineering ceramics. This compound has a high degree of resistance to wear and corrosion. Along with SiC, it has been used as fiber in aluminum matrices to produce composite materials having high strength-to-weight ratios (Ref. 6). The use of ceramic whiskers or particles in matrices is increasing because of significant improvements in structural efficiency. For example, SiC-aluminum composites have an ultimate tensile strength over 690 kPa (100 ksi), a tensile modulus of 117 to 138 GPa (17 to 20 × 10^6 psi), and a density which is not much higher than aluminum (Ref. 7). Graphite and glass also have been used as fiber reinforcements. These reinforced materials are ideal for the reduced weight requirements of aircraft components.

Glass is an amorphous, or noncrystalline, inorganic material composed of silica (SiO_2), lime ($CaCO_3$), and sodium carbonate (Na_2CO_3). In addition to the technical ceramics such as Al_2O_3, the nonsilicate glass systems include phosphate glasses, which are resistant to hydrofluoric acid; borate glasses, formed by the addition of alkali or alkaline earth oxides to diboron trioxide (B_2O_3) (Ref. 8); heat absorbing glasses, which are made with iron monoxide (FeO); and glasses based on oxides of aluminum, vanadium (V), germanium (Ge), and other metals (Ref. 6).

High-silica glasses increase in strength as temperature rises. Increased SiO_2 content also correlates with higher Young's modulus values for certain glass materials (fused silica and 96% silica).

Two glass materials that are used as fiber reinforcements for aerospace composites are S-glass and E-glass. The more frequently used of these materials is E-glass, which is based on lime ($CaCO_3$), Al_2O_3, and borosilicate. S-glass was developed for improved tensile strength, and it consists of silica (SiO_2), Al_2O_3, and magnesia (MgO) (Ref. 9). E- glass has a room-temperature tensile strength of 2413 MPa (350 ksi), equivalent to that of high-strength steel (Ref. 10). The room-temperature tensile strength of S-glass (3450 MPa, or 500 ksi) is significantly greater than that of high-strength steel.

Another type of glass fiber is C-glass, which was developed for acid resistance. C-glass is not used for reinforcement of aerospace composite materials.

Ceramic (vitreous) coatings also are referred to as ceramic or porcelain enamels. These coatings are applied for good protection against high-temperature oxidation and corrosion. Silicon-base ceramic coatings can be used from 540 to 760 °C (1000 to 1400 °F) (Ref. 11). This reflects much greater thermal stability and oxidation resistance compared to organic coatings, which can be used up to approximately 150 °C (300 °F).

Ceramics have been tested for many years for application in gas turbine engines as thermal barrier coatings. The most frequently tested ceramic material has been yttria-stabilized zirconia (Ref. 12-14). Duplex ceramic-metallic coatings offer the best protection against oxidation and thermal diffusion for gas turbine parts. These coatings consist of a ceramic layer (ZrO_2-Y_2O_3) over the metal substrate and a nickel-chromium-aluminum-yttrium coating over the ceramic. The layer between the substrate and outer coating may, alternatively, consist of a cermet (NiCrAlY-Y_2O_3), which has been observed to be very effective in hot corrosion protection (Ref. 15, 16).

High-temperature applications. Ceramic materials under consideration for use in heat engines are Si_3N_4, SiC, ZrO_2, and lithium-aluminum-silicate (Li-Al-SiO_2). Si_3N_4 and SiC are used in components which are subjected to severe thermal and stress-corrosion environments. Both of these ceramics are candidates for rotors, stators, and other engine components (Ref. 17). ZrO_2 is an excellent insulator with good thermal expansion characteristics, suitable for coatings on cylinder liners, piston caps, and intake/exhaust ports. Li-Al-SiO_2 has good thermal properties, but its strength and fatigue life are poor. Applications for Li-Al-SiO_2 are thus limited to nonstructural (non-load-bearing) parts in heat engines.

Plastics and Elastomers

Plastics are highly polymeric materials that will flow when heat and pressure are applied. These versatile materials are composed of repeating organic chemical units called monomers. The structural chain length of a polymer is important in determining many of the properties of a plastic. Increasing chain length produces higher toughness, greater creep resistance, better resistance to SCC, increased melt temperature and melt viscosity, and greater difficulty in processing (Ref. 18). Crystallinity in polymers, such as polyethylene and polypropylene, can significantly increase strength and rigidity (Ref. 6, 18).

Polyethylene is a thermoplastic. A thermoplastic material will soften when heat is applied to it, flow under the influence of stress, and return to its original texture (solid or rubbery) when cooled. However, some thermoplastics, including polytetrafluoroethylene and ultrahigh molecular weight polyethylene, do not flow when heat is applied.

Thermosets are cured, or hardened, into a permanent shape via an irreversible chemical reaction known as crosslinking. Heat is usually applied to carry this reaction to completion (Ref. 10). Therefore, thermosetting plastics cannot be resoftened and remolded, as thermoplastics can.

Elastomers are all of the elastic, or rubberlike, polymers. At room temperature, elastomers can be stretched (under low stress) to at least two times their original length. When the stress is removed, the elastomer returns to its normal configuration. These materials also are referred to as rubbers and thermoplastic elastomers.

The emphasis in current automotive and aerospace design is on weight and cost reduction by the replacement of metallic parts with plastics and with reinforced plastic composite materials. Some of the manufacturing savings that can be attained when plastics are substituted for metallic die castings are:

- Greater design freedom versus die castings, because thinner walls and better detail can be incorporated
- Fewer steps in the assembly process
- Elimination of anticorrosion treatments
- Elimination of expansive automatic machining lines
- Up to 50% weight reduction may be achieved in comparison with the use of a metallic material

Certain plastics can be used in structures exposed to high temperatures. Polyimides and polyamideimides are two classes of thermoplastics that can be used for parts that require high heat resistance (Ref. 19). An example of this type of application is on the wing panels of supersonic reconnaissance planes, which fly at Mach 3 (three times the speed of sound). These wing panels must withstand temperatures up to about 250 °C (480 °F).

Engineering thermoplastics are also candidates for service where resistance to flame propagation is a requirement. An excellent choice for this application is rigid vinyl. The combustion resistance of this material is very high in comparison with other organic construction materials. Once a flame source is removed, rigid vinyl does not continue to burn. Rigid vinyl also does not form flaming droplets that promote fire propagation (Ref. 20).

Acetals. The characteristics of these plastics include high stiffness, resistance to creep, and impact strengths that do not change significantly with temperature. The stiffness of acetal resins is very high among the thermoplastics. Because of their resistance to surface friction and abrasion, acetals are used in moving parts, such as gears and bearings (Ref. 21, 22).

Acrylics. These plastics are dominated by methylmethacrylate, which is of significance because it is transparent. The transparency of colorless acrylic is about the same as the best quality optical glass. Acrylics withstand weathering and are resistant to acids and alkalies. However, acrylics are not resistant to organic solvents. These thermoplastics are also subject to creep, and they have low scratch and abrasion resistance. Acrylics have been combined with PVC to form polymers that are tough, resistant to chemicals and impact, and self-extinguishing. Acrylics are used in the formulation of paints for military equipment. Acrylic paints include the low infrared reflective topcoat conforming to MIL-C-4629 (Ref. 23). An acrylic lacquer meeting the requirements of MIL-L-81352A has also been used for surface protection of military weapon systems (Ref. 24). Another typical application for acrylics is in aircraft canopies (Ref. 21).

Acrylonitrile-butadiene-styrene (ABS) resins. These materials are known for their toughness and chemical resistance. ABS plastics generally have good impact and abrasive resistance. The low temperature limit for the impact resistance of ABS plastics is –40 °C (–40 °F). One problem with ABS materials is weathering. Prolonged exposure to sunlight reduces surface gloss, impact strength, and flex resistance. A potential application for ABS resins is in electrical connectors (Ref. 21).

Alkyds. These thermosetting materials are formed by a condensation reaction between a dibasic acid (or anhydride) and a polyhydric alcohol. Following this initial reaction, a fatty acid (usually long chain) is introduced into the resin molecule (Ref. 25). Alkyds have excellent weatherability and toughness, excellent fungus resistance, and excellent dimensional stability. Alkyd resins also have excellent dielectric strength (good insulating properties). In the radio frequency and ultrahigh frequency ranges, their electrical properties are stable up to 120 to 150 °C (250 to 300 °F). Disadvantages of these materials include their low impact strength and their susceptibility to degradation by strong acids and bases. Alkyds have been used in surface coatings, circuit breaker insulation, and electrical insulation.

Chlorinated polyvinyl chloride (CPVC). This thermoplastic possesses electrical resistance superior to that of rigid PVC. It is an excellent choice for parts subjected to corrosive high-temperature service up to about 105 °C (220 °F). The thermal insulation afforded by CPVC compared with copper and steel is excellent.

Diallyl phthalates (allyls). These high-performance thermosets are important for effective electrical insulation. The high dielectric strength of allyls is maintained up to 200 °C (400 °F) and in high humidity environments (Ref. 25). Other engineering properties of allyls include greater moisture resistance than any of the other thermosets, excellent resistance to corrosion (including microbiological degradation) and chemicals, good resistance to creep at room and elevated temperatures, and ability to be self-extinguishing with addition of appropriate flame retardants. Potential applications for allyls include electrical connectors, insulators, and resistors; nose cones; radomes; and aircraft leading edges.

Epoxies. Most of these thermosetting resins are derived from bisphenol A, which is produced by a reaction between phenol and acetone (Ref. 10). Epoxy-terminated molecules, with two epoxide groups, are produced by reacting bisphenol A with epichlorhydrin. Epoxies have superior thermal resistance and dimensional stability. They are also highly immune to solvents and other chemicals. Epoxy resins are widely used in the encapsulation (casting) of electronic components. As the plastic substrates for reinforcement with glass fibers or graphite fibers, epoxies are being used for reinforced plastic composite materials.

Additional epoxy applications include surface coatings, such as the epoxy based primer that meets MIL-P-23377 (Ref. 26); structural adhesives; and foams. Widely used in electronics applications, epoxy resins have superior thermal resistance and dimensional stability. Epoxies also are highly immune to solvents and other chemicals.

Fluorocarbons and fluorosilicones. Tetrafluoroethylene (TFE), or polytetrafluoroethylene (PTFE), is the most widely known material in this classification. TFE is noted for its excellent electrical and thermal properties. The serviceable temperature range for PTFE is –100 to 260 °C (–150 to 500 °F) (Ref. 27). PTFE has good electrical properties over a wide frequency range, and these favorable electrical characteristics are similar to those of polyethylene. Fluorocarbon plastics and elastomers are highly resistant to chemicals, except gaseous fluorine at high temperatures and pressures, and molten alkali (Ref. 25, 27). These materials also have good impact strength, good weather and abrasion resistance, and relatively high thermal and hydrolytic stability.

Typical applications for TFE include nonlubricated bearings, high-temperature electronic parts, packings, gaskets, seals, and rings (Ref. 28, 29). Chlorotrifluoroethylene (CTFE) is used in fuel sight lenses, electrical insulators, and inserts. Fluorinated ethylene-propylene (FEP) finds use in wire insulation and jacketing, coils, gaskets, electrical terminals, tube sockets, and terminal insulators. Vinylidene fluoride is used in electrical insulation, seals, and gaskets. Applications for ethylene trifluoroethylene (ETFE) include electrical connectors and electrical insulation (Ref. 30). Ethylene-chlorotrifluoroethylene (E-CTFE) is used in wire and cable coatings (for applications requiring high-performance wire) and film for laminates used in aircraft interiors (Ref. 21). Aerospace applications for the fluorosilicone elastomers are O-rings for fuel systems, electrical connectors, fuel seals, and channel sealants and faying surface sealants (Ref. 28).

Furan polymers. These thermosets are products of a condensation reaction initiated by either furfural or furfuryl alcohol. The engineering characteristics of furan polymers include excellent resistance to acids, alkalies, and solvents; good wetting properties; and low viscosity, which permits these resins to penetrate semiporous products and thus add strength, weatherability, and heat resistance. Furan polymers are relatively inexpensive, and they can be used for adhesives (Ref. 25).

Methylmethacrylate has a high strength-to-weight ratio, good dimensional stability, and low water absorption. It also has good weatherability and good electrical insulation characteristics. Methylmethacrylate is easily shaped and has been widely used for aircraft canopies and windows (Ref. 27).

Nylon. This group of polymeric materials is also referred to as the polyamides. Their noteworthy characteristics include high tensile strengths (55 to 76 MPa, or 8 to 11 ksi, in some grades); high melting point; good resistance to abrasion; good toughness and impact strength; excellent high temperature performance; poor resistance to water absorption, making nylon unsuitable in damp environments when good dimensional stability or good electrical properties is required; good resistance to alkalies and organic chemicals; and poor resistance to acids and polar solvents, such as alcohols and glycols (Ref. 21, 25).

Nylon loses its physical properties and oxidizes with prolonged exposure in air above 100 °C (212 °F). Nylon is typically used in gears, cams, and other sliding contact parts; wire insulation, gaskets, high pressure flexible tubing, and rope used by cargo helicopters for lifting and transporting military equipment (Ref. 21, 29).

Phenolic resins. Important characteristics of these plastics include superior heat resistance, good flame resistance, excellent dimensional stability, good resistance to chemical agents, long shelf life, good impact strength, and high flexural modulus (Ref. 9, 25, 30).

Commercial phenolic resins, which are primarily manufactured from phenol and formaldehyde, are referred to as phenolformaldehyde resins. Phenolics have been used in the electrical components of aircraft and guided missiles (Ref. 9, 31, 32).

Polycarbonates are produced by the condensation of bisphenol A with phosgene. Polycarbonates are tough and transparent, resistant to heat and flames, and dimensionally stable. Humidity and short-term exposure to boiling water produce negligible effects on the dimensions and properties of molded polycarbonates. Polycarbonates are also noted for their creep resistance and electrical insulating properties. In general, polycarbonates are not affected by greases, oils, or acids. However, continuous long-term exposure to boiling water produces embrittlement of this resin. Crazing polycarbonates are induced by exposure to aromatic solvents, esters, and ketones.

Polycarbonate resins have been used for ducting and trim in aircraft. Additional polycarbonate applications include protective covers for electrical relays, electrical switch components, and electrical connects (Ref. 25, 32, 33).

Polyesters can be either thermosetting or thermoplastic. The thermosetting polyesters can be either very hard (brittle) and tough (resilient) or soft and flexible. One or more of the following compounds can be added to thermosetting polyesters to confer fire retardance: chlorendic anhydride, tetrabromophthalic anhydride, tetrachlorophthalic anhydride, dibromoneopentyl glycol, and chlorostryrene (Ref. 22). Thermosetting polyesters can be made resistant to chemicals by the addition of the following chemicals: neopentyl glycol, isophthalic acid, hydrogenated bisphenol A, and trimethyl pentanediol. Neopentyl glycol and methylmethacrylate can be added to thermosetting polyesters for increased weathering resistance.

Unreinforced thermoplastic polyesters have good hardness, strength, and toughness; high abrasion resistance; good chemical resistance and low moisture absorption; good resistance to fatigue and SCC; and good electrical characteristics. Typical applications for high-temperature aromatic polyesters include self-lubricating bearings, high-temperature circuit boards, and encapsulation for electronic parts, such as diodes and transistors (Ref. 21, 33). Thermoplastic polyesters can be used in gears and bearings.

Polyimides are thermoplastic resins suitable for relatively high-temperature applications. For continuous service at 260 °C (500 °F), polyimides are capable of maintaining 50 to 60% of their room-temperature strength. These polymers can be used for fabrication of high-temperature foams. Additional applications for polyimides are in gears, bushings, turbofan engine backing rings, piston rings, valve seats, and high-temperature bearings of jet engines (Ref. 34).

Polysulfone. Significant properties of polysulfone resins include a heat-deflection temperature of 45 °C (345 °F) at 1.8 MPa (260 psi), and a long-term-use temperature of 150 to 175 °C (300 to 345 °F). These resins show environmental stress-cracking resistance which is greatly improved by the addition of glass fibers (5 to 10% concentration). In addition, they have high tensile strength (70 MPa, or 10 ksi) at yield; a flexural modulus of elasticity of close to 2800 MPa (400 ksi) at room temperature; total strain (creep) well below 2% at 100 °C (210 °F) and 20 MPa (3 ksi); stable electrical properties up to 175 °C (350 °F); and susceptibility to attack by polar organic solvents such as ketones, chlorinated hydrocarbons, and aromatic hydrocarbons.

Potential applications of polysulfone resins include injection-molded printed circuit boards; housings for meters, switches, and electronic components, aircraft cabin interior parts (because of the self-extinguishing and low smoke density characteristics of polysulfone); and high-temperature curing adhesives (Ref. 21, 25, 34, 35).

Polyurethanes may be classified as either plastics or elastomers. They are extremely tough and resistant to tearing, impact, puncturing, and abrasion. The tensile strength of polyurethanes is relatively high. Polyurethanes are also noted for resistance to oil and chemical solvents. Polyurethane coating systems are used extensively in military and commercial equipment. These coatings generally consist of a mixture of pigmented polyesters and aliphatic isocyanates. Polyurethanes can also be used for the manufacture of very tough adhesives, which can be used from about –270 to 120 °C (–450 to 250 °F) (Ref. 10, 36, 37).

Polyvinyl chloride (PVC) has a high fatigue strength relative to other thermoplastics. See also the section on chlorinated polyvinyl chloride (CPVC).

Silicones have a characteristic polymeric structure containing alternating silicon and oxygen atoms. The Si-O matrix, which constitutes the backbone of silicone polymers, is responsible for the unique characteristics of these materials. The Si-O bond is significantly stronger than a C-C or C-O bond.

Significant design properties of silicones include thermal stability and a mild dependence of physical properties on temperature. For example, general purpose silicone rubber performs between 100 and 260 °C (–150 and 500 °F) without physical property deterioration. In addition, these materials show good resistance to weathering, good dielectric properties and resistance to glow discharges, resistance to ozone and ultraviolet light attack, and resistance to most chemical agents at normal ambient temperatures (Ref. 9, 38, 39). The heat resistance of silicones is superior to that of other plastics elastomers. Up to 200 °C (400 °F), it has been observed that silicone rubber maintains its tensile strength, ductility, and hardness better than other organic rubber materials (Ref. 38).

Reinforced silicones can be used for electronic component encapsulation of such items as transistors, diodes, resistors, and capacitors. Laminated silicones are useful for aircraft radomes (Ref. 30).

Synthetic rubbers. This group of elastomers includes a broad variety of compounds. Some of the most significant of these materials for military applications are described below.

Polychloroprene (neoprene) is noted for its outstanding oil and oxidation resistance. Depending on the particular formulation of neoprene, the mechanical and physical properties of this material will vary. Tensile strengths can be up to 26 MPa (4 ksi) and elongation values up to 900%. Neoprene has about the same degree of resilience and abrasion resistance as natural rubber. Applications for solid neoprene include wire and cable insulation, tube and hose covers, and power transmission belts (Ref. 27, 40, 41).

Nitrile butadiene rubber (NBR) has good resistance to oils and fuels, good tensile and elongation characteristics, heat resistance, and low compression set. The weathering properties of NBR are improved substantially by the addition of anti-ozonants. Ozone oxidizes the carbon atoms of various types of elastomers, producing deep fissures (narrow cracks in the material). This degradation is induced only when the elastomer is under a tensile stress (Ref. 42). Nitrile rubbers have very good resistance to petroleum-based materials, including aircraft fuels. An application for NBR is fuel and hydraulic components, such as hoses (Ref. 41).

Styrene butadiene rubber (SBR), also known as Buna-S, has high elongation and elastic recovery, high tensile strength, and good flexibility down to –80 °C (–120 °F), but SBR materials do soften with increasing temperature. SBRs are impervious to water, alcohol, dilute acids, or dilute alkalies, but they are soluble in ketones, esters, and many hydrocarbons. These materials show excellent stability at temperatures from 120 to 200 °C (250 to 400 °F). Applications for SBR include adhesives, sealants, tire cord, and tubing (Ref. 25, 41).

Acrylic rubbers (AR) are copolymers of acrylic esters and olefins. These rubbers do not include acrylonitrile. AR can tolerate temperatures from –40 to 200 °C (–40 to 400 °F). They are used in seals, O-rings, packings, and adhesives (Ref. 41).

Epichlorohydrin rubber (CO and ECO) has a polyether backbone to which chloromethyl groups are attached. These elastomers have excellent resistance to the diffusive action of gases. CO and ECO can be used where resistance to solvents, fuel, oil, and ozone is required (Ref. 41).

Ethylene-propylene-diene monomer (EPDM) is a fairly new class of elastomers. Although they have poor resistance to oil and hydrocarbons, EPDM elastomers have outstanding resistance to ozone and weathering. EPDM is useful for coverings on electrical wires (Ref. 41, 42).

Fluorocarbon rubbers (FCRs) have especially favorable properties as a result of their high fluorine content. They are resistant to strong acids, oils, chemical solvents, and ozone. FCRs are generally not resistant to strong alkalies. FCRs are also stable up to 220 to 230 °C (425 to 450 °F), and they have good toughness and electrical properties. Applications for FCRs include seals, O-rings, and foam (Ref. 41).

Composite Materials

Composite materials are in use for an increasing number of commercial and military equipment applications. Based on a conservative extrapolation of this growing trend in composites usage, between 35 and 45% of the next generation of military aircraft structures will be made of composite materials.

One major purpose for the selection of composite materials is to obtain desirable structural engineering properties. Among these favorable characteristics are reduced weight (especially for aircraft), higher strength:weight ratios, improved stiffness (modulus):density ratios, and better corrosion resistance. Attainment of these properties will ensure that the equipment system performs with greater mechanical efficiency and durability. Various types of structural composite materials have been developed, primarily for use in aerospace equipment.

Fiber-reinforced plastic (FRP) is any polymeric (resinous) material reinforced with fibers. Reinforcement of plastic materials will improve or modify the properties of these materials. One type of FRP composite material is a fiber-reinforced thermosetting plastic known as sheet molding compound (SMC). SMC does not require processing (such as drying of volatiles or advancement of cure) after manufacture to prepare it for use at the molding press, and it can be molded without producing reaction byproducts. The amount of pressure required to mold SMC is just enough to make the material flow and compact. (High pressure is unnecessary.) SMC has been produced as a polyester-glass fiber (Ref. 43, 44).

Another type of FRP is premix, which also has been called dough molding compound, flow mix, or bulk molding compound. Premix is similar to SMC because both compounds can be molded without excessive pressure or reaction byproducts.

Composite laminates are manufactured by bonding two or more superimposed sheets of reinforcing fibers. Laminates are usually formed by applying heat and pressure to the layered structure (Ref. 25, 45).

Structural sandwich is formed by bonding two thin facings (layers) to a thick core. Nearly all the bending rigidity of the structural sandwich is provided by the facings. The core material separates the facings, transmits shear between the facings, and provides most of the shear rigidity of the sandwich construction. Structural sandwich composites having high stiffness:weight ratios can be manufactured through proper selection of facing and core materials. Lightweight sandwich cores are usually made of low-density material (Ref. 46).

Adhesives are used in sandwich composite construction to bond facings to inserts, such as reinforcing plates and edge strips. Most of these adhesives are formulations of organic resins, which provide high-strength bonds over a wide range of mechanical loading conditions. Inorganic (ceramic) adhesives have also been used in the bonding of composite sandwiches.

Structural sandwich facing material may consist of any thin sheet that is capable of carrying the major applied structural loads. Some of the materials used for facings are metals, including aluminum alloys, steel, titanium alloys, magnesium alloys, nickel-base alloys, cobalt-base alloys, niobium alloys, molybdenum alloys, and beryllium; reinforced plastic materials; and advanced composite materials.

Metal-matrix composite (MMC) materials are advanced composites that consist of a metallic matrix reinforced with high-strength and high-stiffness fibers. MMCs are called advanced composites because they have potential for use in high-performance military equipment.

Current MMCs generally consist of either a reinforced aluminum or magnesium matrix. Aluminum-matrix composites are usually reinforced with either boron (B), silicon carbide (SiC), borsic (Si-coated B), or graphite filaments. Magnesium-matrix composites are usually reinforced with alumina (Al_2O_3). The use of MMCs is minor in comparison to other advanced reinforced plastic composites, particularly graphite and aramid-Kevlar composites (Ref. 47-50).

Fiber-reinforced ceramic composites are currently in the experimental phase of development. The most successful efforts in the manufacture of ceramic composites have been with glass or glass-ceramic matrices and ceramic fiber reinforcements. Recent tests on graphite-fiber, glass-matrix composites have shown that the high strength and fracture toughness of these materials can be retained up to 593 °C (1110 °F). Reinforcing fibers, composed of alumina (Al_2O_3) and SiC, have been effective in improving the strength and toughness of glass matrices. However, ceramic matrices such as Al_2O_3 and magnesium oxide reinforced with carbon fibers have suffered from reduced strength despite their high toughness. Low strengths also have been observed in SiC-reinforced silicon nitride (Si_3N_4) ceramic composites (Ref. 51).

Carbon-carbon (CC) composite materials have been developed for very high-temperature applications up to 2800 °C (5000 °F). Because the mechanical properties of CC composites can be tailored to particular applications, these materials can be manufactured in various forms. CC composites are well suited for high-temperature applications such as the thermal protection system of the space shuttle because these materials do not lose their strength as the temperature increases to about 2200 °C (4000 °F) (Ref. 52-54).

References

1. S. Hayakawa, *Industrial Research and Development*, Vol. 25, No. 2, 1983, pp 142-147
2. M. Sapoff, "Varistor," *The Encyclopedia of Electronics*, C. Susskind, Ed., Reinhold Publishing Corp., 1962, p 911
3. W.C. Nieberding, *Industrial Research and Development*, Vol. 25, No. 9, 1983, pp 148-150
4. D.J. Godfrey, M.W. Lindley, E.R.W. May, and R.L. Brown, *Engineering Ceramics*, Admiralty Materials Laboratory, U.K., 1971
5. L.E. Ferreira, D.D. Briggs, and R.G. Barnhart, "High-Alumina Ceramics," *Source Book on Materials Selection, Vol. II*, R.B. Gunia, Ed., American Society for Metals, 1977, pp 404-408
6. "Materials Reference Issue," *Machine Design*, Vol. 49, No. 6, 1977
7. H.E. Chandler and D.F. Baxter, *Metal Progress*, Vol. 123, No. 1, 1983
8. W.D. Kingery, H.K. Bowen, and D.R. Uhlmann, *Introduction to Ceramics*, 2nd ed., John Wiley and Sons, Inc., 1976
9. Military Handbook MIL-HDBK-17A, *Plastics for Aerospace Vehicles: Part 1: Reinforced Plastics*, Department of Defense, 1971
10. W.E. Driver, *Plastics Chemistry and Technology*, Van Nostrand Reinhold Company, 1977
11. P.E. France, *Chemical Engineering*, Vol. 90, No. 13, 1983, pp 61-63
12. S.K. Anderson, S.K. Lau, R.J. Bratton, S.Y. Lee, K.L. Rieke, J. Allen, and K.E. Munson, "Advanced Ceramic Coating Development for Industrial/Utility Gas Turbine Applications," NASA CR-165619, National Aeronautics and Space Administration, 1982
13. R.J. Keller, "Research and Development for Improved Thermal Barrier Coatings," ER-8216-1, Air Force Systems Command, Aeronautical Systems Division, Wright-Patterson Air Force Base, 1982
14. J.W. Vogan and A.R. Stetson, "Advanced Ceramic Coating Development for Industrial/Utility Gas Turbines," NASA CR 169852, National Aeronautics and Space Administration, 1982
15. D.J. Bak, *Design News*, Vol. 39, No. 1, 1983
16. M.A. Gedwill, T.K. Glasgow, and S.R. Levine, *Thin Solid Films*, Vol. 95, 1982, pp 66-72
17. M.R. Pasucci, "The Role of Ceramics in Engines—An Assessment," *Current Awareness Bulletin*, Metals and Ceramics Information Center, Battelle Columbus Laboratories, Issue 126, August 1983, pp 1-4
18. S.L. Rosen, *Fundamental Principles of Polymeric Materials*, John Wiley and Sons, Inc., NY, 1982, p 42
19. M. Bakker, *Plastics Design Forum*, Vol. 8, No. 6, 1983, pp 83-85
20. "Designing with Thermoplastics," Chemical Group, The BF Goodrich Company, 1980
21. *Plastics Engineering: Handbook of the Society of the Plastics Industry, Inc.*, 4th ed., J. Frados, Ed., Van Nostrand Reinhold Company, 1976, p 29
22. *Handbook of Plastics and Elastomers*, C.A. Harper, Ed., McGraw-Hill, Inc., 1975
23. Military Specification MIL-L-46159A, "Lacquer, Acrylic, Low Reflective, Olive Drab," U.S. Army Materials and Mechanics Research Center, 1977
24. Military Specification MIL-L-81352, "Lacquer, Acrylic (for Naval Weapons Systems)," U.S. Naval Air Systems Command, 1973
25. Military Standardization Handbook MIL-HDBK-700A, *Plastics,* Department of Defense, March 1975
26. Military Specification MIL-P-23377, "Primer Coating, Epoxy Polyamide, Chemical and Solvent Resistant," U.S. Army Materials and Mechanics Research Center, March 1978
27. H.J. Sharp, *Engineering Materials, Selection and Value Analysis*, American Elsevier Publishing Company, Inc., 1966
28. D.A. Strivers, "Fluoroelastomers," *The Vanderbilt Rubber Handbook*, R.O. Babbit, Ed., R.T. Vanderbilt Company, Inc., 1978, pp 244-258
29. R.C. Beercheck, *Machine Design*, Vol. 54, No. 5, 1982, pp 157-161
30. "Mechanical and Physical Properties of Engineering Plastics," *Source Book on Materials Selection, Vol. II*, R.B. Gunia, Ed., 1977, pp 400-403
31. *Plastics Design Forum*, Vol. 8, No. 6, 1983, pp 75-76
32. D.V. Rosato and G. Lubin, "Plastics in Aircraft and Aerospace," *Handbook of Fiberglass and Advanced Plastics Composites*, G. Lubin, Ed., Van Nostrand Reinhold Company, 1969, pp 801, 832
33. H.L. Thomsa, *Industrial Research and Development*, Vol. 25, No. 6, 1983, pp 96-99
34. J.N. Anderson, "Designing with High Performance Plastics in Off the Road and Automotive Equipment Applications," Society of Automotive Engineering, Technical Paper 810-968, 1981
35. L.T. Manzione, *Plastics World*, Vol. 41, No. 11, 1983, pp 57-59
36. Military Handbook MIL-HDBK-17, *Plastics for Aerospace Vehicles, Part II: Transparent Glazing Materials*, Air Force Materials Laboratory, Wright-Patterson Air Force Base, June 1977
37. Military Specification MIL-C-46168B, "Coating, Aliphatic Polyurethane, Chemical Agent Resistant," U.S. Army Mobility Equipment Research and Development Command, DRDME-DS, 1982
38. Military Specification MIL-C-83286B, "Coating, Urethane, Aliphatic Isocyanate, for Aerospace Applications," US Air Force Materials Laboratory, Wright-Patterson Air Force Base, 1980
39. B.B. Hardman and R.W. Shade, *Materials Technology*, General Electric Company, 1980
40. A.R. Mersberg and J.W. Lee, *Materials Performance*, Vol. 19, No. 12, 1980, pp 13-17
41. S.W. Schmitt, "The Neoprenes," *The Vanderbilt Rubber Handbook*, R.O. Babbit, Ed., R.T. Vanderbilt Company, 1978, pp 137-146
42. Military Handbook MIL-HDBK-149A, *Rubber and Rubber-Like Materials*, U.S. Army Materials and Mechanics Research Center, June 1965, pp 63-64
43. D.A. Riegner and J.C. Hau, *Automotive Engineering*, Vol. 90, No. 5, 1982, pp 51-57
44. P.R. Young, "Reinforced Molding Compounds," *Handbook of Fiberglass and Advanced Plastics Composites*, G. Lubin, Ed., Van Nostrand Reinhold Company, 1969, pp 369-420
45. C.A. Harper, Ed., *Handbook of Plastics and Elastomers*, McGraw-Hill Book Co., 1975
46. Military Handbook MIL-HDBK-23A, *Structural Sandwich Composites*, Department of Defense, 1968
47. E.L. Foster, "Technological Development of Metal Matrix Composites for DOD Application Requirements; Part 2: Findings and Recommendations, Including the Proceedings of the Second and Third MMC Workshops," Report P-1177, Defense Advanced Research Projects Agency, February 1977
48. "Advanced Composite Repair Guide," Report NOR 82-60, Air Force Wright Aeronautical Laboratories, Air Force Systems Command, Wright-Patterson Air Force Base, March 1982

49. *Advanced Composites Design Guide, Vol. 1: Design*, Structures Division, Air Force Flight Dynamics Laboratory, Air Force Systems Command, Wright-Patterson Air Force Base, September 1976
50. E.M. Lenoe, "Comments on the Status of Composite Structures Technology," *Fibrous Composites in Structural Design*, E.M. Lenoe, D.W. Oplinger, and J.J. Burke, Ed., Plenum Press, 1980
51. D.K. Shelty, "Ceramic Matrix Composites," *Current Awareness Bulletin*, No. 118, Metals and Ceramics Information Center, Battelle Columbus Laboratories, 1982
52. D.J. Holt, *Aerospace Engineering*, Vol. 3, No. 4, 1983, pp 14-18
53. D.R. Rummler, *Machine Design*, Vol. 55, No. 23, 1983, pp 127-128
54. D.R. Rummler, "Recent Advances in Carbon-Carbon Materials Systems," *Advanced Materials Technology*, NASA Conference Publication 2251, National Aeronautics and Space Administration, Langley Research Center, 1982, pp 293-312

Composites

Fibers, Fillers, and Reinforcements

General

Specific gravity of selected fillers

Fillers	Specific gravity
Barium ferrite	5.4
Barium sulfate	4.5
Barium titanate	6.1
Bronze	8.80
Calcium carbonate	2.7
Carbon fiber (polyacrylonitrile)	1.8
Carbon fiber (pitch)	1.99
Clay	2.6
Glass beads	2.54
Glass fiber	2.54
Graphite powder	2.08
Hydrated alumina	2.4
Iron	7.9
Mica	2.8
MoS_2	4.8
Silica	2.3
Silicone oil	0.98
Talc	2.75
Polytetrafluoroethylene lubricant	2.15
Tungsten	19.35
Wollastonite	2.9

Classification of some fibers

Category	Materials
Metal	Beryllium, molybdenum, steel, tungsten
Glass	Vitreous silica, E-glass, S-glass
Carbonaceous(a)	Polyacrylonitrile (high strength), polyacrylonitrile (high modulus), pitch, rayon (very high modulus),, rayon (high modulus)
Polymer	Aramid, olefin, nylon, rayon
Inorganic	Alumina (monocrystal), alumina (polycrystal), alumina (whisker), alumina silicates, asbestoes, boron (tungsten core), boron nitride, silicon sarbide (carbon core), silicon carbide (polycrystal), silicon carbide (whisker), silicon nitride (whisker), zirconia (polycrystal)

(a) Material refers to starting process

Properties of fiber materials

Material	Density, g/cm^3	Tensile modulus		Tensile strength	
		GPa	10^6 psi	MPa	ksi
Polyester	1.36	13.8	2.0	1100	160
E-glass	2.52	72.3	10.5	3450	500
S-glass	2.49	85.4	12.4	4130	600
Kevlar 49	1.44	124	18.0	2760	400
T-300	1.72	218	31.6	2240	325
VSB-32	1.99	379	55.0	1210	175
FP	3.96	379	55.0	1380	200
Boron	2.35	455	66.0	2070	300
Silicon carbide	3.19	483	70.0	1520	220
GY-70	1.97	531	77.0	1720	250

Typical properties of reinforcing fibers

Fiber	Density, g/cm^3	Tensile strength		Tensile modulus		Elongation, %
		(MPa)	ksi	(GPa)	ksi	
Graphite						
AS4	1.8	3590	521	234	34	1.6
T700	1.8	4902	711	230	33	2.1
Aramid						
Twaron 1055/1056 Kevlar 49	1.44	2896	420	117	17	2.5
Twaron 1010 Kevlar 29	1.44	2896	420	62	9	3.6
Glass						
E-glass	2.54	3448	500	72	10	4.3
S-glass	2.49	4598	667	85	12	4.8
Metal fibers						
Stainless steel nickel-coated	7.9	1965	285	186	27	1 to 2

Properties of natural fibers and typical synthetic fibers

Fiber	Density, g/cm^3	Tensile modulus GPa	Tensile modulus 10^6 psi	Tensile strength GPa	Tensile strength 10^6 psi	Elongation at failure, %
Hemp	1.52	70	10	0.92	0.13	1.7
Jute	1.52	60	9	0.86	0.12	2.0
Flax	1.52	100	15	0.84	0.12	1.8
Cotton	1.52	27	4	0.2 to 0.8	0.03 to 0.12	6 to 12
Silk	1.34	10	2	0.6	0.09	18 to 20
S-glass	2.50	84	12	4.6	0.67	2 to 5
Carbon (type 1)	1.90	380	55	2.0	0.29	1 to 2
Aramid	1.44	133	20	2.8	0.41	2 to 4

Thermal properties of selected fibers

Fiber	Diameter, μm	Heat capacity, kJ/(kg · K)	Thermal conductivity, W/(m · K)	Coefficient of thermal expansion, 10^{-6}/°C
Graphite				
PAN HM	7	0.7	1003	−1.1
PAN HTS (T300)	8	0.7	1003	−1.1
Rayon (T50)	6	0.7	1003	−1.1
Thornel 75 (T75)	5	0.7	1003	−1.1
Pitch (type P)	5 to 10	0.7	1003	−1.1
Pitch UHM	11	0.7	1003	−1.1
Boron on tungsten	102 to 203	1.3	38	5.0
Borsic	102 to 203	1.3	38	5.0
Boron on carbon	102 to 203	1.3	38	5.0
Silicon carbide on tungsten	102 to 203	1.2	16	4.3
Silicon carbide on carbon	102	1.2	16	4.3
Beryllium	127	1.9	150	11.5
Alumina (FP)	20	...	...	8.3
S-glass	9	0.7	13	5.0
E-glass	9	0.7	13	5.0
Molybdenum	127	0.3	145	4.9
Steel	127	0.5	29	13.3
Tantalum	508	0.2	55	6.5
Tungsten	381	0.1	168	4.5
Whisker ceramic Al_2O_3	10 to 25	0.6	24	7.7
Metallic (Fe)	127	0.5	29	13.3

Thermal properties of key reinforcements

Material	Coefficient of thermal expansion				Thermal conductivity				Specific thermal conductivity (axial)	
	Axial		Transverse		Axial		Transverse			
	10^{-6}/K	10^{-6}/°F	10^{-6}/K	10^{-6}/°F	W/m · K	Btu/h · ft · °F	W/m · K	Btu/h · ft · °F	W/m · K	Btu · ft · °F
Carbon fiber (pitch)										
$E = 55 \times 10^6$ psi	−1.3	−0.7	...	...	120	70	...	...	60	35
$E = 75 \times 10^6$ psi	−1.4	−0.8	...	...	185	107	...	...	92	55
$E = 100 \times 10^6$ psi	−1.6	−0.9	18	10	520	300	5	3	236	137
$E = 120 \times 10^6$ psi	−1.6	−0.9	18	10	640	370	5	3	291	168
$E = 130 \times 10^6$ psi	−1.6	−0.9	18	10	1100	610	5	3	500	290
Carbon fiber (PAN)										
$E = 34 \times 10^6$ psi	−0.6	−0.3	...	...	9	5	...	...	5	3
$E = 75 \times 10^6$ psi	−1.1	−0.6	18	10	...	...	...	...	...	...
Carbon fiber (vapor-phase)	...	...	...	...	2000	1160	...	...	1110	642
Boron fiber	4.9	2.7	4.9	2.7	...	...	...	...	...	...
Quartz fiber	0.54	0.3	0.54	0.3	...	...	...	...	...	...
Aramid fiber (Kevlar 49)	−2.0	−1.1	59	33	...	...	...	...	...	...
E-glass fiber	5.0	2.8	5.0	2.8	2.1	1.2	2.1	1.2	0.8	0.5
Silicon carbide particles										
Commercial	4.9	2.7	4.9	2.7	81	47	81	47	28	16
High purity	4.9	2.7	4.9	2.7	490	280	490	280	169	96

Differences between bulk and whisker strengths of various materials

	Tensile strength			
	GPa		10^6 psi	
Material	Bulk	Whisker	Bulk	Whisker
Iron	0.028	13.1	0.004	1.9
Copper	0.0014	2.8	0.0002	0.4
Silicon	0.034	3.79	0.005	0.55
Graphite	0.28	20.7	0.04	3.0
Boron carbide	0.1551	6.653	0.0225	0.965
Alumina	0.55	42.7	0.08	6.2
Silicon carbide	0.21	20.7[a](a)	0.03	3.0[a](a)

(a) Bend test (all other values are tensile data)

Effect of fiber type on selected ultimate properties

	Property									
	Specific	Tensile strength		Tensile modulus		Compressive strength		Thermal conductivity	
Fiber type	gravity	MPa	ksi	GPa	10^6 psi	MPa	ksi	W/m · k	Btu · in./h · ft^2 · °F
Glass(a)	2.0	690	100	40	6	410	60	0.30	2
Carbon(b)	1.65	1000 to 1500	150 to 220	100 to 140	15 to 20	620 to 970	90 to 140	0.85 to 1.4	6 to 10
Aramid(c)	1.28	1400	200	80	12	280	40	0.15	1

(a) E-glass unidirectional rovings. (b) Type AS graphite fibers. (c) DuPont Kevlar 49 fibers

Comparison of long and short fibers (a)

	Reported value for:			
	Polycarbonate matrix reinforced with 40% E-glass		Nylon 6/6 matrix reinforced with 40% E-glass	
Property	Short	Long	Short	Long
Izod impact strength, J/cm (ft · lb/in.):				
Notched	1.2 (2.2)	4.8 (8.9)	1.4 (2.6)	5.94 (11.13)
Unnotched	6.4 (12)	12 (23)	8.0 (15)	17 (31)
Flexural strength, MPa (ksi)	180 (26)	255 (37)	275 (40)	340 (49)
Flexural modulus, GPa (10^6 psi)	9.65 (1.40)	13.4 (1.95)	11.7 (1.70)	11.7 (1.70)
Tensile strength, MPa (ksi)	145 (21)	150 (22)	220 (32)	180 (26)
Tensile modulus, GPa (10^6 psi)	11.7 (1.70)	17.9 (2.60)	13.1 (1.90)	17.9 (2.60)
Compressive strength, MPa (ksi)	150 (22)	220 (32)	160 (23)	255 (37)
Specific gravity	1.50	1.50	1.45	1.45

Short, length-to-diameter aspect ratio of 50. Long, length-to-diameter aspect ratio of 300

Fiber bundle dimensions

	Yield/tow		Filament size	
Material	m/kg	yd/lb	μm	μin.
Graphite (1000 to 12000 filaments per tow)	300 to 1200	150 to 600	5 to 10	200 to 390
Fiberglass (2450 to 12240 filaments per tow)	490 to 2400	245 to 1200	4 to 13	160 to 510
Aramid (800 to 3200 filaments per tow)	2000 to 7850	980 to 3900	12	470

Mechanical properties of key reinforcements

Material	Density		Elastic modulus					
			Axial		Transverse		Specific modulus (axial)	
	g/cm^3	lb/in.^{3}	GPa	10^6 psi	GPa	10^6 psi	GPa	10^6 psi
Carbon fiber (pitch)								
$E = 55 \times 10^6$ psi	2.0	0.072	380	55	...	...	190	28
$E = 75 \times 10^6$ psi	2.0	0.072	520	75	...	...	260	38
$E = 100 \times 10^6$ psi	2.2	0.078	690	100	5	0.7	314	46
$E = 120 \times 10^6$ psi	2.2	0.078	830	120	5	0.7	377	55
$E = 130 \times 10^6$ psi	2.2	0.078	895	130	5	0.7	407	59
Carbon fiber (PAN)								
$E = 34 \times 10^6$ psi	1.8	0.064	230	34	14	2	128	19
$E = 75 \times 10^6$ psi	...	...	520	75	6.9	1.0	...	...
Carbon fiber (vapor-phase)	1.8	0.064	230	34	...	...	128	19
Boron fiber	2.5	0.090	400	58	400	58	160	23
Quartz fiber	2.2	0.079	69	10	69	10	...	...
Aramid fiber (Kevlar 49)	1.44	0.052	124	18	...	...	86	13
E-glass fiber	2.6	0.094	72	10.5	72	10.5	28	4
Silicon carbide particles								
Commercial	2.9	0.106	520	75	520	75	179	26
High purity	2.9	0.106	...	...	...	...	...	...

Mechanical properties of high-modulus fibers

Material	Tensile strength		Elastic modulus		Specific gravity
	MPa	ksi	GPa	10^6 psi	
Boron	3450	500	400	58	2.6
Kevlar 49	2750	400	130	19	1.45
E-glass	1510	220	70	10	0.55
High-modulus graphite	2000	300	380	55	1.94
High-tensile strength graphite	2750	400	260	38	1.76

Some filament properties at room and elevated temperatures

Property	Type of filament				
	Borsic	SiC	Mo (TZM)	Sapphire	Be
Diameter:					
µm	107 to 145	102	102	254	127 to 1525
mils	4.2 to 5.7	4	4	10	5 to 60
Density:					
g/cm^3	2.768	3.460	10.186	3.958	1.855
lb/in.^{3}	0.100	0.125	0.368	0.143	0.067
Modulus, GPa:					
Room temperature	415	450	295	470	295
205 °C (400 °F)	...	450	...	...	...
315 °C (600 °F)	...	450	...	...	...
425 °C (800 °F)	345	450	...	...	290
540 °C (1000 °F)	275	450	...	460	255
Modulus, 10^6 psi:					
Room temperature	60	65	43	68	43
205 °C (400 °F)	...	65	...	...	...
315 °C (600 °F)	...	65	...	...	...
425 °C (800 °F)	50	65	...	...	42
540 °C (1000 °F)	40	65	...	67	37
Tensile strength, MPa:					
Room temperature	2930	2275	2550	2550	690 to 965
205 °C (400 °F)	2860	2070	...	2035	550
315 °C (600 °F)	2655	2040	...	1895	345
425 °C (800 °F)	1795	2025	...	1795	240
540 °C (1000 °F)	1550	2015	...	1795	170
Tensile strength, ksi:					
Room temperature	425	330	370	370	100 to 140
205 °C (400 °F)	415	300	...	295	80
315 °C (600 °F)	385	296	...	275	50
425 °C (800 °F)	260	294	...	260	35
540 °C (1000 °F)	225	292	...	260	25
Coefficient of thermal expansion, room temperature to 540°C (1000°F):					
10^{-6}/°C	5.22	3.24	5.94	7.92	15.3
10^{-6}/°F	2.90	1.80	3.30	4.40	8.50

Mechanical properties and maximum service temperatures of whiskers

	Whisker diameter		Density		Tensile strength		Tensile modulus		Maximum service temperature	
	μm	mils	kg/m³	lb/in.³	MPa	ksi	GPa	10⁶ psi	°C	°F
Al_2O_3	3 to 10	0.11 to 0.39	3958	0.143	20 685	3000	427	62	1650	3000
BeO	10 to 30	0.39 to 1.17	2851	0.103	13 100	1900	345	50	1925	3500
B_4C	...	...	2519	0.091	13 790	2000	483	70	1095	2000
SiC	1 to 3	0.03 to 0.11	3211	0.116	20 685	3000	483	70	600	1110
Si_3N_4	...	...	3183	0.115	13 790	2000	379	55	600	1110
Quartz	...	...	...	...	4 135	600	76	11	...	...
Sapphire	...	...	...	...	11 720	1700	510	74	...	...
Cr	...	...	7197	0.260	8 895	1290	241	35	540	1000
Cu	...	...	8913	0.322	3 275	475	124	18	260	500
Fe	...	...	7833	0.283	13 100	1900	200	29	540	1000
Ni	...	...	8968	0.324	3 860	560	200	29	540	1000
Graphite	...	...	1661	0.060	19 615	2845	703	102	(a)	(a)
Au	...	...	...	...	1 655	240	76	11	...	...
Zirconia	...	...	...	...	4 135	600	427	62	...	...
Silicon	...	...	...	...	3 790	550	158	23	...	...

(a) 316 °C (600 °F) in air; 2760 °C (5000 °F) in inert atmosphere

Some fibers used in plastic

Fiber	Fiber diameter μm	mil	Density g/cm³	lb/in.³	Tensile strength MPa	ksi	Tensile modulus GPa	10⁶ psi	Use limit °C	°F
Continuous										
E-glass	3 to 20	0.12 to 0.79	2.49	0.090	3450	500	72.4	10.5	425	800
S-glass	10 to 20	0.39 to 0.79	2.49	0.090	4590	665	86.9	12.6	425	800
Kevlar	12	0.47	1.44	0.052	2760	400	125	18	425	800
Carbon/graphite-PAN	7.0	0.28	1.72 to 1.80	0.062 to 0.065	2410 to 4830	350 to 700	230 to 395	33 to 57	1650(a)	3000(a)
Carbon/graphite-pitch	5.1 to 12.7	0.20 to 0.50	1.99 to 2.16	0.072 to 0.078	2070	300	380 to 690	55 to 100	1650(a)	3000(a)
Discontinuous										
Processed mineral fiber	4 to 6	0.16 to 0.24	2.68	0.097	830	120	105	15	760	1400
Fiberfrax	2 to 5	0.08 to 0.20	2.60	0.094	1030	150	105	15	1150	2100
Fibermax	3 to 6	0.12 to 0.24	2.99	0.108	860	125	150	22	1760	3200

(a) Oxidation begins at lower temperatures

Some fibers used in metal-matrix composites

Fiber	Fiber diameter μm	mils	Tensile strength MPa	ksi	Modulus GPa	10⁶ psi	Use limit °C	°F
Continuous								
Boron	102 to 103	4.0 to 8.0	3450	500	400	58	540	1000
Carbon/graphite-PAN	7.0	0.28	2410 to 4830	350 to 700	230 to 395	33 to 57	>1650(a)	>3000(a)
Carbon/graphite-pitch	5.1 to 12.7	0.20 to 0.50	2070	300	380 to 690	55 to 100	>1650(a)	>3000(a)
SiC monofilament	140	5.51	4140	600	425	62	930	1700
FP alumina	20	0.79	1380	200	380	55	>1650	>3000
Discontinuous								
Fiberfrax	2 to 5	0.08 to 0.20	1030	150	105	15	1150	2100
Fibermax	3 to 6	0.12 to 0.24	830	120	150	22	1760	3200
Whisker								
SiC	6	0.24	3340	485	485 to 825	70 to 120	930	1700

(a) Oxidation begins at lower temperatures

Critical lengths (L_c) and critical aspect ratios (L_c/d)

		L_c		L_c/d	
Fiber	Matrix	mm	in.	mm	in.
E-glass(a)	Polypropylene	1.78	0.0700	140	6
E-glass(a)	Epoxy	0.43	0.017(b)	34(b)	1.3
E-glass(a)	Polyester	1.27	0.0500(c)	100(c)	4
Carbon (AS1)(d)	Epoxy	0.33	0.013	47	1.9
Carbon (AS4)(d)	Epoxy	0.42	0.017	60	2.4
Carbon (AS4)(d)	Polycarbonate	0.74	0.030	106	4.17
Carbon (XAS)(e)	Polycarbonate	0.35	0.014	50	2
Carbon (XAS)(e)	Epoxy	0.36	0.014	51	2

(a) Proprietary sizing. (b) At 40°C (105°F). (c) At 50°C (120°F). (d) Hercules Inc. (e) Hysol-Grafil Ltd.

Fiber tow characteristics before impregnation

Material	Yield/tow		Filament size	
	m/kg	yd/lb	μm	μin.
Graphite (1000 to 12 000 filaments/tow)	300 to 1200	150 to 600	5 to 10	200 to 390
Fiberglass (2450 to 12 240 filaments/tow)	490 to 2400	245 to 1200	4 to 13	160 to 510
Aramid (800 to 3200 filaments/tow)	2000 to 7850	980 to 3900	12	470

Filament diameter nomenclature

Alphabet	Filament diameter μm	10^{-4} in.
AA	0.8 to 1.2	0.3 to 0.5
A	1.2 to 2.5	0.5 to 1.0
B	2.5 to 3.8	1.0 to 1.5
C	3.8 to 5.0	1.5 to 2.0
D	5.0 to 6.4	2.0 to 2.5
E	6.4 to 7.6	2.5 to 3.0
F	7.6 to 9.0	3.0 to 3.5
G	9.0 to 10.2	3.5 to 4.0
H	10.2 to 11.4	4.0 to 4.5
J	11.4 to 12.7	4.5 to 5.0
K	12.7 to 14.0	5.0 to 5.5
L	14.0 to 15.2	5.5 to 6.0
M	15.2 to 16.5	6.0 to 6.5
N	16.5 to 17.8	6.5 to 7.0
P	17.8 to 19.0	7.0 to 7.5
Q	19.0 to 20.3	7.5 to 8.0
R	20.3 to 21.6	8.0 to 8.5
S	21.6 to 22.9	8.5 to 9.0
T	22.9 to 24.1	9.0 to 9.5
U	24.1 to 25.4	9.5 to 10

Towpreg form parameters

Parameter	Typical range
Strand weight per length, g/m (lb/yd)	0.74 to 1.48 (0.0015 to 0.0030)
Resin content, %	28 to 45
Tow width, cm (in.)	0.16 to 0.64 (0.06 to 0.25)
Package size, kg (lb)	0.25 to 4.5 (0.5 to 10)

Typical woven reinforcement properties

Reinforcement type	Coefficient of thermal expansion, 10^{-6}/K	Dielectric constant, κ, 1 MHz	Styles widely available	Thickness/ply mm	mils
E-glass	5.04	5.80	106 to 7642	0.05 to 0.23	2.0 to 9.0
S-glass	2.80	4.52	1080, 116, 7628	0.06 to 0.18	2.5 to 7.0
D-glass	2.00	3.95	1080, 2116, 7628	0.06 to 0.18	2.5 to 7.0
Quartz	0.54	3.78	503, 525	0.08 to 0.13	3.0 to 5.0
Kevlar 49	–5.20	4.00	120	0.1	4.0
Technora HM-50	–7.50	4.00	120	0.1	4.0

Polymer Fibers

Properties of aramid fibers

Property	Kevlar 49	Kevlar 149
Physical		
Specific gravity	1.44	1.38
Density, g/cm^3 (lb/in.^{3})	1.44 (0.052)	1.42 (0.050)
Filament diameter, μm (mil)	12 (0.47)	...
Mechanical		
Tensile strength, MPa (ksi)	...	1400 (203), 0°, Dir. 1
Tensile modulus, GPa (10^6 psi)	...	105 (15.4) 0°, Dir. 1
Elongation at RT	2.5	1.3, 0°, 120, 90°
Thermal		
Coefficient of expansion, axial and lateral, 10^{-6}/K	...	–268, 0°; 120, 90°
Electrical		
Dielectric constant	...	At 1 kHz, 4.14 dry; at 1 MHz, 3.90 dry
Dissipation factor	...	At 1 kHz, 0.0103 dry; at 1 MHz, 0.0142 dry

Organic fibers for filament winding (in order of ascending modulus of strand, normalized to 100% fiber volume)

	Strand nominal tensile modulus		Strand nominal tensile strength		Maximum	
Type	GPa	10^6 psi	MPa	ksi	number of filaments/strand	Fiber density, g/cm^3
Aramid (medium modulus)	62	9.0	2758	400	1000	1.44
Oriented polyethylene(a)	117	17	2585	375	118	0.97
Aramid(a)	121	17.5	4067	590	...	1.44
Aramid	124	18	3792	550	5000	1.44
Oriented polyethylene(b)	170	24.8	3274	471	...	0.97

(a) Development status. (b) Research and development status

Thermal, physical, and chemical properties of a typical aramid

Heat-deflection temperature at 1.82 MPa, °C	260
Maximum resistance to continuous heat, °C	150
Coefficient of linear thermal expansion, 10^{-5}/°C	2.6
Tensile strength, MPa	120
Elongation, %	5
Specific gravity	1.2
Dielectric constant	3.0
Resistance to chemicals at 25 °C(a):	
Nonoxidizing acids (10% H_2SO_4)	Q
Oxidizing acids (10% HNO_3)	U
Aqueous salt solutions (NaCl)	S
Aqueous alkalis (NaOH)	S
Polar solvents (C_2H_5OH)	Q
Nonpolar solvents (C_6H_6)	S
Water	S

(a) S, satisfactory; Q, questionable; U, unsatisfactory

Mechanical properties and maximum use temperature of Kevlar 49

Tensile strength:	
At room temperature (16 months)	No loss in strength
At 50 °C (120 °F) in air (2 months)	No loss in strength
At 100 °C (210 °F) in air, MPa (ksi)	3170 (460)
At 200 °C (390 °F) in air, MPa (ksi)	2720 (395)
Tensile modulus:	
At room temperature (16 months)	No loss in modulus
At 50 °C (120 °F) in air (2 months)	No loss in modulus
At 100 °C (210 °F) in air, GPa (10^6 psi)	114 (16.5)
At 200 °C (390 °F) in air, GPa (10^6 psi)	110 (16.0)
Long-term use temperature in air, °C (°F)	160 (320)
Decomposition temperature, °C (°F)	500 (930)

Isotropic properties of 20% aramid-fiber-reinforced nylon 6/6 composites

Property	Flow direction	Transverse direction	Difference %
Flexural strength, MPa (ksi)	107 (15.5)	101 (14.7)	5
Flexural modulus, GPa (10^6 psi)	3.6 (0.52)	3.3 (0.48)	8
Izod impact strength, J/cm (ft·lb/in.):			
Notched	0.64 (1.2)	0.64 (1.2)	0
Unnotched	1.6 (3.0)	1.4 (2.7)	11
Coefficient of linear thermal expansion, 10^{-5}/°C (10^{-5}/°F)	4.3 (2.4)	4.7 (2.6)	8

Properties of pultruded unidirectional IPN-Kevlar (58 wt %)

Property	Room temperature	Reported value at: 99 °C (210 °F)	150 °C (300 °F)
Flexural strength, MPa (ksi)	605.8 (87.86)	491.4 (71.28)	331.4 (48.07)
Flexural modulus, GPa (10^6 psi)	62 (9.0)	...	...
Short-beam shear, MPa (ksi)	37.9 (5.49)	35.2 (5.10)	22.1 (3.20)

Properties of Spectra polyethylene fibers

	Spectra 900	Spectra 1000
Density g/cm^3	0.97	0.97
Filament diameter μm (μin.)	38 (1500)	27 (1060)
Tensile modulus GPa (10^6 psi)	117 (17)	172 (25)
Tensile strength GPa (10^6 psi)	2.6 (0.380)	2.9 to 3.3 (0.430 to 0.480)
Tensile elongation %	3.5	0.7
Available yarn counts no. filaments	60 to 120	60 to 120

Kevlar 49 yarn and roving sizes

Denier	Yield m/kg	Yield yd/lb	Number of filaments
55	163636	81175	25
195	46155	22895	134
380	23684	11749	267
1140	7895	3916	768
1420	6388	3144	1000
2130	4225	2097	1000
4560	1973	980	3072
7100	1268	630	5000

Note: All Kevlar 49 yarns have approximately 1.5 denier units/filament, with the exception of the 2130-denier product and the 55-denier product, which have a denier filament ratio of approximately 2.1 and 2.25, respectively

Kevlar 29 yarn and roving sizes

Denier	Yield m/kg	Yield yd/lb	Number of filaments
200	45000	22320	134
400	22500	11160	267
1000	9000	4464	1000
1500	6000	2976	4000
9000	1000	497	4000
15000	600	298	10000

Effect of electron radiation on Kevlar 49

Mrad exposure	Tenacity MPa	Tenacity ksi	Tensile modulus GPa	Tensile modulus 10^6 psi	Elongation, %
0	2860	415	128	18.6	2.4
100	2940	426	130	18.8	2.4
200	3010	436	133	19.3	2.5

(Tenacity, tensile modulus and elongation: single-filament properties)

Using resonant transformer and filament wrapped in aluminum foil over dry ice, conditions of exposure were 1 Mrad every 13.4 s, 0.5 mA, 2MV, and 30-cm (10-in.) distance

Typical properties of Compet RPC fibers

Property	Type 1W69	Type 1W71	Type IR69
Generic name	Polyester	Polyester	Nylon 6
Tensile strength, MPa (ksi)	1105 (160)	1035 (150)	965 (140)
Extension at break, %	14.0	22.0	20.0
Modulus of elasticity, GPa (10^6 psi)	14 (2.0)	9.7 (1.4)	5.2 (0.75)
Shrinkage at 175 °C (350 °F), %	10	2.5	12.0
Diameter, μm (mil)	23 (0.91)	23 (0.91)	23 (0.91)
Specific gravity	1.38	1.38	1.16
Melting point, °C (°F)	253 (488)	253 (488)	220 (428)

Properties of hybrid yarns

Property	PEEK hybrid	PPS hybrid
Density, g/cm^3	1.60	1.61
Total denier, g/10,000 m	3000	2950
Number of filaments(a)	3100	3100
Fiber area, cm^2	0.0019	0.0019
Yield, m/kg	3370	3390

(a) Fabricated with Celion G30-500 3K carbon fibers. Available with up to 61 vol % carbon, with other high-performance carbon fibers, and filament counts

Mechanical properties of aramid, polyamide, polyester, and nylon fibers

Fiber	Density Mg/m^3	Density lb/in.^{3}	Tensile strength MPa	Tensile strength ksi	Tensile modulus GPa	Tensile modulus 10^6 psi	Ultimate elongation, %
Aramid-Kevlar 29	1.44	0.052	3620	525	83	12	4.4
Aramid-Kevlar 49	1.44	0.052	3620	525	124	18	2.9
Polyamide	1.13	0.041	830	120	2.8	0.4	...
Polyester-Dacron Type 68	1.38	0.050	1120	162	4.1	0.6	14.5
Nylon-Du Pont 728(a)	1.13	0.041	990	143	5.5	0.8	18.3
Spectra-900	0.97	0.035	2590	375	117	17	...

(a) Unimpregnated twisted yarn test–ASTM D2256

Properties of PPS fiber

Property	Value
Tenacity, grams per denier (gpd)	3.5
Elongation, %	40
Modulus, gpd at 10% extension	16
Elastic recovery, %	
2% extension	100
5% extension	96
10% extension	86
Dry shrinkage, % at 130 °C (265 °F)	4
Moisture regain, %	0.6
Specific gravity, g/cm^3	1.37

Kevlar 49 fabric and woven roving specifications

Style no.	Weave	Basis weight g/m^2	Basis weight oz/yd^2	Fabric construction ends/cm	Fabric construction ends/in.	Yarn denier	Fabric thickness mm	Fabric thickness 10^{-3} in.
Light weight								
166(a)	Plain	30.6	0.9	37 × 37	94 × 94	55	0.04	1.5
199(a)	Plain	61.13	1.8	24 × 24	60 × 60	55	0.05	2
120	Plain	61.1	1.8	13 × 13	34 × 34	195	0.11	4.5
220	Plain	74.7	2.2	9 × 9	22 × 22	380	0.11	4.5
Medium weight								
181	8-harness satin	169.8	5.0	20 × 20	50 × 50	380	0.23	9
281	Plain	169.8	5.0	7 × 7	17 × 17	1140	0.25	10
285	Crowfoot	169.8	5.0	7 × 7	17 × 17	1140	0.25	10
328	Plain	230.9	6.8	7 × 7	17 × 17	1420	0.33	13
335	Crowfoot	230.9	6.8	7 × 7	17 × 17	1420	0.30	12
500	Plain	169.8	5.0	5 × 5	13 × 13	1420	0.28	11
Unidirectional								
143	Crowfoot	190.2	5.6	39 × 8	100 × 20	380 × 195	0.25	10
243	Crowfoot	227.5	6.7	15 × 7	38 × 18	1140 × 380	0.33	13
Woven roving								
1050	4 × 4 basket	356.6	10.5	11 × 11	28 × 28	1420	0.46	18
1033	8 × 8 basket	509.4	15.0	16 × 16	40 × 40	1420	0.66	26
1350	4 × 4 basket	458.5	13.5	10 × 9	26 × 22	2130	0.64	25

(a) Only available on special order: custom fabric will be woven to specifications

Metals Fibers, Whiskers and Wires

Composite wire properties

	6061-Al (% ROM)(a)	AZ91-Mg (% ROM)
Fiber vol %	50.2 ± 2.0(b)	54.3 ± 2.0
Mean tensile strength, MPa (ksi)	1385 ± 110 (201.0 ± 16)(b)(c)	1240 ± 90 (180.0 ± 13.0)(d)(e)
Mean elastic modulus, GPa (10^6 psi)	420 ± 25 (60.7 ± 3.4)(c)(f)	>390 (>57)

(a) Based on the mean properties of virgin yarn. (b) Average 3200 tests. (c) 100% ROM. (d) Average of 360 tests. (e) 95% ROM. (f) Average of 105 tests

Representative properties of refractory alloy wires

Alloys	Density, g/cm^3	Wire diameter mm	Wire diameter in.	Ultimate tensile strength MPa	Ultimate tensile strength ksi	Stress for 100-h rupture MPa	Stress for 100-h rupture ksi	Stress/density for 100-h rupture cm × 10^3	Stress/density for 100-h rupture in. × 10^3
Tungsten alloys, 1093 °C (2000 °F) data									
218CS	19.1	0.20	0.008	869	126	434	63	234	92
W-1%ThO_2	19.1	0.20	0.008	979	142	531	77	282	111
W-2%ThO_2	18.9	0.38	0.015	1193	173	655	95	356	140
W-3%Re	19.4	0.20	0.008	1475	214	476	69	249	98
W-5%Re-2%ThO_2	19.1	0.20	0.008	1213	176	483	70	254	100
W-24%Re-2%ThO_2	19.4	0.20	0.008	1455	211	345	50	183	72
W-Hf-C	19.4	0.38	0.015	1427	207	1110	161	584	230
W-Re-Hf-C	19.4	0.38	0.015	2165	314	1413	205	744	293
Tungsten alloys, 1204 °C (2200 °F) data									
218CS	19.1	0.20	0.008	745	108	317	46	170	67
W-1%ThO_2	19.1	0.20	0.008	841	122	372	54	198	78
W-2%ThO_2	18.9	0.38	0.015	1034	150	483	70	257	101
W-3%Re	19.4	0.20	0.008	1082	157	317	46	168	66
W-5%Re-2%ThO_2	19.1	0.20	0.008	1020	148	303	44	160	63
W-24%Re-2%ThO_2	19.4	0.20	0.008	1014	147	193	28	102	40
W-Hf-C	19.4	0.38	0.015	1386	201	765	111	404	159
W-Re-Hf-C	19.4	0.38	0.015	1937	281	910	132	480	189

Mechanical properties and melting points of stainless steel and other metallic fibers

Fiber	Fiber diameter μm	Fiber diameter mils	Density g/cm^3	Density lb/in.^{3}	Tensile strength MPa	Tensile strength ksi	Tensile modulus GPa	Tensile modulus 10^6 psi	Melting point °C	Melting point °F
AFC-77	150 to 1270	6 to 50	7.75	0.280	3540 to 4135	528 to 600	207	30.0	1370	2500
Rene 41	25 to 50	1 to 2	8.25	0.298	2000 to 2345	290 to 340	220	31.9	1370	2500
Udimet 700	255	10	7.92	0.286	1515 to 2330	220 to 338	221	32.0	1404	2559
Ribtec-HT(a)	205 to 510	8 to 20	...	...	57	8.3	82.7	12.0	1480 to 1530	2700 to 2790
Ribtec-LR 430(a)	205 to 510	8 to 20	...	...	47	6.8	82.7	12.0	1480 to 1530	2700 to 2790
Ribtec-GR 304(a)	205 to 510	8 to 20	...	...	124	18	124	18.0	1400 to 1455	2550 to 2650
Ribtec-OS 446(a)	205 to 510	8 to 20	...	...	52	7.6	96.5	14.0	1425 to 1510	2600 to 2750
Ribtec-310(a)	205 to 510	8 to 20	...	...	151	22	124	18.0	1400 to 1455	2550 to 2650
Ribtec-OC 330(a)	205 to 510	8 to 20	...	...	193	28	134	19.5	1345 to 1425	2450 to 2600
Steel (wire)	25	1	...	...	3445	500	207	30.0	...	...
Aluminum	...	...	2.68	0.097	620	90	73.1	10.6	...	...

(a) Modulus of elasticity computed at 315 °C (600 °F); tensile strength computed at 870 °C (1600 °F)

Tensile properties of refractory metal wires

Wire material	Wire diameter cm	Wire diameter in.	Test temperature °C	Test temperature °F	Tensile strength MPa	Tensile strength ksi	Elongation in 25 mm (1 in.), %	Reduction in area, %
W-Hf-C, in-process annealed	0.038	0.015	21	70	2700	392	5.4	21.1
			1095	2000	1430	207	...	67.8
			1205	2200	1390	201	...	70.9
W-Hf-C, hard drawn	0.038	0.015	21	70	2250	326	2.8	1.9
			1095	2000	1740	253	...	44.2
			1205	2200	1540	224	...	46.9

(continued)

Tensile properties of refractory metal wires (continued)

Wire material	Wire diameter cm	Wire diameter in.	Test temperature °C	Test temperature °F	Tensile strength MPa	Tensile strength ksi	Elongation in 25 mm (1 in.), %	Reduction in area, %
W-Re-Hf-C, hard drawn	0.038	0.015	21	70	3160	458	4.8	27.5
			1095	2000	2160	314	...	24.7
			1205	2200	1940	281	...	37.6
ASTAR 811C	0.051	0.020	21	70	1700	247	6.9	51.0
			1095	2000	744	108	...	80.8
			1205	2200	490	71	...	89.8
	0.038	0.015	21	70	1740	253	5.3	42.9
			1095	2000	779	113	...	66.4
			1205	2200	550	80	...	66.9
B-88	0.051	0.020	21	70	1480	215	4.8	26.5
			1095	2000	530	77	...	87.4
			1205	2200	350	50	...	97.9
	0.038	0.015	21	70	1620	235	7.7	54.8
			1095	2000	490	71	...	94.5
			1205	2200	310	45	...	95.7
W-2ThO_2	0.038	0.015	21	70	2650	384	5.5	14.2
			1095	2000	1190	173	...	50.2
			1205	2200	1030	150	...	51.0

Mechanical properties and melting points of refractory metallic fibers/wires

Type of filament or wire	Nominal composition	Melting point °C	Melting point °F	Density g/cm^3	Density lb/in.	Tensile strength MPa	Tensile strength ksi
Cr	Cr	1865	3390	7.2	0.26	1586	230
Nb-Sul6	Nb-11W-3Mo-2Hf-0.08C	~2590	~4700	9.27	0.335	889 to 979	129 to 142
Nb-Su31	Nb-17W-3.5Hf-0.12C	~2590	~4700	9.46	0.342	1034 to 1503	150 to 218
Nb, FS85	Nb-28Ta-10W-1Zr-0.005C	2590	4695	10.6	0.383	1510	219
Nb, AS30	Nb-20W-1Zr	~2590	~4700	9.6	0.347	1758	255
Nb, B88	Nb-28W-2Hf-0.06C	~2590	~4700	10.3	0.372	1620	235
Mo	Mo	2610	4730	10.2	0.369	2206	320
Mo + 0.5 Ti	Mo-0.5Ti	2610	4730	10.1	0.367	1793	260
Mo, TZM	Mo-0.5Ti-0.08Zr-0.015C	~2610	~4730	10.2	0.369	1965	285
Mo, TZC	Mo-1.25Ti-0.3Zr-0.015C	~2610	~4730	10.1	0.367	2268	329
Ta, ASTAR 811C	Ta-8W-1Re-0.9Hf-0.03C	~2990	~5400	16.9	0.610	1703	247
W	W	3410	6170	19.3	0.697	1648 to 3268	239 to 474
W, 218CS	W	3410	6170	19.2	0.695	2386 to 2661	346 to 386
W + 1 ThO_2 (NF)	W-1ThO_2	~3410	~6170	19.1	0.691	2255 to 2310	327 to 335
W + 2 ThO_2	W-2ThO_2	~3410	~6170	18.9	0.683	2647 to 2751	384 to 399
W + 3 Re (3D)	W-3Re	~3410	~6170	19.4	0.70	2785	404
W + 5 Re	W-5Re	~3410	~6170	19.4	0.70	1689 to 2647	245 to 384
W + Hf + C	W-0.03Hf-0.036C	~3410	~6170	19.4	0.70	2248 to 2358	326 to 342
W + Re + Hf + C	W-4Re-0.38Hf-0.02C	~3410	~6170	19.4	0.70	3158	458

Type of filament or wire	Specific strength 10^6 cm	Specific strength 10^6 in.	Modulus of elasticity GPa	Modulus of elasticity 10^6 psi	Specific modulus 10^6 cm	Specific modulus 10^6 in.	Typical cross section (diameter) 10^{-3} cm	Typical cross section (diameter) 10^{-3} in.
Cr	2.2	0.89	290	42	411	162	25	10
Nb-Su16	0.98 to 1.1	0.39 to 0.42	121 to 134	17.6 to 19.5	~159 to 177	~63 to 70	25 to 89	10 to 35
Nb-Su31	1.1 to 1.6	0.44 to 0.64	122 to 135	17.7 to 19.6	~160 to 178	~63 to 70	61 to 102	24 to 40
Nb, FS85	1.5	0.57	138	20	132	52	13	5
Nb, AS30	1.9	0.74	~138	~20	~147	~58	13	5
Nb, B88	1.6	0.63	~138	~20	~137	~54	38	15
Mo	2.2	0.87	358	52	358	141	15	6
Mo + 0.5 Ti	1.8	0.71	317	46	318	125	13	5
Mo, TZM	2.0	0.77	317	46	318	125	20 to 25	8 to 10
Mo, TZC	2.3	0.89	317	46	318	125	13	5
Ta, ASTAR 811C	1.0	0.41	200	29	122	48	38 to 51	15 to 20
W	0.9 to 1.7	0.34 to 0.68	406	59	216	85	5.1 to 127	2 to 50
W, 218CS	1.3 to 1.4	0.5 to 0.56	406	59	216	85	20 to 38	8 to 15
W + 1ThO_2(NF)	1.2 to 1.2	0.47 to 0.48	406	59	216	85	20 to 51	8 to 20
W + 2 ThO_2	1.3 to 1.5	0.56 to 0.58	406	59	218	86	20 to 51	8 to 20
W + 3 Re (3D)	1.5	0.58	406	59	213	84	7.6 to 20	3 to 8
W + 5 Re	0.9 to 1.4	0.35 to 0.55	406	59	213	84	25 to 127	10 to 50
W + Hf + C	1.2 to 1.4	0.47 to 0.56	406	59	213	84	38	15
W + Re + Hf + C	1.7	0.65	406	59	213	84	38	15

Recrystallization temperature of tungsten wires in various matrices

System	Ni content of matrix, wt%	Recrystallization temperature(a) °C	°F
W-Ni	100	1150 to 1200	2100 to 2190
W-NiCr	20	1300	2370
W-2%ThO_2 NiCr	20	1250	2280
W-2%ThO_2 Ni	100	1080 to 1130	1980 to 2065
W-2%ThO_2 Inconel 718	52	1175	2150
W-2%ThO_2 Hastelloy X	48	1200	2190
W-2%ThO_2 Kovar	29.5	1250 to 1300	2280 to 2370
W-2%ThO_2 Stainless steel	10	1435	2615
W-2%ThO_2 Stainless steel	10	1465	2670

(a) 50 μm (1950 μin.) in 1 h

Physical properties of metallic wires

Material	Specific gravity	Melting point °C	°F	Tensile strength MPa	ksi	Modulus of elasticity GPa	10 psi	Coefficient of thermal expansion, 10^{-6}/K
Aluminum	2.71	660	1220	290	40	68.9	10.0	23.6
Beryllium	1.85	1350	2460	1100	160	310	45.0	11.6
Copper	8.90	1083	1980	413	60	124	18.0	16.5
Tungsten	19.3	3410	6170	2890	130	345	50.0	4.6
Austenitic stainless steel	7.9	1539	2800	2390	350	200	29.0	8.5
Molybdenum	10.2	2625	4750	2200	320	331	48.0	...

Stress-rupture properties of wire materials

Wire material	Wire diameter cm	in.	Test temperature °C	°F	Stress MPa	ksi	Rupture time, h	Reduction in area, %
Tungsten-base alloys								
W-Hf-C, thermally annealed during drawing	0.038	0.015	1095	2000	1300	189	4.4	44.2
					1290	187	10.3	58.4
					1230	178	21.1	23.2
					1210	175	19.1	35.0
					1150	167	61.5	44.5
					1110	161	108.5	18.0
	0.038	0.015	1205	2200	918	133	28.3	15.3
					841	122	42.9	21.9
					765	111	104.3	11.5
					689	100	188.4	28.5
W-Hf-C, hard drawn	0.038	0.015	1095	2000	1310	190	17.7	44.2
					1240	180	139.4	37.0
					1230	178	88.6	22.6
					1210	175	262.0	57.3
					1100	160	(a)	...
	0.038	0.015	1205	2200	1170	170	6.0	29.4
					1140	165	4.3	30.6
					1100	160	11.1	20.2
					1040	150	22.5	24.9
					965	140	18.6	20.2
					896	130	37.4	16.6
					827	120	63.0	22.6
					793	115	74.3	17.8
					758	110	334.1	50.2
					689	100	329.6	65.8
W-Re-Hf-C, hard drawn	0.038	0.015	1095	2000	1590	230	15.6	15.7
					1520	220	36.2	19.5
					1480	215	42.8	34.0
					1450	210	72.1	27.1
					1380	200	442.6	34.7
					1310	190	104.2	16.7
					1240	180	522.3	37.3

(continued)

Stress-rupture properties of wire materials (continued)

Wire material	Wire diameter cm	Wire diameter in.	Test temperature °C	Test temperature °F	Stress MPa	Stress ksi	Rupture time, h	Reduction in area, %
W-Re-Hf-C, hard drawn (cont'd)	0.038	0.015	1205	2200	1140	165	14.4	32.0
					1040	150	18.4	43.2
					965	140	39.7	23.0
					896	130	49.8	32.8
					862	125	365.5	33.7
					827	120	345.5	44.3
					793	115	342.2	43.2
Tantalum-base alloy								
ASTAR 811C	0.051	0.020	1095	2000	690	100	7.3	17.5
					620	90	68.5	8.7
					590	85	43.0	7.0
					520	75	338.2	3.8
	0.038	0.015	1095	2000	620	90	14.6	29.8
					590	85	94.6	4.2
					570	82	19.1	19.3
					570	82	162.8	6.6
					550	80	338.2	7.4
					550	80	(b)	...
	0.051	0.020	1205	2200	350	50	10.8	8.8
					310	45	28.5	9.7
					280	40	78.3	8.3
					240	35	166.7	7.3
	0.038	0.015	1205	2200	520	75	10.2	15.3
					480	70	14.7	10.3
					410	60	45.4	5.2
					380	55	20.1	6.5
					350	50	62.7	2.8
					310	45	391.9	<1.0
Niobium-base alloy								
B-88	0.051	0.020	1095	2000	380	55	4.1	32.8
					370	53	37.1	15.8
					350	50	101.3	18.6
					310	45	102.1	16.6
	0.038	0.015	1095	2000	350	50	44.8	22.1
					310	45	55.4	23.3
					280	40	199.3	20.7
	0.051	0.020	1205	2200	280	40	2.6	34.7
					240	35	14.4	39.9
					210	30	78.5	20.9
	0.038	0.015	1205	2200	240	35	25.4	32.0
					210	30	86.8	28.6
					170	25	224.1	26.1

(a) Test stopped at 233.4 h. (b) Test stopped at 348.9 h

Carbon and Graphite Fibers

Classification of carbon fibers by modulus and strength

	State-of-the-art				Newer fibers			
	Modulus		Strength		Modulus		Strength	
Fiber type	GPa	10^6 psi	GPa	ksi	GPa	10^6 psi	GPa	ksi
High strength	228	33	3.45	500	241	35	4.14	600
					255	37	5.17	750
Intermediate modulus	...	...	...	...	283	41	4.83	700
High modulus	379	55	2.41	350	...	...	...	...
Very high modulus	517	75	2.07	300	...	...	...	...
Ultrahigh modulus	...	...	...	...	690	100	2.24	325
					827	120	2.41	350

Carbon and graphite fibers (in order of ascending modulus of strand, normalized to 100% fiber volume)

	Strand nominal tensile modulus		Strand nominal tensile strength		Maximum number of	
Class of fiber	MPa	10^6 psi	MPa	ksi	filaments/strand	Density, g/cm^3
High tensile strength	227	33	3102	450	12 000	1.75
High strain	234	34	4100	594	6 000	1.79
Intermediate modulus	275	40	4295	623	12 000	1.74
High modulus	358	52	2482	360	3000	1.81
High modulus pitch	379	55	2068	300	4000	2.0
Ultra-high modulus	517	75	1816	270	384	1.96
Ultra-high modulus, pitch	517	75	2068	300	2000	2.0
Extreme-high modulus, pitch	689	100	2240	325	2000	2.15

Properties of carbon fibers: low-modulus rayon and isotropic pitch precursor fibers

Properties	Rayon precursor		Isotropic pitch precursor	
Axial				
Tensile modulus, GPa (10^6 psi)	41	6	55	8
Tensile strength, GPa (10^6 psi)	1.0	(0.15)	0.7	0.10
Elongation at break, %	2.5	...	1.4	...
Electrical resistivity, μΩ · m	20	...	30	...
Bulk				
Density, g/cm^3	1.6	...	1.6	...
Filament diameter, μm (μin.)	8.5	(330)	10	(390)
Carbon assay, %	99	...	98	...

Thermal, physical, and chemical properties of typical graphite fibers

Tensile strength, MPa	2000
Elongation, %	0.6
Specific gravity	1.63
Modulus of elasticity, GPa	550

Graphite fiber properties

Fiber	Bulk dc resistivity ohm · m × 10^{-4}	Tensile modulus GPa	Tensile modulus 10^6 psi
Carbon			
AS-4	15	231	33.5
T300	18	230	33.4
Graphite			
HMS-4	7.5	341	49.9
P-75	1.8	517	75.0

Properties of carbon fiber types

Fiber type	Density, g/cm^3	Modulus of elasticity GPa	Modulus of elasticity 10^6 psi	Tensile strength GPa	Tensile strength 10^6 psi	Electric resistivity Ω · m	Thermal conductivity W/m·K	Thermal conductivity Btu · in./ft^2 · h · °F
High-strength (PAN)	1.7 to 1.8	230 to 250	33 to 36	2.8 to 4.0	0.41 to 0.58	12 to 30	7 to 10	50 to 70
Ultrahigh strength (PAN)	1.7 to 1.8	260 to 290	38 to 42	4.1 to 5.7	0.59 to 0.83	14 to 20	7 to 9	50 to 60
High-modulus (PAN, mesophase pitch)	1.8 to 2.0	350 to 550	50 to 80	1.7 to 3.5	0.25 to 0.50	5 to 10	60 to 200	420 to 1400
Ultrahigh-modulus (mesophase pitch)	2.0 to 2.2	600 to 900	90 to 130	2.1 to 2.5	0.30 to 0.36	1 to 4	400 to 2500	2800 to 17,300
Low-modulus (rayon, pitch)	1.3 to 1.7	40 to 60	6 to 9	0.6 to 1.0	0.085 to 0.145	30 to 100	7 to 28	50 to 190

Properties of carbon fibers

Properties	Amoco T300	Amoco	Toray	Hercules	FMI
Physical					
Specific gravity	1.77	2.18	1.82	1.88	1.80
Density, g/cm^3 (lb/in.^{3})	1.77 (0.064)	2.18 (0.079)	1.82 (0.066)	...	...
Filament diameter, μm (mils)	7.01 (0.276)	...	...	...	...
Mechanical					
Tensile strength at RT, MPa (ksi)	3650 (530)	2240 (325)	6900 (1000)	3790 (550)	5170 (750)
Tensile modulus at RT, GPa (10^6 psi)	230 (33.5)	830 (120)	290 (42)	425 (62)	260 (38)
Elongation, %	1.4	...	2.4	0.75	1.97
Thermal					
Coefficient of thermal expansion, 10^{-6}/K	–0.54	...	...	...	...
Thermal conductivity, W/m · K (Btu/ft · h · °F)	8.7 (5.0)	...	...	...	...
Specific heat					
At RT, kJ/kg · K (Btu/lb · °F)	0.92 (0.22)	...	...	...	...
From 20 to 1480 °C mean (70 to 2700 °F mean)	1.7 (0.4)	...	...	...	...
Electrical					
Electrical resistivity, μΩ · m	18.0	...	...	...	...

Typical mechanical property values of commercially available carbon fibers

Product name	Manufacturer	Precursor type	Density, g/cm^3	Tensile strength		Tensile modulus	
				GPa	10^6 psi	GPa	10^6 psi
AS-4	Hercules, Inc.	PAN	1.78	4.0	0.580	231	33.5
AS-6	Hercules, Inc.	PAN	1.82	4.5	0.652	245	35.5
IM-6	Hercules, Inc.	PAN	1.74	4.8	0.696	296	42.9
T300	Union Carbide/Toray	PAN	1.75	3.31	0.480	228	32.1
T500	Union Carbide/Toray	PAN	1.78	3.65	0.530	234	33.6
T700	Toray	PAN	1.80	4.48	0.650	248	36.0
T-40	Toray	PAN	1.74	4.50	0.652	296	42.9
Celion	Celanese/ToHo	PAN	1.77	3.55	0.515	234	34.0
Celion ST	Celanese/ToHo	PAN	1.78	4.34	0.630	234	39.0
XAS	Grafil/Hyson	PAN	1.84	3.45	0.500	234	34.0
HMS-4	Hercules, Inc.	PAN	1.78	3.10	0.450	338	49.0
PAN 50	Toray	PAN	1.81	2.41	0.355	393	57.0
HMS	Grafil/Hysol	PAN	1.91	1.52	0.220	341	49.4
G-50	Celanese/ToHo	PAN	1.78	2.48	0.360	359	52.0
GY-70	Celanese	PAN	1.96	1.52	0.220	483	70.0
P-55	Union Carbide	Pitch	2.0	1.73	0.250	379	55.0
P-75	Union Carbide	Pitch	2.0	2.07	0.300	517	75.0
P-100	Union Carbide	Pitch	2.15	2.24	0.325	724	100
HMG-50	Hitco/OCF	Rayon	1.9	2.07	0.300	345	50.0
Thornel 75	Union Carbide	Rayon	1.9	2.52	0.365	517	75.0

Properties of pultruded unidirectional graphite

Property	Reported value, at temperature of:		
	Room temperature	99 °C (210 °F)	150 °C (300 °F)
IPN-graphite (70 wt %)			
Flexural strength, MPa (ksi)	1610 (233)	1300 (188)	855 (124)
Flexural modulus, GPa (10^6 psi)	122 (17.7)	...	...
Short-beam shear, MPa (ksi)	97.9 (14.2)	69 (10)	37.0 (5.37)
Epoxy graphite			
Flexural strength, MPa (ksi)	1790 (260)	...	1140 (165)
Flexural modulus, GPa (10^6 psi)	128 (18.5)	...	103 (15.0)
Short-beam shear, MPa (ksi)	97 (14)	...	52 (7.5)

Specific tensile strength and modulus of carbon fiber relative to other reinforcements

Fiber	Tensile strength/density 10^7 cm	10^6 in.	Tensile modulus/density 10^9 cm	10^9 in.
AS-4	2.25	8.86	1.29	0.508
IM-6	2.76	10.9	1.70	0.669
E-glass	1.33	5.24	0.28	0.11
S-glass	1.73	6.81	0.32	0.13
Kevlar 49	2.50	9.84	0.90	0.35
Boron	1.50	5.91	1.60	0.63
P-75	1.04	4.09	2.58	1.02

Properties of graphite fibers

Property	Hercules HMS PAN	Amoco WYB rayon	Amoco P-120 pitch	Amoco-rayon T50	T-75
Physical					
Specific gravity	1.83	1.32	2.18	1.67	1.80
Density, g/cm^3 (lb/in.^{3})	1.83 (0.066)	1.32 (0.048)	2.18 (0.079)	1.67 (0.060)	1.80 (0.065)
Filament diameter, μm (mil)	8.00 (0.315)	9.4 (0.37)	10 (0.4)	6.6 (0.26)	6.0 (0.24)
Mechanical					
Tensile strength, MPa (ksi)	2200 (320)	620 (90)	2240 (325)	2170 (315)	2620 (380)
Tensile modulus, GPa (10^6 psi)	340 (50)	40 (6)	825 (120)	395 (57)	540 (78)
Elongation, %	0.58	1.5	0.27	0.60	0.50
Thermal					
Coefficient of thermal expansion, 10^{-5}/K:					
Axial	−0.99	...	−1.62	...	...
Lateral	16.8	...	...	...	...
Thermal conductivity, axial, W/m · K (Btu/ft · h · °F	104 (60.0)	...	609 (352)	118 (68)	156 (90)
Specific heat, kJ/kg · K (Btu/lb · °F)	0.71 (0.17)	...	...	0.71 (0.17)	0.71 (0.17)
Electrical					
Electrical resistivity, μΩ · m:					
Axial	13	...	2.2	...	...
Lateral	10	...	...	...	...

Mechanical properties of selected carbon/graphite commercial fibers

Fiber	Fiber diameter μm	mils	Density Mg/m^3	lb/in.^{3}	Tensile strength MPa	ksi	Tensile modulus GPa	10^6 psi	Ultimate elongation, %
Magnamite HMS	8.00	0.315	1.83	0.066	2210	320	345	50	0.58
Magnamite HMU	8.00	0.315	1.85	0.067	2760	400	380	55	0.70
Magnamite chopped fiber	8.00	0.315	1.77	0.064	2480	360	205	30	1.2
Magnamite AS1	8.00	0.315	1.80	0.065	3100	450	230	33	1.32
Magnamite AS2	8.00	0.315	1.80	0.065	2760	400	230	33	1.3(a)
Magnamite AS4	8.00	0.315	1.80	0.065	3590	520	235	34	1.53
Magnamite AS6	...	...	1.82	0.0657	4140	600	243	35.3	1.65(a)
Magnamite IM6	...	...	1.74	0.0627	4380	635	279	40.4	1.50
Microfil 55	4.32	0.170	1.77	0.064	3620	525	380	55	1.00
Microfil 40	4.32	0.170	1.69	0.061	4480	650	275	40	1.65
Celion C-6S	7.11	0.280	1.77	0.064	3790	550	231	33.5	1.64
Celion G-50	6.60	0.260	1.77	0.064	2480	360	360	52	0.7
Celion GY-70	8.38	0.330	1.91 to 1.97	0.069 to 0.071	1520	220	485	70	0.38
Celion GY-80	...	...	1.91 to 1.97	0.069 to 0.071	1520	220	550	80	...
Celion 1000	7.11	0.280	1.77	0.064	3240	470	235	34	1.4
Celion 3000	7.11	0.280	1.77	0.064	3790	550	231	33.5	1.64
Panex 1/4CF-30	7.92	0.312	1.74	0.063	2410	350	205	30	1.2
Panex 30	7.92	0.312	1.74	0.063	2590	375	220	32	1.3
Fortafil 3(C)	7.37	0.290	1.77	0.064	3100	450	230	33	1.4
Fortafil 3(O)	5.33 to 13.97	0.210 to 0.550	1.77	0.064	2760	400	230	33	1.2
Fortafil 5(O)	4.32 to 11.94	1.170 to 0.470	1.77	0.064	3100	450	345	50	0.9
Grafil XA-S (standard)	...	...	1.79	0.0646	3100	450	234	34.0	1.31
Grafil XA-S (high performance)	...	...	1.79	0.0646	3450	500	234	34.0	1.45
Grafil XA-S (high strain)	...	...	1.79	0.0648	3860	560	234	34.0	1.65
Grafil IM-S	...	...	1.76	0.0635	3100	450	290	42.0	1.07

(continued)

Mechanical properties of selected carbon/graphite commercial fibers (continued)

Fiber	Fiber diameter μm	mils	Density Mg/m^3	lb/in.^{3}	Tensile strength MPa	ksi	Tensile modulus GPa	10^6 psi	Ultimate elongation, %
Grafil HM-S/10/K	...	...	1.85	0.067	2480	360	345	50.0	0.73
Grafil HM-S/16/K	...	...	1.85	0.067	2760	400	372	54.0	0.74
Hi-Tex	...	...	1.80	0.065	3100 to 3240	450 to 470	228	33.0	...
Hi-Tex HS	...	...	1.80	0.065	3620 to 3690	525 to 535	234	34.0	...
Thornel P-25W 4K	10.92	0.430	1.91	0.069	1380	200	160	23	0.90
Thornel T-40 12K	5.94	0.234	1.80	0.065	5650	820	275 to 290	40 to 42	2.0
Thornel T-50 3K	6.45	0.254	1.80	0.065	2410	350	395	57	0.70
Thornel 75	5.56	0.219	1.83	0.066	2620	380	545	79	...
Thornel T-300 6K	6.93	0.273	1.77	0.064	3240	470	231	33.5	...
Thornel 400	...	...	1.77	0.064	3100	450	205	30	...
Thornel T-500	6.93	0.273	1.80	0.065	3860	560	241	35.0	1.5
Thornel T-600	...	...	1.80	0.065	4140	600	241	35.0	1.7
Thornel T-700	...	...	1.80	0.065	4480	650	250	36	1.8
Modmor I	7.75	0.305	1.97	0.071	1380	200	380	55	...
Modmor II	8.03	0.316	1.74	0.063	2410	350	240	35	...

(a) Minimum

Properties of carbon fibers: polyacrylonitrile precursor fibers

Properties	Standard grades				New grades					
	Low modulus		High modulus		Low modulus		Intermediate modulus		High modulus	
Axial										
Tensile modulus, GPa (10^6 psi)	230	(30)	390	(55)	230	(35)	270	(40)	320	(45)
Tensile strength, GPa (10^6 psi)	3.3	(0.48)	2.4	(0.35)	4.5	(0.65)	5.3 to 6.8	(0.77 to 0.99)	5.5	(0.80)
Elongation at break, %	1.4	...	0.6	...	2.0	...	2.0 to 2.5	...	1.8	...
Thermal conductivity, W/m · K (Btu · in./h · ft^3 · °F)	8.5	(59)	70	(490)	7	(50)				
Electrical resistivity, μΩ · m (μΩ · cm)	18	(1800)	9.5	(950)	18	(1800)				
Coefficient of thermal expansion at 21 °C (70 °F), 10^{-6}/K	–0.7	...	–0.5	...	...	...	...	...	...	...
Transverse										
Tensile modulus, GPa (10^6 psi)	40	(6)	21	(3)	...	...	...	...	...	...
Coefficient of thermal expansion at 50 °C (120 °F), 10^{-6}/K	10	(2)	7	(1)	...	...	...	...	...	...
Bulk										
Density, g/cm3	1.76	...	1.9	...	1.8	...	1.8	...	1.8	...
Filament diameter, μm (μin.)	7 to 8	(280 to 310)	7	(280)	5 to 6	(200 to 240)	6	(240)	4	(160)
Carbon assay, %	92 to 97	...	100	...	92 to 97	...	96	...	96	...

Properties of carbon fibers: mesophase pitch precursor fibers

Properties	Standard grades						New grades			
	Low modulus		High modulus		Very high modulus		Low modulus		High modulus	
Axial										
Tensile modulus, GPa (10^6 psi)	160	(25)	380	(55)	7.25	(110)	225	(35)	380	(55)
Tensile strength, GPa (10^6 psi)	1.4	(0.20)	1.7	(0.25)	2.2	(0.32)	3.1	(0.45)	3.1	(0.45)
Elongation at break, %	0.9	...	0.4	...	0.3	...	...	...	...	...
Thermal conductivity, W/m · K (Btu · in./h · ft^2 · °F)	...	...	100	(690)	520	(3600)	...	...	...	...
Electrical resistivity, μΩ · m (μΩ · cm)	13	(1300)	7.5	(750)	2.5	(250)	...	...	...	...
Coefficient of thermal expansion at 21 °C (70 °F), 10^{-6}/K	...	...	–0.9	...	–1.6	...	...	...	...	...
Transverse										
Tensile modulus, GPa (10^6 psi)	...	...	21	(3)	...	...	...	...	...	...
Coefficient of thermal expansion at 50 °C (120 °F), 10^{-6}/K	...	...	7.8	...	...	...	...	...	...	...
Bulk										
Density, g/cm3	1.9	...	2.0	...	2.15	...	2.05	...	2.15	...
Filament diameter, μm (μin.)	11	(430)	10	(390)	10	(390)	8	(310)	8	(310)
Carbon assay, %	97+	...	99+	...	99+	...	99+	...	99+	...

Classification of carbon-base fibers

Classification	Carbon content, %	Maximum processing temperature °C	Maximum processing temperature °F	X-ray diffraction crystal structure	Crystallite orientation	Treatment	Modulus of elasticity GPa	Modulus of elasticity 10^6 psi	Tensile strength MPa	Tensile strength ksi
Carbon	>80	<1000	<1830	Crystallites too small to be detected	Amorphous	Carbonization	34	5	690	100
Graphite	>99	>2500	>4530	Crystallites large enough to be detected	Similar to precursor; random	Graphitization	97	14	1035	150
Structural carbon or structural graphite	>99	>2500	>4530	Crystallite number and size greater than in graphite fiber	Preferred orientation of graphite rystallites in a carbon matrix; turbostratic	Combination thermomechanical treatments	>172	>25	>1240	>180

Effect of reinforcement conductivity on conductive plastic resistivity

Material	Fiber conductivity μΩ · m	Surface resistivity Ω/square
PAN carbon fiber	3600	10^4
Nickel-coated carbon fiber	80	10

(a) Fiber concentration of 15% in nylon 6/6 matrix

Ceramic Fibers

General categories of ceramic fibers

Service temperature limit, °C (°F)	Composition, %					
	Al_2O_3	SiO_2	Fe_2O_3	TiO_2	Zr_2O_3	Cr_2O_3
670 to 700 (1260 to 1315)	44 to 51	47 to 53	0.1	1.0 to 1.8	...	...
1425 to 1480 (2600 to 2700) Type A	41 to 48	52 to 55	...	...	...	0.3 to 4
1425 to 1480 (2600 to 2700) Type B	32 to 36	44 to 48	...	...	14 to 21	...
1650 (3000)	96	4	...	...	...	...

Maximum use temperatures of some refractory fibers in oxidizing and nonoxidizing atmospheres

Fiber type	Maximum use temperature			
	Oxidizing atmosphere		Nonoxidizing atmosphere	
	°C	°F	°C	°F
Al_2O_3	1540	2805	1600	2910
ZrO_2	1650	3000	1650	3000
SiO_2	1060	1940	1060	1940
Al_2O_3-SiO_2	1300	2370	1300	2370
Al_2O_3-SiO_2-Cr_2O_3	1427	2600	1427	2600
Al_2O_3-SiO_2-B_2O_3	1427	2600	1427	2600
C	400	750	2500	4530
B	560	1040	1200	2190
BN	700	1290	1650	3000
SiC	1800	3270	1800	3270
Si_3N_4	1300	2370	1800	3270

Characteristics of discontinuous fibers

Material	Form	Diameter		Density, g/cm^3	Ultimate tensile strength		Tensile modulus		Coefficient of thermal expansion, 10^{-6}/K
		μm	μin.		MPa	ksi	GPa	10^6 psi	
Silicon carbide	Crystalline	0.2	8	3.2	800	120	500	75	4.3
Silicon carbide	Crystalline	120	4700	3.2	...	...	...	...	...
Alumina	Crystalline	3	120	3.3	2000	290	300	45	8.1
Mullite	Crystalline	3	120	3.2	690	100	152	20	5.1
Aluminosilicate	Amorphous	2	80	2.7	1730	250	104	15	...
Zirconia	Crystalline	5	200	5.7	...	...	206	30	10.5

Properties of short fibers and whiskers

Material	Density, g/cm^3	Tensile strength		Tensile modulus	
		GPa	10^6 psi	GPa	10^6 psi
Alumina					
Whiskers	4.0	10 to 20	1 to 3	700 to 1500	100 to 220
Sintered fibers	<4.0	0.2 to 0.7	0.030 to 0.10	140 to 300	20 to 40
Boron, thermally formed fibers	2.3	2.75	0.400	400	60
Boron nitride, fibers	1.8 to 2.0	0.3 to 1.4	0.045 to 0.20	28 to 80	4 to 10
Carbon					
Whiskers	>2.0	...	...	700	100
Fibers	1.8 to 2.0	2 to 3	0.30 to 0.45	230 to 550	35 to 80
Silicon nitride, whiskers	3.2	5 to 7	0.75 to 1.0	350 to 380	50 to 55

Mechanical properties and maximum use temperatures of refractory fibers

Fiber	Fiber diameter µm	Fiber diameter mils	Density g/cm^3	Density lb/in.^{3}	Tensile strength MPa	Tensile strength ksi	Tensile modulus GPa	Tensile modulus 10^6 psi	Maximum use temperature °C	Maximum use temperature °F
Avco boron	102	4.0	2.57	0.093	3520	510	400	58	...	...
	142	5.6	2.49	0.090	3520	510	400	58	...	...
	203	8.0	2.46	0.089	3520	510	400	58	...	...
Boron (tungsten core)	51	2.0	3.38	0.122	2760	400	400 to 415	58 to 60	...	...
Boron on carbon	102	4.0	2.24	0.081	3280	475	365	53	315	600
	142	5.6	2.27	0.082	3280	475	380	55	315	600
	203	8.0	2.30	0.083	3170	460	345	50	...	...
SiC-coated boron	107	4.2	2.66	0.096	2410	350	400 to 415	58 to 60	...	...
	145	5.7	2.57	0.093	2410	350	400 to 415	58 to 60	...	...
Boron carbide	102	4.0	2.35	0.085	2690	390	425	62	315	600
Boron nitride	6.9	0.27	1.91	0.069	1380	200	90	13	1095	2000
Titanium boride (TiB_2)	...	...	4.48	0.162	105	15	510	74	2205	4000
TiC	...	...	4.90	0.177	1540	224	450	65	...	...
Zirconium oxide	...	...	4.84	0.175	2070	300	345	50	1925	3500
Nextel 312	9.9 to 11.9	0.39 to 0.47	2.71	0.098	1380 to 1720	200 to 250	150	22	1205	2200
Nextel 440	9.9 to 11.9	0.39 to 0.47	3.10	0.112	1380 to 2070	200 to 300	205 to 240	30 to 35	...	...
Fiberfrax	2 to 12	0.08 to 0.47	2.60	0.094	1030 to 1720	150 to 250	105	15	1795(a)	3260(a)
Al_2O_3 (polycrystalline)	...	...	3.15	0.114	2070	300	170	25	1650	3000
FP Al_2O_3	15.2 to 25.4	0.6 to 1.0	3.71	0.134	1380(b)	200(b)	345	50	2045(a)	3710(a)
SiO_2-coated alumina	...	...	3.71	0.134	1900(b)	275(b)	380	55	...	...
Al_2O_3 monocrystal (sapphire)	...	...	3.96	0.143	2550	370	470	68	2040(a)	3700(a)
Nicalon	...	...	...	...	>2410	>350	180	26	1095	2000
Avco SiC	142	5.6	3.04	0.110	3450	500	425	62	...	...
SiC	102	4.0	3.46	0.125	2280	330	450	65	...	...
SiC (carbon core)	142	5.6	3.29	0.119	3790(b)	550(b)	345	50	...	...
SiC (tungsten core)	142	5.6	3.29	0.119	3790(b)	550(b)	415	60	600	1110
$Al_2O_3 \cdot Cr_2O_3$ monocrystal (ruby)	...	...	3.99	0.144	3450 to 4140	500 to 600	470	68	2040(a)	3700(a)
Borsic (SiC/B/W)	107 to 145	4.2 to 5.7	2.77	0.100	2930	425	415	60	2300(a)	4170(a)
Borsic/C	107 to 145	4.2 to 5.7	2.30	0.083	3170	460	350 to 365	51 to 53	...	...
Saffil alumina, RF grade	...	...	3.29	0.119	2000	290	295	43	1600	2910

(a) Melting or softening point. (b) Ultimate tensile strength

Influence of alumina and aluminosilicate fiber reinforcement on coefficient of thermal expansion of tempered cast aluminum

Matrix	Fiber, vol %	Coefficient of thermal expansion, 10^{-6}/K 0°		90°
332.0 T5	...	...	24.5	...
	Alumina, 5	239	...	23.6
	Mullite, 5	22.3	...	23.8
	Alumina, 15	18.9	...	22.3
339.0 T5	...	...	20.4	...
	Alumina, 10	18.0	...	...
	Alumina, 20	16.4	...	...

Typical properties of Nicalon SiC fiber

Production method	Si- and C-containing polymer spun, cured, and pyrolyzed
Diameter	10 to 20 µm, or 0.39 to 0.79 mil (500 per yarn)
Modulus	180 GPa, or 26.1 × 10^6 psi (420 GPa, or 60.9 × 10^6 psi, for β-SiC)
Strength at 20 °C (68 °F):	
As-produced	2 GPa (0.29 × 10^6 psi)
After treatment at 1400 °C or 2550 °F (argon)	<1 GPa (<0.15 ± 10^6 psi)
Strength at 1400 °C or 2550 °F (oxygen)	<0.5 GPa (<0.07 × 10^6 psi)
Creep strain at 1300 °C (2370 °F), 0.6 GPa (0.09 × 10^6 psi), 20 h	4.5%

Characteristics of ceramic whiskers

Material	Whisker morphology and size	Crystal structure	Properties
SiC	Rod or needle, 3 to 10 µm (120 to 390 µin.) diam, 10 to 1000 µ (40 to 3950 µin.) long	Alpha and/or beta phases	>500 GPa (>70 × 10^6 psi) modulus of elasticity 2 to 7 GPa (0.3 to 1.0 × 10^6 psi) tensile strength
Si_3N_4	Rod or needle, 0.2 to 0.5 µm (8 to 20 µin.) diam, 50 to 300 µm long	Alpha plus beta phase	390 GPa (60 × 10^6 psi) modulus of elasticity up to 1.5 GPa (0.2 × 10^6 psi) tensile strength

Ultimate tensile strengths of tungsten-coated and uncoated SiC filaments at various temperatures

Temperature		Ultimate tensile strength (coated)(a)		Ultimate tensile strength (uncoated)		
°C	°F	GPa	ksi	GPa	ksi	σ/σ_0
RT	RT	1.835	266	2.828	410	0.649
800	1470	1.648	239	2.359	342	0.699
1000	1830	1.531	222	2.083	302	0.735
1200	2190	1.414	205	1.883	273	0.751
1400	2550	1.083	157	1.359	197	0.797
1600	2910	0.855	124	1.062	154	0.805

(a) Mean ultimate tensile strength. Coating thickness, 12.7 μm (0.0005 in.).

Thermal shock response of SiC whisker reinforced alumina as indicated by flexural strength retained after quenching from elevated temperature into boiling water

	Temperature quenched from		Retained fracture strength	
	°C	°F	MPa	ksi
Alumina-20 vol% SiC	No thermal shock		620	89.9
Single thermal shock cycle	400	750	630	91.4
	600	1110	685	99.3
	800	1470	615	89.2
	1000	1830	710	103.0
Ten thermal shock cycles	400	750	610	88.5
	500	930	570	82.7
	800	1470	540	78.3
	1000	1830	545	79.0
Alumina	No thermal shock		315	45.7
Single thermal shock cycle	300	570	250	36.3
	400	750	225	32.6
	500	930	125	18.1

Influence of silicon carbide fiber reinforcement on coefficient of thermal expansion of tempered aluminum

Matrix	Fiber, vol %	Coefficient of thermal expansion, 10^{-6}/K, 0 °		90 °
339.0 T5	...	...	20.4	...
	Silicon carbide, 15	16.9	...	...
356.0 T5	...	...	21.4	...
	Silicon carbide, 14	16.5	...	...
2024 T6	...	...	21.1	...
	Silicon carbide, 25	14.9	...	16.4
	Silicon carbide, 40	13.0	...	...
6061 T7	...	...	21.6	...
(Extruded)	Silicon carbide, 25	12.1	...	...

Influence of fiber reinforcement on thermal conductivity at 200 °C (390 °F)

Matrix	Fiber, vol %	Thermal conductivity, W/m · K (Btu · in./h · ft² · °F), 0 °C				90 °	
332.0 T5	...	...	...	176	1220	...	...
	Alumina, 5	158	1100	...	...	140	970
	Mullite, 5	152	1050	...	...	128	890
	Alumina, 15	134	930	...	...	98	680
	Zirconia, 19	116	800	...	...	87	600
339.0 T5	...	...	...	144	1000	...	...
	Alumina, 10	112	780	...	...	...	...
	Alumina, 15	99	690	...	...	...	...
	SiC, 15	136	940	...	...	...	...
2024 T6	...	...	...	151	1050	...	...
	SiC, 25	102	710	...	...	...	...

Properties of boron and alumina fibers

Fiber	Density, g/cm³	Diameter, μm	Diameter, μin.	Tensile strength, GPa	Tensile strength, 10^6 psi	Tensile modulus, GPa	Tensile modulus, 10^6 psi
Boron-tungsten	2.6	100 to 200	3950 to 7850	5.5 to 7.0	0.80 to 1.0	400	58
Boron-carbon	2.3	100 to 200	3950 to 7850	5.0	0.73	400	58
α-alumina(a)	3.95	20	790	14(b)	2.0	390	57
				19(c)	2.8	390	57

(a) Slurry-spun continuous fiber. (b) Uncoated. (c) Silicon carbide-coated

Commercially available continuous oxide fibers

Composition, wt %	Identification	Company	Forms(a)
>99 Al_2O_3	Fiber FP	E.I. Du Pont de Nemours & Co., Inc.	C, Y, F
85 Al_2O_3, 15 SiO_2	High performance alumina fiber	Sumitomo Chemical Co. Ltd., Japan; distributed by Avco Specialty Materials, Textron, Inc.	C, Y, F
80 Al_2O_3, 20 SiO_2	Long alumina fiber	Denki Kagaku Kogyo K.K., with Nichibi Co. Ltd.	C, Y
70 Al_2O_3, 2 B_2O_3, 28 SiO_2	Nextel 440 and Nextel 480	Minnesota Mining & Manufacturing Co.	C, R, Y, F
62 Al_2O_3, 14 B_2O_3, 24 SiO_2	Nextel 312	Minnesota Mining & Manufacturing Co.	C, R, Y, F
99.95 SiO_2	Astroquartz II	Distributed by J.P. Stevens & Co. Inc.	C, R, Y, F
98 SiO_2, 2 rem	Refrasil	Hitco Materials Div., Armco Inc.	Y, F
98 SiO_2, 2 rem	Siltemp	Ametek, Inc.	F
68 ZrO_2, 32 SiO_2	Nextel Z-11	Minnesota Mining & Manufacturing Co.	C, R, Y, F

(a) C, continuous; Y, yarn; F, fabric; R, roving

Commercially available discontinuous oxide fibers

Composition, wt %	Identification	Company	Forms(a)
95 Al_2O_3, 5 SiO_2	Saffil	Imperial Chemical Industries, PLC Ltd., England; distributed by Babcock & Wilcox	D, D, M, Ch
72 Al_2O_3, 28 SiO_2	Fibermax	Sohio Engineered Materials (formerly Carborundum)	D, B
70 Al_2O_3, 2 B_2O_3, 28 SiO_2	Nextel 440 and Nextel 480 Ultrafiber	Minnesota Mining & Manufacturing Co.	D, M, Ch
60 to 68 Al_2O_3, 4 to 9 B_2O_3, 23 to 32 SiO_2	Staple Fiber	Nichias Corp.	D, B
62 Al_2O_3, 14 B_2O_3, 24 SiO_2	Nextel 312 Ultrafiber	Minnesota Mining & Manufacturing Co.	D, M, Ch
52 Al_2O_3, 48 SiO_2	Fiberfrax	Sohio Engineered Materials	D, B, M, F
49 to 50 Al_2O_3, 50 to 51 SiO_2	Innswool	A.P. Green Refractories	D, B, M
52 to 55 Al_2O_3, 41 to 44 SiO_2	Cer-wool	Combustion Engineering	D, B, M
47 Al_2O_3, 53 SiO_2	Cerafiber	Manville Corp.	D, B, M
42.5 Al_2O_3, 2.5 Cr_2O_3, 55 SiO_2	Cerachrome	Manville Corp.	D, B, M
40 Al_2O_3, 50 SiO_2, 5 CaO, 3.5 MgO, 1.5 TiO_2	Cerawool	Manville Corp.	D, B, M
92 ZrO_2, 8 Y_2O_3	Zircar	Zircar Products, Inc.	D, B, M, F

(a) D, discontinuous; B, bulk; M, mat or blanket; Ch, chopped; F, fabric

Commercially available carbide and nitride fibers

Composition	Identification	Company	Forms(a)
SiC	Nicalon	Nippon Carbon Co. Ltd.	C, Y, Ch, F, M, R
Si-Ti-C	Tyranno	Ube Industries, Ltd.	C
Si-Zr-C	Tyranno	Ube Industries, Ltd.	C
SiC on C core	CVD SiC	Tokai Carbon Co., Ltd.; distributed by Avco Specialty Materials, Textron, Inc.	C
SiC	Tokawhisker	Tokai Carbon Co., Ltd.; distributed by Avco Specialty Materials, Textron, Inc.	D
SiC	Silar	Arco Metals Co.	D
SiC	Tateho	Tateho Chemical Industry Co., Ltd.; distributed by ICD Group Inc.	D
Si_3N_4	Tateho	Tateho Chemical Industry Co., Ltd.; distributed by ICD	D

(a) C, continuous; Y, yarn; Ch, chopped; F, fabric; M, mat or blanket; R, roving; D, discontinuous

Properties of continuous oxide fibers

Fiber	Composition, wt %					Density, g/cm^3	Average diameter		Tensile strength		Tensile modulus of elasticity		Use temperature	
	Al_2O_3	B_2O_3	SiO_2	ZrO_2	Rem		μm	μin.	MPa	ksi	GPa	10^6 psi	°C	°F
Fiber FP(a)	>99	...	...	...	...	3.95	20	790	1380	200	379	55	1320	2410
Sumitomo alumina	85	...	15	...	...	3.2	17	670	1450	210	193	28	1250	2280
Nextel 440	70	2	28	...	...	3.05	11	430	2070	300	193	28	1430	2605
Nextel 480	70	2	28	...	...	3.1	11	430	2240	325	207 to 241	30 to 35	1430	2605
Nextel 312	62	14	24	...	...	2.7	11	430	1720	250	155	22	1200	2190
Nextel Z-11	...	...	32	68	...	3.7	14	550	1310	190	76	11	1000	1830
Astroquartz II	...	...	99.95	...	...	2.2	9	350	3450	500	69	10	1050	1920
Refrasil	...	...	97.9	...	2.1	2.1	...	...	210 to 410	3 to 6	72	10.5	1095	2000
Siltemp	...	...	98	...	2	2.2	...	...	...	...	...	...	...	...
Denki	80	...	20	...	...	...	10	390	...	...	...	...	...	...

(continued)

Properties of continuous oxide fibers (continued)

Fiber	Melt or liquidus temperature °C	Melt or liquidus temperature °F	Coefficient of thermal expansion 10^{-6}/K	Dielectric constant	Resistivity Ω · m at 20 °C	Resistivity Ω · cm at 68 °F	Refractive index
Fiber FP(a)	2045	3710	...	...	...	...	...
Sumitomo alumina	...	...	8.8(b)	...	10^{11}	10^{13}	1.65
Nextel 440	>1800	>3270	5	5.8	...	...	1.617
Nextel 480	>1800	>3270	...	5.8	...	...	>1.617
Nextel 312	1800	3270	3.5	5	...	...	1.572
Nextel Z-11	2000	3630	...	...	...	...	1.75
Astroquartz II	1700	3090	0.54	3.8	10^{16}	10^{18}	1.459
Refrasil	>1760	>3200	...	...	...	...	...
Siltemp	>1760	>3200	...	...	...	...	...
Denki	...	...	...	...	...	...	...

(a) Compressive strength is 6.9 GPa (10^6 psi). (b) At 200 to 400 °C (390 to 750 °F)

Stress-rupture strengths (100-h) of SiC/C and SiC/W filaments

Filament	1093 °C (2000 °F) GPa (ksi)	1093 °C (2000 °F) S/ρ × 10^6 in.	1204 °C (2200 °F) GPa (ksi)	1204 °C (2200 °F) S/ρ × 10^6 in.	1316 °C (2400 °F) GPa (ksi)	1316 °C (2400 °F) S/ρ × 10^6 in.
SiC/C(a)	1.93 (280)	2.3	1.034 (150)	1.4	0.69 (100)	0.96
SiC/W(b)	1.07 (155)	1.3	0.863 (126)	1.1	0.34 (50)	0.43

(a) C is used as a substrate. (b) W is used as a substrate

Properties of discontinuous silicon carbide and silicon nitride whiskers

Fiber	Composition	Max free carbon, wt %	Crystalline species	Whisker content, %	Particle content, %	Density g/cm³	Average diameter μm	Average diameter μin.	Predominant length μm	Predominant length μin.
Silar SC-9	SiC	0.10	α	80 to 90	10 to 20	3.2	0.6	24	10 to 80	390 to 3150
Silar SC-10	SiC	0.20	α	70 to 80	20 to 30	3.2	6.6	260	10 to 80	390 to 3150
Tokawhisker	SiC	Negligible	β	...	<1	3.19	0.1 to 0.5	4 to 20	30 to 100	1200 to 3950
Tateho SiC	SiC	...	β	...	...	3.18	0.05 to 1.5	2 to 59	20 to 200	790 to 7900
Tateho Si_3N_4	Si_3N_4	...	α	...	...	3.18	0.1 to 1.6	4 to 63	20 to 200	790 to 7900

Fiber	Surface area m²/kg	Surface area ft²/lb	Tensile strength MPa	Tensile strength ksi	Tensile modulus of elasticity GPa	Tensile modulus of elasticity 10^6 psi	Use temperature: stability to °C	Use temperature: stability to °F	Electrical conductivity
Silar SC-9	3000	14,600	6900(a)	1000	690	100	1760(b)	3200	...
Silar SC-10	3000	14,600	6900	1000	690	100	1760(b)	3200	...
Tokawhisker	...	...	3000 to 14,000	440 to 2030	400 to 700	58 to 101	1600 (in air)	2910	...
Tateho SiC	...	...	...	...	...	...	...	...	Conductive
Tateho Si_3N_4	...	...	...	...	...	...	...	...	Nonconductive

(a) Estimated. (b) Atmospheric environment not stated

Composition and room-temperature mechanical properties of commercially available polymer-derived monofilament ceramic fibers

Fiber	Composition, wt % Si	N	C	O	Ti	Mechanical properties: Tensile strength MPa	Tensile strength ksi	Elastic modulus GPa	Elastic modulus 10^6 psi	Density g/cm³	Density lb/in.³
CG Nicalon	58.4	0.1	31.2	10.1	...	2970	430	193	28	2.55	0.0921
Tyranno	49.3	0.3	27.9	20.2	1.2	2800	406	193	28	2.40	0.0867
HPZ	59.4	28.9	10.1	3.1	...	2600	377	193	28	2.45	0.0885

Properties of continuous silicon carbide fibers

Fiber	Composition: Si	O	C	Ti	H	N	Rem	Crystalline species	Density, g/cm³	Average diameter: µm	µin.	Tensile strength: MPa	ksi	Tensile modulus of elasticity: GPa	10^6 psi
Nicalon SiC	54.3	11.8	30.0	...	...	...	3.9	β-SiC	2.55	10 to 15(a)	390 to 590	2500 to 3200	360 to 470	180 to 200	26 to 29
Tyranno Si-Ti-C	44.2	12.3	24.5	11.0	0.6	3.4	4.0	β-SiC + TiC	2.3	10 to 15	390 to 590	1990	286	117	17
Avco CVD SiC	...	...	...	...	...	...	...	β on carbon core	...	~140, with 33 · µm core	~5510, with 1300 µin. core	>3400	>500	428	62

Fiber	Use temperature: °C	°F	Coefficient of thermal expansion, 10^{-6}/K	Dielectric constant	Resistivity: Ω · m at 20°C	Ω · m at 68°F	Chemical resistance, % wt reduction in 24 h at 80°C (175°F): 6N HCl	18N H_2SO_4	7N $HClO_3$	30% NaOH aqueous
Nicalon SiC	1200(b) (c)	2190	3.1(d)	6 to 8	10^3	10^4	<1%	<1%	<1%	<1%
Tyranno	1300	2370	...	...	...	...	...	...	...	...
Avco CVD SiC	1100	2010	4.9	...	...	...	...	...	...	...
	1200	2190	...	...	...	...	...	...	...	...
	>1400	2550	...	...	...	...	...	...	...	...

(a) Round cross section. (b) Substantial loss of strength. (c) Loss in strength above ~ 1000°C (1830°F). (d) Along fiber axis, 0 to 200°C (32 to 390°F)

Properties of discontinuous oxide fibers

Fiber	Composition, wt %: Al_2O_3	B_2O_3	CaO	Cr_2O_3	Fe_2O_3	MgO	SiO_2	TiO_2	Y_2O_3	ZrO_2	rem	Density, g/cm³	Average diameter: µm	µin.
Cerachrome	42.5	...	...	2.5	...	...	55	...	...	...	...	...	3.5	138
Cerafiber	47	...	...	...	...	...	52.8	...	...	...	0.2	2.65	3.5	138
Cerawool	40	...	5	...	...	3.5	50	1.5	...	...	...	2.54	3.5	138
Cer-wool	52 to 55	...	...	...	0.1 to 0.2	...	41 to 44	0.1 to 0.2	...	...	1 to 2	...	3.0	118
Fiberfrax	51.9	...	...	...	...	...	47.9	...	...	...	0.1	2.73	2 to 3	79 to 118
Fibermax	72	...	...	...	...	...	28	...	...	...	...	3.0	2 to 3.5	79 to 138
Innswool	49 to 50	...	...	...	...	...	50 to 51	...	...	...	<0.5	...	3 to 5	118 to 197
Kaowool	45	...	...	...	...	...	53	...	...	...	2	2.56	2.8	110
Nextel 312 Ultrafiber	62	14	...	...	...	...	24	...	...	...	...	2.75	3.5	138
Nextel 440 Ultrafiber	70	2	...	...	...	...	28	...	...	...	...	3.1	3.3	130
Nichias	60 to 68	4 to 9	...	...	...	...	23 to 32	...	...	...	...	...	10.5	413
Saffil RF Grade	96 to 97	...	...	...	...	...	3 to 4	...	...	...	...	3.3	3.0	118
Saffil RG Grade	96 to 97	...	...	...	...	...	3 to 4	...	...	...	...	3.3 to 3.5	3.0	118
Zircar	...	...	...	...	...	...	...	...	8	92	...	5.6 to 5.9	4 to 6	157 to 236

Fiber	Tensile strength: MPa	ksi	Tensile modulus of elasticity: GPa	10^6 psi	Use temperature: °C	°F	Melt or liquidus temperature: °C	°F	Specific heat: J/kg · K at 1000 °C	Btu/lb · °F at 1830 °F
Cerachrome	...	...	...	...	1425	2600	>1760	>3200	1148	0.2741
Cerafiber	...	...	...	...	1315	2400	>1760	>3200	1148	0.2741
Cerawool	...	...	...	...	875	1610	>1648	>3000	1148	0.2741
Cer-wool	...	...	...	...	...	...	...	...	...	...
Fiberfrax	1900	276	100	14.6	1260	2300	1790	3255	1130	0.2698
Fibermax	1030	150	150	22	1650	3000	1890	3435	...	...
Innswool	...	...	...	...	1235	2255	1760	3200	...	...
Kaowool	1130	165	84	12.2	1260	2300	1760	3200	1088	0.2598
Nextel 312 Ultrafiber	1720	250	152	22	1200	2190	1800	3270	...	...
Nextel 440 Ultrafiber	1310	190	207 to 241	30 to 35	1430	2605	>1800	>3270	...	...
Nichias	1790	260	...	...	...	...	...	...	...	...
Saffil RF Grade	2000	290	310	45	1600	2910	>2000	>3630	...	...
Saffil RG Grade	1000 to 2000	145 to 290	297	43	1600	2910	>2000	>3630	...	...
Zircar	...	...	...	...	2200	3990	2600	4710	...	...

Glass Fibers

Fiberglass grades

Type	Description
A	Glass of soda-lime composition similar to bottle glass. Poor thermal and chemical properties; not used for fibers
C	Chemically resistant soda-lime-borosilicate glass used for its high corrosion and chemical attack resistance
D	Low-density glass with high electrical resistance
E	Pyrex composition glass. Good electrical properties; good for general-purpose application when a combination of good strength and chemical resistance is observed
S	A high-strength high-modular glass for specific applications. Higher in cost

Compositional ranges for glass fibers used in composite materials

	Composition range, wt %			
Compound	E-glass	S-glass	C-glass	E-CR glass
Silicon dioxide	52 to 56	65	64 to 68	54 to 56
Aluminum oxide	12 to 16	25	3 to 5	9 to 14.5
Boric oxide	5 to 10	...	4 to 6	...
Sodium oxide and potassium oxide	0 to 2	...	7 to 10	0 to 1
Magnesium oxide	0 to 5	10	2 to 4	0 to 4
Calcium oxide	16 to 25	...	11 to 15	17 to 25
Barium oxide	...	...	0 to 1	...
Zinc oxide	...	...	...	0 to 5
Titanium oxide	0 to 1.5	...	...	0 to 4
Zirconium oxide	...	...	...	0 to 0.8
Iron oxide	0 to 0.8	...	0 to 0.8	...
Fluorine	0 to 1	...	...	0 to 0.8

Glass fiber compositions (wt %)

	Grade of glass			
Component	A (high alkali)	C (chemical)	S (high strength)	E (electrical)
Silicon oxide	72.0	64.6	64.2	54.3
Aluminum oxide	0.6	4.1	24.8	15.2
Ferrous oxide	...	...	0.21	...
Calcium oxide	10.0	13.2	0.01	17.2
Magnesium oxide	2.5	3.3	10.27	4.7
Sodium oxide	14.2	7.7	0.27	0.6
Potassium oxide	...	1.7	...	...
Boron oxide	...	4.7	0.01	8.0
Barium oxide	...	0.9	0.2	...
Miscellaneous	0.7	...	...	...

Typical properties for glass fiber types

		Tensile strength										
	Density, bulk annealed,	at −190 °C (−310 °F)		at 23 °C (72 °F)		at 371 °C (700 °F)		at 538 °C (1000 °F)		Modulus of elasticity at 538 °C (1000 °F)		Elongation
Material	g/cm^3	MPa	ksi	MPa	ksi	MPa	ksi	MPa	ksi	GPa	10^6 psi	%
E-glass	2.62	5310	770	3445	500	2620	380	1725	250	72.3	10.5	4.88
S-glass	2.50	8275	1200	4585	665	4445	645	2415	350	88.9	12.9	5.7
C-glass	2.56	5380	780	3310	480	...	...	...	...	...	...	4.8
E-CR glass	2.76	5310	770	3445	500	...	...	...	...	81.3	11.8	...

Typical properties for glass fiber types (continued)

Material	Chemical resistance (percent weight loss): in H_2O 24 h	in H_2O 186 h	10% HCl 24 h	10% HCl 168 h	10% H_2SO_4 24 h	10% H_2SO_4 168 h	1% Na_2CO_3 24 h	1% Na_2CO_3 168 h	10% NaOH 168 h	Relative permittivity at 1 MHz	Relative permittivity at 60 Hz	Dissipation factor at 1 MHz	Dissipation factor at 60 Hz
E-glass	0.7	0.9	42	43	39	42	2.1	2.1	20	6.6	6.7	0.0025	0.0034
S-glass	0.5	0.7	3.8	5.1	4.1	5.7	2.0	2.1	66	5.3	5.4	0.0034	0.0129
C-glass	1.1	2.9	4.1	7.5	2.2	4.9	24	31	...	6.9	...	0.0085	...
E-CR glass	0.7	0.7	5.4	7.7	6.2	10.4	...	1.8	16	6.9	7.2	0.0028	0.0031

Material	Volume resistivity, Ω · m	Surface resistivity, Ω	Dielectric strength KV/cm	Dielectric strength V/mil	Viscosity softening point °C	Viscosity softening point °F	Viscosity annealing point °C	Viscosity annealing point °F	Viscosity strain point °C	Viscosity strain point °F	Thermal expansion, 10^{-6}/K(a)	Specific heat at 23 °C (72 °F) kJ/kg · K (Btu/lbf · °F)	Specific heat at 200 °C (392 °F) kJ/kg · K (Btu/lbf · °F)	Refractive index, bulk annealed
E-glass	0.402×10^{15}	0.42×10^{16}	103	262	846	1555	657	1215	615	1140	5.4	0.810 (0.193)	1.03 (0.247)	1.562
S-glass	0.905×10^{13}	0.886×10^{13}	130	330	970	1778	810	1490	760	1400	1.6	0.737 (0.176)	...	1.525
C-glass	...	...	...	...	750	1382	588	1090	552	1025	6.3	0.787 (0.188)	0.90 (0.215)	1.537
E-CR glass	0.384×10^{15}	0.116×10^{17}	98	250	882	1619	...	...	...	...	...	...	0.97 (0.232)	1.583

(a) From -30°C (-20°F) to 250°C (480°F)

Glass fibers for filament winding (in order of ascending modulus of strand, normalized to 100% fiber volume)

Type	Strand nominal tensile modulus GPa	Strand nominal tensile modulus 10^6 psi	Strand nominal tensile strength MPa	Strand nominal tensile strength ksi	Maximum number of filaments/strand	Fiber density, g/cm^3
E	72.4	10.5	3447	500	4000	2.60
R	86.2	12.5	2068	300	60	2.49
S	86.9	12.6	4585	665	...	2.55

Inherent properties of glass fibers

	Specific gravity	Tensile strength MPa	Tensile strength ksi	Tensile modulus GPa	Tensile modulus 10^6 psi	Coefficient of thermal expansion, 10^{-6}/K	Dielectric constant (a)	Liquidus temperature °C	Liquidus temperature °F
E-glass	2.58	3450	500	72.5	10.5	5.0	6.3	1065	1950
A-glass	2.50	3040	440	69.0	10.0	8.6	6.9	996	1825
ECR-glass	2.62	3625	525	72.5	10.5	5.0	6.5	1204	2200
S-glass	2.48	4590	665	86.0	12.5	5.6	5.1	1454	2650

(a) At 20 °C (72 °F) and 1 MHz.

Properties of quartz fibers

Properties	Astroquartz II
Physical	
Specific gravity	2.20
Density, g/cm^3 (lb/in.^{3})	2.19 (0.079)
Filament diameter, μm (mils in.)	8.9 (0.35)
Mechanical	
Tensile strength at RT, MPa (ksi)	3450 (500)
Tensile modulus at RT, GPa (10^6 psi)	69.0 (10.0)
Elongation, %	5
Thermal (pure silica block)	
Specific heat from –20 to 500 °C (0 ° to 932 °F), kJ/kg · K (Btu/lb · °F)	0.96 (0.23)
Electrical (pure fused silica block)	
Electrical resistivity at RT, Ω · m	10^{16}
Dielectric constant at RT, 1 MHz	3.78

Commercial forms of glass fiber reinforcements

Nominal form	General description	Process	Nominal glass content of typical laminates, %	Typical applications
Rovings	Continuous strands of glass fibers	Filament winding, continuous panel, preforming (matched-die molding), spray-up, pultrusion	25 to 80	Pipe, automobile bodies, rod stock, rocket-motor cases, ordnance
Chopped strands	Strands cut to lengths of 0.125 to 2 in. (3.2 to 50.8 mm)	Premix molding, wet slurry preforming	15 to 40	Electrical and appliance parts, ordnance components
Reinforcing mats	Continuous or chopped strands in random matting	Matched-die molding, hand lay-up, centrifugal casting	20 to 45	Translucent sheets, truck and auto body panels
Surfacing and overlaying mats	Nonreinforcing random mat	Matched-die molding, hand lay-up, filament winding	5 to 15	Where smooth surfaces are required (automobile bodies, some housings)
Yarns	Twisted strands	Weaving, filament winding	60 to 80	Aircraft, marine, electrical laminates
Woven fabrics	Woven cloths from glass fiber yarns	Hand lay-up, vacuum bag, autoclave, high-pressure laminating	45 to 65	Aircraft structures, marine, ordnance hardware, electrical flat sheet and tubing
Woven roving	Woven glass fiber strands (coarser and heavier than fabrics)	Hand lay-up	40 to 70	Marine, large containers

The effect of glass form and amount on mechanical properties

Type of glass fiber reinforcement	Glass content, wt %	Density, g/cm^3	Tensile strength, MPa	Tensile strength, ksi	Tensile modulus, GPa	Tensile modulus, 10^6 psi	Elongation, %	Flexural strength, MPa	Flexural strength, ksi	Flexural modulus, GPa	Flexural modulus, 10^6 psi	Compressive strength, MPa	Compressive strength, ksi
Neat cured resin	0	1.22	59	8.6	5.40	0.783	2.0	88	12.8	3.90	0.565	156	22.6
Chopped strand mat	30	1.50	117	17.0	10.80	1.566	3.5	197	28.6	9.784	1.419	147	21.3
Chopped strand mat	50	1.70	288	41.8	16.70	2.422	3.5	197	28.6	14.49	2.102	160	23.2
Roving fabric	60	1.76	314	45.5	19.50	2.828	3.6	317	46.0	15.00	2.175	192	27.8
Woven glass fabric	70	1.88	331	48.0	25.86	3.750	3.4	403	58.4	17.38	2.520	280	40.6
Unidirectional roving fabric	70	1.96	611	88.6	32.54	4.720	2.8	403	58.4	29.44	4.270	216	31.3

Effect of glass content on mechanical properties

Material	Glass content, wt %	Flexural strength, MPa	Flexural strength, ksi	Flexural modulus, GPa	Flexural modulus, 10^6 psi	Tensile strength, MPa	Tensile strength, ksi	Tensile modulus, GPa	Tensile modulus, 10^6 psi	Compressive strength, MPa	Compressive strength, ksi
Orthophthalic	30	170	25	5.5	0.8	140	20	4.8	0.7	...	...
	40	220	32	6.9	1.0	150	22	5.5	0.8	...	...
Isophthalic	30	190	28	5.5	0.8	150	22	8.3	1.2	...	...
	40	240	35	7.6	1.1	190	28	11.7	1.7	210	30
BPA fumarate	25	120	17	5.1	0.7	80	12	7.6	1.1	170	24
	35	150	22	8.3	1.2	100	14	10.3	1.5	170	24
	40	160	23	9.0	1.3	120	18	11.0	1.6	180	26
Chlorendic	24	120	17	5.9	0.8	80	11	7.6	1.1	140	21
	34	160	23	6.9	1.0	120	18	9.6	1.4	120	18
	40	190	28	9.6	1.4	140	20	9.6	1.4	120	18
Vinyl ester	25	110	16	5.4	0.8	86.2	12.5	7.0	1.0	180	26.5
	35	260	37.3	9.5	1.4	153.4	22.25	10.8	1.6	230	34
	40	220	32	8.9	1.3	160	23	11.0	1.6	210	30

Polymer Matrix Composites

General

Fiber-resin composite properties

Physical	Mechanical	Thermal	Electrical	In-service conditions
Specific gravity	Tensile strength	Coefficient of thermal expansion	Dielectric constant	Service temperature
Density	Tensile modulus	Thermal conductivity	Dielectric strength	TGA
	Poisson's ratio	Specific heat	Dissipation factor	Temperature allowed on all standard loads
	Compressive strength		Volume resistivity	Flammability
	Compressive modulus			EMI/RFI protection
	Poisson's ratio			
	Shear strength			
	Shear modulus			

Thermoset and thermoplastic trade-offs for commercial aircraft composites

Property	Thermosets	Thermoplastics
Resin cost	Low	Low to high
Prepregability	Excellent	Poor (new methods such as emulsions could change this)
Prepreg tack/drape	Excellent	None (revised lay-up techniques are required)
Volatile-free prepreg	Good	Good to excellent
Prepreg shelf life and out-time	Poor	Excellent
Prepreg quality assurance	Fair	Excellent
Prepreg cost	Good	High (new methods needed)
Composite processing	Slow	Slow (unless automated processes are developed)
Shrinkage	Moderate	Low
Composite mechanical properties	Good (room for improvement in damage tolerance)	Good (more data and experience needed)
Interlaminar fracture toughness	Low	High
Resistance to fluids/solvents	Good	Poor to good
Resistance to creep	Good	Currently not known
Crystallinity problems	None	Yes

Comparison of engineering properties of fiberglass reinforced plastics and competitive materials(a)

Material	Mechanical and physical properties										
	Glass fiber content, wt%	Flexural strength, ksi(b)	Flexural modulus, 10^5 psi(c)	Tensile strength at yield, ksi(b)	Tensile modulus, 10^5 psi(c)	Compressive strength, ksi(b)	Ultimate tensile elongation, %	Izod impact strength, ft · lb/in. of notch(d)	Thermal conductivity, Btu · in./ft^2 · h · °F (K value) (e)	Specific heat, Btu/lb · °F	Flammability (UL94) (g)
Glass-fiber-reinforced thermosets											
Sheet molding compound (SMC)	15 to 30	18 to 30	14 to 20	8 to 20	16 to 25	15 to 30	0.3 to 1.5	8 to 22	1.3 to 1.7	0.30 to 0.35	5V
Bulk molding compound (BMC)	15 to 35	10 to 20	14 to 20	4 to 10	16 to 25	20 to 30	0.3 to 5	2 to 10	1.3 to 1.7	0.30 to 0.35	5V
Preform/mat (compression molded)	25 to 50	10 to 40	13 to 18	25 to 30	9 to 20	15 to 30	1 to 2	10 to 20	1.3 to 1.8	0.30 to 0.33	V-0
Cold press molding-polyester	20 to 30	22 to 37	13 to 19	12 to 20	...	...	1 to 2	9 to 12	1.3 to 1.8	0.30 to 0.33	V-0
Spray-up-polyester	30 to 50	16 to 28	10 to 12	9 to 18	8 to 18	15 to 25	1.0 to 1.2	4 to 12	1.2 to 1.6	0.30 to 0.34	V-0
Filament wound-epoxy	30 to 80	100 to 270	50 to 70	80 to 250	40 to 90	45 to 70	1.6 to 2.8	40 to 60	1.92 to 2.28	0.23 to 0.25	V-0
Rod stock-polyester	40 to 80	100 to 180	40 to 60	60 to 180	40 to 60	30 to 70	1.6 to 2.5	45 to 60	1.92 to 2.28	0.22 to 0.25	V-0
Molding compound-phenolic	5 to 25	18 to 24	30	7 to 17	26 to 29	14 to 35	0.25 to 0.6	1 to 8	1.1 to 2.0	0.20 to 0.30	V-0
Glass-fiber-reinforced thermoplastics											
Acetal	20 to 40	15 to 28	8 to 13	9 to 18	8 to 15	11 to 17	2	0.8 to 2.8	...	...	HB
Nylon	6 to 60	7 to 50	2 to 28	13 to 33	2 to 20	13 to 24	2 to 10	0.8 to 4.5	...	0.30 to 0.35	V-0
Polycarbonate	20 to 40	17 to 30	7.5 to 15	12 to 25	7.5 to 17	14 to 24	2	1.5 to 3.5	...	...	V-0
Polyethylene	10 to 40	7 to 12	2.1 to 6	6.5 to 11	4 to 9	4 to 8	1.5 to 3.5	1.2 to 4.0	...	...	V-0
Polypropylene	20 to 40	7 to 11	3.5 to 8.2	5.5 to 10.5	4.5 to 9	6 to 8	1 to 3	1 to 4	...	...	V-0
Polystyrene	20 to 35	10 to 17	8 to 12	10 to 15	8.4 to 12.1	13.5 to 19	1.0 to 1.4	0.4 to 4.5	...	0.23 to 0.35	V-0
Polysulfone	20 to 40	21 to 27	8 to 15	13 to 20	15	21 to 26	2 to 3	1.3 to 2.5	...	...	V-0
ABS (acrylonitrile butadiene styrene)	20 to 40	23 to 26	9.2 to 15	11 to 16	6 to 10	12 to 22	3 to 3.4	1 to 2.4	...	...	V-0
PVC (polyvinyl chloride)	15 to 35	20 to 25	9 to 16	14 to 18	10 to 18	13.4 to 16.8	2 to 4	0.8 to 1.6	...	...	V-0
Polyphenylene oxide (modified)	20 to 40	17 to 31	8 to 15	15 to 22	9.5 to 15	18 to 20	1.7 to 5	1.6 to 2.2	...	...	V-0
SAN (styrene acrylonitrile)	20 to 40	15 to 21	8.0 to 18	13 to 18	9 to 18.5	12 to 23	1.1 to 1.6	0.4 to 2.4	...	...	V-0
Thermoplastic polyester	20 to 35	19 to 29	8.7 to 15	14 to 19	13 to 15.5	16 to 18	1 to 5	1.0 to 2.7	1.3	...	V-0
Unreinforced thermoplastics											
Acetal	N.Ap.	13 to 14	4	8 to 10	4 to 5	5	25 to 60	1.2 to 2.3	...	0.35	HB
Nylon	N.Ap.	5 to 18	2 to 4	9	2 to 5	7 to 10	29	1 to 4	...	0.40	V-0
Polycarbonate	N.Ap.	13	3	9 to 11	3.5	12	100 to 130	16	...	0.30	V-0
Polyethylene (high density)	N.Ap.	...	0.7 to 2.6	4	0.6 to 1.5	2.7 to 3.6	30 to 900	0.6 to 20.0	...	...	V-0
Polypropylene	N.Ap.	5 to 8	1.2 to 2.7	3 to 5	1.2	3.7 to 8	200 to 700	0.5 to 20.0	...	...	V-0
Polystyrene (high impact)	N.Ap.	3 to 10	1 to 5	3 to 5	2 to 4	4 to 9	15 to 30	0.7 to 3.6	...	...	V-0
Polysulfone	N.Ap.	1.5	4	10	3.6	14	50 to 100	1.3	...	...	V-0
ABS (high heat)	N.Ap.	9	3 to 4	6 to 6	2.8 to 4.1	6.8 to 12.5	10 to 20	2.5	...	...	V-0
PVC	N.Ap.	13 to 16	4	6 to 7	4	...	...	2 to 20	...	...	V-0
Polyphenylene oxide (modified)	N.Ap.	15	4	10	3.7	15	50 to 100	1.5 to 1.9	...	...	V-0
SAN	N.Ap.	9.7 to 17.5	5	9 to 11	5	14 to 17	2.5 to 3.7	0.4	...	...	V-0
Metals											
Gray cast iron	N.Ap.	10	N.Av.	15 to 30	120	25	1	4 to 4.4	288 to 408	0.13 to 0.19	N.Ap.
Low-carbon steel (cold rolled)	N.Ap.	28	300	29 to 33	300	28	38 to 39	N.Ap.	260 to 460	0.10 to 0.11	N.Ap.
Stainless steel	N.Ap.	30 to 35	280	30 to 35	280	30	50 to 60	8.5 to 11.0	96 to 185	0.12	N.Ap.
Aluminum, wrought	N.Ap.	20	100	6 to 27	100	N.Av.	30 to 40	N.Ap.	810 to 1620	0.22 to 0.23	N.Ap.
Aluminum, die cast	N.Ap.	8 to 26	100	8 to 26	100	9	6 to 8	N.Ap.	610 to 1100	0.22 to 0.23	N.Ap.
Magnesium, die cast	N.Ap.	14	65	8 to 30	65	10 to 14	4 to 6	3	288 to 960	0.245 to 0.25	N.Ap.
Zinc, die cast	N.Ap.	N.Av.	N.Av.	10 to 25	N.Av.	N.Av.	10	4.3	764 to 792	0.10	N.Ap.
Brass, plain yellow wrought	N.Ap.	14	150	14	150	N.Av.	60 to 65	N.Ap.	804	0.09	N.Ap.

(continued)

Comparison of engineering properties of fiberglass reinforced plastics and competitive materials(a) (continued)

Material	Glass fiber content, %	Hardness, Rockwell	Dielectric strength V/mil(h)	Specific gravity	Density, lb/in.^{3}(j)	Heat distortion at 284 psi, °F (k)	Continuous heat resistance, °F(k)	Coefficient of thermal expansion 10^{-6} psi/°F(m)	Chemical resistance(n): Weak acids	Strong acids	Weak alkalis	Strong alkalis	Organic solvents
Glass-fiber-reinforced thermosets													
Sheet molding compound (SMC)	15 to 30	H50 to 112	300 to 450	1.7 to 2.1	0.061 to 0.075	400 to 500	300 to 400	8 to 12	G to E	F	F	P	G to E
Bulk molding compound (BMC)	15 to 35	H80 to 112	300 to 450	1.8 to 2.1	0.065 to 0.075	400 to 500	300 to 400	8 to 12	G to E	F	F	P	G to E
Preform/mat (compression molded)	25 to 50	H40 to 105	300 to 600	1.5 to 1.7	0.054 to 0.061	350 to 400	150 to 400	10 to 18	G to E	F	F	P	G to E
Cold press molding-polyester	20 to 30	H40 to 105	300 to 600	1.5 to 1.7	0.054 to 0.061	350 to 400	150 to 400	10 to 18	G to E	F	F	P	G to E
Spray-up polyester	30 to 50	H40 to 105	200 to 400	1.4 to 1.6	0.050 to 0.058	350 to 400	150 to 350	12 to 20	G to E	F	F	P	G to E
Filament wound-epoxy	30 to 80	M98 to 120	300 to 400	1.7 to 2.2	0.061 to 0.079	350 to 400	500	2 to 6	E	F	E	G	E
Rod stock-polyester	40 to 80	H80 to 112	200 to 400	1.6 to 2.0	0.058 to 0.072	325 to 375	150 to 500	3 to 8	G to E	F	F	F	G to E
Molding compound-phenolic	5 to 25	M90 to 99	150 to 370	1.7 to 1.9	0.061 to 0.069	400 to 500	325 to 350	4.5 to 9	F	P	F	P	F
Glass-fiber-reinforced thermoplastics													
Acetal	20 to 40	M78 to 94	500 to 600	1.55 to 1.69	...	315 to 335	185 to 220	19 to 35	F	P	F	P	E
Nylon	6 to 60	...	400 to 500	1.47 to 1.7	0.049	300 to 500	300 to 400	11 to 21	G	P	E	F	G
Polycarbonate	20 to 40	M75 to 100	450	1.24 to 152	...	285 to 300	275	12 to 18	E	G(p)	G	F	P(q)
Polyethylene	10 to 40	...	450 to 500	1.16 to 1.28	...	150 to 200	280 to 300	17 to 27	E	G(p)	E	E	G(r)
Polypropylene	20 to 40	R95 to 115	500 to 600	1.04 to 1.22	...	230 to 300	300 to 320	16 to 24	E	G(p)	E	E	G(r)
Polystyrene	20 to 35	M70 to 95	350 to 425	1.20 to 1.29	0.045 to 0.048	200 to 220	180 to 200	17 to 22	E	G(p)	G	G	P(q)
Polysulfone	20 to 40	M85 to 92	...	1.38 to 1.55	...	330 to 350	...	12 to 17	E	E(p)	E	E	G
ABS (acrylonitrile butadiene styrene)	20 to 40	M75 to 102	...	1.23 to 1.38	...	215 to 240	200 to 230	16 to 20	E	G	E	E	P(s)
PVC (polyvinyl chloride)	15 to 35	M80 to 88	500 to 550	1.45 to 1.62	...	155 to 165	...	12	E	G	E	E	P(s)
Polyphenylene oxide (modified)	20 to 40	M95	...	1.20 to 1.38	...	220 to 315	240 to 265	10 to 20	E	F	E	E	G(t)
SAN (styrene acrylonitrile)	20 to 40	M77 to 103	...	1.22 to 1.40	...	210 to 230	200 to 220	16 to 21	G	G(u)	G	G	P(s)
Thermoplastic polyester	20 to 35	R118 to M70	560 to 750	1.45 to 1.61	...	380 to 470	275 to 375	24 to 33	F	P	P	P	F
Unreinforced thermoplastics													
Acetal	N.Ap.	M78 to 94	465 to 500	1.42	0.052	230 to 255	185 to 220	45	F	P	F	P	E
Nylon	N.Ap.	R108 to 118	300 to 470	1.12 to 1.14	0.039 to 0.041	120 to 150	250 to 300	55 to 63	G	P	E	F	G
Polycarbonate	N.Ap.	M70	400 to 425	1.20	0.043	265 to 290	275	39	E	G(p)	G	F	P(q)
Polyethylene (high density)	N.Ap.	...	450 to 500	0.95	...	100 to 130	180 to 230	6	E	G(p)	E	E	P(q)
Polypropylene	N.Ap.	R50 to 110	500 to 600	0.9	...	125 to 140	190 to 240	38	E	G(p)	E	E	P(q)
Polystyrene (high impact)	N.Ap.	M12 to 45	300 to 600	1.05	0.039	175 to 205	150 to 180	22 to 56	E	G(p)	G	G	P(q)
Polysulfone	N.Ap.	M69 to R12.0	425	1.24	...	345	300 to 345	31	E	E	E	E	G
ABS (high heat)	N.Ap.	R113	350 to 500	1.05	...	215 to 245	190 to 230	41 to 52	G	G(u)	G	G	P(s)
PVC	N.Ap.	D80	...	1.4	...	155 to 165	...	...	E	E	E	E	P(s)
Polyphenylene oxide (modified)	N.Ap.	M75	...	1.06	...	375	...	30	E	E	E	E	G(v)
SAN	N.Ap.	M80	400 to 500	1.08	...	190 to 220	140 to 205	36	G	G(u)	G	G	P(s)
Metals													
Gray cast iron	N.Ap.	B93	C	7.19	0.26	N.Ap.	N.Av.	6					
Low-carbon steel (cold rolled)	N.Ap.	B72	C	7.8	0.28	N.Ap.	N.Av.	6 to 8					
Stainless steel	N.Ap.	B90	C	7.92	0.29	N.Ap.	N.Av.	9 to 10					
Aluminum, wrought	N.Ap.	B1 to 85	C	2.6 to 2.8	0.10	N.Ap.	N.Av.	12 to 13					
Aluminum, die cast	N.Ap.	E59	C	2.57 to 2.96	0.09	N.Ap.	N.Av.	12 to 13					
Magnesium, die cast	N.Ap.	E50 to 59	C	1.81	0.07	N.Ap.	N.Av.	14 to 16					
Zinc, die cast	N.Ap.	B44	C	6.6	0.24	N.Ap.	N.Av.	15 to 16					
Brass, plain yellow wrought	N.Ap.	F58 to 64	C	8.5	0.31	N.Ap.	N.Av.	11 to 12					

(a) Data from Owens-Corning Fiberglas. N.Av. = not available; N.Ap. = not applicable. (b) To convert to MPa, multiply listed values by 6.894757. (c) To convert to GPa, multiply listed values by 0.6894757. (d) To convert to J/cm of notch, multiply listed values by 0.5338. (c) To convert to W/m · K, multiply listed values by 0.14423. (f) To convert to kJ/kg · K, multiply listed values by 4.1868. (g) Classification shown is highest obtainable rating. Less-critical applications may permit use of materials with lower classifications. (h) To convert to kV/mm, multiply listed values by 0.03937. (j) To convert to g/cm^3, multiply listed value by 27.68. (k) To convert to °C, subtract 32 from listed values, then divide by 1.8. (m) To convert to 10^{-6}/°C, multiply listed values by 1.8. (n) E, excellent (outstanding); G, good (acceptable); F, fair (test before using); P, poor (not recommended). (p) Attacked by oxidizing acids. (q) Soluble in aromatic and chlorinated hydrocarbons. (r) Below 80 °C (176 °F). (s) Soluble in ketones and esters, and in aromatic and chlorinated hydrocarbons. (t) Soften in some aromatic and chlorinated aliphatics. Resistant to alcohol. (u) Disintegrates in sulfuric acid. (v) Dissolves or swells in some aromatic and chlorinated aliphatics. Resistant to alcohol

Resin-dependent properties of graphite composites

Resin type	Continuous use temperature °C	°F	Maximum use temperature °C	°F	Interlaminar shear strength MPa	ksi	Comments
Thermosets							
Epoxy, 120 °C (250 °F) cure	70	160	105	225	55 to 103	8 to 15	690 kPa (100 psi) mold pressure
Epoxy, 175 °C (350 °F) cure	120	250	150	300	103	15	690 kPa (100 psi) mold pressure
Polyimide, 315 °C (600 °F) cure	290	550	370	700	103	15	690 kPa (100 psi) mold pressure with postcure
Phenolic	260	500	315	600	...	...	...
Thermoplastic							
Polysulfone	150	300	175	350	97	14	Moisture resistant
Polyphenylsulfone	180	360	205	400	97	14	Moisture resistant

Comparative properties of 30% glass reinforced electrical-grade plastics

Property	DAP	PBT	PET	Nylon	PPS
Comparative tracking index	>650	230	260	>600	200
Arc resistance, s	185	126	117	131	34
Heat-deflection temperature, °C (°F)	230 (450)	195 (380)	220 (430)	210 (410)	260 (500)
Tensile creep modulus, GPa (10^6 psi)	3.14 (0.455)	1.14 (0.165)	2.21 (0.320)	1.45 (0.210)	2.59 (0.375)
Flexural modulus, GPa (10^6 psi)	15.9 (2.3)	7.6 (1.1)	9.7 (1.4)	5.5 (0.8)	11.7 (1.7)
High volt track rate	0.0	7.8	0.8	0.3	7.1
High amp ignition	>300	15	77	>200	>200

Typical E-glass laminate thermal properties

Material	Glass transition temperature(a) °C	°F	Solder blister, seconds at 290 °C (550 °F) (a)	Flammability(a)
Fr -4 epoxy	125 to 135	235 to 275	175 to 200	V-0
Polyfunctional FR -4	140 to 150	285 to 300	225 to 250	V-0
High-temperature, one-component epoxy system	170 to 180	340 to 355	300 to 325	V-0
Bismaleimide triazine epoxy	180 to 190	355 to 375	300 to 325	V-0
Polyimide epoxy	250 to 260	480 to 500	300 to 325	V-0
Cyanate ester	240 to 250	465 to 480	750 to 800	V-0
Polyimide	>260	>500	>1200	V-1
PTFE, melting point	327	620	>1200	V-0

(a) 1.6 mm (0.062 in.) laminate, 40% resin content

Typical E-glass laminate chemical properties

Material	Methylene chloride absorption(a), % weight gain	Water absorption(b)(c)
FR-4 epoxy	7.0 to 8.0	1.10 to 1.20
Polyfunctional FR-4	3.0 to 4.0	1.10 to 1.20
High-temperature, one-component epoxy system	5.0 to 6.0	0.40 to 0.60
Bismaleimide triazine epoxy	1.0 to 2.0	0.80 to 0.90
Polyimide epoxy	0.10 to 0.30	1.15 to 1.25
Cyanate ester	0.40 to 0.60	0.60 to 0.70
Polyimide	0.05 to 0.25	1.40 to 1.50
PTFE	Not available	0.20 to 0.30

(a) 0.13 mm (0.005 in.) laminate made with two-ply style. (b) 1.6 mm (0.062 in.) laminate. (c) Conditioning with immersion in distilled water, for 24 h, at 100 °C (212 °F)

Typical E-glass laminate electrical properties

Material	Dielectric constant at 1 MHz(a)(b)	Dissipation factor at 1 MHz(a)(b)	Electric field strength(c) kV/mm	Electric field strength(c) V/mil	Dielectric breakdown(c) kV
FR-4 epoxy	4.10 to 4.20	0.028 to 0.030	48 to 56	1200 to 1400	70 to 75
Polyfunctional FR-4	4.10 to 4.20	0.028 to 0.030	48 to 56	1200 to 1400	70 to 75
High-temperature, one-component epoxy system	4.45 to 4.55	0.020 to 0.022	36 to 44	900 to 1100	70 to 75
Bismaleimide triazine epoxy	3.85 to 3.95	0.011 to 0.013	48 to 56	1200 to 1400	70 to 75
Polyimide epoxy	4.00 to 4.10	0.011 to 0.013	48 to 56	1200 to 1400	70 to 75
Cyanate ester	3.50 to 3.60	0.0045 to 0.0065	32 to 40	800 to 1000	65 to 70
Polyimide	3.95 to 4.05	0.008 to 0.010	48 to 56	1200 to 1400	70 to 75
PTFE	2.45 to 2.55(d)	0.001 to 0.003(d)	32 to 40	800 to 1000	40 to 45

(a) Two-ply laminate at 57% resin content. (b) Conditioning with humidity, for 24 h, at 23 °C (73 °F). (c) Conditioning with immersion in distilled water, for 48 h, at 50 °C (120 °F). (d) Laminate at 73% resin content

Typical E-glass laminate mechanical properties

Material	1-oz copper adhesion, ambient(a) kg/m	1-oz copper adhesion, ambient(a) lb/in.	1-oz copper adhesion(a)(b) kg/m	1-oz copper adhesion(a)(b) lb/in.	Flexural aging(c)
FR-4 epoxy	180 to 195	10.0 to 11.0	150 to 170	8.5 to 9.5	24 to 48
Polyfunctional FR-4	140 to 160	8.0 to 9.0	107 to 125	6.0 to 7.0	48 to 72
High-temperature, one-component epoxy system	160 to 180	9.0 to 10.0	140 to 160	8.0 to 9.0	168 to 192
Bismaleimide triazine epoxy	90 to 107	5.0 to 6.0	80 to 100	4.5 to 5.5	168 to 192
Polyimide epoxy	150 to 170	8.5 to 9.5	125 to 140	7.0 to 8.0	Not available
Cyanate ester	90 to 107	5.0 to 6.0	80 to 100	4.5 to 5.5	288 to 312
Polyimide	140 to 160	8.0 to 9.0	140 to 160	8.0 to 9.0	>6000
PTFE	180 to 215	10.0 to 12.0	140 to 180	8.0 to 10.0	Not available

(a) Tested on multilayer laminates. (b) Conditioning at temperature for 1 h, at 125 °C (255 °F). (c) Hours at 200 °C (390 °F) to lose 50% of original flexural strength

Typical properties of some laminates with potential for high-speed applications

Laminate reinforcement/resin	Supplier	Relative permittivity	Loss tangent	Water uptake mg	Water uptake gr
E-glass/epoxy	Many	4.7	0.021	10	0.15
E-glass/BT-blend	Many	4.5	0.010	15	0.25
E-glass/polyimide	Many	4.5	0.018	25	0.40
E-glass/cyanate ester	Nelco	3.9	0.003		
E-glass/XU71787	Norplex	3.8	0.003	6	0.09
S-glass/XU71787	Norplex	3.6	0.003	6	0.09
Quartz/polyimide	Mica	3.6	0.010	25	0.40
Kevlar/epoxy	DuPont	3.7	0.030	10	0.15
Kevlar/polyimide	DuPont	3.6	0.008	25	0.40
e-PTFE/epoxy	Gore	2.8	0.012	10	0.15
e-PTFE/polyimide	Gore	2.8	0.010	25	0.40
e-PTFE/XU71787	Gore	2.4	0.003	6	0.09
Ceramic/PTFE	Rogers	2.9	0.0012	1	0.015
Nonwoven glass/PTFE	Rogers	2.2	0.0008	1	0.015

Thermoplastics

General

Properties of long-glass-fiber-reinforced thermoplastic composites

Matrix	Reinforcement	Composite efficiency	Tensile strength, MPa	Modulus of elasticity, GPa	Impact strength, J/cm
PET	Twill weave	0.37	220.0	13.4	9.7
PP	Twill weave	0.45	270.0	16.2	11.1
PP	Noncrimp fabric	0.24	275.1	7.6	...
PP	Long chopped	0.19	120.0	6.3	3.0
PP	Swirl mat	0.19	120.0	5.0	7.8

Applications for glass fiber reinforced plastics

Industry	Advantages of FRP	Applications
Automotive	High-volume production; fine finishes, reduced costs	Automobile body components; fender extenders; front ends; headlamp and taillamp housings; hoods; spoilers, instrument panels; shift consoles; under-the-hood components; truck hoods; fenders; cab and body components; insulated tanks; engine covers; housings; fender liners
Agricultural	Ruggedness; corrosion resistance	Farm tractor hoods, grilles, instrument housings, seating, fenders; garden tractor and lawn mower bodies and housings; fertilizer and pesticide tanks and sprayers; feed troughs
Appliances	Ability to produce complex molded parts without fasteners or welds	Room air conditioner cases, base pans, bulkheads; condenser and compressor fans; humidifier cases and blower wheels; dishwasher pump bodies; dryer ducts; home laundry tubs; water softener tanks, controls, piping; fan housings; gears; vacuum cleaner housings; iron handles; soap dispensers; microwave oven cook trays; television swivel stands; sump pump bases
Aviation/aerospace	High strength with light weight	Aircraft interior components for passengers and cargo; wing tips; antenna components; radomes; wing fuel tanks; ducting; rocketmotor cases; nozzles; nose cones; pressure vessels; instrument housings; launch tubes
Business machines	Excellent surface finish and dimensional stability at elevated temperatures; high strength	Machine covers and housings; access panels; keyboard caps; keys; printer heads; gears, cams, and levers; frames; mounting panels; printed circuit boards; fans and blowers
Chemical processing	Corrosion resistance	Chemical and fuel tanks, pipes, and ducting; storage tanks and hoppers; process pump and valve bodies, casings, impellers; pressure vessels; filters; fume-collection hoods and duct systems; scrubbing towers; electroplating racks and handling equipment; photographic processing equipment
Construction	Ruggedness; moderate cost; good appearance	Structural shapes; paneling; siding; skylighting; curtain wall components; glazing panels; patio covers; concrete pouring forms
Electrical/electronic	High dielectric strength with low moisture absorption	Electrical pole line hardware, crossarms, strain insulators, standoffs, brackets; shatterproof street lighting globes; switch control rods; hot sticks; electronic components; housings and backboards; utility line maintenance equipment
House and home	Beauty; low maintenance; low cost	Architectural components; appliance and equipment components; furniture–chairs, tables, lawn furnishings; sinks; bathroom tub/shower units; skylights
Marine	Ease of repair; low maintenance; high performance	Pleasure, commercial, and military boat hulls and superstructures; barge covers; lighters; fuel tanks; water tanks; masts and spars; bulkheads; duct work; ventilation cowls; marker buoys; floating docks; outboard engine shrouds
Materials handling	High strength with light weight	Tote trays and bins; food-processing and delivery trays, boxes, bins; tanks and pipes; conveyor-system components; pallets and skids; cargo-handling equipment
Recreational	Low maintenance; good appearance	Motor homes; travel trailers; truck campers; camping trailers; pickup covers; water and snow skis; surfboards; golf clubs; hockey sticks; lacrosse sticks; archery bows; fishing rods; vaulting poles; recreational water craft, canoes; snowmobile and all-terrain vehicle bodies; golf carts; protective helmets; swimming pools; diving boards; playground equipment
Transportation	Toughness; lightness	Railway passenger and freight car components; transport seating; freight car roofs; hopper car covers; refrigerator car liners; air cargo "igloos"; motor truck and bus components; rapid transit car ends; third-rail covers; barges; truck trailer panels; refrigerated truck bodies

Candidate matrix resins for thermoplastic advanced composites

Candidate resin	Company	Polymer	Glass transition temperature °C	Glass transition temperature °F	Melt temperature °C	Melt temperature °F
Semicrystalline						
Victrex PEEK	ICI	Polyether etherketone	143	290	343	650
Ryton	Phillips	Polyphenylene sulfide	88	190	290	555
HTX	ICI	Poly aromatic ketone	205	400	358	675
Ultra Pek	BASF	Polyether ketone Etherketone ketone	172	340	372	700
Liquid crystal						
Xydar	Dartco	Polyester			415	780
Vectra	Celanese	Polyester			415	780
Amorphous thermoplastic						
Udel	Amoco	Polysulfone	190	375	...	...
Victrex PES	ICI	Polyether sulfone	230	445	...	...
Ultem	GE	Polyetherimide	215	420	...	...
Various	Various	Polycarbonate	140 to 150	285 to 300	...	...
PASII	Phillips	Poly aromatic sulfide	204	400	...	...
Pseudothermoplastics						
Avimid KIII	DuPont	Polyimide	255	490	...	...
Terlon	AMOCO	Polyamideimide	275	525	...	...
LARC-TPI	NASA	Polyimide	255	490	...	...

Properties of selected resins filled with mica(a)

Property	Polypropylene homopolymer(b) Unfilled	Polypropylene homopolymer(b) 45% mica(c)	Nylon 6/6(d) Unfilled	Nylon 6/6(d) 40% mica(e)	PBT(f) Unfilled	PBT(f) 40% mica(g)	PET(h) Unfilled	PET(h) 40% mica(j)
Tensile strength:								
MPa	33.1	46.9	77.9(k)	86.2(k)	50.3	68.3	71.7	79.3
ksi	4.8	6.8	11.3(k)	12.5(k)	7.3	9.9	10.4	11.5
Flexural strength:								
MPa	25.5	71.0	58.6(k)	137.9(k)	62.7	110.3	90.3	117.9
ksi	3.7	10.3	8.5(k)	20.0(k)	9.1	16.0	13.1	17.1
Flexural modulus:								
GPa	1.241	7.584	3.054(k)	9.377(k)	2.392	11.58	3.116	17.31
10^6 psi	0.180	1.100	0.443(k)	1.360(k)	0.347	1.680	0.452	2.510
Heat-distortion temperature at 1.82 MPa (284 psi):								
°C	56	116	67	201	53	169	80	218
°F	133	241	153	394	127	336	176	424
Izod impact resistance at 23 °C (73 °F):								
Unnotched								
J/cm	No break	2.6	...	...	No break	1.7	3.9	1.0
ft · lbf/in.	No break	4.9	...	...	No break	3.1	7.3	1.9
Notched								
J/cm	0.24	0.26	0.43(k)	0.28(k)	...	...	0.28	0.24
ft · lbf/in.	0.45	0.49	0.80(k)	0.53(k)	...	...	0.52	0.45

(a) Data from Marietta Resources International. Micas are all phlogopite grades from Marietta Resources. (b) Profax 6523, Hercules, Inc. (c) Suzorex 200 QX. (d) Zytel 101, Du Pont. (e) Suzorex 325-PO. (f) Valox 310, General Electric. (g) Suzorite 325 HK. (h) Eastman 7352. (j) Suzorite 605. (k) Dry

Comparative properties of carbon- and glass-reinforced engineering thermoplastics

Resin type	Tensile strength MPa	Tensile strength ksi	Flexural modulus GPa	Flexural modulus 10^6 psi	Impact strength, notched/unnotched J/cm	Impact strength, notched/unnotched ft · lbf/in.	Heat-deflection temperature °C	Heat-deflection temperature °F
Amorphous								
Acrylonitrile-butadiene-styrene (ABS):								
30% glass fiber	100	14.5	7.6	1.1	0.75/3.5	1.4/6.5	105	220
30% carbon fiber	130	18.8	12.4	1.8	0.59/2.4	1.1/4.5	105	220
Nylon:								
30% glass fiber	148	21.5	7.9	1.15	0.64/3.7	1.2/7.0	140	285
30% carbon fiber	207	30	15.2	2.2	0.64/4.3	1.2/8.0	145	290
Polycarbonate:								
30% glass fiber	128	18.5	8.3	1.2	2.0/9.34	3.7/17.5	150	300
30% carbon fiber	165	24	13.1	1.9	0.96/5.34	1.8/10.0	150	300

(continued)

Comparative properties of carbon- and glass-reinforced engineering thermoplastics (continued)

Resin type	Tensile strength MPa	ksi	Flexural modulus GPa	10^6 psi	Impact strength, notched/unnotched J/cm	ft · lbf/in.	Heat-deflection temperature °C	°F
Polyetherimide:								
30% glass fiber	197	28.5	8.6	1.25	0.75/5.60	1.4/10.5	215	420
30% carbon fiber	234	34	17.2	2.5	0.75/6.67	1.4/12.5	215	420
Polyphenylene oxide (PPO):								
30% glass fiber	145	21	9.0	1.3	1.2/5.1	2.3/5.1	155	310
30% carbon fiber	159	23	11.7	1.7	0.53/3.0	1.0/5.6	155	310
Polysulfone:								
30% glass fiber	124	18	8.3	1.2	0.96/7.5	1.8/14	185	365
30% carbon fiber	159	23	14.5	2.1	0.64/3.5	1.2/6.5	185	365
Styrene-maleic-anhydride (SMA):								
30% glass fiber	103	15	9.0	1.3	0.59/2.4	1.1/4.5	120	250
Thermoplastic polyurethane:								
30% glass fiber	57	8.2	1.3	0.19	5.1/15	9.5/28	170	340
Crystalline								
Acetal:								
30% glass fiber	134	19.5	9.7	1.4	0.96/4.8	1.8/9.0	165	325
20% carbon fiber	81	11.8	9.3	1.35	0.53/1.6	1.0/3.0	160	320
Nylon 66:								
30% glass fiber	179	26	9.0	1.3	1.5/11	2.9/20	255	490
30% carbon fiber	241	35	20.0	2.9	0.80/6.4	1.5/12	257	495
Polybutylene terphthalate (PBT):								
30% glass fiber	134	19.5	9.7	1.4	1.4/9.1	2.6/17	210	410
30% carbon fiber	152	22	15.9	2.3	0.64/3.5	1.2/6.5	210	410
Polythylene terephthalate (PET):								
30% glass fiber	159	23	9.0	1.3	1.0/···	1.9/···	225	435
Polyphenylene sulfide (PPS):								
30% glass fiber	138	20	11.0	1.6	0.75/4.5	1.4/8.5	260	500
30% carbon fiber	186	27	16.9	2.45	0.59/2.9	1.1/5.5	265	505

Interlaminar fracture toughness of thermoplastic composites as determined by the double-cantilever-beam test

Material	Process conditions °C/kPa	°F/psi	Interlaminar fracture toughness), G_{Ic} J/m²	ft · lbf/ft²
Polysulfone (Udel)	340/4150	650/600	1175	80
Polyetherimide (Ultem)	400/4850	750/700	950	65
Polyamideimide (Torlon)	340/1400	650/200	1050	70
Polyphenylene sulfide (Ryton)	340/1400	650/200	720	50
Polyether etherketone (PEEK)	400/700	750/100	1600	110

Design properties of carbon-fiber-reinforced thermoplastic composites(a)

Base resin	PAN carbon fiber content, %	Physical properties: Specific gravity(b)	Water absorption,(c) %	Mold shrinkage(d), in./in.	Mechanical properties: Tensile strength ksi(e)	Shear strength ksi(f)	Flexural strength ksi(g)	Flexural modulus 10^6 psi(h)	Izod impact strength(j), ft · lbf/in.(k) Notched	Unnotched
Nylon 6/6	10	1.18	0.80	0.0040	20.0	...	30.0	1.00	1.0	6.0
	20	1.23	0.60	0.0025	28.0	12.0	42.0	2.40	1.1	8.0
	30(x)	1.28	0.50	0.0020	35.0	13.0	51.0	2.90	1.5	12.0
	40	1.34	0.40	0.0020	40.0	14.0	60.0	3.40	1.6	10.0
	30(y)	1.38	0.48	0.0025	30.5	12.5	43.0	2.30	1.4	13.0
	30(z)	1.36	0.45	0.0025	27.0	11.5	36.0	1.75	1.4	11.0
	30(aa)	1.27	0.80	0.0020	32.0	12.5	46.5	2.50	2.1	14.0
Nylon 6	30	1.28	0.80	0.0020	32.0	12.5	46.0	2.40	1.8	13.0
Nylon 6/12	30	1.22	0.15	0.0020	29.0	12.0	42.0	2.30	1.8	13.0
Nylon 6/10	30	1.23	0.12	0.0020	28.0	12.0	40.5	2.20	1.8	14.0
Supertough nylon	30	1.22	0.35	0.0025	24.0	11.0	34.0	2.00	3.0	19.0
Amorphous nylon	30	1.27	0.12	0.0015	30.0	12.5	47.5	2.20	1.2	8.0
Polycarbonate	30	1.33	0.08	0.0015	24.0	10.0	36.0	1.90	1.8	10.0
Polysulfone	30	1.37	0.15	0.0015	23.0	9.5	32.0	2.05	1.2	6.5
Polyethersulfone	30	1.48	0.30	0.0015	26.0	10.5	37.0	2.05	1.0	5.5
	30(y)	1.57	0.20	0.0020	23.5	9.3	33.5	1.90	1.0	4.8
PEEK	15(y)	1.46	0.10	0.0020	19.5	7.8	24.0	1.50	0.9	9.0
	30	1.39	0.10	0.0010	31.0	12.4	36.0	2.20	1.2	12.0
Acetal	20	1.46	0.50	0.0050	11.8	8.2	13.7	1.35	1.0	3.0
Polyester (PBT)	30	1.41	0.04	0.0020	22.0	8.0	29.0	2.30	1.2	6.5
Polyphenylene sulfide	30	1.45	0.04	0.0010	27.0	9.5	34.0	2.45	1.1	5.5
	30(y)	1.57	0.03	0.0015	25.5	8.0	28.0	2.40	0.8	4.5
ETFE(bb)	20	1.72	0.02	0.0025	12.0	6.5	16.0	1.20	5.5	7.5
PVDF	15	1.77	0.02	0.0060	13.5	7.5	18.0	1.15	1.0	7.0

(continued)

Design properties of carbon-fiber-reinforced thermoplastic composites(a) (continued)

Base resin	PAN carbon fiber content	Thermal properties: Heat deflection temperature °F(m)	Thermal conductivity Btu · in./ft^2 · h · °F(n)	Coefficient of linear thermal expansion 10^{-5}/°F(p)	Flammability (UL 94)
Nylon 6/6	10	485	4.0	2.0	HB
	20	495	5.5	1.4	HB
	30(x)	495	7.0	1.1	HB
	40	500	8.5	0.8	HB
	30(y)	490	7.0	1.1	HB
	30(z)	485	7.0	1.1	HB
	30(aa)	495	6.8	1.1	HB
Nylon 6	30	425	7.0	1.0	HB
Nylon 6/12	30	420	6.5	0.9	HB
Nylon 6/10	30	425	6.5	0.9	HB
Supertough nylon	30	410	5.8	1.2	HB
Amorphous nylon	30	290	7.0	1.1	HB
Polycarbonate	30	300	4.9	0.9	VI
Polysulfone	30	365	5.5	0.6	V-0
Polyethersulfone	30	415	6.0	0.8	V-0
	30(y)	410	5.7	0.8	V-0
PEEK	15(y)	600	4.5	1.8	V-0
	30	600	7.1	0.7	V-0
Acetal	20	320	4.6	2.2	HB
Polyester (PBT)	30	430	4.6	0.5	HB
Polyphenylene sulfide	30	505	5.2	0.6	V-0
	30(y)	500	5.4	0.8	V-0
ETFE(bb)	20	435	6.0	1.0	V-0
PVDF	15	300	2.2	2.5	V-0

Base resin	PAN carbon fiber content, %	Tribological properties: Coefficient of friction, Static	Coefficient of friction, Dynamic	Wear factor 10^{-10} in.^{3} · min/ft · lbf · h	Limiting PV, 10 fpm	Limiting PV, 100 fpm	Limiting PV, 1000 fpm	Electrical: surface resistivity(q), Ω/in.^{2}	Chemical resistance (r,σ): Acids	Bases	Solvents
Nylon 6/6	10	0.17	0.21	60	10000	16000	5000	10^7	F	G	E
	20	0.16	0.20	40	19000	25000	7000	1300	F	G	E
	30(t)	0.16	0.20	20	21000	27000	8000	150	F	G	E
	40	0.13	0.18	14	22000	27500	8500	75	F	G	E
	30(u)	0.11	0.15	10	29000	42000	19000	200	F	G	E
	30(v)	0.10	0.11	6	29000	43000	20000	1000	F	G	E
	30(w)	0.18	0.22	22	20000	26500	7000	150	F	G	E
Nylon 6	30	0.18	0.21	30	18000	22000	7500	150	F	G	E
Nylon 6/12	30	0.19	0.23	25	18000	20000	17000	250	F	G	E
Nylon 6/10	30	0.20	0.25	25	18000	21000	7500	250	F	G	E
Supertough nylon	30	0.20	0.25	25	20000	26000	7500	1100	F	G	E
Amorphous nylon	30	0.19	0.24	90	10000	11000	6000	150	F	F	F
Polycarbonate	30	0.18	0.17	85	8000	8500	5500	3500	F	F	P
Polysulfone	30	0.17	0.14	75	8500	8500	6000	200	G	E	P
Polyethersulfone	30	0.17	0.15	80	10000	10000	7000	100	G	E	P
	30(u)	0.13	0.17	40	35000	33000	16000	100	G	E	P
PEEK	15(u)	0.18	0.20	60	42000	40000	22000	5000	E	E	G
	30	0.19	0.13	60	...	...	...	120	E	E	G
Acetal	20	0.11	0.14	40	13000	20000	15000	2000	F	G	E
Polyester (PBT)	30	0.12	0.15	24	18000	22000	10000	500	F-G	P-F	E
Polyphenylene sulfide	30(u)	0.23	0.20	160	12000	20000	10000	250	E	E	E
	30(u)	0.13	0.15	75	27000	35000	30000	150	E	E	E
ETFE(x)	20	0.16	0.18	28	...	...	...	1200	E	E	E
PVDF	15	0.25	0.25	14	15000	11000	<5000	700	E	E	E

(a) Data from LNP Corp. This information is based on experience and is intended for use as a guide only. (b) ASTM D 792. (c) ASTM D 570. (d) ASTM D 955. (e) ASTM D 638. (f) To convert to MPa, multiply listed values by 6.894757. (g) ASTM D 790. (h) To convert to GPa, multiply listed values by 6.894757. (j) ASTM D 256. (k) To convert to J/cm, multiply listed values by 0.5338. (m) To convert to °C, subtract 32 from listed values, then divide by 1.8 (n) To convert to W/m · K, multiply listed values by 0.14423. (p) To convert to 10^{-5}/°C, multiply listed values by 1.8. (q) To convert to Ω/m^2, multiply listed values by 1550. (r) ASTM D 257. (s) E, excellent; G, good; F, fair; P, poor. (t) Grade RC 1006. (u) Plus 15% PTFE lubricant. (v) Plus 15% PTFE/silicone lubricant. (w) Grade RD 1006 HI. (x) Based on Du Pont Tefzel fluoropolymer

Properties of nylon 6/6, nylon 6, and PBT reinforced with glass/mica mixture

Property	Nylon 6/6	Nylon 6	PBT
Resin content, wt %	60	60	60
Mica content, wt %	25	25	25
Glass-fiber content, wt %	15	15	15
Tensile strength, MPa (ksi)	141 (20.5)	141 (20.4)	100 (14.5)
Tensile elongation, %	3.4	3.8	3.1
Flexural strength, MPa (ksi)	197 (28.5)	252 (36.6)	148 (21.5)
Flexural modulus, GPa (10^6 psi)	9.997 (1.450)	11.34 (1.645)	9.653 (1.400)
Impact strength, J/cm (ft · lbf/in.):			
Unnotched	0.43 (0.8)	0.48 (0.9)	0.43 (0.8)
Notched	8.01 (15.0)	6.14 (11.5)	5.87 (11.0)
Heat-distortion temperature at 1.82 MPa (264 psi), °C (°F)	225 (490)	200 (395)	215 (420)
Mold shrinkage, mm/mm (in./in.)	0.006	0.0055	0.0065
Coefficient of thermal expansion, 10^{-5}/ °C (10^{-5}/ °F)	3.8 (2.1)	3.6 (2.0)	3.4 (1.9)

Properties of pitch carbon-fiber-reinforced composites(a)

Property	ASTM test method	Nylon 20 wt% chopped fiber	Nylon 30 wt% chopped fiber	Nylon 40 wt% chopped fiber	Nylon 30 wt% chopped fiber(f)	Polycarbonate 25 wt% chopped fiber	Polypropylene 40 wt% chopped fiber
Mechanical							
Tensile strength	D 638						
MPa		89.6	107	121	138	72.3	31.7
ksi		13.0	25.5	17.5	20.0	10.5	4.60
Elongation, %	D 638	2.5	2.0	1.5	1.9	2.0	1.1
Tensile modulus	D 638						
GPa		10.3	13.8	17.9	17.2	11.4	...
10^5 psi		15.0	20.0	26.0	15.0	16.5	...
Flexural strength	D 790						
MPa		152	179	193	227	68.9	23.4
ksi		22.0	26.0	28.0	33.0	10.0	3.4
Flexural modulus	D 790						
GPa		6.9	10.3	13.8	13.1	...	7.6
10^5 psi		10.0	15.0	20.0	19.0	...	11.0
Izod impact strength, notched(b)	D 256						
J/m		32.0	37.4	42.7	64.1	58.7	37.4
ft · lbf/in.		0.6	0.7	0.8	1.2	1.1	0.7
Izod impact strength, unnotched(b)	D 256						
J/m		...	694	...	...	320	214
ft · lbf/in.		...	13.0	...	...	6.0	4.0
Compressive strength	D 695						
MPa		138	138	158	172	...	...
ksi		20.0	20.0	23.0	25.0	...	...
Shear strength	D 732						
MPa		65.5	65.5	68.9	68.9	...	...
ksi		9.5	9.5	10.0	10.0	...	...
Hardness, Rockwell E	D 785	44	53	54	55	...	...
Deformation under load, %(c)	D 621	0.54	0.38	0.2	...	...	...
Physical							
Specific gravity	D 792	1.24	1.30	1.36	1.40	1.35	1.15
Water absorption in 24 h, %	D 570	0.75	0.6	0.5	0.5	...	0.03
Linear mold shrinkage in 3 mm (1/8 in.), %	...	0.4	0.3	0.2	0.2	0.1	0.1
Thermal							
Coefficient of linear thermal expansion	D 696						
10^{-5}/ °C		2.3	1.6	0.9	1.3	...	...
10^{-5}/ °F		1.3	0.9	0.5	0.7	...	...
Deflection temperature under load(d)	D 648						
°C		232	241	246	243	143	127
°F		459	465	475	470	290	260
Deflection temperature under load(e)	D 648						
°C		254	254	259	254	...	...
°F		490	490	498	490	...	...
Flammability at minimum thickness (UL 94)		HB	HB	HB	HB	V-0	HB
Electrical							
Volume resistivity, Ω · cm	D 257	10^6	10^4	10^3	10^4	10^5	10^2
Surface resistivity, Ω /square	D 257	10^6	10^4	10^3	10^4	10^5	10^2

(a) All tests conducted at 23 °C (73 °F) unless otherwise noted. (b) Izod impact test bars 6.35 × 12.7 mm (1/4 × 1/2 in.). (c) Deformation at 27.6 MPa and 50 °C (4 ksi and 122 °F). (d) At 1.82 MPa (264 psi). (e) At 0.45 MPa (66 psi). (f) Contains 15 wt % glass fiber

Properties of thermoplastic matrix composites

Properties	Aromatic copolyester Resin	Aromatic copolyester 40% glass fiber-resin	PBT Resin	PBT 15 to 40% glass fiber-resin	PET Resin	PET 30 to 45% glass fiber-resin
Heat deflection temperature at 1800 kPa (264 psi), °C (°F)	355 (671)	...	55 (130)	205 (400)	...	225 (435)
UL in-service temperature rating, °C (°F)	240 (464)	...	120 (248)	140 (284)	140 (284)	150 to 180 (302 to 356)
Processing melt temperature, °C (°F)	400 to 450 (750 to 840)	...	270 (520)	...	290 (550)	...
Specific gravity	1.35	1.70	1.31	1.53	...	1.56 to 1.69
Density, g/cm^3 (lb/in.^{3})	1.35 (0.049)	1.70 (0.061)	1.31 (0.047)	1.53 (0.055)	...	156 to 1.69 (0.056 to 0.061)
UL flammability	94 V-0	94 V-0	94 HB/94 V-0	94 HB/94 V-0	94 HB/94 V-0	94 HB/94 V-0

Tensile strengths of filled and unfilled heat-resistant resins after thermal aging at 260 °C (500 °F)(a)

Base resin	Fiberglass content, wt %	Tensile strength, sik (MPa), after aging for: 0 h	100 h	250 h	500 h	750 h	1000 h	1500 h
ETFE	20	11.3 (77.9)	11.5 (79.3)	10.0 (69.0)	7.0 (48.3)	5.0 (34.5)	3.8 (26.2)	2.3 (15.9)
FEP	20	5.0 (34.5)	5.1 (35.2)	4.8 (33.1)	4.7 (32.4)	4.7 (32.4)	4.6 (31.7)	4.5 (31.0)
Polyphenylene sulfide	40	23.2 (160)	16.4 (113)	16.0 (110)	15.5 (107)	15.0 (103)	14.5 (100)	13.8 (95.2)
Polyethersulfone	40	22.7 (157)	15.6 (108)	14.8 (102)	14.3 (98.6)	13.7 (94.5)	12.2 (84.1)	10.5 (72.4)
Polyimide	30	13.0 (89.6)	15.0 (103)	14.3 (98.6)	13.4 (92.4)	12.8 (88.3)	12.0 (82.7)	11.2 (77.2)
Polyamide-imide	0	27.4 (189)	27.2 (188)	26.6 (183)	26.0 (179)	24.5 (169)	23.5 (162)	22.0 (152)
Polyarylsulfone	0	13.1 (90.3)	11.5 (79.3)	10.5 (72.4)	10.0 (69.0)	9.5 (65.5)	8.4 (57.9)	7.6 (52.4)
Poly-*p*-oxybenzoate	0	23.0 (159)	18.0 (124)	16.3 (112)	16.0 (110)	15.5 (107)	15.1 (104)	13.0 (89.6)
Nylon 6/6	50	31.0 (214)	17.6 (121)	10.3 (11.0)	9.4 (64.8)	...	...	...
Polyester	40	22.1 (152)	Melted	...	...	...	...	...
Polysulfone	40	20.3 (140)	Melted	...	...	...	...	...

(a) Tested at 23 °C (73 °F)

Comparison of fatigue behavior of various resins

Base resin	Fiber type(a)	Cyclic failure stress MPa	ksi
SAN	Glass	45	6.5
Styrene	Glass	41	6.0
Polycarbonate	Glass	38	5.5
ETFE copolymer	Glass	24	3.5
	Carbon	42	6.1
Polysulfone	Glass	34	5.0
Polyethersulfone	Carbon	55	8.0
Acetal copolymer	Glass	48	7.0
Polypropylene	Glass	31	4.5
Polyphenylene sulfide	Carbon	66	9.5
Nylon 6/6c(b)	Glass	41	6.0
	Carbon	55	8.0
	Bead	27	3.9
Polyester (PBT)	Glass	39	5.6
	Carbon	51	7.4
Modified PPO	Glass	34	4.9
PEEK	Carbon	121	17.5

(a) Fiber content, 30%. (b) The "c" indicates moisture conditioned at 50% RH

Flexural strength as a function of fiber orientation in filled materials

		Flexural strength			
		Bar 1		Bar 2	
Basin	Filler	MPa	ksi	MPa	ksi
Nylon 6/6	...	110	16.0	115	16.8
Nylon 6/6	15% glass fiber	180	26.6	150	22.2
Nylon 6/6	30% glass fiber	240	34.6	175	25.6
Nylon 6/6	40% glass beads	130	19.5	140	20.3
Nylon 6/6	40% carbon fiber	300	43.0	190	27.9
Polycarbonate	...	100	14.5	105	15.4
Polycarbonate	30% glass fiber	190	27.6	145	21.1

Effect of filled and unfilled resins on weld line integrity

			Tensile strength				
			Single gate		Double gate		Percent
Resin	Reinforced	Filler	MPa	ksi	MPa	ksi	retained
PSU	None	...	65	9.6	65	9.6	100
PSU	30% glass fiber	...	115	16.8	70	10.4	62
SAN	None	...	80	11.3	60	9.0	80
SAN	30% glass fiber	...	110	16.2	45	6.5	40
PP	None	...	35	5.4	35	4.7	86
PP	20% glass fiber	...	65	9.1	30	4.3	47
PP	15% glass fiber	15% glass beads	45	6.5	19	2.7	42
PP	30% glass fiber	10% PTFE	67	9.7	20	2.8	29
PPS	None	...	60	8.8	50	7.3	83
PPS	10% glass fiber	...	70	10.3	25	3.9	38
PPS	40% glass fiber	...	140	20.5	28	4.1	20

Wear and frictional properties of long and short carbon fiber reinforced thermoplastic composites

		Wear factor,	Coefficient of friction	
Matrix resin	Fiber type and content, %	10–10 in.^{3} min./ft · lbf · h	Static 40 psi	Dynamic 40 psi, 50 fpm
Nylon 6/6	20% short carbon fiber	40	0.16	0.20
Nylon 6/6	30% short carbon fiber	20	0.16	0.20
Nylon 6/6	40% short carbon fiber	14	0.13	0.18
Nylon 6/6	40% long carbon fiber	8	0.09	0.09
Polycarbonate	40% short carbon fiber	26	0.16	0.13
Polycarbonate	40% long carbon fiber	17	0.13	0.13

Specific

Physical properties of random glass reinforced acrylamate composites

Property	Value: 30% glass reinforced	40% glass reinforced
Mechanical		
Tensile strength, MPa (ksi)	114 (16.6)	154 (22.3)
Tensile modulus, GPa (10^6 psi)	7.65 (1.11)	8.69 (1.26)
Tensile elongation, %	2.1	2.3
Flexural strength, MPa (ksi)	165 (23.9)	248 (36.0)
Flexural modulus, GPa (10^6 psi)	6.89 (1.00)	8.62 (36.0)
Compressive strength, MPa (ksi)	165 (23.9)	211 (30.6)
Compressive modulus, GPa (10^6 psi)	7.1 (1.03)	11 (1.6)
Izod impact, notched, J/m (ft · lbf/in.)	646 (12.1)	747 (14.0)
Izod impact, unnotched, J/m (ft · lbf/in.)	779 (14.6)	1.12 (21.0)
Shear strength, MPa (ksi)	81 (11.7)	88 (12.8)
Barcol hardness	28	30
Poisson's ratio	0.33	0.34
Thermal		
Deflection temperature, at 1.82 MPa (0.264 ksi), °C (°F)	240 (464)	240 (464)
Coefficient of thermal expansion, 10^{-6}/K	29.3	27.0
Thermal conductivity, W/m · K (Btu · in./h · ft^2 · °F)	0.195 (1.35)	...
Flammability		
Vertical burn, UL 94	HB(a)	HB(a)
Other		
Specific gravity	1.41	1.47
Mold shrinkage	0.002	0.001
Water absorption, %		
At 24 h, 23 °C (74 °F)	0.25	0.25
At equilibrium, 23 °C (74 °F)	0.91	0.80

(a) Fire-retardant versions are available

Physical properties of oriented glass and carbon reinforced acrylamate composites

Property	50% bidirectional glass cloth reinforced	35% carbon 10% glass reinforced
Tensile strength, MPa (ksi)	203 (29.4)	255 (37)
Tensile modulus, GPa (10^6 psi)	14.8 (2.14)	27.4 (3.98)
Tensile elongation, %1.6	1.1	
Flexural strength, MPa (ksi)	350 (50.8)	345 (50.1)
Flexural modulus, GPa (10^6 psi)	13 (1.89	31.2 (4.52)
Izod impact, notched J/m (ft · lbf/in.)	710 (13.3)	523 (9.8)
Izod impact, unnotched J/m (ft · lbf/in.)	977 (18.3)	822 (15.4)
Specific gravity	1.57	1.35

Physical properties of chopped fiber reinforced acrylamate composites

Property	32% reinforcement 0°	32% reinforcement 90°	21% reinforcement 0°	21% reinforcement 90°
Notched Izod impact strength, J/m (ft · lbf/in.)	800 (15)	960 (18)	640 (12)	530 (10)
Heat-deflection temperature at 1.82 MPa (0.264 ksi), °C (°F)	238 (460)	227 (440)	204 (400)	162 (320)
Tensile strength, MPa (ksi)	86.9 (12.6)	82.7 (12.0)	51 (7.4)	47.6 (6.9)
Tensile modulus, GPa (10^6 psi)	7.54 (1.09)	7.17 (1.04)	5.75 (0.834)	5.93 (0.860)
Elongation, %	2.46	1.87	2.09	2.15
Flexural strength, MPa (ksi)	160 (23.2)	144 (20.9)	108 (15.6)	100 (14.4)
Flexural modulus, GPa (10^6 psi)	6.67 (0.967)	5.82 (0.844)	4.78 (0.693)	4.80 (0.696)

Physical properties of acrylamate prepreg composites

Property	43% 120-style glass cloth	60% Kevlar 285	60% unidirectional AS-4 graphite
Flexural strength, MPa (ksi)	530 (76.5)	303 (43.9)	1390 (201)
Flexural modulus, GPa (10^6 psi)	2.14 (0.31)	2.30 (0.335)	11.2 (1.63)
Tensile strength, MPa (ksi)	350 (50.8)	530 (77.1)	...
Tensile modulus, GPa (10^6 psi)	2.30 (0.335)	2.62 (0.38)	...
Compressive strength, MPa (ksi)	514 (74.5)	87.6 (12.7)	...
Compressive modulus, GPa (10^6 psi)	2.28 (0.33)	2.41 (0.35)	...

Environmental resistance of 40% random glass reinforced acrylamate composites

Fluid(a)	Tensile strength retention, % 3 weeks	6 weeks
Water	75	72
10% saltwater	80	70
50% antifreeze	96	98
Motor oil	110	108
Brake fluid	102	97
Automatic transmission fluid	102	109
Methyl ethyl ketone	76	75
Windshield washer solution(b)	101	103
Water(c)	32	...

(a) Temperature of fluid is 65 °C (150 °F) unless otherwise noted. (b) At 20 °C (70 °F). (c) At 100 °C (212 °F)

Mechanical and physical properties of carbon-base filled and reinforced nylon 6/6

Property	Carbon black	10% PAN carbon fiber	20% pitch carbon fiber	15% nickel-coated carbon fiber
Tensile strength, MPa (ksi)	44.1 (6.4)	138 (20.0)	89.6 (13.0)	96.5 (14.0)
Tensile elongation, %	2.0	4.0	2.5	1.6
Flexural strength, MPa (ksi)	77.2 (11.2)	207 (30.0)	152 (22.0)	145 (21.0)
Flexural modulus, GPa (10^6 psi)	2.9 (0.42)	6.8 (0.99)	6.9 (1.0)	6.9 (1.0)
Izod impact, notched, J/m (ft · lbf/ft)	21.4 (4.81)	42.7 (9.60)	32 (7.19)	26.7 (6.00)
Heat-deflection temperature under load, at 1.8 MPa (0.264 ksi), °C (°F)	74 (165)	249 (480)	232 (450)	238 (460)
Specific gravity	1.16	1.18	1.24	1.2
Surface resistance, Ω/square	10^6	10^6	10^6	10
Volume resistance, Ω · m	...	10^4	10^4	10^{-2}
Static decay rate, s	<0.05	<0.05	<0.05	<0.05

Mechanical properties of long and short 50 wt % glass fiber reinforced nylon 6/6 composites

Property	RF-100-10 short glass	Verton RF-700-10 long glass
Tensile strength, ksi	32	38
Tensile modulus, ksi	2500	2900
Flexural strength, ksi	46.5	58
Flexural modulus, ksi	2200	2400
Izod impact strength, ft · lbf/in.		
Notched	2.6	7.0
Unnotched	20.0	35.0
Falling dart impact strength, ft · lbf	6.2	9.2

Mechanical properties of long and short 50 wt % glass fiber nylon 6 composites

Property	PF-100-10HI short glass	Verton® PF-700-10HI long glass
Tensile strength, ksi	28	30.5
Flexural strength, ksi	35	50
Flexural modulus, ksi	1400	2000
Izod impact strength, notched, ft · lbf/in.	4.0	7.0

Instantaneous strength properties of nylon 6/6 composites

Reinforcement	Ultimate tensile strength, ksi	Ultimate tensile elongation, %	Elastic modulus, ksi
None	11.8	60.0	410
33% short glass fiber	27	4.0	1300

Thermochemical properties of nylon-composite samples(a) 3.2-mm (1/8-in.) laminates with various fillers

Composite(b)	Linear mold shrinkage								Coefficient of linear thermal expansion							
	Longitudinal				Transverse at 12.7-mm (0.5-in.) width				Longitudinal				Transverse at 12.7-mm (0.5-in.) width			
	3.2-mm (0.125-in.) sample		6.4-mm (0.25-in.) sample		3.2-mm (0.125-in.) sample		6.4-mm (0.25-in.) sample		3.2-mm (0.125-in.) sample		6.4-mm (0.25-in.) sample		3.2-mm (0.125-in.) sample		6.4-mm (0.25-in.) sample	
	mm	in.	mm	in.	mm	in.	mm	in.	10^{-5}/°C	10^{-5}/°F	10^{-5}/°C	10^{-5}/°F	10^{-5}/°C	10^{-5}/°F	10^{-5}/°C	10^{-5}/°F
100% nylon 6/6	0.3683	0.0145	0.6706	0.0264	0.7010	0.0276	0.8661	0.0341	7.2	4.0	11.2	6.2	14	7.8	14	7.8
10% GF/90% nylon	0.1270	0.0050	0.2616	0.0103	0.3759	0.0148	0.6299	0.0248	5.7	3.2	6.1	3.4	20	11.1	19	10.6
10% CF/90% nylon	0.0559	0.0022	0.1168	0.0046	0.4064	0.0160	0.6655	0.0262	3.5	1.9	2.5	1.4	19	10.6	19	10.6
20% GF/80% nylon	0.0686	0.0027	0.1321	0.0052	0.4013	0.0158	0.6200	0.0244	2.8	1.6	2.9	1.6	21	11.7	19	10.6
20% CF/80% nylon	0.0076	0.0003	0.0610	0.0024	0.3912	0.0154	0.5588	0.0220	1.6	0.9	1.6	0.9	19	10.6	17	9.4
30% GF/70% nylon	0.0356	0.0014	0.0940	0.0037	0.4064	0.0160	0.5385	0.0212	2.4	1.3	2.1	1.2	18	10.0	18	10.0
30% CF/70% nylon	0.0076	0.0003	0.1780	0.0007	0.3810	0.0150	0.5334	0.0210	1.2	0.7	1.1	0.6	22	12.2	15	8.3

(a) Measurement vs. flow direction. (b) GF, glass fibers; CF, carbon fibers

Properties of long and short glass fiber reinforced nylon 6/6 at 300 °F

Property	Reinforcement			
	35% short glass fiber	35% long glass fiber	50% short glass fiber	50% long glass fiber
Tensile strength, ksi	12.9	15.1	13.8	20.3
Tensile elongation, %	9.3	5.3	7.8	5.6
Flexural strength, ksi	13.8	18	14.5	23.7
Flexural modulus, ksi	465	580	527	890

Properties of long and short glass fiber reinforced nylon 6/6 at 400 °F

Property	Reinforcement			
	35% short glass fiber	35% long glass fiber	50% short glass fiber	50% long glass fiber
Tensile strength, ksi	6.8	8.2	7.3	8.8
Tensile elongation, %	8.6	6.2	9.5	6.8
Flexural strength, ksi	7	9.2	7.4	10
Flexural modulus, ksi	426	550	480	750

Surface finish of long and short glass fiber reinforced nylon 6/6 composites(a)

Glass fiber content % short	% long	60-degree gloss measurement	Average roughness, Ra A	B
35	...	40.0	55.3	56.6
...	35	80.2 - 91.0(b)	13.8 - 18.6(b)	12.7 - 17.9(b)
50	...	35.6	88.2	86.4
...	50	68.6 - 83.6(b)	18.6 - 28.6(b)	16.8 - 27.2(b)

(a) All material molded with 200 °F mold temperature. (b) Range due to different grades of nylon 6/6 resins

Static decay rate in s (to complete decay) versus concentration of PAN and pitch carbon fibers in nylon 6/6 matrix

Carbon fiber type	Carbon fiber, wt%					
	5	10	15	20	25	40
PAN	>100	0.05	0.05	0.05	0.05	0.05
Pitch	>100	>100	>100	0.05	0.04	0.05

Mechanical and physical properties of metallic-base filled and reinforced polycarbonate

Property	40% aluminum flake	25% metallized glass fiber	10% stainless steel fiber	10% glass fiber
Tensile strength, MPa (ksi)	44.1 (6.40)	82.7 (12.0)	75.8 (11.0)	66.2 (9.6)
Tensile elongation, %	2.0	2.0	4.0	9.0
Flexural strength, MPa (ksi)	86.1 (12.5)	130 (18.8)	117 (17.0)	103 (14.9)
Flexural modulus, GPa (10^6 psi)	6.5 (0.94)	6.2 (0.90)	3.4 (0.49)	3.4 (0.49)
Izod impact, notched, J/m (ft · lbf/ft)	69.4 (15.6)	80.1 (18.0)	74.8 (16.8)	80.1 (18.0)
Heat-deflection temperature under load, at 1.8 MPa (0.264 ksi), °C (°F)	143 (290)	143 (290)	141 (285)	142 (288)
Specific gravity	1.54	1.4	1.35	1.25
Surface resistance, Ω/square	10	...	...	10^{13}
Volume resistance, Ω · m	10^{-1}	10^4	10^{-2}	...
Static decay rate, s	<0.05	<0.05	<0.05	<0.05

Comparison of injection-moldable polycarbonates for EMI shielding

Property	Unfilled	Proprietary system(a)	25% carbon fiber	25% metallized glass	30% Al flake
Tensile strength					
MPa	62	55	138	82.7	41
ksi	9.0	8.0	20.0	12.0	6.0
Elongation, %	6 to 8	7	1.1	2.0	2.9
Izod impact					
J/cm	1.3	0.9	1.1	0.9	1.1
ft · lbf/in.	2.4	1.7	2.0	1.6	2.0
Flexural strength					
MPa	89.6	93.1	193	145	75.8
ksi	13.0	13.5	28.0	21.0	11.0
Flexural modulus					
GPa	2.3	2.3	13.8	6.8	4.4
10^6 psi	0.33	0.33	2.00	0.98	0.64
Deflection temperature under load(b)					
°C	132	142	146	143	141
°F	270	288	295	290	285
Specific gravity	1.20	1.27	1.31	1.40	1.44
Attenuation at 1.0 GHz, dB	0	40	30	20	30

(a) Conductive additives include stainless steel fibers. (b) At 1.82 MPa (264 psi)

Electrical properties of isophthalic polyester(a) 3.2-mm (1/8-in.) laminates with various fillers

Material	Dielectric strength short time V/μm	Dielectric strength short time V/mil	Volume resistivity, 10^{-13} Ω · m	Dielectric constant, 1 MHz	Dissipation factor, 1 MHz	Dielectric constant, 1 kHz	Dissipation factor, 1 kHz	Dielectric constant, 60 Hz	Dissipation factor, 60 Hz	Arc resistance avg	Arc resistance max	Arc resistance min	Track resistance, V	Dielectric breakdown short time, kV	Dielectric breakdown step-by-step, kV
Calcium carbonate	15.0	380	7.8	4.10	0.007	4.18	0.005	4.19	0.003	157	181	140	840	58	61
Gypsum $CaSO_4$	14.4	365	2.1	3.69	0.011	4.04	0.023	4.19	0.027	153	184	141	840	70	55
Alumina trihydrate	15.4	390	2.6	3.67	0.009	3.81	0.010	3.89	0.011	183.5	184	183	860	67	51
Clay	14.4	365	6.4	4.08	0.018	4.61	0.040	5.10	0.057	182.5	183	182	840	59	57

(a) Vinyl toluene monomer

Electric properties of BPA fumarate polyester(a) 3.2-mm (1/8-in.) laminates with various fillers

Material	Dielectric strength short time V/μm	Dielectric strength short time V/mil	Volume resistivity, 10^{-13} μ · m	Dielectric constant, 1 MHz	Dissipation factor, 1 MHz	Dielectric constant , 1 kHz	Dissipation factor, 1 kHz	Dielectric constant, 60 Hz	Dissipation factor, 60 Hz	Arc resistance avg	Arc resistance max	Arc resistance min	Track resistance, V	Dielectric breakdown short time, kV	Dielectric breakdown step-by-step, kV
Calcium carbonate	6.1	155	1.6	3.94	0.005	4.00	0.004	4.03	0.004	140	143	133	840	58	52
Gypsum $CaSO_4$	5.9	150	3.3	3.72	0.009	4.03	0.024	4.24	0.029	144	151	137	820	50	40
Alumina trihydrate	11.8	300	3.3	3.64	0.008	3.81	0.015	3.93	0.025	182	184	181	820	55	52
Clay	12.6	320	3.5	4.08	0.023	4.68	0.043	5.11	0.053	183	184	181	840	61	43

(a) Vinyl toluene monomer

Electrical properties of fiberglass-polyester composites

Property	Value
Volume resistivity, 50% relative humidity, Ω · m	10^{10} to 10^{13}
Dielectric strength	
Short time, 3.2 mm, or 1/8 in., kV/mm (kV/in.)	13.6 to 16.5 (345 to 420)
Step-by-step, 3.2 mm, or 1/8 in., kV/mm (kV/in.)	10.8 to 15.4 (275 to 390)
Dielectric constant	
at 60 Hz	5.3 to 7.3
at 1 kHz	4.68
at 1 MHz	5.2 to 6.4
Dissipation factor	
at 60 Hz	0.011 to 0.41
at 1 MHz	0.008 to 0.022
Arc resistance, s	120 to 200

Effect of filler and glass-fiber reinforcement on mechanical properties of polyester resins

Physical property	Neat resin casting (A material)	80 pph A, 20 pph $CaCo_3$ (B material)	74 pph B, 26 pph 2.5 cm long chopped strand
Flexural strength, MPa (ksi)	129 (18.7)	109 (15.8)	183 (26.5)
Flexural modulus, GPa (10^6 psi)	3.6 (0.52)	4.3 (0.62)	6.1 (0.88)
Tensile strength, MPa (ksi)	69.6 (10.1)	52 (7.5)	116 (16.8)
Tensile modulus, GPa (10^6 psi)	3.5 (0.51)	5.6 (0.81)	9.7 (1.4)
Tensile elongation, %	3.0	1.26	1.72
Barcol hardness	30	45	45
Heat deflection temperature (for 0.25 mm, or 10 mil deflection at 1.82 MPa, or 0.264 ksi), °C (°F)	58 (135)	67 (150)	>260 (500)

pph, parts per hundred

Effect of glass type and amount on unsaturated polyester

Type of glass fiber reinforcement	Wt%	Tensile strength MPa	Tensile strength ksi	Tensile modulus GPa	Tensile modulus 10^6 psi	Elongation, %	Flexural strength MPa	Flexural strength ksi	Flexural modulus GPa	Flexural modulus 10^6 psi	Compressive strength MPa	Compressive strength ksi	Density, g/cm^3
Neat cured resin	0	59	8.5	5.4	0.78	2.0	87.6	12.7	3.9	0.56	156	22.6	1.22
Chopped-strand mat	30	117	16.9	10.8	1.56	3.5	197	28.5	9.72	1.41	147	21.3	1.50
Chopped-strand mat	50	288	41.7	16.7	2.42	3.5	197	28.5	14.5	2.10	160	23.2	1.70
Roving fabric	60	314	45.5	19.4	2.82	3.6	316	45.9	15.0	2.17	192	27.8	1.76
Woven glass fabric	70	330	47.9	25.9	3.75	3.4	403	58.4	17.4	2.52	280	40.6	1.88
Unidirectional roving fabric	70	610	88.5	32.5	4.72	2.8	403	58.4	29.4	4.27	216	31.3	1.96

Effect of glass content on mechanical properties of fiberglass-polyester resin composites

Polyester resin	Glass content, %	Flexural strength MPa	Flexural strength ksi	Flexural modulus GPa	Flexural modulus 10^6 psi	Tensile strength MPa	Tensile strength ksi	Tensile modulus GPa	Tensile modulus 10^6 psi	Compressive strength MPa	Compressive strength ksi
Orthophthalic	30	172	25	5.5	0.80	138	20	4.8	0.70	...	...
	40	221	32	6.89	1.00	152	22	5.5	0.80	...	...
Isophthalic	30	193	28	5.5	0.80	152	22	8.27	1.20	...	...
	40	241	35	7.58	1.10	193	28	11.7	1.70	210	30
BPA fumarate	25	117	17	5.1	0.74	83	12	7.58	1.10	170	24
	35	152	22	8.27	1.20	97	14	10.3	1.50	170	24
	40	159	23	8.96	1.30	124	18	11.0	1.60	180	26
Chlorendic	24	117	17	5.9	0.85	76	11	7.58	1.10	140	21
	34	159	23	6.89	1.00	124	18	9.65	1.40	120	18
	40	193	28	9.65	1.40	138	20	9.65	1.40	120	18

Properties of fiberglass-reinforced polyester resins

Property	ASTM test method	Woven cloth	Chopped roving	Sheet molding compound
Mechanical				
Specific gravity	D 792	1.5 to 2.1	1.35 to 2.30	1.65 to 2.60
Flexural yield strength	D 790			
MPa		276 to 552	68.9 to 276	68.9 to 248
ksi		40 to 80	10 to 40	10 to 36
Flexural modulus of elasticity	D 790			
GPa		...	6.9 to 20.7	6.9 to 15.2
10^5 psi		...	10 to 30	10 to 22
Compressive strength	D 695			
MPa		172 to 345	103 to 207	103 to 207
ksi		25 to 50	15 to 30	15 to 30
Tensile strength	D 638			
MPa		207 to 345	103 to 207	55.2 to 138
ksi		30 to 50	15 to 30	8 to 20
Elongation, %	D 638	0.5 to 2.0	0.5 to 5.0	...
Tensile modulus of elasticity	D 638			
GPa		10.3 to 31.0	5.5 to 13.8	...
10^5 psi		15.0 to 45.0	8.0 to 20.0	...
Izod impact strength, notched(a)	D 256			
J/m		267 to 1600	107 to 1070	374 to 1175
ft · lbf/in.		5.0 to 30.0	2.0 to 20.0	7.0 to 22.0
Barcol hardness		60 to 80	50 to 80	50 to 70
Electrical				
Volume resistivity(b), Ω · cm	D 257	10^{14}	10^{14}	10^{14} to 10^{15}
Dielectric strength(c), V/mil	D 149			
Short-time test		350 to 500	350 to 500	380 to 450
Step-by-step test		300 to 400	300 to 450	350 to 400
Dielectric constant	D 150			
At 60 Hz		4.1 to 5.5	3.8 to 6.0	4.4 to 6.3
At 10^3 Hz		4.2 to 6.0	4.0 to 6.0	4.4 to 6.1
At 10^6 Hz		4.0 to 5.5	3.5 to 5.5	4.2 to 5.8
Dissipation (power) factor	D 150			
At 60 Hz		0.01 to 0.04	0.01 to 0.04	0.007 to 0.021
At 10^3 Hz		0.01 to 0.06	0.01 to 0.05	0.007 to 0.015
At 10^6 Hz		0.01 to 0.03	0.01 to 0.03	0.016 to 0.024
Arc resistance, s	D 495	60 to 120	120 to 180	120 to 200
Resistance characteristics				
Heat resistance (continuous)				
°C		149 to 177	149 to 177	149 to 204
°F		300 to 350	300 to 350	300 to 400
Water absorption(d), in 24 h, %	D 570	0.05 to 0.50	0.1 to 1.0	0.10 to 0.15
Effect of sunlight		Slight	Slight	Slight

(a) Test bar 13 × 13 mm (0.5 × 0.5 in.). (b) At 50% RH and 23 °C (73 °F). (c) Material thickness, 3.2 mm (0.125 in.). (d) Material thickness, 3.1 mm (0.125 in.)

Mechanical properties of fiberglass-polyester resin composites

Property	Orthophthalic	Isophthalic	BPA fumarate	Chlorendic	Dicyclopentadiene
Glass content, %	40	40	40	40	34
Barcol hardness	...	45	40	40	...
Tensile strength, MPa (ksi)	152 (22)	193 (28)	124 (18)	138 (20)	96 (14)
Tensile modulus, GPa (10^6 psi)	5.5 (0.8)	11.7 (1.7)	11.0 (1.6)	9.7 (1.4)	7.6 (1.1)
Tensile elongation, %	1.7	2.0	1.2	1.4	1.9
Flexural strength, MPa (ksi)	220 (32)	240 (35)	160 (23)	190 (28)	160 (23)
Flexural modulus, GPa (10^6 psi)	6.9 (1.0)	7.6 (1.1)	9.0 (1.3)	9.7 (1.4)	6.2 (0.9)
Compressive strength, MPa (ksi)	...	205 (30)	180 (26)	125 (18)	...
Izod impact, J/m (ft · lbf/in.)	...	571 (10.7)	640 (12)	374 (7.0)	...

Corrosion resistance of glass fiber-polyester resin composites

Resin	75% H_2SO_4 80 °C (175 °F)	15% NaOH 65 °C (150 °F)	5 1/4% NaOCl 65 °C (150 °F)	Xylene ambient	Deionized water 100 °C (210 °F)	Seawater 80 °C (180 °F)
Isophthalic	-	-	-	+	-	-
Chlorendic	+	-	-	+	+	+
BPA fumarate	-	+	-	+	-	+
Vinyl ester	-	-	+	-	-	+

Performance of selected polyester composites in fire tests

	System I	System II	System III
Material			
Resin	100(a)	100(b)	100(c)
Alumina trihydrate	100	100	100
Antimony oxide	...	5	...
Ferrous oxide	...	...	5
Test method and property			
ASTM E 162			
Flame-spread index	75	7	7
ASTM E 84			
Flame spread	64	23	25
Smoke emission	608	270	268
NBS chamber			
Flaming mode			
Max density	203	433	264
90-s density	2.5	18	11
240-s density	162	245	128
Nonflaming mode			
Max density	481	400	350
90-s density	1	1	5
240-s density	16	45	50

(a) Orthophthalic resin. (b) HET acid resin A, 26% Cl. (c) HET acid resin B, 26% Cl

Selected process factors for carbon-PEEK composite 3.2-mm (1/8-in.) laminates with various fibers

	Area/width limits	Weight/thickness	Properties	Economics	Advantages/ disadvantages
Preimpregnated					
Woven fabric (APC-2 woven)	460 mm (18 in.) only Wider very difficult	216 g/m² max	~APC-2	Difficult to prepreg	Thin skins, good props, limited widths, no cold drape
Woven prepreg tow (APC-2 tow 12K)	1780 mm (70 in.)	300 g/m²	APC-2	Simple weaving	Large area, thin skins, good props, easy consolidation, no cold drape
Postimpregnated					
Film stacked (PEEK film/woven carbon fiber fabric)	Large area limited only by weaving loom and consolidation press size	200 to 300 g/m²	Poorer	Simple weaving and film	Straightforward, large area, difficult impregnation, flat sheet only, no drape
Slit film (cowoven slit PEEK film)			↑		
Cowoven (PEEK monofill or multifill)			90% push and pull 70% bend 50% impact	Slow weaving	Large area, limited drape, difficult impregnation
Comingled and woven (PEEK multifill)			↑	Expensive slit tape Simple weaving	Large area, limited drape with monofill, difficult impregnation
				Four-step process	Large area, good drape, improved impregnation
Powder impregnated (fine PEEK powder trapped in tow or fabric)		200 to 500 g/m² (0.7 to 1.6 oz/ft²)	Poor		Better impregnation; poor polymer weight, function, and control

Properties of reinforced PET grades

Properties	ASTM test method	Product type(s): 15% GF	30% GF	45% GF	35% GF-mineral	30% GF flame retardant
Mechanical						
Tensile strength at yield, MPa (ksi)	D 638	115 (16.8)	150 (22.0)	180 (26.0)	105 (15.0)	130 (19.0)
Ultimate elongation, %	D 638	3.4	2.2	1.6	1.5	2.0
Flexural strength, MPa (ksi)	D 790	175 (25.2)	235 (34.0)	285 (41.5)	170 (25.0)	190 (27.5)
Flexural modulus, GPa (10^6 psi)	D 790	5.79 (0.840)	8.96 (0.130)	13.8 (2.00)	10.3 (1.50)	9.93 (1.44)
Notched Izod impact strength, J/m (ft · lbf/in.)	D 256	70 (1.3)	95 (1.8)	110 (2.0)	55 (1.0)	80 (1.5)
Heat-deflection temperature at 1.82 MPa (0.264 ksi), °C (°F)	D 648	211 (412)	225 (437)	225 (437)	210 (410)	220 (430)
Mold shrinkage, (0.318 cm bar), mm/mm	...	0.005	0.003	0.002	0.003	0.003
Electrical						
Volume resistivity, dry, $\Omega \cdot m$	D 257	10^{13}	10^{13}	10^{13}	10^{13}	10^{13}
Surface resistivity, dry, Ω	D 257	10^{15}	10^{15}	10^{15}	10^{15}	10^{15}
Dielectric strength, short-time (0.318 cm), kV/mm (V/mil)	D 149	19 (490)	22 (570)	22 (570)	21 (540)	21 (550)

(continued)

Properties of reinforced PET grades (continued)

Properties	ASTM test method	15% GF	30% GF	45% GF	35% GF-mineral	30% GF flame retardant
		Product type(s)				
Electrical (continued)						
Dielectric strength, step-by-step (0.318 cm), kV/mm (V/mil)	D 149	17 (410)	16 (400)	16 (400)	17 (410)	16 (400)
Dielectric constant at 10^6 Hz	D 150	3.5	3.5	3.8	3.4	3.5
Dissipation factor at 10^6 Hz	D 150	0.022	0.021	0.022	0.022	0.022
Physical						
Specific gravity	D 792	1.39	1.68	1.70	1.61	1.68
Melting point, °C (°F)	D 789	245 (473)	245 (473)	245 (473)	245 (473)	245 (473)
Coefficient of linear thermal expansion, 10^{-6}/K	D 696	3.0 (1.7)	2.5 (1.4)	2.2 (1.2)	2.2 (1.2)	2.5 (1.4)
Flammability, (0.08 cm)	UL 94(b)	HB	HB	HB	HB	V-0

GF, glass fiber. (a) Tested at 23°C, or 73°F, dry as molded. (b) Underwriters Laboratories test method

Physical properties of reinforced polypropylene

Property	ASTM test method	40% talc: Homopolymer polypropylene	40% talc: Copolymer polypropylene	40% mica: Homopolymer polypropylene	40% mica: Copolymer polypropylene	40% calcium carbonate: Homopolymer polypropylene	40% calcium carbonate: Copolymer polypropylene
Tensile strength, MPa (ksi)	D 638	33 (4.8)	21.4 (3.1)	37.2 (5.4)	18.6 (2.7)	23.4 (3.4)	16.6 (2.4)
Elongation, %	D 638	8.0	10.0	2.0	3.0	15.0	9.0
Flexural modulus, GPa (10^6 psi)	D 790	3.45 (0.500)	2.35 (0.340)	4.76 (0.690)	2.76 (0.400)	2.48 (0.360)	1.65 (0.240)
Notched Izod impact strength, J/m (ft · lbf/in.)	D 256	26.7 (0.5)	203 (4.0)	19.5 (0.36)	90.8 (1.7)	43 (0.8)	673 (12.4)
Deflection temperature under load, °C (°F)	D 648						
At 1.82 MPa (0.264 ksi)		80 (180)	74 (165)	99 (210)	65.5 (150)	71 (160)	63 (145)
At 0.45 MPa (0.066 ksi)		130 (270)	102 (215)	143 (290)	107 (225)	110 (230)	94 (201)

Properties of Kynol-fiber reinforced polypropylene

Property	4	18(a)	18(b)	33	Powder, 20 wt %	Polypropylene(c)
	Fiber, wt %					
Density, g/cm^3	0.91	0.94	0.94	0.98	0.94	0.903
Izod impact strength, J/cm (ft · lbf/in.)						
Unnotched	7.10 (13.3)	5.50 (10.3)	4.8 (9)	4.3 (8)	2.0 (3.8)	12 (22)
Notched	0.37 (0.7)	0.62 (1.17)	0.37 (0.7)	0.64 (1.2)	0.32 (0.6)	0.27 (0.5)
Elongation, %	>50	25	17	7	6	>100
Tensile strength, MPa (ksi)	30 (4.3)	28 (4.1)	35 (5.1)	30 (4.4)	26 (3.7)	34 (4.9)
Heat-deflection temperature, °C (°F)						
At 1.82 MPa (264 ksi)	65 (149)	77 (171)	84 (183)	100 (212)	67 (153)	58 (136)
At 455 kPa (66 psi)	...	131 (268)	136 (277)	...	...	98 (208)
Tensile modulus, GPa (10^6 psi)	1.59 (0.23)	1.86 (0.27)	1.86 (0.27)	1.93 (0.28)	1.59 (0.23)	1.38 (0.20)

(a) Without coupling agent. (b) With coupling agent. (c) Profax 6523, Hercules

Powder blending vs. melt compounding of mica/polypropylene composites

Property	Unfilled	40 wt % fine mica(a): Powder blend	40 wt % fine mica(a): Melt compound	50 wt % coarser mica(b): Powder blend	50 wt % coarser mica(b): Melt compound
Tensile strength, MPa (ksi)	28 (4.1)	34 (4.9)	35 (5.1)	36 (5.2)	29 (4.2)
Flexural modulus, GPa (10^6 psi)	1.372 (0.199)	7.033 (1.020)	6.640 (0.963)	11.38 (1.650)	9.928 (1.440)
Flexural strength, MPa (ksi)	29 (4.2)	57 (8.3)	59 (8.5)	63 (9.2)	52 (7.5)

(a) Suzorite 325-H, phlogopite. (b) Suzorite 60-S, phlogopite

Effects of coupling agents on glass fiber reinforcement in homopolymer polypropylene

Property	ASTM test method	30% glass, uncoupled	30% glass, coupled
Tensile strength, MPa (ksi)	D 638	48 (7.0)	75 (11.0)
Elongation, %	D 638	2.5	2.0
Flexural modulus, GPa (10^6 psi)	D 790	4.14 (0.600)	4.83 (0.700)
Notched Izod impact strength, J/m (ft · lbf/in.)	D 256	54.3 (1.0)	65.1 (1.2)
Deflection temperature under load, °C (°F)	D 648		
At 1.82 MPa (0.264 ksi)		129 (265)	138 (280)
At 0.45 MPa (0.066 ksi)		138 (280)	151 (305)

Nominal properties of PPS composites

Properties	Stampable sheet	Prepreg/ laminates
Fiber reinforcement	Glass, carbon	Glass, carbon, aramid
Fiber form	Random mat	Unidirectional, fabric
Fiber loading, wt%	20 to 40	40 to 70
Tensile strength, MPa (ksi)	90 to 206 (13 to 30)	200 to 1800 (29 to 260)
Tensile modulus, GPa (10^6 psi)	6 to 14 (0.87 to 2.0)	14 to 140 (2.0 to 20)
Impact strength, J/m (ft · lbf/in.)	530 to 1325	635 to 1590
Compressive strength, MPa (ksi)	260 (40)	620 to 900 (90 to 130)

Nominal properties of PPS compounds

Injection molded using a 135 °C (275 °F) mold

Property	ASTM test method	Unfilled	40 wt% glass filled	Glass/mineral filled(a)
Specific gravity, g/cm^3	D 1505	1.35	1.6	1.8 to 2.0
Tensile strength, MPa (ksi)	D 638	65.5 (9.5)	138 (20)	96 (14)
Elongation, %	D 638	1.6	1.25	0.7
Flexural modulus, GPa (10^6 psi)	D 790	3.8 (0.55)	12 (1.7)	17 (2.5)
Flexural strength, MPa (ksi)	D 790	96 (13.9)	160.0 (23.2)	158 (22.9)
Compressive strength, MPa (ksi)	D 695	110 (16)	145 (21)	158 (22.9)
Izod impact, J/m (ft · lbf/in.)	D256			
Notched		16 (0.30)	58 (1.09)	53 (1.0)
Unnotched		100 (1.87)	175 (3.28)	185 (3.5)
Heat-deflection temperature at 1.8 MPa (0.264 ksi), °C (°F)	D 648	135 (275)	243 (469)(c)	244 (471)(b)
Specific heat, J/g · K (Btu/lb · °F)		1.09 (0.260)	1.05 (0.251)	
Hardness Rockwell, F	D 785	120	123	121
Thermal conductivity, W/m · K (Btu · in./h · ft^2 · °F)(d)	...	0.288 (2.00)	0.288 (2.00)	0.6 (4.16)
Coefficient of linear thermal expansion, 10^{-5}/K	...	4.9	4	2.8
Flammability	UL 94(d)	V-0	V-0/5V	V-0/5V
Oxygen index	D 2863	44	46.5	53

(a) Properties vary slightly, depending on filler-glass levels. (b) Greater than 260°C (500°F) after heat treating for 4 h at 260°C (500°F). (c) –30°C to + 30°C determined on thermomechanical analyzer instrument. (d) Underwriters Laboratories test method

Nominal electrical properties for PPS compounds

Property	ASTM test method	40 wt% glass filled	Glass/mineral filled
Dielectric strength(a), kV/mm (V/mil)	ASTM D 149	17.7 (450)	13.4 to 15.7 (340 to 400)
Dielectric constant at 25 °C (77 °F)			
At 1 kHz		3.9	4.6 to 6.6
At 1 MHz		3.8	4.3 to 6.1
Dissipation factor, at 25 °C (77 °F)	ASTM D 150		
At 1 kHz		0.0014	0.01 to 0.08
At 1 MHz		0.0014	0.01 to 0.02
Volume resistivity, Ω · m	ASTM D 257	4.5×10^{14}	1 to 3×10^{13}
Insulation resistance, at 95% RH	ASTM D 257	5×10^9	1×10^{10}
Arc resistance, s	ASTM D 495	34.0	116 to 182
Arc track rate, mm/s (in./min.)	UL 746A(b)	8.4 (20.0)	0 to 2 (0 to 4.7)
Comparative track rate, V	UL 746A(b)	180.0	160 to 235
UL temperature index(c), °C (°F)	UL 746B(b)	200 to 220 (390 to 430)	200 to 240 (390 to 465)

(a) Transformer oil, 500 V/s rate of increase, 1.6 to 3.2 mm (0.6 to 0.13 in.) thickness. (b) Underwriters Laboratories test method. (c) Indexes vary depending on requirement and compound composition.

Thermal aging of PPS compounds

Exposure time, h	Tensile strength retained, %(a), at 175 and 230 °C (350 and 445 °F) 40% glass filled 175	230	Glass/mineral filled 175	230
0	100	100	100	100
250	97	78	93	100
500	97	78	93	100
1 000	86	73	99	89
2 500	84	73	99	90
5 000	79	63	87	81
7 500	47	55	85	82
10 000	55	47	72	90

(a) Specimens molded in 135 °C (275 °F) mold and annealed 2 h at testing temperature. Properties determined at room temperature after aging

Properties of polysulfone resin and fiber-resin composites

Properties	Neat resin	30% glass fiber-resin	30% carbon fiber-resin
Specific gravity	1.24	1.45	1.37
Density, g/cm^3 (lb/in.^{3})	1.24 (0.045)	1.45 (0.052)	1.37 (0.049)
UL electrical rating, °C (°F)	160 (320)	...	...
Deflection temperature at 1800 kPa (264 psi), °C (°F)	175 (345	185 (365)	185 (365)
In-service temperature, °C (°F)	150 to 205 (300 to 400)	...	...
UL flammability rating	94 V-0	94 V-0	94 V-0

Thermosets

General

Properties of unidirectional advanced composites

Property	Boron/ epoxy	Boron/ polymide	S-glass/ epoxy	High-modulus graphite/ epoxy	High-modulus graphite/ polyimide	High-strength graphite/ epoxy(a)	Aramid/ epoxy(b)	High-strength graphite/ epoxy(c)
Reinforcement content, vol %	50	49	72	45	45	70	54	60
Density, g/cm^3 (lb/in.^{3})	2.02 (0.073)	1.99 (0.072)	2.13 (0.077)	1.55 (0.056)	1.55 (0.056)	1.61 (0.058)	1.36 (0.049)	1.58 (0.057)
Tensile strength, MPa (ksi)								
Longitudinal	1370 (199)	1040 (151)	1290 (187)	840 (122)	807 (117)	1500 (218)	1190 (172)	1520 (220)
Transverse	56 (8.1)	11 (1.6)	46 (6.7)	42 (6.1)	15 (2.2)	40 (5.8)	11 (1.6)	55 (8.0)
Tensile modulus, GPa (10^6 psi)								
Longitudinal	201 (29.2)	221 (32.1)	61 (8.8)	190 (27.5)	216 (31.3)	145 (21.0)	84 (12.2)	110 (16.0)
Transverse	22 (3.2)	14 (2.1)	25 (3.6)	6.9 (1.0)	5.0 (0.72)	10 (1.5)	4.8 (0.70)	15 (2.2)
Compressive strength, MPa (ksi)								
Longitudinal	1600 (232)	1090 (158)	820 (119)	883 (128)	652 (94.5)	1700 (247)	290 (42)	1240 (180)
Transverse	123 (17.9)	63 (9.1)	162 (23.5)	197 (28.5)	70 (10.2)	246 (35.7)	65 (9.4)	248 (36.0)
Shear modulus, GPa (10^6 psi)	5.38 (0.78)	7.65 (1.11)	12.0 (1.74)	6.2 (0.9)	4.48 (0.65)	6.9 (1.0)	2.83 (0.41)	4.96 (0.72)
Intralaminar shear strength, MPa (ksi)	63 (9.1)	26 (3.8)	45 (6.5)	61 (8.9)	22 (3.2)	68 (9.8)	28 (4.0)	69 (10.0)
Poisson's ratio								
Major	0.17	0.16	0.23	0.10	0.25	0.28	0.32	0.25
Minor	0.02	0.02	0.09	...	0.02	0.01	0.02	0.034
Moisture coefficient, 10^{-2} mm (10^{-2} in.)								
Longitudinal	0.0762 (0.003)	0.0762 (0.003)	0.3556 (0.014)	0.0762 (0.003)	0.0762 (0.003)	0.1524 (0.006)	0.2032 (0.008)	0.1524 (0.006)
Transverse	4.267 (0.168)	4.267 (0.168)	3.251 (0.128)	3.277 (0.129)	3.277 (0.129)	3.277 (0.129)	3.835 (0.151)	3.277 (0.129)
Coefficient of thermal expansion, 10^{-6}/ °C (10^{-6}/ °F)								
Longitudinal	6.1 (3.4)	4.9 (2.7)	3.8 (2.1)	...	0.0 (0.0)	0.02 (0.01)	-2.88 (-1.60)	0.72 (0.40)
Transverse	30.4 (16.9)	28.4 (15.8)	16.7 (9.3)	33.3 (18.5)	25.4 (14.1)	22.5 (12.5)	56.3 (31.3)	29.5 (16.4)

(a) Union Carbide Thornel 300 fibers. (b) Du Pont Kevlar 49. (c) Hercules AS fibers

Properties of low-temperature thermoset matrix composites

Properties	Neat resin	10 to 40 wt% glass fiber-resin
Specific gravity	1.2 to 1.4	1.6 to 1.9
Density, g/cm^3 (lb/in.^{3})	1.2 to 1.4 (0.043 to 0.051)	1.6 to 1.9 (0.058 to 0.069)
In-service temperature °C (°F)	120 to 150 (250 to 300)	120 to 205 (250 to 400)
Heat deflection temperature at 1820 kPa (264 psi) °C (°F)	50 to 205 (120 to 400)	190 to 205 (375 to 400)
UL rated in-service temperature °C (°F)	180 (356)	...
UL flammability	...	...

Interlaminar fracture toughness of carbon fiber-thermoset resins as determined by double-cantilever-beam specimen

Material(a)	Interlaminar fracture toughness, G_{Ir} J/m^2	ft · lbf/ft^2
5208	80 to 90	5 to 6
3502	120 to 150	8 to 10
3501-6	150 to 214	10 to 15
1504	95 to 123	7 to 8
2220-1	256	18
914	220 to 250	15 to 17
BP907	324 to 397	22 to 27
HST-7	543	37
V378A(b)	72 to 86	5 to 6
5245	134 to 141	9 to 10

(a) Composites fabricated at 175°C (350°F). (b) Postcured at 205°C (400°F)

Specific

Typical amino molding compound properties

Property	Urea-formaldehyde, alpha cellulose	Melamine-formaldehyde: Alpha cellulose	Melamine-formaldehyde: Glass fiber	Melamine-formaldehyde: Mineral
Shrinkage, mm/mm	0.008 to 0.010	0.008 to 0.009	0.002 to 0.004	0.003 to 0.005
Specific gravity	1.5	1.5	1.80 to 1.85	1.72
Water absorption, 24 h at 23 °C, or 73 °F, %	0.4 to 0.8	0.3 to 0.5	0.15	0.2
Impact strength, notched, J/m (ft · lbf/in.)	14 to 18 (0.27 to 0.34)	16 to 19 (0.30 to 0.35)	32 (0.60)	24 (0.45)
Tensile strength, MPa (ksi)	38 to 48 (5.5 to 7.0)	48 to 55 (7 to 8)	41 to 62 (6 to 9)	35 to 48 (5 to 7)
Flexural strength, MPa (ksi)	75 to 110 (11 to 16)	75 to 110 (11 to 16)	95 to 125 (14 to 18)	55 to 70 (8 to 10)
Compressive strength, MPa (ksi)	210 to 260 (30 to 38)	280 to 310 (40 to 45)	240 to 290 (35 to 42)	180 to 210 (26 to 30)
Thermal conductivity, W/m · K (Btu · in./h · ft^2)	0.414 (2.87)	0.414 (2.87)	0.455 (3.16)	0.621 (4.31)
Arc resistance, s	100 to 135	125 to 136	>180	>180
Dielectric constant, 1 MHz	6.8	8.1	6.0	5.6
Dissipation factor, 1 MHz	0.030	0.028	0.020	0.030

Typical physical properties of DAP molding compounds

Property	ASTM test method	Dacron-filled DAP	Short glass fiber filled DAP	Short glass fiber filled DAIP	Long glass fiber filled DAP	Long glass fiber filled DAIP
Flexural strength, MPa (ksi)	D 790	76 (11)	90 to 131 (13 to 19)	90 to 117 (13 to 17)	117 to 124 (17 to 18)	117 to 124 (17 to 18)
Flexural modulus, GPa (10^6 psi)	D 790	3.4 to 4.1 (0.5 to 0.6)	9.0 to 13.8 (1.3 to 2.0)	9.0 to 13.8 (1.3 to 2.0)	9.7 to 13.1 (1.4 to 1.9)	9.7 to 13.1 (1.4 to 1.9)
Izod impact strength, J/m (ft · lbf/in.), notch	D 256	240 to 270 (4.5 to 5.0)	25 to 65 (0.5 to 1.2)	32 to 65 (0.6 to 1.2)	210 to 530 (4.0 to 10.0)	210 to 640 (4.0 to 12.0)
Compressive strength, MPa (ksi)	D 695	210 (30)	170 to 220 (25 to 32)	180 to 220 (26 to 32.5)	180 to 220 (26 to 32)	180 to 240 (26 to 35)
Tensile strength, MPa (ksi)	D 638	32 to 43 (4.6 to 6.2)	55 to 76 (8 to 11)	55 to 76 (8 to 11)	55 to 76 (8 to 11)	65 to 76 (9.5 to 11)
Mold shrinkage, mm/mm	D 955	0.009 to 0.010	0.002 to 0.005	0.002 to 0.005	0.001 to 0.004	0.002 to 0.004
Postmold shrinkage, 480 h at 125 °C (257 °F) mm/mm	...	<1	1	1	1	1
Water absorption, 48 h, at 50 °C (120 °F), %	D 570	0.2	0.2 to 0.3	0.2 to 0.3	0.2 to 0.3	0.2 to 0.3
Heat deflection temperature, °C (°F)	D 648	145 (290)	175 to 260 (350 to 500)	230 to 280 (450 to 540)	200 to 260 (390 to 500)	255 to 280 (490 to 540)
Continuous heat resistance, °C (°F)	D 794	175 to 205 (350 to 400)	175 to 230 (350 to 450)	220 to 260 (425 to 500)	175 to 230 (350 to 450)	220 to 260 (425 to 500)

Electrical properties of DAP and DAIP molding compounds

Property	ASTM test method	Dacron-filled DAP	Short glass filled DAP	Short glass filled DAIP	Long glass filled DAP	Long glass filled DAIP
Dielectric strength, MV/m (V/mil)						
Short time	D 149	0.0157 (400)	0.0156 to >0.0157 (395 to >400)	0.0150 to 0.0157 (380 to 400)	0.0156 (395)	0.0150 (380)
Step-by-step	D 149	0.0161 (410)	0.0156 to 0.0157 (395 to 400)	0.0154 to 0.0157 (390 to 400)	0.0148 to (375)	0.0148 (375)
Volume resistivity, Ω · m	D 257	$>10^{11}$	$>10^{11}$	$>10^{14}$	$>10^{11}$	$>10^{14}$
Dielectric constant	D 150					
At 1 kHz		3.7 to 3.8	4.1 to 4.4	4.1 to 4.4	4.2 to 4.4	4.1 to 4.2
At 1 MHz		3.5 to 3.6	3.5 to 4.5	3.4 to 4.4	3.5 to 4.3	3.4 to 4.0
Dissipation factor	D 150					
At 1 kHz		0.004 to 0.008	0.004 to 0.008	0.004 to 0.009	0.004 to 0.007	0.004 to 0.008
At 1 MHz		0.012 to 0.016	0.009 to 0.012	0.010 to 0.014	0.010 to 0.013	0.010 to 0.012
Arc resistance, s	D 495	125	125 to 140	130 to 180	125 to 130	125 to 180

Thermal properties of fiber-epoxy composite materials

Type of material	Coefficient of thermal expansion, 10^{-6}/K: 0 °	Coefficient of thermal expansion, 10^{-6}/K: Quasi-isotropic	Thermal conductivity, W/m · K (Btu · ft/h · ft^2 °F): 0 °	Thermal conductivity, W/m · K (Btu · ft/h · ft^2 °F): Quasi-isotropic
AS graphite-epoxy	−0.1	0.6	7 to 9 (4 to 5)	5 to 7 (3 to 4)
HMS graphite-epoxy	−0.2	0.3	10 to 55 (30 to 32)	28 to 30 (16 to 18)
Kevlar 49-epoxy	−1.1	−0.63	0.17 (0.101)	...
S-glass-epoxy	1.9	3.3	3.5 (2)	0.35 (0.2)

Epoxy resin and fiber-resin properties

Property	DGEBA	Epoxy resin systems: Phenolic-novolac	Cycloaliphatic	Glass-resin fiber	Chopped glass fabric resin	Continuous fiber-fabric epoxy resin systems: S-2 filament winding glass fiber-resin	Kevlar-resin	Carbon-resin	Graphite fiber-resin	Lay-up quartz fabric-resin
Specific gravity	1.15	1.24	1.22	1.72	1.79	1.86	1.25	1.46	1.58	1.75
Density, g/cm^3 (lb/in.^{3})	1.15 (0.042)	1.24 (0.045)	1.22 (0.044)	1.72 (0.062)	1.79 (0.065)	1.86 (0.067)	1.25 (0.045)	1.46 (0.053)	1.58 (0.057)	1.75 (0.063)
Service temperature, °C (°F)	80 to 88 (175 to 190)	230 (450)	230 to 260 (450 to 500)	150 (300)	150 (300)	150 (300)	150 (300) for 1 to 2 min.	150 (300) (300)	150 (300)	150 (300)
Heat-deflection temperature at 1.82 MPa (264 psi), °C (°F)	110 (230)	150 to 205 (300 to 400)	150 to 275 (300 to 525)	NA ...	NA ...	NA ...	NA ...	NA ...	NA ...	NA ...
TGA, stability temperature, °C (°F) 0% wt loss	200 (390)	...	...	NA	...	...	...	...	...	...
Reinforcement volume	NA	NA	NA	35	49.3	60	62.9	56	54	65 wt %

Typical properties of molded epoxy composite materials

Fiber: Type	Product code	Specific gravity, g/cm^3	Fiber content: wt %	vol %	Tensile strength: MPa	ksi	Flexural strength: MPa	ksi	Flexural modulus: GPa	10^6	Compressive strength: MPa	ksi	Impact strength: J/mm	ft · lbf/in.
Fiberglass	Type E	1.88	63	46	190	27	470	68	28	4.1	290	42	1.6	30
Fiberglass	Type S-2	1.85	63	46	210	30	430	62	30	4.3	260	38	1.7	32
PAN-based carbon	High-strength	1.48	58	49	140	20	330	48	38	5.5	190	28	0.55	10
PAN-based carbon	High-modulus	1.51	58	48	170	25	340	50	55	8.0	210	30	0.70	13
Aramid	Kevlar 49	1.34	53	49	160	23	290	42	21	3.0	150	22	1.8	34
Aramid	Kevlar 29	1.33	53	49	110	16	270	39	19	2.8	130	19	2.1	40

Properties of typical fiber-epoxy composites and structural metals

Material	Density, g/cm^3	Modulus of elasticity, Longitudinal: GPa	10^6 psi	Modulus of elasticity, Transverse: GPa	10^6 psi	Shear modulus: GPa	10^6 psi	Poisson's ratio	Tensile strength, Longitudinal: MPa	ksi	Tensile strength, Transverse: MPa	ksi	Compressive strength, Longitudinal: MPa	ksi	Compressive strength, Transverse: MPa	ksi	Shear strength: MPa	ksi
Unidirectional composites ($V_{f=0.6}$)																		
E-glass	1.94	45	6.5	12	1.7	4.4	0.64	0.25	1000	150	34	5	550	80	140	20	40	6
Kevlar-49	1.30	76	11.0	5.5	0.8	2.1	0.3	0.34	1380	200	28	4	280	40	140	20	55	8
T-300	1.47	132	19.2	10.3	1.5	6.5	0.95	0.25	1240	180	45	6.5	830	120	140	20	62	9
VSB-32	1.63	229	33.2	6.9	1.0	5.5	0.8	0.25	1170	170	41	6	690	100	140	20	75	11
Boron	1.86	274	39.8	15	2.2	52	7.5	0.25	1310	190	34	5	2480	360	310	45	100	15
GY-70	1.61	320	46.4	5.5	0.8	4.1	0.6	0.25	690	100	41	6	620	90	140	20	96	14
Metals																		
2024-T3	2.77	72.3	10.5	72.3	10.5	27.6	4.0	0.31	462	67	455	66	345	50	345	50	276	40
7075-T6	2.80	71.0	10.3	71.0	10.3	27.6	4.0	0.31	544	79	530	77	475	69	475	69	324	47
4130	7.84	207	30.0	207	30.0	82.7	12.0	0.25	655	95	655	95	1100	160	1100	160	380	55

Typical in-plane stiffness properties of epoxy matrix composite materials

Material	E_{11}: GPa	10^6 psi	E_{22}: GPa	10^6 psi	ν_{12}	G_{12}: GPa	10^6 psi
Carbon-epoxy prepreg (AS4/3501-6)	131.0	19.0	11.2	1.63	0.28	6.52	0.945
Carbon-epoxy prepreg (T300/5208)	153	22.1	11.2	1.63	0.33	7.10	1.03
Boron-epoxy	204	29.6	18.5	2.68	0.23	5.59	0.810
Kevlar 49-epoxy	76.0	11.0	5.50	0.80	0.34	2.30	0.33
E-glass-epoxy	38.6	5.60	8.27	1.20	0.26	4.14	0.60

Behavior of carbon-epoxy versus metals under various conditions

Condition	Carbon-epoxy behavior relative to metals
Stress-strain relationship	More linear strain to failure
Notch sensitivity	
Static	Greater sensitivity
Fatigue	Less sensitivity
Transverse properties	Weaker
Mechanical properties variability	Higher
Sensitivity to aircraft hygrothermal environment	Greater
Damage growth mechanism	In-plane delamination instead of through-the-thickness cracks

Typical properties of boron/epoxy laminates(a)

Temperature °C	Temperature °F	0 ° tensile strength MPa	0 ° tensile strength ksi	0 ° tensile modulus GPa	0 ° tensile modulus 10^6 psi	0 ° failure strain %	90 ° tensile strength MPa	90 ° tensile strength ksi	0 ° compressive strength MPa	0 ° compressive strength ksi
Unidirectional construction (0 °)										
-55	-67	1240 to 1380	180 to 200	205 to 220	30 to 32	0.59 to 0.64	68 to 90	10 to 13	3105 to 3310	450 to 480
24	75	1275 to 1450	185 to 210	205 to 215	30 to 31	0.62 to 0.69	55 to 83	8 to 12	2585 to 3070	375 to 445
125	260	1275 to 1345	185 to 195	200 to 205	29 to 30	0.63 to 0.66	41 to 55	6 to 8	1860 to 2070	270 to 300
190	375	1105 to 1170	160 to 170	195 to 200	28 to 29	0.56 to 0.61	21 to 28	3 to 4	585 to 1000	85 to 145
Cross-plied construction (0 °/90 °)										
-55	-67	690 to 760	100 to 110	130	18.9	0.54 to 0.56	690 to 760	100 to 110	...	...
24	75	690 to 760	100 to 110	125	18.0	0.57 to 0.61	690 to 760	100 to 110	...	...
125	260	655 to 690	95 to 100	120	17.5	0.58	655 to 690	95 to 100	...	...
190	375	580 to 655	84 to 95	105 to 115	15.5 to 16.5	0.56 to 0.58	580 to 655	84 to 95	...	...

(a) Nominal fiber content, 55 vol %

Properties of a high-modulus graphite-epoxy lamina

$E_1 = 170$ GPa (25.0×10^6 psi)
$E_2 = 12$ GPa (1.7×10^6 psi)
$G_{12} = 4.5$ GPa (0.65×10^6 psi)
$\upsilon_{12} = 0.30$
$\rho = 0.056$
$\alpha_1 = -0.54 \times 10^{-6}$/K
$\alpha_2 = 35.1 \times 10^{-6}$/K
$\sigma_L^{tu} = 758$ MPa (110.0 ksi)
$\sigma_T^{tu} = 28$ MPa (4.0 ksi)
$\sigma_L^{u} = 62$ MPa (9.0 ksi)
$\sigma_L^{cu} = 758$ MPa (110.0 ksi)
$\sigma_T^{cu} = 138$ MPa (20.0 ksi)
$V_f = 0.6$

Note: Ply thickness, 0.13 mm (0.0052 in.)

Typical mechanical properties of graphite/epoxy composites with various fiber orientations

Fiber type	Ultimate tensile strength MPa	Ultimate tensile strength ksi	Tensile modulus GPa	Tensile modulus 10^6 psi	Ultimate compressive strength MPa	Ultimate compressive strength ksi	Compressive modulus GPa	Compressive modulus 10^6 psi
Chopped fiber:								
Molding compound	352	51	108	15.7	469	68	...	...
Unidirectional (nonwoven):								
High-strength fiber (0 °)	1625	236	138	20.0	993	144	113	16.4
High-strength fiber (0 °/±45 °)	496	72	57	8.3	503	73	50	7.3
Medium-strength fiber (0 °)	1455	211	147	21.3	1405	204	131	19.0
Medium-strength fiber (0 °/±45 °)	503	73	65	9.4	503	73	64	9.3
High-modulus fiber (0 °)	1240	180	215	31.2	758	110	177	25.6
Fabric (woven):								
Medium-strength fiber	510	74	70	10.2	510	74	63	9.2

Comparison of 2D and 3D models of Kevlar/epoxy composites in Charpy impact test

	Impact energy-Y			Impact energy-Z		
Model	J	ft · lbf	Standard deviation	J	ft · lbf	Standard deviation
2D	14.57	10.75	1.06	17.83	13.15	0.92
3D	15.25	11.25	0.35	19.0	14.0	0.71

Comparison of 2D (with and without a circular hole) and 3D models of Kevlar/epoxy composites in four-point bending test

	Maximum flexural strength		Fracture toughness	
Model	MPa	ksi	J/cm^3	in. · lbf/in.^{3}
2D without hole	30.9	4.48	506	273
2D with hole	26.1	3.79	136	73.4

Prepreg characteristics, cure conditions, and mechanical properties of SiC/epoxy laminates(a)

Property	Value
Prepreg characteristics	
Resin content	26 wt %
Volatile content	1.0%
Tack (at 23 °C, or 73 °F)	Good
Width tolerance	±0.51 mm (±0.02 in.)
Filament count (nominal)	5.51/mm (140/in.)
Storage temperature, °C (°F)	−18 (0)
Typical cure conditions	
Temperature °C (°F)	177 (350)
Pressure, kPa (psi)	345 to 585 (50 to 85)
Time, min.	90(b)
Mechanical	
Tensile strength, MPa (ksi)	
At room temperature	1579 (229)
At 127 °C (260 °F)	1309 (190)
Tensile modulus, GPa (10^6 psi)	
At room temperature	227 (33)
At 127 °C (260 °F)	227 (33)
Compressive strength, MPa (ksi)	
At room temperature	2248 (326)
At 127 °C (260 °F)	1600 (232)
Horizontal shear strength, MPa (ksi)	
At room temperature	103 (314)
At 127 °C (260 °F)	62 (9.0)
Flexural strength, MPa (ksi)	
At room temperature	2165 (314)
At 127 °C (260 °F)	2179 (316)
Flexural modulus, GPa (10^6 psi)	
At room temperature	221 (32.0)
At 1276C (260 °F)	207 (30.0)
Density, g/cm^3 (lb/in.^{3})	2.325 (0.084)

(a) Laminates were fabricated and tested by the University of Dayton under AFML Contract (F33615-78-C-5172). (b) Postcure, 4 h at 190 °C (375 °F)

Physical, thermal, mechanical, and electrical properties of reinforced phenolic grades

Property	Cellulose	Mineral	Glass
Specific gravity	1.35 to 1.45	1.50 to 1.70	1.75 to 2.10
Shrinkage, mm/mm	0.006 to 0.008	0.003 to 0.006	0.001 to 0.003
Water absorption, %	0.50 to 0.70	0.15 to 0.40	0.05 to 0.20
UL index, °C (°F)	150 (300)	150 to 170 (300 to 340)	150 to 180 (300 to 360)
Heat-deflection temperature at 1.82 MPa (0.264 ksi), °C (°F)	165 to 205 (330 to 400)	190 to 230 (375 to 450)	175 to 260 (350 to 500)
Coefficient of thermal expansion, 10^{-6}/K	35 to 45	25 to 35	15 to 20
Coefficient of thermal conductivity, W/m^2 · K (Btu · in./s · ft^2 · °F)	29 to 39 (2.0 to 2.7)	39 to 53 (2.7 to 3.7)	49 to 70 (3.4 to 4.9)
Creep modulus at room temperature and 14 MPA (2 ksi), MPa (ksi)	>14 (>2)	>14 (>2)	>17 (>2.5)
Hardness, Rockwell M	100 to 110	105 to 115	110 to 120
Dimensional stability, MIL-M-14, %	0.50 to 0.70	0.45 to 0.55	0.10 to 0.25
Compressive strength, MPa (ksi)	170 to 205 (25 to 30)	170 to 205 (25 to 30)	205 to 280 (30 to 40)
Flexural strength, MPa (ksi)	70 to 90 (10 to 13)	70 to 90 (10 to 13)	80 to 140 (12 to 20)
Elastic modulus, GPa (10^6 psi)	6.9 to 9.7 (1.0 to 1.4)	10.3 to 17.2 (1.5 to 2.5)	17.2 to 20.7 (2.5 to 3.0)
Izod impact strength, MV/m (V/mil)	16 to 58.7 (0.30 to 1.10)	16 to 24 (0.30 to 0.45)	21 to 800 (0.40 to 15.0)
Dielectric strength, J/m (ft · lbf/in.)			
Short-time	13.8 to 15.7 (350 to 400)	15.7 to 16.7 (350 to 425)	14.8 to 16.7 (375 to 425)
Step-by-step	10.8 to 13.8 (275 to 350)	11.8 to 15.7 (300 to 350)	11.8 to 14.8 (300 to 375)
Volume resistivity, Ω · m	10^{10} to 10^{11}	10^{10} to 10^{11}	10^{10} to 10^{11}

Phenolic glass fiber-resin and fiber-resin properties

Property	Phenolic resin	Molding compound glass fiber-resin	Fiber-resin composites: Tape wrapping fabric-resin: E-glass fabric-resin	Carbon fabric-resin	Graphite fabric-resin
Specific gravity	1.28	1.95	1.90	1.45	1.42
Density, g/cm^3 (lb/in.^{3})	1.28 (0.046)	1.95 (0.070)	1.90 (0.069)	1.45 (0.052)	1.42 (0.051)
Service temperature, °C (°F)	150 to 230 (300 to 450)	150 (300)	538 (1000)	3038 (5500)	3038 (5500)
Heat deflection temperature at 1820 MPa (264 psi), °C (°F)	120 to 315 (250 to 600)	NA	NA	NA	NA
UL flammability	94 V-1	94 V-0	94 V-0	...	...
TGA, stability temperature at 5 to 10% wt loss, °C (°F)	230 (450)	NA	...	...	...

Mechanical properties of rubber-toughened and unmodified phenolic composites

Property	Phenolic matrix: Rubber-toughened	Unmodified
Flexural strength, MPa	200	159
Flexural modulus, GPa	18.6	9.7
Flexural strain at failure, %	1.1	1.9
Izod impact strength, notched, J/m	48.1	37.2
Izod impact strength, unnotched, J/m	107.0	133.3
Drop weight crack initiation energy, J	0.33	0.48

Flexural properties of experimental long fiber phenolics

Property	Carbon fiber: 3.2 mm	Carbon fiber: 6.4 mm	Glass fiber: 3.2 mm	Glass fiber: 6.4 mm	Glass fiber: 12.7 mm
Flexural strength, MPa	144	181	172	176	174
Flexural modulus, GPa	26.9	29.2	19.3	18.5	18.3
Failure strain, %	0.73	0.74	1.00	1.01	1.07

Fracture energy of 40% glass-phenolic composites

Material	Fracture energy, G_{Ic}, kJ/m^2: Resin	Composite
Phenolic	0.07	5.52
Phenolic/15% thermoplastic	0.28	9.43

Properties of polyimide thermoset resin and fiber-resin composites

Property	Resin	50% glass fiber-resin
Heat deflection temperature at 1820 kPa (264 psi), °C (°F)	305 to 360 (582 to 680)	350 (660)
In-service temperature, °C (°F)	260 to 370 (500 to 700)	260 (500)
Specific gravity	1.43	1.65
Density, g/cm^3 (lb/in.^{3})	1.43 (0.052)	1.65 (0.060)

Properties of fiberglass-filled S/MA resins

Property	ASTM test method	Melt precompounded products: High modulus	High toughness	High toughness	High temperature	Dry blends: High modulus	High temperature
Mechanical							
Tensile strength, MPa (ksi)	D 638	88.9 (12.9)	51.0 (7.4)	77.9 (11.3)	70.3 (10.2)	87.5 (12.7)	78.5 (11.4)
Flexural strength, MPa (ksi)	D 790	144 (20.9)	91.7 (13.3)	132 (19.1)	122 (17.7)	152 (22.0)	137 (19.9)
Flexural modulus, GPa (10^6 psi)	D 790	6.72 (0.975)	3.55 (0.515)	5.86 (0.850)	5.45 (0.790)	6.82 (0.990)	6.13 (0.890)
Izod impact, J/m (ft · lbf/in.)	D 256	69 (1.3)	91 (1.7)	101 (1.9)	101 (1.9)	140 (2.7)	160 (3.0)
Mold shrinkage, cm/cm or in./in.	D 955	0.003	0.004	0.003	0.003	0.003	0.003
Water absorption, %	D 570	0.11	0.12	0.12	0.19	0.19	0.19

(continued)

Properties of fiberglass-filled S/MA resins (continued)

Property	ASTM test method	Melt precompounded products: High modulus	Melt precompounded products: High toughness	Melt precompounded products: High toughness	Melt precompounded products: High temperature	Dry blends: High modulus	Dry blends: High temperature
Thermal							
Vicat softening point, °C (°F)	D 1525	123 (253)	124 (255)	124 (255)	135 (275)	123 (253)	135 (275)
Deflection temperature under load, 1.82 MPa (0.264 ksi), 3.2 mm (1/8 in.) specimen, °C (°F)	D 648	114 (238)	107 (224)	111 (231)	118 (245)	115 (239)	122 (252)
Coefficient of linear thermal expansion, 10^{-5}/K	D 696	3.6	4.5	3.6	3.6	3.6	3.6
Burn rate, cm/min. (in./min.)	MVSS-302(a)	2.3 (0.9)	5.1 (2.0)	1.8 (0.7)	5.1 (2.0)	5.1 (2.0)	5.1 (2.0)
Other							
Specific gravity, 23 °C (73 °F), g/cm^3	D 792	1.22	1.13	1.20	1.21	1.23	1.22
Melt flow, g/10 min. condition L	D 1238	0.9	0.6	0.35	0.4	...	...
Glass fiber content, %	TGA(b)	20	10	20	20	20	20

(a) MVSS, Motor Vehicle Safety Standard. (b) TGA, thermogravimetric analysis

Performance comparison for reinforced S/MA with random glass orientation

Property(a)	Reinforcement: Chopped Glass	Reinforcement: Long Glass
Flexural strength, MPa		
Parallel	170	175
Diagonal	135	163
Perpendicular	117	173
Flexural modulus, GPa		
Parallel	10.3	8.9
Diagonal	7.8	7.4
Perpendicular	6.4	7.3
Flexural toughness, MPa		
Parallel	0.20	0.22
Diagonal	0.17	0.28
Perpendicular	0.17	0.35
Izod impact strength, notched, J/m		
Parallel	75	200
Diagonal	75	290
Perpendicular	80	370

(a) Test axis is referenced to the flow direction

Performance comparison for reinforced S/MA with preferred glass orientation

Property	Reinforcement: Chopped Glass	Reinforcement: Long Glass
Flexural strength, MPa	200	250
Flexural modulus, GPa	12.0	12.6
Flexural toughness, MPa	0.2	0.4
Izod impact strength, notched, J/m	90	420
Penetration impact, maximum load, N	1290	2000

Mechanical properties of glass fiber reinforced vinyl ester

Property	Glass content, wt %: 40	Glass content, wt %: 52
Tensile strength, MPa (ksi)	143 (21)	261 (40)
Tensile modulus, GPa (10^6 psi)	12 (1.7)	15 (2.2)
Flexural strength, MPa (ksi)	204 (30)	342 (50)
Flexural modulus, GPa (10^6 psi)	7 (1.0)	14 (2.0)

Corrosion resistance of fiber-reinforced vinyl ester resins versus competitive materials in various chemical environments

Materials	Sulfuric acid, dilute	Sulfuric acid, concentrate	Hydrochloric acid, dilute	Hydrochloric acid, concentrate	Hydro-fluoric acid	Phosphoric acid, dilute	Phosphoric acid, concentrate	Sodium hydroxide, dilute	Sodium hydroxide, concentrate	Acid chloride salts	Bleach	Wet chlorine	Nitric acid
Fiber-reinforced vinyl ester	R	R to 75%	R	R	R	R	R	R	R to 50%	R	R	R	R to 40%
Teflon-lined fiber-rein forced vinyl ester	R	R	R	R	R	R	R	R	R	R	R	R	R
Carbon steel (1020)	NR	R above 85%	NR	NR	NR	NR	NR	R	R	NR	NR	NR	NR
316 stainless	R below 5%	R above 85%	NR	NR	NR	R	R	R	NR	NR	NR	NR	R
Hastelloy C	R	R	R	R	R	R	R	R	R	R	R	R	R
Aluminum	NR	NR	NR	NR	NR	NR	NR	NR	NR	NR	NR	NR	NR

Note: R, resistant; NR, not resistant

Maximum service temperature versus chemical environment for corrosion-resistant reinforced vinyl ester resins

	Vinyl ester types			
	BPA		Novolac	
Chemical environment	°C	°F	°C	°F
Hydrochloric acid, 37%	65	150	80	180
Hydrofluoric acid, 10%	65	150	65	150
Sulfuric acid, 75%	40	100	50	120
Sulfuric acid, 70%	80	180	80	180
Phosphoric acid, 100%	100	210	100	210
Sodium hydroxide, 25%	80	180	80	180
Toluene, 100%	25	80	50	120
Chlorine dioxide, all concentrations	80	180	80	180
Chlorine water, saturated	80	180	100	210
Gasoline, unleaded	80	180	80	180

Processing

Process comparisons for glass fiber composites

	Resin transfer molding	Injection molding	Pultrusion	RRIM	Hand lay-up spray-up
Factor limiting maximum size of part	Machine size	Machine size	Materials	Metering equipment	Mold size: part transport
Maximum size to date, m^2 (ft^2)	9.3 (100)	9.3 (100)	1.6 mm to 1.22 m × 2.44 m (1/16 to 48 × 96 in.)	4.6 (50)	279 (3 000)
Shape limitations	Moldable	Moldable	Round, rectangular	Moldable	None
Usual useful production	Medium	High	Medium	Medium-high	Low-medium
Volume, parts/year	1000 to 20 000	50 000 to 1 000 000	3050 m (10 000 ft)	15 000 to 100 000	0 to 1 000
Production cycle time	10 to 20 min.	15 s to 15 min.	10 to 30 min.	1 to 2 min.	3 min. to 24 h
Typical glass content, %	15 to 25	20 to 40	30 to 75	5 to 25	20 to 35
Strength orientation	Random	Random	Highly oriented	With flow	Random (usually)
Strength category	Low to medium	Low	High	Low	Medium
Wall thickness:					
Minimum, mm (in.)	0.76 (0.03)	0.76 (0.03)	1.6 (1/16)	2.03 (0.080)	0.76 (0.030)
Maximum, mm (in.)	25.4 (1.0)	12.7 to 25.4 (1/2 to 1)	12.7 (1/2)	12.7 (0.500)	38 and up (1.5 and up)
Tolerance, mm (in.)	...	...	+0.25 ± 25.4 (+0.01 ± 1.0)	±0.05 (±0.002)	+0.51 (+0.02)
Variations	Uniform	Uniform	Uniform	Uniform	As desired
Minimum draft:					
To depth 150-mm (6-in.)	1 °	1 °	0 to 2 °	1 ° to 3 °	0 to 2 °
Over depth 150-mm (6-in.)	1 °	1 ° +	0 to 2 °	3 °+	0 to 2 °
Minimum inside radius, mm (in.)	1/2 part depth	1/2 part depth	1.5 (0.06)	1/2 part depth	6.4 (0.25)
Ribs	Yes	Yes	No	Yes	Yes
Bosses	Yes	Yes	No	Yes	Yes
Undercuts	Possible	Possible	No	Yes, with proper	Avoid
Holes:					
Parallel	Yes	Yes	No	Yes	Yes
Perpendicular	Yes	Yes	No	Yes	Yes
Built-in cores	Yes	Yes	No	Yes	Possible
Metal inserts	Yes	Yes	No	Yes	Yes
Metal edge stiffeners	Yes	Yes	No	Yes	Yes
Surface finish:					
Number of finished surfaces	All	2	2	2	1
Quality of surface	Excellent	Excellent	Fair to good	Excellent	Excellent
Gel-coat surface	...	Yes	No	No	Yes
Surfacing mat	Yes	Yes	No	No	Yes
Combination with thermoplastic liner	Yes	Yes	No	No	Yes
Trim in mold	No	No	No	Yes	No
Molded-in labels	Yes	Yes	No	No	Yes
Raised numbers	Yes	Yes	No	Yes	Yes
Translucency	Yes	Yes	No	No	Yes
Tool cost	High	High	Low	Low-medium	Low
Capital equipment cost	High	High	Low	Low-medium	Low

Process comparisons for glass fiber composites

	Filament winding	Compression: sheet molding compound	Compression: bulk molding compound	Perform molding
Factor limiting maximum size of part	Winding machine	Press rating and size	Press rating and size	Press rating and size
Maximum size to date, m^2 (ft^2)	93 (1 000)	4.6 (50)	4.6 (50)	18.6 (200)
Shape limitations	Surface of revolution	Moldable	Moldable	Moldable
Usual useful production	Low-medium	High	High	High
Volume, parts/year	0 to 1 000	1 000 to 1 000 000	1 000 to 1 000 000	1 000 to 1 000 000
Production cycle time	5 min.	1 1/2 to 5 min.	1 1/2 to 5 min.	1 1/2 to 5 min.
Typical glass content, %	65 to 90	15 to 35	15 to 35	24 to 45
Strength orientation	Highly oriented	Random	Random	Random
Strength category	Very high	Low-medium	Low-medium	Medium-high
Wall thickness:				
Minimum, mm (in.)	0.25 (0.01)	0.76 (0.03)	1.5 (0.06)	0.76 (0.03)
Maximum, mm (in.)	51 (2.0)	6.4 (0.25)	25.4 (1.0)	6.4 (0.25)
Tolerance, mm (in.)	+0.25 (+0.01)	+0.20 (+0.008)	+0.13 (+0.005)	+0.13 (+0.005)
Variations	As desired	Uniform desirable <3:1	As desired	Uniform desirable <2:1
Minimum draft:				
To depth 150-mm (6-in.)	3 °	1 ° to 3 °	1 ° to 3 °	1 ° to 3 °
Over depth 150-mm (6-in.)	3 °+	3 °+	3 °+	3 °+
Minimum inside radius, mm (in.)	3.18 (0.125)	1.5 (0.06)	1.5 (0.06)	3.18 (0.125)
Ribs	No	Yes	Yes	Not recommended

(continued)

Process comparisons for glass fiber composites (continued)

	Filament winding	Compression: sheet molding compound	Compression: bulk molding compound	Perform molding
Bosses	No	Yes	Yes	Not recommended
Undercuts	No	Avoid	No	No
Holes:				
Parallel	Yes	Yes	Yes	Not recommended
Perpendicular	...	Undesirable	Undesirable	Undesirable
Built-in cores	Possible	Possible	Possible	Possible
Metal inserts	Yes	Yes	Yes	Not recommended
Metal edge stiffeners	No	No	No	Yes
Surface finish:				
Number of finished surfaces	1	2	All	2
Quality of surface	Excellent	Very good	Excellent	Very good
Gel-coat surface	Yes	No	No	Yes
Surfacing mat	Yes	No	No	Yes
Combination with thermoplastic liner	Yes	No	No	No
Trim in mold	Yes	Yes	Yes	Yes
Molded-in labels	Yes	Difficult	Difficult	Difficult
Raised numbers	...	Yes	Yes	Yes
Translucency	Yes	No	No	Yes
Tool cost	Low	High	High	Medium

Process reinforcement capabilities and selected properties

Process	Reinforcement Type	%	Limitations to reinforcement	Properties: Surface finish (++ to −−)	Flexural modulus GPa	Flexural modulus 10^6 psi	Temperature resistance (trend)	Tendency to warp (high to low)
Thermoplastics								
Injection	None	...	Longer than 2 mm (0.08 in.) is difficult;	++	2.0	0.3	Low	High
	Short glass	40	glass increases stiffness but lowers	+	4.0	0.6	to	High
	Long glass	50	impact	−−	7.0	1.0	high	Very high
Injection compression	None	...	Can handle glass up to 50 mm (2 in.), but	++	2.0	0.3	Low	Slight
	Short glass	40	is better with 10 mm (0.4 in.) or less	+	4.0	0.6	to	Slight
	Long glass	60		−	7.5	1.2	high	High
Hollow injection	None	...	Single large gate allows use of up to 10	++	2.0	0.3	Low	Slight
	Short glass	40	mm (0.4 in.), but easier with 2 mm	+	4.0	0.6	to	Slight
	Long glass	50	(0.08 in.)	−−	7.0	1.0	medium	High
Foam injection	None	...	Glass tends to increase stiffness but	−	2.5	0.4	Low	Slight
	Short glass	20	spoils cell structure, surface, impact	−−	5.0	0.7	to	High
	Long glass	40		−−	8.0	1.2	medium	High
Sandwich molding	Glass in core	40	<10 mm (<0.4 in.) is acceptable with development	++	3.5	0.5	Low	High
Compression	Very long glass	40	Glass orientation in ribs	−−	5.0	0.7	Medium	Slight
Stamping	None	...	Single thickness only, but up to 70%	+	0.5	0.1	Low	High
	Very long glass	40	if cloth used	−	9.5	1.4	Medium	High
	Continuous	70		−	14.0	2.0	High	High
Blow molding	None	...	Very difficult	+ to −−	1.0	0.2	Low	High
Thermoforming	None	...	Very difficult	+	1.0	0.2	Low	Very high
Filament winding	Continuous	70	Shape limited	−−	40.0	6.0	High	Low
Rotational casting	None	...	Difficult	+	1.0	0.2	Low	Slight
Thermosets								
Compression	Long glass (SMC)	30	Very long glass limits material flow and	+	7.0	1.0	High	Very low
	+ very long (HMC)	50	surface finish	−	15.0	2.1	High	Low
Injection	Long glass (BMC)	20	Difficult to handle if longer than 30 mm	++	5.5	0.8	High	Very low
	Long glass (ZMC)	30	(1.2 in.); glass breakage	++	7.0	1.0	High	Very low
Stamping	Very long glass	70	No ribs or bosses	−	15.0 to 35.0	2.2 to 5.0	High	Slight
Reaction injection molding (RIM)	None	...	Reduced impact as glass loading	++	0.5	0.1	Low	Slight
	Short or flake	20	increases	+	2.0	0.3	Medium	High
Resinject (resin transfer molding)	Very long glass	55	Glass bulk limits maximum	+	12.5 to 27.5	1.8 to 4.0	High	Low
Foam polyurethane	None	...	Simple shapes and only very open glass	+ to −−	0.3	<0.1	Low	Slight
	Very long glass	30	structures	−−	7.0	1.0	Low	Slight
Filament winding	Continuous	70	Limited cross strength	−−	40.0	6.0	High	Low
Hand lay-up/vacuum bagging	Very long glass/cloth	30	...	++ to −−	7.0	1.0	High	Slight

Property comparison by process

Process	Reinforcement, wt%	Tensile strength, MPa	Tensile strength, ksi	Tensile modulus, GPa	Tensile modulus, 10^6 psi	Flexural strength, MPa	Flexural strength, ksi	Compressive strength, MPa	Compressive strength, ksi	Impact strength, J/m	Impact strength, ft · lbf/ft	Thermal conductivity, W/m · K	Thermal conductivity, Btu · in./h · ft^2 · °F	Heat distortion at 1.8 MPa, °C	Heat distortion at 1.8 MPa, °F	Dielectric strength, kV/cm	Dielectric strength, kV/in.
Spray	30 to 50 glass-polyester	60 to 120	9 to 18	5.5 to 12	0.8 to 1.8	110 to 190	16 to 28	100 to 170	15 to 25	210 to 640	48 to 144	0.17 to 0.23	1.2 to 1.6	175 to 205	350 to 400	80 to 160	200 to 400
Compression	15 to30 glass-SMC	55 to 140	8 to 20	11 to 17	1.6 to 2.5	120 to 210	18 to 30	100 to 210	15 to 30	430 to 1150	96 to 264	0.19 to 0.25	1.3 to 1.7	205 to 260	400 to 500	120 to 180	300 to 450
Compression	25 to50 glass mat-polyester	170 to 210	25 to 30	6.2 to 14	0.9 to 2.0	70 to 280	10 to 40	100 to 210	15 to 30	530 to 1050	120 to 240	0.19 to 0.26	1.3 to 1.8	175 to 205	350 to 400	120 to 240	300 to 600
Filament winding	30 to80 glass-epoxy	550 to 1700	80 to 250	28 to 62	4.0 to 9.0	690 to 1850	100 to 270	310 to 480	45 to 70	2150 to 3200	480 to 720	0.27 to 0.33	1.9 to 2.3	175 to 205	350 to 400	120 to 160	300 to 400
Pultrusion	40 to80 glass mat-polyester	410 to 1050	60 to 150	28 to 41	4.0 to 6.0	690 to 1050	100 to 150	210 to 480	30 to 70	2400 to 3200	540 to 720	0.27 to 0.33	1.9 to 2.3	205 to 260	400 to 500	80 to 160	200 to 400
Pultrusion	30 to50 glass mat-polyester	80 to 210	12 to 30	6.9 to 17	1.0 to 2.5	170 to 210	25 to 30	210 to 340	30 to 50	530 to 1350	120 to 300	0.22 to 0.27	1.5 to 1.85	95 to 150	200 to 300	80 to 120	200 to 300
Pultrusion	30 to55 glass mat and roving-vinyl ester resin	70 to 280	10 to 40	6.9 to 21	1.0 to 3.0	100 to 280	15 to 40	140 to 340	20 to 50	270 to 1600	60 to 360	0.22 to 0.33	1.5 to 2.3	175 to 230	350 to 450	80 to 130	200 to 325
Pultrusion	30 to55 glass mat and roving-polyester resin	50 to 240	7 to 35	5.5 to 17	0.8 to 2.5	70 to 210	10 to 30	100 to 280	15 to 40	210 to 1350	48 to 300	0.22 to 0.33	1.5 to 2.3	175 to 205	350 to 400	80 to 120	200 to 300

The range of values reflects transverse and axial testing directions as well as percent of reinforcement

Equivalent-thickness factor of FRP pultrusion relative to traditional structural materials

		Steel			Aluminum			Wood		
Pultrusion construction	Specific gravity	Tensile strength	Rigidity	Flexural strength	Tensile strength	Rigidity	Flexural strength	Tensile strength	Rigidity	Flexural strength
50% mat and roving	1.85	2.5	2.15	1.82	1.0	1.49	1.16	0.25	0.79	0.45
70% roving only	2.00	1.0	1.71	1.12	0.4	1.19	0.71	0.10	0.63	0.27

Metal Matrix Composites

General

Classification of most composite system interfaces

Class I(a)	Class II(b)	Class III(c)
Copper/tungsten	Copper (chromium)/ tungsten	Copper (titanium)/ tungsten
Copper/alumina	Eutectics	Aluminum/carbon (<700 °C)
Silver/alumina	Niobium/tungsten	Titanium/alumina
Aluminum/BN-coated boron	Nickel/carbon	Titanium/boron
Magnesium/boron	Nickel/tungsten(e)	Titanium/silicon carbide
Aluminum/boron(d)		Aluminum/silica
Aluminum/stainless steel(d)		
Aluminum/silicon carbide(d)		

Coefficients of thermal expansion of matrices and reinforcing fibers

Matrix	Expansion, 10^{-6}/°C	Filament	Expansion, 10^{-6}/°C
Aluminum	23.9	Boron	6.3
Titanium	8.4	Borsic	6.3
Iron	11.7	SiC	4.0
Nickel	13.3	Alumina	8.3
Silicon	2.5	Carbon(a)	~0
Copper	17	SiO_2	~0
Silver	19	Tungsten	4.5
Magnesium	26	Al_2O_3	8
		Beryllium	12

(a) Parallel to basal plane

Some fiber/metal combinations

Matrix	C	B	Al_2O_3	SiC	Steel	Be	W	Ni_3Nb
Ti		+	+	+		+		
Al	+	+	+	+	+	+		
Ni	+		+	+			+	+

Comparison of fiber-matrix reactions for various matrix metals

Annealing temperature °C	°F	Matrix	No. of compositions investigated	Relative number of cases, %: Recrystallization	Intermetallic compound	Diffusion penetration	No recrystallization	No reaction
1200	2190	Nickel-based	27	93	55	...	7	4
		Cobalt-based	29	10	83	12	90	10
		Iron-based	30	3	30	...	97	70
1300	2370	Nickel-based	27	96	63	...	4	4
		Cobalt-based	19	21	84	...	79	10
		Iron-based	30	20	80	3	80	13

Strength properties of several types of fiber-reinforced metals

Filament	Matrix	Fabrication technique(a)	Test temperature, °C	V_f	L/d_f	S_c MPa	S_c ksi	$\bar{S}_f$ MPa	$\bar{S}_f$ ksi	$\bar{S}_f^*$ MPa	$\bar{S}_f^*$ ksi	β
Metallic filements												
Stainless steel	Al	HPF	RT	0.11	250	179	26.9	1515	220	1405	204.0	0.93
W	Cu	VI	RT	0.63	∞	772	112.0	1240	180	1215	176.0	0.98
W	Cu	VI	250	0.69	∞	855	124.0	1240	180	1225	178.0	0.99
W	Cu	VI	250	0.56	40	558	81.0	1240	180	972	141.0	0.78
W	Cu	VI	250	0.51	20	490	71.0	1240	180	938	136.0	0.76
W	Cu	VI	250	0.57	10	386	56.0	1240	180	655	95.0	0.53
W	Cu	VI	RT	0.768	∞	1755	254.9	2255	327	2250	326.0	~1.00
W	Cu	VI	RT	0.357	75	827	120.0	2255	327	2220	322.0	0.99
Oxide filaments												
Al-coated E-glass	Al	HP	RT	0.50	∞	310	45.0	896	130	593	86.0	0.66
Al-coated E-glass	Al	HP	480	0.50	∞	178	28.85	731	106	342	49.6	0.47
Al-coated E-glass	Al	HP	540	0.50	∞	135	19.63	710	103	263	38.1	0.37
Al-coated E-glass	Al	VI	RT	0.50	∞	121	17.5	896	130	230	33.4	0.26
Al-coated E-glass	Al	VI	480	0.50	∞	96	13.89	731	106	189	27.48	0.26
Al-coated SiO_2	Al	HP	RT	0.48	∞	871	126.3	3035	440	1725	250.0	0.57
Al-coated SiO_2	Al	HP	100	0.48	∞	958	139.0	3035	440	1935	281.0	0.65
Al-coated SiO_2	Al	HP	500	0.48	∞	288	41.8	1795	260	593	86.0	0.33
Pb-coated E-glass	Pb	VI	RT	0.08	∞	103	15.0	1515	220	1055	153.0	0.70

(a) HPF, hot pressed filaments on powdered matrix; VI, vacuum infiltration; HP, hot pressed metal-coated filaments. (b) S_c, composite tensile strength; $\bar{S}_f$, average strength of fibers tested individually; $\bar{S}_f^*$, average strength of fibers in the matrix during fracture; β, effective fiber strength factor = $\bar{S}_f^*/\bar{S}_f$

Room-temperature tensile and specific strengths of various fiber-reinforced composites(a)

Composition Matrix	Composition Fiber	Density, lb/in.^{3} Matrix	Density, lb/in.^{3} Fiber	Fiber content(b), vol %	Composite Tensile strength(c), ksi	Composite Strength/density, 10^3 in.	Source
Al	B/W	0.097	0.095	50	110	1150	United Aircraft Research Laboratory
	B	...	...	10	43	453	GE-AETD
	Steel	0.097	0.282	25	173	1210	Harvey Aluminum Co.
	Be	0.097	0.067	40	80	830	North American
	SiO_2	0.097	0.079	48	118	1340	Rolls-Royce Ltd.
	Al_2O_3	0.097	0.143	35	161	1425	GE-SSL
	B_4C	0.097	0.091	10	29	302	GE-SSL
	$CuAl_2$	0.097	0.157	50	39	307	United Aircraft Research Laboratory
	Al_3Ni	0.097	0.143	10	48	470	United Aircraft Research Laboratory
Nb	Nb_2C	0.299	0.286	31	172	570	United Aircraft Research Laboratory
Ni	B/W	0.320	0.095	75	384	1470	General Technology, Inc.
	W	0.320	0.695	9.4	61.4	173	Battelle N.W.
	W	...	...	40	161	344	NGTE (England)
	Al_2O_3	0.320	0.143	19	171	600	GE-SSL
	C	0.320	0.054	56	80	467	Union Carbide Corp.
Al-10.2Si	Al_2O_3	0.096	0.143	15(d)	40.7	395	Melpar, Inc.
Ni-20Cr	Al_2O_3	0.308	0.143	9	255	870	Horizons, Inc.
NiCr	W	...	...	22	73	185	Clevite Corp.
Fe	Al_2O_3	0.284	0.143	36	237	1017	Horizons, Inc.
Ta	Ta_2C	0.598	0.544	29	155	267	United Aircraft Research Laboratory, twice solidified
	Ta_2C	0.598	0.544	29	118	203	United Aircraft Research Laboratory, cold swaged, 67% reduction
Ag	Si_3N_4	0.378	0.115	15(d)	40	119	Explosive Research Development Est. (England)
	Al_2O_3	0.378	0.143	24	232	720	GE-SSL
	Steel	...	...	44	65	191	MIT
Ti	Mo(e)	...	...	20	96	457	Clevite Corp.
Cu	W	...	...	77	255	420	NASA-Lewis
Co	W	...	...	30	107	244	Clevite Corp.
	Mo	...	...	17	52	156	Clevite Corp.
316 stainless steel	W	...	...	18	58.6	175	Clevite Corp.

(a) Early and developmental metal-matrix composite work. (b) Unidirectional fiber orientation except where otherwise noted. (c) Highest reported values. (d) Random fiber orientation. (e) Short, discontinuous fibers and whiskers

Typical mechanical properties of some metal-matrix composites

Fiber	Matrix	Reinforcement, vol%	Density, g/cm^3	Longitudinal tensile strength, MPa	Longitudinal modulus, GPa	Transverse tensile strength, MPa	Transverse modulus, GPa
G T50	201 Al	30	2.380	620	170	50	30
G T50	201 Al	49	...	1120	160	...	...
G GY70	201 Al	34	2.380	660	210	30	30
G GY70	201 Al	30	2.436	550	160	70	40
G HM pitch	6061 Al	41	2.436	620	320	...	...
G HM pitch	AZ31 Mg	38	1.827	510	300	...	...
B on W, 142-μm fiber	6061 Al	50	2.491	1380	230	140	160
Borsic	Ti	45	3.681	1270	220	460	190
G T75	Pb	41	7.474	720	200	...	...
G T75	Cu	39	6.090	290	240	...	...
FP	201 Al	50	3.598	1170	210	(140)	140
SiC	6061 Al	50	2.934	1480	230	(140)	140
SiC	Ti	35	3.931	1210	260	520	210
SiC whisker	Al	20	2.796	340	100	340	100
B_4C on B	Ti	38	3.737	1480	230	>340	>140
G T75	Mg	42	1.799	450	190	...	...
G HM	Pb	35	7.750	500	120	...	...
G T75	Al-7% Zn	38	2.408	870	190	...	...
G T75	Zinc	35	5.287	770	120	...	...
G T50	Nickel	50	5.295	790	240	...	...
G T75	Nickel	50	5.342	828	310	30	40
G (81.3 μm)	2024 Al	50	2.436	760	140	...	...
G (142 μm)	2024 Al	60	2.436	1100	180	...	...
Superhybrid	Grafitic	60	2.048	860	120	220	60
Superhybrid	S-glass	60	2.159	740	80	190	30
Superhybrid	Kevlar	60	1.799	700	80	190	10

Factors affecting the realizable strengths of fiber-reinforced metals

Factor	Ideal case
Fiber parameter–strength	
Initial strength	Should be at a maximum
Scatter in strength	Variation in strength between individual fibers should be minimized
Strength retention	Should be maximized during handling, fabrication, etc.
Fracture statistics	All fibers should be loaded evenly. Also depends on points discussed above for initial strength and scatter in strength
Fiber parameter–shape and size	
Taper	Should be minimized
Diameter	Should be minimized. Fiber strength usually increases with decreasing diameter. Statistically, more fibers can be packed in per unit volume for small diameters
Length	Should be optimized in terms of length greater than critical length. Note: strength also increases with decreasing length
Variation in fiber diameter	Should be as uniform as possible
Cross section	Not all fibers are circular, hence cross-sectional shape will affect maximum packing density and stress distribution
Fiber parameters–relation to matrix	
Orientation	Fibers should be parallel to tensile axis. However, properties can be tailor-made by varying orientation
Distribution	Fibers should be uniformly distributed in the matrix
Overlap	Short fibers must have a minimum overlap distance; otherwise, crack in matrix will propagate through regions between fiber ends
Fiber-fiber	Fibers must be separated by matrix; otherwise, contact points serve as stress concentrators
Matrix parameters	
Mechanical	Rate of work hardening should be optimized, properties relative to fibers–i.e., E_m, E_f
Chemical	Should wet fibers, but not react to weaken fibers, particularly at high temperatures; oxidation and corrosion resistance are important
Thermal	Coefficient of expansion should be similar to that of fiber, recrystallization, temperature
Fabrication parameters	
Fiber production	Factors include cost to obtain optimum values cited under "Fiber parameters"
Fiber handling	Technique depends on length and length -to-diameter ratio, and on properties desired in composite (see "Fiber parameters–relation to matrix"); surface treatment (metallizing)
Incorporation into matrix	Many techniques–powder metallurgy, infiltration, impact forming, etc. Fibers must be wetted and bonded to matrix. Must form composite of theoretical density
Composite shaping	Extrusion, forging, pressing–depends on fiber brittleness (and size)
Joining	Chemical, pressure, electron welding, and diffusion bonding techniques are more effective than mechanical joining

Methods used for fabrication of metallic-matrix whisker composites

Method used to combine constituents	Matrix	Whisker(a) Composition	Alignment	Coating	Consolidation process(b)
Deposition (molecular)					
Chemical vapor deposition	Ni or NiCr	Al_2O_3, SiC	R	C	HP or LPHP
Electrodeposition	Ni	Al_2O_3	R	C	HP
	Ni	Al_2O_3, SiC	R	C, U	
	Ni	Al_2O_3	A	C, U	As deposited or hot forged
	Ni	Al_2O_3	A	C	HP
Electroplate	Ni	Al_2O_3, SiC	A, R	C	Cold and hot rolling, HP
Electroform	Ni	Al_2O_3	A	C	...
Liquid (matrix)					
Alloy (eutectic)	Al, Nb	$NiAl_3$, Nb_2C	A	U	Unidirectional solidification
	Resin	Al_2O_3	A	C	Polymerize matrix
	Ag	Al_2O_3	A	C	Solidification of matrix
	Cu, Ni-alloys	Al_2O_3	A	C	Solidification of matrix
Infiltration	Al	B_4C	A	C	Solidification of matrix
	Al-alloys	Al_2O_3	A	C	Solidification of matrix
Melting of powdered matrix	Fe, NiCr	Al_2O_3	R	C, U	LPHP
Casting of melt and whiskers	Co-alloys	Al_2O_3, SiC	R	C, U	Solidification of melt
	Ag, Fe, Ni	Si_3N_4	A	U	Dry, burn off organic carrier, HP
Spin, extrude, draw slurry of whisker matrix powder, and carrier solution	Al	SiC	A	U	Dry, burn off organic carrier, LPHP
	Al-alloy	SiC	A	U	Dry, burn off organic carrier, HP
	Cu, Al, Mg	SiC	A	U	Dry, burn off, sinter, HP
	Al-alloy	SiC	A	U	Dry, burn off, HP
	Ag-alloy	Si_3N_4	R	U	LPHP
Filter slurry or settle out whiskers and matrix	Cu, Al-alloy	Al_2O_3, SiC	R	U	HP
	NiCr, Al-alloy	SiC	A	C, U	LPHP
	NiCr, Al-alloy	Al_2O_3, SiC	A, R	C, U	LPHP
	Mg-alloy	Al_2O_3, SiC	A, R	C, U	LPHP
Impregnation into whisker strands and tapes	Resin	SiC	A	U	HP and squeeze out excess resin
	Resin	Al_2O_3	A	U	HP and squeeze out excess resin
	Resin	Si_3N_4	A	U	HP and squeeze out excess resin
Solid state					
Powder matrix and whiskers	Ni	Al_2O_3	R,	C	HP
	Al	Si_3N_4	A	U	Extrusion
	Ni, Ti	Al_2O_3, SiC	R	U	High-energy-rate forming
Deposit whiskers on Al-foil	Al	SiC	A	U	Diffusion bonding

Method used to combine constituents	Matrix	Forming or shaping process	Remarks
Deposition (molecular)			
Chemical vapor deposition	Ni or NiCr	Hot roll	SiC whiskers reacted chemically with matrix
Electrodeposition	Ni	Hot roll	Problem of voids and low whisker content
	Ni	Hot roll	Problem of voids and low whisker content
	Ni	Hot roll	Align whiskers via flow in electroplating bath
	Ni	Hot roll	Problem of consolidation and whisker breakage
Electroplate	Ni	Hot roll	SiC whiskers reacted with matrix, also whisker breakage
Electroform	Ni	Hot roll	Excellent properties, limited to very small specimens
Liquid (matrix)			
Alloy (eutectic)	Al, Nb	Extrude, roll	Whisker content fixed by alloy composition
	Resin	Extrude, roll	Achieved high-strength composites
	Ag	Extrude, roll	Achieved high-strength composites
	Cu, Ni-alloys	Extrude, roll	Achieved varying degrees of success; coating stability a problem
Infiltration	Al	Extrude, roll	Achieved varying degrees of success; coating stability a problem
	Al-alloys	Extrude, roll	Achieved varying degrees of success; coating stability a problem
Melting of powdered matrix	Fe, NiCr	Extrude, roll	Problems of interfacial reactions and whisker segregation
Casting of melt and whiskers	Co-alloys	Extrude, roll	Whiskers concentrated at grain boundaries, low whisker content
	Ag, Fe, Ni	Extrude, roll	Excellent whisker alignment and packing
Spin, extrude, draw slurry of whisker matrix powder, and carrier solution	Al	Extrude, roll	Excellent whisker alignment and packing
	Al-alloy	Extrude, roll	Excellent whisker alignment, fabrication of complex shapes
	Cu, Al, Mg	Extrude, roll	Problem of matrix porosity and wetting
	Al-alloy	Extrude, roll	Excellent whisker alignment
	Ag-alloy	Extrude, roll	Excellent whisker alignment, problem carbon burn-off
Filter slurry or settle out whiskers and matrix	Cu, Al-alloy	Hot roll	Little whisker alignment, much whisker breakage
	NiCr, Al-alloy	Clad and roll	Ni-coated whiskers were aligned magnetically during settling in slurry
	NiCr, Al-alloy	Clad and roll	Ni-coated whiskers were aligned magnetically during settling in slurry
	Mg-alloy	Hot extrusion	Successfully rolled composites with aligned whiskers

(continued)

Methods used for fabrication of metallic-matrix whisker composites (continued)

Method used to combine constituents	Matrix	Forming or shaping process	Remarks
Impregnation into whisker strands and tapes	Resin	Hot extrusion	Achieved excellent alignment, little whisker breakage
	Resin	Hot extrusion	Achieved excellent alignment, little whisker breakage
	Resin	Hot extrusion	Achieved excellent alignment, little whisker breakage
Solid state			
Powder matrix and whiskers	Ni	Hot extrusion	Many whiskers broken
	Al	Hot extrusion	Extended long rods, but many whiskers broken
	Ni, Ti	Hot extrusion	Chemical reactions minimized, but whiskers still broken
Deposit whiskers on Al-foil	Al	Hot extrusion	Achieved excellent alignment of whiskers

(a) R, random whisker alignment; A, whiskers unidirectionally aligned; C, coated whiskers; U, uncoated whiskers. (b) HP, hot pressing; LPHP, liquid phase hot pressing

Carbon/Metal

Typical properties of 55MSI graphite/6061 aluminum composites

Density, g/cm³	Density, lb/in.³	Reinforcement content, vol %	Fiber orientation	Modulus of elasticity, GPa	Modulus of elasticity, 10^6 psi
2.35	0.085	34	0°	182.2 ± 6.6	26.43 ± 0.95
2.35	0.085	34	90°	33	4.8

Comparative properties of various graphite fiber reinforced metals

Composite	Fiber content, vol %	Strength, ksi	Modulus of elasticity, 10^6 psi	Density, lb/in.³	Strength/density, 10^6 in.	Modulus/density, 10^6 in.
Graphite(a)/lead	41	104	29.0	0.270	0.385	107.0
Graphite(b)/lead	35	72	17.4	0.280	0.260	62.3
Graphite(a)/zinc	35	110.9	16.9	0.191	0.580	88.5
Graphite(a)/magnesium	42	65	26.6	0.064	1.016	393.7

a) Thornel 75 fiber. (b) Courtaulds HM fiber

Typical properties of graphite-magnesium castings(a)

Fiber type	Fiber content/orientation	Casting	Fiber preform method	Tensile strength, 0°, GPa	Tensile strength, 0°, 10^6 psi	Modulus of elasticity, 0°, GPa	Modulus of elasticity, 0°, 10^6 psi	Tensile strength, 90°, GPa	Tensile strength, 90°, 10^6 psi	Modulus of elasticity, 90°, GPa	Modulus of elasticity, 90°, 10^6 psi	Coefficient of thermal expansion, 10^{-6}/K
P55	40%/0°	Rod	Filament wound	0.72	0.105	172	25	...	...	...	...	...
P100	35%/0°	Rod	Filament wound	0.72	0.105	248	36	...	...	...	...	...
P75	40%/± 16° plus 9%/90°	Hollow cylinder	Filament wound	0.45	0.065	179	26	0.061	0.089	86	12.5	1.3
P100	40%/± 16°	Hollow cylinder	Filament wound	0.56	0.081	228	33	0.38	0.055	30	4.4	–0.07
P55	40%/0°	Plate	Prepreg	0.48(b)	0.070(b)	159	23	0.02	0.003	21	3	2.3
P55	30%/0° plus 10%/90°	Plate	Prepreg	0.28	0.04	83	12	0.010	0.015	34	5	4.5
P55	20%/0° plus 20%/90°	Plate	Prepreg	8.45(b)	1.225(b)	90	13	0.24	0.035	90	13	...

(a) All materials contain pitch-base fibers. (b) Equivalent 0° tensile strength at 400 °C (750 °F)

Tensile strengths and moduli for two graphite/aluminum composites

Composite	Fiber loading, vol %	Tensile strength, MPa	Tensile strength, ksi	Tensile modulus, GPa	Tensile modulus, 10^6 psi	Wire diameter, mm	Wire diameter, in.
VS0054/201 Al	48 to 52	1035 to 1070	150 to 155	345	50	0.64(a)	0.025(a)
GY70SE/201 Al	37 to 38	793 to 827	115 to 120	207	30	0.71(b)	0.028(b)

(a) Two strand. (b) Eight strand

Property data for a commercial graphite/aluminum composite

Fiber(a)	Modulus of elasticity, Longitudinal, GPa	Modulus of elasticity, Longitudinal, 10^6 psi	Modulus of elasticity, Transverse, GPa	Modulus of elasticity, Transverse, 10^6 psi	Tensile strength, Longitudinal, MPa	Tensile strength, Longitudinal, ksi	Tensile strength, Transverse, MPa	Tensile strength, Transverse, ksi
P55	207 to 221	30 to 32	28 to 41	4 to 6	517 to 621	75 to 90	28 to 48	4 to 7
P75	276 to 296	40 to 43	28 to 41	4 to 6	621 to 724	90 to 105	28 to 48	4 to 7
P100	379 to 414	55 to 60	28 to 41	4 to 6	552 to 834	80 to 121	28 to 48(b)	4 to 7(b)

(a) Union Carbide Thornel fibers. (b) Estimated

Silicon Carbide/Metal

Tensile strength of SCS-2-Al

Fiber orientation	No. of plies	Tensile strength		Tensile modulus		Total strain	Poisson's ratio	Coefficient of thermal expansion, 10^{-6}/K
		MPa	ksi	GPa	10^6 psi			
0°	6, 8, 12	1462	212	204.1	29.6	0.89	0.268	6.6
90°	6, 12, 40	86.2	12.5	118.0	17.1	0.08	0.124	21.3
$[0°/90°/0°/90°]_s$	8	673	97.6	136.5	19.8	0.90	...	...
$[0_2°/90°/0°]_s$	8	1144	166.0	180.0	26.1	0.92	...	...
$[90_2/0°/90°]_s$	8	341.3	49.5	96.5	14.0	1.01	...	...
±45°	8, 12, 40	309.5	44.9	94.5	13.7	10.6	0.395	...
$[0°/\pm45°/0°]_{s+2s}$	8, 16	800.0	116	146.2	21.2	0.86	...	...
$[0°/\pm45°/90°]_s$	8	572.3	83.0	127.0	18.4	1.0	...	...

Data on investment-cast SCS-Al

Fiber orientation	Fiber, vol %	Ultimate tensile strength		Range of measurement %	Tensile modulus		ROM, %	Ultimate compressive strength		Compressive modulus	
		MPa	ksi		GPa	10^6 psi		MPa	ksi	GPa	10^6 psi
$0_3°/90_6°/0_3°$	33	458.5	66.5	75	122.0	17.7	107	1378.9	200	...	...
$90_3°/0_6°/90_3°$	33	584.0	84.7	95	124.8	18.1	110	1378.9	200	...	...
0°	34	1034.2	150	85	172.4	25	100	1896.1	275	186.2	27.0

Room-temperature tensile strength of silicon carbide-aluminum alloy composites

Material	Fiber, vol %	Ultimate tensile strength			
		Base		Reinforced	
		MPa	ksi	MPa	ksi
Pure aluminum	11	59	8.6	235	34.1
6061-T6	16	300	43.5	441	64.0
2024-T4	20	470	68.2	565	81.9

Reactivity of metals with SiC fibers

Metal	Melting temperature, °C	Reactivity(a) in 1 h in H_2 gas at temperature, °C, of:									
		450	530	620	650	750	850	950	1000	1100	1200
Al	660	O	O	O							
Ag	961				O	O	O				
Cu	1083					O	O	O			
Ni	1453								O	×	×
Co	1495								O	Δ	×
Fe	1537								O	O	O
Ti	1668								Δ	Δ	Δ
Cr	1875								O	O	Δ
Mo	2610								O	O	O

(a) O, no reaction; Δ, slight reaction; ×, significant reaction

Tensile data for SiC-whisker-reinforced aluminum alloy

Fiber content, vol %	Yield strength (0.2%) MPa	ksi	Standard deviation	Range of measurement	Tensile strength MPa	ksi	Standard deviation	Range of measurement	Modulus of elasticity GPa	10^6 psi	Standard deviation	Range of measurement
0	210	30.5	3.8	9.5	297	43.1	1.8	3.5	71.9	10.4	4.5	13
0.12	266.5	38.7	4.2	10.6	359	52.1	33.6	85.6	95.3	13.8	1.6	6
0.16	264.5	38.4	0.6	1.6	374	54.2	8.0	23.0	90.0	13.1	3.7	9
0.20	298	43.2	4.0	10.2	383.6	55.6	15.2	38.8	111.0	16.1	5.0	13

Yield strength and ultimate tensile strength of aluminum alloy reinforced with SiC whiskers at different temperatures

Fiber, vol %	350 °C (660 °F) Yield strength MPa	ksi	Ultimate tensile strength MPa	ksi	300 °C (570 °F) Yield strength MPa	ksi	Ultimate tensile strength MPa	ksi	250 °C (480 °F) Yield strength MPa	ksi	Ultimate tensile strength MPa	ksi
Polycrystalline alumina												
0	35	5.1	55	8.0	...	...	70	10.2	70	10.2	115	16.7
0.05	54	7.8	63	9.1	79	11.5	88	12.8	112	16.2	134	19.4
0.12	68	9.9	74	10.7	...	...	...	...	...	...	...	...
0.20	110	16.0	112	16.2	154	22.3	155	22.5	186	27.0	198	28.7
SiC whiskers												
0	35	5.1	55	8.0	...	...	70	10.2	70	10.2	115	16.7
0.12	94	13.6	124	18.0	153	22.2	180	26.1	197	28.6	226	32.8
0.16	120	17.4	147	21.3	...	...	...	...	...	...	...	...
0.20	163	23.6	184	26.3	207	30.0	235	34.1	268	38.9	284	41.2

Compression strength of SCS-2-Al

Direction	Plies	Load N	lb	Stress MPa	ksi	Tensile modulus GPa	10^6 psi	Poisson's ratio
0°	12	36 000	8 100	2 647	383.9	...	...	...
		38 250	8 600	2 708	392.7	...	...	...
		38 700	8 700	2 739	397.3	...	...	...
		40 500	9 100	2 878	417.4	...	...	...
		48 900	11 000	3 296	478.0	212.4	30.8	0.241
		53 100	11 940	3 689	535.0	222.7	32.3	...
90°	12	4 220	948	294.4	42.7	104.8	15.2	...
		4 380	985	300.6	43.6	116.5	16.9	0.174
		4 270	960	294.4	42.7	...	...	...
		4 230	950	292.3	42.4	113.1	16.4	0.173
		3 960	890	273.0	39.6	115.8	16.8	...
		3 780	850	259.2	37.6	124.1	18.0	...
90°	40	13 480	3 030	293.7	42.6	...	...	...
		14 610	3 285	294.4	42.7	131.7	19.1	0.136
		13 280	2 985	290.0	42.0	102.7	14.9	...
		13 430	3 020	287.5	41.7	108.9	15.8	...
		13 520	3 040	294.4	42.7	115.1	16.7	...
		13 680	3 075	297.2	43.1	142.0	20.6	0.158

Shear strength of SCS-2-Al

Test temperature °C (°F)	Failure stress MPa	ksi	Shear strength MPa	ksi	Shear modulus MPa	ksi
Room temperature	455.7	66.1	113.8	16.5	42.5	6.17
	452.3	65.6	113.1	16.4	39.5	5.73
	479.2	69.5	120.0	17.4	39.8	5.77
	422.6	61.3	105.5	15.3	40.3	5.85
Average	**452.5**	**65.6**	**113.1**	**16.4**	**40.5**	**5.88**
75 (165)	437.1	63.4	109.6	15.9	40.2	5.83
	434.4	63.0	108.9	15.8	43.2	6.27
	424.7	61.6	106.2	15.4	41.7	6.05
Average	**432.1**	**62.6**	**108.2**	**15.7**	**41.7**	**6.05**
−55 (−65)	501.2	72.7	125.5	18.2	44.5	6.46
	482.6	70.0	120.7	17.5	39.6	5.75
	453.0	65.7	113.1	16.4	39.6	5.75
Average	**479.0**	**69.4**	**119.8**	**17.3**	**41.3**	**5.98**

Typical properties of SiC whisker-reinforced aluminum alloy sheet

Sheet thickness, mm	Sheet thickness, in.	Test specimen orientation	Ultimate tensile strength, MPa	Ultimate tensile strength, ksi	Yield strength(a), MPa	Yield strength(a), ksi	Elongation, %	Modulus of elasticity, GPa	Modulus of elasticity, 10^6 psi	Fracture toughness, K_c, MPa$\sqrt{m}$	Fracture toughness, K_c, ksi$\sqrt{in}$
2.54	0.100	Longitudinal (Along roll direction)	718	104	573	83.1	5.3	114	16.5	55	50
2.54	0.100	Transverse (90° to roll direction)	559	81.0	386	56.4	8.5	95	14	59	54

(a) 0.2% offset

Notched strengths of SCS-2-Al

Specimen	Average gross stress, RT, MPa	Average gross stress, RT, ksi	Average net stress, RT, MPa	Average net stress, RT, ksi	Notch factor, RT	Notch factor, 75 °C (165 °F)	Notch factor, –55 °C (–65 °F)
Double-edge notch, 0°	814.8	118.2	1269.5	184.1	0.92	0.85	0.80
Center hole, 1.6-mm (1/16-in.) diam, 0°	1125.9	163.3	1163.8	168.8	0.84	...	...
Center hole, 3.2-mm (1/8-in.) diam, 0°	991.7	143.8	1061.6	154.0	0.77	0.75	0.72
	898.0(a)	130.2	956.1(a)	138.7	0.70(a)	...	...
Center hole, 6.4-mm (1/4-in.) diam, 0°	842.1	122.1	966.4	140.2	0.70	...	...
	800.9(a)	116.2	911.3(a)	132.2	0.66(a)	...	...
Center hole, 3.2-mm (1/8-in.) diam, 0°/±45°	728.1	015.6	777.4	112.8	0.90	...	...
Center hole, 6.4-mm (1/4-in.) diam, 0°/±45°	620.5	90.0	710.8	103.1	0.83	...	...
Center hole, 3.2-mm (1/8-in.) diam, 0°/±45°/90°	437.1	63.4	467.5	67.8	0.90	...	...
Center hole, 6.4-mm (1/4-in.) diam, 0°/±45°/90°	400.6	58.1	460.6	66.8	0.89	...	...
Center hole, 2.4-mm (3/32-in.) diam, ±45°	244.7	35.5	256.6	37.2	0.85	...	...
Center crack, 6.4-mm (1/4-in.) EDM slot, 0°	822.0	119.2	944.2	137.0	0.68	...	...
Center crack, 12.7-mm (1/2-in.) EDM slot, 0°	659.9	95.7	886.9	128.6	0.64	...	...
	621.6(a)	90.20	819.8(a)	118.9	0.60(a)	...	...

(a) 40-ply material

Typical properties of SiC/2024-T6 Al billet and extruded plate showing the effects of SiC whisker alignment

MMC material form	Test specimen orientation	Ultimate tensile strength, MPa	Ultimate tensile strength, ksi	Yield strength(a), MPa	Yield strength(a), ksi	Coefficient of thermal expansion, 10^{-6}/K	Density, g/cm^3
12-in.-diam cylindrical billet	Longitudinal (axial)	496	71.9	351	50.9	16.1	2.86
1/2-in. by 5-in. extrusion	Longitudinal	737	107	448	64.9	13.0	2.86
12-in.-diam cylindrical billet	Transverse	503	72.9	358	51.9	16.4	2.86
1/2-in. by 5-in. extrusion	Transverse (long)	462	67.0	379	54.9	19.6	2.86

(a) 0.2% offset

Tensile properties of a forged SiC/6061-T6 Al turbine wheel(a)

Test-specimen orientation	Tensile strength, MPa	Tensile strength, ksi	Yield strength(b), MPa	Yield strength(b), ksi	Elongation, %	Modulus of elasticity, GPa	Modulus of elasticity, 10^6 psi
Radial	518	75.1	386	55.9	4.7	111	16.1
Transverse (circumferential)	490	71.0	379	54.9	6.1	108	15.7

(a) SiC whisker content, 20 vol %. (b) 0.2% offset

Properties of aluminum/SiC$_p$ composites(a)

Property	Al 2124-T6/30 vol % SiC$_p$	Al 2124-T6/40 vol % SiC$_p$
Microyield strength, MPa (ksi)	177 (17)	...
Coefficient of thermal expansion, 10^{-6}/K (10^{-6}/°F)	12.4 (6.9)	10.8 (6.0)
Modulus of elasticity, GPa (10^6 psi)	117 (17)	145 (21)
Density, g/cm^3 (lb/in.^{3})	2.91 (0.105)	2.96 (0.107)
Specific modulus, 10^6 m (10^6 in.)	4.11 (162)	4.98 (196)
Thermal conductivity, W/m · K (Btu/ft · h · °F)	125 (72)	116 (67)
Specific heat, kJ/kg · K (Btu/lb · °F)	0.80 (0.19)	0.67 (0.16)

(a) SiC$_p$ stands for SiC particulate

Typical mechanical properties of SiC particulate reinforced aluminum alloy composites

	Modulus of elasticity		Yield strength		Ultimate tensile strength		
Alloy and vol %	GPa	10^6 psi	MPa	ksi	MPa	ksi	Ductility, %
6061							
Wrought	68.9	10	275.8	40	310.3	45	12
15	96.5	14	400.0	58	455.1	66	7.5
20	103.4	15	413.7	60	496.4	72	5.5
25	113.8	16.5	427.5	62	517.1	75	4.5
30	120.7	17.5	434.3	63	551.6	80	3.0
35	134.5	19.5	455.1	66	551.6	80	2.7
40	144.8	21	448.2	65	586.1	85	2.0
2124							
Wrought	71.0	10.3	420.6	61	455.1	66	9
15	...	...	...	...	...	...	...
20	103.4	15	400.0	58	551.6	80	7.0
25	113.8	16.5	413.7	60	565.4	82	5.6
30	120.7	17.5	441.3	64	593.0	86	4.5
35	...	...	...	...	...	...	...
40	151.7	22	517.1	75	689.5	100	1.1
7090							
Wrought	72.4	10.5	586.1	85	634.3	92	8
15	...	...	...	...	...	...	...
20	103.4	15	655.0	95	724.0	105	2.5
25	115.1	16.7	675.7	98	792.9	115	2.0
30	127.6	18.5	703.3	102	772.2	112	1.2
35	131.0	19	710.2	103	724.0	105	0.90
40	144.8	21	689.5	100	710.2	103	0.90
7091							
Wrought	72.4	10.5	537.8	78	586.1	85	10
15	96.5	14	579.2	84	689.5	100	5.0
20	103.4	15	620.6	90	724.0	105	4.5
25	113.8	16.5	620.6	90	724.0	105	3.0
30	127.6	18.5	675.7	98	765.3	111	2.0
35	...	...	...	...	...	...	...
40	139.3	20.2	620.6	90	655.0	95	1.2

Pin bearing strengths of SiC particulate reinforced aluminum

Composite	Edge distance (multiplied by pin diameter from edge)	Bearing yield strength MPa	ksi	Bearing ultimate strength MPa	ksi
20 vol% SiC-7091 (T-6)	1.5	689.5	100	724.0	105
	2.0	1000.0	145	1310.0	190
	3.0	1000.0	145	1448.0	210
25 vol% SiC-2124	2.0	827.4	120	...	...

Property comparison of aluminum/SiC$_p$ composites

Property	Al 2124-T6	Al 2124-T6/30 vol % SiC$_p$(a)	Al 2124-T6/40 vol % SiC$_p$(a)	Al 6061-T6/35 vol % SiC$_p$(a)
Coefficient of thermal expansion(b), 10^{-6}/K (10^{-6}/°F)	23.4 (13.0)	12.4 (6.9)	10.8 (6.0)	12.1 (6.7)
Microyield strength(c), MPa (ksi)	117 (17)	117 (17)	124 (18)	124 (18)
Modulus of elasticity, GPa (10^6 psi)	72 (10.5)	117 (17)	145 (21)	131 (19)
Density, g/cm^3 (lb/in.^{3})	2.78 (0.100)	2.91 (0.105)	2.96 (0.107)	2.88 (0.104)

(a) SiC mean particle size of 3.5 μm (1.38×10^{-4} in.) used in instrument-grade MMC materials. (b) Average value over a temperature range of 222 to 366 K (–60 to +200 °F). (c) Microyield strength, also known as precision elastic limit, is the stress required to cause 10^{-6} m/m (1 μin./in.) of residual (plastic) strain

Mechanical data on SCS-Mg cast rod

Sample	Exposure time, min	Ultimate tensile strength MPa	ksi	Strain to failure, %	Elastic modulus GPa	10^6 psi	Fiber, vol %
VIR 67	5	1000	145	0.83	169.6	24.6	34
VIR 69	10	1524	221	0.88	209.6	30.4	46
VIR 72	10	1331	193	0.78	230.3	33.4	50
VIR 77	10	1379	200	0.95	180.6	26.2	37

Shear strength of SiC particulate reinforced aluminum composites

Composite	Shear strength MPa	ksi
25 vol% SiC-6061 (T-6)	277.9	40.3
30 vol% SiC-6061 (T-6)	289.6	42.0
25 vol% SiC-7091 (T-6)	279.2	55.0
30 vol% SiC-7090	430.9	62.5
25 vol% SiC-2124 (T-4)	344.8	50.0

Abrasive wear testing of 20% silicon carbide reinforced 2024-20-T6 Al alloy

	Unreinforced 2024	Silicon carbide-2024
6000 cycles		
Wt loss, g (oz)	0.12 (0.0040)	zero
Scar depth, μm (mil)	15-25 (0.6-1.0)	zero
20 000 cycles		
Wt loss, g (oz)	0.34 (0.0120)	0.06 (0.0020)
Scar depth, μm (mil)	35-60 (1.4-2.4)	1-15 (0.4-0.6)

Mechanical data on SCS-Cu

Panel	Fiber, vol %	Axial ultimate tensile stress MPa	ksi	Axial modulus GPa	10^6 psi
84-014	0.23	690	100	172.4	25.0
84-153	0.33	965	140	202	29.3
84-377	0.33	900	130	187.5	27.2

Data on SCS-6-Ti (sample size, 62 panels)

	Tensile strength MPa	ksi	Elastic modulus GPa	10^6 psi	Strain to failure, %
Mechanical properties of SiC-Ti-6Al-4V (35 vol %)					
As fabricated:					
Mean	1690	245	186.2	27.0	0.96
Standard deviation	119.3	17.3	7.58	1.1	0.091
After heating 7 h at 905 °C (1660 °F)					
Mean	1434	208	190.3	27.6	0.86
Standard deviation	108.9	15.8	8.3	1.2	0.087
Mechanical properties of SiC-Ti-15V-3Sn-3Cr-3Al (38 to 41 vol %)					
As fabricated:					
Mean	1572	228	197.9	28.7	...
Standard deviation	138	20	6.21	0.9	...
After heat treating 16 h at 480 °C (900 °F), 13 samples:					
Mean	1951	283	213.0	30.9	...
Standard deviation	96.5	14	4.83	0.7	...

Alumina/Metal

Adhesion of metals to polycrystalline Al_2O_3

Metal	Sintering temperature, °C	Sintering atmosphere	Adhesion, erg/cm^2
Cu	450	H_2	815
Ni	1000	H_2	435
Au	1000	Air	530
Fe	1000	H_2	810
Ni	1400	Ar	518
Ag	700	H_2	435

Coefficients of linear thermal expansion of polycrystalline alumina and alumina/chromium composites

Material	Method	Temperature °C	Temperature °F	Coefficient of linear thermal expansion 10^{-6}/K	Coefficient of linear thermal expansion 10^{-6}/°F
Al_2O_3	(a)	27 to 799	80 to 1470	7.90	4.39
		27 to 1315	80 to 2400	9.40	5.22
$70Al_2O_3/30Cr$	Telemicroscope	25 to 800	77 to 1472	8.6	4.8
		22 to 1315	72 to 2400	9.45	5.25
$34Al_2O_3/66(Cr, Mo)$	Telemicroscope	24 to 802	75 to 1475	7.96	4.42
		24 to 1315	75 to 2400	10.5	5.82
$28Al_2O_3/72Cr$	Telemicroscope	25 to 800	77 to 1472	8.62	4.79
		22 to 1315	72 to 2400	10.4	5.75
$23Al_2O_3/77Cr$ (LT-1)	...	25 to 1000	77 to 1832	8.91	4.95

(a) Average of many reported values obtained by a variety of methods

Yield strength and ultimate tensile strength of aluminum alloy reinforced with polycrystalline alumina at different temperatures

	350 °C (660 °F)				300 ° (570 °F)				250 °C (480 °F)			
	Yield strength		Ultimate tensile strength		Yield strength		Ultimate tensile strength		Yield strength		Ultimate tensile strength	
Fiber, vol %	MPa	ksi	MPa	ksi	MPa	ksi	MPa	ksi	MPa	ksi	MPa	ksi
Polycrystalline alumina												
0	35	5.1	55	8.0	...	...	70	10.2	70	10.2	115	16.7
0.05	54	7.8	63	9.1	79	11.5	88	12.8	112	16.2	134	19.4
0.12	68	9.9	74	10.7	...	...	...	...	...	...	...	...
0.20	110	16.0	112	16.2	154	22.3	155	22.5	186	27.0	198	28.7
SiC whiskers												
0	35	5.1	55	8.0	...	...	70	10.2	70	10.2	115	16.7
0.12	94	13.6	124	18.0	153	22.2	180	26.1	197	28.6	226	32.8
0.16	120	17.4	147	21.3	...	...	...	...	...	...	...	...
0.20	163	23.6	184	26.3	207	30.0	235	34.1	268	38.9	284	41.2

Tensile data for polycrystalline-alumina-reinforced aluminum alloy

	Yield strength (0.2%)				Tensile strength				Modulus of elasticity			
Fiber content, vol %	MPa	ksi	Standard deviation	Range of measurement	MPa	ksi	Standard deviation	Range of measurement	GPa	10^6 psi	Standard deviation	Range of measurement
0	210	30.5	3.8	9.5	297	43.1	1.8	3.5	71.9	10.4	4.5	13
0.05	232	33.6	4.2	10.4	282	40.9	6.5	15.1	78.4	11.4	2.3	6
0.12	251.5	36.5	14.6	38.3	273	40.0	19.6	49.6	83.0	12.0	7.8	21
0.20	282.5	41.0	11.3	25.2	312	45.3	16.0	42.3	95.2	13.8	2.7	7

Typical compositions and physical properties of alumina/chromium composites

Typical composition			Melting point	
wt %	Approximate vol %	Density, g/cm^3	°C	°F
70Al_2O_3/30Cr	81Al_2O_3/19Cr	4.60 to 4.65	>1705(a)	>3100(a)
34Al_2O_3/66(Cr, Mo)(b)	49.9Al_2O_3/50.1(Cr, Mo)	5.82	>1730(a)	>3150(a)
28Al_2O_3/72Cr	42Al_2O_3/58Cr	5.9	>1730(a)	>3150(a)
23Al_2O_3/77Cr	35Al_2O_3/65Cr	5.9 to 6.0	1850(c)	3360(c)
Al_2O_3		3.98	2050	3720

(a) Temperature given is sintering temperature at which composite was prepared. (b) Second phase added as 80Cr-20Mo alloy. (c) Maximum service temperature: 1315 °C (2400 °F) for long term; 1650 °C (3000 °F) for short term

Wetting angles in various alumina/liquid metal systems

System	Wetting angle	Temperature, °C, atmosphere
Al_2O_3/Ni	>90°	1450, air
Al_2O_3/Cu	138°	1200, vacuum
Al_2O_3/Fe-Cr	~40°	1650, reducing
Al_2O_3/Cr	1 to 10°	1950, reducing
Al_2O_3/Fe	139°	1550, N_2

Hardness at 25 °C of Al-9Si-3Cu alloy reinforced with Al_2O_3 fibers(a)

Volume fraction	Vickers hardness
0	131
0.12	179
0.18	190
0.24	212

(a) "Saffil" fiber, RF grade

Coefficients of thermal expansion for Al_2O_3 fiber/Al alloy composites

Volume fraction (V_f)	Coefficient of thermal expansion(a), 10^{-5}/°C In-plane	Normal
0	2.03	2.03
0.12	1.66	1.76
0.18	1.54	1.66
0.24	1.55	1.57

(a) Coefficients (at 20 to 200 °C) of composites containing "Saffil" fiber, RF grade, in Al-9Si-3Cu alloy measured parallel and normal to planes of fiber orientation

Boron/Metal

Reactivity of boron with various metals

	Definite reaction			Little or no reaction		
	Temperature			Temperature		
Metal	°C	°F	Time(a), h	°C	°F	Time(a), h
Fe	700	1292	50	600	1112	(b)
	900	1652	1	...	...	...
Co	700	1292	50	600	1112	(b)
	900	1652	1	...	...	...
Al	700(c)	1292(c)	0.2	600	1112	1
Mg	...	...	...	600	1112	100
	...	...	...	700(c)	1292(c)	0.2
Be	...	...	...	1000	1832	24
Ti	...	...	...	600	1112	100
Cr	900	1652	1	...	...	...
Ag	...	...	...	900	1652	2
Ni	600	1112	100	...	...	...
	700	1292	100	...	...	...
	900	1652	1	...	...	...

(a) Time required for consolidation of metal powders and fibers by hot pressing. (b) Reaction time so short that the reaction is finished as soon as fibers have been hot pressed. (c) Molten

Wetting angles of borides by iron-group melts

	Iron			Nickel			Cobalt		
	Wetting angle		Work of adhesion, erg/cm^2	Wetting angle		Work of adhesion, erg/cm^2	Wetting angle		Work of adhesion, erg/cm^2
Boride	Vacuum	Argon		Vacuum	Argon		Vacuum	Argon	
TiB_2	39	94	1650	25	72	2220	...	64	2590
ZrB_2	97	102	1410	72	78	2020	...	81	2100
HfB_2	...	98	1530	...	99	1430	...	...	...
NbB_2	...	0	...	...	24	...	...	22	3460
TaB_2	...	0	...	...	...	...	...	...	...
CrB_2	...	0	...	20	21	3310	...	0	...
Mo_2B_5	...	0	...	...	0	...	...	22	3460
W_2B_5	...	...	...	...	...	...	...	94	1680

Typical properties of eutectoid bonded boron-aluminum(a)

	Tensile strength				Ultimate strain 10^{-6} m/m (10^{-6} in./in.)		Initial modulus			
	Room temperature		315 °C (600 °F)		Room	315 °C	Room temperature		315 °C (600 °F)	
Type of loading	MPa	ksi	MPa	ksi	temperature	600 °F	GPa	10^6 psi	GPa	10^6 psi
Longitudinal tension:										
Coupon	1100	160	1070	155	6500	5800	205	29.8	194	28.1
Beam	1340	194	...	...	7200	...	198	28.7	...	...
Longitudinal compression, beam	2360(b)	343(b)	...	...	10500	...	254	36.8	...	...
Transverse tension, coupon	115	16.7	27	3.9	4200	6500	134	19.5	97	14.0
Transverse compression, beam	259	37.5	66	9.6	2400	10700	138	20.0	120	17.4
In-plane shear, rail shear	69	10	26	3.7	16000	13000	69	10.0	54	7.8

(a) All tests conducted on 42 to 45 vol % B/Al. (b) Failure occurred in honeycomb core – no failure in B/Al face sheet

Wetting angles of refractory borides with various metal melts

Boride	Wetting angle for metal melt and temperature of:								
	Cu 1130 °C	Al 900 °C	Ga 800 °C	In 250 °C	Si 1500 °C	Ge 1000 °C	Sn 300 °C	Pb 400 °C	Bi 320 °C
TiB_2	143	98	115	124	15	...	114	106	141
ZrB_2	135	106	117	114	...	102	110	...	...
HfB_2	...	134	...	114	...	140	...	...	...
NbB_2	109	125	101	133	0	60	102	125	110
TaB_2	...	138	...	117	...	...	...	...	...
CrB_2	26	107	123	97	...	126	100	124	128
Mo_2B_5	0	134	...	...	...	28	100	...	...
W_2B_5	104	...	...	130	22	128	100	...	...

Mechanical properties of boron/aluminum composites(a)

Matrix	Fiber orientation	Tensile strength		Modulus of elasticity	
		MPa	ksi	GPa	10^6 psi
Al-6061	0°	1515	220	207	30
	90°	138	20	138	20
Al-2024	0°	1550	225	207	30
	90°	214	31	145	21

(a) These samples contain 48% Avco 5.6 mil (142-μm) boron. The longitudinal tensile specimens are 6 in. (152 mm) by 0.3125 in. (7.9 mm) by 6 ply, and the transverse tensile bars are 6 in. (152 mm) by 0.5 in. (12.7 mm) by 6 ply

Room-temperature properties of boron-aluminum (0°) with 50 vol % filament

Property	Reported value
Tensile strength, MPa (ksi):	
Longitudinal	1100 (160.0)
Transverse	110 (16.0)
Compressive strength, MPa (ksi):	
Longitudinal	1215 (176.0)
Transverse	159 (23.0)
Shear strength, MPa (ksi):	
In-plane	69 (10.0)
Interlaminar	126 (18.3)
Strain, μm/m (μin./in.):	
Longitudinal	5000-6000
Transverse	6000-12,000
Tensile modulus, GPa (10^6 psi):	
Longitudinal	235 (34.0)
Transverse	138 (20.0)
Compressive modulus, GPa (10^6 psi):	
Longitudinal	207 (30.0)
Transverse	131 (19.0)
In-plane shear modulus, GPa (10^6 psi)	66 (9.5)
Poisson's ratio:	
Longitudinal	0.23
Transverse	0.17
Density, g/cm^3	2.7
Coefficient of thermal expansion, 10^{-6}/°C (10^{-6}/°F):	
Longitudinal	5.8 (3.2)
Transverse	19.1 (10.6)

Average tensile and shear strengths of soldered boron/aluminum tee sections

Test mode	Test temperature		Failure stress	
	K	°F	MPa	ksi
Tension	294	700	39	5.70
	366	200	38	5.50
Shear	294	70	73	10.60
	366	200	75	10.70

Typical mechanical properties of 50 vol % unidirectional reinforced B/Al 6061 composites

Tensile strength, MPa (ksi):	
Longitudinal	1490 (216)
Transverse	138 (20)
Tensile modulus, GPa (10^6 psi):	
Longitudinal	214 (31)
Transverse	138 (20)
Poisson's ratio:	
Longitudinal	0.23
Transverse	0.13
Compressive strength, MPa (ksi):	
Longitudinal	1725 (250)
Transverse	207 (30)
Compressive modulus, GPa (10^6 psi):	
Longitudinal	221 (32)
Transverse	138 (20)
Longitudinal shear strength, MPa (ksi)	159 (23)
Longitudinal shear modulus, GPa (10^6 psi)	41 (6)
Longitudinal bearing strength, MPa (ksi)	827 (120)
Unnotched fatigue strength at runout (10^7 cycles), MPa (ksi):	
Longitudinal	1035 (150)
Transverse	41 (6)
Creep at 370 °C (700 °F), MPa (ksi)	At 1105 (160) total strain averages 0.06% in 100 h

Room-temperature properties of cross-ply boron-aluminum (0°/90°) with 50 vol % filament

Property	Reported value
Tensile strength, MPa (ksi):	
Longitudinal	483 (70)
Transverse	483 (70)
Compressive strength, MPa (ksi):	
Longitudinal	607 (88)
Transverse	607 (88)
Shear strength, MPa (ksi):	
In-plane	103 (15)
Interlaminar	69 (10)
Strain, longitudinal, μm/m (μin./in.)	6700
Tensile modulus, GPa (10^6 psi):	
Longitudinal	145 (21)
Transverse	145 (21)
Compressive modulus, GPa (10^6 psi):	
Longitudinal	145 (21)
Transverse	145 (21)
Density, g/cm^3	2.7

Axial tensile strength of 140-μm (5.6-mil) boron-aluminum at various fiber levels

Matrix	Boron content, %	Tensile strength MPa	ksi	Modulus of elasticity GPa	10^6 psi	Strain to failure, %
2024, as fabricated	45	1287.5	186.7	202.1	29.3	0.775
	47	1420.7	206.0	222.1	32.2	0.795
	52	1721.0	249.6	...	...	...
	54	1798.6	260.8	...	...	...
	64	1527.6	221.5	275.9	40.0	0.72
	66	1739.2	251.6	...	...	...
	70	1927.6	279.5	...	...	...
2024-T6	46	1458.7	211.5	220.7	32.0	0.810
	64	1924.1	279.0	275.9	40.0	0.755
6061, as fabricated	48	1489.7	216.0	...	...	...
	50	1343.4	194.8	217.2	31.5	0.695
6061-T6	51	1417.2	205.5	231.7	33.6	0.735

Room-temperature longitudinal tensile strengths of thermally exposed and unexposed boron-aluminum

Exposure condition	Average monolayer strength MPa	ksi	Average filament-bundle strength MPa	ksi	Degradation, %
Unexposed controls	1275	185	3040	441	...
30 min. at 550 °C (1020 °F)	1035	150	2510	364	17.6
7 min. at 555 °C (1030 °F)	1150	167	2680	389	11.9
15 min. at 555 °C (1030 °F)	1095	159	2470	358	19.0
7 min. at 570 °C (1060 °F)	1080	157	2580	374	15.5
15 min. at 570 °C (1060 °F)	915	133	2185	317	28.3
35 min. at 570 °C (1060 °F)	850	123	1995	289	34.6
7 min. at 595 °C (1100 °F)	930	135	2180	316	28.6

Impact energy of full-size notched Charpy specimens containing 55% boron fibers

Fiber diameter, μm	Matrix	Total impact energy, J
200	Aluminum 1100	95
145	Aluminum 1100	55
145	Al alloy 6061	35
Unreinforced	Ti-6Al-4V	25
Unreinforced	Al alloy 6061-T6	15

Refractory Metal Fiber/Metal Matrix Composites

Elevated-temperature tensile properties of W/Fe-Cr-Al-Y

Test temperature °C	°F	Specimen	Filament orientation	Ultimate tensile strength MPa	ksi	Total elongation, %
648	1200	P34-15-2	±15°	776	112.6	25.2
		P36-15-2	±15°	717	104.0	12.8
		P34-45-2	±45°	564	81.8	35.5
		P36-45-2	±45°	539	78.2	24.6
		P34-90-2	90°	189	27.4	3.5
		P36-90-2	90°	180	26.2	3.2
760	1400	P34-15-1	±15°	571	82.8	12.0
		P36-15-1	±15°	534	77.4	14.2
		P34-45-1	±45°	185	26.9	23 + (a)
		P36-45-1	±45°	163	23.6	24.6
		P36-90-1	90°	111	16.1	6.5
648	1200	P21-1-1(b)	0°	737	106.9	3.0
		P21-1-2(c)	0°	768	111.4	2.9

(a) Test terminated before failure. (b) Creep specimen after 1077-h creep test at 1037 °C (1900 °F). (c) Creep specimen after 990-h creep test at 1093 °C (2000 °F)

Room-temperature tensile properties of tungsten fiber/copper alloy composites

Binder material	Maximum solubility of alloying element in tungsten	Alloying element content wt %	at. %	Fiber content, vol %	Tensile strength MPa	ksi	Reduction in area, %	Type of fracture
Pure copper	Insoluble	0	0	65	1556	225.7	...	Ductile
				70.2	1641	238.0	...	Ductile
				75.4	1722	249.8	...	Ductile
Copper-nickel	0.3	5	5.4	79	1700	246.6	34	Ductile
				78.4	1724	250.0	37	Ductile
				76	1509	218.9	32	Ductile
		10	10.9	74.1	908	131.7	Nil	Brittle
				75.5	750	108.8	Nil	Brittle
				79.5	354	51.3	Nil	Brittle
Copper-cobalt	0.3	1	1.1	77.3	1513	219.4	...	Semiductile
		5	5.4	76	1470	213.2	1.5	Semiductile
				74.8	1581	229.3	2.3	Ductile
				74.7	1015	147.2	...	Brittle
				74.9	1187	172.1	...	Brittle
Copper-aluminum	2.6	5	11.3	63.4	682	98.9	Nil	Brittle
				72.4	1060	153.8	Nil	Semiductile
				76.1	1065	154.5	Nil	Semiductile
		10	20.8	76.7	955	138.5	...	Brittle
Copper-titanium	8	10	12.8	78.2	1542	223.7	...	Semiductile
				71.7	1518	220.1	10	Semiductile
		25	30.7	76.3	1287	186.7	...	Brittle
Copper-zirconium	3	10	7.2	72.8	1489	216.0	Nil	Brittle
				78.5	1760	255.3	Nil	Ductile
				75.6	1564	226.9	Nil	Semiductile
				64.7	1190	172.6	Nil	Brittle
				64.3	1349	195.7	Nil	Semiductile
		33	25.5	75.9	736	106.7	Nil	Brittle
Copper-chromium	Complete solid solubility (miscibility gap)	1	1.2	78.7	1541	223.5	7.4	Semiductile
				77.5	1572	228.0	25.8	Ductile
				77.2	1558	225.9	7.5	Semiductile
		2	2.4	76.4	1666	241.7	16.4	Ductile
Copper-niobium	Complete solid solubility	1	0.6	75.4	1635	237.1	20.6	Ductile
				75.1	1538	223.1	24.7	Ductile

Rupture strengths for tungsten alloy wire reinforced composites

Alloy, wt %	Wire	Wire diam mm	Wire diam in.	Vol %	Density, g/cm^3	100-h rupture strength MPa	100-h rupture strength ksi	Strength to density for 100-h rupture cm × 10^3	Strength to density for 100-h rupture in. × 10^3
100-h rupture strength at 1100 °C (2010 °F)									
Ni-12.5Cr-7W-4.8Mo-5Al-2.5Ti(ZhS6)	VRN Tungsten	0.3-0.5	0.012-0.020	40	12.5	138	20	112.5	44.3
Ni-11W-6Al-6Cr-2Mo-1.5Nb(EPD-16)	No reinforcement	...	...	0	8.3	51	7.4	63.5	25.0
	Tungsten	0.25	0.010	40	12.7	131	19	104	40.9
Ni-12.5Cr-2.5Fe-2Nb-4Mo-6Al-1Ti(Nimocast 713C)	No reinforcement	...	...	0	8.0	48	7	61.3	24.1
	Tungsten	1.27	0.050	20	10.3	93	13.5	92.7	36.5
Co-21.5Cr-25W-10Ni-3.5Ta-0.8Ti(Mar-M322E)	No reinforcement	...	...	0	...	48	7	...	...
	W-2%ThO_2	0.08	0.003	40	...	207	30	...	...
Ni-25W-15Cr-2Al-2Ti	No reinforcement	...	...	0	9.15	23	3.3	25.4	10
	218CS (Tungsten)	0.38	0.015	40	13.3	138	20	105.8	41.7
	W-2%ThO_2	0.38	0.015	40	13.0	193	28	151.3	59.6
	W-Hf-C	0.38	0.015	40	13.3	324	47	249.1	98.0
Fe-24Cr-5Al-1Y	W-1%ThO_2	0.38	0.015	56	12.5	242(a)	35	195.7	76.8
Fe-24Cr-5Al-1Y	W-Hf-C	0.38	0.015	35	11.3	242	35	214.7	84.5

(a) 831-h rupture strength

Effect of thermal cycling on residual room-temperature tensile properties of W-1ThO_2/Fe-Cr-Al-Y

Exposure	Observations	Ultimate tensile strength MPa	Ultimate tensile strength ksi	Modulus of elasticity GPa	Modulus of elasticity 10^6 psi	Strain to failure μin./in.
As-fabricated	...	655	95.0	179	26.0	3400
As-fabricated	...	581	84.2	201	29.2	3700
100 cycles, 29 to 1095 °C (85 to 2000 °F)	No visual change	563	81.7	259	37.6	2400
100 cycles, 29 to 1095 °C (85 to 2000 °F)	No visual change	618	89.6	219	31.7	4300
1000 cycles, 29 to 1095 °C (85 to 2000 °F)	Surface roughening	590	85.5	258	37.4	3200
1000 cycles, 29 to 1095 °C (85 to 2000 °F)	Surface roughening	557	80.8	177	25.6	3300
100 cycles, 29 to 1205 °C (85 to 2200 °F)	No visual change	624	90.5	250	36.3	3100
100 cycles, 29 to 1205 °C (85 to 2200 °F)	No visual change	587	85.2	228	33.1	3600
1000 cycles, 29 to 1205 °C (85 to 2200 °F)	Surface roughening	503	72.9	158	22.9	3200
1000 cycles, 29 to 1205 °C (85 to 2200 °F)	Surface roughening	487	70.6	170	24.6	3000

Cermets

Properties of simultaneously pressed and sintered Al_2O_3/TiC cermet(a)

Property	Value
Mechanical	
Transverse rupture strength(b), MPa (ksi)	760 (110)
Fracture toughness, MPa $\sqrt{m}$ (ksi $\sqrt{in.}$)	5.9 (5.3)
Elastic constants:	
Elastic modulus, GPa (10^6 psi)	
At 25 °C (77 °F)	376 (54.5)
At 500 °C (930 °F)	353 (51.2)
At 1000 °C (1830 °F)	324 (47.1)
Shear modulus, GPa (10^6 psi)	
At 25 °C (77 °F)	154 (22.4)
At 500 °C (930 °F)	144 (21.0)
At 1000 °C (1830 °F)	134 (19.4)
Poisson's ratio	
At 25 °C (77 °F)	0.22
At 500 °C (930 °F)	0.22
At 1000 °C (1830 °F)	0.21
Microhardness, Vickers DPH, GPa (kg/mm^2)	
100-g load	23.0 (2350)
500-g load	18.6 (1900)
Thermal	
Linear thermal expansion, 10^{-6}/°C	
From 25 to 300 °C (77 to 570 °F)	7.0
From 25 to 500 °C (77 to 930 °F)	7.6
From 25 to 800 °C (77 to 1470 °F)	8.3
Thermal conductivity, W/cm · K (cal/cm · s · °C)	
At 25 °C (77 °F)	0.13 (0.030)
At 500 °C (930 °F)	0.13 (0.030)
At 1000 °C (1830 °F)	0.11 (0.027)
Thermal diffusivity, cm^2/s	
At 25 °C (77 °F)	0.042
At 500 °C (930 °F)	0.029
At 1000 °C (1830 °F)	0.025
Specific heat (C_p), J/g · K (cal/g · °C)	
At 25 °C (77 °F)	0.58 (0.14)
At 500 °C (930 °F)	0.84 (0.20)
At 1000 °C (1830 °F)	0.88 (0.21)
Electrical	
Volume resistivity, dc, $10^{-4}\Omega \cdot m$	
At 25 °C (77 °F)	2
At 100 °C (212 °F)	2
At 300 °C (570 °F)	3

(a) Composite is electrically conductive. Density, 5.16 g/cm^3. Porosity, vacuum tight. Water absorption, none. Grain size, ~2 μm. Color, black. (b) In four-point bending

General property data for titanium carbide cermets

Property	Typical values reported at 24 °C (75 °F)(a): TiC/metal composite	TiC
Density, g/cm^3	5.5 to 6.8	4.65 to 4.92
Melting point:		
°C	1455(b)	3065
°F	2650(b)	5550
Specific heat:		
kJ/kg · K	0.46	0.50
Btu/lb · °F	0.11	0.12
Thermal conductivity:		
W/m · K	28 to 35	26 to 35
Btu/ft · h · °F	16 to 20	15 to 20
Coefficient of linear thermal expansion (21-980 °C or 70-1800 °F):		
10^{-6}/°C	2.8 to 3.6	2.4
10^{-6}/°F	5.0 to 6.5	4.3
Bend strength:		
MPa	689 to 1380	414 to 827
ksi	100 to 200	60 to 120
Compressive strength:		
MPa	2760 to 3450(c)	1380 to 2760
ksi	400 to 500(c)	200 to 400
Tensile strength:		
MPa	689 to 965(d)	241 to 276
ksi	100 to 140(d)	35 to 40
Impact strength:		
J	5.42 to 21.47	<1.36
in. · lb	48 to 190	<12
Modulus of elasticity:		
GPa	310 to 379	276 to 448
10^6 psi	45 to 55	40 to 65
Hardness, Rockwell A	84 to 89	93
Oxidation resistance (20 h at 870 to 980 °C or 1600 to 1800 °F), % weight gain	<1	<1

(a) Data are for composite materials containing up to 50 wt % metal; values were obtained from various sources. (b) Melting point for nickel or cobalt metal component. (c) Compressive yield stress. (d) Yield of 0.2 to 0.3% for indicated stresses

Components of typical carbide/metal cermets

Carbide components(a): Major	Minor	Metal components(a): Major	Minor
TiC	...	Ni or Co	...
TiC	...	Ni or Co	Cr, Mo, W, Al, or Fe
TiC (NbC, TaC)(b)	...	Ni or Co	(c)
TiC	Cr_3C_2	Ni or Co	...
TiC	(d)	Ni or Co	(c)
Cr_xC_y(e)	(d)	Ni	(c)
WC	(d)	Co	(c)

(a) Composite range of about 30 to 70 wt % of either major component. (b) A solid-solution component. (c) With or without other metal additions, such as Cr, Mo, W, Al, or Fe. (d) With or without other additions of transition-metal carbides. (e) As Cr_3C_2 or mixed carbides of chromium

Ceramic Matrix Composites

General

Some ceramic/metal combinations

Oxide matrix	Metal	Oxide matrix	Metal
Binary			
Cr_2O_3	Cr, Mo, W, Re	$Nd_2O_3(CeO)$	Nb
$(Cr, Al)_2O_3$	Cr, Mo, W	TiO_2	Cr, Nb, Ta
Gd_2O_3	Mo, W	UO_2	Mo, Ta, W
$Gd_2O_3(CeO2)$	Mo	$UO_2(ThO_2)$	W
$HfO_2(CaO)$	Mo, W	$Y_2O_3(CeO_2)$	Mo, W
$HfO_2(Y_2O_3)$	W	ZrO_2	Ta
La_2O_3	Mo, W	$ZrO_2(CaO)$	W
Nd_2O_3	Mo, W	$ZrO_2(Y_2O_3)$	W
$LaCrO_3$	Cr, Mo, W	SiO_2	Cr
$YCrO_3$	Cr, Mo, W	TiC	Mo, Fe, Ni, Co
SiC	Ag, Co, Cr	Al2O3	Al, Co, Fe, Cr
WC	Co		
Ternary			
UO_2-MgO	W	La_2O_3-$LaCr_2O_3$	W
MgO-ZrO_2	W	CaO-$CaCr_2O_4$	W
MgO(Cr_2O_3)-ZrO_2	Mo	Cr_2O_3-ZrO_2	W, Mo
Cr_2O_3-$LaCrO_3$	W	Cr_2O_3-HfO_2	Mo

Fabrication techniques for ceramic composites

Architectures	Matrix densification
Filament winding	Infiltration
Chopped fiber	Glass
Braiding	Polymer precursor
Fabric lay-up	Sol gel
1D, 2D, 3D	Si
Whiskers	CVD
Particle dispersion	Hot pressing
	Sintering
	Reaction sintering
	Plasma spraying

Flaw size, particle or fiber spacings, and strengths of ceramic composites

Material	Flaw size(a), μm	Particle or fiber Material	Particle or fiber Diameter, μm	Spacing, μm, for volume fraction of: 10%	20%	30%
Al_2O_3	65	Random carbon fibers	8	32	21	14
Glass	23	Al_2O_3 particles	60	45	23	12
			15	11	5.7	3.1
Glass	30	Aligned carbon fibers	8	14	8	5
Glass	30	Random carbon fibers	8	32	21	14
Glass	72	Al_2O_3 particles	3.5	2.6	1.4	0.7
			11	8	4.2	2.2
			44	33	17	9
MgO	82(b)	Random Ni fibers	89	360	240	160
	G(c)	Co, Fe, or Ni particles	~40	30	15	8
Si_3N_4	62	SiC	5	3.7	1.9	1.0
			9	6.7	3.4	1.8
			32	24	12	6.6

(a) Flaw size calculated from *S*, *E*, and γ of matrix. (b) Assuming uniform spacing of equal spherical or cylindrical particles, for random fibers, only one-third were assumed to be oriented to significantly affect crack propagation. (c) Grain size (*G*), 10 μm or less and $0.03 < P < 0.07$

Ceramic matrix composite toughening concepts

Concept	Basic requirements	Status of verification and modeling
Modulus transfer of load from matrix to fibers	$E_f > E_m$, preferably by a factor of >2	Verified, reasonable modeling
Prestressing of fibers and matrix	$\alpha_f > \alpha_m$ so axial tensile stresses in fibers < their fracture stress to give reasonable compressive axial stress in matrix	Not verified; basic modeling not expected to be difficult
Crack-impeding second phases	Fracture toughness of fibers (or particles) > local matrix so crack is either arrested or bows out, *i.e.*, gives line tension effects between fibers or particles	Arrest impractical; line tension modeling, but uncertain verification
Fiber pull-out	Fiber (or elongated particles) have high enough transverse fracture toughness so failure occurs along fiber matrix interface	Limited verification and modeling
Crack deflection or multiplication	Sufficiently weak fiber (or particle) matrix interfaces, or appropriate mismatch of properties, especially thermal expansions between matrix and particles (fibers) and use of appropriate particles (fiber sizes)	Limited verification: no modeling. Some verification: possible modeling developing

Theoretical maximum temperatures for some fiber-reinforced ceramics (based on melting or softening points)

System	Maximum temperature, °C
Carbon–Pyrex glass	700 to 800
Carbon–glass-ceramic	1300
Silicon carbide–glass	>650
FP alumina–glass	>650
Silicon carbide–silicon	1410
Silicon carbide–silicon nitride	1900
Carbon–carbon	3550

Fabrication methods for metal fiber/ceramic matrix composites

Fibers	Matrix	Fabrication method
Cr Mo	Al_2O_3-Cr_2O_3 CeO_2-doped Gd_2O_3	Hot pressing of grains of previously directionally solidified eutectic composites
V, Nb, Ta	Cr_2O_3	Directional solidification
Cr, Nb, Ta Ta	TiO_2 ZrO_2	Hot pressing of grains of previously grown eutectic
Cr	Fe_3O_4, Al_2O_3, Cr_2O3, and mixtures	Directional solidification
Ta	Unstabilized HfO_2	Directional solidification
W, Mo	Stabilized HfO_2	Hot pressing
Ni, Fe, Co	MgO	Hot pressing
W	Fused SiO_2	Hot pressing
Ta, Mo, Nb	UO_2	Directional solidification
Stainless steel	Wüstite	Hot pressing
Cu, Cu-Be, Be	Be_4B, Be_2B	Hot pressing, plasma spraying, or vapor deposition
Ti, Cr	SiC	Whisker formation *in situ*
Ta, W	Si_3N_4	Hot pressing
W, Mo	Si_3N_4	Flame spraying of silicon and heating in nitriding atmosphere
Mo, Ta, W	Sialon, Si_3N_4, Si_3N_4-C, TaC	Hot pressing
Ta	TaC	Hot pressing
W, W-Re	TaC	Hot pressing
Nb	$MoSi_2$	Hot pressing
Nb	Borosilicate glass	Hot pressing
Ni	Glass-ceramic	Hot pressing
W, Mo, stainless steel, or carbon steel	Glass, glass-ceramic	Fusing of glass-coated fibers together using pressure
Stainless steel	PbO glass	Hot pressing, vacuum injection, or pultrusion

Cyclic fatigue results for some ceramic composites

Material	Environment	Mean strength, MPa	Maximum stress, MPa	Number of cycles of failure
30 vol % BN/mullite	Liquid N_2	...	320 to 360	2 to 100
30 vol % BN/alumina	Liquid N_2	...	410	3
30 vol % BN/Si_3N_4	Silicone oil	330	275	10 to 100
Woven carbon/carbon	Silicone oil	97	80 to 110	2 to 70

Property data of ceramic composite materials

Material	Density, g/cm^3	Softening point °C	Softening point °F	Elastic modulus GPa	Elastic modulus 10^6 psi	Coefficient of thermal expansion, 10^{-6}/K
Corning 6810 soda-zinc glass	2.65	770	1420	6.6	0.96	6.9
Al_2O_3	3.91	...	...	385	55.8	8.85
ZrO_2	5.65	...	...	142	20.6	5.47

Some potential matrix, fiber, and dispersion material options for ceramic composites

Matrix materials
- Si_3N_4
- SiC
- ZrO_2, HfO_2
- Al_2O_3
- Glass
- Glass-ceramic
- Mullite
- Cordierite

Fiber materials
- SiC
- BN
- Si_3N_4
- $Al_2O_3 \cdot B_2O_3 \cdot SiO_2$
- Graphite
- Coatings
- Mullite
- Al_2O_3

Dispersion materials
- SiC
- ZrO_2
- TiC
- BN

Ceramic

Calculated abrasion wear resistance parameters for Si_3N_4-base composites compared with some Al_2O_3 systems

Material	Abrasion wear resistance parameter, $*H^{1/2} \cdot K_{Ic}^{3/4}$	Fracture toughness (K_{Ic}) MPa$\sqrt{m}$
Si_3N_4 + 6 wt % Y_2O_3	11.87	4.8
Si_3N_4 + 10 wt % TiC	11.91	4.8
Si_3N_4 + 30 wt % TiC	11.84	4.4
Si_3N_4 + 50 wt % TiC	8.48	2.7
Si_3N_4 + 30 wt % (W, Ti)C	9.57	3.5
Si_3N_4 + 30 wt % WC	13.06	5.2
Si_3N_4 + 30 wt % TaC	11.15	4.6
Si_3N_4 + 30 wt % HfC	9.81	3.6
Si_3N_4 + 30 wt % SiC	9.74	3.6
Al_2O_3	7.38	2.3
Al_2O_3 + ZrO_2	9.32	3.2
Al_2O_3 + TiC	9.92	3.2
Sialon	9.87	4.0

* Calculated using wear resistance index $\left(K_{Ic}^{3/4}H^{1/2}\right)$ value

Hardness and fracture toughness of Al_2O_3 ceramic cutting-tool materials compared with those of Sialon and $Si_3N_4 + Y_2O_3$

Material	Knoop hardness, GPa	Fracture toughness (K_{Ic}) MPa$\sqrt{m}$
Al_2O_3	15.6	2.3
Al_2O_3 + TiC	17.2	3.2
Al_2O_3 + ZrO_2	15.2	3.2
Sialon	12.2	4.0
Si_3N_4 + Y_2O_3	13.4	4.8

Examples of directed metal oxidation ceramic-matrix systems

Parent metal	Reaction product
Aluminum	Oxide, nitride, boride, titanate
Silicon	Nitride, boride, carbide
Titanium	Nitride, boride, carbide
Zirconium	Nitride, boride, carbide
Hafnium	Boride, carbide
Tin	Oxide

Fabrication methods for ceramic fiber/ceramic matrix composites

Fibers	Matrix	Fabrication method
Al_2O_3	Al_2O_3	Sintering
AlN	Al_2O_3	Tape casting, aligning of AlN needles, and sintering
AlN, Si_3N_4	AlN, Si_3N_4	Hot pressing
Al_2O_3, C, ZrO_2	$Mg_3(PO_4)_2$	Hot pressing
Al_2O_3, C, B, SiC, SiO_2	Al_2O_3, $3Al_2O_3 \cdot 2SiO_2$	Hot pressing
Al_2O_3, C, B, BN, SiC	Si_3N_4	Reaction sintering
$3Al_2O_3 \cdot 2SiO_2$	$3Al_2O_3 \cdot 2SiO_2 \cdot Al_2O_3$	Slip casting and firing
BN	BN	Hot pressing
BN	BN	Chemical vapor deposition
BN	BN	Firing of B_2O_3-containing composite in nitriding atmosphere
C	Al_2O_3, $3Al_2O_3 \cdot 2SiO_2$	Coating of fibers with LiC; sintering
C	Al_2O_3	Hot pressing
C	Carbides, borides, silicides, oxides	Hot pressing
C	Pyrolytic materials	Chemical vapor deposition
C	C-SiC, TiC	Chemical vapor deposition
C	Si_3N_4	Hot pressing, reaction sintering; coating of fibers–e.g., with SiC–to improve compatibility
C	Sialon, Si_3N_4, Si_3N_4-C, TaC	Hot pressing
C	TaC	Vacuum impregnation with precursor solution and pyrolyzing
C	C-TaC	Hot pressing of Ta-coated fibers
C	ZrB_2-Si-C	Hot pressing
C, fused SiO_2	Powdered ceramic	Application of aqueous slurry and drying
MgO	Cubic ZrO_2	Directional solidification
SiC	Si	Impregnation of carbon fiber preform with molten silicon
SiC	Si	Heating of a mixture of carbon fibers and silicon powder
SiC	Si	Infiltration of SiC fibers with molten silicon
SiC	SiC	Chemical vapor deposition
SiC	SiC, Si_3N_4, AlN, BN	Hot pressing or sintering
SiC	Si_3N_4	Reaction sintering
Si_3N_4	Si_3N_4	Hot pressing
ZrO_2	Al_2O_3	Directional solidification
ZrO_2	CaO-ZrO_2	Directional solidification
ZrO_2	MgO	Hot pressing
ZrO_2	ZrO_2	Hot pressing
ZrO_2	ZrO_2	Impregnation

Manufactured fiber-reinforced ceramic composites and their corresponding processes

Processing	Composite (fiber-matrix)	Comments
Hot pressing	W-glass, Ni-glass, Mo-thoria, Mo-alumina, W-ceramic, stainless steel-alumina, C-glass, C-glass ceramic, C-MgO, C-Al_2O_3, ZrO_2-MgO, ZrO_2-ZrO_2, SiC-glass, SiC-glass-ceramics, Al_2O_3-glass, C-Si_3N_4, Ta-Si_3N_4	Fibers and matrix powder are mixed together and hot pressed to produce low-porosity composites, with uncracked matrices, if thermal expansion coefficients are matched. Aligned continuous fiber composites can have very high strengths
Cold pressing and sintering	C-glass, metal fiber-ceramic	Fibers and matrix are mixed, cold pressed, and sintered. Disappointing results because the large shrinkage produces cracked composites
Devitrification	C-glass-ceramic, SiC-glass-ceramic	Fibers and glass powder are hot pressed at relatively low temperatures to give a reinforced glass. Further high-temperature heat treatment is used to devitrify the glass to a glass-ceramic. The C-glass system gives disappointingly low strengths, probably because of volume changes during devitrification. The SiC glass-ceramic system is reported to have good properties
Reaction bonding	Reinforced Si_3N_4	Fibers are incorporated into flame-sprayed silicon that is subsequently reaction-sintered in nitrogen
Slip-casting	Ceramic fiber-fused silica	Ceramic fibers are incorporated into slips of finely divided fused silica and fired. Increased porosity due to the presence of fiber usually results in a degradation of properties
Plasma-spraying	Mo-Al_2O_3, W-Al_2O_3	Alumina powder is plasma-sprayed. Processing is believed to be slow
Chemical vapor infiltration and deposition	SiC fibers in SiC, C fibers in SiC	Fiber integrity can be preserved by lack of mechanical movement and relatively low process temperatures

Izod impact strength after elevated-temperature exposure for a glass-reinforced ceramic composite(a)

Temperature (2-h exposure)		Izod impact strength	
°C	°F	J/cm	ft · lbf/in.
315	600	17.1	32.0
425	800	9.0	16.9

(a) S-944, style 181 fiberglass fabric in matrix of a modified aluminum phosphate

Fabrication methods for some ceramic whisker/ceramic matrix composites

Whiskers	Matrix	Fabrication method
$3Al_2O_3 \cdot 2SiO_2$, αAl_2O_3, ZrO_2	Oxides and nitrides	Hot pressing
$3Al_2O_3 \cdot 2SiO_2$, αAl_2O_3, SiC, Si_3N_4, ZnO	TiO_2	Hot pressing
$3Al_2O_3 \cdot SiO_2$	Al_2O_3, Al_2O_3-Mo, Cr_2O_3, ZrO_2, Al_2O_3-Cr, AlN, BN, Si_3N_4, V_2O_3, TiN, SiO_2	Hot pressing
αAl_2O_3, AlN, SiC	$3Al_2O_3 \cdot 2SiO2$, Al_2O_3	Hot pressing
Si_3N_4	Si_3N_4	Sintering
SiC, BN, C	Si_3N_4, AlN	Sintering or hot pressing
ZrO_2	Stabilized ZrO_2	Hot pressing
ZrO_2	MgO	Hot pressing
Ground whiskers	Several oxides	Powder metallurgy techniques

Formulation and fabrication parameters of multidirectional continuous fiber-reinforced ceramic-ceramic composites

Property/composite	Three-directional silica-silica	Four-directional silica-silica	Three-directional alumina-alumina	Three-directional alumina-silica	Three-directional boron nitride/boron nitride
Preform reinforcement type	Three-directional orthogonal	Four-directional cubic braided	Three-directional orthogonal	Three-directional orthogonal	Three-directional orthogonal
Fiber type	Fused quartz continuous	Fused quartz continuous	Polycrystalline alumina staple	Polycrystalline alumina staple	Continuous polycrystalline boron nitride
Fiber electrical resistivity, ohm · m	10^{16} (at 293 K) 2×10^5 (at 1073 K)	10^{16} (at 293 K) 2×10^5 (at 1073 K)	10^{10} (estimated)	10^{10} (estimated)	10^8 (at 1075 K)
Fiber strength, GPa (10^6 psi)	0.7(0.10)	0.7 (0.10)	1.4 (0.20)	1.4 (0.20)	0.35 to 0.7 (0.05 to 0.10)
Fiber volume fraction	50%	50%	30%	30%	40%
Matrix/densification	Colloidal silica	Colloidal silica	Colloidal alumina	Colloidal silica	Nitrided boric oxide
Process or stabilization temperature, K	923 to 1013	923	110	923	2073
Composite bulk density, g/cm^3	1.6	1.6	1.9	2.0	1.6

Performance properties of multidirectional continuous fiber-reinforced ceramic-ceramic composite materials at 300 K

Property	Three-directional silica-silica	Four-directional silica-silica	Three-directional alumina-alumina	Three-directional alumina-silica	Three-directional boron nitride-boron nitride
Tensile					
Modulus of elasticity, GPa (10^6 psi)	15.6 (2.26)	9.7 to 13.1 (1.4 to 1.90)(a)	36.3 to 5.26	33.8 (4.90)	15.4 (2.23)
Tensile strength, MPa (ksi)	26.7 (3.87)	20.4 to 26.6 (2.96 to 3.86)(a)	71.1 (10.3)	74.8 (10.8)	24.8 (3.60)
Tensile strain, %	0.2	1.2	0.2	0.2	0.2
Compressive					
Compressive modulus, GPa (10^6 psi)	21.9 (3.18)	8.6 (1.3)	31.4 (4.55)	...	29.2 (4.23)
Compressive strength, MPa (ksi)	144.8 (21.0)	70.6 (10.2)	224.7 (32.59)	...	36.5 (5.29)
Compressive strain, %	1.6	1.5	>0.6	...	0.2
Shear					
Shear modulus, GPa (10^6 psi)	1.5 (0.22)	4.4 (0.64)	3.4 (0.49)	1.7 (0.25)	...
Poisson's ratio	0.09	...	...	...	...
Mean coefficient of thermal expansion, to 600 K, 10^{-6}/K	0.54	0.47	6.4	6.4	2.7
Thermal conductivity at 300 K, W/m · K (Btu · in./h · ft^2 · °F)	0.66 (4.6)	0.65 (4.5)	1.62 (11.2)	0.68 (4.7)	9.0 (62.4)
Dielectric constant(b)	3.5	2.9	3.7	3.8(c)	3.0
Loss tangent(b)	0.0009	0.001	0.0045	0.004(c)	0.002

(a) Lower value for "water-desensitized": version corresponding to dielectric properties. (b) At 8.5 to 9.4 GHz, or as indicated. (c) At 35 GHz

Mechanical properties of Si_3N_4 composites containing 30 vol % of metal carbide dispersoid (2 μm average particle diameter)

Matrix	Dispersed phase	Density, g/cm^3	Knoop hardness, GPa	Fracture toughness (K_{Ic}), MPa$\sqrt{m}$	Modulus of rupture, MPa: RT	1000 °C	1200 °C
Si_3N_4 + 6 wt % Y_2O_3	None	3.26	13.4 ± 0.3	4.8 ± 0.3	110.9 ± 1.6	88.3 ± 3.5	49.2 ± 5.0
Si_3N_4 + 6 wt % Y_2O_3	TiC	3.81	15.21 ± 0.3	4.4 ± 0.5	80.6 ± 5.9	120.4 ± 12.2	64.4 ± 2.9
	(Ti, W) C	4.55	14.06 ± 0.3	3.5 ± 0.3	75.5 ± 3.2	86 ± 0	52.9 ± 0.5
	WC	7.70	14.4 ± 0.4	5.2 ± 0.4	89.1 ± 31.8	136.4 ± 1.6	55.7 ± 0.5
	TaC	6.87	12.6 ± 0.2	4.6 ± 0.4	86.2 ± 7.3	124.5 ± 16.0	43.2 ± 2.0
	HfC	5.74	14.1 ± 0.4	3.6 ± 0.2	86 ± 0.8	...	68.6 ± 0.5
	SiC	3.24	13.6 ± 0.2	3.65 ± 0.5	97.6 ± 8.5	94.0 ± 4.9	52.3 ± 3.2
Al_2O_3	TiC	4.28	17.2 ± 0.2	3.2 ± 0.4	72.2 ± 13.0	69.4 ± 4.3	57.0 ± 4.1

Selected properties of BN/Al_2O_3 and ZrO_2/Al_2O_3 composites

Material	Bend strength, MPa	Modulus of elasticity, GPa	Impact energy, J/m^2	Critical temperature, °C	Fracture toughness, MPa $\sqrt{m}$	Fracture strength, σ_f, MPa	Specific strength, σ_τ/σ_i
30 vol % BN-1/Al_2O_3, ⊥ HPA(a)	170	120	60	700 to 850(b)	3.8	...	...
30 vol % BN-2/Al_2O_3, ⊥ HPA(a)	400	190	40 > 100	450(b)	3.9 to 9.0	...	...
30 vol % BN-2/Al_2O_3, ∥ HPA(a)	160	140	25	...	2.6	...	...
30 vol % BN/Al_2O_3, ⊥ HPA(a)(c)	200	200	3	400 to 800(b)	1.1	...	...
0.5 vol % ZrO_2/Al_2O_3	...	400	20	600	...	410	0.35
4.0 vol % ZrO_2/Al_2O_3	...	410	25	600	...	420	0.4
9.0 vol % ZrO_2/Al_2O_3	...	380	55	950	...	700	...
11.5 vol % ZrO_2/Al_2O_3	...	380	60	1150	...	780	0.6 to 0.75
14.0 vol % ZrO_2/Al_2O_3	...	375	35	800	...	680	0.65
19.0 vol % ZrO_2/Al_2O_3	...	350	35	800	...	350	0.65

(a) ⊥ HPA = stress direction perpendicular to hot pressing axis. ∥ HPA = stress direction parallel to hot pressing axis. (b) Determined by 22 °C (72 °F) water quench test using 3-mm-(0.19-in.) square bars. (c) Calculated treating BN as pseudoporosity

Mechanical properties of a two-dimensional Nicalon™ fiber-reinforced Al_2O_3-matrix composite

Test temperature, °C	°F	4-point flexural strength, MPa	ksi	Weibull Modulus	Chevron notch toughness, MPa $\sqrt{m}$	ksi $\sqrt{in.}$
21	70	461 ± 28	67	15.5	27.8	25.3
1200	2190	488 ± 22	71	20.6	23.3	21.2
1300	2370	400 ± 12	58		19.2	17.5
1400	2550	340 ± 11	49		15.6	14.2

Thermal and physical properties of a glass-reinforced ceramic composite(a)

Specific gravity	1.8 to 1.9
Specific heat:	
kJ/kg · K	0.84
Btu/lb · °F	0.2
Coefficient of thermal expansion:	
Parallel ti reinforcement	
10^{-6}/K	3.6
10^{-6}/°F	2.0
Perpendicular to reinforcement	
10^{-6}/K	1.1
10^{-6}/°F	0.6
Thermal conductivity:	
W/m · K	0.58
Btu · in./ft^2 · h · °F	4.0
Total normal emissivity:	
At 425 °C (800 °F)	0.73
At 650 °C (1200 °F)	0.76

(a) S-944, style 181 fiberglass fabric in matrix of a modified aluminum phosphate

SiC/Ceramic

Types and properties of Si/SiC

Designation	SiC content, vol %	Bend strength MPa	Bend strength ksi	Modulus of elasticity GPa	Modulus of elasticity 10^6 psi	Density g/cm^3	Density lb/in.^{3}
		Properties at 25 °C (77 °F)(a)					
G.E.-reported data							
Type TH(b)	80 to 85	483	70	393	57	3.30	0.012
Type THL	38 to 40	276	40	303	44	2.70	0.098
Type F(c)	20 to 25	172	25	200	29	2.60	0.094
Fansteel-reported data							
Tensile strength at room temperature:							
MPa							827
ksi							120
Charpy impact strength:							
J							13.6
ft · lbf							10
Oxidation resistance							Excellent
Thermal shock resistance							Excellent

(a) Measured on 2.54 × 2.54 × 15.8 mm (0.1 × 0.1 × 0.625 in.) specimens tested in three-point bending. (b) Unidirectional orientation. (c) Omnidirectional orientation

Physical properties of fiber-grain CVI SiC-matrix composites

Material system	Density g/cm^3	Density lb/in.^{3}	Modulus of rupture MPa	Modulus of rupture ksi	Modulus of elasticity GPa	Modulus of elasticity 10^6 psi	Strain, %
Carbon fibers/SiC grain	2.06	0.074	40.0	5.8	20	2.9	0.57
Mullite fibers/SiC grain	1.80	0.065	35.9	5.2	15.9	2.3	0.56
Silica fibers/SiC grain	1.95	0.070	33.1	4.8	22.1	3.2	0.46

Properties of silicon carbide composites

Property	Silicon nitride-bonded SiC(a)	Silicon oxynitride-bonded SiC(b)	SiC bonded graphite aggregate(c)	Self-bonded SiC(d)
Density, g/cm^3	2.5 to 2.8	~2.7	2.3 to 2.8	3.10
Porosity, %	13 to 15	~18	8 to 26	~5
Maximum recommended working temperature:				
Oxidizing environment				
°C	1650	1650	815 to 1540	1650
°F	3000	3000	1500 to 2800	3000
Neutral environment				
°C	2205	2205	2205	2315
°F	4000	4000	4000	4200
Specific heat (27 to 1370 °C; 80 to 2500 °F):				
kJ/kg · K	0.84 to 1.55	0.84 to 1.55	0.71 to 1.47	0.71 to 1.38
Btu/lb · °F	0.2 to 0.37	0.2 to 0.37	0.17 to 0.35	0.17 to 0.33
Thermal conductivity (980 °C; 1800 °F):				
W/m · K	16	16	48 to 59	42
Btu/ft · h · °F	9.5	9.5	28 to 34	24
Coefficient of linear thermal expansion (21 to 1370 °C; 70 to 2500 °F):				
10^{-6}/°C	1.4	1.4	1.5(e)	1.51
10^{-6}/°F	2.6	2.6	2.7(e)	2.72
Bend strength:				
24 °C (75 °F)				
MPa	41 to 69	41 to 55	34 to 69	165
ksi	6 to 10	6 to 8	5 to 10	24
1095 °C (2000 °F)				
MPa	48 to 69	55 to 62	34 to 69	172
ksi	7 to 10	8 to 9	5 to 10	25
1500 °C (2730 °F)				
MPa	21	~21	48 to 83	124
ksi	3	~3	7 to 12	18

(continued)

Properties of silicon carbide composites (continued)

Property	Silicon nitride-bonded SiC(a)	Silicon oxynitride-bonded SiC(b)	SiC bonded graphite aggregate(c)	Self-bonded SiC(d)
Compressive strength (24 °C; 75 °F):				
MPa	138	~138	434(e)	1380
ksi	20	~20	63(e)	200
Impact strength (21 °C; 70 °F):				
J	<0.113	...	<0.113	<0.113
in. · lbf	<1	...	<1	<1
Modulus of elasticity:				
24 °C (75 °F)				
GPa	117		~241(e)	379
10^6 psi	17	17	~35	55
1000 °C (1830 °F)				
GPa	69	69	207(e)	365
10^6 psi	10	10	30(e)	53
Bend-creep resistance–conditions to promote a tensile creep strain of 0.125%:				
Time, h	>300 (1095 °C; 2000 °F)	...	>630 (1205 °C; 2200 °F)	>1000 (1205 °C; 2200 °F)
Temperature (300 h)				
°C	>1095	...	>1205	>1300
°F	>2000	...	>2200	>2370
Bend stress (100 h)				
MPa	28 to 41 (1000 °C)	...	28 to 41 (1000 °C)	159 (1205 °C)
ksi	4 to 6 (1830 °F)	...	4 to 6 (1830 °F)	23 (2200 °F)

(a) Composition is approximately 80 wt % SiC and 20 wt % Si_3N_4 (b) Composition is approximately 80 wt % SiC and 20 wt % Si_2ON_2. (c) Variable graphite content; compositions reportedly range from 78 to 50 wt % SiC (dense bond phase), 20 to 46 wt % graphite aggregate, and 1 to 4 wt % silicon. (d) Fine-grain, high-density material. (e) Value for low-graphite-content material

Mean matrix crack spacing and first cracking stress for SiC/RBSN composites

Fiber fraction, %	Matrix crack spacing, mm	Composite stress at which matrix first cracked, MPa
23 ± 3	2.0 ± 0.3(a)	237 ± 25(b)
40 ± 2	0.9 ± 0.2	293 + 15

(a) Standard deviation for 30 cracks on five bend specimens. (b) Standard deviation for 5 specimens measured in 3-point bend

Density and porosity data for SiC/RBSN composites

Volume fraction of fibers, %	Before nitriding: Density, g/cm^3	Before nitriding: Matrix porosity %	After nitriding: Density, g/cm^3	After nitriding: Matrix porosity %
0	1.56	35	1.98	37
23 ± 3	1.70	54(a)	2.19	39(a)
40 ± 2	1.90	51(a)	2.36	40(a)

(a) Matrix porosity calculated from composite density and from theoretical density for CVD SiC fiber (3.0 g/cm^3) and from density for silicon (2.4 g/cm^3) or for Si_3N_4 (3.2 g/cm^3)

Room-temperature strength of RBSN and SiC/RBSN composites

	Axial strength, MPa		
Test	0% fiber	23 ± 3% fiber	40 ± 2% fiber
4-point bend (L/H = 15)(a)	107 ± 26(b)	539 ± 48(b)	616 ± 36(b)
4-point bend (L/H = 45)(a)	...	675 ± 42	868 ± 32
3-point bend (L/H = 35)(a)	...	717 ± 80	958 ± 45
Tensile(c)	...	352 ± 73	536 ± 20

(a) L/H refers to span-to-height ratio of test specimen (H ≈ 1.2 mm). (b) 50-mm gage length. (c) Standard deviation for five tests

Physical properties of various ceramic- and graphite-reinforced CVI SiC composites

Material system	Density g/cm^3	Density lb/in.^{3}	Modulus of rupture MPa	Modulus of rupture ksi	Modulus of elasticity GPa	Modulus of elasticity 10^6 psi	Strain, %
PAN 8HS/SWB	1.74	0.063	229	33.2	40.7	5.9	3.0
PAN knit/KFB	1.95	0.070	215	31.2	43.4	6.3	3.2
Nicalon/8HS	2.09	0.076	262(a)	38.0(a)	53.1	7.7	0.64
Saffil/Al_2O_3 paper	...	...	296	43.0	208.3	30.2	1.0
Saffil/Al_2O_3 paper	...	...	125	18.2	94.5	13.7	1.0
Saffil/Al_2O_3 paper	...	...	101	14.6	17.9	2.6	0.8

(a) Ceramic-grade fiber; results for nonceramic-grade fiber were 108 MPa (15.6 ksi) high and 37 MPa (5.4 ksi) low

Thermal properties of SiC-whisker-reinforced ceramics

Composite	Thermal conductivity at 22 °C (70 °F) W/m · K	Thermal conductivity at 22 °C (70 °F) Btu · in./h · ft^2 · °F	Thermal conductivity at 600 °C (1110 °F) W/m · K	Thermal conductivity at 600 °C (1110 °F) Btu · in./h · ft^2 · °F	Linear coefficient of thermal expansion at 22 to 1100 °C (70 to 2010 °F), 10^{-6}/K
Alumina	36 ± 5	250 ± 35	12 ± 3	85 ± 20	7.8 to 8.2
20 vol% SiC whiskers	32	220	16	110	7.35
30 vol% SiC whiskers	...	...	...	...	6.70
60 vol% SiC whiskers	...	...	...	...	5.82
SiC	95	660	50	350	4.8
20 vol% SiC whiskers-mullite	7.2	50	...	...	5.60

Shock strengths of SiC-whisker-reinforced ceramics

Matrix	Whisker content, vol %	Aid/wt %	Temperature °C	Temperature °F	Pressure MPa	Pressure ksi	Time, min	Density g/cm^3	Density lb/in.^{3}	Charpy shock strength, J (in. · lbf) Room temperature	Charpy shock strength, J (in. · lbf) 1095 °C (2000 °F)	Charpy shock strength, J (in. · lbf) 1315 °C (2400 °F)
Si_3N_4	5	MgO/1	1700	3092	28	4	65	3.15	0.114	0.150 (1.33)	0.160 (1.40)	0.15 (1.31)
	10	MgO/4	1700	3092	28	4	65	3.12	0.113	0.195 (1.73)	0.165 (1.46)	0.11 (0.99)
	38	MgO/1	1700	3092	28	4	75	3.00	0.108	0.094 (0.83)	0.080 (0.73)	0.21 (1.85)
	10	MgO/1	1700	3092	28	4	95	2.79	0.101	0.069 (0.61)	0.230 (2.03)	0.146 (1.29)
SiC	38	Al_2O_3/3; C/2	2140	3884	28	4	180	3.12	0.113	0.094 (0.83)	0.076 (0.67)	0.069 (0.61)
	50	Al_2O_3/3	2170	3938	28	4	180	3.11	0.112	0.072 (0.64)	0.079 (0.70)	0.069 (0.61)
	65	Al_2O_3/3	2140	3884	28	4	180	2.93	0.106	0.067 (0.59)	0.060 (0.54)	0.11 (0.96)

Graphite/Ceramic

Thermal-strain parameters (at 20 °C) and measured works of fracture for 20 vol % carbon-fiber composites

Material	Work of fracture, J/m^2	10^{-3} ($\alpha_m - \alpha_f$) ΔT
MgO	10	...
CM	110	6.6
Al_2O_3	38 to 66	...
CA	40	0.4
Pyrex	4	...
CP	344	−2.3
Glass-ceramic	4	...
CGC	100	−6.0

Properties of graphite fiber/$Li_2O \cdot Al_2O_3 \cdot 8SiO_2$ composites(a) tested perpendicular and parallel to hot pressing direction

	Modulus of rupture		Average modulus of rupture		Standard deviation			Modulus of elasticity		Average modulus of elasticity		Standard deviation		
Test direction	MPa	ksi	MPa	ksi	MPa	ksi	Variance	GPa	10^6 psi	GPa	10^6 psi	GPa	10^6 psi	Variance
Perpendicular to hot pressing direction	804.6	116.7	793.2	115.0	57	8.2	7.1	146	21.2	144	20.9	11.7	1.7	8.1
	756.4	109.7						149	21.6					
	777.7	112.8						143	20.7					
	766.7	111.2						123	17.9					
	721.9	104.7						134	19.5					
	892.9	129.5						153	22.2					
	832.2	120.7						158	22.9					
Parallel to hot pressing direction	837.7	121.5	860.7	124.8	41	5.9	4.7	159	23.0	159	23.0	4.8	0.7	3.1
	897.7	130.2						155	22.5					
	815.0	118.2						152	22.0					
	877.0	127.2						160	23.2					
	912.2	132.3						166	24.1					
	824.6	119.6						159	23.0					

(a) Fiber content, 39.6 vol %. Bulk density, 2.06 g/cm^3

Properties of graphite fiber/$Li_2O \cdot Al_2O_3 \cdot 8SiO_2$ composites(a)

Fiber	Bulk density,	Modulus of rupture		Average modulus of rupture		Standard deviation		
content, vol %	g/cm^3	MPa	ksi	MPa	ksi	MPa	ksi	Variance
29.0	2.19	637	92.4	633	91.8	37	5.4	5.9
		641	93.0					
		689	99.9					
		601	87.1					
		597	86.6					
33.1	2.16	698.4	101.3	782.1	113.4	48	7.0	6.2
		799.1	115.9					
		821.2	119.1					
		806.7	117.0					
		785.3	113.9					
35.1	2.17	845.3	122.6	834.4	121.0	58	8.4	7.0
		878.4	127.4					
		830.8	120.5					
		737.7	107.0					
		879.8	127.6					
36.3	2.15	804.6	116.7	878.4	127.4	51	7.4	5.8
		839.1	121.7					
		896.3	130.0					
		951.5	138.0					
		892.9	129.5					
		886.0	128.5					

(a) Hot pressed for 5 min. at 1375 °C (2510 °F) and 6.9 MPa (1.0 ksi)

Mechanical properties of graphite fiber-reinforced ceramic composites(a)

Property(b)	RT	650 °C (1200 °F)
Flexural strength:		
MPa	153	194
ksi	22.2	28.2
Flexural modulus:		
GPa	86.2	80.7
10^6 psi	12.5	11.7
Compressive strength:		
MPa	205	247
ksi	29.8	35.8
Compressive modulus:		
GPa	106	...
10^6 psi	15.4	...
Tensile strength(c):		
MPa	303	...
ksi	44.0	...
Tensile modulus:		
GPa	68.7	...
10^6 psi	9.96	...
Interlaminar shear strength:		
MPa	13.0	13.2
ksi	1.89	1.92

(a) Data from Acurex Corp. Ceramic matrix is Chemceram, a modified aluminum phosphate. (b) Flexural specimens failed in shear. Specific gravity, 1.77 g/cm^3 (0.064 lb/in.^{3}). Load applied once temperature achieved on specimen. (c) 228 MPa (33.0 ksi) at 1260 °C (2300 °F); 210 MPa (30.4 ksi) at 1650 °C (3000 °F)

Izod impact energy of HMS graphite fiber/$Li_2O \cdot Al_2O_3 \cdot 8SiO_2$ composites

Density, g/cm^3	Fiber content, vol %	Notched	Izod impact energy J/cm^2	ft · lbf/in.^{2}
2.19	31.6	No	15.9	75.5
2.19	31.6	Yes	5.94	28.3
2.18	34.2	Yes	4.20	20.0
2.18	34.4	Yes	4.68	22.3

Modulus and strength fractions of rule-of-mixture values for some carbon-fiber-reinforced ceramics(a)

Matrix	Modulus fraction	Strength fraction
Pyrex glass	0.86	0.81
Glass-ceramic	0.71 to 0.80	0.81
Carbon	1.00	0.50

(a) Fibers are continuous and aligned, with V_f between 0.4 and 0.5

Typical properties of a carbon/carbon composite compared with polycrystalline bulk graphite

Property	Carbon/carbon composite(a)	Bulk graphite(b)
Density, g/cm^3	1.65	1.83
Tensile strength, MPa (ksi):		
Room temperature	103 (15.0)	35.2 (5.1)
2480 °C (4500 °F)	68.9 (10.0)(c)	53.1 (7.7)
Tensile modulus, GPa (10^6 psi):		
Room temperature	41.4 (6.0)	11.7 (1.7)
2480 °C (4500 °F)	10.3 (1.5)	11.0 (1.6)
Compressive strength, MPa (ksi):		
Room temperature	68.9 (10.0)	80.7 (11.7)
2480 °C (4500 °F)	159 (23.0)	...
Compressive modulus, GPa (10^6 psi):		
Room temperature	22.8 (3.3)	...
2480 °C (4500 °F)	10.3 (1.5)	...
Flexural strength, MPa (ksi):		
Room temperature	96.5 (14.0)	42.0 (6.1)
1925 °C (3500 °F)	103 (15.0)	72.4 (10.5)
Flexural modulus, GPa (10^6 psi):		
Room temperature	27.6 (4.0)	...
1925 °C (3500 °F)	17.2 (2.5)	...
Coefficient of thermal expansion from RT to 1095 °C (2000 °F), 10^{-6}/mm · K (10^{-6} in./in. · °F)	0.54 (0.3)	3.24 (1.8)
Thermal conductivity, W/m · K (Btu/ft · h · °F):		
260 °C (500 °F)	55.4 (32)	114 (66)
2480 °C (4500 °F)	24.2 (14)	46.7 (27)

(a) Properties measured in z direction. (b) Properties are with grain. (c) Shear failure

Physical properties of a graphite-fiber-reinforced graphite composite(a)

Property	Standard product	Special grade
Density:		
g/cm^3	1.6	2.0
lb/ft.^{3}	100	125
Tensile strength:		
MPa	11.7	24.1
ksi	1.7	3.5
Tensile modulus:		
GPa	8.3	17.9
10^6 psi	1.2	2.6
Compressive strength:		
MPa	44.8	82.7
ksi	6.5	12.0
Compressive modulus:		
GPa	6.9	...
10^6 psi	1.0	...
Shear modulus:		
GPa	3.4	...
10^6 psi	0.5	...
Flexural strength:		
MPa	...	34.5
ksi	...	5.0
Thermal conductivity at 538 °C (1000 °F):		
W/m · K	86.4	144
Btu · in./ft^2 · h · °F	600	1000
Coefficient of thermal expansion at 538 °C (1000 °F):		
10^{-6}	0.58	0.97
10^{-6}	0.32	0.54
Open porosity, %	...	0.2

(a) All properties measured in the plane panel direction

Properties of some commercial carbon/carbon composites

Material	Bond	Fiber content, vol %	Fiber orientation	Density, g/cm^3	Tensile strength MPa	Tensile strength ksi	Flexural strength MPa	Flexural strength ksi	Interlaminar strength MPa	Interlaminar strength ksi
Ag Carb 101	Resin char	66	Flat	1.45	75.8	11.0	96.5	14.0	5.2	0.75
PTE	Pitch char	>60	Flat	1.57	74.5	10.8	75.8	11.0	...	...
Carbitax 530	...	66	Flat	...	75.8	11.0	110	16.0	14	2.0
Carbitax 513	Resin char	66	Filament wound	1.45	103 to 138	15.0 to 20.0	...	...	...	...
Carbitax 515	...	66	Unidirectional	...	276	40.0	...	...	12	1.8
Pyrolarex	Resin char	...	Chopped	1.00	8.674	1.258	17.47	2.534	...	...
Pyrolarex 350	Resin char	>60	Flat	1.20	43	6.2	86.2	12.5	6.9	1.0
Pyrocarb 400	Char/CVD	>60	Flat	1.30	96.5	14.0	138	20.0	6.6	0.95
Haveg 41 G(L)	Char	>60	Flat	1.45	40.3	5.84	44.1	6.39	14.6	2.12
LTV-CG	Char/CVD	>60	Flat	1.38	64.0	9.28	65	9.4	15	2.2
RPG	CVD	5 to 7	Felt	1.75	43	6.2	82.7	12.0	41	6.0
Pyro-Bond	CVD	50	Filament wound; circular wrap	1.34	827	120	...	...	...	...
Pyco-Bond	CVD	34	Flat	1.65	...	...	169	24.5	31	4.5
Pyco-Bond	CVD	44	Filament wound; 75° wrap	1.45	To 391	To 56.7	...	...	...	...

Effect of heat treatment temperature on CVD/felt carbon/carbon composites with various fiber content

Fiber content, vol %	Heat treatment temperature, °C	Strength MPa	Strength ksi	Modulus GPa	Modulus 10^6 psi	d_{002}, Å	Macroporosity, vol %
9	1100	61.3	8.9	21.4	3.1	3.49	15
	2630	62.0	9.0	16.5	2.4	3.45	
	3000	52.4	7.6	11.0	1.6	3.41	
25	1100	95.2	13.8	22.8	3.3	3.49	10.7
	2630	83.5	12.1	21.4	3.1	3.46	
	3000	76.5	11.1	17.2	2.5	3.41	
36	1100	111.0	16.1	24.1	3.5	3.49	8.8
	2630	91.0	13.2	23.4	3.4	3.47	
	3000	89.6	13.0	17.9	2.6	3.42	
46	1100	130.2	18.9	26.2	3.8	3.49	5.5
	2630	108.9	15.8	23.4	3.4	3.47	
	3000	97.9	14.2	17.2	2.5	3.42	

Oxidation of carbon/carbon composites

Designation	Material	d_{002}, Å	Fiber content, vol %	Heat treatment temperature, °C	Apparent energy, kcal/mol		
GC-30	Glassy carbon	...	...	3000	52 ± 2		
PG-30	Pyrolytic graphite	3.364	...	3000	54 ± 1		
EG-5H	Commercial graphite	3.367	...	...	55 ± 1 (0 to 8%), 57 ± 1 (>8%)		
UD-50-28	Carbon fiber/glassy carbon composite	3.360	50	2800	50 ± 2, 53 ± 2(a)	...	...
UD-10-24	Carbon fiber/glassy carbon composite	3.363	10	2400	49 ± 2(a)	50 ± 2(b)	48 ± 3(c)
UD-50-20	Carbon fiber/glassy carbon composite	...	50	2000	...	46 ± 2(b)	48 ± 2(c)
UD-50-18	Carbon fiber/glassy carbon composite	...	50	1800	...	44 ± 2(b)	46 ± 1(c)
UD-50-15	Carbon fiber/glassy carbon composite	...	50	1500	...	38 ± 1(b)	40 ± 2(c)

(a) Apparent activation energy of oxidation for graphite. (b) Apparent activation energy of oxidation for anisotropic area. (c) Apparent activation energy of oxidation for isotropic area

Modulus of rupture vs thermal shock cycles between room temperature and 1200 °C (2190 °F) for graphite fiber/$Li_2 \cdot Al_2O_3 \cdot 8SiO_2$ composites

Thermal shock cycles	Modulus of rupture MPa	Modulus of rupture ksi	Average modulus of rupture MPa	Average modulus of rupture ksi
0	951.5	138.0	857.7	124.4
	839.8	121.8		
	781.9	113.4		
1	897.7	130.2	893.6	129.6
	889.4	129.0		
5	838.4	121.6	861.8	125.0
	885.3	128.4		

Glass Matrix

Coefficient of thermal expansion for borosilicate 7740 glass matrix composites

Filament	Orientation	Coefficient of thermal expansion(a), 10^{-6}/°C
35 vol % SiC monofilament	0°	4.20
	90°	4.60
40 vol % SiC yarn	0°	3.25
	90°	2.70

(a) From 22 to 500 °C

Coefficient of thermal expansion for graphite fiber unidirectionally reinforced borosilicate matrix composites

Fiber type	Fiber elastic modulus GPa	Fiber content, vol %	Coefficient of thermal expansion(a), 10^{-6}/°C 0°	90°
Thornel 300	234	54	–0.10	+4.6
HM	350	70	–0.50	+6.5
P-100	654	50	–1.0	+4.4
Chopped Cel 6000	234	30	1.7	1.7(4.2)
Borosilicate glass	...	0	3.25	3.25

(a) From 22 to 500 °C

Selected glass-matrix composites and applications

Glass composition type	Crystal components	Application
Lead borosilicate	Pd/Au, Pt/Au	Conductor
	Pd/Au	Conductor
Proprietary glass	Pd/Ag	Conductor
CdBi lead borosilicate	Ba/Nd titanate	Capacitor
B_2O_3, lead borate	Ba titanate	Capacitor
B_2O_3	Ba ferrite	Magnets
Lead borate	Ag flake	Die attach
	Ag powder	Conductor (bus bars)
Proprietary glass	AlN	Substrate

Properties of SiC-fiber-reinforced borosilicate 7740 glass

Property	Monofilament		Yarn
Fiber content, vol %	35	65	40
Density, g/cm^3	2.6	2.9	2.4
Axial flexural strength, MPa:			
At 22 °C	650	830	290
At 350 °C	...	930	360
At 600 °C	825	1240	520
Axial elastic modulus, GPa, at 22 °C	185	290	120
Axial fracture toughness, MN$\sqrt{m}$:			
At 22 °C	18.8	...	11.5
At 600 °C	14.3	...	7.0

Three-point flexural strengths and elastic moduli of SiC-yarn-reinforced 7930 glass composites

Test temperature, °C	Specimens fabricated at 1500 °C Flexural strength, MPa	Flexural modulus, GPa	Specimens fabricated at 1600 °C Flexural strength, MPa	Flexural modulus, GPa
22	467	111	509	109
	413	107	482	97.1
	447	112	527	101
950	643	108	724	101
	565	88	657	102
1050	...	...	742	83
	...	...	668	87
1150	541	64	...	...
	498	65	...	...
1200	419	60	...	...
1250	243	48	...	...
	280	52	...	...

Fracture toughness of high-modulus-fiber-reinforced borosilicate glass matrix composites

Test temperature, °C	Test speed, cm/s	Fracture toughness, MPa $\sqrt{m}$
22	330.0	21.4
22	0.002	22.4
600	330.0	15.8
650	330.0	19.0

Coefficient of thermal expansion for some graphite-reinforced borosilicate glass composites

Composite type	Coefficient of thermal expansion, 10^{-6}/K(a) Longitudinal	Transverse
60 vol % GY-70/7740	−0.29	7.6
60 vol % HMS/7740	−1.0	3.6
60 vol % T-300/7740	0.38	4.3

(a) From 295 to 423 K

Comparison of properties of selected matrix and reinforcement materials used in glass matrix composites

Material	Modulus of elasticity GPa	10^6 psi	Fracture strength MPa	ksi	Coefficient of thermal expansion, 10^{-6}/K	Density, g/cm^3
Matrix						
Borosilicate glass	60	8.7	100	14.5	3.5	2.3
Soda-lime glass	60	8.7	100	14.5	8.9	2.5
Lithium aluminosilicate glass-ceramic	100	14.5	100 to 150	14.5 to 21.8	1.5	2.0
Magnesium aluminosilicate glass-ceramic	100	14.5	110 to 170	16 to 24.7	2.5 to 5.5	2.6 to 2.8
Particulate						
Alumina	360 to 400	52 to 58	250 to 350	36.3 to 50.8	8.5	3.9 to 4.0
Zirconia	200	29	200 to 500	29 to 73	8	5.6
Mullite	220	32	200 to 300	29 to 44	5.3	3.1
Cordierite	132	19	245	36	1.4	2.5
Silicon carbide	400 to 410	58 to 59.5	310	45	4.8	3.2
Tungsten	340 to 410	49 to 59.5	3800(a)	550(a)	4.8	19.3
Stainless steel	150 to 210	22 to 30.5	2100(a)	305(a)	...	7.7 to 8.0
Kovar	135	19.6	...	...	5.6	8.2

(a) Tensile yield point

Design, Tooling and Manufacturing

Composite Design Considerations

Physical properties relating to reinforced plastic design considerations

Mechanical properties
Tensile properties
Compressive properties
Flexural properties
Shear properties
Impact strength
Properties at high rates of loading (dynamic properties)
Bearing strength
Surface hardness
Creep properties (creep-rupture and stress-relaxation)
Fatigue (cyclic properties)
Poisson's ratio
Notch sensitivity
Shatterproofness
Shockproofness
Tear resistance

Thermal properties
Thermal conductivity
Thermal expansion
Specific heat
Flow temperature
Flammability (flame resistance)
Heat-distortion (deflection temperature under load)
Thermal shrinkage
Maximum safe operating temperature
Ignition properties
Brittleness temperature

Electrical properties
Arc resistance
Electrical resistance (insulation resistance–volume and surface)
Dielectric strength and dielectric breakdown voltage
Dielectric constant and power factor

Optical properties
Index of refraction (refractive index)
Light diffusion
Crazing resistance
Spectral transmission (haze)
Internal stress (transparent plastics)
Surface stability, optical
Optical uniformity and distortion

Chemical and permanence properties
Water absorption
Water vapor permeability (diffusion)–gas transmission rate
Sunlight and weather exposure (aging)
Resistance to chemical reagents
Effects of radiation
Toxicity
Volatile loss (outgassing)
Stress-crazing
Impact sensitivity (LOX)
Accelerated service (temperature and humidity)

Characteristics of three-directional woven preforms

Material	Bulk density, g/cm^3	No. of yarn bundles			Center-to-center bundle spacing				Fiber, vol %(a)		
					X, Y		*Z*				
		X	*Y*	*Z*	mm	in.	mm	in.	*X*	*Y*	*Z*
Thornel 50(b)	0.64	1	1	1	0.56	0.022	0.58	0.023	0.14	0.14	0.13
	0.75	1	1	2	0.71	0.028	0.58	0.023	0.11	0.11	0.23
	0.68	2	2	1	1.02	0.040	0.58	0.023	0.14	0.14	0.12
	0.80	2	2	6	0.69	0.027	1.02	0.040	0.12	0.12	0.24
Thornel 75(b)	0.70	1	1	2	0.56	0.022	0.58	0.023	0.09	0.09	0.17
	0.65	2	2	1	0.84	0.033	0.58	0.023	0.12	0.12	0.09
	0.72	2	2	2	1.07	0.042	0.58	0.023	0.09	0.09	0.18

(a) Volume fraction of total preform volume occupied by fiber in each orthogonal direction. (b) Center-to-center bundle spacing

Design guides for fiberglass composites

Characteristic	Compression molding: Sheet molding compound	Compression molding: Bulk molding compound	Compression molding: Preform molding	Injection molding (thermoplastics)	Cold press molding	Spray-up and hand lay-up
Minimum inside radius, mm (in.)	1.59 (1/16)	1.59 (1/16)	3.18 (1/8)	1.59 (1/16)	6.35 (1/4)	6.35 (1/4)
Molded-in holes	Yes(a)	Yes(a)	Yes(a)	Yes(a)	No	Large
Trimmed in mold	Yes	Yes	Yes	No	Yes	No
Core pull and slides	Yes	Yes	No	Yes	No	No

(continued)

Design guides for fiberglass composites (continued)

Characteristic		Compression molding: Sheet molding compound	Compression molding: Bulk molding compound	Compression molding: Preform molding	Injection molding (thermoplastics)	Cold press molding	Spray-up and hand lay-up
Undercuts		Yes	Yes(b)	No	Yes	No	Yes(b)
Minimum recommended draft, °/in.		Depths of 6.35 to 152 mm ($\frac{1}{4}$ to 6 in.), 1 to 3°; depths of 152 mm (6 in.) and over, 3° or as required				2° 3°	0°
Minimum practical thickness, mm (in.)		1.27 (0.50)	1.52 (0.060)	0.76 (0.030)	0.89 (0.035)	2.03 (0.080)	1.52 (0.060)
Maximum practical thickness, mm (in.)		25.4 (1)	25.4 (1)	6.35 (0.250)	12.7 (0.500)	12.7 (0.500)	No limit
Normal thickness variation, mm (in.)		±0.13 (±0.005)	±0.13 (±0.005)	±0.20 (±0.008)	±0.13 (±0.005)	±0.25 (±0.010)	±0.50 (±0.020)
Maximum thickness build-up, heavy build-up and increased cycle		As required	As required	2-to-1 maximum	As required	2-to-1 maximum	As required
Corrugated sections		Yes	Yes	Yes	Yes	Yes	Yes
Metal inserts		Yes	Yes	NR(c)	Yes	No	Yes
Bosses		Yes	Yes	Yes	Yes	NR(c)	Yes
Ribs		As required	Yes	NR(c)	Yes	NR(c)	Yes
Molded-in labels		Yes	Yes	Yes	No	Yes	Yes
Raised numbers		Yes	Yes	Yes	Yes	Yes	Yes
Finished surfaces (reproduces mold surface)		Two	Two	Two	Two	Two	One

(a) Parallel or perpendicular to ram action only. (b) With slides in tooling, or split mold. (c) Not recommended

Reinforcement efficiencies of selected composites

Matrix	Reinforcement	Composite efficiency (Φ)	Tensile strength (σ), MPa	Strain (∈)
Epoxy	Twill weave	0.39	251.3	0.26
Epoxy	Noncrimp fabric	0.42	427.2	0.41
Epoxy	Random mat	0.43	250.4	0.22
PP(a)	Short chopped	0.19	103.4	0.17
PP(b)	Swirl mat	0.19	120.0	0.20
SMC	Long chopped	0.35	158.6	0.13

(a) Injection molded. (b) Azdel

Reinforcement efficiencies of thermoplastic composites

Matrix	Reinforcement	Composite efficiency (Φ)	Tensile strength (σ), MPa	Strain (ε)
PP	Long chopped	0.19	120.0	0.20
PP	Noncrimp fabric	0.24	275.1	0.44
PP	Twill weave	0.33	236.5	0.27
PP	Twill weave	0.45	270.0	0.23
PET	Twill weave	0.37	237.1	0.21
PBT	Twill weave	0.28	208.4	0.24
PBT	Long chopped	0.16	120.6	0.20
Nylon 12	Atochem weave	0.33	209.9	0.24
PEEK	Random mat	0.31	227.5	0.21

Typical fabric styles and composite properties

Weave	Yarns/in. Warp Fill	Weight kg/m²	Weight oz/yd²	Thickness at 25 kPa (3.4 psi) mm	Thickness at 25 kPa (3.4 psi) in.	Yarn (carbon)
Eight-harness satin	24 × 23	0.370	10.9	0.46	0.018	Thornel 300 3K
Eight-harness satin	24 × 23	0.370	10.9	0.48	0.019	Celion 3K
Plain	12½ × 12½	0.190	5.6	0.30	0.012	Thornel 300 3K, Kevlar aramid tracers
Five-harness satin	24 × 24	0.125	3.7	0.20	0.008	Thornel 300 1K
CFS	24 × 12	0.20	6.0	0.23	0.009	Celion 3K warp 150 1/o glass fill
Plain	11½ × 11½	0.19	5.7	0.25	0.010	Magnamite AS-4 3K
Five-harness satin	11 × 11	0.370	10.9	0.50	0.020	Magnamite AS-4 6K
Plain	8 × 8	0.525	15.5	0.81	0.032	Celion 12K
Eight-harness satin	10½ × 10½	0.755	22.2	1.0	0.040	Thornel 300 15K
Plain	10 × 10	0.345	10.2	0.48	0.019	75 1/o glass warp, Grafil E/XA-S 12K fill
8HS	21 × 21	0.393	11.6	0.38	0.015	HITEX 3K

Typical composite properties (balanced weave)

Property	Value
Tensile strength, MPa (ksi)	620 to 690 (90 to 100)
Tensile modulus, GPa (10^6 psi)	69 to 76 (10 to 11)
Flexural strength, MPa (ksi)	690 to 900 (100 to 130)
Flexural modulus, GPa (10^6 psi)	62 to 69 (9 to 10)
Compressive strength, MPa (ksi)	620 to 690 (90 to 100)
Compressive modulus, GPa (10^6 psi)	62 to 69 (9 to 10)
Short beam shear strength, kPa (psi)	55 to 69 (8 to 10)
Specific gravity	1.6

Methods for determining composite density

Given	Unknown	Equation
% Fiber by volume	% Fiber by weight	$\frac{va}{va+(1-v)b}$
Fiber density = a Fiber volume = v Fiber weight = $v \cdot a$	% Resin by weight	$\frac{(1-v)b}{va+(1-v)b}$
Resin density = b Resin volume = $1 - v$	Composite density	$va + (1 - v)b$
Resin weight = $(1 - v) \cdot b$	Bulk factor	$1/v$
% Resin by volume	% Fiber by volume	$\frac{(1-w)/a}{(1-w)/a+w/b}$
Fiber density = a Fiber weight = $1 - w$ Fiber volume = $(1 - w)/a$	% Resin by volume	$\frac{w/b}{(1-w)/a+w/b}$
Resin density = b Resin weight = w Resin volume = w/b	Composite density	$\frac{1}{(1-w)/a+w/b}$
	Bulk factor	$1+\frac{wa}{(1-w)b}$
Bulk factor = BE	% Fiber by weight	$\frac{a}{a+(BF-1)b}$
Fiber density = a Fiber volume = 1 Fiber weight = a	% Resin by weight	$\frac{(BF-1)b}{a+(BF-1)b}$
Resin density = b Resin volume = $BF - 1$	% Fiber by volume	$1/BF$
Resin weight = $(BF - 1)b$	% Resin by volume	$(BF - 1)/BF$
	Composite density	$\frac{a+(BF-1)\,b}{BF}$

Properties of woven cloth in epoxy

Property	100% "S2" glass	100% Kevlar	100% carbon
Tensil strength (0°), MPa (ksi)	495 (72)	565 (82)	565 (82)
Tensile modulus (0°), GPa (10^6 psi)	30 (4.3)	42 (6.1)	71.0 (10.3)
Flexural strength, (0°), MPa (ksi)	670 (97)	475 (69)	600 (87)
Flexural modulus (0°), GPa (10^6 psi)	26 (3.8)	33 (4.8)	61 (8.9)
Short-beam shear strength, MPa (ksi)	52 (7.5)	28 (4.0)	59 (8.5)

Effects of weave pattern on fiberglass composite mechanical properties

Fabric style(a)	Plies, number	Resin content, wt %	Thickness		Flexural strength		Flexural modulus		Compressive strength		Tensile strength	
			mm	in.	MPa	ksi	GPa	10^6 psi	MPa	ksi	MPa	ksi
7628	18	37.1	3.15	0.124	371	53.8	26.8	3.89	177	25.7	317	45.9
76281(b)	18	36.7	3.07	0.121	584	84.7	23.5	3.41	393	57.0	408	59.2
16-149	12	36.5	3.05	0.120	436	63.2	26.3	3.81	331	48.0	405	58.7
7781(c)	12	37.6	3.05	0.120	600	87.0	22.3	3.24	443	64.3	414	60.0

(a) Each material is polyester compatible. (b) 5-end satin weave version of style 7781. (c) 8-end satin weave

Typical composite efficiencies attained in reinforced plastics

Fiber configuration	Fiber length	Total fiber content (V_f), vol %	F_{long}(a), ksi (MPa) F_{theor}(b)	F_{test}(c)	Composite efficiency(d) %
Filament-wound	Continuous	0.77	310 (2140)	180 (1240)	58.0
Cross-laminated fibers	Continuous	0.48	197 (1360)	72.5 (500)	36.8
Cloth-laminated fibers	Continuous	0.48	197 (1360)	43.0 (296)	21.8
Mat-laminated fibers	Continuous	0.48	197 (1360)	57.2 (394)	29.0
Chopped fiber systems (random)	Noncontinuous	0.13	60.7 (418)	15.0 (103)	24.7
Glass flake composites	Noncontinuous	0.70	165.5 (1141)	20.0 (138)	12.1

(a) F_{long}, tensile strength in direction of greatest fiber content (longitudinal), if there is one. (b) Theoretical strength based on "rule of mixtures": $F_{theor} = V_f S_f + (1-V_f) S_m$ = 400 ksi (2758 MPa)–typical boron or carbon fiber strength, and S_m = 10 ksi (69 MPa)–typical resin strength. (c) F_{test}, typical experimental strength values. (d) Composite efficiency = $(F_{test}/F_{theor}) \times 100$

Composite Tooling and Machining

Properties of typical composite tooling materials

Material	Coefficient of thermal expansion, 10^{-6}/K	Thermal conductivity, W/m · K	Thermal conductivity, Btu · in./h · ft² °F
Fiberglass-epoxy	7.9	...	...
Graphite-epoxy	–0.9	0.222	0.15
Aluminum	23.0	0.221	1.53
Steel	13.9	0.048	0.33
Electroless nickel	13.3	0.035	0.24

Parameters for composite trimming

Operation	Equipment	Cutter type	Speed, m/s	Speed, ft/min.
Straight-line cuts	Pneumatic saw	Diamond-coated circular saw(a)	60 to 90	12 000 to 18 000
	Hand router	Diamond router(b) Carbide router(c)	10 to 15	2000 to 3000
Irregular outline	Hand router	Diamond router(b) Carbide router(c)	10 to 15	2000 to 3000
Chamfer, deburr	Hand	Abrasive drum(d)	NA	NA
Finish operations	Hand drill motor	Abrasive disc(d)	5 to 60	1000 to 12 000
	Hand	Abrasive cloth(d)	NA	NA

(a) Diamond circular saw, 0.050 kerf, 36/44 grit. (b) Diamond routers, 36/44 grit roughing, 80/100 grit finishing. (c) Carbide diamond-shaped, chisel-cut routers. (d) 80 grit (rough), 220 grit (finish)

Parameters for peck drilling

Material	Speed, rpm	Feed rate per revolution, mm	Feed rate per revolution, in.	Peck cycle, in.
Titanium	550	0.050	0.002	60 min
Graphite	550	0.10	0.004	30 min
Aluminum	550	0.10	0.004	30 min

Performance of tooling materials

Material	Coefficient of thermal expansion 10^{-6}/K	Thermal correction required, 0.9-m (3-ft) part, mm	0.9-m (3-ft) part, in.	9-m (30-ft) part, mm	9-m (30-ft) part, in.	Springback of square corners, degrees(b)
Graphite-epoxy(a)	3.6	0.28	0.011	2.79	0.110	1°15'
Glass-epoxy(a)	11.7 to 13.1	0.76	0.030	9.14	0.360	...
Steel	12.1	0.76	0.030	9.14	0.360	1°30'
Nickel electroform	13.3	0.76	0.030	8.89	0.350	...
Fiberglass wet lay-up	14.4 to 18.0	...	...	...	...	...
Aluminum	22.5	1.52	0.060	13.97	0.550	2°0'

(a) MXG-7620 resin. (b) Angle on tool must be larger than engineering call-out

Thermal characteristics of tooling materials

Material	Tool thickness, mm	Tool thickness, in.	Tool rise time(a) (ambient to 110 °C, or 230 °F) min.	Overshoot (degrees over ambient), °C	Overshoot (degrees over ambient), °F
Aluminum	6.4	0.25	41.6	13	23
Steel	6.4	0.25	(b)	9	15
Carbon-epoxy	6.4	0.25	45.8	14	24
Aluminum	25.4	1.00	48.0	(b)	(b)
Steel	12.7	0.50	48.6	7	12
Carbon-epoxy	12.7	0.50	51.7	13	23
Steel	38.1	1.50	73.8	2	3
Aluminum	101.6	4.00	86.3	0	0

(a) In response to 2 °C/min. (4 °F/min.) ramp from ambient to 240 °F. (b) Testing difficulties prevented accurate determination of values

Approved fastener materials

Materials being joined	Preferred	Fastener material Recommended with barrier coatings	Not recommended
Graphite-epoxy(a) to aluminum(b)	Titanium pin with aluminum collars or nuts bearing on aluminum	Multiphase alloys, Inco 718, or austenitic stainless steel, with aluminum or stainless steel collars or nuts bearing on aluminum, or titanium-columbium systems	Copper, brass, aluminum, low-alloy steel, martensitic stainless steels, or cadmium-plated fastener systems
Graphite-epoxy to titanium, A286, austenitic stainless steel, or graphite-epoxy	Multiphase alloys, Monel fastening systems, Inconel, or titanium, with stainless steel, Inconel, or titanium collars or nuts	Stainless steel pin or screw with aluminum collars or nuts bearing on stainless steel or titanium. Stainless steel collars or nuts bearing on graphite-epoxy, titanium or stainless steel, or titanium-columbium systems	Aluminum collars or nuts bearing on graphite-epoxy, copper, brass, aluminum, or low-alloy steel elements, or cadmium-plated fastener systems

(a) Graphite is used to mean either graphite or carbon reinforcement. (b) These materials are incompatible at the faying surface without a barrier coating

Parameters for graphite drilling

Hole diam, max			Feed rate per revolution	
mm	in.	Speed, rpm	mm	in.
3.967	0.1562	2800	0.025 to 0.040	0.0010 to 0.0015
4.763	0.1875	2800	0.025 to 0.040	0.0010 to 0.0015
6.350	0.2500	2800	0.025 to 0.040	0.0010 to 0.0015
7.938	0.3125	1800	0.045 to 0.055	0.0017 to 0.0022
9.525	0.3750	1800	0.045 to 0.055	0.0017 to 0.0022

Beam press versus roller press

	Beam	Roller
Tonnage	70 to 200	Unlimited
Cutting speed	125 mm/s (5 in./s)	430 mm/s (17 in./s)
Material ply height	12.7 mm ($\frac{1}{2}$ in.) or more, depending on capacity of die and press stroke	3.2 to 4.8 mm ($\frac{1}{8}$ to $\frac{3}{16}$ in.) max
Die maintenance	Up and down hydraulic action, easy on dies	Mechanical pincer action can distort die knives. More die repairs
Press maintenance	Minimal	Minimal
Cutting pad	Fixed-in head	Rides on top of die and tends to curl because of the roller action. Disposable pads required
Cuttable materials	All composite unidirectional tapes and fabrics	Same

Faying-surface sealant categories

Material type	Use	Specifications(a)
Polysulfide	General-purpose corrosion-inhibiting sealant	MIL-S-81733C(b)
	Fuel tank sealant (also for general-purpose nonfuel areas)	MIL-S-8802E
	General-purpose low-adhesion sealant	MIL-S-8784B
	Low-adhesion sealant for fuel tank areas	AMS 3267(c)
	High-temperature sealant (to 180 °C, or 360 °F, peak)	MIL-S-83430A
Silicone	High temperature sealing applications (-60 to 205 °C, or -80 to 400 °F)(d)	AMS 3373

(a) Some of these specifications have several types, grades, and classes. A careful review of these categories and the recommended use for each must be made before sealant selection. (b) Material conforming to this specification is recommended for permanent graphite composite assemblies because of the corrosion-inhibiting formulation and also because of the long pot life, which permits a long assembly time for complex structures. (c) Material conforming to this specification is recommended for removable graphite composite assemblies because of the corrosion-inhibiting formulation. (d) A two-part silicone sealant, with catalytic curing agents, must be used. One-part silicones require moisture (from exposure to air) for cure and thus cannot be used on a faying surface with more than 25-mm (1-in.) width

Cutting speeds versus materials and thicknesses

Material	Thickness mm	in.	Cutting speed m/min.	in./min.
Abrasive cutting at 240 MPa (35 ksi), 100-grit garnet, 20 hp				
Resin-impregnated graphite, aramid, glass fibers	3.2	0.125	1.60	63
	6.4	0.250	0.75	30
	12.7	0.500	0.46	18
	19.1	0.750	0.30	12
	25.4	1.000	0.13	5

(continued)

Cutting speeds versus materials and thicknesses (continued)

Material	Thickness mm	Thickness in.	Cutting speed m/min.	Cutting speed in./min.
Ferrous metal, stainless, Hastelloy, Inconel, mild steel, high carbon, 60 HRC	1.6	0.063	0.50	20
	3.2	0.125	0.28	11
	6.4	0.250	0.18	7
	12.7	0.500	0.075	3
	25.4	1.000	0.019	0.75
Aluminum, magnesium titanium, brass alloys	1.6	0.063	1.47	58
	3.2	0.125	0.74	29
	6.4	0.250	0.41	16
	12.7	0.500	0.25	10
	25.4	1.000	0.075	3
Nonabrasive cutting at 360 MPa (52 ksi), 20 hp				
Balsa wood, light foam rubber styrofoam	1.6	0.063	15	600+
	12.7	0.500	12	450
	25.4	1.000	7	275
Polyurethane, rubber compounds, polyethlene(a) (30 + Durometer)	6.4	0.250	6	225
	12.7	0.500	3	100
	25.4	1.000	1	40
Paper, fabric, corrugated board, rubber(a) (30 - Durometer)	0.13	0.005	15	600+
	0.38	0.015	8	300+
	0.81	0.032	8	300+
	1.6	0.063	8	300+
	3.2	0.125	8	300+
	6.4	0.250	8	300+

(a) These materials may require additional support on a table

Factors for drilling of composites

Material	Speed sfm	Feed (equivalent) mm/rev	Feed (equivalent) ipr	Tool recommendation	Drilling time, min	Tool life, holes
Boron-epoxy, 2.03 mm (0.08 in.)	300 to 600	0.013	0.0005	Core with end-set diamonds	0.1	300 to 400
Boron-epoxy, 25.4 mm (1.00 in.)	300 to 600	0.013	0.0005	Core with end-set diamonds	1.0	75 to 100
Boron-epoxy/titanium multilayer, 12.7 mm (0.50 in.) total	200 to 500	0.0013	0.00005	Core	4.0	30 to 50

Composite Manufacturing Methods

Factors affecting prepreg form selection

10 = highest cost or worst case; 1 = lowest cost or best case

Material form and fabrication process	Facility cost	Production rate	Importance of operator's skill	Part complexity possible	Part reproducibility	Material cost	Material use efficiency
Unidirectional tape, hand lay-up	1	10	10	5	10	3	7
Machine-cut, hand lay-up	5	5	5	5	5	3	4
Machine lay-down	10	1	1	5	1	6	2
Multidirectional tape, hand lay-up	1	5	9	7	8	8	5
Machine-cut, hand lay-up	5	3	7	7	4	8	4
Fabrics, hand lay-up	1	10	8	1	8	5	7
Machine-cut, hand lay-up	5	5	4	1	4	5	7
Towpreg	8	5	7	3	5	2	3

Sizing classifications and functions

Type	Purpose	Example	Remarks
film-forming organics and polymers	To protect the reinforcement during processing	Polyvinyl alcohol (PVA), polyvinyl acetate (PVAc)	The polymer is formulated to wet-spread to form a uniform coating that is applied to aid processing but later may be removed by washing or heat cleaning (e.g., fugitive sizing)
Adhesion promotors	To improve composite mechanical properties and/or moisture resistance	Silane coupling agents	Principally used on inorganic reinforcement (e.g., glass fiber)
Interlayer	To enhance composite properties by creating an interphase between matrix and reinforcement	Elastomeric coating	Not in commercial use
Chemical modifiers	React to form protective coating	Silicon carbide on boron fibers	

Commercial designations and sources of processing materials

Material	Purpose	Commercial designation	Source	Comments
Peel ply	Provides a bondable surface	Burlease 51789 (nylon-6)	Burlington Mills	Not suitable for phenolics
		Burlease 60 000 (polyester)	Burlington Mills	Suitable for phenolics
		Bleeder-lease E (fiberglass with release agent)	Airtech International	
Separator (release)	Separates cured laminates from other process materials without damage	A5000 Teflon FEP film	Richmond Technology	
		A4000 halogen release film	Airtech International	
		Wrightlon 4500 (halocarbon film)	Airtech International	
		104 TFE Teflon-coated fiberglass-style fabric	Various	
Bleeder	Absorbs excess resin	120-style fiberglass fabric	Commercial	
		181-style fiberglass fabric	Commercial	
		Organic fiber felts	Commercial	
Barrier	Limits resin flow to only bleeder plies	See Separator		
		Unperforated TFE film		
Breather	Evacuates the vacuum bag so that the desired autoclave pressure is applied	Airweave N-10, Airweave N-4	Airtech International	Organic fiber, stretchable felt
		A-3000	Richmond Technology	Organic fiber, stretchable felt
		Airweave HP	Airtech International	1.4-mm (0.055-in.) thick fiberglass fabric
		Burflo 4819	Burlington Mills	Polyester nonwoven fabric
Dam	Prevents resin flow from edges	Rubber neoprene cork, Rubatex 886	Groendyke	Tape with pressure-sensitive adhesive
Vacuum bag	Applies autoclave pressure	Vac-Pak	Richmond Technology	Various types for different cure temperatures
		Wrightlon, IPPLON, Wrightcast, Vacalloy	Airtech International	Various types for different cure temperatures
		Silicone rubbers	Various	Semipermanent

Effect of resin matrices on properties of type E fiberglass composite materials

Resin type	Specific gravity, g/cm³	Fiber wt %	Fiber vol %	Tensile strength MPa	Tensile strength ksi	Flexural strength MPa	Flexural strength ksi	Flexural modulus GPa	Flexural modulus 10^6 psi	Compressive strength MPa	Compressive strength ksi	Impact strength J/mm	Impact strength ft · lbf/in.
Epoxy	1.88	63	46	190	27	470	68	28	4.1	290	42	1.6	30
Polyimide	1.95	63	47	140	21	260	37	21	3.1	220	32	1.2	22
Phenolic	1.78	56	34	110	16	240	35	21	3.0	340	35	1.1	20
Polyester	1.98	55	39	80	12	170	25	17	2.5	180	26	0.8	15
Silicone	2.02	46	34	30	4	70	10	14	2.0	80	11	0.25	5

(All values are typical values.)

Manufacturing processes for polymer matrix composites

Process	Tooling material	Tool life, number of parts: Low	Tool life, number of parts: High	Typical materials used	Advantages	Limitations	Applications
Structural foam (thermoplastic and RIM)	Steel Machine aluminum Cast aluminum; Kirksite Cast aluminum-filled epoxy Cast epoxy (100%)	200 000 5 000 500 10 2	1 000 000 250 000 50 000 500 25	Polyurethanes, polycarbonates, polyphenylene oxide, phenylene ether copolymers, polybutylene terephthalate, ABS, polyethylene, polystyrene, polypropylene	Large detailed parts, low production costs, rigidity, complex shapes, parts consolidation. High strength-to-weight ratio. Low density. Molded-in inserts. Low pressure molding due to low viscosity. Wide material selection. Minimizing or eliminating of sink marks. Low molded-in stresses. Improved chemical resistance	Surfaces usually need secondary finishing and/or painting. Some sacrifice of physical properties relative to base resin. Longer cycle times than injection-molded parts	Business machine housings, automotive fascias, medical and electronic cabinetry, furniture, materials-handling equipment
Injection molding	Steel Machined aluminum	200 000 5 000	1 000 000 250 000	Broad range of thermoplastics and thermosets	High-volume production runs, close tolerances, molded-in color, low part cost, large material selection, parts consolidation	High tooling investment. Long tooling lead times	Wide range of applications in all industries
Compression molding	Steel	200 000	1 000 000	Thermosets (polyesters, sheet molding compound, alkyds, ureas, phenolics, epoxies, dialyl phthalates)	High-strength, heat-resistant parts. High modulus. Complex shapes. Parts consolidation. Excellent surface finish. Used for molding large parts, such as with polyester, as well as full range of sizes. Excellent fatigue resistance	High tooling costs. Deflashing needed. Labor intensive. Parts generally need painting	Automotive parts, electrical connectors, business-machine parts, power-tool housings
Vacuum forming	Machined aluminum Cast aluminum; Kirksite Cast aluminum-filled epoxy Cast epoxy (100%)	5 000 500 10 2	250 000 50 000 500 25	ABS and ABS alloys, PVC, PPO, acrylic, polystyrene, polyethylene, polycarbonate, polypropylene	Excellent for complex contours, with minimum internal details. Large or small parts	Often limited by large radii, shallow depths, large draft angles, loose tolerances. Exposed edges must be trimmed and buffed or milled	Signage, business-machine housings, furniture, medical cabinetry, recreational products (boats, campers), transportation (interior and exterior parts), packaging (cups, plates)
Hand lay-up sprayed glass fiber	Machined aluminum Cast aluminum; Kirksite Wooden pattern	5 000 500 10	250 000 50 000 1 000	Thermoset polyesters	Large parts, basically shells, can be molded with complex curves, excellent surface finish, and high rigidity	Internal details (bosses, ribs) must be manually layed into inside wall and then overlayed with glass fiber. Labor intensive	Recreational boating, materials handling, furniture, construction, transportation (truck hoods, bus seats)

(continued)

Manufacturing processes for polymer matrix composites (continued)

Process	Tooling material	Tool life, number of parts Low	High	Typical materials used	Advantages	Limitations	Applications
Die casting	Steel	200 000	1 000 000	Zinc, magnesium, aluminum alloys	Complex shapes, intricate details, tight tolerances, excellent surface finishes direct from die. High-strength fatigue-resistant parts	Usually requires costly secondary operations, such as deburring, deflashing, tapping, painting	Small appliances, hardware, automotive parts, motors, power tools
Matched metal die forming	Steel	200 000	1 000 000	Mild steel, aluminum	Effective for production of compound-surfaced parts which require high strength, rigidity, heat resistance	High tooling costs due to need for progressive dies. Deburring required. Painting necessary	Automotive body panels, major appliances, housings, large containers
Sandcasting	Wooden pattern	10	1 000	Iron, bronze, brass, copper, aluminum	Complex shapes with low capital investment. Very rigid parts. High heat resistance and strength. Good surface finish	Limited to bulky parts. Frequently require machining. Labor intensive. Secondary finishing operations can be extensive	Wide range of industrial uses. Transportation components, materials-handling equipment, large generator parts
Sheetmetal/brake formed	Fixtures			Steel, aluminum, brass	Excellent for punching and bending sheetmetal	Compound curves, internal ribs should be avoided. Assembly intensive. Painting	Electronic cabinetry, ducting, furniture, construction

Hand, mechanized, and fully automated lay-up processes

Process	Ply generation	Placement on tool	Forming of prepreg to tool shape
Hand lay-up	Lay tape on tool or on lay-up templates Cut fabric to shape using templates	Place manually using tooling to coordinate the partial plies to the tool: Lay-up templates Bank rails at edges	Use vacuum bag and localized heating to soften the prepreg
Mechanized	Cut from wide goods with the Gerber cutter Cut from wide goods with Clicker press and steel rule dies	Stack on tool manually as above, but use an optical/layer system to locate plies on the mold	Use vacuum bag plus localized heat as above. Use mechanical devices to seat plies on the webs of sine wave spars (rollers) Use devices to apply vacuum/pressure to seat plies on male tools by means of a diaphragm Use devices to heat the prepreg so it can be formed easily by hand
	Lay-up plies in the flat with a tape layer	Use robotics or other mechanized system to transfer plies to the tool	
Fully automated	For wind and tail skins and parts of similar gentle contour, use contour tape layer to lay up the plies directly on the mold. Applications limited by the capability of the tape layer For automated laminating cell, cut plies with a Gerber cutter Use a robot to pick up the plies from the Gerber table and place them on the tool		

Comparison of filament-winding impregnation methods

	Prepreg	Wet winding	Wet rerolled
Cleanliness	Best	Worst	Almost equal to prepreg, mess is away from winder
Fiber availability	Poor. Not all fibers are available; many necessitate special order	Best. Any fiber that system will handle	Best. All fibers
Control of resin content	Best. Constant speed and viscosity	Poor. Speed of mandrel varies, viscosity of resin may vary	Better. Process is away from winding and is faster; little viscosity change
Quality assurance	Highest. Can be done far ahead	Worst. Imposes quality control procedures onto factory floor and can lead to errors	Good. Can be done ahead
Ability to use complex resin systems	Yes. Hot melts available	Very difficult. Requires complex impregnators to remove solvents or liquify hot melts	Difficult. Still requires complex impregnators
Large data base resin systems	Yes	Commercial resins generally not available as liquids; the wet systems with large data bases may be proprietary	Same as wet winding
Graphite fibers encapsulated (to prevent electronic shortouts)	Yes	No	Graphite fibers not released at winder
Storage	Must be refrigerated and storage records maintained	Easy mix at winder; dry fibers have long shelf life	Must be stored like prepreg, but shorter storage life records must be kept

(continued)

Comparison of filament-winding impregnation methods (continued)

	Prepreg	Wet winding	Wet rerolled
Fiber damage	Depends on impregnator; fiber is handled twice	May require special equipment; less damage potential because of less handling	All handling of fiber is under control of user
Cost	Highest	Lowest	Slightly above wet but also requires capital investment for impregnation equipment
Large roving package	Depends on impregnator	Whatever is available dry from fiber manufacturers	Whatever is available dry from fiber manufacturers
Room-temperature cure	Not possible	Possible	Possible
Simple resin formulation	Possible	Necessary	Necessary
Winding speed	Can be highest. Resin throw from fiber is minimized	Lowest speed	Intermediate. Resin can be staged to lower resin throw
Stability on nongeodesic path	Highest possible	Lowest. Wet resin may cause slippage	Intermediate. Resin can be staged to increase tack

Property comparison by process(a)

Process	Reinforcement, wt %	Tensile strength MPa	Tensile strength ksi	Tensile modulus GPa	Tensile modulus 10^6 ksi	Flexural strength MPa	Flexural strength ksi
Spray	30 to 50 glass-polyester	60 to 120	9 to 18	5.5 to 12	0.8 to 1.8	110 to 190	16 to 28
Compression	15 to 30 glass-SMC	55 to 140	8 to 20	11 to 17	1.6 to 2.5	120 to 210	18 to 30
Compression	25 to 50 glass mat-polyester	170 to 210	25 to 30	6.2 to 14	0.9 to 2.0	70 to 280	10 to 40
Filament winding	30 to 80 glass-epoxy	550 to 1700	80 to 250	28 to 62	4.0 to 9.0	690 to 1850	100 to 270
Pultrusion	40 to 80 glass mat-polyester	410 to 1050	60 to 150	28 to 41	4.0 to 6.0	690 to 1050	100 to 150
Pultrusion	30 to 50 glass mat-polyester	80 to 120	12 to 30	6.9 to 17	1.0 to 2.5	170 to 210	25 to 30
Pultrusion	30 to 55 glass mat and roving vinyl ester resin	70 to 280	10 to 40	6.9 to 21	1.0 to 3.0	100 to 280	15 to 40
Pultrusion	30 to 55 glass mat and roving vinyl ester resin	50 to 240	7 to 35	5.5 to 17	0.8 to 2.5	70 to 210	10 to 30

Process	Compressive strength MPa	Compressive strength ksi	Impact strength J/m	Impact strength ft · lbf/ft	Thermal conductivity W/m · K	Thermal conductivity Btu · in./h · ft^2 · °F
Spray	100 to 170	15 to 25	210 to 640	48 to 144	0.17 to 0.23	1.2 to 1.6
Compression	100 to 210	15 to 30	430 to 1150	96 to 264	0.19 to 0.25	1.3 to 1.7
Compression	100 to 210	15 to 30	530 to 1050	120 to 240	0.19 to 0.26	1.3 to 1.8
Filament winding	310 to 480	45 to 70	2150 to 3200	480 to 720	0.27 to 0.33	1.9 to 2.3
Pultrusion	210 to 480	30 to 70	2400 to 3200	540 to 720	0.27 to 0.33	1.9 to 2.3
Pultrusion	210 to 340	30 to 50	530 to 1350	120 to 300	0.22 to 0.27	1.5 to 1.85
Pultrusion	140 to 340	20 to 50	270 to 1600	60 to 360	0.22 to 0.33	1.5 to 2.3
Pultrusion	100 to 280	15 to 40	210 to 1350	48 to 300	0.22 to 0.33	1.5 to 2.3

Process	Heat distortion at 1.8 MPa °C	Heat distortion at 1.8 MPa °F	Dielectric strength kV/cm	Dielectric strength kV/in.
Spray	175 to 205	350 to 400	80 to 160	200 to 400
Compression	205 to 260	400 to 500	120 to 180	300 to 450
Compression	175 to 205	350 to 400	120 to 240	300 to 600
Filament winding	175 to 205	350 to 400	120 to 180	300 to 400
Pultrusion	205 to 260	400 to 500	80 to 160	200 to 400
Pultrusion	95 to 150	200 to 300	80 to 120	200 to 300
Pultrusion	175 to 230	350 to 450	80 to 130	200 to 325
Pultrusion	175 to 205	350 to 400	80 to 120	200 to 300

(a) Range of values reflects transverse and axial testing directions as well as percent reinforcement

Comparison of tape and fabric prepregs for manual lay-up

Advantages	Disadvantages
Tape	
Better strength/stiffness control	More plies required
Lower resin content	Longer time to cut patterns and lay up
Can be spliced parallel to fibers	...
Lower coefficient of thermal expansion	...
Fabric	
Fewer plies required	Lower mechanical strength (due to higher resin content)
Less time to lay up; easier to form over large curved areas	Difficulties in splicing large parts
Lower cost	Higher void content

CVD densification of multidirectional preform (typical)

Process conditions	
Temperature, °C (°F)	1100 (2012)
Pressure, Pa (torr)	1350 (10)
Hydrocarbon	Natural gas

Process steps

1. Process three-directional preform to 1.2 g/cm^3 density
2. Machine preform surfaces
3. Process for 640 h to a density of 1.6 g/cm^3

Effect of matrix precursor on composite modulus

	Heat treatment At 1000 °C (1830 °F)	Heat treatment At 2600 °C (4710 °F)
Phenolic resin	110%	140%
Pitch	130%	210%

Note: Fiber stiffening factor assuming all stiffness comes from HM fiber

Ceramics

General

Standards and General Applications

Relevant ASTM standards for ceramic-metal seals

Thermal expansion

Standard Practice for Making and Testing Reference Glass-Metal Bead-Seals to Determine Degree of Mismatch, ASTM F-14 (10.04), 1990
Standard Practice for Making and Testing Reference Glass-Metal Butt-Seals to Determine Degree of Mismatch, ASTM F-140 (10.04), 1990
Standard Practice for Making and Testing Reference Glass-Metal Sandwich-Seals to Determine Degree of Mismatch, ASTM F-144 (10.04), 1990
Standard Test for Linear Thermal Expansion of Porcelain Enamel Frit by the Interferometric Method, ASTM C-539 (02.05, 15.02), 1990
Standard Test for Linear Thermal Expansion of Porcelain Enamels, Glaze Frits and Fired Ceramic Whiteware Products by the Dilatometry Method, ASTM C-372 (15.02), 1990
Standard Test for Linear Thermal Expansion of Refractories Under Load, ASTM C-832 (15.01), 1990
Standard Test for Linear Thermal Expansion of Rigid Solids with Interferometry, ASTM E-289 (03.01, 14.02), 1990
Linear Thermal Expansion of Solid Materials by Thermomechanical Analysis, ASTM E-831 (08.03, 14.02), 1990
Standard Test for Linear Thermal Expansion of Solid Materials with Vitreous Silica Dilatometer, ASTM E-228 (03.01, 14.02), 1990
Standard Test for Linear Thermal Expansion of Vitreous Glass Enamels and Glass Color Frits by Dilatometry, ASTM C-824 (15.02), 1990

Thermal conductivity/diffusivity

Standard Test for Thermal Diffusivity of Carbon/Graphite by the Thermal Pulse Method, ASTM C-714 (15.01), 1990
Standard Test for the Thermal Conductivity of Ceramic Whitewares, ASTM C-408 (15.02), 1990
Standard Test for Thermal Conductivity of Electrical Grade Magnesium Oxide, ASTM D-2858 (10.02), 1990
Standard Test for Thermal Conductivity of Solids by Guarded-Comparative-Longitudinal Heat Flow Technique, ASTM E-1225 (14.02), 1990
Standard Practice, Requirements/Guidelines for Thermal Transmission Properties Calculated from Steady-State Heat Flux Measurements, ASTM C-1045 (04.06), 1990
Standard Test for Thermal Conductivity of Brick other than Insulating Firebrick, ASTM C-202 (15.01), 1990
Standard Test for Thermal Conductivity of Carbon and Carbon-Bearing Refractories, ASTM C-767 (15.01), 1990
Standard Test for Thermal Conductivity of Insulating Firebrick, ASTM C-182 (15.01), 1990
Standard Test for Thermal Conductivity of Unfired Monolithic Refractories, ASTM C-417 (15.01), 1990
Standard Test for Thermal Transmission Properties of Insulating Specimens by heat Flow meter, ASTM C-518 (04.06, 14.01), 1990
Standard Practice for Thermal Transmission Properties of Insulating Specimens Using a Guarded Hot Plate in One-Sided Mode, ASTM C-1044 (04.06), 1990
Standard Test for Thermal Transmission Properties of Insulating Specimens by Guarded Hot Plate Apparatus, ASTM C-177 (04.06, 08.01, 14.01), 1990

Specific heat capacity

Standard Test for Specific Heat Capacity of Materials by Differential Scanning Calorimetry, ASTM E-1269 (14.02), 1990
Standard Test for Specific Heat of Liquids/Solids, ASTM D-2766 (05.02), 1990

Emittance

Standard Test for Total Normal Emittance by Inspection-Meter Technique, ASTM E-408 (15.03), 1990
Standard Test for Spectral Normal Emittance at Elevated Temperatures, ASTM E-307 (15.03), 1990
Standard Test for Spectral Normal Emittance of Nonconducting Specimens at Elevated Temperatures, ASTM E-423 (15.03), 1990
Standard Test for Total Hemispherical Emittance by Calorimetric Determination, ASTM E-424 (15.03), 1990
Standard Test for Total Hemispherical Emittance of Metal/Coated-Metal Surfaces from 20° to 1400 °C, ASTM C-835 (04.06), 1990

General

Standard Definitions of Terms Relating to Thermophysical Properties, ASTM E-1142, 1990

Strong ceramics for engineering

Main compound	Fabrication method	Minor components, %	Typical mean strength(a), MPa	Property advantages
Al_2O_3	Sintering	~0.05 MgO	200 to 400	Refractory, high-purity types, can be translucent
		0 to 2 MgO, 0 to 4 SiO_2	200 to 250	Acid-resistant, debased type
		0 to 3 CaO, 0 to 10 SiO_2	200 to 300	General-purpose electrical and mechanical types
		10 to 20 ZrO_2	200 to 500	Transformation-toughened type
	Hot pressing	20 to 40 TiC	200 to 500	High-strength machine-tool tips
ZrO_2	Sintering	4 to 8 CaO, MgO, Y_2O_3	200	Stabilized type, refractory, heat-insulating but not strong or thermal shock resistant
	Sintering plus aging	2 to 4 CaO, MgO, Y_2O_3	500	Partially stabilized type, transformation-toughened but not refractory
	Sintering	1 to 2 Y_2O_3	1000	All-tetragonal (TZP) type, transformation-toughened but not refractory
SiC	Reaction bonding	8 to 20 Si	400	Hard, wear-resistant, quite strong, thermal shock resistant, refractory, acid-resistant
	Hot pressing	1 to 2 Al_2O_3	500	Stronger but less dimensional flexibility
	Sintering	~1 B	300	Acid- and alkali-resistant
Si_3N_4	Reaction bonding		200	Thermal shock resistant, moderate strength, refractory, but porous
	Hot pressing	1 to 5 MgO, Y_2O_3	800	Harder, less refractory, impermeable
	Sintering	5 to 15 MgO, Y_2O_3	600	High strength, more shape flexibility
Sialons	Sintering	5 to 15 MgO, Y_2O_3	600	Strong, hard, wear-resistant, and some are quite refractory
B_4C	Hot pressing	~1 B	400	Very hard and wear-resistant, but oxidizes above 800 °C

(a) Typical flexural strength as determined from small test bars tested in bending

Applications of hot isostatic pressed ceramics

Materials			Applications	
Al_2O_3	ThO_2	ZrB_2	Let down valves for coal liquefaction	Drill-bit inserts
$Al_2O_3 + ZrO_2$	TiC	HfB_2	Turbine disks	Nuclear waste consolidation
$Al_2O_3 + TiC$	B_4C	$BaTiO_3$	Turbine blades	Nuclear waste containers
$Al_2O_3 + TiN$	SiC	$BaFeO_4$	Turbine vanes	Bearings
BeO	$SiC + Si_3N_4$	$MnZnFeO_4$	Stirling heater heads	Radomes and infrared domes
UO_2	Si_3N_4	$NiZnFeO_4$	Cutting tools	β-alumina tubes
ZrO_2	Sialon	$Pb(ZrTi)O_3$	Orthopedic implants and dental ceramics	Nuclear reactor core insulators
ZnO_2	TiN	$(PbBi)(ZrTi)O_3$	Magnetic tape heads	Fusion reaction insulators
SiO_2	BN	$(PbLa)(ZrTi)O_3$	Transducers	Special dies
T_2O_3	TiB_2		Laser windows	Multilayer capacitors
			Sputtering targets	

Present uses of bioceramics

Application	Material(s) used
Orthopedic load-bearing applications	Al_2O_3
Coatings for chemical bonding (orthopedic, dental and maxillary prosthetics)	HA, surface-active glasses and glass-ceramics
Dental implants	Al_2O_3, HA, surface-active glasses
Alveolar ridge augmentations	Al_2O_3, HA, HA-autogenous bone composite, HA-PLA composite, surface-active glasses
Otolaryngological applications	Al_2O_3, HA, surface-active glasses and glass-ceramics
Artificial tendons and ligaments	PLA-carbon fiber composites
Coatings for tissue ingrowth (cardiovascular, orthopedic, dental, and maxillofacial prosthetics)	Al_2O_3
Temporary bone space fillers	Trisodium phosphate, calcium and phosphate salts
Periodontal pocket obliteration	HA, HA-PLA composites, trisodium phosphates, calcium and phosphate salts, surface-active glass
Maxillofacial reconstruction	Al_2O_3, HA, HA-PLA composites, surface-active glasses
Percutaneous access devices	Bioactive glass-ceramics
Orthopedic fixation devices	PLA-carbon fibers, PLA-calcium/phosphorus-base glass fibers

Typical materials used for ceramic-metal seals and important properties

Material	Coefficient of thermal expansion, 10^{-6}/K	Temperature range °C	Temperature range °F	Yield strength MPa	Yield strength ksi	Tensile strength MPa	Tensile strength ksi	Elongation modulus, %	Elastic modulus GPa	Elastic modulus 10^6 psi
Alumina	7.6	At 500	At 930	...	...	~5000	~725	0	390	57
Mullite	5.1 to 5.8	At 1000	At 1830	...	...	~4000	~580	0	145	21
Beryllia	7.6	At 500	At 930	...	...	~4000	~580	0	380	55
PSZ (2 wt% Y_2O_3)	7.5 to 13	20 to 1000	68 to 1830	...	...	~4000	~580	0	207	30
Alloy 42 (UNS 94100)	7.9	20 to 500	68 to 930	295	45	550	80	43.7	145	21
Kovar	5.2	20 to 200	68 to 390	410	60	534	77	25	138	20
Nickel (Nickel 200)	13.2	25 to 100	77 to 212	148	20	462	67	27	204	30
Titanium (Grade 1)	10.1	20 to 815	68 to 1500	170 to 241	25 to 35	240 to 331	35 to 48	30	102.7	15
Copper (UNS C10100)	17.7	20 to 300	68 to 570	69	10	220	32	45	115	17
TZM (UNS R03630)	4.9	20 to 40	68 to 105	860	125	965	140	10	315	46
Steel (AISI-ASE 1010)	15	20 to 700	68 to 1290	115	17	310 to 380	45 to 55	33	205	30
410 stainless steel (UNS S41000)	11.6	0 to 538	32 to 1000	205	30	450	65	22	200	29

Properties and applications of ceramic and cermet coatings commonly used in molten particle deposition

Classification	Plasma spray material	Maximum service temperature(a) °C	°F	Applications	Properties
Cermets	WC-W_2C with Co or Ni (7, 12, 17, 20 wt%)	450	840	Wear resistance	Metal matrix lowers wear resistance but increases resistance to mechanical or thermal shocks
					Erosion (cavitation) or abrasion depends on porosity, carbon content, and mean diameter of WC grains
					Sliding wear resistance generally poor except if WC decomposed
	Cr_3C_2-NiCr (80:20 wt%)	≤800	≤1470	Same use as WC-Co but at higher temperatures (mostly in aeronautics)	...
Carbides	TiC-Cr_3C_2	...	...	...	Good oxidation resistance up to 800 to 900 °C (1470 to 1650 °F)
	NbC-TaC	...	...	...	High corrosion resistance at high temperature
					High tenacity (oxidation resistance limited to 600 to 700 °C, or 1110 to 1290 °F)
Borides	TiB_2	...	...	Wear resistance	Low reactivity with molten metals
					High neutron-capture cross section
	CrB_2	...	...	Wear resistance	Low reactivity with molten metals
					High neutron-capture cross section
	HfB_2	...	...	Wear resistance	Low reactivity with molten metals
					High neutron-capture cross section
	TaB_2	...	...	Wear resistance	Low reactivity with molten metals
					High neutron-capture cross section
Oxides	ZrO_2 stabilized with CaO or MgO	<1000	<1830	Thermal barriers	Good wear resistance
					Good mechanical properties
	ZrO_2 stabilized with Y_2O_3 or CeO_2	<1350	<2460	Thermal barriers	Good wear resistance
					Good mechanical properties
	Al_2O_3	≤950	≤1740	...	Good wear resistance
					Dielectric
					High reactivity with molten salts
					Poor resistance to thermal shocks
	$Al_2O_3 + xTiO_2$ ($x = 3, 13, 40$ wt%)	≤800	≤1470	...	Addition of TiO_2 lowers Al_2O_3 porosity, improves adhesion, gives an excellent surface finish and good wear resistance
	Cr_2O_3	1200	2190	...	Excellent wear resistance
					Low porosity
					Excellent finish
					Used on sliding surfaces
	$ZrSiO_4$	1100	2010	...	Good resistance to liquid glass and combustion gases
					Not wetted by liquid metals: casting
	TiO_2	≤800	≤1470	...	Excellent surface finish
					Different colors (from white to black) due to oxygen losses

(a) Related to the coating detached from its substrate and assuming no severe thermal shock upon heating

Thermodynamic and Solid State Data

Transformation temperatures and heats of transformation of various ceramic compounds

Compound	Molecular weight	Solid-state (SS) transformations: SS Transformation temperature (T_{tr}), K	SS Latent heat (ΔL_{tr}), J/mole	Melting temperature (T_m), K	Latent heat of melting (ΔL_m), J/mole
Elements					
Al	26.98	...	...	933	10,711
C	12.01	...	...	[4070](a)	...
Mo	95.94	...	...	2897	39,099
Si	28.08	...	...	1685	50,208
W	183.85	...	...	3680	35,397
Single oxides					
Al_2O_3	101.96	...	...	2327	111,085
CaO	56.08	...	...	3200	79,496
Cr_2O_3	151.99	...	...	2603	129,704
Fe_2O_3	159.69	906	0	*1730*(b)	...
Fe_3O_4	231.54	850	0	1870	138,072
MgO	40.30	...	...	3105	77,822
MnO	70.94	...	...	2115	43,932
MoO_3	143.94	...	...	1075	48,911
NiO	74.69	525	0	2228	54,392
		565	0	...	...
½P_4O_{10}	141.94	...	...	[632](a)	53,000
SiO_2	60.08	847 and 1079	728 and 1996	1996	9566
TiO_2	79.88	...	...	2130	66,944
ZnO	81.39	...	...	2248	54,392
ZrO_2	123.22	1478	5941	2950	87,027
Mixed oxides					
$3Al_2O_3 \cdot 2SiO_2$ (mullite)	426.05	...	...	2023	...
$BaTiO_3$	233.20	395 and 1733	201 and 0	1889	...
$CaSiO_3$ (pseudowollastonite)	116.16	1398	5400	1817	56,066
Ca_2SiO_4 (olivine)	172.24	1121 and 1712	13,987 and 14,188	2403	71,128
$Ca_3(PO_4)_2$	310.18	1423 and 1743	18,828 and 0	2083	167,360
$CaO \cdot Al_2O_3 \cdot 2SiO_2$ (anorthite)	278.21	...	...	1823	81,170
$CaO \cdot TiO_2 \cdot SiO_2$ (sphene)	196.04	...	...	1673	123,846
$CaWO_4$	287.93	...	...	1853	...
$CuFe_2O_4$	239.24	675 and 795	753 and 0	1358	13,054
Fe_2SiO_4 (fayalite)	203.78	...	...	1490	92,174
$LiAlO_2$	65.92	...	...	1883	87,864
$LiAlSiO_4$ (eucryptite)	126.01	1300	1255	...	...
$MgAl_2O_4$	142.27	...	...	2408	196,648
$MgCr_2O_4$	192.30	...	...	2623	...
$MgSO_4$	120.37	...	...	1400	14,644
Mn_2SiO_4	201.96	...	...	1618	89,621
$MnTiO_3$	150.82	...	...	1677	...
$NaAlSiO_4$ (nepheline)	142.05	467 and 1180	0 and 0	...	...
Na_2CO_3	105.99	723	690	1123	29,665
$NiAl_2O_4$	176.65	...	...	2383	...
$PbTiO_3$	303.08	763	5648	1559	...
Zn_2SiO_4	222.86	...	...	1785	83,680
Zirconates					
$CaZrO_3$	179.30	...	...	2623	...
$SrZrO_3$	226.84	1003, 1133, and 1408	...	3023	...
$ZrSiO_4$	183.31	*1811*(b)	...	...	...
Borides					
AlB_{12}	156.71	...	...	2423	...
CoB	69.74	...	...	1733	...
CrB	62.81	...	...	2343	...
CrB_2	73.62	...	...	2437	...
FeB	66.66	...	...	1923	...
MnB_2	76.56	...	...	2261	...
Ni_4B_3	267.19	...	...	1853	...
TaB_2	202.57	...	...	3373	83,680
TiB_2	69.50	...	...	3193	100,416
VB_2	72.56	...	...	3020	...
ZrB_2	112.85	...	...	[*4466*](a)(b)	...

(continued)

Transformation temperatures and heats of transformation of various ceramic compounds (continued)

Compound	Molecular weight	Solid-state (SS) transformations: SS Transformation temperature (T_{tr}), K	SS Latent heat (ΔL_{tr}), J/mole	Melting temperature (T_m), K	Latent heat of melting (ΔL_m), J/mole
Carbides					
Al_4C_3	143.96	...	...	2500	...
B_4C	55.26	...	...	2743	104,600
Be_2C	30.04	...	...	2400	75,312
CaC_2	64.10	720	5523	...	...
$Cr_{23}C_6$	1267.98	...	...	*1793*(b)	...
Fe_3C	179.55	485	7029	1500	51,463
HfC	190.50	...	...	4103	...
Nb_2C	197.82	1503 and 2723	...	*3259*(b)	...
SiC	40.10	...	...	*3245*(b)	...
TiC	59.89	...	...	3290	71,128
ZrC	103.24	...	...	3805	79,496
Nitrides					
AlN	40.99	...	...	[2790](a)	...
BN	24.82	...	...	[*2600*](a)(b)	...
Be_3N_2	55.05	...	...	2473	129,286
Cr_2N	118.00	...	...	[*1810*](a)(b)	...
Fe_4N	237.40	753	8368	...	...
Li_3N	34.83	...	...	1086	...
Mg_3N_2	100.93	823 and 1061	920 and 1088	...	...
NbN	106.91	1643	4184	2323	46,024
Si_3N_4	140.28	...	...	[*2151*[(a)(b)	...
TiN	61.89	...	...	3223	62,760
VN	64.95	...	...	[*2619*](a)(b)	...
ZrN	105.23	...	...	3225	67,362
Halides					
$AlCl_3$	133.34	...	...	466	35,350
CaF_2	78.08	1424	4770	1691	29,706
$CaCl_2$	110.98	...	...	1045	28,543
CsI	259.81	...	...	900	25,606
KBr	119.00	...	...	1007	25,522
LiF	25.94	...	...	1121	27,087
MgF_2	62.30	...	...	1536	58,702

(a) Sublimation reactions are distinguished from melting reactions by the use of brackets. (b) Temperatures of incongruent melting or solid-state decomposition are italicized to distinguish them from congruent transformations

Coefficients for calculation of heat capacity

C_p (in J/mole) $= a + b \times 10^{-3}\,T + c \times 10^{5}\,T^{-2}$

Compound	Temperature range(a), K	*a*	*b*	*c*
Elements				
Al (solid)	298 to 933	19.212	14.451	0.728
Al (liquid)	933 to 2791	31.748	...	...
C (graphite)	298 to 2000	14.719	6.410	−7.209
	2000 to 4100	23.601	1.121	−30.123
C (diamond)	298 to 1200	8.033	7.627	−146.015
Mo (solid)	298 to 2897	18.312	9.916	2.477
Mo (liquid)	2897 to 4978	40.350	...	...
N_2 (gas)	298 to 2000	27.266	4.929	0.331
	2000 to 5000(a)	36.169	0.469	−45.687
O_2 (gas)	299 to 2000	30.249	4.209	−1.891
	2000 to 5000(a)	34.893	1.749	−26.358
Si (solid)	298 to 1685	23.760	2.937	−4.121
Si (liquid)	1685 to 3505	27.196	...	...
W (solid)	298 to 3680	24.329	3.104	−0.879
W (liquid)	3680 to 5931	35.654	...	...
Oxides				
Al_2O_3 (α, solid)	298 to 2327	115.493	11.539	−35.487
Al_2O_3 (γ, solid)	298 to 2290	112.415	22.383	−32.324
Al_2O_3 (liquid)	2327 to 2500	46.000	...	...
CO (gas)	298 to 2000	28.065	4.627	−0.259
	2000 to 3000(a)	34.207	1.038	...

(continued)

Coefficients for calculation of heat capacity (continued)

C_p (in J/mole) $= a + b \times 10^{-3}\,T + c \times 10^{5}\,T^{-2}$

Compound	Temperature range(a), K	a	b	c
Oxides (con't)				
CO_2 (gas)	298 to 2000	45.365	8.686	–9.619
	2000 to 3000(a)	61.381	0.619	–90.563
CaO (liquid)	298 to 3200	50.741	3.682	–8.702
CaO (liquid)	3200 to 3500	(62.76)(b)	...	...
Cr_2O_3 (solid)	298 to 2603	109.645	15.455	...
Cr_2O_3 (liquid)	2603 to 3000	(156.90)(b)	...	...
Fe_2O_3 (solid)	298 to 960	98.194	80.613	–16.430
	960 to 1800	142.417	...	...
Fe_3O_4 (solid)	298 to 850	79.756	225.390	34.014
	850 to 1870	49.850	72.514	855.382
H_2O (liquid)	298 to 373	75.480	...	...
H_2O (gas)	373 to 3000	304.246	10.426	–0.393
MgO (solid)	298 to 3105	47.515	4.309	–10.347
MgO (liquid)	3105 to 4000	(66.94)(b)	...	...
MnO (solid)	298 to 2115	46.921	7.774	–4.569
MnO (liquid)	2115 to 2500	(60.67)(b)	...	...
MoO_3 (solid)	298 to 1075	75.505	32.081	–8.974
MoO_3 (liquid)	1075 to 1400	126.231	...	...
NiO (solid)	298 to 525	–6.322	131.229	10.221
	525 to 565	–34.249	168.440	...
	565 to 2228	–39.913	12.367	21.885
NiO (liquid)	2228 to 2500	(54.39)(b)	...	...
½P_4O_{10} (solid)	298 to 632	70.275	169.754	13.374
½P_4O_{10} (gas)	632 to 2500	158.131	4.157	90.238
PbO (solid)	298 to 762	43.578	15.204	–2.042
	762 to 1159	45.202	12.526	–1.916
PbO (liquid)	1159 to 1500	57.887	...	...
SiO_2 (gas)	298 to 2000	47.377	6.661	–8.439
	2000 to 3000(a)	58.331	0.577	–50.147
SiO_2 (solid)	298 to 847	40.495	44.599	08.322
	847 to 1079	67.589	42.577	–1.364
				–23.731
	1079 to 1996	67.639	4.192	
SiO_2 (liquid)	1996 to 3000	(85.77)(b)	...	...
TiO_2 (solid)	298 to 2130	70.765	5.104	–15.317
TiO_2 (liquid)	2130 to 3000	(100.41)(b)	...	...
ZnO (solid)	298 to 2000	45.336	7.288	–5.771
ZrO_2 (solid)	298 to 1478	70.116	7.020	–14.242
	1478 to 2500	74.745	...	...
ZrO_2 (liquid)	2950 to 3300	(87.86)(b)	...	...
Mixed oxides				
$3Al_2O_3 \cdot 2SiO_2$	298 to 2023	498.60	31.664	–162.326
Al_2SiO_5 (kyanite) (solid)	298 to 2000	168.073	28.139	...
$BaTiO_3$ (solid)	298 to 395	61.938	136.566	–0.058
				–26.501
	395 to 1733	130.633	5.523	
	1773 to 1889	(134.93)(b)	...	...
$CaSiO_3$ (solid)	298 to 1817	103.748	18.529	...
$CaSiO_3$ (liquid)	1817 to 2200	146.44	...	...
Ca_2SiO_4 (solid)	298 to 1121	133.122	51.547	–19.413
	1121 to 1712	126.621	51.224	...
	1712 to 2403	205.016	...	...
Ca_2SiO_4 (liquid)	2403 to 2800	(209.20)(b)	...	...
$Ca_3(PO_4)_2$ (solid)	298 to 1423	183.657	148.055	...
	1423 to 1743	(376.56)(b)	...	...
	1743 to 2083	(376.56)(b)	...	...
$Ca_3(PO_4)_2$ (liquid)	2083 to 2500	(380.74)(b)	...	...
$CaAl_2Si_2O_8$ (solid)	298 to 1823	(260.54)(b)	(63.43)(b)	...
$CaAl_2Si_2O_8$ (liquid)	1823 to 2100	380.74	...	...
$CaO \cdot TiO_2 \cdot SiO_2$ (solid)	298 to 1673	177.35	23.18	–40.29
$CaO \cdot TiO_2 \cdot SiO_2$ (liquid)	1673 to 2000	279.491	...	...
$CaWO_4$ (solid)	298 to 1000	110.787	45.771	...
$CuFe_2O_4$ (solid)	298 to 675	139.621	117.779	–23.431
	675 to 795	227.191	...	...
	795 to 1358	166.021	41.003	...
$CuFe_2O_4$ (liquid)	1358 to 1500	225.936	...	...
Fe_2SiO_4 (solid)	298 to 1490	152.751	39.160	–28.031
Fe_2SiO_4 (liquid)	1490 to 1700	240.580	...	...
$LiAlO_2$ (solid)	298 to 1883	92.337	12.163	–25.007
$LiAlO_2$ (liquid)	1883 to 2500	133.89	...	...

(continued)

Coefficients for calculation of heat capacity (continued)

C_p (in J/mole) $= a + b \times 10^{-3}\,T + c \times 10^{5}\,T^{-2}$

Compound	Temperature range(a), K	a	b	c
Mixed oxides (cont'd)				
$LiAlSiO_4$ (solid)	298 to 1300	154.337	28.449	−43.934
	1300 to 1600	129.701	50.21	...
$MgAl_2O_4$ (solid)	298 to 2408	153.964	26.776	40.918
$MgAl_2O_4$ (liquid)	2408 to 2500	230.12	...	...
$MgCr_2O_4$ (solid)	298 to 2623	167.436	14.894	40.081
$MgSO_4$ (solid)	298 to 1400	112.449	41.076	−25.330
$MgSO_4$ (liquid)	1400 to 2000	158.992	...	...
$2MgO \cdot 2Al_2O_3 \cdot 5SiO_2$ (solid)	298 to 1700	(612.19)(b)	(109.97)(b)	...
Mn_2SiO_4 (solid)	298 to 1618	159.074	19.500	−31.128
Mn_2SiO_4 (liquid)	1618 to 3000	243.09	...	...
$MnTiO_3$ (solid)	298 to 1677	(121.97)(b)	(9.29)(b)	(−21.88)(b)
Na_2CO_3 (solid)	298 to 723	29.234	226.225	...
	723 to 1123	50.087	129.072	...
Na_2CO_3 (liquid)	1123 to 2000	189.54	...	...
$NaAlSiO_4$ (solid)	298 to 467	27.739	0.295	...
	467 to 1180	112.087	67.113	...
	1180 to 1525	171.997	5.528	...
$NiAl_2O_4$ (solid)	298 to 2000	(159.20)(b)	(23.35)(b)	(−30.75)(b)
$PbTiO_3$ (solid)	298 to 763	119.534	17.911	−18.198
	763 to 1559	109.075	22.804	−13.337
$ZnFe_2O_4$ (solid)	298 to 1000	(136.48)(b)	(54.15)(b)	...
Zn_2SiO_4 (solid)	298 to 1785	(144.88)(b)	(36.94)(b)	(−30.29)(b)
Zn_2SiO_4 (liquid)	1785 to 2000	213.384	...	...
Zirconates				
$BaZrO_3$ (solid)	298 to 2000	(122.80)(b)	(8.79)(b)	(−21.96)(b)
				−22.945
$CaZrO_3$ (solid)	298 to 2623	119.194	10.702	...
$Li_2Zr_2O_7$ (solid)	298 to 1500	(139.20)(b)	(25.39)(b)	(−33.12)(b)
$Nd_2Zr_2O_7$ (solid)	298 to 1400	(255.22)(b)	(44.77)(b)	(−40.17)(b)
$SrZrO_4$ (solid)	298 to 2000	(124.06)(b)	(12.22)(b)	(−21.61)(b)
$Y_2Zr_2O_7$ (solid)	298 to 1500	(273.45)(b)	(9.54)(b)	(−54.45)(b)
$ZrSiO_4$ (solid)	298 to 2000	131.706	17.656	−33.805
Borides				
AlB_{12} (solid)	298 to 2423	(211.29)(b)	(115.06)(b)	(−85.35)(b)
CoB (solid)	298 to 1300	(42.43)(b)	(14.64)(b)	(−11.26)(b)
CrB (solid)	298 to 1200	42.342	16.024	−10.041
CrB_2 (solid)	298 to 1200	9.63	10.7	...
FeB (solid)	298 to 1923	(48.29)(b)	(6.44)(b)	...
MnB_2 (solid)	298 to 1200	(40.25)(b)	(44.85)(b)	...
Ni_4B_3 (solid)	298 to 1400	(37.28)(b)	(11.74)(b)	(−9.03)(b)
TaB_2 (solid)	298 to 3373	59.452	18.785	−15.062
TaB_2 (liquid)	3373 to 3500	125.52	...	...
TiB_2 (solid)	298 to 3193	70.724	10.884	−26.511
TiB_2 (liquid)	3193 to 3500	108.784	...	...
VB_2 (solid)	298 to 2500	50.578	28.784	−10.828
ZrB_2 (solid)	298 to 3323	66.062	17.656	−14.685
ZrB_2 (liquid)	3323 to 3800	96.232	...	...
Carbides				
Al_4C_3 (solid)	298 to 2500	154.675	28.726	−41.922
B_4C (solid)	298 to 2743	96.186	22.593	−44.850
B_4C (liquid)	2743 to 3500	135.980	...	...
Be_2C (solid)	298 to 2400	36.894	21.364	...
Be_2C (liquid)	2400 to 3000	(92.05)(b)	...	...
CaC_2 (solid)	298 to 720	68.614	11.882	−8.660
	720 to 1200	66.571	6.520	...
$Cr_{23}C_6$ (solid)	298 to 1793	638.474	240.968	−73.072
Fe_3C (solid)	298 to 485	(82.18)(b)	(83.68)(b)	...
	485 to 1500	107.189	12.551	...
Fe_3C (liquid)	1500 to 2000	(121.34)(b)	...	...
HfC (solid)	298 to 3900	47.448	5.439	−13.021
Nb_2C (solid)	298 to 1800	(66.44)(b)	(12.55)(b)	(−8.58)(b)
SiC (solid)	298 to 3245	49.628	2.238	−20.728
TiC (solid)	298 to 3290	49.494	3.347	−14.978
TiC (liquid)	3290 to 3500	(62.76)(b)	...	...
ZrC (solid)	298 to 3805	50.905	3.386	−124.547
ZrC (liquid)	3805 to 4000	(62.76)(b)	...	...
Nitrides				
AlN (solid)	298 to 2000	(34.39)(b)	(16.94)(b)	(−8.37)(b)
BN (solid)	298 to 1900	33.889	14.727	−23.053
Be_3N_2 (solid)	298 to 2473	132.005	−10.470	−59.824

(continued)

Coefficients for calculation of heat capacity (continued)

C_p (in J/mole) $= a + b \times 10^{-3} T + c \times 10^{5} T^{-2}$

Compound	Temperature range(a), K	*a*	*b*	*c*
Nitrides (cont'd)				
Be_3N_2 (liquid)	2473 to 3000	(133.89)(b)	...	...
Cr_2N (solid)	298 to 1810	63.761	28.450	...
Fe_4N (solid)	298 to 753	(112.29)(b)	(34.14)(b)	...
	753 to 953	(138.07)(b)	...	...
Li_3N (solid)	298 to 1086	57.565	84.028	−6.527
Mg_3N_2 (solid)	298 to 823	95.432	30.542	...
	823 to 1061	123.840		
	1061 to 2000	123.595		
NbN (solid)	298 to 1643	46.674	7.443	−8.803
	1643 to 2323	(62.76)(b)	...	...
NbN (liquid)	2323 to 3000	(62.76)(b)	...	...
Si_3N_4 (solid)	298 to 2200	70.539	98.738	...
TiN (solid)	298 to 3323	49.829	3.933	12.384
TiN (liquid)	3323 to 3500	(66.94)(b)	...	...
VN (solid)	298 to 2619	45.771	8.786	−9.246
ZrN (solid)	298 to 3225	46.940	7.029	−7.196
ZrN (liquid)	3225 to 3500	(58.58)(b)	...	...
Halides				
$AlCl_3$ (solid)	298 to 466	64.933	87.860	...
$AlCl_3$ (liquid)	466 to 1000	125.520	...	...
$CaCl_2$ (solid)	298 to 1045	69.836	15.388	−1.594
$CaCl_2$ (liquid)	1045 to 2206	122.263	−14.903	0.703
CaF_2 (solid)	298 to 1424	41.056	55.460	8.497
				−729.609
	1424 to 1691	154.098	0.000	
CaF_2 (liquid)	1691 to 2804	99.914	...	...
CsI (solid)	298 to 743	52.820	7.439	−2.297
	743 to 900	66.230	...	...
CsI (liquid)	900 to 1400	74.221	...	...
KBr (solid)	298 to 1007	46.808	16.267	0.364
KBr (liquid)	1007 to 1671	69.873	...	...
LiF (solid)	298 to 1121	43.281	16.727	−5.711
LiF (liquid)	1121 to 2000	64.347	...	...
MgF_2 (solid)	298 to 1536	73.137	8.878	−12.660
MgF_2 (liquid)	1536 to 2600	94.922	...	...
$SiCl_4$ (gas)	298 to 2000	105.260	1.699	−13.786
SiF_4 (gas)	298 to 2000	96.232	6.786	−21.902

(a) For some gases, separate coefficients are listed above 2000 K (3140 °F) to more accurately reflect actual values over a wide temperature range. (b) Parentheses indicate estimated values

Heats and free energies of formation at 298 K (25 °C)

Compound(a)	ΔH^0_{298} kJ/mole	Δ^0_{298}, kJ/mole
Elements		
C (diamond)	1.90	2.90
Oxides		
Al_2O_3 (α)	−1675.69	−1582.27
Al_2O_3 (γ)	−1656.86	−1563.85
CO (gas)	−110.54	−137.18
CO_2 (gas)	−393.50	−393.36
CaO	−635.09	−603.51
Cr_2O_3	−1139.70	−1058.07
Fe_2O_3	−824.25	−742.29
Fe_3O_4	−1118.38	−1015.23
H_2O (liquid)	−285.83	−237.14
H_2O (gas)	−241.83	−228.62
MgO	−601.24	−568.94
MnO	−385.22	−362.90
MoO_3	−745.09	−667.99
NiO	−239.70	−211.54
½P_4O_{10}	−1504.97	−1361.67
PbO	−219.40	−189.28
SO_2 (gas)	−296.81	−300.10
SiO_2	−910.86	−856.44
TiO_2	−944.75	−889.41
ZnO	−350.46	−320.48
ZrO_2	−1097.46	−1039.72

Compound(a)	ΔH^0_{298} kJ/mole	Δ^0_{298}, kJ/mole
Mixed oxides		
$3Al_2O_3 \cdot 2SiO_2$	−6820.46	−6443.05
Al_2SiO_5 (kyanite)	−2594.08	−2443.88
$BaTiO_3$	−1659.80	−1572.44
$CaSiO_3$	−1628.40	−1544.74
Ca_2SiO_4	−2315.22	−2198.59
$Ca_3(PO_4)_2$	−4120.80	−3884.97
$CaAl_2Si_2O_8$	−4227.89	−4002.22
$CaO \cdot TiO_2 \cdot SiO_2$	−2603.30	−2461.78
$CaWO_4$	−1645.15	−1538.42
$CuFe_2O_4$	−967.97	−863.24
Fe_2SiO_4	−1479.90	−1378.98
$LiAlO_2$	−1188.67	−1126.31
$LiAlSiO_4$	−2124.20	−2010.11
$MgAl_2O_4$	−2299.90	−2175.01
$MgCr_2O_4$	−1784.06	−1660.50
$MgSO_4$	−1284.90	−1170.58
$MgO \cdot 2Al_2O_3 \cdot 5SiO_2$	−9161.70	−8651.34
Mn_2SiO_4	−1730.50	−1632.13
$MnTiO_3$	−1358.55	−1279.38
Na_2CO_3	−1130.77	−1048.00
$NaAlSiO_4$	−2094.66	−1980.01
$NiAl_2O_4$	−1915.90	−1797.12
$PbTiO_3$	−1198.72	−1111.85

(continued)

Heats and free energies of formation at 298 K (25 °C) (continued)

Compound(a)	ΔH^0_{298} kJ/mole	$\Delta ^0_{298}$, kJ/mole
$ZnFe_2O_4$	–1171.58	–1065.79
$ZnSi_2O_4$	–1262.57	–1179.50
Zirconates		
$BaZrO_3$	–1779.46	–1694.68
$CaZrO_3$	–1766.90	–1681.06
Li_2ZrO_3	–1760.20	–1666.84
$Nd_2Zr_2O_7$	–4046.92	–3844.62
$SrZrO_3$	–1767.30	–1681.68
$Y_2Zr_2O_7$	–4121.99	–3917.87
$ZrSiO_4$	–2023.80	–1909.32
Borides		
AlB_{12}	–266.10	–272.24
CoB	–94.14	–92.55
CrB	–75.31	–77.00
CrB_2	–94.14	–95.22
FeB	–71.13	–69.52
Borides		
MnB_2	–94.14	–91.41
Ni_4B_3	–311.71	–305.05
TaB_2	–209.20	–206.57
TiB_2	–323.80	–319.65
VB_2	–203.76	–200.65
ZrB_2	–322.59	–318.24
Carbides		
Al_4C_3	–208.80	–196.46
B_4C	–71.13	–70.55
Be_2C	–116.98	–114.51
CaC_2	–59.80	–64.89
$Cr_{23}C_6$	–328.44	–338.57
Fe_3C	25.10	20.03
HfC	–251.04	–248.63
Nb_2C	–190.00	–185.66
SiC	–73.22	–70.85
TiC	–184.50	–180.84
ZrC	–196.65	–193.28
Nitrides		
AlN	–317.98	–287.00
BN	–254.39	–228.50
Be_3N_2	–588.27	–532.87
Cr_2N	–125.52	–102.20
Fe_4N	–11.09	3.72
Li_3N	–164.56	–128.64
Mg_3N_2	–460.66	–400.50
NbN	–235.10	–205.97
Si_3N_4	–744.75	–647.34
TiN	–337.86	–309.16
VN	–217.15	–191.08
ZrN	–365.26	–336.70
Halides		
$AlCl_3$	–705.63	–630.00
$CaCl_2$	–795.79	–748.11
CaF_2	–1225.91	–1173.54
CsI	–346.60	–340.59
KBr	–393.80	–380.43
LiF	–616.93	–588.66
MgF_2	–1124.24	–1071.11
$SiCl_4$ (gas)	–662.75	–622.76
SiF_4 (gas)	–1614.94	–1572.70

(a) Room-temperature values of the compound in the solid state, unless specified otherwise as either as gas or liquid

Summary of ceramic crystal structures

Class	Name	Anion lattice	Coordination	Cation occupation
AX	Rock salt	Cubic close-packed	6:6	All octahedral sites
AX	Nickel arsenide	Hexagonal close-packed	6:6	All octahedral sites
AX	Cesium chloride	Simple cubic	8:8	All cubic sites
AX	Wurtzite	Hexagonal close-packed	4:4	$\frac{1}{2}$ octahedral sites
AX	Zinc blende	Cubic close-packed	4:4	$\frac{1}{2}$ tetrahedral sites
AX_2	Fluorite	Simple cubic	8:4	$\frac{1}{2}$ cubic sites
AX_2	Rutile	Tetragonal close-packed	6:3	$\frac{1}{2}$ octahedral sites
AX_2	Silica	Connected tetrahedra	4:2	
A_2X	Antifluorite	Cubic close-packed	4:8	All tetrahedral sites
A_2X_3	Corundum	Hexagonal close-packed	6:4	$\frac{2}{3}$ octahedral sites
ABX_3	Perovskite	Cubic close-packed	12:6:6	$\frac{1}{4}$ octahedral sites
ABX_3	Ilmenite	Hexagonal close-packed	6:6:4	$\frac{2}{3}$ octahedral sites
AB_2O_4	Spinel	Cubic close-packed	4:6:4	$\frac{1}{8}$ tetrahedral sites (*A*) and $\frac{1}{2}$ octahedral sites (*B*)
$B(AB)O_4$	Inverse spinel	Cubic close-packed	4:6:4	$\frac{1}{8}$ tetrahedral sites (*B*) and $\frac{1}{2}$ octahedral sites (*A*, *B*)
AB_2O_4	Olivine	Hexagonal close-packed	6:4:4	$\frac{1}{2}$ octahedral sites (*A*) and $\frac{1}{8}$ tetrahedral sites (*B*)

Refractive indices of selected crystalline materials

Material	Chemical formula	Average refractive index	Birefringence
Silicon chloride	$SiCl_4$	1.412	...
Lithium fluoride	LiF	1.392	...
Sodium fluoride	NaF	1.326	...
Calcium fluoride	CaF_2	1.434	...
Corundum	Al_2O_3	1.76	0.008
Periclase	MgO	1.74	...
Quartz	SiO_2	1.55	0.009
Spinel	$MgAl_2O_4$	1.72	...
Zircon	$ZiSiO_4$	1.95	0.055
Orthoclase	$KAlSi_3O_8$	1.525	0.007
Albite	$NaAlSi_3O_8$	1.529	0.008
Anorthite	$CaAl_2Si_2O_8$	1.585	0.008
Sillimanite	$Al_2O_3 \cdot SiO_2$	1.65	0.021
Mullite	$3Al_2O_3 \cdot 2SiO_2$	1.64	0.010
Rutile	TiO_2	2.71	0.287
Silicon carbide	SiC	2.68	0.043
Litharge	PbO	2.61	...
Galena	PbS	3.912	...
Calcite	$CaCO_3$	1.65	0.17
Silicon	Si	3.49	...
Cadmium telluride	CdTe	2.74	...
Cadmium sulfide	CdS	2.50	...
Strontium titanate	$SrTiO_3$	2.49	...
Lithium niobate	$LiNbO_3$	2.31	...
Yttrium oxide	Y_2O_3	1.92	...
Zinc selenide	ZnSe	2.62	...
Barium titanate	$BaTiO_3$	2.40	...

Defects in crystalline ceramics

Defect	Cause	Relative scale
Impurity	Foreign atom or ion introduced by substitution or interstitially; vacancies may be created to balance valance charges	Atomic in three dimensions
Vacancy	Missing positive or negative ion, or atom	Atomic in three dimensions
Vacancy pair	Missing positive ion coupled with missing negative ion	Atomic in three dimensions
Vacancy cluster	Aggregation of vacancies	Microscopic (large clusters can be resolved optically)
Color center	Anion vacancy plus electron or cation vacancy plus electron hole	Atomic in two dimensions, microscopic in length
Dislocation	Structural misfit, linear character	Atomic in two dimensions, microscopic in length
Subgrain boundary	Structural misfit (at small angles), surface character	Atomic in one dimension, microscopic in two dimensions
Grain boundary	Structural misfit (at large angles), surface character	Atomic to microscopic in one dimension, microscopic to macroscopic in two dimensions
Phase boundary	Compositional or crystallographic discontinuity, surface character	Atomic to microscopic in one dimension, microscopic to macroscopic in two dimensions

Physical and Mechanical Properties

Properties of aluminas and other ceramic materials(a)

Property	ASTM method	Units	Test or material condition	Refractory (alumina-mullite)
Specific gravity	C 20	...	...	2.84
Hardness	E 18	Rockwell 45N	...	...
	(1000-g load)	Knoop (GPa)		
Surface finish	Profilometer	μm (μin.)	As-fired	...
	(0.75-mm cutoff)		Ground	...
			Polished	...
Crystal size	...	μm (μin.)	Range	...
			Average	...
Water absorption	C 373	...	...	...
Gas permeability(d)	...	...	...	...
Color	...	...	...	White
Compressive strength	C 773	MPa (ksi)	25 °C	83 (12)
			1000 °C	...
Flexural strength	F 417	MPa (ksi)	25 °C: typical	...
			min (e)	...
			1000 °C: typical	...
			min (e)	...
Tensile strength	ACMA Test #4	MPa (ksi)	25 °C	...
			1000 °C	...
Modulus of elasticity	C 623	GPa (10^6 psi)	...	...
Shear modulus	C 623	GPa (10^6 psi)	...	...
Bulk modulus	C 623	GPa (10^6 psi)	...	...
Transverse sonic velocity	C 623	m/s (ft/s)	...	...
Poisson's ratio	C 623		...	...
Maximum-use temperature		°C (°F)	No load	1750 (3180)
Coefficient of linear thermal expansion	C 371	10^{-6}/°C (10^{-6}/°F)	–200 to 25 °C	...
			25 to 200 °C	...
			25 to 500 °C	...
			25 to 800 °C	...
			25 to 1000 °C	...
			25 to 1200 °C	6.0 (3.4)
Thermal conductivity	C 408	W/m · K (cal/cm · s · °C)	20 °C	...
			100 °C	...
			400 °C	...
			800 °C	2.1 (0.005)
Specific heat	C 351	J/kg · K (cal/g · °C)	100 °C	630 (0.15)
Dielectric strength	D 116	kV/mm (V/mil)	6.35 mm(f)	...
			3.18 mm(f)	...
			1.27 mm(f)	...
			0.64 mm(f)	...
			0.25 mm(f)	...
Dielectric constant	D 150,	...	1 kHz	...
(at 25 °C)	D 2520		1 MHz	...
			100 MHz	...
Dissipation factor	D 150,	...	1 kHz	...
	D 2520		1 MHz	...
			100 MHz	...
Loss index	D 150,	...	1 kHz	...
	D 2520		1 MHz	...
			100 MHz	...
Volume resistivity	D 1829	Ω · cm^2/cm	25 °C	...
			300 °C	...
			500 °C	...
			700 °C	...
			1000 °C	...
T_c value(g)	...	°C (°F)	...	...

Zirconia(b)	90%(c) Al_2O_3 (opaque)	94%(c) Al_2O_3	Alumina 96%(c) Al_2O_3	99.5%(c) Al_2O_3	99.9%(c) Al_2O_3
5.4	3.69	3.62	3.72	3.89	3.96
68	75	78	78	83	90
...	10.8	11.1	11.1	14.7	15.2
...	1.6 (63)	1.6 (63)	1.6 (63)	0.9 (35)	0.5 (20)
...	1.0 (39)	1.3 (51)	1.3 (51)	0.5 (20)	0.9 (35)
0.4 (14)	0.1 (3.9)	0.3 (12)	0.3 (12)	0.1 (3.9)	<0.03 (<1.2)
...	2 to 40 (79 to 1575)	2 to 25 (79 to 985)	2 to 20 (79 to 788)	5 to 50 (197 to 1970)	1 to 6 (39 to 236)

(continued)

Properties of aluminas and other ceramic materials(a) (continued)

Zirconia(b)	90%(c) Al_2O_3 (opaque)	94%(c) Al_2O_3	Alumina 96%(c) Al_2O_3	99.5%(c) Al_2O_3	99.9%(c) Al_2O_3
55 (2170)	6 (2336)	12 (473)	11 (433)	17 (670)	3 (118)
None	None	None	None	None	None
None	None	None	None	None	None
Ivory	Black	White	White	Ivory	Ivory
...	2413 (350)	2103 (305)	2068 (300)	2620 (380)	3792 (550)
...	...	345 (50)	...	...	1930 (280)
414 (60)	365 (53)	352 (51)	358 (52)	379 (55)	552 (80)
...	324 (47)	317 (46)	324 (47)	...	517 (75)
...	...	138 (20)	712 (25)	...	414 (60)
...	...	117 (17)	138 (20)	...	379 (55)
...	228 (33)	193 (28)	193 (28)	262 (38)	310 (45)
...	...	103 (15)	96 (14)	...	221 (32)
...	308 (45)	283 (41)	303 (44)	372 (54)	386 (56)
...	124 (18)	117 (17)	124 (18)	152 (22)	158 (23)
...	...	165 (24)	172 (25)	228 (33)	228 (33)
...	$9.3 (31) \times 10^3$	$8.9 (29) \times 10^3$	$9.1 (30) \times 10^3$	$9.8 (32) \times 10^3$	$9.9 (32) \times 10^3$
...	0.24	0.21	0.21	0.22	0.22
...	1500 (2730)	1700 (3100)	1700 (3100)	1750 (3180)	1900 (3450)
...	4.4 (2.4)	3.4 (1.9)	3.4 (1.9)	3.4 (1.9)	3.4 (1.9)
...	6.4 (3.6)	6.3 (3.5)	6.0 (3.4)	7.1 (4.0)	6.5 (3.6)
...	7.3 (4.0)	7.1 (4.0)	7.4 (4.1)	7.6 (4.3)	7.4 (4.1)
...	8.0 (4.4)	7.6 (4.3)	8.0 (4.5)	8.0 (4.5)	7.8 (4.4)
4.9 (2.7)	8.4 (4.7)	7.9 (4.4)	8.2 (4.6)	8.3 (4.6)	8.0 (4.5)
...	...	8.1 (4.5)	8.4 (4.7)	...	8.3 (4.6)
...	12.6 (0.030)	18.0 (0.043)	24.7 (0.059)	35.6 (0.085)	38.9 (0.093)
...	11.3 (0.027)	14.2 (0.035)	18.8 (0.045)	25.9 (0.062)	27.6 (0.066)
...	7.5 (0.018)	7.9 (0.017)	10.0 (0.024)	12.1 (0.028)	13.4 (0.032)
...	...	5.0 (0.010)	5.4 (0.013)	6.3 (0.015)	6.3 (0.015)
...	1045 (0.25)	880 (0.21)	880 (0.21)	880 (0.21)	880 (0.21)
...	5.3 (135)	8.7 (220)	8.3 (210)	8.7 (220)	9.4 (240)
...	8.8 (225)	11.8 (300)	10.8 (275)	11.4 (290)	12.8 (325)
...	16.3 (415)	16.7 (425)	14.6 (370)	16.9 (430)	18.1 (460)
...	21.2 (540)	21.6 (550)	17.7 (450)	22.8 (580)	23.2 (590)
...	28.3 (720)	28.3 (720)	22.8 (580)	33.1 (840)	31.5 (800)
...	22.0	8.9	9.0	9.8	9.9
...	9.8	8.9	9.0	9.7	9.8
...	...	8.9	9.0	...	...
...	0.3000	0.0002	0.0011	0.0002	0.0020
...	0.0200	0.0001	0.0001	0.0003	0.0002
...	...	0.0005	0.0002	...	...
...	6.6	0.002	0.010	0.002	0.020
...	0.200	0.001	0.001	0.003	0.002
...	...	0.004	0.002	...	...
...	$>10^{14}$	$>10^{14}$	$>10^{14}$	$>10^{14}$	$>10^{15}$
...	5.5×10^{10}	9.0×10^{11}	3.1×10^{11}	...	1.0×10^{15}
...	5.5×10^7	2.5×10^9	4.0×10^9	...	3.3×10^{12}
...	1.7×10^6	5.0×10^7	1.0×10^8	...	9.0×10^9
...	4.0×10^4	5.0×10^5	1.0×10^6	...	1.1×10^7
...	...	950 (1742)	1000 (1832)	...	1170 (2138)

(a) All data measurements are typical and made at room temperature unless otherwise noted. Ceramic property values vary somewhat with method of manufacture, size, and shape of part. Ceramic compositions are controlled using modern chemical, spectrographic, and x-ray fluorescent methods. (b) Partially stabilized with MgO. (c) Nominal. (d) No helium leak through a plate 25.4 mm in diameter by 0.25 mm thick measured at a vacuum of 3×10^{-7} torr vs. approximately one atmosphere of helium pressure for 15 s at room temperature. (e) Minimum flexural strength is a minimum mean for a sample of ten specimens. (f) Specimen thickness. (g) Temperature at which resistivity is 1 MΩ · cm

Properties of some ceramic materials

Material	Melting point, °C	Limit of application, °C	Hardness, Mohs scale	Density, g/cm³	Specific heat (mean), J/kg · °C, at 25 to 1000 °C	Linear coefficient of expansion, 10^{-6}/°C, at 25 to 800 °C	Thermal conductivity, W/m · °C, at temperature, °C	Electrical resistivity, Ω · cm, at temperature °C
Alumina (Al_2O_3)	2050	1950	9	3.96	1050	8.0	4 at 1315	10^6 at 1100
Beryllia (BeO)	2550	2400	9	3.0	2180	7.5	29 at 1000	4×10^8 at 600; 8×10^{12} at 2100
Magnesia (MgO)	2850	2400	6	3.60	1170	13.5	59 at 1100	2×10^8 at 850
Thoria (ThO_2)	3220(a)	2700	7	9.5 to 9.9	290	9.5	3 at 1000	2.6×10^7 at 550; 1.5×10^4 at 1200
Zirconia (ZrO_2)	2700	2400	6.5	5.5 to 5.8	590	7.5	3 at 1315	10^6 at 385; 3.6×10^2 at 1200

(continued)

Properties of some ceramic materials (continued)

Material	Melting point, °C	Limit of application, °C	Hardness, Mohs scale	Density, g/cm^3	Specific heat (mean), J/kg · °C, at 25 to 1000 °C	Linear coefficient of expansion, 10^{-6}/°C, at 25 to 800 °C	Thermal conductivity, W/m · °C, at temperature, °C	Electrical resistivity, Ω · cm, at temperature °C
Zircon ($ZrO_2 \cdot SiO_2$)	2500(a)	1870	7.5	4.5 to 4.7	630	4.5	4 at 1200	High
Spinel ($MgO \cdot Al_2O_3$)	2130	1900	8	3.60	1050	8.5	2 at 1315	2.8×10^7 at 500; 2.0×10^5 at 1100
Mullite ($3Al_2O_3 \cdot 2SiO_2$)	1850	1800	...	2.8	840	5.0	4 at 1200	10^5–10^3 at 815-1370
Sillimanite ($Al_2O_3 \cdot SiO_2$)	1800(a)	1800	6.5	3.2	840	5.0	2 at 1300	10^4–10^5 at 815-1370
Silicon carbide (SiC)	2200 to 2700(b)	1400 to 1700(c)	9	3.2	840	4.5	13 at 1100	7420-745 at 1000 – 1500
Silicon nitride (Si_3N_4)	1900(d)	1400 in air; 1850 in inert atmosphere	9	3.18	1050	$\alpha = 2.9$; $\beta = 2.3$	9.5 at 1200	10^{13} at 25; 10^{10} at 480
Carbon graphite (C)(e)	3600(d)	...	0.5 to 1.0	2.2	1600	2.2	147 at 50; 63 at 900	10^{-3}
Quartzite (SiO_2)	1400	1090	7	2.65	1170	8.6	2.06 at 1200	10^{14} at 20; 5×10^3 at 1300
Boron carbide (B_4C)	2350	540 in air; 2260 in inert atmosphere	9.3	2.5	2090	5.7	17.3 at 800	...
Titanium carbide (TiC)	3140	1500(c)	9 to 10	6.5	1050	6.9	40 at 1100	...
Tungsten carbide (WC)	2780	...	9 to 10	14.3	300	6.3	43.3 at 1100	...
Boron nitride (BN)	2721	650(c)	2	2.1	1570	7.5(∥); 0.77(⊥)	26 at 900	1.7×10^{13} at 25(∥); 2.3×10^{10} at 480(∥)

(a) Approximate. (b) Decomposes. (c) Oxidizes. (d) Sublimes. (e) Properties vary with type

Property comparison of various ceramic materials and abrasive-resistant steel

Product	Density g/cm³	Density lb/ft.³	Color	Modulus of rupture MPa	Modulus of rupture ksi	Fracture toughness, MPa	Knoop hardness, kg/mm²	Elastic modulus GPa	Elastic modulus 10^6 psi	Thermal expansion from 25 to 1000 °C (77 to 1830 °F) ppm/°C	Thermal expansion ppm/°F	Maximum thermal shock Δ °C	Maximum thermal shock Δ °F	Dry erosion resistance test(a); volume lost, cm³/h	Slurry erosion resistance(b)	Sliding abrasion(c), mm³/h
Sintered alumina, 85%(d)	3.41	213	White	317	46	3 to 4	960	221	32	7.2	4.0	300	540	0.65	0.85	16,000
Sintered alumina, 96%(d)	3.72	232	White	358	52	3 to 4	1088	303	44	8.2	4.6	250	450	0.50	1.0	14,000
Partially stabilized zirconia(e)	5.74	358	Ivory	820	118	8 to 12	1120	205	30	10.2	5.7	375	675	1.33	9.0	6200
Zirconia-toughened alumina(f)	4.15	259	Ivory	585	85	5.5	1380	335	49	8.0	4.4	445	800	0.70	1.8	10,200
Silicon nitride-bonded silicon carbide(g)	2.54	159	Gray	48	7	1.5	620	152	22	3.9	2.2	400	720	2.48	9.0	3700
Reaction-bonded silicon carbide(g)	3.09	193	Black	280	40	4.93	1880	380	55	5.0	2.8	350	630	0.10	40.0	1600
Sintered silicon carbide(g)	3.10	193	Black	460	67	4.6	2800	410	59	4.0	2.2	325	585	0.10	91.0	1800
Silicon carbide																
ceramic/metal	3.26	203	Black	90	13	5.5	800	313	45	5.4	3.0	600	1080	0.20	45.0	1500
composites(h)	3.28	205	Black	140	20	6.0	800	313	45	5.4	3.0	600	1080	0.30	25.0	2300
AZS fused cast(i)	3.60	225	Ivory	110	16	2.0	1000	...	...	7.5	4.2	300	540	...	...	...
Basalt fused cast(j)	2.90	181	Brick	64	9	1.5	880	104	15	10.1	5.6	100	180	...	...	...
NiHard IV abrasive-resistant steel(k)	7.4	462	Gray	650	94	2.7	650	104	15	7.3	4.1	>600	1130	9.62	4.5	12,200

(a) Grit blast testing with a stream of 150 to 300 μm silica particles at a pressure of 275 kPa (40 psi) directed at a stationary specimen at 30 degrees for five min. (b) Values listed are a ratio of volume loss for a reference pin (96% alumina) divided by volume loss of test specimen. Test pins rotated in an aqueous slurry of 40 wt% silica (300 to 600 μm) at 1750 rpm for 20 h. (c) Miller test. *Mechanical data values from product literature of:* (d) Coors Ceramic Company. (e) Nilcra Ceramics Inc. (f) Diamonite Products. (g) The Carborundum Company. (h) Alanx Products L.P. (i) Cohart Refractories Corporation. (j) Abresist Corporation. (k) Steel Casting Companies

Properties of ceramics

Material	Crystal structure	Theoretical density, g/cm³	Knoop or Vickers hardness GPa	Knoop or Vickers hardness 10^6 psi	Transverse rupture strength MPa	Transverse rupture strength ksi	Fracture toughness MPa$\sqrt{m}$	Fracture toughness ksi$\sqrt{in.}$	Modulus of elasticity GPa	Modulus of elasticity 10^6 psi	Poisson's ratio	Thermal expansion, 10^{-6}/K	Thermal conductivity, W/m · K
Glass-ceramics	Variable	2.4 to 5.9	6 to 7	0.9 to 1.0	70 to 350	10 to 51	2.4	2.2	83 to 138	12 to 20	0.24	5 to 17	2.0 to 5.4(a) 2.7 to 3.0(b)
Pyrex glass	Amorphous	2.52	5	0.7	69	10	0.75	0.7	70	10	0.2	4.6	1.3(a) 1.7(c)
TiO_2	Rutile tetragonal	4.25	7 to 11	1.0 to 1.6	69 to 103	10 to 15	2.5	2.3	283	41	0.28	9.4	8.8(a)
	Anatase tetragonal	3.84	...	...	...	...	...	...	...	...	...	...	3.3(d)
	Brookite orthorhombic	4.17	...	...	...	...	...	...	...	...	...	...	...
Al_2O_3	Hexagonal	3.97	18 to 23	2.6 to 3.3	276 to 1034	40 to 150	2.7 to 4.2	2.5 to 3.8	380	55	0.26	7.2 to 8.6	27.2(a) 5.8(d)
Cr_2O_3	Hexagonal	5.21	29	4.2	>262	>38	3.9	3.5	>103	>15		7.5	10 to 33(e)
Mullite	Orthorhombic	2.8	...	...	185	27	2.2	2.0	145	21	0.25	5.7	5.2(a) 3.3(d)
Partially stabilized ZrO_2	Cubic, monoclinic, tetragonal	5.70 to 5.75	10 to 11	1.5 to 1.6	600 to 700	87 to 102	(f)	(f)	205	30	0.23	8.9 to 10.6	1.8 to 2.2
Fully stabilized ZrO_2	Cubic	5.56 to 6.1	10 to 15	1.5 to 2.2	245	36	2.8	2.5	97 to 207	14 to 30	0.23 to 0.32	13.5	1.7(a) 1.9(g)
Plasma-sprayed ZrO_2	Cubic, monoclinic, tetragonal	5.6 to 5.7	...	...	6 to 80	0.9 to 12	1.3 to 3.2	1.2 to 2.9	48(h)	7	0.25	7.6 to 10.5	0.69 to 2.4
CeO_2	Cubic	7.28	...	...	...	...	...	...	172	25	0.27 to 0.31	13	9.6(a) 1.2(d)

(continued)

Properties of ceramics (continued)

Material	Crystal structure	Theoretical density, g/cm³	Knoop or Vickers hardness GPa	Knoop or Vickers hardness 10^6 psi	Transverse rupture strength MPa	Transverse rupture strength ksi	Fracture toughness MPa$\sqrt{m}$	Fracture toughness ksi$\sqrt{in.}$	Modulus of elasticity GPa	Modulus of elasticity 10^6 psi	Poisson's ratio	Thermal expansion, 10^{-6}/K	Thermal conductivity, W/m · K
TiB_2	Hexagonal	4.5 to 4.54	15 to 45	1.5 to 6.5	700 to 1000	102 to 145	6 to 8	5.5 to 7.3	514 to 574	75 to 83	0.09 to 0.13	8.1	65 to 120(i) 33 to 80(j) 54 to 122(k)
TiC	Cubic	4.92	28 to 35	4.0 to 5.1	241 to 276	35 to 40	...	...	430	62	0.19	7.4 to 8.6	33(a) 43(d)
TaC	Cubic	14.4 to 14.5	16 to 24	2.3 to 3.5	97 to 290	14 to 42	...	...	285	41	0.24	6.7	32(a) 40(d)
Cr_3C_2	Orthorhombic	6.70	10 to 18	1.5 to 2.6	49	7.1	...	...	373	54		9.8	19
Cemented carbides	Variable	5.8 to 15.2	8 to 20	1.2 to 2.9	758 to 3275	110 to 475	5 to 18	4.6 to 16.4	396 to 654	57 to 95	0.2 to 0.29	4.0 to 8.3	16.3 to 119
SiC	α, hexagonal	3.21	20 to 30	2.9 to 4.4	(l)	(l)	(m)	(m)	207 to 483	30 to 70	0.19	4.3 to 5.6	63 to 155(a) 21 to 33(d)
	β, cubic	3.21	...	...	...	...	...	...	...	...	...	...	...
SiC (CVD)	β, cubic	3.21	28 to 44	4.1 to 6.4	(n)	(n)	5 to 7	4.6 to 6.4	415 to 441	60 to 64		5.5	121(a) 34.6(g)
Si_3N_4	α, hexagonal	3.18	8 to 19	1.2 to 2.8	(o)	(o)	(p)	(p)	304	44	0.24	3.0	9 to 30(a)
	β, hexagonal	3.19	...	...	...	...	...	...	...	...	...	...	...
TiN	Cubic	5.43 to 5.44	16 to 20	2.3 to 2.9	...	...	...	...	251	36	...	8.0	24(a) 67.8(q) 56.9(r)

(a) At 400 K. (b) At 1200 K. (c) At 800 K. (d) At 1400 K. (e) At 350 K. (f) 8 to 9 (7.3 to 8.2) at 293 K, 6 to 6.5 (5.5 to 5.9) at 723 K, and 5 (4.6) at 1073 K, in units of MPa$\sqrt{m}$ (ksi$\sqrt{in.}$). (g) At 1600 K. (h) 21 GPa (3×10^6 psi) at 1373 K. (i) At 300 K. (j) At 1100 K. (k) At 2300 K. (l) Sintered: 96 to 520 (14 to 75) at 300 K, and 250 (36) at 1273 K. Hot pressed: 230 to 825 (33 to 120) at 300 K, and 398 to 743 (58 to 108) at 1273 K, MPa (ksi). (m) Sintered: 4.8 (4.4) at 300 K, and 2.6 to 5.0 (2.4 to 4.6) at 1273 K. Hot pressed: 4.8 to 6.1 (4.4 to 5.6) at 300 K, and 4.1 to 5.0 (3.7 to 4.6) at 1273 K, MPa $\sqrt{m}$ (ksi$\sqrt{in.}$). (n) 1034 to 1380 (150 to 200) at 300 K, and 2060 to 2400 (300 to 350) at 1473 K, MPa (ksi). (o) Sintered: 414 to 650 (60 to 94). Hot pressed: 700 to 1000 (100 to 145). Reaction bonded: 250 to 345 (36 to 50), MPa (ksi). (p) Sintered: 5.3 (4.8). Hot pressed: 4.1 to 6.0 (3.7 to 5.5). Reaction bonded: 3.6 (3.3), MPa $\sqrt{m}$ (ksi$\sqrt{in.}$). (q) At 1773 K. (r) At 2573 K

Typical properties of hot pressed ceramics

Ceramic	Density, g/cm³	Modulus of rupture(a), MPa	Modulus of rupture(a), ksi	Modulus of elasticity, GPa	Modulus of elasticity, 10^6 psi	Fracture toughness, MPa$\sqrt{m}$	Fracture toughness, ksi$\sqrt{in.}$	Poisson's ratio	Knoop hardness, GPa	Knoop hardness, 10^6 psi	Coefficient of thermal expansion, 10^{-6}/K	Conductivity, W/m · K	Conductivity, (Btu · in./h · ft² · °F)	Thermal shock parameter, ΔT_c
Si_3N_4	3.30	850	125	310	45	6.1	5.6	0.27	14.7	2.13	3.4	29.5	205	700
Al_2O_3-SiC_w	3.64	640	95	390	56	7.7	7.0	0.23	17.6	2.55	5.9	35	245	900
Al_2O_3-TiC_p	4.26	760	110	395	57	4.5	4.1	0.22	18.1	2.63	8.0	40	275	300
SiC	3.18	610	90	430	62	5.2	4.7	0.17	24.5	3.55	4.5	110	760	440
TiB_2	4.43	345	50	540	78	5.0	4.6	0.11	26.5	3.84	8.2	75	520	200
B_4C	2.48	350	51	470	68	3.5	3.2	0.17	28.4	4.12	5.6	16	110	200
Al_2O_3	3.98	550	80	390	56	4.5	4.1	0.23	15.2	2.20	8.1	29	200	150
AlN	3.25	300	45	318	46	4.5	4.1	0.25	13.5	1.96	5.6	180	1245	550

(a) Measured in strongest direction

Properties of technical ceramics and tool steel

Material	Vickers hardness, GPa	Vickers hardness, ksi	Fracture toughness, MPa$\sqrt{m}$	Fracture toughness, ksi$\sqrt{in.}$	Thermal conductivity, W/m · K	Thermal conductivity, Btu/ft · h · °F	Thermal expansion, 10^{-6}/°C	Flexural strength, MPa	Flexural strength, ksi
Al_2O_3	14	2030	4	3.6	16-39	9.3-22.5	8	50	7.2
Zr_2O_3	12	1740	11	10.0	2	1.2	10.5	90	13.0
SiC	25	3626	3	2.7	110	63.6	4.3	65	9.4
WC	18	2610	20	18.2	100	57.8	5.1	300	43.5
Tool steel	10	1450	80	72.8	25	14.5	6	550	79.7

Comparison of properties important for structural use of advanced ceramics

Material(a)	Strength, MPa	Toughness, MPa$\sqrt{m}$	Thermal expansion, 10^{-6}/°C
Silicon nitrides			
RBSN	300	3.6	3.3
HPSN	1100	6.6	3.5
GPSSN	440	2.9	3.5
Silicon carbides			
α SSC	420	2 to 3	4.1
β SSC	533	2.4	4.1
HPSC	800	3.9	4.2
Transformation-toughened ceramics			
PSZ	700+	8+	10.2
TTA	900	8	7
Ceramic-ceramic composite			
SiC-LAS	620	15	1 to 4

(a) RBSN, reaction-bonded silicon nitride; HPSN, hot pressed silicon nitride; GPSSN, gas-pressure-sintered silicon nitride; α SSC, alpha-phase-sintered silicon carbide; β SSC, beta-phase-sintered silicon carbide; HPSC, hot pressed silicon carbide; PSZ, partially stabilized zirconia; TTA, transformation-toughened alumina; SiC-LAS, silicon carbide fibers in lithium aluminosilicate glass

Comparison of properties for tungsten carbide and hot pressed silicon nitride ceramics

Material	Density, g/cm³	Density, lb/in.³	Hardness, HK	Strength RT, MPa	Strength RT, ksi	Strength 800 °C (1470 °F), MPa	Strength 800 °C (1470 °F), ksi	Strength 1000 °C (2010 °F), MPa	Strength 1000 °C (2010 °F), ksi	Relative thermal shock resistance(a), RT	Relative thermal shock resistance(a), 800 °C (1470 °F)	Relative thermal shock resistance(a), 1000 °C (2010 °F)	Thermal conductivity at 20 °C (70 °F), W/m · K
WC	12.5 to 15.1	0.452 to 0.546	1700	1550	225	1000	145	414	60.03	0.33	0.23	0.07	121
HPSN	3.2 to 3.25	0.116 to 0.117	2000	793	115	793	115	758	110	1.0	0.65	0.65	32

(a) Obtained by normalizing the values with reference to the property of HPSN at room temperature

Physical properties of ceramic substrate materials

	96% Al_2O_3	99.5% Al_2O_3	BeO	AlN	Mullite	Various glass-ceramic materials
Density, g/cm^3	3.75	3.90	2.85	3.25	2.82	2.5 to 2.8
Flexural strength, MPa (ksi)	400 (58)	522 (80)	207 (30)	345 (50)	186 (27)	138 (20)
Thermal expansion from 25 to 500 °C (77 to 930 °F), 10^{-6}/°C	7.4	7.5	7.5	4.4	3.7	3.0 to 4.5
Thermal conductivity at 20 °C (68 °F), W/m · °C	26	35	260	140 to 220	4	4 to 5
Dielectric constant at 1 MHz	9.5	9.9	6.7	8.8	5.4	4 to 8
Dielectric loss at 1 MHz (tan δ)	0.0004	0.0002	0.0003	0.001 to 0.0002	0.003	>0.002

Property comparison of technical ceramics

Property	Silicon nitride	Silicon carbide	Zirconia
Bulk density, g/cm^3	3.2	3.0	6.0
Flexural strength, MPa (ksi)			
at room temperature	676 (98)	490(71)	980 (142)
at 1000 °C (1830 °F)	598 (87)	490 (71)	274 (40)
at 1200 °C (2190 °F)	392 (57)	490 (71)	167 (24)
Fracture toughness, MPa$\sqrt{m}$ (ksi$\sqrt{in.}$)	5.9 (5.4)	3.4 (3.1)	6.7 (6.1)
Thermal shock resistance, water quench, ΔT, °C (°F)	670 (1240)	350 (660)	300 (570)
Thermal conductivity at room temperature, W/m · K (Btu · in./h · ft^2 · °F)	25 (175)	71 (490)	4 (30)

Electrooptic properties of ceramic and single-crystal materials

Material	κ(a)	*n* (b)	r_c (c), 10^{10} m/V	*R* (d), 10^{16} m^2/V^2
Ceramic				
PLZT 8.5/65/35	5000	2.50	...	38.6
PLZT 9/65/35	5700	2.50	...	3.8
PLZT 9.5/65/35	5500	2.50	...	1.5
PLZT 8/70/30	5400	2.48	...	11.7
PLZT 8/40/60	980	2.57	1.02	...
PLZT 12/40/60	1300	2.57	1.20	...
PLZT 14/30/70	1025	2.59	1.12	...
Single-crystal				
$LiNbO_3$ (r_{33})	37	2.20	0.32	...
$LiNbO_3$ (r_{13})	37	2.29	0.10	...
$BaTiO_3$ (r_{33})	373	2.36	0.28	...
$BaTiO_3$ (r_{51})	372	2.38	8.20	...
$KNbO_3$ (r_{33})	30	2.17	0.64	...
$KNbO_3$ (r_{42})	137	2.25	3.80	...
SBN (T = 560 K)(e)	119	2.22	0.56	...
SBN (T = 300 K)(e)	3400	2.30	13.40	...
$Ba_2NaNb_5O_{15}$	86	2.22	0.56	...

(a) κ is the dielectric constant. (b) *n* is the index of refraction at 633 nm. (c) r_c is the linear electrooptic coefficient. (d) *R* is the quadratic electrooptic coefficient. (e) SBN, strontium barium niobate

Comparison of basic properties of selected compounds used in advanced ceramics

Compound	Density, g/cm^3	Hardness, kg/mm^2	Melting point, °C	Thermal conductivity, cal/cm · s · °C
Aluminum oxide	3.98	2100	2050	0.07
Zirconium oxide(a)	6.27	1200	2715	0.005
Silicon carbide	3.22	2500	2220(b)	0.16
Silicon nitride	3.17	2400	1900	0.04
Silica glass	2.20	...	...	0.002

(a) Stabilized in cubic form. (b) Decomposes

Thermal conductivity of selected ceramics

Material	Thermal conductivity, cal/cm · s · °C, at: 100 °C	1000 °C
Al_2O_3	0.072	0.015
BeO	0.525	0.049
MgO	0.090	0.017
$MgAl_2O_4$	0.036	0.014
ThO_2	0.025	0.007
Mullite	0.014	0.009
$UO_{2.00}$	0.024	0.008
Graphite	0.43	0.15
ZrO_2 (stabilized)	0.0047	0.0055
Fused silica glass	0.0048	0.006
Soda-lime-silica glass	0.004	
TiC	0.060	0.014
Porcelain	0.004	0.0045
Fire-clay refractory	0.0027	0.0037
TiC cermet	0.08	0.02

Physical properties of ceramic materials

Material	Bending strength, MPa, at: 800 K	1400 K	Density, g/cm^3	Modulus of elasticity, GPa at 1260 K	Thermal expansion, 10^{-6}/K at 300 to 1260 K	Thermal conductivity, W/m · K at 1260 K
S-Si_3N_4	530	300	3.1	300	3.2	12
RB-Si_3N_4	300	300	2.6	180	3.0	9
S-SiC	450	450	3.15	400	4.5	40
MAS	70	20	2.2	12	0.6	1
ZrO_2	600	300	5.7	200	9.8	2.5
Al_2O_3-TiO_2	40	20	3.2	23	3.0	2
Inco 713 C	900	200	7.9	170	15.0	25

Moduli of elasticity for selected ceramics

Material	Modulus of elasticity GPa	10^6 psi
Aluminum oxide crystals	380	55
Sintered alumina(a)	365	53
Alumina porcelain(b)	365	53
Sintered beryllia(a)	310	45
Hot pressed boron nitride(a)	83	12
Hot pressed boron carbide(a)	290	42
Graphite(c)	9.0	1.3
Sintered magnesia(a)	210	30.5
Sintered molybdenum silicide(a)	405	59
Sintered spinel(a)	238	34.5
Dense silicon carbide(a)	470	68
Sintered titanium carbide(a)	310	45
Sintered stabilized zirconia(a)	150	22
Silica glass	72.4	10.5
Vycor glass	72.4	10.5
Pyrex glass	69	10
Mullite porcelain	69	10
Steatite porcelain	69	10
Superduty fire-clay brick	97	14
Magnesite brick	170	25
Bonded silicon carbide(c)	345	50

(a) Approximately 5% porosity. (b) 90 to 95% Al_2O_3. (c) Approximately 20% porosity

Mechanical properties of selected ceramics and glasses

Material	Knoop hardness, GPa	Transverse rupture strength MPa	ksi	Modulus of elasticity GPa	10^6 psi	Fracture toughness $MPa\sqrt{m}$	$ksi\sqrt{in.}$	Modulus of resilience 10^3 (J/mm^3)	$lbf \cdot in./in.^3$
Al_2O_3	10 to 20	276 to 700	40 to 102	380	55	2.7 to 4.2	2.5 to 3.8	0.1 to 0.6	15 to 87
Reaction-bonded Si_3N_4	8 to 19	250 to 345	36 to 50	304	44	3.6	3.3	0.1 to 0.2	15 to 29
Hot pressed Si_3N_4	8 to 19	700 to 1000	100 to 150	304	44	4.1 to 6.0	3.7 to 5.5	0.8 to 1.6	116 to 233
Sintered Si_3N_4	8 to 19	414 to 650	60 to 94	304	44	5.3	4.8	0.03 to 0.68	4.4 to 99
Partially stabilized ZrO_2 (PSZ)	10 to 11	60 to 700	87 to 102	205	30	8 to 9	7.3 to 8.2	0.87 to 1.19	127 to 173
Fully stabilized ZrO_2	10 to 15	245	35.5	97 to 207	14 to 30	2 to 8	1.8 to 7.3	0.2	29
Hot pressed SiC	20 to 30	230 to 825	33 to 120	207 to 483	30 to 70	4.8 to 6.1	4.4 to 5.6	0.08 to 0.98	12 to 143
Sintered SiC	20 to 30	96 to 520	14 to 75	207 to 483	30 to 70	4.8	4.4	0.013 to 0.39	1.9 to 57
TiB_2	15 to 36	700 to 1000	102 to 145	514 to 574	74.5 to 83.2	6 to 8	5.5 to 7.3	0.45 to 0.92	65 to 134
Pyrex glass	5	60	8.7	70	10	0.75	0.68	0.026	3.8
Glass ceramics	6 to 7	70 to 350	10 to 51	83 to 138	12 to 20	2.4	2.2	0.22 to 0.55	32 to 80
Quartz	8 to 10	70	10	74	11	...	...	0.033	4.8
Cemented carbide (16% Co)	18	1860(a)	270(a)	524	76	...	...	3.3	480

(a) Ultimate tensile strength

Mechanical properties of SiC-whisker-reinforced ceramics

Composite	Whisker control, vol %	Fracture toughness, $MPa\sqrt{m}$	$ksi\sqrt{in.}$	Flexural strength, MPa	ksi
Corning 1723 glass	0	<1	<0.9	...	...
	25	2.1 to 3.4	1.9 to 3.1	200 to 340	30 to 50
Barium osumilite	25	4.5	4.1	360 to 400	50 to 60
Si_3N_4	0	5 to 7	4.6 to 6.4	400 to 650	60 to 95
	10	6.5 to 9.5	5.9 to 8.6	400 to 500	60 to 75
	30	7.5 to 10	6.8 to 9.1	350 to 450	50 to 65
Spinel	30	...	...	415	60
Mullite	0	1.8 to 2.2	1.6 to 2.0	250	40
	20	4.6	4.2	440	65
ZrO_2	0	6.2	5.6	1080	160
Toughened alumina (TTA)	20	8.5 to 13.5	7.7 to 12.3	700 to 880	100 to 130
Cordierite	0	2.2	2.0	180	25
	20	3.7	3.4	260	40
$MoSi_2$	0	5.3	4.8	150	20
	20	8.2	7.5	310	45

Average strengths of ceramics(a)

Material	Compressive strength MPa	ksi	Tensile strength MPa	ksi	Flexural strength MPa	ksi
Alumina: 85%	1620	235	125	18	293	42.5
90%	2410	350	140	20	315	46
95%	2410	350	195	28	340	49
99%	2590	375	205	30	345	50
Alumina silicate	275	40	17	2.5	62	9
$ZrO_2 \cdot Al_2O_3$	2410	350	...	...	...	...
3% ½O_3 PSZ(b)	2960	430	...	...	1170	170
TTZ(c)	1760	255	350	51	635	92
9% MgO PSZ(b)	1860	270	...	...	690	100
Slip cast Si_3N_4	140	20	24	3.5	69	10
Reaction-bonded SiC	690	100	140	20	255	37
Pressureless-sintered SiC	3860	560	170	25	550	80
Sintered SiC with free silicon	1030	150	165	24	325	47
Sintered SiC with graphite	415	60	35	5	55	8
Reaction-bonded Si_3N_4	770	112	...	...	205	30
Hot pressed Si_3N_4	3450	500	...	...	860	125

(a) Strength is dependent on test method, sample preparation, and sample size. Data are from a variety of sources. (b) Partially stabilized zirconia. (c) Transformation-toughened zirconia

Fracture toughness values achieved to date utilizing various toughening mechanisms

Toughening mechanism	Highest toughness values achieved MPa√m	ksi√in.	Exemplary material-composite systems
Microcracking	~10	~9.1	Al_2O_3-ZrO_2
Transformation	~20	~18.2	ZrO_2 (MgO)
Particle	~8	~7.3	Si_3N_4-SiC
Platelet	~8	~7.3	HIPRBSN-SiC
	~14	~12.7	Si_3N_4-SiC
Whisker	~8.5	~7.7	Al_2O_3-SiC
	~11	~10.0	Si_3N_4-SiC
Fibers	>30	>27.3	Glass-SiC
	>25	>22.8	Glass-ceramic/SiC
	>30	>27.3	SiC-SiC
	~16	~14.6	Si_3N_4-SiC
Metal dispersion	~25	~22.8	Al_2O_3-Al, Al_2O_3-Ni

Fracture toughness of ceramics

Material	Condition	Fracture toughness(a), MPa√m
Soda-lime glass(b)	Amorphous	0.74 DCB
Aluminosilicate glass	Amorphous	0.91 DCB
ZnSe	Vapor deposited	0.9
WC	Co-bonded	13.0
ZnS	Vapor deposited	1.0
Si_3N_4	Hot pressed	5.0
Al_2O_3	MgO-doped	4.0
Al_2O_3 (sapphire)	Monocrystal	2.1
SiC	Hot pressed	4.0
SiC-ZrO_2	Hot pressed(c)	5.0
MgF_2	Hot pressed	0.9
MgO	Hot pressed	1.2
B_4C	Hot pressed	6.0
Si	Monocrystal	0.6
ZrO_2	Ca-stabilized	7.6 DCB

(a) Double-torsion measurement technique, except where double-cantilever-beam test (DCB) indicated. (b) Commercial sheet glass. (c) 20% ZrO_2, 14 wt % mullite. ZrO_2 present in monoclinic form; no transformation toughening

Moduli of rupture for selected ceramics

Material	Modulus of rupture MPa	ksi
Aluminum oxide crystals	345 to 1034	50 to 150
Sintered alumina(a)	207 to 345	30 to 50
Alumina porcelain(b)	345	50
Sintered beryllia(a)	138 to 276	20 to 40
Hot pressed boron nitride(a)	48 to 103	7 to 15
Hot pressed boron carbide(a)	345	50
Sintered magnesia(a)	103	15
Sintered molybdenum silicide(a)	690	100
Sintered spinel(a)	90	13
Dense silicon carbide(a)	172	25
Sintered titanium carbide(a)	1100	160
Sintered stabilized zirconia(a)	83	12
Silica glass	107	15.5
Vycor glass	69	10
Pyrex glass	69	10
Mullite porcelain	69	10
Steatite porcelain	138	20
Superduty fire-clay brick	5.2	0.75
Magnesite brick	27.6	4.0
Bonded silicon carbide(c)	13.8	2.0
1090 °C insulating firebrick(d)	0.28	0.04
1430 °C insulating firebrick(e)	1.17	0.17
1650 °C insulating firebrick(f)	2.0	0.3

(a) Approximately 5% porosity. (b) 90 to 95% Al_2O_3. (c) Approximately 20% porosity. (d) 80 to 85% porosity. (e) Approximately 75% porosity. (f) Approximately 60% porosity

High-Temperature Applications

Advantages and disadvantages of ceramics as engine components

+ Abundant raw materials	– More efficient fuel burn – Reduced emissions
+ Higher operating temperatures	– Brittleness – High-temperature lubrication
+ Lower density/lighter weight	– Lower inertia for moving parts – Lower overall engine weight

Summary of ceramics applications for adiabatic engines

	Desire characteristics							
Adiabatic components	Low friction	Light weight	Insulation	Wear resistance	Heat resistance	Corrosion resistance	Expansion coefficient	High-technology ceramics(a)
Piston		X	X		X	X	X	Si_3N_4, PSZ, TTA
Piston ring				X	X			SSN, PSZ, coating
Cylinder liner	X			X	X	X	X	Si_3N_4, PSZ, coating
Prechamber			X		X	X		PSZ, Si_3N_4, SSN, PSZ,
Valve		X	X	X	X	X		composite
Valve seat insert			X	X	X	X		PSZ, SSN
Valve guides	X		X		X	X		PSZ, SSN, SiC
Exhaust/intake ports			X		X	X		ZrO_2, Si_3N_4, $TiO_2Al_2O_3$
Manifolds			X		X	X		ZrO_2, Si_3N_4, $TiO_2Al_2O_3$
Tappets		X		X		X		PSZ, SiC, Si_3N_4
Mechanical seals	X			X		X		SiC, Si_3N_4, PSZ
Turbocharger								
Turbine rotor		X	X		X	X	X	Si_3N_4, SiC
Turbine housing			X		X	X	X	LAS
Heat shield			X		X	X	X	ZrO_2, LAS
Ceramic bearings	X	X		X	X	X	X	SSN

(a) SSN, sintered silicon nitride; PSZ, partially stabilized zirconia; LAS, lithium alumina silicate; TTA, transformation toughened alumina

Applications of ceramics in internal-combustion engines

Component	Material properties needed(a)	Material types being considered(b)	Reasons for substituting for metals
Turbines			
Flame can	1, 2, 3	A, B	Metal parts limited to 1100 °C surface temperature; ceramic parts may not
Volute	1, 2,	A, C	need parasitic losses of air cooling to cope with higher temperatures
Stator, nozzle rings	1, 2, 3	C, D, E, F	Lower mass; no cooling required
Rotor	1, 2, 3	D, E, F	Bearings running hot are needed to cope with higher temperatures
Bearings	1, 4	D, E, F	
Heat exchanger	2	H	Required to improve efficiency of land-vehicle engines
Diesels			
Cylinder liners	2, 4, 5	A, G, G′	Reduce heat losses through cylinder wall (adiabatic diesel); may no longer
Cylinder block, head	2, 4, 5	A, C	need cooling system, saving weight and space; may reduce emissions
Valve heads	1, 2	G′	Reduce oxidation at higher running temperatures
Valve seats	1, 2, 3, 4	A, B, D, E, F	More corrosion resistant in hot exhaust stream
Pistons or piston caps, swirl chambers	2, 3, 5	C, D, G, I, G′	Reduce heat losses through piston
Precombustion chambers	1, 2, 3	C, E	Cheaper than nickel alloy types
Exhaust ports	1, 2, 5	C, G′	Maintain high temperature to maximize turbocharger efficiency
Turbocharger rotors	1, 2, 3	D, E, F	Cheaper and lower inertia than nickel alloy types
Wear parts:			
Cam followers	3, 4,	E, F, G	Reduce inertia and friction giving less wear, λεσσ φρεθυεντ αδφυστμεντ,
Valve guides	3, 4	E, F, G	ανδ σμαλλερ παρασιτιχ λοσσεσ
Tappet faces	3, 4	E, F, G, G′	
Gudgeon pins	3, 4	D, E, F	Lower thermal expansion to match that of ceramic piston
Gasoline engines			
For gasoline engines which have temperature limitations on fuel injection, consideration is mostly being given to wear parts such as those listed above for diesels			

(a) 1, High temperature capability (>1000 °C); 2, thermal shock resistance; 3, high strength; 4, hard or wear-resistant; 5, low thermal conductivity. (b) A, reaction-bonded silicon carbide; B, sintered silicon carbide; C, reaction-bonded silicon nitride; D, hot pressed silicon nitride; E, sintered silicon nitride; F, sintered sialons; G, zirconia ceramics; H, aluminous keatite; I, aluminum titanate; ′, coating on a metal

Selected ceramics used for heat exchanger applications

Material	Processing method	Manufacturer	Brand	Density, g/cm^3	Strength MPa	Strength ksi
SiC-Si	Extruded, slip cast	Norton	NC-430/CS101-K	3.07	160(a)	23(a)
SiC (100%α)	Extruded, slip cast	Carborundum	Hexaloy SA	3.10	320(b)	46(b)
					246(c)	35.7(c)
SiC	Extruded, slip cast	Norton	CS-101	2.7	91(a)	13(a)
					100 to 200(b)	14.5 to 29(b)
Si_3N_4-bonded SiC	Extruded	Norton	CX-589	2.51	84(a)	12.2(a)
SiC	Extruded, slip cast	Coors	SC-2	3.10	525(d)	76(d)
					175(c)	25(c)
SiC	Slip cast	Coors	RBSC-205	3.05	287(d)	41.5(d)
SiC	Chemical vapor infiltration	Amercom	...	2.7 to 3.2	300	43.5
Al_2O_3/ZrO_2	Sol-gel winding	B&W	...	3.7	350(c)	50(c)
SiC whisker-reinforced alumina	DIMOX(e)	Lanxide	...	...	460(c)	66.5(c)

(a) O-ring strength test. (b) Four-point bending flexural strength. (c) C-ring strength test. (d) Three-point bending flexural strength. (e) Proprietary process

Property requirements for ceramic engine materials

Thermal-property requirements	Low thermal conductivity, low specific heat, high thermal shock resistance
Mechanical-property requirements	High strength, high fracture toughness, good wear resistance, low coefficient of friction
Chemical-property requirement	Chemical inertness for high resistance to corrosion and erosion

Typical bulk properties of heat-engine ceramics

Material	Modulus of rupture, MPa, at: Room temperature	1000 °C	1375 °C	Modulus of elasticity(a), GPa	Linear coefficient of expansion, $10^{-6}/°C$	Thermal conductivity, W/m · °C
Aluminosilicates						
Lithium aluminosilicate (LAS)	83	69	...	...	450 ppm(b)	...
Magnesium aluminosilicate (MAS)	140	140	...	160	2.2	3.5
Aluminosilicate (AS)(c)	...	...	...	...	...	...
Silicon nitrides						
Hot-pressed (MgO additive)	690	620	330	315	3.0	30-15
Sintered (Y_2O_3 additive)	655	585	275	275	3.2	28-12
Reaction-bonded (2.45 g/cm^3)	205	345	380	165	2.8	6-3
Silicon carbides						
Hot-pressed (Al_2O_3 additive)	655	585	515	450	4.5	85-35
Sintered (α phase)	310	310	310	405	4.8	100-50
Reaction-sintered (20 vol % free Si)	380	415	275	345	4.4	100-50
CVD (lower values)	415	550	550	415	...	...

(a) At room temperature. (b) Total thermal excursion in ppm from 0 to 1000 °C. (c) Properties available only in matrix form

Properties of various materials used in adiabatic engines

Material	Silicon nitride	Silicon carbide	Zirconia	TiC	Alumina	Plasma-sprayed alumina	Plasma-sprayed Cr_2O_3	Cr_2O_3 densified coating	Hard chrome plate	Cast iron (flake graphite)
Bulk density, g/cm^3	3.2	3.1 to 3.2	6.0	6.0	3.8	3.5	5.0	3.3	7.0	7
Hardness, HV	1500	2000 to 2800	1128 to 1250	1630	1170	850	1260	1800	900	23
Linear thermal expansion coefficient, $10^{-6}/°C$ ($10^{-6}/°F$)	3.2 (1.8)	4.0 to 4.4 (2.2 to 2.4)	10 to 11 (5.5 to 6.0)	7.4 (4.1)	8.0 (4.4)	7.4 (4.1)	7.4 (4.1)	13.3 (7.4)	9.4 (5.2)	11 (6.1)
Thermal conductivity, W/m · K	20.9	62.8 to 75.4	2.9 to 3.8	16.7	20.9	2.7	2.6	3.7	19.2	33
Specific heat, kJ/kg · K	0.67	0.63	0.46 to 0.50	0.71	0.92	...	...	0.59	0.46	0.5
Melting point °C (°F)	1900 (3450)	2350 to 2700 (4260 to 4890)	1600 to 2677	...	2400 (4350)	2035 (3695)	2400 (4350)	1610 (2930)	1850 (3360)	...
Thermal shock resistance, Δ °C (Δ °F)	600 (1080)	380 to 400 (685 to 720)	300 to 350 (540 to 630)	350 (630)	150 (270)	...	...	800 (1440)	...	...

Properties of ceramic materials for heat engines

Material	Density, g/cm³	Properties
Zirconia	5.5	+ Low coefficient of friction, good wear and corrosion resistance, low thermal conductivity, relatively high thermal expansion, good thermal shock resistance, high fracture toughness – Some uncertainty about high-temperature performance
Sintered silicon nitride	3.3	+ Good thermal shock resistance, good mechanical properties, significantly higher strength and lower thermal conductivity than SiC at high temperatures – Poor thermal insulator
Sintered silicon carbide	3.1	+ Low coefficient of friction, good wear resistance, good corrosion resistance, good thermal shock resistance, retains strength at high temperatures
Reaction-bonded silicon nitride	2.8	+ <1% shrinkage during firing, retains accurate dimensions – Lower density than sintered Si_3N_4
Lithium aluminum silicate	2.3	+ Very low thermal expansion, very low thermal conductivity, good thermal shock resistance, good thermal insulator –Very poor strength and fatigue life

Properties of selected high-temperature ceramics

Property	Alumina (Al_2O_3) 96%	Alumina (Al_2O_3) 99%	Alumina (Al_2O_3) 99.5%	Silicon nitride (Si_3N_4) Reaction bonded	Silicon nitride (Si_3N_4) Hot pressed	Titanium diboride (TiB_2)	Boron nitride (BN), 96%
General							
Color	White	White	Ivory	Gray	Black	Gray	White
Particle size, μm	9	9	13	...	...	...	...
Maximum continuous temperature (no load):							
°F	3000	3092	3092	2372	2372	2192	5027
°C	1650	1700	1700	1300	1300	1200	2775
Physical							
Thermal conductivity:							
Btu · in./ft² · h · °F	220	205	225	95	173.6	175.79	13.3
W/m · K	31	29	32	13.7	25	25.35	1.9
Coefficient of thermal expansion:							
10^{-6}/°F	3.5	4.1	3.5	1.4 to 1.8	2	4.5	(a)
10^{-6}/°C	6.4	7.4	6.4	2.6 to 3.2	3.6	8.1	(b)
Specific heat:							
Btu/lb · °F	0.19	0.20	0.20	0.17	...	0.23	...
J/kg · K	795	837	837	712	...	963	...
Specific gravity	3.76	3.85	3.94	0.75 to 2.7	3.19	4.52	2.08
Water absorption (porosity), %	0.00	0.00	0.00	(76 to 15)	(0.00)	...	1.1
Mechanical							
Compressive strength:							
ksi	340	375	380	77 to 112	500	192	(c)
MPa	2344	2585	2620	531 to 772	3448	1324	(d)
Tensile strength:							
ksi	25	30	30	...	...	...	...
MPa	172	207	207	...	...	...	...
Flexural strength:							
ksi	46	50	50	30	125	35	(e)
MPa	317	345	345	206.9	861.9	241	(f)
Modulus of elasticity:							
10^6 psi	45	50	55	24	45	62	(g)
GPa	310	345	379	165	310	427.5	(h)
Impact resistance:							
in. · lbf Charpy	7	7	6	...	...	...	...
N · m	0.79	0.79	0.68	...	...	...	...
Hardness:							
Mohs scale	9	9	9	9	...	...	...
Vickers	...	...	...	...	1800	...	...
Electrical							
T_c value:							
°F	1832	2012	1472	...	...	...	...
°C	1000	1100	800	...	...	...	...
Dielectric strength:							
V/mil	225	230	230	...	...	...	950
kV/mm	8.9	9.0	9.0	...	...	...	374
Electrical resistivity, Ω · cm	$>10^{15}$	$>10^{14}$	$>10^{14}$	6.6×10^{10}	10^{12}	9 to 15×10^{16}	$>2 \times 10^{14}$
Dissipation factor, MHz	0.00019	0.0001	0.0001	...	...	...	0.00034
Loss factor, MHz	0.0018	0.0009	0.0009	...	...	...	0.00139
Dielectric constant, MHz	9.3	9.3	9.5	...	...	...	4.08

(a) WG, 2.3; AG, 1.1. (b) WG, 4.1; AG, 2. (c) WG, 45; AG, 34. (d) WG, 310; AG, 234. (e) WG, 11.7; AG, 13.9. (f) WG, 80.7; AG, 95.8 (g) WG, 9.4; AG, 6. (h) WG, 64.8; AG, 41.4; WG = with grain; AG = against grain

Friction, Wear, and Abrasion

Friction behavior of glasses and glass-ceramics

Materials	Static coefficient of friction
Glass on brass	0.18
Glass on chromium	0.16
Glass on tin plate	0.29
Brass on glass	0.24
Steel on glass	0.18
Glass-ceramic on glass-ceramic:	
Lithium-zinc silicate (low ZnO)	0.19
Lithium-zinc silicate (high ZnO)	0.09
Lithium aluminosilicate	0.16

Scuffing temperatures (T_s) and coefficients of friction (μ_f) between ring and cylinder liner materials

	Piston ring								
Cylinder liner	Plasma-sprayed alumina	Silicon carbide	Zirconia	TiC cermet	Cast iron	Silicon nitride	Plasma-sprayed Cr_2O_3	Hard chrome plate	Densified Cr_2O_3
Silicon nitride									
μ_f	0.014	0.022	0.005	0.008	0.019	0.010	0.007	0.004	0.009
T_s, °C	166	160	177	200	240	242	260	268	286
Zirconia									
μ_f	...	0.026	0.012	0.012	...	0.012	...	0.004	0.006
T_s, °C	...	127	177	255	...	223	...	250	250
Cast iron									
μ_f	...	0.034	0.013	0.039	0.038	0.023	0.065	0.013	0.013
T_s, °C	...	195	160	165	137	138	179	155	222
Cr_2O_3 densified									
μ_f	...	0.037	...	...	...	0.007	...	0.007	0.007
T_s, °C	...	199	...	...	...	250	...	238	288

Factors affecting wear of ceramics

Wear factor	Variables affecting wear factors	
Nature of the abrasive	Size	Wear increases with increasing particle size
	Shape	Angular particles cause about twice the wear of rounded ones
	Density	Dense materials create greater wear since kinetic energy of the particles in higher
	Hardness	Wear increases rapidly when the particle hardness exceeds that of the surface being abraded
	Concentration	In slurry systems, wear increases with concentration
Nature of contact	Velocity	The rate of wear increases rapidly with increasing velocity
	Impact angle	Wear increases when the impingement angle is increased towards 90°
	Load	Increased throughput causes an increase in wear rate
	Corrosivity	Complex interaction, but greater severity corrosion, lower wear resistance
Wear material	Fracture toughness	Greater wear resistance is usually obtained with higher fracture toughness
	Corrosion resistance	Very complex, but wear best for inert materials
	Porosity	Wear resistance best for materials with zero porosity
	Coefficient of friction	Lower coefficient of friction generally gives better wear resistance
	Grain size	For many sintered ceramics, fine grain size usually gives better wear resistance. However with SNBSC, SCC/MC, and other multi-phase composites, the coarser grain size gives better wear, expecially in abrasive slurries
	Hardness	Higher hardness ceramics usually have better wear resistance

Wear behavior of ceramic materials

MS-grade PSZ against		Value of wear coefficient, k, at various temperatures(a) 20 °C (70 °F)	205 °C (400 °F)	425 °C (795 °F)
MS-grade PSZ	k pin	0.003	11.5	...
	k disc	...	25.5	222.0
Si_3N_4	k pin	9.3	17.8	95.1
	k disc	71.3	25.2	96.9
SiC	k pin	4.9	2.3	...
	k disc	19.1	0.77 to 3.70	29.0
Al_2O_3	k pin	67.6	36.3	107.2
	k disc	83.8	193.0	26.4

(a) k = wear coefficient = volume material removed/force × distance

Wear behavior of ceramics under abrasive conditions

Material	ASTM test G 65 procedure A, volume loss, mm^3
Standard tungsten carbide	5 to 20
Nilcra-PSZ-MS grade	10 to 12
Alumina	8 to 85
Stellite 1016	18 to 23
D2 tool steel	32 to 38
316 stainless steel	100 to 250
Carbon steel	200 to 250

Properties of wear-resistant ceramics

Property	Alumina 85%	90%	92%	95%	96%	99%	99.5%	99.8%	Alumina silicate
Modulus of elasticity:									
GPa	220	270	290	295 to 315	310	345	380	385	55
10^6 psi	32	39	42	43 to 46	45	50	56	8	...
Impact resistance:									
J	0.71 to 0.73	0.73	0.73 to 0.77	0.73 to 0.86	0.8	0.8	0.7	0.8	0.37
in. · lbf Charpy	6.3 to 6.5	6.5	6.5 to 6.8	6.5 to 7.6	7	7	6	7	3.3
Hardness:									
Vickers	...	...	...	...	...	...	...	...	...
Mohs	9	9	9	9	9	9	9	9	9
Knoop	...	...	...	...	...	...	...	...	...
Rockwell	...	...	...	...	...	...	...	93.5 A	...

Property	$ZrO_2 \cdot Al_2O_3$	3% $½O_3$ PSZ	TTZ	9% MgO PSZ	SiC composite graphite	Cast Si_3N_4 bonded SiC	Reaction-bonded SiC	Pressureless-sintered SiC
Modulus of elasticity:								
GPa	260	...	200	205	14	115	385	405
10^6 psi	38	...	29	30	2	17	56	59
Impact resistance:								
J	...	...	...	...	...	Up to 0.09	...	...
in. · lbf Charpy	...	...	...	...	...	Up to 0.8	...	...
Hardness:								
Vickers	...	...	...	...	...	...	...	...
Mohs	...	...	...	...	8+	...	...	...
Knoop	1470	...	...	1080-1520	...	...	2700	2800
Rockwell	...	91.5A	74 to 79 C	...	...	...	...	...

Property	Sintered SiC with free silicon	Sintered SiC with graphite	Reaction-bonded Si_3N_4	Hot pressed Si_3N_4
Modulus of elasticity:				
GPa	380	205	165	310
10^6 psi	55	30	24	45
Impact resistance:				
J	...	Similar to silicon carbide		
in. · lbf Charpy	...	...		
Hardness:				
Vickers	...	...	...	1800
Mohs	...	...	>9	...
Knoop	1900	1000	...	...
Rockwell	...	...	...	75 to 80 C

Conventional abrasives and superabrasives applied to the machining of selected ceramic and glass components

Material	Bond type	Applications
Conventional		
Silicon carbide	Vitreous	Grind soft ceramics
		Flame spray ceramic coatings
		Glass beveling
	Rubber	Cutoff glass tubing
	Coated	Seaming plate glass
		Finishing picture tubes
		Scientific stemware
Aluminum oxide	Vitreous	Glass beveling
Zirconia alumina	Coated	Seaming plate glass
Ceramic alumina	...	...
Superabrasives		
Diamond	Metal	Precision pencil edging and beveling of tempered plate and safety glass
		Optical lens production
		Glass/quartz tube cutoff
		General precision grinding of all types of glass
		Production grinding of aluminas, quartz, and refractories
	Resin	All production and finish grinding of silicon carbide, titanium nitride, silicon nitride, and partially stabilized zirconia, and other technical ceramics
Cubic boron nitride	...	...

Typical physical and thermal properties of selected abrasives

Abrasive	Chemical composition	Density g/cm^3	Density lb/in.^{3}	Relative thermal conductivity	Coefficient of thermal expansion, 10^{-6}/K	Threshold temperature for degradation °C	Threshold temperature for degradation °F
Diamond	C	3.52	0.127	100 to 350	4.8	800	1470
Aluminum oxide	Al_2O_3	3.92	0.142	1	8.6	1750	3180
Silicon carbide	SiC	3.21	0.116	10	4.5	1500	2730

Mechanical and thermal properties of selected conventional abrasives and superabrasives

Material	Knoop hardness, GPa	Elastic modulus GPa	Elastic modulus 10^6 psi	Comprehensive strength MPa	Comprehensive strength ksi	Thermal conductivity W/m · K	Thermal conductivity Btu/ft · h · °F
Superabrasive							
Diamond	70 to 100	1,075	156	10,444	1,515	2,092	1,209
Cubic boron nitride	45	662	96	7,061	1,024	1,300	750
Conventional							
SiC	27	207	30	569	82.5	41.5	24.0
Al_2O_3	25	308	45	2,942	426.6	33.5	19.4

Hardness of several common abrasives

Material	Knoop hardness (100 g), kg/mm^2
Sapphire (alumina)	2000 to 2050
Tungsten carbide (WC)	2050 to 2150
Silicon carbide (SiC)	2150 to 2950
Boron carbide (B_4C)	2900 to 3100
Cubic boron nitride (CBN)	4500 to 4600
Diamond (C)	8000 to 8500

Typical properties of cemented carbide, sintered alumina, and Sialon cutting-tool materials

Property	Cemented carbide	Sintered alumina	Alumina-titanium carbide composite	Sialon
Hardness:				
GPa	12.3 to 15.1	15.3 to 15.9	17.0 to 17.4	12.2 to 15.2
HRA	91 to 92	93.2 to 93.6	94.0 to 94.4	90.8 to 93.5
Melting point:				
K	1673	2273	3413 (TiC)	Decomposes at 2200 K
°F	2552	3632	5684 (TiC)	
Coefficient of thermal expansion, 10^{-6}/K	4.5 to 7.2	7.5	7.6	3.2
Modulus of elasticity:				
GPa	520 to 660	440	420	300
10^6 psi	75 to 96	64	61	44
Transverse rupture strength at 25 °C:				
MPa	1000 to 2400	700 to 840	840 to 940	830
ksi	145 to 350	100 to 122	122 to 136	120
Fracture toughness,				
$MPa\sqrt{m}$	...	2.2 to 2.5	3.1 to 3.5	3.6 to 5.2
Density, g/cm^3	12.0 to 15.1	3.80 to 3.90	4.20 to 4.30	3.35

Comparative cutting performances of hard metal, alumina, and Sialon

Cutting parameter	Cast iron	Hardened steel En31	Incoloy 901
Hard metal			
Cutting speed:			
m/min.	245	4.6	21
ft/min.	800	15	70
Depth of cut:			
mm	6.4	...	...
in.	0.25	...	...
Feed rate:			
mm/rev	0.51	...	...
in/rev	0.020	...	...
Alumina			
Cutting speed:			
m/min.	610	(a)	90
ft/min.	2000	(a)	300
Depth of cut:			
mm	6.4	(a)	(b)
in.	0.25	(a)	(b)
Feed rate:			
mm/rev	0.25	(a)	(b)
in./rev	0.010	(a)	(b)
Sialon			
Cutting speed:			
m/min.	1070	120	305
ft/min.	3500	400	1000
Depth of cut:			
mm	9.6	0.51	2.03
in.	0.38	0.020	0.080
Feed rate:			
mm/rev	0.51	0.25	0.25
in./rev	0.020	0.010	0.010

(a) Impossible to cut. (b) No second entry

Environmental Effects

Results of corrosion tests of selected ceramics in liquids

Test environment(a), wt %	Temperature °C	Temperature °F	Corrosive weight loss(b), mg/cm²·yr: Si/SiC composites (12% Si)	Tungsten carbide (6% Co)	Aluminum oxide (99%)	Silicon carbide (no free Si)
98% H_2SO_4	100	212	55.0	>1000	65.0	1.8
50% NaOH	100	212	>1000	5.0	75.0	2.5
53% HF	25	77	7.9	8.0	20.0	<0.2
85% H_3PO_4	100	212	8.8	55.0	>1000	<0.2
70% HNO_3	100	212	0.5	>1000	7.0	<0.2
45% KOH	100	212	>1000	3.0	60.0	<0.2
25% HCl	70	158	0.9	85.0	72.0	<0.2
10% HF + 57% HNO_3	25	77	>1000	>1000	16.0	<0.2

(a) Test time, 125 to 300 h of submersive testing (with continuous stirring). (b) Corrosion weight loss guide: >1000 mg/cm^2 · yr, specimen completely destroyed within days; 100 to 999, not recommended for service longer than one month; 50 to 100, not recommended for service long than one year; 10 to 49, caution recommended, based on the specific application; 0.3 to 9.9, recommended for long-term service; <0.2, recommended for long-term service (no corrosion, other than as a result of surface cleaning, was evidenced)

Ceramic corrosion in the presence of combustion products

Material	Corrosion depth data(a) µm	Corrosion depth data(a) mils	Projected corrosion µm/y	Projected corrosion mils/y
Cordierite	9.4	0.37	58.0	2.3
Mullite	0.85	0.034	8.2	0.32
Reaction-bonded SiC	0.20	0.0080	1.9	0.08
	0.16	0.0063	1.5	0.06
	0.18	0.0069	1.7	0.07
Sintered αSiC	0.37	0.014	3.5	0.14
Reaction-bonded Si_3N_4	0.90	0.035	8.6	0.34
Glass-enamel	1.4	0.054	13.3	0.52

(a) Data are based on a heat condensing application where products were exposed to combustion products (H_2SO_4, HCl, and nitrous oxides) under cyclic conditions up to 290 °C (550 °F) for 900 h

Thermal etching of selected ceramics

Material	Atmosphere	Time	Temperature °C	Temperature °F
αAl_2O_3 (<96 wt% Al_2O_3)	Air	0.5 to 4.5 h	1250 to 1500	2280 to 2730
βAl_2O_3	Air	1 to 5 min.	1470(a)	2680(a)
Strontium and barium ferrites	Air	1 h	1050 to 1150	1920 to 2100
TiO_2	Air	1 h	1350	2460
Si_3N_4	Vacuum	15 min.	1250	2280
	Nitrogen	5 h	1600	2910
SiC	Vacuum	1 to 3 h	1300 to 1500	2370 to 2730

(a) Thermal etching should be performed 200 °C (360 °F) below the sintering temperature

Examples of chemical etching of selected ceramics

Material	Etchant	Temperature °C	Temperature °F	Time
Silicate ceramics	2 to 5% hydrofluoric acid (HF)	20	70	1 to 20 min.
Calcia and magnesia ceramics	Conc HCl	20	70	3 s to 6 min.
Barium titanate	1/3 HF/conc HCl	20	70	7 min. to 2 h
Silicon nitride	NaOH melt	400 to 450	750 to 840	1 to 10 min.
Silicon carbide	Modified Murakami's reagent (3 g KOH or NaOH, 30 g K-hexacyanoferrate (III), 60 mL water)	Boiling	Boiling	3 to 30 min.
Zirconia ceramics	Concentrated phosphoric acid	Boiling	Boiling	5 to 10 min.

Conc, concentrated

Manufacturing and Production

Pressure densification techniques and parameters for various ceramic materials

Technique and material	Pressure MPa	Pressure ksi	Temperature °C	Temperature °F
Uniaxial hot pressing				
Al_2O_3	12 to 40	1.7 to 5.8	1150 to 1350	2100 to 2460
SiO_2	7 to 17	1.0 to 2.5	1100 to 1200	2010 to 2190
NiO	4 to 130	0.6 to 18.8	800 to 1100	1470 to 2010
CoO	75	11	950 to 1100	1740 to 2010
MgO	5 to 30	0.7 to 4.4	1440 to 1790	2625 to 3255
$YBa_2Cu_3O_x$	10 to 40	1.5 to 5.8	850 to 950	1560 to 1740
Hot isostatic pressing				
TiN	100 to 200	15 to 30	1000 to 1600	1830 to 2910
TaC; WC	70 to 100	10 to 15	1500 to 1760	2730 to 3200
SiC	138	20	1850 to 2100	3360 to 3810
Gas pressure sintering(a)				
Si_3N_4 (Be_3N_2)	2/7	0.3/1.0	2050/1950	3720/3540
Al_2O_3	Vacuum/100	Vacuum/15	1650	3000
Si_3N_4 (Y_2O_3/Al_2O_3)	0.1/2	0.015/0.3	1780	3235
PLZT; $BaTiO_3$	0.1/20	0.015/3	1170	2140

(a) Values separated by virgule indicate first step/second step

Ceramic materials produced by chemical vapor deposition

Coating	Chemical mixture	Deposition temperature °C	Deposition temperature °F	Method	Application
Carbides					
TiC	$TiCl_4$-CH_4-H_2	900 to 1000	1650 to 1830	CVD	W
	$TiCl_4$-CH_4(C_2H_2)-H_2	400 to 600	750 to 1110	PACVD	E
HfC	$HfCl_x$-CH_4-H_2	900 to 1000	1650 to 1830	CVD	W, C/O
ZrC	$ZrCl_4$-CH_4H_2	900 to 1000	1650 to 1830	CVD	W, C/O
	$ZrBr_4$-CH_4-H_2	>900	>1650	CVD	W, C/O
SiC	CH_3SiCl_3-H_2	1000 to 1400	1830 to 2550	CVD	W, C/O
	SiH_4-C_xH_y	200 to 500	390 to 930	PACVD	E, C
B_4C	BCl_3-CH_4-H_2	1200 to 1400	2190 to 2550	CVD	W
B_xC	B_2H_6-CH_4	400	750	PACVD	W, E, C
W_3C	WF_6-CH_3OCH_3-H_2	350 to 500	660 to 930	CVD	W
W_2C	WF_6-CH_4-H_2	400 to 700	750 to 1290	CVD	W
Cr_7C_3	$CrCl_2$-CH_4-H_2	1000 to 1200	1830 to 2190	CVD	W
Cr_3C_2	$Cr(CO)_6$-CH_4-H_2	1000 to 1200	1830 to 2190	CVD	W
TaC	$TaCl_5$-CH_4-H_2	1000 to 1200	1830 to 2190	CVD	W, E
VC	VCl_2-CH_4-H_2	1000 to 1200	1830 to 2190	CVD	W
NbC	$NbCl_5$-CCl_4-H_2	1500 to 1900	2730 to 3450	CVD	W
Nitrides					
TiN	$TiCl_4$-N_2-H_2	900 to 1000	1650 to 1830	CVD	W
	$TiCl_4$-N_2-H_2	250 to 1000	480 to 1830	PACVD	E
HfN	$HfCl_x$-N_2-H_2	900 to 1000	1650 to 1830	CVD	W, C/O
	HfI_4-NH_3-H_2	>800	>1470	CVD	W, C/O
Si_3N_4	$SiCl_4$-NH_3-H_2	1000 to 1400	1830 to 2550	CVD	W, C/O
	SiH_4-NH_3-H_2	250 to 500	480 to 930	PACVD	E, C/O
	SiH_4-N_2-H_2	300 to 400	570 to 750	PACVD	E
BN	BCl_3-NH_3-H_2	1000 to 1400	1830 to 2550	CVD	W
	BCl_3-NH_3-H_2	25 to 1000	77 to 1830	PACVD	E
	$BH_3N(C_2H_5)_3$-Ar	25 to 1000	77 to 1830	PACVD	E
	$B_3N_3H_6$-Ar	400 to 700	750 to 1290	CVD	W, W
	BF_3-NH_3-H_2	1000 to 1300	1830 to 2370	CVD	W
	B_2H_6-NH_3-H_2	400 to 700	750 to 1290	PACVD	E
ZrN	$ZrCl_4$-N_2-H_2	1100 to 1200	2010 to 2190	CVD	W, C/O
	$ZrBr_4$-NH_3-H_2	>800	>1470	CVD	W, C/O
TaN	$TaCl_5$-N_2-H_2	800 to 1500	1470 to 2730	CVD	W
AlN	$AlCl_3$-NH_3-H_2	800 to 1200	1470 to 2190	CVD	W
	$AlBr_3$-NH_3-H_2	800 to 1200	1470 to 2190	CVD	W
	$AlBr_3$-NH_3-H_2	200 to 800	390 to 1470	PACVD	E, W
	$Al(CH_3)_3$-NH_3-H_2	900 to 1100	1650 to 2010	CVD	E, W

(continued)

Ceramic materials produced by chemical vapor deposition (continued)

Coating	Chemical mixture	Deposition temperature °C	°F	Method	Application
VN	VCl_4-N_2-H_2	900 to 1200	1650 to 2190	CVD	W
NbN	$NbCl_5$-N_2-H_2	900 to 1300	1650 to 2370	CVD	W, E
Oxides					
Al_2O_3	$AlCl_3$-CO_2-H_2	900 to 1100	1650 to 2010	CVD	W, C/O
	$Al(CH_3)_3$-O_2	300 to 500	570 to 930	CVD	E, C
	$Al[OCH(CH_3)_2]_3$-O_2	300 to 500	570 to 930	CVD	E, C
	$Al(OC_2H_5)_3$-O_2	300 to 500	570 to 930	CVD	E, C
SiO_2	SiH_4-CO_2-H_2	200 to 600	390 to 1110	PACVD	E, C
	SiH_4-N_2O	200 to 600	390 to 1110	PACVD	E
SiO_2	$Si(OEt)_4$-O_2	650 to 750	1200 to 1380	CVD	E
SiO_2	SiH_2Et_2-O_2	350 to 450	660 to 840	CVD	E
TiO_2	$TiCl_4$-H_2O	800 to 1000	1470 to 1830	CVD	W, C
	$TiCl_4$-O_2	25 to 700	77 to 1290	PACVD	E
	$Ti[OCH(CH_3)_2]_4$-O_2	25 to 700	77 to 1290	PACVD	E
ZrO_2	$ZrCl_4$-CO_2-H_2	900 to 1200	1650 to 2190	CVD	W, C/O
Ta_2O_5	$TaCl_5$-O_2-H_2	600 to 1000	1110 to 1830	CVD	W, C, E
Cr_2O_3	$Cr(CO)_6$-O_2	400 to 600	750 to 1110	CVD	W
Borides					
TiB_2	$TiCl_4$-BCl_3-H_2	800 to 1000	1470 to 1830	CVD	W, C, E
MoB	$MoCl_5$-BBr_3	1400 to 1600	2550 to 2910	CVD	W, C
WB	WCl_6-BBr_3-H_2	1400 to 1600	2550 to 2910	CVD	W, C
NbB_2	$NbCl_5$-BCl_3-H_2	900 to 1200	1650 to 2190	CVD	W, C
TaB_2	$TaBr_5$-BBr_3	1200 to 1600	2190 to 2910	CVD	W, C
ZrB_2	$ZrCl_4$-BCl_3-H_2	1000 to 1500	1830 to 2730	CVD	W, C, E
HfB_2	$HfCl_x$-BCl_3-H_2	1000 to 1600	1830 to 2910	CVD	W, C

CVD, conventional CVD; PACVD, plasma-assisted CVD; W, wear-resistant coating; E, electronics; C, corrosion-resistant coatings; O, oxidation-resistant coatings

Raw materials for advanced structural and magnetic (ferrite) ceramics

End product	Raw materials	Key product properties	Applications
Al_2O_3	Bayer process alumina derived from bauxite	Low permittivity Hardness Wear resistance	Substrates, insulators, spark plugs, wear parts, milling media, thread guides, armor, radomes
ZrO_2	Zirconia derived from zircon by chemical processes	Ionic conductivity Electronic conductivity Wear resistance	Oxygen sensors, fuel cells (potential), high-temperature heater, milling media
CBN	High temperature and pressure transformation of hexagonal form of BN (HBN) (formed by reacting B_2O_3 and urea in nitrogen)	High thermal conductivity High electrical resistivity High hardness	Substrates in electronics Machining of ferrous metals
BeO	Beryllia powder from beryl or bertrandite ores	High thermal conductivity High electrical resistivity	Substrates (heatsinks) in electronics
SiC	Acheson process: $SiO_2 + C \rightarrow SiC + CO$ Pyrolysis of polycarbosilanes	Extreme hardness Resistance to thermal shock	Wear parts As a fiber whisker or particle in MMCs and CMCs
$Al_2O_3 \cdot ZrO_2$	High-quality alumina High-quality zirconia	Improvement in strength and toughness over Al_2O_3	Wear parts
Si_3N_4	Silicon nitride powder derived from silicon and nitrogen	Hardness Resistance to thermal shock	Wear parts
Sialons	Silicon nitride, alumina, 21*R* polytype (AlN), yttria	Hardness Toughness Resistance to thermal shock	Wear parts Extrusion dies Cutting tools
AlN	AlN powder, prepared by carbothermal reduction of Al_2O_3 in nitrogen, or direct nitridation of Al	Low permittivity High thermal conductivity	Electronic substrates
SnO_2	High-purity tin dissolved in nitric acid and coprecipitated with other oxides	Surface controlled conductivity	Sensors
PZT (lead zirconium titanate)	High-purity oxides Coprecipitated oxides	High piezoelectric coefficients Change of polarization with temperature	Transducers Actuators Pyroelectrics
PMN (lead magnesium niobate)	High-purity oxides Coprecipitated oxides	High permittivity and breakdown voltage Controlled deformation in an applied field	Capacitors Actuators
PLZT (lead lanthanum zirconium titanate)	High-purity oxides	Change of birefringence with applied field	Electro-optics, head-up displays, flash goggles
	Coprecipitated oxides	Controlled deformation in an applied field	
	Metal alkoxide-base coatings	Change of birefringence with applied field	Actuators
ZTS (zirconium titanium stannate)	High-purity oxides	Stable permittivity at high frequencies over a wide temperature range and very low dielectric and insertion losses	Microwave resonators and filters

(continued)

Raw materials for advanced structural and magnetic (ferrite) ceramics (continued)

End product	Raw materials	Key product properties	Applications
PBNT ($PbO \cdot BaO \cdot Nd_2O_3 \cdot TiO_2$)	High-purity oxides	Stable permittivity at high frequencies over a wide temperature range and very low dielectric and insertion losses	Microwave resonators and filters
ZnO	High-purity oxides (derived from metal smelting) plus praeseodymium or bismuth oxides	Change of resistivity with applied field	Varistors
$YBa_2Cu_3O_{7-\delta}$	Barium, strontium, calcium salts (chlorides, carbonates, nitrates, and peroxides)	Superconductivity	Demonstration devices
$Bi_2Sr_2CaCu_2O_8$	Barium, strontium, calcium salts (chlorides, carbonates, nitrates, and peroxides	Superconductivity Very low insertion losses	Microwave filters
$Bi_2Sr_2Ca_2Cu_3O_{10}$ stabilized with PbO	High-purity oxides Coprecipitation, sol-gel, metal alkoxides, sputtering, chemical vapor deposition	Superconductivity conductor in very high magnetic fields	Nuclear magnetic resonance (NMR) imagers
TiB_2	Powder made by carbothermal reduction of TiO_2 with B_2O_3	Electrical conductivity Resistance to molten aluminum coupled with complete wetting of surface	Potential cathode material in primary aluminum production Evaporator boats (with BN)
B_4C	Carbothermal reduction of B_2O_3	Very hard Abrasion resistant Absorbs thermal neutrons	Shot blast nozzles, bearings, armor, and nuclear energy industry uses
Ferrites			
Hard ferrites			
$SrFe_{12}O_{19}$	$SrCO_3/Fe_2O_3$	High residual flux density High coercive force	Permanent magnets
$BaFe_{12}O_{19}$	$BaCO_3/Fe_2O_3$	High coercive force High residual flux	Motors
Soft ferrites			
$MnZnFe_4O_8$	Mixed oxide, iron oxide derived from thermal hydrolysis of ferric chloride Mixed oxide, iron oxide derived from thermal hydrolysis of ferric chloride	High initial permeability Low loss High initial permeability High saturation flux density	Wide band/pulse, transformers Inductors, telecommunications Wide band/pulse, transformers Power transformers, magnetic recording heads
Microwave			
$Y_3Fe_5O_{12}$	Coprecipitation or milling of pure oxides	Narrow line widths and extremely low losses at microwave frequencies	Elements in microwave circuitry
$BaTiO_3$	Barium carbonate and titania	High permittivity	Capacitors

Traditional Ceramics

Whitewares and Clay-Based Products

Ceramic whiteware classes and products

Class and type	Absorption, %	Products
Earthenware		
Natural	+15	Tableware, artware, tiles
Fine	10 to 15	Tableware, artware, kitchenware, tiles
Talc	10 to 20	Artware, tiles, ovenware
Semivitreous	4 to 9	Tableware, artware
Stoneware		
Natural	0 to 5	Drain pipe, kitchenware, artware, tiles
Fine	0 to 5	Cookware, artware, tableware
Technical vitreous	0 to 0.2	Chemical ware
Jasper	0 to 1	Artware
Basalt	0 to 1	Artware
China		
Hotel	0.1 to 0.3	Tableware
Bone	0.3 to 2	Tableware, artware
Frit	0 to 0.5	Tableware, artware
Vitreous plumbing fixtures	0.1 to 0.3	Lavatories, closet bowls, flush tanks, urinals
Cookware	1 to 5	Ovenware, stoveware
Porcelain		
Hard	0 to 0.5	Tableware, artware
Technical vitreous	0 to 0.2	Chemical ware, ball mill balls and linings
Triaxial electrical	0 to 0.2	Low-frequency insulators
High-strength electrical	0 to 0.2	Low-frequency insulators
Dental	0 to 0.1	Dentures
Technical ceramics		
Steatite	0 to 0.05	High-frequency insulators, low-loss dielectrics
Electrical porcelains	0 to 0.2	Low-frequency insulators

Structural ceramic products

Product	ASTM standard
Facing materials	
Face brick	C 216
Terra cotta	...
Thin brick veneer	C 1088
Sculptured brick	...
Special shapes	...
Glazed brick	C 126
Glazed structural tile	C 126
Load-bearing units	
Building brick	C 62
Hollow brick	C 652
Face brick	C 216
Structural tile facing	C 212
Structural tile floor	C 57
Structural tile wall	C 34
Paving units	
Light traffic paves	C 902
Quarry tile	...
Paving brick	...
Chemically resistant units	
Chemically resistant masonry units	C 279
Industrial floor brick	C 410
Chemical stoneware	...
Sewer pipe	C 700
Chimney brick	C 980
Filter block	C 159
Roofing tile	...
Miscellaneous	
Flue lining	C 315
Lightweight aggregate	C 331
Drain tile	C 4

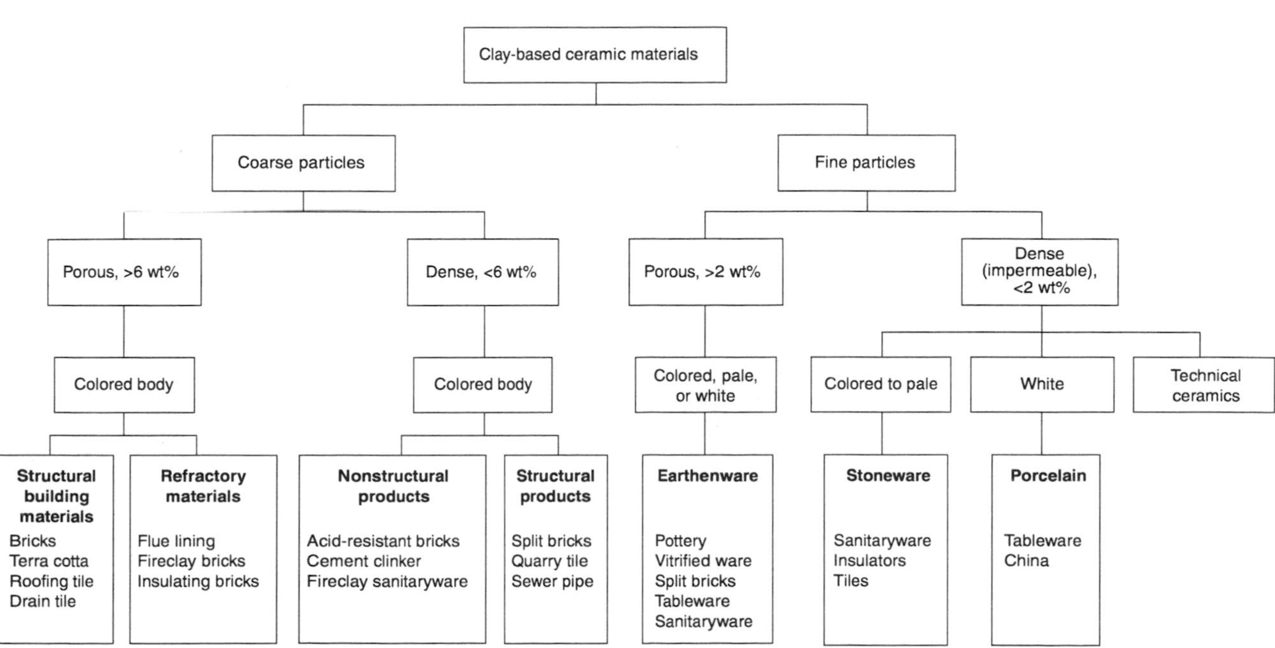

Fig. 1 Classification of clay-based ceramics. Water percent values represent water absorptivity

A summary of important materials properties for structural ceramics

Property	Components				
	Heat exchangers	Valve components	Turbine engines	Bearings	Cutting tools
Thermal					
Conductivity	X	X	X	X	X
Diffusivity	X	X	X	X	X
Expansion	X	X	X	X	
Specific heat	X	X	X	X	X
Shock resistance	X	X	X	X	X
Emissivity	X				
Melting point	X	X	X		X
Maximum service temperature	X	X	X	X	X
Mechanical					
Elastic modulus	X	X	X	X	X
Poisson's ratio	X	X	X	X	X
Shear (torsion) modulus			X		
Tensile strength	X	X	X	X	X
Flexural strength	X	X	X	X	X
Modulus of rupture	X	X	X	X	X
Weibull modulus	X	X	X	X	X
Hardness		X		X	X
Toughness	X	X	X	X	X
Creep	X	X	X	X	X
Impact resistance		X		X	X
Coefficient of friction		X		X	X
Wear resistance		X		X	X
Chemical					
Corrosion products	X	X	X	X	X
Corrosion rate	X	X	X	X	X
Oxidation rate	X	X	X	X	X

Typical ball clay compositions

Material	Location/content, %					
	Kentucky (US)	Kentucky (US)	Tennessee (US)	North Devon (UK)	South Devon (UK)	Dorset (UK)
Compound						
SiO_2	55.6	51.6	61.0	63.0	52.0	54.0
Al_2O_3	28.6	28.6	24.5	24.0	30.0	30.0
Fe_2O_3	1.0	0.9	1.0	0.9	1.2	1.4
TiO_2	1.8	1.7	1.3	1.5	1.0	1.2
CaO	0.1	0.1	0.1	0.2	0.2	0.3
MgO	0.4	0.5	0.1	0.4	0.4	0.4
K_2O	1.1	0.5	1.7	2.4	2.1	0.4
Na_2O	0.1	0.2	0.4	0.5	0.3	0.5
Ignition loss, %	11.4	15.9	9.7	6.8	12.2	8.8
Mineral						
Kaolinite	...	...	...	45.0	64.0	57.0
Mica	...	...	...	28.0	22.0	32.0
Quartz	...	...	...	26.0	11.0	11.0
Feldspar	...	...	...	...	...	...
Other	...	...	...	1.0	3.0	...

Typical china clay and kaolin compositions

Material	Location/content, %					
	Georgia (US)	North Carolina (US)	South Carolina (US)	Florida (US)	Cornwall (UK)	Cornwall (UK)
Compound						
SiO_2	44.4	47.3	44.6	46.1	48.0	50.2
Al_2O_3	39.6	36.4	39.5	38.8	36.9	34.1
Fe_2O_3	0.4	1.0	0.5	0.7	0.7	0.8
TiO_2	1.7	0.1	1.4	0.4	...	0.1
CaO	...	...	0.1	0.2	0.1	...
MgO	...	...	0.1	...	0.3	0.3
K_2O	0.1	1.3	0.5	0.3	1.6	4.1
Na_2O	0.1	0.3	...	...	0.1	0.2
Ignition loss, %	13.7	13.6	13.3	13.5	12.3	10.2

(continued)

Typical china clay and kaolin compositions (continued)

Material	Georgia (US)	North Carolina (US)	South Carolina (US)	Florida (US)	Cornwall (UK)	Cornwall (UK)
	Location/content, %					
Mineral						
Kaolinite	...	...	...	...	84.0	66.0
Mica	...	...	...	...	13.0	23.0
Quartz	...	...	...	...	...	...
Feldspar	...	...	...	...	1.0	9.0
Other	...	...	...	...	2.0	2.0

Characterization of two common clay-based ceramics

Properties	Vitreous sanitaryware	Vitreous china
Compound, wt%		
SiO_2	65.0	69.4
Al_2O_3	23.1	19.5
Fe_2O_3	0.44	0.30
TiO_2	0.28	0.14
CaO	0.33	1.33
MgO	0.13	0.11
K_2O	2.68	1.45
Na_2O	2.41	1.14
Ignition loss	5.67	6.46
Mole of flux(a)	0.0766	0.0604
Minerals, wt%		
Smectite	3.7	3.0
Kaolin group	32.7	33.3
Mica	8.8	5.8
Free quartz	23.7	39.6
Organic	0.46	0.23
Auxiliary flux	...	2.0
Particle size, %		
<20 μm (<800 μin.)	76	76
<5 μm (<200 μin.)	47	45
<2 μm (<80 μin.)	33	36
<1 μm (<40 μin.)	19	21
Surface area, methylene blue index, meq/100 g(b)	3.3	2.7

(a) Mole of flux is the sum of the percentages of CaO, MgO, K_2O and Na_2O, divided by their respective molecular masses. (b) Measured in terms of methylene blue index (MBI); the milliequivalents of methylene blue cation (chlorine salt) absorbed per 100 g of clay

Characteristics of commercial clays found in the United States

Properties	Coarse kaolin, sedimentary(a)	Fine kaolin, sedimentary(b)	Dark, fine ball clay(c)	Light, coarse ball clay(d)
Compound, wt%				
SiO_2	45.7	46.7	50.5	60.4
Al_2O_3	38.3	38.2	28.7	27.0
Fe_2O_3	0.41	0.60	0.91	0.93
TiO_2	1.55	1.42	1.48	1.62
CaO	0.08	0.12	0.40	0.28
MgO	0.06	0.20	0.30	0.26
K_2O	0.06	0.15	0.89	1.70
Na_2O	0.14	0.03	0.18	0.50
Ignition loss	13.65	13.79	16.58	7.59
Minerals, wt%				
Montmorillonite	nil	3	8	7
Kaolin group	96	93	58	44
Mica	2	2	10	21
Free quartz	trace	1	14	26
Organic	trace	trace	8	0.5
Particle size, %				
<20 μm (<800 μin.)	95	99	99	98
<5 μm (<200 μin.)	69	88	95	79
<2 μm (<80 μin.)	52	72	82	61
<1 μm (<40 μin.)	35	56	69	43
<0.5 μm (<20 μin.)	28	41	51	29
Surface area, methylene blue index, meq/100 g	1.6	10.5	12.1	5.6

(a) Washington County, Georgia. (b) Wilkinson County, Georgia. (c) Graves County, Kentucky. (d) Weakley County, Tennessee

Typical oxide compositions of raw materials, wt%

Oxide	Silica sand	Clay	Feldspar	Borate	Talc	Zircon	Calcite	Dolomite
SiO_2	85 to 99	40 to 65	60 to 75	1 to 10	50 to 65	30 to 35	0 to 1	0 to 1
Al_2O_3	0 to 10	10 to 35	10 to 20	0 to 5	0 to 3	0 to 2	0 to 1	0 to 1
B_2O_3	<0.1	<0.1	<0.1	20 to 40	<0.1	<0.1	<0.1	<0.1
Fe_2O_3	0 to 1	1 to 10	0 to 2	0 to 1	0 to 3	<0.5	<0.5	<0.5
Na_2O	0 to 1	0 to 1	0 to 10	0 to 20	0 to 1	<0.2	0 to 1	0 to 1
K_2O	0 to 5	0 to 5	0 to 15	0 to 1	0 to 5	<0.5	0 to 1	0 to 1
CaO	0 to 5	0 to 15	0 to 3	0 to 25	0 to 5	<0.5	50 to 55	25 to 35
MgO	0 to 1	0 to 5	0 to 2	0 to 5	20 to 30	<0.5	0 to 5	15 to 25
TiO_2	0 to 1	0 to 4	0 to 2	0 to 1	0 to 1	<0.5	0 to 1	0 to 1
ZrO_2	<0.1	<0.1	<0.1	<0.1	<0.1	60 to 65	<0.1	<0.1
Other	0 to 1	0 to 1	0 to 1	0 to 1	0 to 1	0 to 1	0 to 1	0 to 1

Composition and physical properties of raw materials used for whiteware applications

Raw material	Composition: Frit materials	Composition: Raw glaze materials	Composition: Constituents	Typical properties of fired ware
Body materials				
Bone china	...	...	Bone ash, china clay, quartz, feldspars	Pure white, translucent, high strength, zero water absorption
Porcelain	...	...	China clay, feldspars, quartz	Pure, bluish, or off-white, translucent, high strength, zero water absorption
Hotelware	...	...	Quartz, ball clay, china clay, feldspars	Off-white, good physical and chemical durability, zero water absorption
Earthenware	...	...	Quartz/flint, china clay, ball clay, feldspars	Ivory to pure white, lower strength, water absorption 4 to 8%
Sanitaryware	...	...	China clay, ball clay, fire clay, quartz, feldspars, nepheline syenite	...
Wall/floor tiles	...	...	Ball clay, fire clay, quartz, feldspars, limestone	...
Glazes	<1150 °C (2100 °F): Quartz, sodium borate, boric acid, limestone, feldspars, china clay, lead oxides, zircon, zinc oxide, alkali metal, carbonates, and nitrates	...	...	Transparent to opaque, colored or colorless, glossy, matte, vellum, or textured
Pigments	...	>1150 °C (2100 °F): Feldspars, quartz, clays, nepheline syenite, limestone, dolomite, zinc oxide, zircon	...	Transparent to opaque, colored or colorless, glossy, matte, vellum or textured
Zircon-based	...	...	Zirconia, quartz	...
Spinel-based or oxide colorants	...	...	Iron, chromium, zinc, nickel, copper oxides or compounds	...
Sphene-based	...	...	Quartz, limestone, tin oxide	...

Typical oxide compositions of various ceramic bodies, wt%

Oxide	Earthenware	Porcelain	Stoneware	Brick	Refractory
SiO_2	30 to 60	40 to 60	45 to 65	45 to 65	20 to 60
Al_2O_3	5 to 20	15 to 30	10 to 20	10 to 20	25 to 90
Na_2O	0 to 1	0 to 1	0 to 1	0 to 1	0 to 2
K_2O	0 to 1	0 to 1	0 to 1	0 to 2	0 to 2
CaO	0 to 20	0 to 2	0 to 2	0 to 10	0 to 5
MgO	0 to 10	0 to 1	0 to 1	0 to 2	0 to 3
Fe_2O_3	0 to 1	0 to 1	0 to 1	0 to 5	0 to 2
TiO_2	0 to 1	0 to 2	0 to 2	0 to 3	0 to 3
Other	0 to 1	0 to 1	0 to 1	0 to 1	0 to 3

Typical ceramic body compositions, %

	Fine whiteware A	Fine whiteware B	Earthenware, wall tile	Vitreous floor tile	Chemical stoneware	Tender porcelain	Hard porcelain	Electrical porcelain C	Electrical porcelain D	Steatite porcelain	Bone china	Hotel china E	Hotel china F	Cookware	Vitreous china, sanitaryware
China clay	30.6	18	22	32	55	30	50.4	20	17	7	37.5	36.8	34.7	62	22.3
Ball clay	14.4	38	25			...	...	25	35	...	...	6.2	7.1	...	26.7
Feldspar	27.5	32	3	58	30	30	17.1	35	32	6	20	21.7	18.3	10	33.7
Flint	27.5	12	25	8	14	35 to 37	32.3	...	16	...	5	35.2	27.5	7	17.3
Bone ash $[Ca_3(PO_4)]_2$	...	...	...	...	...	...	...	...	...	...	37.5	...	...	...	...
Whiting ($CaCO_3$)	...	...	...	...	1	3 to 5	...	...	...	...	...	...	...	...	...
Alumina	...	...	...	...	...	...	...	20	...	...	...	...	12	...	...
$MgCo_3$	...	...	...	...	...	...	...	...	...	...	...	...	...	21	...
Talc	...	...	10	2	...	...	...	...	...	87	...	...	...	...	...
Pyrophyllite	...	...	15	...	...	...	...	...	...	...	...	...	...	...	...
Firing range (PCE)	8	9	5 to 6	9	14	8 to 9	14	10 to 11	10 to 11	10	8 to 10	10	10 to 12	14	9

A, white body, low plasticity; B, ivory body, high workability; C, high voltage; D, low voltage; E, standard; F, high-strength body

Properties of structural ceramic materials

Material	Elastic modulus GPa	Elastic modulus 10^6 psi	Poisson's ratio	Thermal conductivity, W/m · K RT	Thermal conductivity, W/m · K 600 °C (1110 °F)	Thermal expansion, 10^{-6}/K	Specific heat, J/g · °C	Density, g/cm³	Strength, MPa (ksi) RT	Strength, MPa (ksi) 600 °C (1110 °F)	Weibull modulus, at RT	Maximum use temperature °C	Maximum use temperature °F
Silicon nitride													
Hot-pressed (HPSN)	290	42.1	0.3	29	22	2.7	0.75	3.3	830 (120)	805 (117)	7	1400	2550
Sintered (SSN)	290	42.1	0.28	33	18	3.1	1.1	3.3	800 (116)	725 (105)	13	1400	2550
Reaction-bonded (RBSN)	200	29.0	0.22	10	10	3.1	0.87	2.7	295 (43)	295 (43)	10	1400	2550
Silicon carbide													
Hot-pressed (HPSC)	430	62.4	0.17	80	51	4.6	0.67	3.3	550 (80)	520 (75)	10	1500	2730
Sintered (SSC)	390	56.6	0.16	71	48	4.2	0.59	3.2	490 (71)	490 (71)	9	1500	2730
Reaction-bonded (RBSC)	413	59.9	0.24	225	70	4.3	1.0	3.1	390 (57)	390 (57)	10	1300	2370
Partially stabilized zirconia (PSZ)	205	29.7	0.30	2.9	2.9	10.5	0.5	5.9	1020 (148)	580 (84)	14	950	1740
Lithium-aluminum-silicate	68	9.9	0.27	1.4	1.9	0.5(a)	0.78	2.3	96 (14)	96 (14)	10	1200	2190
Aluminum-titanate	11	1.6	0.22 to 0.26	2	5	1.0	0.88	3.0	41 (6)	...	15	1200	2190
Common metals (Reference)													
Cast iron	170	24.7	0.28	49	40	12	0.45	7.1	620 (90)	100 (14.5)	...	500	930
Steel	200	29.0	0.28	38	...	14	0.45	7.8	1500 (218)	140 (20)	...	600	1110
Aluminum	70	10.2	0.33	160	...	22.4	0.96	2.7	370 (54)	0	...	350	660

These properties are typical of these classes of materials, but in many cases, large variations may exist between various formulations. Strengths are for four-point bending spans of 19.05/9.525 mm (0.745/0.375 in.) and bar cross sections of 6.35 × 3.175 mm (0.25 × 0.125 in.). (a) Maximum excursion from 350 to 800 °C (660 to 1470 °F), initial expansion is negative

Selected properties of structural clay product types

Product	Compressive strength MPa	ksi	Water absorption, %
Fancy brick	4 to 137	0.58 to 20	2 to 25
Paving brick	17 to 137	2.50 to 20	1 to 6
Structural tile	3.5 to 69	0.51 to 10	2 to 28
	17.5 to 146	2.53 to 21	1 to 8
Drain tile for pipe	10 to 73	1.45 to 11	5 to 16

Physical properties of whitewares

Property	Earthenware	Hard porcelain	Bone china	Hotel china	Normal electrical porcelain	High-strength electrical porcelain
Water absorption, %	6 to 8	0.0 to 0.5	0.0 to 1.0	0.1 to 0.3	0.0	0.0
Specific gravity	2.6	...	2.75	2.6	2.4	2.8
Bulk density, g/m^3	2200	2400	2700	2600	2400	2770
Compressive strength, MPa (ksi)	...	400 (58)	...	...	700 (102)	700 (102)
Tensile strength, MPa (ksi)	...	23 to 34 (3 to 5)	...	...	35 (5)	56 (8)
Modulus of rupture, MPa (ksi)	55 to 72 (8 to 10)	39 to 69 (5.5 to 10)	97 to 111 (14 to 16)	82 to 96 (12 to 14)	105 (15)	175 (25)
Modulus of elasticity, GPa (10^6 psi)	55 (8)	69 to 79 (5.5 to 11.5)	96 (13.9)	82 (12)	...	...
Linear thermal expansion (μm/m · K)						
20 to 500 °C	7.3 to 8.3	...	8.4	7.3 to 8.3	5.7	6.7
20 to 1000 °C	...	3.5 to 4.5	...	...	...	...
Thermal conductivity, W/m · K						
20 to 100 °C	1.26	...	1.26	...	...	...
Dielectric constant at 1 MHz	...	...	...	...	5.6	6.9
Power factor at 1 MHz	...	...	...	...	0.8	0.7

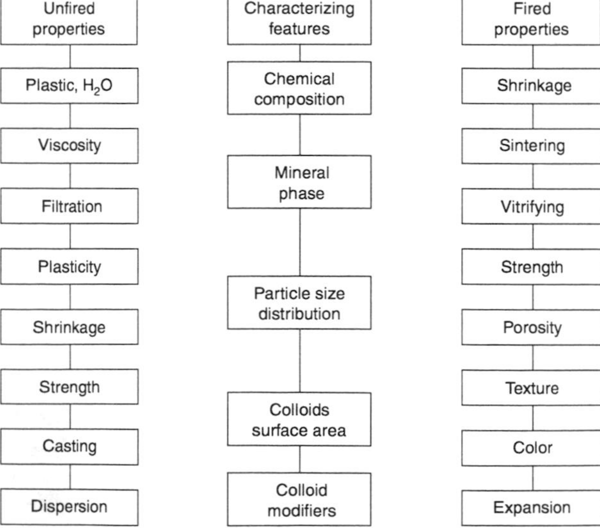

Fig. 2 Characteristic features of traditional ceramics and the properties that canbe measured in unfired and fired states

Reactions that can occur when firing clay-based triaxial bodies

Temperature °C	°F	Reaction
≤100	≤212	Loss of free moisture
100 to 200	212 to 390	Loss of adsorbed water
200 to 450	390 to 840	Crystal structure of clay minerals altered by removal of OH-groups; pyrophyllite shows a marked expansion
400 to 700	750 to 1290	Organic matter in the form of lignite is oxidized; pyrophyllite expands further
573	1065	Quartz inverts to high-temperature polymorph
700 to 950	1290 to 1740	Pyrophyllite reaches maximum expansion; metakaolin converts to spinel in clay
950 to 1100	1740 to 2010	Mica structure is destroyed. Talc decomposes to protoenstatite and glass. Mullite forms from spinel. Pyrophyllite converts to mullite and glass
1100 to 1200	2010 to 2190	Feldspars melt; clay and cristobalite dissolve. Vitrification begins, porosity decreases, shrinkage increases
>1200	>2190	Protoenstatite from talc converts to clinoenstatite. Mica breaks down into alumina and glass. Glass content increases, mullite needles grow; only closed porosity remains

Moisture content and pressure ranges for shaping clay-based bodies

Forming process	Superimposed pressure MPa	ksi	Liquid content, %
Soft plastic	0.1 to 0.75	0.015 to 0.109	20 to 30
Rigid plastic			
Piston	3 to 15	0.45 to 2.18	12 to 16
Auger	20 to 50	2.9 to 7.3	15 to 20
Dry press	100 to 200	14.5 to 29	5 to 15
Dust press	150 to 250	21.8 to 36.3	0 to 2
Isostatic press	200 to 400	14.5 to 58.0	0 to 15
Slip cast	0.1 to 3	0.015 to 0.45	20 to 35

Tile and Brick

Properties of ceramic tiles measured according to ASTM methods

Property	Tile type: Flat glazed wall tile	Flat ceramic mosaic tile	Flat quarry tile	Flat paver tile
Non-destructive tests				
Thickness	X	X	X	X
Facial dimension	X	X	X	X
Spacers	X			
Warpage	X	X	X	X
Wedging	X	X	X	X
Color uniformity	X			
Electrical properties		X		
Destructive tests				
Water absorption	X	X	X	X
Crazing	X	X	X	X
Thermal shock	X	X	X	X
Bond strength	X	X	X	X
Breaking strength	X	X	X	X
Abrasive hardness		X	X	X

European (EN) standards for ceramic tiles

Standard type	Standard no.	Description/classification
General	EN 87	Definition, classification
	EN 163	Sampling
Testing method	EN 98	Format, appearance
	EN 99	Water absorption
	EN 100	Bending strength
	EN 101	Mohs hardness
	EN 102	Wear resistance of unglazed tiles
	EN 103	Linear thermal expansion
	EN 104	Thermal shock resistance
	EN 105	Crazing resistance
	EN 106	Resistance of unglazed tiles to chemicals
	EN 122	Resistance of unglazed tiles to chemicals
	EN 154	Wear resistance of glazed tiles
	EN 155	Steam expansion
	EN 202	Frost resistance
Product requirements	EN 121	Group AI
	EN 159	Group BIII
	EN 176	Group BI
	EN 177	Group BIIa
	EN 178	Group BIIb
	EN 186	Group AIIa
	EN 187	Group AIIb

Classification of ceramic tiles in compliance with EN 87

Tile type	Tile classification/water absorption, E, wt%: $E \leq 3$ (Group I)	$3 < E \leq 6$ (Group IIa)	$6 < E \leq 10$ (Group IIb)	$E > 10$ (Group III)
Extruded tiles(a)	AI	AIIa	AIIb	AIII
Pressed tiles(b)	BI (d)	BIIa	BIIb	BIII
Cast tiles(c)	CI	CIIa	CIIb	CIII

(a) Includes split tiles and individually extruded tiles. (b) Includes dry-pressed floor and wall tiles. (c) Includes cast floor and wall tiles. (d) Water absorption for BI unglazed tiles is <1.5 wt%. For unglazed and very vitrified tiles, E<0.5

Selected technical features of vitrified, half-vitrified, and porous ceramic tiles

Test method standard	Property	Tile type/product requirement standard: BI (EN 176)	BIIa (EN 177)	BIIb (EN 178)	BIII (EN 159)
EN 100	Bending strength, MPa (ksi)	≥27 (≥3.9)	≥20 (≥2.9)	≥16 (≥2.3)	≥12 to 15 (≥1.7 to 2.2)
EN 101	Scratch hardness (Mohs scale)				
	Glazed tiles	≥5	≥5	≥5	...
	Unglazed tiles	≥6	≥6	≥6	...
	Wall tile	...	...	...	≥3
	Floor tile	...	...	...	≥5
EN 102	Abrasion resistance, mm^3(a)	≤205	≤345	≤540	...
EN 103	Coefficient of thermal expansion at 20 to 100 °C, 10^{-6}/K	≤9	≤9	≤9	≤9
EN 155	Steam expansion, mm/m	...	...	≤0.6	...

(a) The abrasion resistance for unglazed tiles. For glazed tiles, the abrasion resistance is specified by the manufacturer

Property comparisons of several types of ceramic wall tiles

Test method standard	Property	Tile type			
		Group BIII (EN 159)	Porous single-fired	Cottoforte	Majolica
EN 99	Water absorption, *E*, wt%	>10	13 to 18	16 to 19	18 to 24
EN 100	Bending strength, MPa (ksi)	≥12 (≥1.7)	≥20 (≥2.9)	≥15 (≥2.2)	≥13 (≥1.9)
EN 101	Scratch hardness (Mohs scale)				
	Wall tiles	≥3	4	4	4
	Floor tiles	≥5	5	5	5
EN 154	Abrasion resistance(a)		Specified by the manufacturer		
EN 103	Coefficient of thermal expansion at 20 to 100 °C, 10^{-6}/°C	≤9	7 to 7.5	6.5	7 to 7.5

(a) Glazed tile requirement

Selected properties of porcelain tile

Standard	Property	Value prescribed by the standards	Real value of the products
EN 99	Water absorption, *E*, wt%	≤0.5	<0.2
EN 100	Bending strength, MPa (ksi)	≥27 (≥3.9)	≥50 (≥7.25)
EN 101	Scratch hardness (Mohs scale)	≥6	718
EN 102	Abrasion resistance(a), mm^3	≤205	<130
EN 103	Coefficient of thermal expansion at 20 to 100 °C, 10^{-6}/°C	≤9	<7

(a) Unglazed tile requirement

The influence of brick and mortar strength on wall strengths

			Type M mortar						Type N mortar					
	Brick compressive strength		Masonry cube strength		Wall strength		Mortar strength		Masonry cube strength		Wall strength		Mortar strength	
Sample no.	MPa	psi	MPa	psi	MPa	psi	MPa	psi	MPa	psi	MPa	psi	MPa	psi
B3	96	14,000	35.5	5,147	20.0	2,900	14.8	2,140	30.6	4,440	14.2	2,053	4.8	709
C3	90	13,000	37.9	5,507	25.0	3,633	11.7	1,703	38.6	5,592	17.3	2,513	3.8	553
A3	83	12,000	32.6	4,730	21.3	3,090	12.4	1,795	30.6	4,440	15.6	2,260	5.3	775
F3	52	7,500	20.3	2,940	15.7	2,280	18.6	2,693	14.6	2,120	11.0	1,603	5.8	840
E3	48	7,000	19.5	2,833	16.1	2,340	13.1	1,900	17.8	2,583	10.9	1,583	5.4	784
E16	45	6,500	24.0	3,483	15.3	2,213	11.1	1,615	20.2	2,933	9.7	1,412	5.0	725
G14	31	4,500	12.7	1,848	13.2	1,915	16.7	2,427	10.0	1,452	8.9	1,293	4.9	720
G5	31	4,500	14.9	2,160	12.8	1,860	14.7	2,135	14.9	2,165	10.3	1,487	6.4	919

The general range of property values for commercial brick used in the United States

Property	Value
Compressive strength:	
Extruded brick, MPa (psi)	41 to 110 (6,000 to 16,000)
Molded brick, MPa (psi)	21 to 45 (3,000 to 6,500)
Modulus of rupture, MPa (psi)	4.8 to 27.6 (700 to 4,000)
Modulus of elasticity, GPa (10^6 psi)	9.7 to 34.5 (1.4 to 5.0)
Bulk density, g/cm^3	1.65 to 2.08
Absorption for room-temperature water, %	0.5 to 10.0
Saturation coefficient	0.6 to 0.9
Initial rate of absorption, $kg/m^2 \cdot$ min. (g/min. · 30 $in.^2$)	0.05 to 3.6 (1 to 70)
Thermal expansion coefficient, μm/m · K (μin./in. · °F)	4.5 to 9.0 (2.5 to 4.5)
Moisture expansion, %	0.02 to 0.09
Shrinkage in service, %	0.0
Corrosion resistance	Can contain all acids hot or cold except hydrofluoric; resistant to all alkalies
Thermal conductivity, W/m · K (Btu · in./ft^2 · h · °F)	0.43 to 1.44 (3 to 10)

Cements and Concretes

Uses of portland cement mortars and concretes and some alternative materials

Concrete products	Alternative materials
Factory-fabricated concrete products	
Unreinforced products:	
Block	Clay brick, adobe brick, stone
Brick	Clay brick, adobe brick, stone
Tile	Clay tile, slate, wood shingles, fiber-reinforced cement
Steel-reinforced products:	
Pipe	Steel, fiber-reinforced cement, plastic, clay
Panels	Clay brick, fiber-reinforced concrete, plywood
Beams	Steel, lumber
Railroad ties	Wood
Boat hulls	Steel, wood, fiber-reinforced plastic
Other products:	
Fiber-reinforced products	Aluminum, steel, wood, plastic, glass, fired clay
Extruded products	Aluminum, steel, wood, plastic
Field-cast concrete structures	
Predominantly unreinforced concrete:	
Highway pavements	Bituminous concrete
Canal linings, dams	Earth
Foundations	Steel, wood
Steel-reinforced concrete:	
Columns	Steel
Floor slabs	
Tunnel linings	Steel
Bridge decks	Bituminous concrete
Marine structures	Steel
Nuclear pressure vessels	Steel
Railroad structures	Steel
Other:	
Terazzo	Resin formulations
Stucco	Resin formulations
Masonry mortar	Resin formulations
Oil-well grouts	
Concrete patching	Resin formulations, bituminous materials
Roof decks	Steel, plywood

Chemical constituents for various types of portland cement

	$3CaO \cdot SiO_2$	$2CaO \cdot SiO_2$	$3CaO \cdot Al_2O_3$	$4CaO \cdot Al_2O_3 \cdot Fe_2O_3$	$CaSO_4$	MgO	Free CaO
Type I	45	27	11	8	3.1	2.9	0.5
Type II	44	31	5	13	2.8	2.5	0.4
Type III	53	19	11	9	4	2	0.7
Type IV	28	49	4	12	3.2	1.8	1.9
Type V	38	43	4	9	2.7	1.9	0.5

Approximate composition and fineness ranges for ASTM standard types of portland cement

	Composition, wt%				
Compound/property	Type I	Type II	Type III	Type IV	Type V
C_3S	42 to 65	35 to 60	45 to 70	20 to 30	40 to 60
C_2S	10 to 30	15 to 35	10 to 30	50 to 55	15 to 40
C_3A	0 to 17	0 to 8	0 to 15	3 to 6	0 to 5
C_4AF	6 to 18	6 to 18	6 to 18	8 to 15	10 to 18
$C\bar{S}H_2$	3 to 6	3 to 6	3 to 6	3 to 6	3 to 6
Specific surface area, m^2/kg (ft^2/lb)	300 to 400 (1465 to 1955)	280 to 380 (1365 to 1855)	450 to 600 (2195 to 2930)	280 to 320 (1365 to 1560)	290 to 350 (1415 to 1710)

Compressive strength data for typical cements versus time

Cement type	1-day strength MPa	ksi	3-day strength MPa	ksi	7-day strength MPa	ksi	28-day strength MPa	ksi	91-day strength MPa	ksi
Type I	9.3	1.3	22.5	3.3	32	4.6	42	6.1	50.5	7.3
Type II	14	2.0	27	3.9	36.6	5.3	46.3	6.7	52.5	7.6
Type III	21	3.0	37.5	5.4	44.2	6.4	52.3	7.6	56.0	8.1
Type IV	...	...	9.6	1.4	13.9	2.0	34.3	5.0	...	...
Type V	...	...	22.1	3.2	29.5	4.3	41.3	6.0	...	...
White portland	...	...	26.5	3.8	36.2	5.3	46.5	6.7	...	...
Portland/blast-furnace	8.6	1.2	13.1	1.9	25.3	3.7	45.0	6.5	53.3	7.7

Typical ranges of compressive and tensile strengths of moist-cured concretes

Property	Strength range MPa	psi
Compressive:		
1 day	2 to 40	300 to 6000
3 days	6 to 60	900 to 9000
7 days	10 to 65	1500 to 9500
28 days	20 to 70	3000 to 10,000
Tensile:		
1 day	0.2 to 4	30 to 600
3 days	0.6 to 6	60 to 900
7 days	1 to 6	150 to 900
28 days	2 to 7	300 to 1000

Ingredients other than portland-cement clinker used in blended cements or added separately to concrete as mineral admixtures or supplementary cementing materials

Ingredient	Composition, wt% (major oxides only) CaO	SiO_2	Al_2O_3	Fe_2O_3	Typical % used(a)
Pozzolans:					
Natural pozzolan	0.7 to 9.3	45 to 89	3 to 19	2.5 to 10	15 to 25
Fly ash	1 to 13	35 to 62	12 to 23	3 to 45	15 to 25
Silica fume	0.1 to 0.6	86 to 98	0.1 to 0.6	0 to 1	10 to 15
Ground, granulated blast-furnace slag(b)	37 to 45	32 to 37	10 to 16	0.3 to 9.5 (FeO)	15 to 50

(a) By mass of the portland cement. (b) Slag also contains 3.5 to 8.5% MgO

Characteristics of concretes and mortars in which improvements could bring benefits to specific applications

Application	Compressive strength	Flexural or tensile strength	Bond strength	Volume stability	Controlled expansion	Uniform appearance	Color	Low density	Flow properties	Young's modulus	Impact resistance	Ductility	Energy absorption	Fracture toughness	Early strength
Block	X			X		X	X	X			X				X
Brick	X			X		X					X				
Pipe	X	X		X							X			X	X
Panels		X		X		X	X	X			X			X	X
Beams	X	X		X						X		X		X	X
Tile		X				X		X	X		X				X
Extruded products		X		X		X			X		X		X	X	X
Fiber-reinforced products		X		X		X			X		X	X	X	X	X
Boats		X		X				X			X				
Railroad ties		X									X	X	X	X	X
Foundations	X	X													
Missile silos	X	X		X							X		X	X	
Columns	X	X				X				X					
Slabs		X		X				X							
Highways		X		X					X						

(continued)

Characteristics of concretes and mortars in which improvements could bring benefits to specific applications (continued)

Application	Compressive strength	Flexural or tensile strength	Bond strength	Volume stability	Controlled expansion	Uniform appearance	Color	Low density	Flow properties	Young's modulus	Impact resistance	Ductility	Energy absorption	Fracture toughness	Early strength
Canal linings		X		X					X					X	
Tunnel linings	X	X		X					X		X	X			X
Bridge decks		X		X											
Desalinization plants				X											
Dams	X			X											
Marine construction	X	X		X											
Nuclear pressure vessels	X	X		X					X						
Terazzo		X	X	X		X									X
Stucco		X	X	X		X			X		X				
Masonry mortar	X		X	X					X						
Oil well grouts			X	X					X						
Concrete patching			X	X											X
Refractory linings			X	X				X	X					X	X
Roofing		X	X	X		X		X	X		X				
Elevated railroad structures	X	X	X			X					X	X		X	
Hardened MX missile sites	X	X	X								X	X	X	X	

Application	Quick setting	Low heat liberation	Low permeability	Freeze-thaw resistance	Sulfate and salt resistance	Low thermal expansion	Abrasion resistance	Stain resistance	Low thermal conductivity	High temperature resistance	Low cost	Estimate of fraction of total quantity of cement use
Block			X						X			4
Brick									X			0.2
Pipe	X		X		X		X					2
Panels						X		X				2
Beams	X					X						2
Tile	X		X	X				X				0.5
Extruded products	X		X									0.2
Fiber-reinforced products	X		X		X	X						2
Boats			X	X	X	X						0.1
Railroad ties	X			X	X		X				X	0.5
Foundations		X			X						X	40
Missile silos		X	X	X	X							0.2
Columns												8
Slabs			X	X	X							15
Highways			X	X	X	X	X				X	15
Canal linings			X	X	X	X	X				X	2.5
Tunnel linings	X		X		X	X	X				X	2.0
Bridge decks				X	X	X	X		X			1.5
Desalinization plants			X		X	X	X		X			0.1
Dams		X		X	X	X					X	0.7
Marine construction			X	X	X	X	X					1.1
Nuclear pressure vessels		X				X						0.1
Terazzo	X		X			X	X	X				0.1
Stucco			X					X				0.5
Masonry mortar				X		X						4.5
Oil well grouts					X					X		1.4
Concrete patching	X			X	X	X	X					0.1
Refractory linings	X	X				X	X		X	X		0.1
Roofing				X		X						0.1
Elevated railroad structures				X	X							0.5
Hardened MX missile sites			X	X							X	Large

Refractories

Chemical analysis of some typical refractory raw materials

Constituent	Refractory material composition, wt%							
	Magnesia(a)	Chromite(b)	Dolomite(c)	Bauxite(d)	Alumina(e)	Kyanite(c)	Flint clay(f)	Quartzite
Silica (SiO_2)	0.8	5.1	0.7	6.5	0.06	40.0	51.0	99.5
Alumina (Al_2O_3)	0.4	29.9	0.3	88.3	≥99.5	56.6	44.0	0.14
Titania (TiO_2)	trace	trace	trace	3.2	trace	1.0	2.5	0.03
Iron oxide (Fe_2O_3)	1.4	13.0	0.9	1.8	0.06	1.0	0.9	0.17
Lime (CaO)	0.9	0.5	57.7	...	...	0.5	0.2	...
Magnesia (MgO)	96.3	18.7	40.4	...	trace	0.5	0.3	0.01
Chromia (Cr_2O_3)	ND(g)	32.2	ND	ND	ND	ND	ND	ND
Alkalis ($Na_2O + K_2O + Li_2O$)	ND	ND	ND	0.02	0.05	0.2	0.6	0.03
Loss on ignition	...	0.3	47.0	0.25	...	14.2	13.2	...

(a) Seawater, U.K. (b) Philippine lump. (c) United States. (d) Refractory grade, Guyana. (e) Tabular. (f) Missouri. (g) No data

Physical properties of some typical fired refractory brick

	Magnesia (95% MgO)	Chrome (30% Cr_2O_3)	90% alumina	70% alumina	Zircon	Fireclay (Missouri superduty)	Silicon carbide	Silica (superduty)
Bulk density, g/cm^3	2.805 to 2.950	3.060 to 3.140	2.900 to 2.965	2.530 to 2.600	3.605 to 3.720	2.310 to 2.370	2.565 to 2.660	1.780 to 1.875
Porosity, %	15 to 19	16 to 20	14 to 18	17.5 to 21.5	19 to 23	11 to 14	11 to 15	20 to 24
Cold crushing strength, MPa (ksi)	48 to 70 (7 to 10)	35 to 55 (5 to 8)	62 to 95 (9 to 14)	27 to 48 (4 to 7)	48 to 76 (7 to 11)	12 to 21 (1.7 to 3)	69 to 83 (10 to 12)	27 to 41 (4 to 6)
Modulus of rupture, MPa (ksi)	17 to 24 (2.5 to 3.5)	14 to 21 (2 to 3)	17 to 21 (2.5 to 3)	7.6 to 11 (1.1 to 1.6)	15 to 23 (2.2 to 3.3)	4.8 to 6.9 (0.7 to 1)	21 to 24 (3 to 3.5)	4.1 to 6.9 (0.6 to 1)
Reheat test, % permanent linear change after heating to:								
1600 °C (2910 °F)	...	...	...	+3.5 to 6.0	...	0.0 to 0.9	...	...
1650 °C (3000 °F)	...	...	...	...	0.0	...	...	...
1725 °C (3135 °F)	−0.2 to 1.0	...	+1.0 to 1.0	...	...	...	-0.1 to 0.1	...
Load test at 170 kPa (25 ksi); withstands load to temperature, °C (°F)	1620 (2950)	1400 (2550)	1760 (3200)	1450 (2640)	1600 (2910)	1450 (2640)	1650 (3000)	1680 (3055)

Mean specific heats of refractory materials between 0 °C (32 °F) and indicated temperature

Temperature		Specific heat, kJ/kg · K				
°C	°F	Fireclay brick	Silica brick	Magnesia brick	Chrome brick	99% alumina brick
0	32	0.808	0.708	0.871	0.712	0.716
200	390	0.863	0.883	0.971	0.762	0.896
400	750	0.913	0.988	1.043	0.808	0.976
600	1110	0.976	1.051	1.097	0.850	1.034
800	1470	1.022	1.080	1.130	0.879	1.063
1000	1830	1.063	1.110	1.168	0.909	1.093
1200	2190	1.097	1.139	1.206	0.930	1.118
1400	2550	1.122	1.164	1.239	0.942	1.139
1500	2730	1.143	1.177	1.256	0.950	1.164

Composition and selected properties of basic refractory materials

Type	Composition	Maximum use temperature in oxygen °C	Maximum use temperature in oxygen °F	Thermal conductivity, kcal/min. · °C At 300 °C (570 °F)	At 800 °C (1470 °F)	At 1200 °C (2190 °F)	Refractoriness under load of 197 kPa (28.5 psi) °C	°F
Silica	93 to 96% SiO_3	1700	3090	0.8 to 1.0	1.2 to 1.4	1.6 to 1.8	1650 to 1700	3000 to 3090
Fireclay	15 to 45% Al_2O_3 55 to 80% SiO_2	1300 to 1450	2370 to 2640	0.9 to 0.9	1.0 to 1.2	2.5 to 2.8	1250 to 1450	2280 to 2640
Magnesite	80 to 95% MgO Fe_2O_3, Al_2O_3	1800	3270	3.8 to 9.7	2.8 to 4.7	2.5 to 2.8	1500 to 1700	2730 to 3090
Chromite	30 to 45% Cr_2O_3 14 to 19% MgO 10 to 17% Fe_2O_3	1700	3090	1.3	1.6	1.8	1400 to 1450	2550 to 2640

Composition and selected properties of high-duty refractory oxides

Type	Composition	Melting point °C	Melting point °F	Maximum use temperature in oxygen °C	°F	Thermal conductivity, kcal/min. · °C At 100 °C (212 °F)	At 500 °C (930 °F)	At 1000 °C (1830 °F)	At 1500 °C (2730 °F)	Refractoriness under load of 196 kPa (28.4 psi) °C	°F
Aluminum oxide	100% Al_2O_3	2015	3660	1950	3540	26.0	9.4	5.3	5.0	2000	3630
Beryllium oxide	100% BeO	2550	4620	2400	4350	189.0	56.3	17.5	13.5	2000	3630
Magnesium oxide	100% MgO	2800	5070	2400	4350	31.0	12.0	6.0	5.4	2000	3630
Silicon dioxide	100% SiO_2	...	...	1200	2190	0.8	1.4	1.8	...	...	...
Mullite	72% Al_2O_3 28% SiO_2	1830(a)	3325(a)	1850	3362	5.3	3.8	3.4	...	...	...

(a) Incongruent

Composition, applications, and origin of selected body raw materials, bonding materials, and special additives used to produce refractory products

Raw material	Typical composition	Applications	Source of supply
Body materials			
Silica	>95% SiO_2 silica rock	Production of fired silica brick for use in: coke ovens, glass tanks, roofs	Worldwide supplies of suitable material
	Silica sand	Production of sand molds for metal castings	
Chammotte	42 to 44% Al_2O_3 calcined aluminosilicate Low Fe_2O_3 (typically <1%) Low alkalis Low alkaline earths	Production of fired 42/44 Al_2O_3 aluminosilicate bricks for use in: blast furnaces; cement kilns; production of 42/44 aluminosilicate general purpose castable	Reserve of high-quality clay limited. Main supplies from South Africa, United States, China, and France
Andalusite	58 to 60% Al_2O_3 aluminosilicate mineral liberated from host rock Low Fe_2O_3 (typically <1.5%) Low alkalis Low alkaline earths	Production of fired 60% Al_2O_3 aluminosilicate bricks for use in: blast furnaces; steel ladles; torpedo ladles; and aluminum anode baking furnace	Reserve of high-quality andalusite limited. Main supplies from South Africa and France. China now entering the market
Bauxite	85 to 88% Al_2O_3 calcined aluminosilicate <1.5% Fe_2O_3, <3.5% TiO_2 Low alkalis Low alkaline earths	Production of fired 85 to 88% Al_2O_3 aluminosilicate bricks for use in: steel ladles, torpedo ladles, aluminum holding vessels Production of phosphate-bonded brick for aluminum remelt furnaces Production of castables	Reserves of high-quality ore limited to Guyana (gibbsite $Al_2O_3 \cdot 3H_2O$) and China (diaspore $Al_2O_3 \cdot H_2O$)
Doloma	40% MgO, <1.5% Fe_2O_3, <1% SiO_2, <1% Al_2O_3, ~ 56 to 60% CaO	Production of fired doloma brick for use in: cement kiln hot zones; argon oxygen decarburization (AOD) vessels for stainless steel; and steel ladles	Reserves of dolomite for dead burning available in United States, Belgium, Germany, and United Kingdom
Magnesia	High-quality dead burnt magnesia Low porosity (bulk density >3.40 g/cm^3) Controlled CaO:SiO_2 ratio: >2.5:1 0.5% SiO_2, <0.2% Fe_2O_3, <0.1% Al_2O_3, <0.05% B_2O_3, <96% MgO >90% MgO	Used in the production of magnesia-carbon brick for basic oxygen steelmaking, electric arc furnaces, and steel ladles Used in the production of fired brick for glass tank regenerators, magnesia-spinel bricks for cement kilns, and slide gate plates High-quality magnesia chrome for use in secondary steel making Used in the production of magnesia chrome, chrome-magnesia fired bricks for cement kilns, nonferrous (copper), and secondary steel making Also as a gumming repair material	High-quality material sources from brine sea water and beneficiated natural magnesite found in Israel, Holland, Ireland, United Kingdom, United States, Greece, and Japan Low-quality material sources from sea water and natural magnesite found in Greece, China, Brazil, and Czechoslovakia

(continued)

Composition, applications, and origin of selected body raw materials, bonding materials, and special additives used to produce refractory products (continued)

Raw material	Typical composition	Applications	Source of supply
Chrome ore	Iron chromite mineral: a mixed spinel of FeO, MgO, Fe_2O_3, Cr_2O_3, and Al_2O_3 Cr_2O_3 levels: 32 to 56%	Used in the production of fired magnesia-chrome and chrome-magnesia bricks	Supplied after separation from impurities from Philippines, South Africa, and Zimbabwe
Graphite	Flake graphite mineral is separated from host rock Carbon level 85 to 95% Remainder aluminosilicate impurities	Used in the production of magnesia-carbon bricks and alumina-carbon, zirconia-carbon continuous casting products	Supplied from China, Norway, Sri Lanka, and Malagasy
Carbon	Electrocalcined or gas calcined anthracite or petroleum coke Carbon level >95% Remainder aluminosilicate impurities	Used in the production of fired carbon for blast furnaces and as the cathode in aluminum reduction cells	Supplied locally to suit market needs in United States, Germany, United Kingdom, and Poland
Alumina	Fused or tabular alumina: >99% Al_2O_3	Used in alumina carbon continuous casting products Fired alumina slide gates High-quality monolithics for the petrochemical industry and for blast furnace application	Produced by fusing or calcining Bayer grade alumina in United States, Holland, United Kingdom, and Japan
Aluminosilicate fiber	42 to 44% Al_2O_3 fiber produced from high-quality kaolinitic clay	Used as a fiber insulation in kilns, glass tanks, and so on	Produced in United States, Europe, and Japan
Mullite	Sintered or fused 72% Al_2O_3 aluminosilicate, $3Al_2O_3 \cdot 2SiO_2$	Creep-resistant refractory used as fired shapes in glass tanks, tunnel kilns, continuous casting	Produced in United States, United Kingdom, and Japan
Spinel	Sintered or fused magnesia aluminate, $MgO \cdot Al_2O_3$	Addition to fired magnesia bricks to improve thermal shock resistance for application in cement kilns and glass tanks	Produced in United States, United Kingdom, Germany, and Japan
Olivine	Naturally occurring magnesium silicate	Used as a monolithic coating in tundishes	Scandinavia
Silicon carbide	SiC produced by the Acheson process: >98% SiC Impurities: Si; SiO_2; C; and Fe_2O_3	Used as a nitride Sialon or self-bonded product in blast furnaces and aluminum reduction cells As a silicate-bonded product in incinerators and power plants	Norway and China
Zircon	Naturally occurring zirconium silicate sand, $ZrO_2 \cdot SiO_2$, containing small quantities of HfO_2, Al_2O_3, TiO_2, and Fe_2O_3	Used in investment casting, foundries, as a glass contact refractory, an aluminum contact refractory Fluctuating demand as a ladle refractory Raw material for the production of zirconia and zirconia mullite	Australia, South Africa, and United States
Zirconia	Fused zirconia with >96% ZrO_2 Main impurities: HfO_2; SiO_2 (<0.5%); Al_2O_3 (<0.5%); Fe_2O_3 (<0.1%); TiO_2 (0.2%)	Used in zirconia carbon continuous casting refractories as fired zirconia shapes, nozzles, kiln furniture, and so on, and as an addition to cement kiln doloma refractories to improve thermal shock resistance	Naturally occurring ZrO_2 (baddeleyite) found in South Africa Zirconia derived from zircon in United States, United Kingdom, and Germany
Zirconia mullite	Fused or sintered impurities in zircon diluted by high-purity mullite ($ZrO_2 \cdot 3Al_2O_3 \cdot 2SiO_2$)	Used in continuous casting refractories	United Kingdom, United States, and Japan
Bonding materials			
Lignosulfonates	...	Temporary green bond in fired refractories	By-product of the paper industry
Plastic clay	...	Permanent green bond in fired aluminosilicate refractories and for plastic bond in rammable monolithics	Widespread
Calcium aluminate cements	...	Permanent cementitious bond for castables	Japan, United States, France, and United Kingdom
Phosphoric acid	...	Permanent bond for low-temperature fired (400 °C, or 750 °F) refractories	Commodity chemical
Sodium silicate	...	Air setting bond for mortars	Commodity chemical
Tars/pitches	...	Bonding systems for carbon-containing products Materials carbonize on firing in reducing atmosphere	By-product of coke plants and petrochemical refineries
Phenol formaldehyde resins, and furane resins	...	Bonding systems for magnesia graphite, alumina graphite, and zirconia graphite	Commodity chemicals
Special additives			
Volatilized silica	...	Used as an addition to low cement castables to improve rheological properties and sintering	By-product of the silicon industry Norway
Calcined alumina	...	Used as an addition to castables and high-quality alumina products to improve sintering	Produced from the Bayer process
Chromium oxide	...	Used as an addition to improve corrosion resistance in the presence of siliceous slags	Minor use of pigment grade Cr_2O_3
Silicon, aluminum	...	Used as powdered additives in carbon-bonded refractories to improve strength and improve oxidation resistance	Commodity materials
Proprietary phosphates	...	Used as dispersing aids in monolithic refractories	Commodity materials
Ethyl silicate	...	Used as a bond in special precast refractories	Commodity materials

Glazes, Enamels, and Coatings

Common raw materials used as sources of oxides for selected ceramic coatings

Oxide coating desired	Possible raw materials	Other oxides introduced
Al_2O_3	Clay	SiO_2
	Corundum	...
	Feldspars	SiO_2, Na_2O, K_2O
	Nepheline syenite	SiO_2, Na_2O, K_2O
BaO	Barium carbonate	...
CaO	Dolomite	MgO
	Whiting	...
	Wollastonite	SiO_2
Li_2O	Spodumene	SiO_2, Al_2O_3
MgO	Dolomite	CaO
	Heavy magnesium oxide	...
	Talc	SiO_2, CaO
Na_2O	Feldspars	SiO_2, Al_2O_3, K_2O
	Nepheline syenite	SiO_2, Al_2O_3, K_2O
PbO	Lead bisilicate(a)	SiO_2
K_2O	Feldspars	SiO_2, Al_2O_3, Na_2O
	Nepheline syenite	SiO_2, Al_2O_3, Na_2O
SiO_2	Clay	Al_2O_3
	Feldspars	Al_2O_3, Na_2O, K_2O
	Nepheline syenite	Al_2O_3, Na_2O, K_2O
	Quartz sand	...
	Talc	MgO, CaO
	Wollastonite	CaO
	Zircon	ZrO_2
SrO	Strontium carbonate	...
ZnO	Zinc oxide	...
ZrO_2	Zircon	SiO_2

(a) Lead bisilicate, as sold commercially, is a frit rather than a crystalline raw material

Composition and application of selected frits used for glazes

Frit	Cone classifier	Composition, wt%														Glaze application
		Al_2O_3	B_2O_3	BaO	CaO	F	K_2O	Li_2O	MgO	Na_2O	PbO	SiO_2	SrO	ZnO	ZrO_2	
A	5	5.4	12.3	27.7	...	...	...	...	...	5.5	...	49.1	...	...	...	Partially fritted matte
B	4	5.1	9.2	...	12.1	...	3.6	...	...	1.7	...	59.9	8.4	...	...	All-fritted leadless
C	4	6.2	8.0	...	10.4	...	2.7	...	...	1.8	17.1	53.8	...	...	...	All-fritted lead
D	01	8.0	18.2	...	1.3	...	1.6	0.5	0.3	15.2	...	53.4	...	1.3	...	Partially fritted tile
E	01	12.1	12.7	...	0.4	2.3	5.0	...	...	11.9	...	43.6	...	...	12.1	High opacity tile
F	06	...	23.3	...	20.0	...	...	...	...	10.4	...	46.4	...	...	...	Partially fritted, alkali borosilicate flux
G	06	...	21.3	...	...	...	...	...	...	...	44.7	34.0	...	...	...	All-fritted lead for artware

Typical oxide compositions of frits and glazes, wt%

Oxide	Transparent	Opaque	Lead	Borax	Mat(b)
SiO_2	50 to 60	50 to 60	30 to 40	10 to 40	50 to 60
Al_2O_3	4 to 10	4 to 10	0 to 5	0 to 5	5 to 10
B_2O_3	5 to 15	5 to 10	0 to 20	20 to 40	5 to 10
Na_2O	0 to 10	0 to 5	0 to 10	0 to 5	0 to 5
K_2O	0 to 10	0 to 5	0 to 5	0 to 5	0 to 5
CaO	0 to 10	0 to 15	0 to 5	0 to 5	5 to 20
PbO	0 to 10	0 to 5	30 to 60	20 to 50	0 to 5
ZnO	0 to 10	0 to 10	0 to 10	0 to 10	5 to 20
ZrO_2	0 to 3	5 to 15	0 to 5	0 to 5	0 to 5
Other(a)	0 to 10	0 to 10	0 to 10	0 to 10	0 to 10

(a) SrO, Fe_2O_3, SnO_2, Sb_2O_3, TiO_2, Li_2O, and others. (b) Met glazes or frits

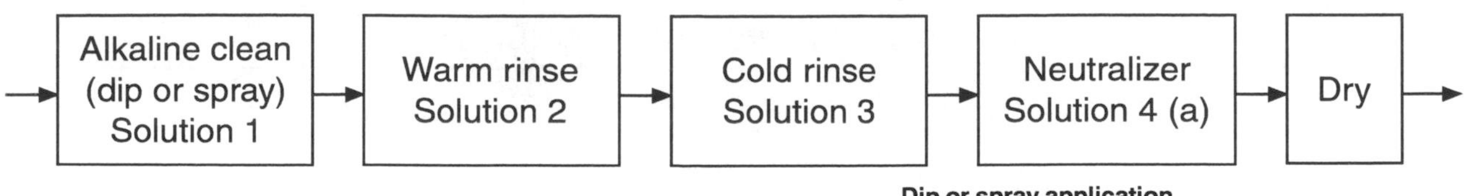

			Dip or spray application			
			Temperature		Cycle time, min.	
No.	Solution	Composition	°C	°F	Dip	Spray
1	Alkaline cleaner	Cleaner, 3.7-60 g/L (½-8 oz/gal)	Ambient to 100	Ambient to 212	6 to 12	1 to 3
2	Warm rinse	Water	49 to 60	120 to 140	½ to 4	½ to 1
3	Cold rinse	Water	Ambient	Ambient	½ to 4	½ to 1
4	Neutralize(a)	2/3 Na_2CO_3 and 1/3 borax, 0.60-2.10 g/L (0.08-0.28 oz/gal) as Na_2O	49 to 71	120 to 160	3 to 6	½ to 1

(a) Some systems may not require a neutralizer

Fig. 3 Process for preparing sheet steel for porcelain enameling using the "clean only" system

Applications in which porcelain enamels are used for resistance to corrosive environments

	Corrosive environment			
	Temperature			
Application	°C	°F	pH	Corrosive media
Bathtubs	≤ 49	≤ 120	5 to 9	Water, cleansers
Chemical ware	≤ 100	≤ 212	12	Alkaline solutions
	≤ 100	≤ 212	1 to 2	All acids except hydrofluoric
	175 to 230	350 to 450	1 to 2	Concentrated sulfuric acid, nitric acid, and hydrochloric acid
Home laundry equipment	≤ 71	≤ 160	11	Water, detergents, and bleach
Dishwashers	≤ 82	≤ 180	8 to 12	Water, strong detergents
Range exteriors	21 to 66	70 to 150	2 to 10	Food acids, cleaners
Range oven liners:				
Conventional	20 to 315	70 to 600	2 to 10	Food acids, cleaners
Pyrolytic	20 to 540	70 to 1000	2 to 10	Food acids, cleaners
Range burner grates	20 to 595	70 to 1100	2 to 10	Food acids, cleaners
Refrigerators	-18 to 20	0 to 70	2 to 10	Food acids, cleaners
Kitchen sinks, lavatories	≤ 71	≤ 160	2 to 10	Food acids, water, and cleansers
Water heaters	≤ 71	≤ 160	5 to 8	Water

Typical composition ranges of cast irons used in porcelain enameling

Constituent	Amount, %
Total carbon	3.20 to 3.60
Silicon	2.30 to 3.00
Manganese	0.30 to 0.60
Sulfur	0.05 to 0.12
Phosphorus	0.40 to 0.80

Service temperature limits for porcelain enamels

Service Temperature °C	°F	Limiting conditions
425	800	Usual limit for enamels maturing at about 815 °C (1500 °F)
540	1000	Maximum for enamels maturing at about 815 °C (1500 °F) without reboil
760	1400	Operating limit for special high-temperature enamels
1095	2000	Refractory enamels useful for short periods for protection of stainless steels and special alloys

Processing

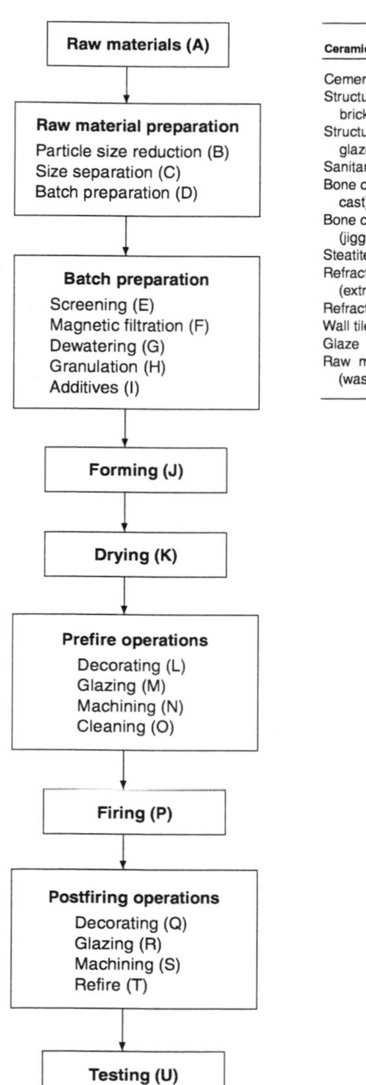

Ceramic type	Sequence of unit operations to produce ceramic type
Cements	A-D/P/L/C/I/U
Structural clay bricks	A-D/J/K/P/U
Structural clay glazed pipe	A-D/J/K/P/R/T/U
Sanitaryware	A/D/E/F/O/M/P/U
Bone china (slip cast)	A/D-F/J/K/O/P/R/T/Q/T/U
Bone china (jiggered)	A/D-G/J/K/O/P/R/T/Q/T/U
Steatite porcelain	A/ D-F/J/K/N/M/P/Q/T/U
Refractory, fireclay (extruded)	A-D/J/K/P/U
Refractory, MgO	A-D/J/K/P/U
Wall tile	A/I/D-F/H/E/J/M/K/P/U
Glaze	A/I/D-F/M
Raw materials (washed clay)	A/C-F/K/C/U

Fig. 3 Fabrication processes for various ceramic types

Factors influencing ceramic forming process selection

Forming process	Component size	Component shape	Production volume
Slip casting	Large	Complex	Low
Tape casting	Thin sheets	Simple	High
Extrusion	Wide range	Constant cross section	High
Injection molding	Small	Complex	High
Dry pressing	Small to medium	Simple, low aspect ratio	High

Note: Labels in table reflect the most preferred condition for the stated process. In practice most forming processes are used over a range of conditions

Characteristics of ceramic starting powders

Physical characteristics	Chemical characteristics
Grains (primary particles)	Bulk composition
Size	Major elements (1 to 100%)
Shape	Minor elements (10 ppm to 1%)
Agglomerates (secondary particles)	Trace elements (<10 ppm)
Size	Inorganic species (for example, sulfates and nitrates)
Shape	Organic species
Porosity	Water and other volatiles
Amount (see density)	Phases
Size	
Shape	
Particle contact	
Coordination	
Strength	
Density	
Specific surface area	Surface composition
Permeability	Major elements
Compactibility	Minor elements
Flowability	Trace elements
	Inorganic species
	Organic species
	Water
	Phases

Viscoelastic properties required by various ceramic-forming processes

Process	Degree of saturation	$y = \frac{V_L}{V_{void}}$	Forming pressure MPa	Forming pressure ksi	Flow property	Viscosity range, Pa · s	Shear stress rate, 1/s	Comment
Dry(a)								
Dry pressing	0.01 to 0.05	~0.3	20 to 200	2.9 to 29	...	...	0 to 10^3(b)	Stability of binder against hydration and degradation essential during storage
Isostatic pressing		...	...	...	Newtonian or elastic	...	0 to 10^3(b)	
Semi-wet plastic forming								
Extrusion, pressing, and jiggering	>0.9	~1	1 to 20	0.15 to 2.9	Newtonian	100 to 1000	10^2 to 10^3	Lubricants used
Injection molding	>0.9	~1	1 to 20	0.15 to 2.9	Pseudoplastic	<1000	10^2 to 10^3	Values measured at 200 °C (390 °F); both soluble and solid lubricants used
Wet								
Slip casting	>1	1.5 to 2.0	~1	~0.15	Pseudoplastic	0.1	1 to 10^3(c)	Thixotropic behavior sometimes desirable
Tape casting	>1	1.5 to 2.0	~1	~0.15	Pseudoplastic	0.1	>10^3	Flexibility after solvent removal required
Screen or thick-film printing	>1	1.5 to 2.0	~1	~0.15	Pseudoplastic	0.1	>10^4(d)	Thixotropy must be carefully controlled
Spray-dry granulation	>1	1.5 to 2.0	>1	>0.15	Newtonian	0.1	1 to 10^3(c)	Values measured at ~115 °C (240 °F)

(a) Based on granulated feedstock. (b) 0 represents static storage of feedstock. (c) highest during mixing. (d) During some application methods

Selected chemical additives used to optimize powder treatment and green forming of ceramics

Material	Application or function
Polyvinyl alcohol	Binder for advanced ceramics
Polyethylene glycol	Binder for advanced ceramics
Sodium polyacrylate	Deflocculant for slip casting
Tertiary amide polymer	Binder for dry pressing
Starch blended with dry colloidal aluminosilicate	Binder for vacuum forming
Cationic alumina plus organic flocculant	Binder for vacuum forming
Pregelled, cationic corn starch	Flocculant for colloidal silica and alumina binder
High-purity sodium carboxymethylcel -lulose	Binder
Inorganic colloidal magnesium aluminum silicate	Suspending agent
Medium-viscosity sodium carboxy-methylcellulose added to Veegum	Suspending agent, viscosity stabilizer
Ammonium polyelectrolyte	Dispersing agent for casting slips for electronic ceramics
Sodium polyelectrolyte	Dispersing agent binder for spray-dried bodies
Microcrystalline cellulose and sodium carboxymethylcellulose	Thickening agent
Polysilazane	Processing aid, binder, and precursor for advanced ceramics

Solvents commonly used in ceramics processing (room temperature)

Solvent	Dielectric constant	Surface tension, mN/m	Viscosity, Pa · s
Polar, aqueous			
Water	80	73	1.0
Polar, organic			
Alcohols			
Methyl alcohol (CH_3OH)	33	23	0.6
Isopropyl alcohol (C_3H_7OH)	18	22	2.4
n-octyl alcohol ($C_8H_{17}OH$)	10	28	10.6
Benzyl alcohol (C_7H_7OH)	13.1	35.5	5.8
Ethylene glycol ($C_2H_6O_2$)	37	48	20
Furfuryl alcohol [2-(CH_4H_3O)CH_2OH]	...	38	4.6
Acids			
Propionic acid ($CH_3CH_2CO_2H$)	3.3	26.7	1.1
Octonoic acid [$CH_3(CH_2)_6CO_2H$]	2.45	28	4.6
Aldehydes			
Octanal ($C_8H_{17}OH$)	10	28	10.6
Benzaldehyde (C_6H_5CHO)	16.8	38.5	1.32
Esters			
Ethyl acetate ($C_4H_8O_2$)	6.15	23.52	0.42
n-butyl *n*-butyrate	5	23.72	0.98
Ethers			
Isopentyl ether	2.82	22.8	0.4
Tetrahydrofuran (C_4H_8O)	7.85	27.4	0.53
Ketones			
Acetone (C_3H_6O)	20.7	25.1	0.32
Heptatone [$CH_3(CH_2)_4COCH_3$]	9.8	26.7	0.76
Methyl ethyl ketone (C_4H_8O)	18	25	0.4
Halogenated hydrocarbons			
Trichloroethylene (C_2HCl_3)	3	...	...
Nonpolar, organic			
Benzene (C_6H_6)	2.3	30.22	0.65
Toluene ($C_6H_5CH_3$)	2.4	28.5	0.6
n-hexane (C_6H_{14})	1.89	18.4	0.33

Single Oxides

Representative crystal structures of single oxides

A/O ratio	Coordination number	Type of crystal structure	Examples of oxides
A_2O	2	Cuprite	Cu_2O, Ag_2O, Pb_2O
	4	Antifluorite	Li_2O, K_2O, Na_2O
AO	4	Wurtzite	ZnO, BeO, CdO
	6	Rock-salt	MgO, CaO, SrO, BaO FeO, CoO, NiO, VO
A_3O_4	4, 6	Spinel	Fe_3O_4, Mn_3O_4, Co_3O_4
	6	Corundum	αAl_2O_3, αFe_2O_3, Cr_2O_3
	6	C-type rare-earth oxide	Y_2O_3, In_2O_3, Sc_2O_3
	7	A-type rare-earth oxide	La_2O_3, Nd_2O_3, Pr_2O_3
AO_2	4	Cristobalite	SiO_2
		Tridymite	SiO_2
		Quartz	SiO_2
	6	Rutile	TiO_2, SnO_2, GeO_2, VO_2
	8	Fluorite	ZrO_2, CeO_2, ThO_2, HfO_2
A_2O_5	...	...	V_2O_5, Nb_2O_5, Ta_2O_5
AO_3	...	...	WO_3, MoO_3, ReO_3

Melting points of *AO*-type oxides

Oxide	*Tm*, K
BeO	2843
MgO	3098
CaO	2887
SrO	2727
BaO	2196
TiO	2023
MnO	2115
FeO	~1640
CoO	2208
NiO	2257
CuO	1300
ZnO	Evaporate
CdO	Evaporate
PbO	1159

Property-value comparison of the principal single oxides

Property	Typical range of values at: Room temperature	1095 °C (2000 °F)	Exceptions (room-temperature values)
Physical			
Crystal system	(a)	...	...
Theoretical density, g/cm^3	3 to 11.5	...	...
Melting point, °C (°F)	2040 to 2870 (3700 to 5200)(b)	...	ThO_2 (3220 °C); TiO_2 (1840 °C)(b)
Thermal			
Specific heat, kJ/kg · K (Btu/lb · °F)	0.4 to 0.8 (0.1 to 0.2)	0.4 to 1.3 (0.1 to 0.3)	BaO; ThO_2; UO_2 (all 0.25 kJ/kg · K)
Thermal conductivity, W/m · K (Btu/ft · h · °F)	1.7 to 17 (1 to 10)	1.7 to 6.9 (1 to 4)	BeO (24 W/m · K); MgO (52 W/m · K); Al_2O_3 (35 W/m · K)
Linear thermal expansion, %	0.2 to 0.3(c)	0.8 to 1.5	SrO (0.4); CaO (0.4); Cr_2O_3 (0.1)(c)
Mechanical			
Bend strength, MPa (ksi)	140 to 275 (20 to 40)	69 to 140 (10 to 20)	Al_2O_3 (450 MPa); CaO (69 MPa)
Compressive strength, MPa (ksi)	2070 (300)	550 to 895 (80 to 130)	MgO (825 MPa); UO_2 (690 MPa)
Tensile strength, MPa (ksi)	69 to 140 (10 to 20)	34 to 105 (5 to 15)	Al_2O_3 (275 MPa)
Impact strength, J (in. · lb)	0.09 to 0.12 (0.8 to 1.1)	...	(d)
Modulus of elasticity, GPa (10^6 psi)	205 to 415 (30 to 60)	205 to 345 (30 to 50)	...
Shear modulus, GPa (10^6 psi)	69 to 140 (10 to 20)	69 to 140 (10 to 20)	...
Bulk modulus, GPa (10^6 psi)	0 to 170 (0 to 25)	...	...
Poisson's ratio	0.2 to 0.5	(e)	...
Creep rate, 1/h	...	(f)	...
Hardness, Mohs	6 to 9	...	BaO; CaO; SrO (all 3 to 4)
Microhardness, kg/mm^2	600 to 1000	...	Al_2O_3 (3000)
Thermal-stress resistance	...	(f)	...
Oxidation and corrosion resistance	...	(f)	...

(a) The most common or useful phases of the majority of the principal single oxides crystallize in the cubic or hexagonal systems. (b) "Room temperature" heading not applicable. (c) At 315 °C (600 °F); "Room temperature" heading not applicable. (d) Data available only for Al_2O_3 and MgO. (e) Values generally higher at elevated temperatures. (f) Varies widely; direct comparison difficult

Physical characteristics of Al_2O_3 bioceramics

	High-alumina ceramics	ISO Standard 6474
Alumina content, %	<99.8	≥99.50
Density, g/cm^3	>3.93	≥3.90
Average grain size, μm	3 to 6	<7
Surface roughness (R_a), μm	0.02	...
Hardness, Vickers	2300	>2000
Compressive strength, MPa (ksi)	4500 (653)	...
Bending strength, MPa (ksi) (after testing in Ringer's solution)	550 (80)	400 (58)
Modulus of elasticity, GPa (10^6 psi)	380 (55.2)	...
Fracture toughness (K_{Ic}), MPa $\sqrt{m}$ (ksi $\sqrt{in.}$)	5 to 6 (4.5 to 5.5)	...

Electrical properties of high-alumina ceramics

Property	Frequency, Hz	vitreous 85% 25 °C (77 °F)	vitreous 85% 500 °C (930 °F)	vitreous 95% 25 °C (77 °F)	vitreous 95% 500 °C (930 °F)	vitreous 99.5+%	porous 99.5+%
Dielectric constant	60	8.4	...	9.2	...	...	...
	10^3	7.65 to 8.75	13.86	8.84 to 10.51	13.3	10	...
	10^6	7.4 to 8.95	8.87	8.81 to 9.60	9.03	...	5.5
	10^8	8.10 to 8.95	...	8.80 to 9.60	...	...	5.3
	10^9	...	...	8.60	...	...	...
	3×10^9	8.14	...	8.80	...	...	...
	10^{10}	8.08 to 8.77	8.26	8.40 to 9.36	9.03	...	7.07
Power factor (tan δ)	60	0.0013 to 0.0015	...	0.0005	...	...	...
	10^3	0.0002 to 0.0014	0.580	0.00007 to 0.0006	1.1	...	...
	10^6	0.0007 to 0.0012	0.024	0.00035 to 0.0035	0.012	...	0.0005
	10^8	0.0009	...	0.00035 to 0.00040	...	...	0.0005
	10^9	...	...	0.0006	...	...	...
	3×10^9	0.0014	...	0.0010	...	...	...
	10^{10}	0.0027	0.0033	0.0008 to 0.0015	0.0021	...	...
Loss factor	60	0.011 to 0.013	...	...	...	...	...
	10^3	0.00175 to 0.0115	8.0	0.0008 to 0.0053	14.6	...	...
	10^6	0.0018 to 0.0078	0.21	0.0014 to 0.0035	0.108	...	0.003
	10^8	0.006 to 0.0074	...	0.0031 to 0.0040	...	...	0.003
	10^9	0.0076	...	0.0038	...	...	...
	3×10^9	0.0114	...	0.0038	...	...	...
	10^{10}	0.013 to 0.218	0.027	0.0067 to 0.0140	0.019	...	0.00075

Engineering properties of αAl_2O_3

Property	Single crystal	99.9%	99.9%	99.5%	96%	90%
Physical						
Density, g/cm^3	...	3.99	3.96	3.87	3.72	3.60
Grain size, μm (mil)	...	15 to 45 (0.6 to 1.8)	1 to 6 (0.04 to 0.24)	5 to 50 (0.2 to 2)	2 to 20 (0.08 to 8)	2 to 10 (0.08 to 0.4)
Surface finish, (authentic average), μm (mil)	...	62 (2.5)	50 (2)	87 (3.5)	162 (6.5)	162 (6.5)
Color	...	Translucent white	Ivory	Ivory	White	White
Coefficient of thermal expansion, 10^{-6}/°C						
25 to 400 °C (77 to 750 °F)	...	7.4	7.4	7.6	7.4	7.0
25 to 1000 °C (77 to 1830 °F)	...	8.3	8.3	8.3	8.2	8.1
Thermal conductivity, W/cm · K						
At 20 °C (68 °F)	0.43	0.39	0.39	0.35	0.24	0.16
At 100 °C (212 °F)	...	0.28	0.27	0.26	0.19	0.13
At 400 °C (750 °F)	...	0.13	0.13	0.12	0.10	0.08
Dielectric constant at 25 °C (77 °F)						
At 1 kHz	...	10.1	9.9	9.8	9.0	8.8
At 1 MHz	...	10.1	9.8	9.7	9.0	8.8
At 10 GHz	...	10.1	9.8	9.7	8.9	8.7
Dissipation factor at 25 °C (77 °F)						
At 1 kHz	...	0.00050	0.0020	0.0002	0.0011	0.0006
At 1 MHz	...	0.00004	0.0002	0.0003	0.0001	0.0004
At 10 GHz	...	0.00009	0.0050	0.0002	0.0006	0.0009
Loss factor at 25 °C (77 °F)						
At 1 kHz	...	0.0050	0.020	0.002	0.010	0.005
At 1 MHz	...	0.0004	0.002	0.003	0.001	0.004
At 10 GHz	...	0.0010	0.005	0.002	0.005	0.008
Dielectric strength, AC, kV/cm (average RMS values at 60 Hz AC)						
0.63 cm thick	...	90.5	94.5	86.6	82.6	92.5
0.13 cm thick	...	200.7	181.1	169.3	145.6	177.1
0.02 cm thick	...	...	314.9	330.7	228.3	299.2
Volume resistivity, Ω · cm^2/cm						
At 25 °C (77 °F)	...	...	$>10^{15}$	$>10^{14}$	$>10^{14}$	$>10^{14}$
At 500 °C (930 °F)	...	...	3.3×10^{12}	...	4.0×10^9	2.8×10^8
At 1000 °C (1830 °F)	...	...	1.1×10^7	...	1.0×10^6	8.6×10^5
Mechanical						
Modulus of elasticity, GPa (10^6 psi)	434 (63)	393 (57)	366 (53)	372 (54)	303 (44)	275 (40)
Modulus of rigidity, GPa (10^6 psi)	...	162 (24)	158 (23)	151 (22)	124 (18)	117 (17)
Bulk modulus, GPa (10^6 psi)	...	234 (34)	227 (33)	227 (33)	172 (25)	158 (23)
Poisson's ratio	...	0.22	0.22	0.22	0.21	0.22
Flexural strength, MPa (ksi)						
At 25 °C (77 °F)	634 (92)	282 (41)	551 (80)	379 (55)	358 (52)	337 (49)
At 1000 °C (1830 °F)	413 (60)	172 (25)	413 (60)	...	172 (25)	...

(continued)

Engineering properties of αAl_2O_3 (continued)

Property	Single crystal	99.9%	99.9%	99.5%	96%	90%
Compressive strength, MPa (ksi)						
At 25 °C (77 °F)	...	2549 (370)	3790 (550)	2618 (380)	2067 (300)	2480 (360)
At 1000 °C (1830 °F)	...	482 (70)	1929 (280)	...	...	...
Tensile strength, MPa (ksi)						
At 25 °C (77 °F)	...	206 (30)	310 (45)	262 (38)	193 (28)	220 (32)
At 1000 °C (1830 °F)	...	103 (15)	220 (32)	...	96 (14)	108 (16)
Transverse sonic velocity, 10^3 m/s	...	9.9	9.9	9.8	9.1	8.8
Hardness (R45N)	...	85	90	83	78	79

Properties of dense alumina

Property	Temperature °C	°F	94% Al_2O_3	96% Al_2O_3	97.6% Al_2O_3	99.5% Al_2O_3
Flexural strength, MPa (ksi)	25	77	345 (50)	365 (53)	296 (43)	310 (45)
Compressive strength, MPa (ksi)	25	77	>2070 (>300)	>2070 (>300)	>1720 (>250)	>2070 (>300)
Density, g/cm^3 (lb/in.^{3})	25	77	3.67 (0.132)	3.72 (0.134)	3.76 (0.134)	3.86 (0.139)
Porosity, % water absorption	...	...	0.00(a)	0.00(a)	0.00(a)	0.00(a)
Color	...	...	White	White	White	White
Hardness, HR45N	...	...	78	79	75	81
Thermal conductivity, W/m · K (Btu/ft · h · °F)	25	77	20.5 (11.9)	25.6 (14.8)	26.8 (15.5)	29.3 (16.9)
Coefficient of linear thermal expansion, 10^{-6}/°C (10^{-6}/°F)	25 to 200	77 to 390	6.3 (3.5)	6.4 (3.6)	6.9 (3.8)	6.9 (3.8)
	200 to 400	390 to 750	7.5 (4.2)	7.6 (4.2)	7.8 (4.3)	7.8 (4.3)
	400 to 600	750 to 1110	8.0 (4.4)	8.2 (4.6)	8.5 (4.7)	8.3 (4.6)
	600 to 800	1110 to 1470	8.6 (4.8)	8.7 (4.8)	8.8 (4.9)	9.0 (5.0)
	800 to 1000	1470 to 1830	9.1 (5.1)	9.0 (5.0)	9.0 (5.0)	9.4 (5.2)
Maximum working temperature, °C (°F)	...	...	1600 (2910)	1620 (2950)	1650 (3000)	1725 (3150)
Dielectric strength (b), dc-kV/mm (dc-V/mil)	25	77	25.6 (650)	26.6 (675)	43.3 (1100)	31.5 (800)
T_c value, °C (°F)	...	...	>950 (>1740)	>950 (>1740)	>1000 (>1800)	>975 (>1790)
Volume resistivity, Ω · cm	25	77	>10^{14}	>10^{14}	>10^{14}	>10^{14}
	300	570	2.0×10^{12}	2.0×10^{12}	2.0×10^{12}	2.0×10^{12}
	600	1110	4.6×10^{8}	5.2×10^{8}	2.3×10^{10}	6.0×10^{8}
	900	1650	3.5×10^{6}	4.1×10^{6}	5.0×10^{8}	2.5×10^{6}
Dielectric constant (κ)						
10 MHz	25	77	9.07	9.30	9.53	9.58
	300	570	9.53	9.65	9.91	9.92
	500	930	9.91	10.10	10.14	10.20
1000 MHz	25	77	9.04	9.20	9.00	9.30
	300	570	...	...	...	...
	500	930	...	...	...	...
8500 MHz	25	77	8.98	9.16	9.04	9.37
	300	570	9.26	9.30	9.32	9.61
	500	930	9.40	9.45	9.54	9.82
Dissipation factor (tan δ)						
10 MHz	25	77	0.00026	0.00030	0.00004	0.00003
	300	570	0.00028	0.00061	0.00016	0.00009
	500	930	0.00341	0.00330	0.00052	0.00040
1000 MHz	25	77	0.00062	0.00044	0.00030	0.00014
	300	570	...	...	...	...
	500	930	...	...	...	...
8500 MHz	25	77	0.00078	0.00062	0.00045	0.00009
	300	570	0.00155	0.00085	0.00040	0.00014
	500	930	0.00155	0.00121	0.00072	0.00025
Loss factor (κ tan δ)						
10 MHz	25	77	0.00236	0.00279	0.00038	0.00029
	300	570	0.00267	0.00588	0.00158	0.00089
	500	930	0.03369	0.03333	0.00527	0.00408
1000 MHz	25	77	0.00560	0.00405	0.00270	0.00130
	300	570	...	...	...	...
	500	930	...	...	...	...
8500 MHz	25	77	0.00700	0.00568	0.00407	0.00084
	300	570	0.01165	0.00719	0.00373	0.000135
	500	930	0.01457	0.01143	0.00687	0.00245

(a) Vacuum tight. (b) For material 2.54 mm (0.100 in.) thick under oil

Properties of high-alumina ceramics

Property	Temperature °C	Temperature °F	85% Al_2O_3 vitreous body	95% Al_2O_3 vitreous body	99.5% Al_2O_3 vitreous body	99.5% Al_2O_3 porous body
Tensile strength, MPa (ksi)	...	...	115 to 160 (17 to 23)	170 to 240 (25 to 35)	259 to 262 (37.5 to 38)	...
Compressive strength, MPa (ksi)	...	...	965 to 2760 (140 to 400)	1720 to 2760 (250 to 400)	2940 (427)	69 to 860 (10 to 125)
Flexural strength, MPa (ksi)	...	...	205 to 310 (30 to 45)	310 to 345 (45 to 50)	295 to 325 (42.7 to 47)	69 to 150 (10 to 22)
Modulus of elasticity GPa (10^6 psi)	...	...	215 to 240 (31 to 35)	270 to 295 (39 to 43)	360 (52)	...
Impact resistance, J (in. · lbf Charpy)	...	...	0.66 to 0.79 (5.8 to 7.0)	0.70 to 0.86 (6.2 to 7.6)	...	0.34 (3.0)
Specific gravity	...	...	3.40 to 3.53	3.61 to 3.75	3.7 to 3.97	2.4 to 3.40
Water absorption, %	...	...	0.00 to 0.02	0.00	0.00	7 to 8
Porosity, %	...	...	<1(a)	<1(a)	<1	7.25
Hardness:						
Mohs scale	...	...	8.5 to 9	9	9	...
Knoop	...	...	1450	1720	...	...
Maximum working temperature, °C (°F)	...	...	1200 to 1400 (2200 to 2550)	1600 to 1700 (2910 to 3100)	0.51 (0.22)	...
Pore size, μm (mil)	...	...	...	...	...	2 to 3 (0.08 to 0.12)
Heat capacity, kJ/kg (Btu/lbf)	...	...	0.419 (0.180)	0.437 to 0.442 (0.188 to 0.190)	1950 (3542)	1400 to 1800 (2550 to 3270)
Thermal conductivity, W/m · K (Btu · in./ft^2 · h · °F)	38	100	13 to 17 (90 to 116)	19 to 22 (130 to 150)	19 (135)	17 (116)
	425	800	26 (180)	35 (240)	...	...
	870	1600	33 (230)	43 (300)	...	...
Coefficient of thermal expansion, 10^{-6}/°C (10^{-6}/°F)	25 to 200	77 to 390	5.47 to 5.68 (3.04 to 3.16)	5.7 to 6.67 (3.2 to 3.70)	...	5.1 (2.8)
	25 to 600	77 to 1110	6.55 to 6.96 (3.64 to 3.87)	6.7 to 7.65 (3.7 to 4.25)	...	...
	25 to 700	77 to 1290	7.6 to 7.9 (4.2 to 4.4)	8.07 (4.48)	7.7 (4.3)	...
	25 to 800	77 to 1470	7.33 (4.07)	7.6 (4.2)	...	...
	25 to 1000	77 to 1830	7.67 to 7.89 (4.26 to 4.38)	8.45 to 9.14 (4.69 to 5.08)	8.4 (4.7)	...
Thermal shock resistance	...	...	Fair	Good	Good	Good
Dielectric strength, ac-kV/mm (ac-V/mil)	25	77	8.07 to 13.8 (205 to 350)	9.84 to 15.7 (250 to 400)	15.0 (380)	1.97 (2.8)
	500	930	...	3.94 to 4.72 (100 to 120)	...	...
	1000	1830	...	0.79 to 1.18 (20 to 30)	...	...
Volume resistivity, Ω · cm^2/cm	25	77	1 to 3.6×10^{14}	10^{16}	...	10^{14}
	100	212	2 to 7.5×10^{13}	9.0×10^{14}	...	8.5×10^{13} to 1×10^{14}
	200	390	...	10×10^{13}	...	...
	300	570	1 to 5.0×10^{10}	5.3×10^{12}	1.2×10^{13}	1×10^{10} to 1.5×10^{11}
	400	750	...	10×10^{10}	...	...
	500	930	1×10^{8} to 7.5×10^{9}	1.2 to 4.5×10^{10}	1.3×10^{11}	7.5×10^{7} to 1.0×10^{9}
	600	1110	...	10^{8}	...	...
	700	1290	3 to 7.0×10^{6}	6.0×10^{8}	...	3.6×10^{6} to 3.0×10^{7}
	800	1470	...	...	3.5×10^{8}	...
	900	1650	4 to 5.0×10^{5}	...	...	5.6×10^{5}
T_c value, °C (°F)	...	...	750 to 1000 (1380 to 1830)	800 to 1100 (1470 to 2010)	1100 (2010)	835 to 1100 (1535 to 2010)

(a) Gas tight (helium mass spectrometer test on 0.254-mm, or 0.010-in., sections)

Typical properties of opaque alumina (92% Al_2O_3)

Property	Value
Density, g/cm^3	3.72
Flexural strength, MPa (ksi)	324 (47)
Thermal expansion, 10^{-6}/°C 25 to 500 °C (77 to 930 °F)	7.2
Thermal conductivity (at 20 °C, W/m · °C)	15
Dielectric constant at 1 MHz	9.6
Dielectric loss at 1 MHz	0.002

Typical chemical composition of thick-film alumina substrates

Substrate	Composition, wt%
Al_2O_3	96
SiO_2	2.5 to 3.0
MgO	0.75 to 1.0
CaO	0.10 to 0.25
Na_2O	0.05 to 0.10 max
Fe_2O_3	0.03 to 0.05
ZrO_2	0 to 0.5

Chemical etching of alumina-containing ceramics (94 to 99.8 wt% Al_2O_3)

Etchant	Temperature °C	Temperature °F	Time
10% HF	20	70	15 min.
Conc H_2SO_4	230	450	2 to 10 min.
Conc H_3PO_4	250	480	1 to 10 min.
$K_2S_2O_4$ melt	650	1200	1 to 10 min.
V_2O_5 melt	900	1650	1 min.
Borax melt	900	1650	15 to 45 s

Standard as-fired dimensional tolerances of thick-film alumina substrates

Dimension	Tolerance
Length, width, and hole centers	Not less than ±1% or ±0.08 mm (0.003 in.)
Thickness	Not less than ±10% or ±0.38 mm (0.0015 in.)
Camber	0.003 mm/mm (0.003 in./in.)
Hole diameters of:	
0.38 to 0.74 mm (0.015 to 0.029 in.)	±0.05 mm (0.002 in.)
0.76 to 2.54 mm (0.030 to 0.1 in.)	±0.08 mm (0.003 in.)
>2.54 mm (0.1 in.)	Not less than ±1%, or ±0.13 mm (0.005 in.)
Perpendicularity (squareness)	≤0.004 mm/mm (0.004 in./in.)
Edge straightness	≤0.003 mm/mm (0.003 in./in.)

Typical properties of thick-film alumina (96% Al_2O_3)

Property	Conditions	ASTM test method	Property value
Surface finish (CLA), μm (μin.)	As-fired	Profilometer 0.762 mm (0.030 in.) cutoff	<0.6 (25)
Flexural strength, MPa (ksi)	21 °C (70 °F)	F 417	400 (58)
Thermal conductivity, W/m · °C	20 °C (68 °F)	C 408	26
	100 °C (212 °F)		20
	400 °C (750 °F)		12
Thermal coefficient of expansion, 10^{-6}/°C	25 to 200 °C (77 to 390 °F)	C 372	6.3
	25 to 500 °C (77 to 930 °F)		7.1
	25 to 800 °C (77 to 1470 °F)		7.6
	25 to 1000 °C (77 to 1830 °F)		8.0
Volume resistivity, Ω · cm	25 °C (77 °F)	D 1829	$>10^{14}$
	300 °C (570 °F)		5.0×10^{10}
	700 °C (1290 °F)		4.0×10^{7}
Temperature, °C (°F), at which resistivity is 1 MΩ · cm	...	D 1829	1000 (1830)
Dielectric strength (60 cycle average), kV_{rms}/mm (V_{rms}/mil)	0.635 mm (0.025 in.) thick specimens	D 116	23.6 (600)
Dielectric constant (relative permittivity) at 25 °C (77 °F)	1 kHz	D 150	9.5
	1 MHz		9.5
	100 MHz		9.5
Loss tangent (dissipation factor)	1 kHz	D 150	0.0010
	1 MHz		0.0004
	100 MHz		0.0004
Loss index	1 kHz	D 150	0.009
	1 MHz	D 150	0.004
	100 MHz	D 2520	0.004

Typical material characteristics of thin-film alumina substrates

Property	ASTM test method	99.5% Al_2O_3	99.6% Al_2O_3
Bulk density, g/cm^3	C 373		
Typical		3.89	3.88
Range		3.86 to 3.92	3.86 to 3.90
Hardness, HR45N	E 18	87	87
Surface finish (as-fired center line average), μm (μin.)	Profilometer: cut-off: 0.75mm (0.030 in.), stylus diam: 0.01 mm (0.0004 in.)	≤0.15 (6)	≤0.075 (3)
Average grain size, μm (mil)	Intercept method	≤2.2 (0.087)	≤1.2 (0.047)
Water absorption, %	C 373	nil	nil
Flexural strength, MPa (ksi)	F 394	570 (83)	570 (83)
Modulus of elasticity, GPa (10^6 psi)	C 623	370 (54)	370 (54)
Poisson's ratio	C 623	0.20	0.20
Coefficient of linear thermal expansion, 10^{-6}/°C, from 25 °C (77 °F) to:			
300 °C (570 °F)	C 372	6.7	6.7
600 °C (1110 °F)		7.5	7.5
800 °C (1470 °F)		8.0	8.0
1000 °C (1830 °F)		8.3	8.3
Thermal conductivity, W/m · °C, at:			
20 °C (68 °F)	Various	34.7	34.7
100 °C (212 °F)		25.5	25.5
400 °C (750 °F)		12.6	12.6
Average rms dielectric strength (60 cycle ac), kV/mm (V/mil):			
0.635 mm (0.025 in.) thick specimen	D 116	670	670
1.3 mm (0.050 in.) thick specimen		450	450
Dielectric constant (relative permittivity) at:			
1 KHz	D 150	9.9 (±2%)	9.9 (±1%)
1 MHz		9.9 (±2%)	9.9 (±1%)
Dissipation factor (loss tangent) at:			
1 KHz	D 150	0.005	0.005
1 MHz		0.002	0.002
Loss index (loss factor) at:			
1 KHz	D 150	0.005	0.005
1 MHz		0.002	0.002
Volume resistivity, Ω · cm, at:			
25 °C (77 °F)	D 1829	$>10^{14}$	$>10^{14}$
300 °C (570 °F)		$>10^{12}$	$>10^{12}$
500 °C (950 °F)		$>10^{9}$	$>10^{8}$
700 °C (1290 °F)		$>10^{8}$	$>10^{8}$

Quality attributes for two grades of thin-film alumina substrates

Visual attribute	Precision-resistor-grade acceptance level	Conductor-grade acceptance level
Burrs	>0.013 mm (0.0005 in.) high, none >0.13 mm (0.005 in.) diam	>0.025 mm (0.001 in.) high, none >0.25 mm (0.010 in.) diam
Pits, holes, and pocks	<0.075 mm (0.003 in.) diam NIF, none >0.13 mm (0.005 in.) diam(a)	<0.075 mm (0.003 in.) diam NIF, None >0.25 mm (0.010 in.) diam(a)
Stains, spots, and contamination	None	None
Blisters	None	None
Scratches	None >0.005 mm (0.0002 in.) deep and >0.635 mm (0.025 in.) length	>0.18 mm (0.007 in.) deep and >0.635 mm (0.250 in.) length
Fins and ridges	None	None
Chips (width only to be considered)	0.75% of substrate length that is not less than 0.13 mm (0.005 in.)	1% of substrate length that is not less than 0.25 mm (0.010 in.)
Cracks	None	None
Allowable density of defects	1/in.^{2}, noncumulative	1/in.^{2}, noncumulative
Inspection level	100%	100%
Camber	≤0.002 mm/mm (in./in.)	≤0.003 mm/mm (in./in.)
Surface finish, μm (μin.)		
99.5% alumina	≤0.15 (6)	≤0.15 (6)
99.6% alumina	≤0.075 (3)	≤0.075 (3)

(a) NIF, not inspected for

Effect of ceramic surface conditions on peel strength

Ceramic surface treatment	Surface roughness, μm	Surface roughness, μin.	Peel strength, N	Peel strength, lbf
As-sintered	1.17	46	111	25
As-ground	1.04	41	58	13
Ground and lapped	0.10	4	111	25
Ground and resintered	0.94	37	107	24

Note: Ceramic, 99.5% Al_2O_3; Metal, Kovar (6.5 × 75 × 0.25 mm, or 0.25 × 3 × 0.01 in.); Brazing filler metal, Cusil ABA (50 μm, or 2 mils, thick); Brazing temperature, 825 °C (1515 °F); Brazing timer, 10 min.; Peel rate, 50 mm/min. (2 in./min.)

Mechanical properties of SiC-whisker-reinforced alumina ceramics

Whisker control, vol %(a)	Temperature, °C	Temperature, °F	Fracture strength, MPa	Fracture strength, ksi	Fracture toughness, MPa $\sqrt{m}$	Fracture toughness, ksi $\sqrt{in.}$
0	22	70	...	...	4.5	4.1
	700	1290	...	...	4.0	3.6
	1000	1830	...	...	3.8	3.5
10	22	70	455 ± 55	65 ± 8	7.1	6.5
	1000	1830	320 ± 36	45 ± 5	...	...
20	22	70	655 ± 135	95 ± 20	7.5 to 9.0	6.8 to 8.2
	700	1290	535 ± 35	80 ± 5	...	...
	1000	1830	570 ± 20	85 ± 3	7.0 to 8.0	6.8 to 8.2
40	22	70	850 ± 130	120 ± 20	6.0	5.5
	700	1290	740 ± 61	110 ± 9	...	...
	1000	1830	665 ± 88	96 ± 13	6.2	5.6

(a) Hot pressed mixture of alumina powder and SiC whiskers

Physical properties of beryllia

Modification	Crystal system	Theoretical density, g/cm^3	Melting point(a), K	Melting point(a), °F
αBeO	Hexagonal	3.008	2843	4658
βBeO(b)	Tetragonal	2.69	2843	4658

(a) Most frequently reported value. Other values range from 2793 to 2923 K (4568 to 4801 °F). (b) α-β transition at 2322 ± 50 K (3720 ± 90 °F)

Typical properties of 99.5% beryllia

Property	Value
Density, g/cm^3	2.85
Flexural strength, MPa (ksi)	241 (35)
Thermal expansion, 10^{-6}/°C, from 25 °C (77 °F) to:	
200 °C (390 °F)	6.4
500 °C (930 °F)	7.2
Thermal conductivity, W/m · °C, at:	
20 °C (68 °F)	248
100 °C (212 °F)	188
Dielectric constant at 1 MHz	6.5
Dielectric loss tangent at 1 MHz	0.0004

Values of bulk modulus and Poisson's ratio for polycrystalline beryllia

Property	Temperature, K	Temperature, °F	Reported values
Bulk modulus, GPa (10^6 psi)	294	70	375 (54.5)
			291 (42.1)
Poisson's ratio	294 to 1273	70 to 1830	0.34
	>1273	>1830	>0.34
	294	70	0.38
			0.30 ± 0.05

Dielectric constants and loss tangents for microwave-absorbing ceramics

Composition, wt%	Dielectric constant (κ′) 8.6 GHz	10.5 GHz	12 GHz	Loss tangent (tan δ) 8.6 GHz	10.5 GHz	12 GHz
98% MgO - 2% SiC	11.22	11.13	10.94	0.15	0.06	0.03
95% MgO - 5% SiC	12.80	12.72	12.64	0.20	0.12	0.10
90% MgO - 10% SiC	17.12	16.84	16.61	0.30	0.19	0.19
80% MgO - 20% SiC	26.58	26.08	25.73	0.40	0.28	0.30
60% MgO - 40% SiC	54.24	53.81	50.46	0.92	0.70	0.73
98% BeO - 2% SiC	8.17	7.40	7.13	0.001	0.039	0.017
95% BeO - 5% SiC	10.60	8.95	8.54	0.032	0.075	0.043
90% BeO - 10% SiC	13.65	11.81	11.53	0.21	0.15	0.14
80% BeO - 20% SiC	20.26	18.70	17.81	0.33	0.24	0.22
60% BeO - 40% SiC	67.68	54.55	49.54	0.73	0.73	0.72

Properties of microwave-absorbing ceramics

Composition	Fracture energy J/m^2	Fracture energy ft · lbf./ft^2	Modulus of rupture MPa	Modulus of rupture ksi	Modulus of elasticity GPa	Modulus of elasticity 10^6 psi
90 BeO - 10% SiC	36	2.47	525	76	385	56
80% BeO - 20% SiC	24	1.64	450	65	395	57
60% BeO - 40% SiC	14	0.96	505	73	405	59
40% BeO - 60% SiC	30	2.05	745	108	415	60
99% MgO - 1% SiC	...	...	230	33	285	41
95% MgO - 5% SiC	28	1.92	...	...	290	42
90% MgO - 10% SiC	9	0.62	125	18	295	43
80% MgO - 20% SiC	13	0.89	195	28	315	46
70% MgO - 30% SiC	16	1.10	215	31	330	48
60% MgO - 40% SiC	15	1.03	205	30	345	50
40% MgO - 60% SiC	44	3.01	310	45	380	55

Physical properties of barium, calcium, and strontium oxides

Oxide	Crystal system(a)	Theoretical density, g/cm^3	Melting temperature K	Melting temperature °F
BaO	Cubic	5.72	2196	3493
CaO	Cubic	3.32(b)	2887(c)	4737(c)
SrO	Cubic	4.7	2727 ± 22	4450 ± 40

(a) Hexagonal forms have also been reported, particularly as sublimation products. Amorphous forms also exist, which convert to the cubic crystalline form at temperatures of about 672 to 700 K (750 to 800 °F). (b) Other reported values are 3.37 and 3.40/g/cm^3. (c) Most reliable value; other reported values range from about 2805 to 2894 K (4590 to 4750 °F)

Physical properties of titanium and chromium oxides

Oxide	Crystal system	Theoretical density, g/cm^3	Melting point K	Melting point °F
αTiO	Monoclinic	4.93	(b)	(b)
βTiO	Cubic	...	2023	3180
Ti_2O_3	Hexagonal	4.6	2128(c)	3770(c)
Ti_3O_5	Monoclinic	4.24	(d)	(d)
TiO_2 (anatase)	Tetragonal	3.84	(e)	(e)
TiO_2 (rutile)	Tetragonal	4.25	(f)	(f)
TiO_2 (brookite)	Orthorhombic	4.17	2109 ± 17(g)	3340 ± 30(g)
Cr_2O_3(h)	Hexagonal	5.21	2299 ± 122(g)	4170 ± 220(g)

(a) Some values may be experimental rather than theoretical. (b) Transforms to a β-phase at 1263 K (1815 °F). (c) Transforms to a β-phase at 472 K (390 °F). (d) Transforms to a β-phase at 450 K (350 °F). (e) Reported to exist in three forms. Transformation temperatures are given as 915 K (1188 °F) for anatase I→anatase II, 1188 K (1679 °F) for anatase II→rutile, and 1323 K (1922 °F) for anatase III→rutile. (f) Transforms to brookite at 1572 K (2370 °F). (g) Range of values most frequently reported. (h) Other oxides of chromium exist but are not stable at elevated temperatures

Physical properties of vanadium, niobium, and tantalum oxides

Oxide	Crystal system	Theoretical density, g/cm^3	Melting point K	Melting point °F
VO	Cubic	5.23	2322	3720
V_2O_3	Hexagonal	4.87	2243 to 2273	3578 to 3632
VO_2	Tetragonal	4.65	1818 to 1911	2813 to 2980
V_2O_5	Orthorhombic	3.35	943	1238
NbO	Cubic	7.30	2218	3533
Nb_2O_3	...	...	2045 to 2050	3222 to 3230
NbO_2	Tetragonal	5.90	2188	3479
Nb_2O_5	Orthorhombic(a)	4.46	1761 to 1783	2710 to 2750
Ta_2O_5	Orthorhombic	8.02	2155	3420

(a) Nb_2O_5 transforms to a monoclinic form at about 1373 K (2012 °F)

Moduli of elasticity for niobium oxides

Oxide	Temperature K	Temperature °F	Modulus of elasticity(a) GPa	Modulus of elasticity(a) 10^6 psi
Nb_2O_5				
Sintered	293	68	40	5.8
	1273	1830	100	14.5
Hot pressed	293	68	150	21.8
	1273	1830	125	18.2
Nb_2O_3				
Hot pressed	293	68	161	23.4

(a) All values approximate

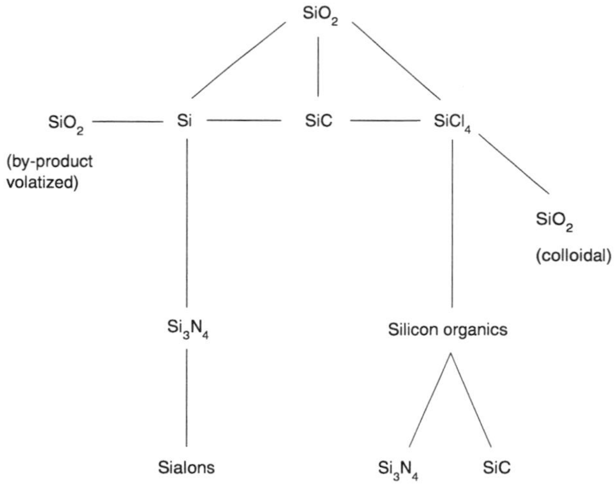

Use of silica as a precursor for ceramics

Density of silica at various pressures

Pressure MPa	Pressure ksi	Radius nm	Radius μin.	Intruded volume, cm^3	Sample volume, cm^3	Apparent density, g/cm^3
0.10	0.015	7260	290	0	1.115	0.967
34	5	21.3	0.838	0.024	1.091	0.988
69	10	10.7	0.421	0.050	1.065	1.01
103	15	7.11	0.280	0.080	1.035	1.04
138	20	5.34	0.210	0.117	0.998	1.08
172	25	4.27	0.168	0.160	0.955	1.13
207	30	3.56	0.140	0.203	0.912	1.18
241	35	3.05	0.120	0.245	0.870	1.24
276	40	2.67	0.105	0.285	0.830	1.30
310	45	2.37	0.093	0.323	0.792	1.36
345	50	2.13	0.084	0.357	0.758	1.42
379	55	1.94	0.076	0.385	0.730	1.48

Crystalline phases of SiO_2 at atmospheric pressure

Classification of phases	Stability range, K	Crystal system	Theoretical density, g/cm³	Melting point, K	Remarks
Quartz, low	<845	Trigonal	2.65	...	Most common crystalline phase
Quartz, high	845 to 1140	Hexagonal	2.52	1743	High-temperature phase
Tridymite S	1140 to 1743	Orthorhombic	2.32	1933	Characteristic phase in refractories
S-I	<337				Polymorphic modification
S-II	337 to 391				Polymorphic modification
S-III	391 to 436				Polymorphic modification
S-IV	436 to 483				Polymorphic modification
S-V	483 to 748				Polymorphic modification
S-VI	748 to 1743				Polymorphic modification
Tridymite M	...	Hexagonal	2.30	...	Unstable, converts to Tridymite S
M-I	<391				Polymorphic modification
M-II	391 to 436				Polymorphic modification
M-III	>436				Polymorphic modification
Cristobalite, low	<545	Tetragonal	2.32	...	Low-temperature phase
Cristobalite, high	545 to 1998	Cubic	2.27	1998	Appears first when above phases are heated to 1672 K

Representative strength values of fused silica and quartz at room temperature

Type of material	Percent of theoretical density	Bend strength MPa	Bend strength ksi	Compressive strength MPa	Compressive strength ksi	Tensile strength MPa	Tensile strength ksi
Bulk fused silica							
clear	100	100(a)	16(a)	690 to 1380	100 to 200	69 (a)	10(a)
	...	55(b)	8(b)	...	...	...	...
	100	48	7	1035	150	...	...
translucent	100	45(a)	6.5(a)	275	40	21	3
Polygranular fused silica							
hot pressed	~95	...	...	690 to 1380	100 to 200	...	...
slip cast, sintered(c)	~34 to 98	3.4 to 120	0.5 to 17.5	17 to 490	2.5 to 7.1	4.8 to 32	0.7 to 4.7
foamed (open pores)	13.6 to 35	0.69 to 2.76	0.10 to 0.40	2.76 to 8.62	0.40 to 1.25	...	...
foamed (closed pores)	7 to 10	0.69 to 1.03	0.10 to 0.15	0.90 to 1.38	0.13 to 0.20	...	...
Single-crystal quartz(d)							
parallel to *c*-axis	100	...	...	~2070	~300	110 to 275	16 to 40
perpendicular to *c*-axis	...	...	...	<2070	<300	...	...

(a) Polished surfaces. (b) Abraded surfaces. (c) Fracture energy of slip cast fused silica is reported to be about 1 kg/cm (6 lb/in.). (d) Natural crystal

Physical properties of rare-earth oxides

Oxide	Crystal system	Theoretical density g/cm³	Melting point K	Melting point °F
Sc_2O_3	Cubic(a)	3.841	2660	4330 ± 158(b)
Y_2O_3	Cubic	5.03	~2700	~4400
La_2O_3	Hexagonal	6.57	2539	4110 ± 90
Ce_2O_3	Hexagonal	6.87	2448	3949 ± 61(b)
CeO_2	Cubic	7.28 ± 0.1	2614	4245 ± 740(b)
PrO_2	Cubic	...	...	...
Pr_6O_{11}	Cubic	5.47	2315	3710 ± 54(b)
Pr_2O_3	Cubic	6.32	2485	4013 ± 153(b)
	Hexagonal(c)	...	...	...
Nd_2O_3	Hexagonal(d)	7.28	2378	3821 ± 369(b)
Pm_2O_3	(e)	...	...	...
SmO	Cubic	...	...	...
Sm_2O_3	Monoclinic	7.43	...	(f)
Sm_2O_3	Cubic(g)	7.62	2576	4177 ± 84(b)
EuO	Cubic	8.16	2131	3376 ± 284(b)
Eu_2O_3	Cubic	7.3 ± 0.04	...	(f)
Eu_2O_3	Monoclinic(h)	7.95	2439	3931 ± 295(b)
Eu_3O_4	Orthorhombic	8.07	2273	3632 ± 180(j)
$Eu_{16}O_{21}$	Orthorhombic	6.74	...	...
Gd_2O_3	Cubic	7.65 ± 0.02	...	(f)
βGd_2O_3	Monoclinic	...	2629	4273 ± 70(b)
Tb_2O_3	Cubic(k)	7.74 ± 0.06	2612	4242 ± 94(b)

(continued)

Physical properties of rare-earth oxides (continued)

Oxide	Crystal system	Theoretical density g/cm^3	Melting point K	Melting point °F
Tb_4O_7	Cubic	...	...	...
Dy_2O_3	Cubic(m)	8.17	2583	4190 ± 148(b)
Ho_2O_3	Cubic	8.41 ± 0.6	2636	4285 ± 59(b)
Er_2O_3	Cubic	8.64	2618	4253 ± 99(b)
Tm_2O_3	Cubic	8.83 ± 0.13	2619	4255 ± 81(b)
Yb_2O_3	Cubic	9.25 ± 0.05	2608	4235 ± 135(b)
Lu_2O_3	Cubic	9.423	2740	4473(n)

(a) Cubic form transforms to monoclinic at an elevated temperature. (b) Range of reported values. (c) Cubic form reportedly transforms to hexagonal at about 1153 K (1616 °F). (d) Cubic form reportedly transforms to hexagonal at about 1113 K (1544 °F). (e) Three types: hexagonal, monoclinic, cubic. (f) Transforms to β modification irreversibly, the rate being a function of time and temperature. (g) Cubic form reportedly transforms to monoclinic at about 1253 K (1796 °F). Monoclinic form reportedly begins transformation to hexagonal at 2573 K (4142 °F). (h) An apparently irreversible transformation from cubic to monoclinic takes place at approximately 1348 K (1967 °F). At higher temperatures, transformations to three other polymorphic forms occur, two of which are hexagonal. (j) Estimated from experimental data. (k) Cubic form reportedly transforms to monoclinic at about 2123 K (3362 °F), and monoclinic form becomes hexagonal at a higher temperature. (m) Cubic form reportedly transforms to monoclinic at about 2413 K (3884 °F). (n) Value most often reported.

Mechanical properties of rare-earth oxides

Oxide	Temperature K	Temperature °F	Property values	
Bend strength			**MPa**	**ksi**
Y_2O_3	293	68	123	17.92
	1273	1830	133	19.34
	1573	2370	122	17.78
	293	68	104	15.07
	1273	1830	139	20.19
	1773	2730	125	18.20
	2023	3180	129	18.77
Sc_2O_3	293	68	213	31.00
	293	68	176	25.60
	973	1290	186	27.00
	1273	1830	181	26.30
	1623	2460	127	18.50
CeO_2	293	68	139	20.19
	1073	1470	95	13.79
	1273	1830	71	10.38
	1473	2190	51	7.39
	1623	2460	23	3.41
Gd_2O_3	293 to 1673	68 to 2550	108 to 137	15.64 to 19.91
Er_2O_3	293 to 1673	68 to 2550	98 to 127	14.22 to 18.49
Dy_2O_3	293	68	78	11.38
	1073	1470	88	12.80
	1473	2190	137	19.91
Yb_2O_3	293	68	73	10.66
	1073	1470	73	10.66
	1473	2190	78	11.38
Compressive strength			**MPa**	**ksi**
Y_2O_3	293	68	392	57.00
CeO_2	293	68	588	85.40
Nd_2O_3	293	68	107 to 134	15.57 to 19.42
Gd_2O_3	293	68	166 to 215	24.17 to 31.28
Er_2O_3	293	68	420 to 430	61.15 to 62.57
Dy_2O_3	293	68	381	55.46
Yb_2O_3	293	68	225	32.71
Bulk modulus			**GPa**	**10^6 psi**
Tm_2O_3	293	68	130	18.86
	791	964	123	17.91
	1269	1825	114	16.56
Lu_2O_3	293	68	139	20.25
	790	962	132	19.24
	1269	1825	125	18.24
Y_2O_3	293	68	135	19.69
Dy_2O_3	293	68	150	21.84
Ho_2O_3	293	68	134	19.53
Er_2O_3	293	68	140	20.40

(continued)

Mechanical properties of rare-earth oxides (continued)

Oxide	Temperature K	Temperature °F	Property values
Poisson's ratio			
Yb_2O_3	293	68	0.284
Gd_2O_3	293	68	0.276
	1273	1832	0.267
Sm_2O_3	293	68	0.32
Tm_2O_3	293	68	0.292
	791	964	0.289
	1269	1832	0.285
Lu_2O_3	293	68	0.287
	790	962	0.288
	1269	1832	0.289
Y_2O_3	293	68	0.298
Dy_2O_3	293	68	0.313
Ho_2O_3	293	68	0.290
Er_2O_3	293	68	0.292
Y_2O_3	293	68	0.295
CeO_3	293	68	0.275 to 0.315
Hardness			**kg/mm²**
Sc_2O_3	293	68	790 to 910
La_2O_3	293	68	300 to 380
CeO_3	293	68	5 to 6 Mohs
Pr_2O_3	293	68	370 to 380
Nd_2O_3	293	68	~380 to 650
Sm_2O_3	293	68	380 to 480
Hardness (con't)			**kg/mm²**
Eu_2O_3	293	68	140 to 500
Gd_2O_3	293	68	380 to 550
Tb_2O_3	293	68	380
Dy_2O_3	293	68	400 to 700
Ho_2O_3	293	68	380
Er_2O_3	293	68	380 to 700
Tm_2O_3	293	68	380
Yb_2O_3	293	68	650
Lu_2O_3	293	68	650 to 830

Thermal and physical properties of polycrystal TiO_2

Property	Rutile	Anatase	Brookite
Density, g/cm³	4.25	3.89	4.14
Melting point, °C (°F)	1855 (3370)	Transformed to rutile	
Coefficient of thermal expansion, 10^{-6}/K	7.14	10.2	11.0
Thermal conductivity, W/m · K	10.4(a)	...	...
	7.4(b)	...	...
At 300 °C	8.4	...	...
Magnetic susceptibility, cm²/g	7.4×10^{-8}	...	...
Refractive index			
At 25 °C (77 °F) and 598 nm	n_ω 2.612	n_ω 2.561	n_α 2.583
	n_ε 2.899	n_ε 2.488	n_β 2.584
			n_γ 2.700
Reflectance, %			
At 400 °C (750 °F)	47 to 50	88 to 90	...
At 500 °C (930 °F)	95 to 96	94 to 95	...
Absorption of UV, at 360 µm %	90	67	...

(a) Parallel to *c*-axis, single crystal. (b) Perpendicular to *c*-axis, single crystal

Hardnesses of thoria, urania, and plutonia

Oxide	Theoretical density, %	Load, g	Knoop hardness, kg/mm²
ThO_2	...	500	640
ThO_2 + ½% CaO	98.3	500	700
UO_2	~100	100	662(a)
UO_2	...	...	666
UO_2	93.3	10	355
UO_2	93.3	50	585
UO_2	93.3	100	625

(continued)

Hardnesses of thoria, urania, and plutonia (continued)

Oxide	Theoretical density, %	Load, g	Knoop hardness, kg/mm^2
UO_2	93.3	500	600
UO_2	93.3	2000	520
UO_2	97	...	600 Vickers
	80	...	~260 Vickers
	60	...	~45 Vickers
$UO_{2.00}$	...	500	640
$UO_{2.02}$	...	500	787
$UO_{2.145}$	...	500	880
UO_2	...	...	6 to 7 (Mohs)
PuO_2	~90	...	~440 Vickers

(a) Average value for annealed single crystal, independent of orientation

Physical properties of thoria, urania, and plutonia

			Melting point	
Oxide	Crystal system	Theoretical density, g/cm^3	K	°F
ThO_2	Cubic	10.00 ± 0.01	3493	5830 ± 180(a)
UO_2	Cubic	10.96 ± 0.01	3113	5144 ± 36
U_3O_8	Orthorhombic(b)	8.39	...	...
UO_3	Orthorhombic(c)	8.34	...	(d)
PuO	Cubic	13.9	...	...
Pu_2O_3	Cubic	10.2	2484	4020 ± 80
	Hexagonal	11.2	2484	4020 ± 80
PuO_2	Cubic	11.46	2633	4334 ± 36

(a) The value 5970 °F (3572 K) is most frequently reported. (b) Becomes "hexagonal" above 673 K (752 °F). (c) As many as five allotropic forms may exist, including an amorphous form. (d) Decomposes to U_3O_8 in the range 772 to 972 K (930 to 1290 °F)

Bend and compressive strengths of thoria

	Average grain size		Temperature		Strength values	
Theoretical density, %	μm	mils	K	°F	MPa	ksi
Bend strength						
80.2	~4	~0.16	297	75	89.4	13.0
91.4	16	0.63	297	75	128.6	18.7
68.9	43	1.70	297	75	31.6	4.6
Compressive strength						
80.2	~4	~0.16	297	75	1444.8	210
91.4	16	0.63	297	75	1561.8	227
68.9	43	1.70	297	75	217.4	31.6
92	...	...	300	80	1472.3	214
			672	750	1073.3	156
			1072	1470	488.5	71
			1472	2190	195.4	28
			1772	2730	<10	<2

Five polymorphs of WO_3

Crystal system	Space group	*a* nm	*b* nm	*c* nm	α, β, γ	Temperature region, K
Tetragonal	*P4/nmn*	0.525	...	0.391	...	>1013
Orthorhombic	...	0.7384	0.7512	0.3846	...	623 to 1013
Monoclinic	$P2_1/n$	0.7297	0.7539	0.7688	β = 90° 91′	290 to 623
Triclinic	$P2_1/n$	0.730	0.752	0.769	α = 88° 50′	233 to 290
		...	...	...	β = 90° 55′	...
		...	...	...	γ = 90° 56′	...
Monoclinic	...	0.527	0.516	0.767	β = 91° 43′	<233

Physical properties of zirconia and hafnia

Oxide	Crystal system	Theoretical density, g/cm^3	Melting temperature K	Melting temperature °F
ZrO_2	Monoclinic	5.56	(a)	(a)
	Tetragonal	6.10	3037 ± 83(b)	5008 ± 150(b)
	Cubic (stabilized ZrO_2)	(c)	(d)	(d)
HfO_2	Monoclinic	9.68	(e)	(e)
	Tetragonal	10.01	3117 ± 55(f)	5152 ± 10(f)
	Cubic (stabilized HfO_2)	(g)	(g)	(g)

(a) Monoclinic ZrO_2 transforms to tetragonal ZrO_2 at 1222 to 1494 K (1740 to 2230 °F), the exact temperature depending on purity. (b) Range of most reliable values reported. (c) The theoretical density of stabilized, cubic ZrO_2 will depend on the amount and kind of stabilizer used; approximate theoretical densities for 7, 5, and 3 wt % CaO-stabilized ZrO_2 are reported as 5.5, 5.75, and 5.9 g/cm^3, respectively. For the same percentage additives, MgO-stabilized ZrO_2 will have nearly identical densities, whereas Y_2O_3-stabilized ZrO_2 will have slightly higher densities than those of CaO-stabilized ZrO_2. (d) Stabilization also leads to lower melting points. For 7 and 5 wt % CaO-stabilized ZrO_2, respective melting points are reported as 2772 and 2872 K (4530 and 4710 °F). (e) Monoclinic HfO_2 transforms to tetragonal HfO_2 at 1883 to 2072 K (2930 to 3270 °F). (f) Range of reported values. (g) For cubic-stabilized HfO_2, values of theoretical density and melting point will depend on the amount and kind of stabilizer used

Crystallographic and physical properties of zirconia

Property	Value
Polymorphism, K	
Monoclinic to tetragonal	1273 to 1473
Tetragonal to cubic	2643
Cubic to liquid	2953
Crystallography, monoclinic	
a	5.1454 Å
b	5.2075 Å
c	5.3107 Å
β	99° 14′
Space group	$P2_1/c$
Crystallography, tetragonal	
a	3.64 Å
c	5.27 Å
Space group	$P4_2/mmc$
Crystallography, cubic	
a	5.065Å
Space group	$Fm3m$
Density, g/cm^3	
Monoclinic	5.68
Tetragonal	6.10
Thermal expansion coefficient, 10^{-6}/K	
Monoclinic	7
Tetragonal	12
Heat of formation, kJ/mol	−1096.73
Boiling point, K	4548
Thermal conductivity, W/m · K	
At 100 °C	1.675
At 1300 °C	2.094
Mohs hardness	6.5
Refractive index	2.15

Hardness of zirconia

Stabilizer(a)	Hardness, Mohs	Micro-hardness, kg/mm^2
8% CaO	...	1180
10% CaO	>6	1470
12% CaO	...	1130
8% MgO	...	1080
10% MgO	...	1520
12% MgO	...	1430
15% Y_2O_3(b)	>6	1360

(a) Mole percent. (b) Single crystal: Knoop 500-g load

Several properties of polycrystalline tetragonal zirconia (3 mol% Y_2O_3)

Property	Value
Color	Ivory
Density, g/cm^3	6 to 6.05
Particle size, (tetragonal) μm (μin.)	0.2 to 1 (8 to 40)
Mechanical properties	
Flexural strength, MPa (ksi)	800 to 1500 (116 to 218)(a)
Compressive strength, MPa (ksi)	>2900 (420)(a)
Fracture toughness, MPa $\sqrt{m}$ (ksi $\sqrt{in.}$)	7 to 12 (6.4 to 10.9)(a)
Vickers hardness, GPa (10^6 psi)	12 to 13 (1.7 to 1.9)
Modulus of elasticity, GPa (10^6 psi)	200 to 210 (29 to 30.5)
Poisson's ratio	0.31
Thermal shock (DT), K	360
Creep rupture, MPa (ksi)	300 (45)(b)
Thermal properties	
Specific heat, cal/g · K	0.11 to 0.12
Thermal expansion, 10^{-6}/K	9.6(c) 10.4(d)
Thermal conductivity, W/m · K	2.0 to 3.3
Electrical properties	
Ionic conductivity, S/cm	~10^{-12}(a) 3×10^{-4} (e)
Dielectric constant	15 to 32
Loss tangent	5 to 7×10^{-3}

(a) Room temperature. (b) At 800 °C (1470 °F), 1000 h. (c) From 20 to 400 °C (68 to 750 °F). (d) From 20 to 1000 °C (68 to 1830 °F). (e) At 500 °C (930 °F)

Mechanical properties of various zirconia materials

Property	Temperature K	Temperature °F	Reported values MPa	Reported values ksi
Yttria-stabilized ZrO_2				
Bend strength	294	70	364 to 577(a)	52.9 to 83.8(a)
	294	70	334 to 527(b)	48.6 to 76.6(b)
	294	70	654 to 663(c)	95.1 to 96.3(c)
	294	70	151	22
	773	932	138	20
	1273	1832	117	17
	1473	2192	69	10
Magnesia-stabilized ZrO_2				
Bend strength	293	68	290	42.2
	1323	1922	217	31.6
Compressive strength	294	70	2064	300
	773	932	1569	228
	1273	1830	1176	171
	1673	2552	127	18.5
	1773	2732	19	2.8
	293	68	450	65.4
	1323	1922	137	19.9
Calcia-stabilized ZrO_2				
Bend strength	294	70	239	34.7
	1273	1830	169	24.5
	1573	2372	113	16.4
Compressive strength	294	70	1085	303
	1473	2192	689	100
Tensile strength	294	70	30.0 to 42.2	4.36 to 6.13
	294	70	79.7	11.6
	294	70	144.5	21
	1472	2190	82.5	12
	1811	2800	~13.8	~2

(a) Dry pressed and fired to a density of about 6.0 g/cm^3. (b) Wet pressed and fired to a density of about 6.0 g/cm^3. (c) Pressure cast and fired to a density of about 6.0 g/cm^3

Typical properties of transformation-toughened zirconia

Property	ASTM test method	Temperature °C	Temperature °F	TTZ
Stabilizer	...	...	...	MgO
Density	C 20	...	...	5.75
Water absorption	C 373	...	...	None
Hardness, HR45N	...	...	...	74 to 79
Flexural strength, MPa (ksi)	F 417	25	77	634 (92)
		500	930	414 (60)
		1000	1830	290 (42)
Tensile strength, MPa (ksi)	ALMA Test #4	25	77	352 (51)
Compressive strength, MPa (ksi)	C 773	25	77	1758 (255)
Modulus of elasticity, GPa (10^6 psi)	C 623	...	...	200 (29)
Shear modulus, GPa (10^6 psi)	C 623	...	...	69 (10)
Bulk modulus, GPa (10^6 psi)	C 623	...	...	373 (54)
Poisson's ratio	...	...	...	0.22
Coefficient of thermal expansion, 10^{-6}/°C (10^{-6}/°F)	C 372	25 to 1000	77 to 1830	10.1 (5.6)
Fracture toughness (K_{Ic}), MPa$\sqrt{m}$ (ksi$\sqrt{in.}$)	SENB	...	...	8 to 12 (7 to 11)
Weibull modulus, m (ft)	Four-point bend	...	...	20 (66)

Mixed Oxides

Data on the properties of various oxide ceramics

Ceramic	Flexural strength MPa	Flexural strength ksi	Fracture toughness $MPa\sqrt{m}$	Fracture toughness $ksi\sqrt{in.}$	Coefficient of thermal expansion, $10^{-6}/°C$	Thermal conductivity, W/m · K
Al_2O_3	200 to 600	29 to 87	3 to 4	2.7 to 3.6	7 to 9	15 to 40
PSZ (CaO, MgO)	600 to 1000	87 to 145	8 to 16	7.3 to 14.6	10	2.2
TZP (Y_2O_3, CeO_2)	800 to 2000	116 to 290	7 to 8	6.4 to 7.3	10	2.2
Al_2O_3-ZrO_2	1500 to 2000	218 to 290	8 to 13	7.3 to 11.8	...	...
$3Al_2O_3 \cdot 2SiO_2$	200	29	2	1.8	5.3	3.5
$2MgO_2 \cdot 2Al_2O_3 \cdot 5SiO_2$	150	22	1 to 2	0.9 to 1.8	2	3

Characteristics of mixed oxides as surface acoustic wave filter substrate materials

Parameters	$LiTaO_3$	$LiNbO_3$	PZT(a)	ZnO
SAW velocity, m/s (10^3 ft/s)	3295 (10.8)	3960 (13.0)	2430 (8.0)	3150 (10.3)
Coupling factor (k^2), %	0.7	6.0	2.9	0.6
Temperature coefficient of frequency, 10^{-6}/K	−31	−78	−17	−15
Permittivity, ε_2	47.9	67.2	350	8.84
Curie point (T_c), °C (°F)	618 (1145)	1210 (2210)	300 (570)	1200 (2190)

(a) $Pb(Mn_{1/3}Nb_{2/3})O_3$–$PbZrO_3$–$PbTiO_3$

Examples of perovskites including unit cell parameters

Compound	Unit cell parameters, Å a	b	c	Remarks
$KNbO_3$	3.9714	5.6946	5.7203	Orthorhombic
$KTaO_3$	3.9885	...	...	Cubic
$NaNbO_3$	5.512	5.577	3.885	Orthorhombic
$NaTaO_3$	3.8851	5.4778	5.5239	Orthorhombic
$TlIO_3$	4.510	...	...	$\alpha = 89.34°$ rhombohedral
$A^{2+}B^{4+}O_3$				
$BaFeO_3$	3.98	...	4.01	Tetragonal
$BaTiO_3$	3.989	...	4.029	Tetragonal
$BaZrO_3$	4.192	...	...	Cubic
$CaTiO_3$	5.381	7.645	5.443	Orthorhombic
$PbSnO_3$	7.86	...	8.13	Tetragonal
$PbTiO_3$	3.896	...	4.136	Tetragonal
PbZrO3	9.28	...	...	Pseudocubic, orthorhombic
$SrCeO_3$	5.986	8.531	6.125	Orthorhombic
$A^{3+}B^{3+}O_3$				
$BiAlO_3$	7.61	...	7.94	Tetragonal
$BiCrO_3$	3.90	3.87	3.90	$\alpha = \gamma = 90°13'$ $\beta = 89°10'$
$DyFeO_3$	5.30	5.60	7.62	Orthorhombic
$DyMnO_3$	3.70	...	...	Cubic
$GdAlO_3$	5.247	5.304	7.447	$GdFeO_3$ structure
$LaTiO_3$	3.92	...	...	...
$NdAlO_3$	3.752	...	...	Rhombohedral
$NdGaO_3$	5.426	5.502	7.706	$GeFeO_3$ structure
$PrCoO_3$	3.787	...	...	$\alpha = 90°13'$; rhombohedral
$SmVO_3$	3.89	...	...	
A_xBO_3 and ABO_{3-x}				
$Ce_{0.33}NbO_3$	3.89	3.91	7.86	Orthorhombic
$Ca_{0.5}TaO_3$	11.068	7.505	5.378	Orthorhombic
Li_xWO_3	($x = 1$) 3.72	...	...	Cubic $x = 0.35$ to 0.57
$A(B'_{0.67}B''_{0.33})O_3$				
$Ba(Dy_{0.67}W_{0.33})O_3$	8.386	...	...	$(NH_4)_3FeF_6$ structure
$Ba(In_{0.67}U_{0.33})O_3$	8.512	...	...	$(NH_4)_3FeF_6$ structure
$Ba(Nd_{0.67}W_{0.33})O_3$	8.513	...	...	$(NH_4)_3FeF_6$ structure
$A^{2+}(B^{2+}_{0.33}B^{5+}_{0.67})O_3$				
$Ba(Cd_{0.33}Nb_{0.67})O_3$	4.168	...	...	...
$Ba(Fe_{0.33}Ta_{0.67})O_3$	4.10	...	...	...
$Ba(Sr_{0.33}Ta_{0.67})O_3$	5.95	...	7.47	Hexagonal ordered

(continued)

Examples of perovskites including unit cell parameters (continued)

Compound	Unit cell parameters, Å: a	b	c	Remarks
$A^{2+}(B^{3+}_{0.5}B^{5+}_{0.5})O_3$				
$Ba(Bi_{0.5}Nb_{0.5})O_3$	8.630	...	...	$(NH_4)_3FeF_6$ structure
$Ba(Cu_{0.5}W_{0.5})O_3$	7.88	...	8.61	Tetragonal
$Ba(Ni_{0.5}Nb_{0.5})O_3$	4.1	...	...	...
$Ba(Sc_{0.5}Re_{0.5})O_3$	8.163	...	...	$(NH_4)_3FeF_6$ structure
$Ca(Gd_{0.5}Nb_{0.5})O_3$	4.03	4.04	4.03	β = 92°42′ monoclinic
$A^{2+}(B^{2+}_{0.5}B^{6+}_{0.5})O_3$				
$Ba(Ca_{0.5}W_{0.5})O_3$	8.39	...	...	$(NH_4)_3FeF_6$ structure
$Ba(Cu_{0.5}U_{0.5})O_3$	8.18	...	8.84	Tetragonal
$Ba(Mg_{0.5}W_{0.5})O_3$	8.099	...	...	$(NH_4)_3FeF_6$ structure

Specific heats of aluminates at room temperature

Aluminate	Specific heat: kJ/kg · K	Btu/lb · °F	Material and test conditions
$CaO \cdot 2Al_2O_3$	0.75	0.18	Powder samples: "high-purity", essentially stoichiometric compositions; 135-to-150-g samples; Nernst method, adiabatic copper calorimeter; 51 to 298 K
$CaO \cdot Al_2O_3$	0.75	0.18	
$3CaO \cdot Al_2O3$	0.75	0.18	
$FeO \cdot Al_2O_3$	0.71	0.17	Powder samples; "high-purity", essentially stoichiometric compositions; 217-g sample; Nernst method, adiabatic copper calorimeter; 51 to 298 K

Physical properties of aluminates

Aluminate	Crystal system	Theoretical density, g/cm^3	Melting point(a): K	°F
$BaO \cdot Al_2O_3$	Cubic/hexagonal	3.99(b)	2270	3630
$BaO \cdot 6Al_2O_3$	Hexagonal	3.64(b)	2135	3380
$BeO \cdot Al_2O_3$	Orthorhombic	3.76	2155	3420
$CaO \cdot 2Al_2O_3$	Monoclinic	2.90	2035(c)	3200(c)
$3CaO \cdot 5Al_2O_3$	Orthorhombic	...	1995	3130
$CaO \cdot Al_2O_3$	...	2.98(b)	1870	2910
$3CaO \cdot Al_2O_3$	Cubic	3.00	1810(c)	2800(c)
$CeO \cdot Al_2O_3$	Cubic	6.17	2345 ± 28	3765 ± 50
$CoO \cdot Al_2O_3$	Cubic	4.37(b)	2235	3560
$Dy_2O_3 \cdot 2Al_2O_3$	Cubic	6.05	2090	3300
$Gd_2O_3 \cdot Al_2O_3$	Cubic	...	2255	3600
$FeO \cdot Al_2O_3$	Cubic	4.35	1710(c)	2620(c)
$K_2O \cdot Al_2O_3$	Cubic	...	>1920	>3000
$Li_2O \cdot Al_2O_3$	...	2.55(b)	2170 to 2270	3450 to 3630
$Li_2O \cdot 5Al_2O_3$	Cubic(d)	3.60(b)	...	...
$MgO \cdot Al_2O_3$	Cubic	3.59 ± 0.01	2410	3875
$MnO \cdot Al_2O_3$	Cubic	4.12	1835(c)	2840(c)
$Na_2O \cdot Al_2O_3$	...	...	>1970	>3090
$NiO \cdot Al_2O_3$	Cubic	4.45	2300	3685
$Sm_2O_3 \cdot Al_2O_3$	Cubic	...	2255	3600
$SrO \cdot Al_2O_3$	...	...	2285	3650
$SrO \cdot 2Al_2O_3$	Monoclinic	3.03	2045	3220
$2YrO_3 \cdot Al_2O_3$	Cubic	...	2310	3700
$3YrO_3 \cdot 5Al_2O_3$	Cubic	...	2255	3600
$Y_2O_3 \cdot Al_2O_3$	Cubic	5.50	...	...
$ZnO \cdot Al_2O_3$	Cubic	4.58(b)	2220	3540

(a) Approximate values. (b) Measured density. (c) Incongruent melting. (d) Reportedly exists in many forms. Other stable spinel phase formed above 1560 K (2350 °F)

Effect of solid solution substitutions on the ferroelectric phase transitions of $BaTiO_3$ ceramics

Additive	Solid solution limit (mol%)	Change in transition temperature, Δ °C/mol%, for: ΔT_c(a)	ΔT_1(b)	ΔT_2(c)
$PbTiO_3$	100	+3.7	−9.5	−6.0
$SrTiO_3$	100	−3.7	−2.0	0
$CaTiO_3$	21	0	−6.7	−6.0
$BaZrO_3$	100	−5.3	+7.0	+18
$BaSnO_3$	100	−8.0	+5.0	+16
$KNbO_3$	100	−9.0	+12	+35
$Ba(Fe_{0.5}Ta_{0.5})O_3$	100	−15	−2	−6
$Ba(Co_{0.5}W_{0.5})O_3$	50	−30	...	...
$(K_{0.5}Nd_{0.5})TiO_3$	15	−10	−8	−6
$(K_{0.5}La_{0.5})TiO_3$	15	−15	...	...
$La_{0.67}TiO_3$	15	−18	...	...
$Ba_{0.5}NbO_3$	14	−26	+12	+6
$Ba_{0.5}TiO_3$	14	−29	...	...

(a) T_c, tetragonal/cubic transition temperature. (b) T_1, orthorhombic/tetragonal transition temperature. (c) T_2, rhombohedral/orthorhombic transition temperature

Designation and composition of several hexagonal ferrites

Ferrite designation	Chemical composition
M (Magnetoplumbite)	$BaO \cdot 6Fe_2O_3$
W	$BeO \cdot 2MO \cdot 8Fe_2O_3$
S (spinel)	$MO \cdot Fe_2O_3$
Z	$3BaO \cdot 2MO \cdot 12Fe_2O_3$
Y	$2BaO \cdot 2MO \cdot 6Fe_2O_3$

Note: W, Z, and Y are common designations for hexagonal ferrite types

Selected properties of PLZT compositions

PLZT type(a)	Density, g/cm³	T_t(b), °C	T_m(c), °C	κ(d)	tanδ(e), %	PR(f), μC/cm²	E_c(g), kV/cm	k_p(h)	d_{33}(i), C/N × 10^{12}	g_{33}(j), V · m/N × 10^3	Q_{11}(k), m^4/C^2	Q_{12}(l), m^4/C^2	$s_{11}E$(m), $m^2/N \times 10^{12}$
8/90/10	7.83	...	...	303	0.4	0	0	0	0	0	...	...	...
7.4/70/30	7.84	52	...	4500	4.8	29	3.7	0.61	...	...	...	...	...
7.6/70/30	7.84	40	100	4940	5.4	28	3.6	0.65	...	...	...	...	...
7.8/70/30	7.83	29	...	5545	5.6	23	2.5	0.59	520	...	...	...	...
8/70/30	7.83	20	85	5100	4.7	0	0	0	0	0	0.010	−0.008	...
2/65/35	7.98	320	325	650	2.5	40	13.7	0.45	150	23	...	...	...
7/65/35	7.84	95	150	1850	1.8	34	5.3	0.62	400	22	0.022	−0.012	13.5
8/65/35	7.82	55	110	3400	3.0	30	3.6	0.65	682	20	0.018	−0.008	12.4
9/65/35	7.81	5	80	5700	6.0	0	0	0	0	0	0.020	−0.010	...
9.5/65/35	7.79	−10	75	5500	5.5	0	0	0	0	0	0.021	−0.009	...
10/65/35	7.78	−25	70	5100	5.4	0	0	0	0	0	...	...	...
12/65/35	7.74	...	60	2200	4.6	0	0	0	0	0	...	...	...
8/40/60	7.84	240	245	980	1.2	28	17.7	0.34	...	...	...	...	...
12/40/60	7.74	140	145	1300	1.3	25	12.5	0.47	235	12	...	...	7.5
8/10/90	7.84	...	355	360	0.4	29	37.5	0.21	...	...	...	...	...
30/0/100	7.41	10	...	2140	0.1	0	0	0	0	0	...	...	...

(a) La/Zr/Ti compositions with lanthanum given in at %. (b) T_t, temperature at which there is a loss of permanent polarization. (c) T_m, temperature at the maximum dielectric constant. (d) κ, dielectric constant. (e) tan δ, dielectric dissipation or the deviation from an ideal capacitor. (f) P_R remanent polarization. (g) E_c, coercive filed. (h) k_p, planar coupling coefficient. (i) d_{33}, longitudinal coupling coefficient. (j) g_{33}, longitudinal open circuit voltage. (k) Q_{11}, longitudinal electrostrictive coefficient. (l) Q_{12}, lateral electrostrictive coefficient. (m) $s_{11}E$, compliance at constant electric field

Piezoelectric, elastic, and dielectric constants of several lead titanate zirconate compositions

	$Pb(Ti_{0.48}Zr_{0.52})O_3$	$Pb_{0.94}Sr_{0.06})(Ti_{0.47}Zr_{0.53})O_3$	$Pb_{0.988}(Ti_{0.48}Zr_{0.52})_{0.976}Nb_{0.024}O_3$
Curie point, °C (°F)	386 (727)	328 (622)	365 (690)
Dielectric constants (unitless)			
κ_1^T	1180	1475	1730
κ_1^S	612	730	916
κ_3^T	730	1300	1700
κ_3^S	399	635	830
Dissipation factor	0.004	0.004	0.02
Piezoelectric coupling factors (unitless)			
k_p	0.52_9	0.58	0.60
k_{31}	0.31_3	0.33_4	0.34_4

(continued)

Piezoelectric, elastic, and dielectric constants of several lead titanate zirconate compositions (continued)

	$Pb(Ti_{0.48}Zr_{0.52})O_3$	$Pb_{0.94}Sr_{0.06}(Ti_{0.47}Zr_{0.53})O_3$	$Pb_{0.988}(Ti_{0.48}Zr_{0.52})_{0.976}Nb_{0.024}O_3$
Piezoelectric coupling factors (unitless)(cont'd)			
k_{33}	0.67_0	0.70	0.70_5
k_{15}	0.69_4	0.71	0.68_5
Piezoelectric constants (units indicated)			
d_{31}, 10^{-12} C/N	–93.5	–123	–171
d_{33}	223	289	374
d_{15}	494	496	584
g_{31}, 10^{-3} Vm/N	–14.5	–11.1	–11.4
g_{33}	34.5	26.1	24.8
g_{15}	47.2	39.4	38.2
h_{31}, 10^8 V/m	...	–9.2	–7.3
h_{33}	...	26.8	21.5
h_{15}	...	19.7	15.2
e_{31}, C/m^2	...	–5.2	–5.4
e_{33}	...	15.1	15.8
e_{15}	...	12.7	12.3
Elastic compliances, 10^{-12} m^2/N			
s_{11}^E	13.8	12.3	16.4
s_{33}^E	17.1	15.5	18.8
s_{12}^E	–4.07	–4.05	–5.74
s_{13}^E	–5.80	–5.31	–7.22
s_{44}^E	48.2	39.0	47.5
s_{66}^E	38.4	32.7	44.3
s_{11}^D	12.4	10.9	14.4
s_{33}^D	9.35	7.90	9.46
s_{12}^D	–5.38	–5.42	–7.71
s_{13}^D	–2.56	–2.10	–2.98
s_{44}^D	25.0	19.3	25.2
Elastic stiffness (constants), GPa			
c_{11}^E	...	139	121
c_{12}^E	...	77.8	7
c_{13}^E	...	74.3	75.2
c_{33}^E	...	115	111
c_{44}^E	...	25.6	21.1
c_{66}^E	...	30.6	22.6
c_{11}^D	...	145	126
c_{12}^D	...	83.9	80.9
c_{13}^D	...	60.9	65.2
c_{33}^D	134	159	147
c_{44}^D	...	518	39.7
Other elastic constants			
Q_M	500	500	75

Mechanical properties of niobates

Niobate	Temperature °C	°F	Bend strength MPa	ksi	Modulus of elasticity GPa	10^6 psi	Material and test conditions
$MgO \cdot Nb_2O_5$	24	75	37	5.4	...	...	Pressed and sintered; 91 to 92% dense
	800	1470	78	11.3	...	...	
	1000	1830	90	13.0	...	...	
	1200	2190	51	7.4	...	...	
$TiO_2 \cdot Nb_2O_5$	24	75	29	4.2	41	6	Pressed and sintered; 91% pure; 86 to 88% dense; Modulus of elasticity determined sonically
	800	1470	34	4.9	41	6	
	1000	1830	48	6.9	41	6	
	1200	2190	36	5.3	...	...	

Physical properties of silicates

Silicate	Crystal system	Density(a), g/cm^3	Melting point K	°F
$3Al_2O_3 \cdot 2SiO_2$	Orthorhombic	3.13 to 3.36	2120	3360
$2BaO \cdot SiO_2$	Orthorhombic	5.20	>2025	>3190
$BaO \cdot SiO_2$	Monoclinic	4.40	1875	2920
$2BaO \cdot 3SiO_2$	...	3.93	1720	2640
$BaO \cdot 2SiO_2$	Orthorhombic	3.73	1690	2585
$BaO \cdot Al_2O_3 \cdot 2SiO_2$	Hexagonal	3.21 to 3.30	1990	3120

(continued)

Physical properties of silicates (continued)

Silicate	Crystal system	Density(a), g/cm^3	Melting point K	Melting point °F
$BaO \cdot 2CaO \cdot 3SiO_2$	Hexagonal	...	1595(b)	2410(b)
$BaO \cdot TiO_2 \cdot SiO_2$	...	...	1670	2550
$BaO \cdot TiO_2 \cdot 2SiO_2$	...	...	1520	2280
$2BeO \cdot SiO_2$	Hexagonal	2.99	1835(c)	2840(c)
$3CaO \cdot SiO_2$	Monoclinic	...	2170(c)	3450(c)
$2CaO \cdot SiO_2$	Monoclinic	3.28	2400	3865
$3CaO \cdot 2SiO_2$	Orthorhombic	...	1740(b)	2670(b)
$CaO \cdot SiO_2$	Triclinic	2.9	1810	2800
$CaO \cdot Al_2O_3 \cdot 2SiO_2$	Triclinic	2.77	1820	2820
$2CaO \cdot Al_2O_3 \cdot 2SiO_2$	Tetragonal	3.04	1865	2895
$CaO \cdot MgO \cdot SiO_2$	Orthorhombic	3.2	1770(b)	2730(b)
$CaO \cdot MgO \cdot 2SiO_2$	Monoclinic	3.28	1665	2535
$2CaO \cdot MgO \cdot 2SiO_2$	Tetragonal	2.94	1735	2660
$3CaO \cdot MgO \cdot 2SiO_2$	Monoclinic	3.15	1845(b)	2865(b)
$CaO \cdot K_2O \cdot SiO_2$	...	...	1905	2970
$CaO \cdot TiO_2 \cdot SiO_2$	Monoclinic	3.5	1655	2520
$2CaO \cdot ZnO \cdot 2SiO_2$	Tetragonal	...	1700	2600
$CaO \cdot ZrO_2 \cdot SiO_2$	...	...	1855	2880
$2CoO \cdot SiO_2$	Orthorhombic	4.68	1695	2590
$Dy_2O_3 \cdot SiO_2$	...	...	2200	3505
$2Dy_2O_3 \cdot 3SiO_2$	...	...	2195	3490
$Dy_2O_3 \cdot 2SiO_2$	...	...	1995(b)	3130(b)
$Er_2O_3 \cdot SiO_2$	...	6.80	2250	3595
$2Er_2O_3 \cdot 3SiO_2$	...	6.22	2170	3450
$Er_2O_3 \cdot 2SiO_2$	...	6.10	2070	3270
$Gd_2O_3 \cdot SiO_2$	...	6.55	2170	3450
$2Gd_2O_3 \cdot 3SiO_2$	Hexagonal	6.29	2220	3450
$Gd_2O_3 \cdot 2SiO_2$	...	5.34	1993	3128
$H{:}O_2 \cdot SiO_2$	Tetragonal	...	...	...
$2FeO \cdot SiO_2$	Orthorhombic	4.24	1470	2190
$La_2O_3 \cdot SiO_2$	Monoclinic	5.72	2200	3505
$2La_2O_3 \cdot 3SiO_2$	Hexagonal	5.30	2245	3580
$La_2O_3 \cdot 2SiO_2$	Monoclinic	4.85	2020(b)	3180(b)
$Li_2O \cdot SiO_2$	Orthorhombic	2.48	1475	2200
$2Li_2O \cdot SiO_2$	Orthorhombic	2.33	1525(b)	2290(b)
$Li_2O \cdot Al_2O_3 \cdot 2SiO_2$	Orthorhombic	2.36	1670(b)	2550(b)
$Li_2O \cdot Al_2O_3 \cdot 4SiO_2$	Monoclinic	...	1700	2600
$Li_2O \cdot Al_2O_3 \cdot 6SiO_2$	...	2.41	1455	2160
$4Li_2O \cdot 3ZrO_2 \cdot 5SiO_2$	...	4.02	1425	2110
$MgO \cdot SiO_2$(f)	...	...	1825	2850
$2MgO \cdot SiO_2$	Orthorhombic	3.22	2185	3470
$2MgO \cdot 2Al_2O_3 \cdot 5SiO_2$	Orthorhombic	2.51	1745(b)	2680(b)
$4MgO \cdot 5Al_2O_3 \cdot 2SiO_2$	...	...	~1725(b)	~2650(b)
$2MnO \cdot SiO_2$	Orthorhombic	4.05	1615(b)	2445(b)
$MnO \cdot SiO_2$	Triclinic	3.71	1570(b)	2370(b)
$3MnO \cdot Al_2O_3 \cdot 3SiO_2$	Cubic	4.18	1470	2190
$2MnO \cdot 2Al_2O_3 \cdot 5SiO_2$	...	...	1470(b)	2190(b)
$Nd_2O_3 \cdot SiO_2$	Hexagonal	...	...	...
$Nd_2O_3 \cdot 2SiO_2$	Monoclinic	...	...	...
$K_2O \cdot Al_2O_3 \cdot SiO_2$	Cubic	...	...	...
$K_2O \cdot Al_2O_3 \cdot 2SiO_2$	Hexagonal	2.6	2020	3180
$K_2O \cdot Al_2O_3 \cdot 4SiO_2$	Cubic	2.47	1960	3070
$K_2O \cdot Al_2O_3 \cdot 6SiO_2$	Triclinic	2.56	1420	2100
$K_2O \cdot ZnO \cdot SiO_2$	Cubic	...	1570(b)	2370(b)
$Pr_2O_3 \cdot SiO_2$	...	...	1670(c)	2550(c)
$Sm_2O_3 \cdot SiO_2$	Monoclinic	6.36	2215	3525
$2Sm_2O_3 \cdot 3SiO_2$	...	5.77	2195	3490
$Sm_2O_3 \cdot 2SiO_2$	Monoclinic	5.20	2050(b)	3230(b)
$Sc_2O_3 \cdot SiO_2$	...	3.49	2223	3542
$S_2O_3 \cdot 2SiO_2$	...	3.39	2133	3380
$2SrO \cdot SiO_2$	Monoclinic	3.84	>1975	>3100
$SrO \cdot SiO_2$	Monoclinic	3.65	1850	2975
$SrO \cdot Al_2O_3 \cdot 2SiO_2$	Triclinic	3.12	1935	3020
$ThO \cdot SiO_2$	Monoclinic	5.3	2250(b)	3595(b)
$Yb_2O_3 \cdot SiO_2$	...	...	2220	3540
$2Yb_2O_3 \cdot 3SiO_2$	...	...	2195	3490
$Yb_2O_3 \cdot 2SiO_2$	...	...	2120	3360
$Y_2O_3 \cdot SiO_2$	Monoclinic	4.49	2250	3595
$2Y_2O_3 \cdot 3SiO_2$	Hexagonal	4.39	2200	3540
$Y_2O_3 \cdot 2SiO_2$	Monoclinic	4.06	~2050	~3230
$ZnO \cdot SiO_2$	Hexagonal	4.1	1785	2750
$ZrO_2 \cdot SiO_2$	Tetragonal	4.68	1815(c)	2805(c)

(a) Values either are calculated from lattice parameters or are pycnometric measurements. (b) Incongruent melting. (c) Decomposition

Typical strengths of commercial silicate ceramics at room temperature(a)

Major phase	Bend strength MPa	Bend strength ksi	Tensile strength MPa	Tensile strength ksi	Compressive strength MPa	Compressive strength ksi
$3Al_2O_3 \cdot 2SiO_2$ (mullite)	170	25	110	16	690 to 1310	100 to 190
$ZrO_2 \cdot SiO_2$ (zircon)	140	20	75	11	480 to 690	70 to 100
$2MgO \cdot SiO_2$ (forsterite)	140	20	65	9.5	550	80
MgO · SiO2 (steatite)	130	19	64	9.3	550	80
$CaO \cdot SiO_2$ (wollastonite)	130	19	55	8	...	...
$2MgO \cdot 2Al_2O_3 \cdot 5SiO_2$ (cordierite)	110	16	54	7.8	345	50

(a) All values approximate

Primary phase wollastonite or diopside commercial slag-ceramics

Product type	Manufacture	Density, g/cm^3	Coefficient of thermal expansion, 10^{-6}/K	Modulus of rupture MPa	Modulus of rupture ksi	Composition, wt% SiO_2	Al_2O_3	Na_2O/K_2O	MgO/CaO	ZnO	MnO	Fe_2O_3	S
Slagsital (wollastonite)	USSR	~2.6	7.2 to 9	80 to 120	12 to 17	55.2	8.3	5.4/0.6	2.2/24.8	1.4	0.9	0.3	0.4
Sigran	USSR	2.6 to 2.8	8 to 8.5	28 to 70	4 to 10	...	...	...	...	...	...	...	...
Minelbite (diopside)	Hungary	...	...	...	...	60.9	14.2	3.2/1.9	5.7/9.0	...	2.0	2.5	0.6
Neoparies	Japan	2.7	6.2	50	7.3	...	...	...	...	...	...	...	...

Primary phase cordierite

Product type	Manufacturer	Density, g/cm^3	Coefficient of thermal expansion, 10^{-6}/K	Modulus of rupture MPa	Modulus of rupture ksi	Composition, wt% SiO_2	Al_2O_3	MgO	CaO	As_2O_3	Fe_2O_3	TiO_2	Nucleant
Commercial													
9606	Corning	2.6 to 2.612	6.5	123 to 370(a)	18 to 54(a)	56.1	19.8	14.7	0.1	0.3	0.1	8.9	...
"Cordierite type"	...	2.61	1.1 to 5.7	240 to 350(b)	35 to 51(b)	...	...	...	...	...	...	...	...
Noncommercial													
M1	...	...	5.3	350	51	(c)	20	15	...	...	...	...	TiO_2
M2	...	...	3.2	162	24	(c)	26	12	...	...	...	...	TiO_2
M3	...	...	2.4	127	18	(c)	30	14	...	...	...	...	TiO_2
M4	...	...	3.1	270	39	(c)	26	12	...	...	...	...	TiO_2, ZrO_2

(a) Strength dependent on surface finish. (b) Higher if chemically strengthened. (c) Balance is SiO_2

Thermal conductivities of titanates

Titanate	Thermal conductivity(a) W/m · K	Thermal conductivity(a) Btu/ft · h · °F
$SrO \cdot TiO_2$	5.9 to 5.5 (43 to 140 °C)	3.4 to 3.2 (110 to 280 °F)
	6.2 to 4.8 (49 to 150 °C)	3.6 to 2.8 (120 to 300 °F)
$CaO \cdot TiO_2$	4.7 to 4.3 (43 to 130 °C)	2.7 to 2.5 (110 to 270 °F)
$BaO \cdot TiO_2$	3.3 to 2.6 (43 to 230 °C)	1.9 to 1.5 (110 to 440 °F)
	2.9 to 2.6 (49 to 230 °C)	1.7 to 1.5 (120 to 390 °F)

(a) Reported values; no material or test details available

Physical properties of titanates

Titanate	Crystal system	Density, g/cm^3	Melting point K	Melting point °F
$\alpha Al_2O_3 \cdot TiO_2$	Orthorhombic	3.68	2135	3380
$\beta Al_2O_3 \cdot TiO_2$	...	...	(a)	(a)
$2BaO \cdot TiO_2$	...	5.3(b)	2135	3380
$BaO \cdot TiO_2$	Hexagonal(c)	5.9(b)	1890 ± 5.5	2940 ± 10
$BaO \cdot 2TiO_2$	...	...	1595(d)	2410(d)
$BaO \cdot 3TiO_2$	...	4.7(b)	1630	2475
$BaO \cdot 4TiO_2$	...	4.6(b)	1700	2600

(continued)

Physical properties of titanates

Titanate	Crystal system	Density, g/cm³	Melting point K	Melting point °F
$CaO \cdot TiO_2$	Cubic	4.10	2245	3580
$3CaO \cdot 2TiO_2$	...	...	2020(d)	3180(d)
$HfO_2 \cdot TiO_2$	Orthorhombic	7.21	...	...
$2MgO \cdot TiO_2$	Cubic	3.52	2005	3150
$MgO \cdot TiO_2$	Hexagonal	~4.00	1905	2970
$MgO \cdot 2TiO_2$	...	3.66	1920	3000
$2MnO \cdot TiO_2$	Cubic	4.49	1725	2650
$MnO \cdot TiO_2$	Hexagonal	4.54	1635(c)	2480(c)
$SrO \cdot TiO_2$	Cubic	5.11	2310	3700
$2ZnO \cdot TiO_2$	Cubic	5.12	1820	2820

(a) Unstable between 1020 and 1570 K (1380 and 2370 °F); converts to α phase at 2090 K (3300 °F). (b) Pycnometric measurements. (c) Three polymorphic forms reported: tetragonal (295 to 395 K, or 70 to 250 °F); cubic (395 to 1735 K, or 250 to 2660 °F); and hexagonal (above 1735 K, or 2660 °F). (d) Incongruent melting

Mechanical properties of titanates

Titanate	Test temperature °C	Test temperature °F	Bend strength MPa	Bend strength ksi	Compressive strength MPa	Compressive strength ksi	Modulus of elasticity GPa	Modulus of elasticity 10^6 psi
$Al_2O_3 \cdot TiO_2$	21	70	21 to 76(a)	3 to 11(a)	...	...	...	...
$BaO \cdot TiO_2$	21	70	...	...	...	...	110(b)	16(b)
	21	70	66 to 108(c)	9.6 to 15.6(c)	...	...	...	...
	21	70	83 to 163(d)	12.0 to 23.6(d)	...	...	...	...
$CaO \cdot TiO_2$	21	70	...	...	48 to 131(e)	7 to 19(e)	...	...
$3CaO \cdot 2TiO_2$	21	70	...	...	69(f)	10(f)	...	...
$MgO \cdot 2TiO_2$	21	70	12 to 14(g)	1.8 to 2.0(g)	...	...	7.6 to 10(g)	1.1 to 1.5(g)
	400	750	14 to 17	2.0 to 2.5	...	...	9.0 to 23	1.3 to 3.3
	600	1110	18 to 24	2.6 to 3.5	...	...	12 to 41	1.8 to 6.0
	800	1470	25 to 41	3.6 to 5.9	...	...	26 to 94	38 to 13.7
	1000	1830	44 to 58	6.4 to 8.4	...	...	43	6.3

(a) Pressed and sintered; 88 to 90% dense. (b) Pressed and sintered; 95% dense; 95% pure; dynamic method. (c) Commercial material, as received; average value, 80.7 MPa (11.7 ksi). (d) Same material described in footnote (c), thermally conditioned at 1290 °C (2350 °F); average value, 101 MPa (14.6 ksi). (e) Pressed and sintered; 75 to 91% dense. (f) pressed and sintered; 89% dense. (g) Pressed and sintered; 95.5% dense; strength and modulus determined in flexure (three-point loading)

Physical properties of zirconates

Zirconate	Crystal system	Theoretical density, g/cm³	Melting point K	Melting point °F
$CaO \cdot ZrO_2$	Monoclinic(a)	4.76 ± 0.02	~2615	~4250
$SrO \cdot ZrO_2$	Cubic(b)	5.48(c)	>2970(d)	>4890(d)
$BaO \cdot ZrO_2$	Cubic	6.26	~2920	~4800

(a) Pseudocubic also reported. (b) Also reported as orthorhombic or distorted cubic. (c) Also reported as 5.52 g/cm³. (d) Melting point of about 3115 K (about 5150 °F) also reported

Typical properties of TZP

Property	Value
Density, g/cm³	6.05
Hardness, HR45N	83
Modulus of rupture, MPa (ksi)(a)	900 (130)
Fracture toughness (K_{Ic}), MPa$\sqrt{m}$ (ksi$\sqrt{in.}$)(b)	14 (12.7)
Elastic modulus, GPa (10^6 psi)	200 (30)
Weibull modulus, *m*	14
Thermal conductivity, W/m · K	2

(a) Four-point bend test. (b) Single-edge notched beam test. Source: Coors Ceramics

Typical properties of Mg-PSZ (3 wt% MgO)

Property	Value
Density, g/cm³	5.75
Hardness, HR45N	74 to 79
Flexural strength, MPa (ksi)	
at 25 °C	634 (92)
at 500 °C	414 (60)
at 1000 °C	290 (42)
Tensile strength, at 25 °C, MPa (ksi)	352 (51)
Compressive strength, at 25 °C, MPa (ksi)	1758 (255)
Elastic modulus, GPa (10^6 psi)	200 (29)
Shear modulus, GPa (10^6 psi)	69 (10)
Bulk modulus, GPa (10^6 psi)	373 (54)
Poisson's ratio	0.22
Thermal expansion, 25 to 1000 °C, 10^{-6}/°C	10.1
Fracture toughness (K_{Ic}), MPa$\sqrt{m}$ (ksi$\sqrt{in.}$)	8 to 12 (7 to 11)
Weibull modulus, *m*	20

Carbides

Physical properties of aluminum and boron carbides

Carbide	Crystal system	Theoretical density, g/cm³	Melting point K	Melting point °F
Al_4C_3	Rhombohedral	2.99	2977 ± 55(a)	~4900 ± 100(a)
B_4C	Rhombohedral	2.52	2700(b)	4400(b)

(a) Decomposition obscures melting-point determinations; estimated value. (b) Congruent melting

Typical properties of boron carbide (B_4C)

Property	Value
General	
Density, g/cm³	2.51
Boron content, %	78.28
Carbon content, %	21.72
Structure	Rhombohedral
Lattice spacings of hexagonal cell, Å	a = 5.60
	c = 12.10
Color	Black
Thermal	
Melting point, °C (°F)	2450 (4440)
Boiling point, °C (°F)	3500 (6330)
Specific heat, cal/g	0.226
Heat of formation, kcal/mol	-13.8
Entropy, kcal/mol · °C	6.5
Thermal expansion, 10^{-6}/°C, from room temperature to:	
500 °C (930 °F)	4.78
800 °C (1470 °F)	4.5 to 5
1000 °C (1830 °F)	5.54
1500 °C (2730 °F)	6.02
2000 °C (3630 °F)	6.53
Thermal conductivity, W/m · K, at:	
20 °C (68 °F)	27 to 29
27 °C (80 °F)	28
425 °C (800 °F)	83
975 °C (1790 °F)	16
Mechanical(a)	
Elastic modulus, GPa (10^6 psi)	445 (64.5)
Shear modulus, GPa (10^6 psi)	186.5 (27)
Bulk modulus, GPa (10^6 psi)	254 (37)
Speed of sound, km/s (mile/s)	14.5 (9)
Flexural strength, MPa (ksi), at:	
room temperature	303 to 480 (44 to 70)
650 °C (1200 °F)	290 (42)
1100 °C (2000 °F)	241 (35)
Tensile strength at 980 °C (1800 °F), MPa (ksi)	155 (22.5)
Compressive strength, MPa (ksi)	2855 (414)
Poisson's ratio	0.17 to 0.207
Hardness numbers, kg/mm²	
Knoop (100 g load)	2800 to 3060
Knoop (1000 g load)	2230
Vickers	3700
Electrical and optical	
Resistivity, Ω · cm, at:	
room temperature	0.1 to 10
600 °C (1100 °F)	0.038
1000 °C (1830 °F)	0.030
2000 °C (3630 °F)	0.022
Coefficient of thermal emf, uV/°C	80
Band gap, eV	1.64
Emissivity of 0.65 μm energy at:	
880 °C (1615 °F)	0.76
1880 °C (3415 °F)	0.56
Total emissivity	0.85
Magnetic nature	Nonmagnetic
Chemical stability(b)	
In helium	Stable to 2250 °C (4080 °F)
In O_2	Stable to 540 °C (1000 °F)
In sulfur, 1200 °C (2200 °F)	Stable
In phosphorus, 1200 °C (2200 °F)	Stable
In iodine, 1200 °C (2200 °F)	Stable
In CO	Stable
In Cl_2, 600 °C (1100 °F)	Reacts
In metal oxides, >1500 °C (2730 °F)	Reacts(c)
Corrosion	
Air, maximum-use temperature, °C (°F)	1000 (1830)
General corrosion weight loss (mg/cm²) in:	
Concentrated HCl	0.095(d)
Concentrated H_2SO_4	0.130(d)
Concentrated HNO_3	0.269(d)
Concentrated H_3PO_4	0.086(d)
40% HF	0.095(d)
40% HF + concentrated HNO_3 + concentrated H_2SO_4	0.139(d)
25% NaOH	0.069(d)

(a) Unless otherwise noted, room-temperature mechanical properties are given. (b) Stability of all carbide powders is a function of their surface area. (c) Boron carbide reacts with many metal oxides at temperatures greater then 1500 °C (2730 °F). (d) General corrosion rate for a hot pressed product (2.49 g/cm³ density) in the specified environment at 200 °C (390 °F) for 8 h

Mechanical properties of boron carbide

Property	Reported values(a)
Bend strength, MPa (ksi)	323 to 346 (46.9 to 50.3)
	242 to 243 (35.2 to 35.3)(b)
	205 to 238 (29.8 to 34.6)(c)
	192 to 205 (27.9 to 29.8)(d)
	199 (28.9)(e)
	241 (35.0)
Compressive strength, MPa (ksi)	2752 (400)
Impact strength, J (in. · lb):	
Unnotched	0.026 to 0.031 (0.23 to 0.28)
Notched	0.0028 to 0.003 (0.025 to 0.028)
Modulus of elasticity, GPa (10^6 psi)	289 to 454 (42 to 66)
Modulus of elasticity, GPa (10^6 psi)(cont'd)	362 to 400 (52.6 to 58.2)
	322 to 384 (46.8 to 55.8)(f)
	277 to 356 (40.2 to 51.8)(g)
	228 to 265 (33.2 to 38.5)(h)
Shear modulus, GPa (10^6 psi)	165 to 206 (24 to 30)
Bulk modulus, GPa (10^6 psi)	193 to 255 (28 to 37)
Poisson's ratio	0.19
Microhardness, kg/mm²:	
Knoop (100-g load)	2800
Vickers (100-g load)	3200 to 5000

(a) Tested at room temperature unless otherwise noted. (b) At 1144 K (1600 °F). (c) At 1366 K (2000 °F). (d) At 1589 K (2400 °F). (e) At 1700 K (2600 °F). (f) At 672 K (750 °F). (g) At 1072 K (1470 °F). (h) At 1472 K (2190 °F)

Mechanical properties versus product condition of boron carbide

Atomic formula	Carbon content, wt%	Fabrication method	Density, g/cm^3	Flexural strength(a) MPa	Flexural strength(a) ksi	Modulus of elasticity, GPa	Modulus of elasticity, 10^6 psi	Shear modulus GPa	Shear modulus 10^6 psi	Poisson's ratio
B_4C	21.7	Hot pressed	2.51	480 ± 40	70 ± 6	441	64	188	27	0.17
$B_4C + 1C$	22.5	Sintered + HIP	2.51	400 ± 20	58 ± 3	433	63	183	26.5	0.18
$B_4C + 3C$	24.8	Sintered + HIP	2.50	430 ± 63	62 ± 9	405	59	171	25	0.16

(a) Four-point bending

General thermal properties of beryllium carbide (Be_2C)

Property	Temperature K	Temperature °F	Reported value(a)
Specific heat, kJ/kg · K (Btu/lb · °F)	293	68	1.632 (0.39)
	302 to 372	85 to 210	1.397 (0.33)
Thermal conductivity, W/m · K	298	77	42.56 (24.6)
Btu/ft · h · °F	422	300	23.35 (13.5)
Coefficient of linear thermal expansion, 10^{-6}/K (10^{-6}/°F)	298 to 872	77 to 1110	10.1 (5.6)

(a) These data are for Be_2C bodies which contain 2 to 12 wt% of other materials as impurities or additions for facilitating fabrication, and which are greater than 90% dense

Mechanical properties of beryllium carbide (Be_2C)

Property	Reported value(a)
Bend strength, MPa (ksi)	89.4 (13)
Compressive strength, MPa (ksi)	722.4 (105)
Modulus of elasticity, GPa (10^6 psi):	
Compression	316 (46)(b)
	241 (35)(c)
	206 (30)(d)
Flexure	344 (50)
Poisson's ratio	0.10
Microhardness, Knoop, kg/mm^2	2700

(a) These data are for hot-pressed Be_2C, 90 to 96% pure and about 92% dense. Tested at room temperature unless otherwise noted. (b) At 811 K (1000 °F). (c) At 1089 K (1500 °F). (d) At 1366 K (2000 °F)

Representative compositions and proprietary designations of various cemented carbide grades

Representative composition, % WC	Co	TaC	TiC	Ni	Cr	Mo_2C	Proprietary designations: Adamas Carbide	Anderson Strathclyde	Carbidie	Carmet	Danit	General Carbide	General Electric Carboloy	GTE Valenite
96.5	3.0	0.5	...	...	...	...	GU2	CA	CD20	CA8	K04	GC003	999	VC3
94.0	5.5	0.5	...	...	...	...	PWX	CF	CD24	CA306	K10	GC005	895	VC2
94.0	6.0	...	...	...	...	...	A	CG	CD30	CA4	K20	GC106	883	...
91.0	9.0	...	...	...	...	...	B	...	CD35F	CA12	K30	GC009	...	VC152
87.0	13.0	...	...	...	...	...	BB	...	CD40	CA10	...	GC313	258	VC11
81.5	18.0	0.5	...	...	...	...	GU1	...	CD650	...	...	...	...	...
79.0	12.0	9.0	...	...	...	...	474	...	...	...	...	...	...	VC047
...	...	...	74.0	12.5	...	13.5	Titan 80	...	...	CA100	...	...	...	VC83
...	...	...	70.5	17.5	1.0	11.0	Titan 60	...	...	...	...	...	...	...
...	...	...	66.5	22.5	1.0	10.0	Titan 50	...	...	...	...	...	...	...
89.0	...	...	1.0	10.0	...	...	...	...	...	R10	...	...	...	VC099
85.0	15.0	...	...	...	...	...	HD15	CPM	CD50	CA11	DG30	GC315	268	VC12
80.0	20.0	...	...	...	...	...	HD20	...	...	CA20	DG40	GC320	...	VC13
75.0	25.0	...	...	...	...	...	HD25	...	...	CA225	DG50	GC325	190	VC14
75.0	20.0	5.0	...	...	...	...	HD20T	...	CD60	...	...	...	...	...
70.0	25.0	5.0	...	...	...	...	HD25T	...	CD70	...	...	...	...	...
94.0	6.0	...	...	...	...	...	575	CR	...	CA3	B030	GC206	44A	...
90.0	10.0	...	...	...	...	...	569	CM	...	2102	B050	GC410	90	...
89.0	11.0	...	...	...	...	...	783	...	CD337	CA411	B055	GC411	115	...
88.0	12.0	...	...	...	...	...	502	CT	...	CA412	B060	GC412	120	...

Representative composition, % WC	Co	TaC	TiC	Ni	Cr	Mo_2C	Proprietary designations: Kennametal	Krupp Widia	Mefasa	Metallwerk Plansee	Mitsubishi	Sandvik	Sumitomo	Teledyne Firth Sterling
96.5	3.0	0.5	...	...	...	...	K11	THF	...	H03T	...	CS05	...	HF
94.0	5.5	0.5	...	...	...	...	K68	GT05	K1	H10T	GTi05	CS10	H1	HA
94.0	6.0	...	...	...	...	...	K6	GT10	K2	H16T	GTi10	HML	G10E	H6
91.0	9.0	...	...	...	...	...	K9	GT15	MK30	H30T	GTi15	H10F	G3	H8
87.0	13.0	...	...	...	...	...	...	GT3H	...	H40T	GTi20	R4	G5	H81
81.5	18.0	0.5	...	...	...	...	...	...	...	...	GTi40	...	...	...
...	...	...	74.0	12.5	...	13.5	K165	...	...	F05T	NX33	CN02	...	...

(continued)

Representative compositions and proprietary designations of various cemented carbide grades (continued)

Representeative composition, %							Proprietary designations							
WC	Co	TaC	TiC	Ni	Cr	Mo_2C	Kennametal	Krupp widia	Mefasa	Metallwerk Plansee	Mitsubishi	Sandvik	Sumitomo	Teledyne Firth Sterling
...	...	...	70.5	17.5	1.0	11.0	...	...	...	F10T	NX55	...	T12A	...
...	...	...	66.5	22.5	1.0	10.0	...	TTF	...	...	...	...	T12B	...
88.2	1.8	10.0	...	...	...	...	K602	...	...	...	...	...	...	...
85.8	10.1	...	...	...	4.1	...	K701	...	...	...	...	...	...	...
93.3	5.8	...	...	...	0.9	...	K703	...	...	...	...	...	...	...
88.4	6.1	4.5	1.0	...	...	...	K714	...	...	...	...	...	...	...
93.7	...	0.3	...	6.0	...	...	K801	...	...	WC6Ni	...	...	...	...
89.0	...	...	1.0	10.0	...	...	K803	...	...	TCR30	...	...	...	...
85.0	15.0	...	...	...	...	...	SP212	BT40	G3	B50T	...	CT60	G6	MPD160
80.0	20.0	...	...	...	...	...	...	...	G4	H60T	GTi40	CT75	G7	...
75.0	25.0	...	...	...	...	...	...	GT55	G5	H70T	...	CT85	G8	...
75.0	20.0	5.0	...	...	...	...	K91	...	...	...	...	...	...	ND20
70.0	25.0	5.0	...	...	...	...	K90	...	...	...	...	...	...	ND25
94.0	6.0	...	...	...	...	...	K3404	BT10	K3	B10T	...	CT30	...	HAN6
90.0	10.0	...	...	...	...	...	K3070	BT25	MK35	B30T	...	CT45	G3	MPD10
89.0	11.0	...	...	...	...	...	K3047	...	MK40	B36T	...	CT50	...	MPD11
88.0	12.0	...	...	...	...	...	K3030	BT30	...	B40T	...	...	G5	...

Note: This table lists approximate compositions and proprietary designations for a number of corrosion-resistant grades. These grades from 16 manufacturers worldwide are meant to be representative only in a general sense. There are well over 100 manufacturers throughout the world (over 25 in the United States alone); therefore, it is not feasible to include all. In addition, cross comparisons are not precisely possible. For example, a grade listed with an approximate composition of WC-25Co may be cross referenced with a comparable grade that contains only 24% Co. Manufacturers can be consulted for more information

Properties of representative grades of cemented carbide

Cemented carbide	Room-temperature hardness, HV	Modulus of elasticity, GPa	Transverse rupture strength, MPa	Coefficient of thermal expansion, 10^{-6}/K	Thermal conductivity, W/m · K	Density, g/cm^3
Iron-bonded TiC	1000	305	2070	7.83	17	6.60
WC-20 wt% Co	1050	490	2850	6.4	100	13.55
WC-10 wt% Co	1625	580	2280	5.5	110	14.50
WC-3 wt% Co	1900	673	1600	5.0	110	15.25
WC-10 wt% Co-22 wt% (Ti, Ta, Nb)C	1500	510	2000	6.1	40	11.40
(Ti, Mo)C-Ni	1900	460	1100	7.5	17	5.50

Properties of representative cobalt-bonded cemented carbides

Nominal composition	Grain size	Hardness, HRA	Density		Transverse strength		Compressive strength		Modulus of elasticity		Relative abrasion resistance (a)	Coefficient of thermal expansion, 10^{-6}/°C		Thermal conductivity, W/m · K
			g/cm^3	oz./in.^{3}	MPa	ksi	MPa	ksi	GPa	10^6 psi		at 200 °C (390 °F)	at 1000 °C (1830 °F)	
97WC-3Co	Medium	92.5 to 93.2	15.3	8.85	1590	230	5860	850	641	93	100	4.0	...	121
94WC-6Co	Fine	92.5 to 93.1	15.0	8.67	1790	260	5930	860	614	89	100	4.3	5.9	...
	Medium	91.7 to 92.2	15.0	8.67	2000	290	5450	790	648	94	58	4.3	5.4	100
	Coarse	90.5 to 91.5	15.0	8.67	2210	320	5170	750	641	93	25	4.3	5.6	121
90WC-10Co	Fine	90.7 to 91.3	14.6	8.44	3100	450	5170	750	620	90	22	...	...	...
	Coarse	87.4 to 88.2	14.5	8.38	2760	400	4000	580	552	80	7	5.2	...	112
84WC-16Co	Fine	89	13.9	8.04	3380	490	4070	590	524	76	5	...	...	...
	Coarse	86.0 to 87.5	13.9	8.04	2900	420	3860	560	524	76	5	5.8	7.0	88
75WC-25Co	Medium	83 to 85	13.0	7.52	2550	370	3100	450	483	70	3	6.3	...	71
71WC-12.5TiC-12TaC-4.5Co	Medium	92.1 to 92.8	12.0	6.94	1380	200	5790	840	565	82	11	5.2	6.5	35
72WC-8TiC-11.5TaC-8.5Co	Medium	90.7 to 91.5	12.6	7.29	1720	250	5170	750	558	81	13	5.8	6.8	50

(a) Based on a value of 100 for the most abrasion-resistant material

Properties of selected interstitial carbides

Carbide	Hardness, HV (50 kg)	Crystal structure	Melting point		Theoretical density, g/cm^3	Modulus of elasticity		Coefficient of thermal expansion, $10^{-6}/°C$	Resistivity, $\mu\Omega \cdot cm$
			°C	°F		GPa	10^6 psi		
TiC	3000	Cubic	3100	5600	4.94	451	65.4	7.7	68
VC	2900	Cubic	2700	4900	5.71	422	61.2	7.2	150
HfC	2600	Cubic	3900	7050	12.76	352	51.1	6.6	109
ZrC	2700	Cubic	3400	6150	6.56	348	50.5	6.7	63
NbC	2000	Cubic	3600	6500	7.80	338	49.0	6.7	74
Cr_3C_2	1400	Orthorhombic	1800(a)	3250	6.66	373	54.1	10.3	...
WC	(0001) 2200	Hexagonal	~2800(a)	5050	15.7	696	101	(0001) 5.2	53
	$(10\bar{1}0)$ 1300							$(10\bar{1}0)$ 7.3	...
Mo_2C	1500	Hexagonal	2500	4550	9.18	533	77.3	7.8	97
TaC	1800	Cubic	3800	6850	14.50	285	41.3	6.3	30
W_2C	2150(b)	Hexagonal	2785	5045	17.2	420	60.9	4 to 4.7(c)	...

(a) Not congruently melting, dissociation temperature. (b) Knoop hardness, 100 g load. (c) Expansion from room temperature to 1500 and 2000 °C (2730 and 3630 °F), respectively

Physical properties of chromium, molybdenum, and tungsten carbides

Carbide	Crystal system	Theoretical density, g/cm^3	Melting point(a)	
			K	°F
$Cr_{23}C_6$	Cubic	6.99	1794	2770
Cr_7C_3	Hexagonal	6.92	2053	3235
Cr_3C_2	Orthorhombic	6.68	2166	3440
Mo_2C	Hexagonal(b)	9.12	2761 to 2794(c)	4510 to 4570(c)
MoC_{1-x}	Hexagonal(d)	...	2825(e)	4625(e)
MoC	Hexagonal	~8.8	2866(f)	4700(f)
W_2C	Hexagonal	~17.3	3050(g)	5030(g)
WC_{1-x}	Cubic	...	3016(h)	4970(h)
WC	Hexagonal	~15.8	3050(k)	5030(k)

(a) Approximate. (b) An orthorhombic form (αMo_2C) reportedly exists below 1473 K (2192 °F). (c) Alpha and beta phases indicated for C:Mo atomic ratios of 0.30 to 0.34; αMo_2C melts congruently; βMo_2C melts congruently but is not stable below 1746 K (2685 °F). (d) Pseudocubic lattice; see also footnote (e). (e) Phase indicated for C:Mo atomic ratios of 0.38 to 0.39; congruent melting; not stable below 1927 K (3010 °F). (f) Phase indicated for C:Mo atomic ratios of 0.41 to 0.50; congruent melting; not stable below 2283 K (3650 °F). (g) Alpha and beta phases indicated for C:W atomic ratios of 0.29 to 0.35; αW_2C melts congruently but is not stable below 1522 K (2280 °F); βW_2C decomposes at about 2794 K (4570 °F) and is not stable below 2722 K (4440 °F). (h) Phase indicated for C:W atomic ratios of 0.375 to 0.40; congruent melting; not stable below 2802 K (4585 °F). (k) Phase at stoichiometric composition; decomposes into melt and graphite

Mechanical properties of chromium, molybdenum, and tungsten carbides

Property	Reported values(a) for:		
	Cr_3O_2	Mo_2C	WC
Bend strength, MPa (ksi)	241 to 323 (35 to 47)	...	...
	~571 (~83)(b)	...	...
	~138 (~20)(c)	...	...
	...	...	482 to 832 (70 to 121)
	...	...	78 to 195 (11.4 to 28.4)(d)
Compressive strength, MPa (ksi)	1039 (151)	...	2683 (390)
	929 (135)(e)	...	1404 (204)(e)
	564 (82)(f)	...	...
	413 (60)(g)	...	...
	1101 (160)	901 (131)	...
	...	...	2958 (430)
	...	...	3523 (512)
Tensile strength, MPa (ksi)	...	...	344 (50)
Modulus of elasticity, GPa (10^6 psi)	385 (56)	...	...
	...	227 (33)	~688 (~100)
	...	...	667 (97)
	...	533 (77.5)	...
Hardness:			
Microhardness(h), kg/mm^2	1350 to 2280	1500 to 1800	1700 to 2400
Rockwell A	~91	~89	~90

(a) Tested at room temperature unless otherwise noted. (b) At 1255 to 1366 K (1800 to 2000 °F). (c) At 1589 K (2400 °F). (d) At 1589 to 2366 K (2400 to 3800 °F). (e) At 1273 K (1830 °F). (f) At 1477 K (2200 °F). (g) At 1672 K (2550 °F). (h) Essentially consistent ranges of microhardness values have also been reported from various sources for $Cr_{23}C_6$, Cr_7C_3, and W_2C. Representative ranges for these compounds, in kg/mm^2, are 970 to 1650, 1336 to 2200, and 1500 to 2060, respectively

Crystal properties of tungsten carbide

Compound	Crystal structure	Structure type	Pearson symbol	Space group	Lattice parameters, nm
Monotungsten carbide (WC)	Face-centered cubic	WC	*hP*2	$P\bar{6}m2$	$a = 0.29063$
Ditungsten monocarbide (W_2C)	Hexagonal(a)	CdI_2	*hP*3	$P\bar{3}m1$	$a = 0.300$ $c = 0.4730$

(a) Low-temperature stable to 2753 K (4500 °F)

Diamond polytype parameters

Crystal system	Crystallographic characteristics						
Ramsdell notation	3*C*	2*H*	4*H*	6*H*	8*H*	15*R*	21*R*
Jagodzinski notation	$(k)_3$	$(h)_2$	$(hk)_2$	$(hkk)_2$	$(hkkk)_2$	$(hkhkk)_3$	$(hkkhkkk)_3$
Zhdanov notation	(Infinity)	(11)	(22)	(33)	(44)	$(23)_3$	$(34)_3$
New *ABC*	*ABC*	AA′	*AA′C′C*	*AA′C′B′BC*	*AA′C′B′A′ABC*	*AA′C′CABB′ A′ABCC′B′BC*	*AA′C′B′BCABB′A′ C′CABCC′B′A′ABC*
Space group	$Fd3m$	$P6_3/mmc$	$P6_3/mmc$	$P6_3/mmc$	$P6_3/mmc$	$R\bar{3}m$	$R\bar{3}m$
Factor group	O_h	$D6_h$	$D6_h$	$D6_h$	$D6_h$	$D3_d$	$D3_d$
Hexagonal, %	0	100	50	33	25	40	29
Lattice parameter, nm	$a = 0.35597$ ...	$a = 0.25221$ $c = 0.41186$	$a = 0.25221$ $c = 0.82371$	$a = 0.25221$ $c = 1.23557$	$a = 0.25221$ $c = 1.64743$	... ...	... ...
Interatomic distance, nm	0.154	0.154	0.154	0.154	0.154	0.154	0.154
Atom positions	$(0,0,0)\left(\frac{1}{2},\frac{1}{2},0\right)$ $\left(0,\frac{1}{2},\frac{1}{2}\right)\left(\frac{1}{2},0,\frac{1}{2}\right)$ $\left(\frac{1}{4},\frac{1}{4},\frac{1}{4}\right)\left(\frac{1}{4},\frac{3}{4},\frac{3}{4}\right)$ $\left(\frac{3}{4},\frac{3}{4},\frac{1}{4}\right)\left(\frac{3}{4},\frac{1}{4},\frac{3}{4}\right)$	4 on $f(C_{3v})$	4 on $e(C_{3v})$ 4 on $f(C_{3v})$	4 on $e(C_{3v})$ 8 on $f(C_{3v})$	4 on $e(C_{3v})$ 12 on $f(C_{3v})$	30 on $c(C_{3v})$	42 on $c(C_{3v})$
Sources	Natural, HTHP(a), explosive, thin film	Natural, explosive, lonsdaleite, meteorites, bort	Thin film, explosive	Thin film, explosive	Thin film, explosive	Thin film, explosive	Thin film, explosive

(a) HTHP, high temperature, high pressure

Thermal properties of diamond

Properties/source	Natural diamond	Synthetic diamond	Polycrystalline diamond	Thin film diamond
Thermal conductivity, W/m · K at 298 K	...	...	543	...
Thermal diffusivity, cm^2/s	...	...	...	0.3 to 8.1
Type 11a at 77 K	4800	3300	...	...
Type 11a at 298 K	10	10	...	...
Coefficient of thermal expansion, 10^{-6}/K, at 298 to 373 K	...	...	1.5 to 3.8	...
Specific heat, J/mol · K, at 300 K	...	...	...	33.2 to 41.8
Heat of sublimation, kJ/mol	669	...	...	...
Debye temperature, K	...	...	...	...
Thermal shock resistance parameter, 10^8 W/m	3.8	...	...	1.9
Oxidation rate, $g/cm^2 \cdot s$	...	...	...	0.356 to 0.879
(111) plane	2.38	...	...	...
(100) plane	0.167	...	...	...
Activation energy of oxidation, kJ/mol · K				
823 to 1023 K	172 to 184	...	...	...
923 to 1023 K	...	...	...	303 to 318

Electrical properties of diamond

Properties/Source	Natural diamond	Synthetic diamond	Polycrystalline diamond	Thin film diamond
Electrical resistivity, Ω · cm, at 298 K				
Type 1a	10^3 to 10^{16}	...	...	...
Type 1b	10^3 to 10^{16}	...	...	...
Type 11a	10^3 to 10^{16}	...	...	...
Type 11b	10 to 10^7	...	...	...

(continued)

Electrical properties of diamond (continued)

Properties/Source	Natural diamond	Synthetic diamond	Polycrystalline diamond	Thin film diamond
Dielectric constant (298 K, 1 MHz)	5.5 to 5.7	5.7	...	3.5 to 5.7
Breakdown strength (V/cm)	$>10^7$	...	...	...
Activation energy, eV				
Type 11b (90 to 290 K)	0.29 to 0.37	0.015 to 0.37	0.42	...
B-doped (270 to 710 K)	...	0.0029 to 0.087	...	...
Al-doped (270 to 710 K)	...	0.32	...	...
B-doped (270 to 710 K)	...	0.17 to 0.18	...	...
Be-doped (270 to 710 K)	...	0.20 to 0.35	...	...
Band gap energy, eV at 298 K	5.2 to 5.6	...	...	...
Effective mass				
Holes (m_h/m_o) Type 11b	<1.0	...	...	...
Electrons (m_e/m_o) Type 11b	0.2	...	...	...
Carrier mobility, $cm^2/V \cdot s$				
Hole	1200 to 1600	...	...	...
Electron	1800 to 2200	...	...	...
Dissipation factor	0.0002	...	...	0.0140
Carrier lifetime, s	10^{-10}	...	...	...
Electron velocity, cm/s	2.7×10^7	...	...	...

Mechanical properties of diamond

Properties	Natural diamond	Synthetic diamond	Polycrystalline diamond	Thin film diamond
Density, g/cm^3	3.51 to 3.52	3.20 to 3.52	3.00 to 4.00	1.80 to 3.50
Elastic properties				
Elastic constants, GPa	C_{11} = 950 to 1079	...	...	...
	C_{12} = 120 to 330	...	...	...
	C_{44} = 430 to 578	...	...	...
Modulus of elasticity, GPa	700 to 1200	800 to 925	749 to 953	536 to 1035
Bulk modulus, GPa	440 to 590	...	290 to 372(a)	...
Poisson's ratio	0.10 to 0.29	0.20	0.07 to 0.20(a)	...
Strength and brittleness	16.4 to 32.4	21.6 to 32.4	0.34 to 1.50(a)	...
Tensile strength, GPa	...	...	...	...
Flexural strength, MPa	1050	800 to 1400	389 to 1550(a)	...
Compressive strength, GPa	8.68 to 16.53	4.5 to 5.8	1.9 to 6.9(a)	...
Fracture toughness, $MPa\sqrt{m}$	3.4	6.0 to 10.7	6.0 to 8.8(a)	...
Cleavage energy, J/m^2	12(b) 10 to 20(d)	10 to 18	26 to 50(c)	...
Knoop hardness, GPa	...	54 to 84	49 to 78	65
(001) face	56 to 102	...	...	...
(110), (111) faces	58 to 88	...	...	...
Vickers hardness, GPa	...	95 to 131	25 to 98	29 to 118
(001) face	88 to 147	...	...	...
(111) face	98(e)	...	...	...
Type 1b	...	88 to 108	...	...
Type 11a	...	108 to 145	...	...
Sliding coefficient of friction (f)				
In air	0.05 to 0.10	0.05 to 0.15	0.02 to 0.40	0.15 to 0.45
In vacuum	0.9	...	...	...
Sound velocity, m/s	...	...	16,330	16,200
(100) face	17,500	...	...	...
(110) face	18,200	...	...	...

(a) Values dependent upon grain size and cobalt content. (b) Experimental values. (c) Fracture energy. (d) Theoretical values vary with planes. (e) Maximum value. (f) Varies with face, direction, surface roughness, and atmosphere

Physical properties of titanium, zirconium, and hafnium carbides

			Melting point(a)	
Carbide	Crystal system	Theoretical density, g/cm^3	K	°F
TiC	Face-centered cubic	4.92	3340 ± 15(b)	5550 ± 27(b)
ZrC	Face-centered cubic	6.56	3693 ± 20(c)	6188 ± 36(c)
HfC	Face-centered cubic	12.67	4203 ± 20(d)	7106 ± 36(d)

(a) Representative values; congruent melting. (b) Maximum melting point at about 44 at. % carbon. (c) Maximum melting point at about 45 at. % carbon. (d) Maximum melting point at about 18.5 at. % carbon

Room-temperature mechanical properties of hafnium carbide

Property	Reported values
Bend strength, MPa (ksi)	234 to 241 (34 to 35)
Modulus of elasticity, GPa (10^6 psi)	316 to 461 (46 to 67)
Shear modulus, GPa (10^6 psi)	179 to 193 (26 to 28)
Bulk modulus, GPa (10^6 psi)	241 (35)
Poisson's ratio	0.17
	0.18
Microhardness, kg/mm^2:	
Knoop (50-g load)	2000 to 2500
Knoop (100-g load)	2260 to 3050
	1800 to 2500(a)
Vickers (200-g load)	1900 to 2100(a)

(a) Single crystal

Mechanical properties of zirconium carbide

Property	Reported values(a)
Bend strength, MPa (ksi)	103 to 206 (15 to 30)
Compressive strength, MPa (ksi)	826 to 2958 (120 to 430)
	1637 (238)
	488 (71)(b)
	255 (37)(c)
Tensile strength, MPa (ksi)	103 (15)
	83 to 96 (12 to 14)(d)
	89 to 110 (13 to 16)(e)
	69 to 110 (10 to 16)(f)
	14 to 48 (2 to 7)(g)
Modulus of elasticity, GPa (10^6 psi)	358 (52)
	550 (80)
	385 to 406 (56 to 59)(h)
	345 to 400 (50 to 58) (j)
Shear modulus, GPa (10^6 psi)	172 (25)(h)
	165 to 200 (24 to 29)
	162 (23.5)
Bulk modulus, GPa (10^6 psi)	220 (32)(h)
	206 (30)
Poisson's ratio	0.19
	0.20 (h)
Hardness:	
Microhardness, kg/mm^2:	
Knoop (100-g load)	1830
Vickers (100-g load)	2600
Vickers (50-g load)	2310 to 2520
Rockwell A	92

(a) Tested at room temperature unless otherwise noted. (b) At 1273 K (1830 °F). (c) At 1472 K (2190 °F). (d) At 1255 K (1800 °F). (e) At 1477 K (2200 °F). (f) At 1866 K (2900 °F). (g) At 1999 K (3630 °F). (h) Single crystal. (j) At 294 to 1505 K (70 to 2250 °F)

Mechanical properties of titanium carbide

Property	Reported values(a)
Bend strength, MPa (ksi)	282 to 667 (41 to 97)
	688 to 5504 (100 to 800)(b)
Compressive strength, MPa (ksi)	757 to 2958 (110 to 430)
	227 (33)(c)
	89 (13(d)
Tensile strength, MPa (ksi)	241 to 275 (35 to 40)
	110 to 117 (16 to 17)(e)
	55 to 62 (8 to 9)(F)
Modulus of elasticity, GPa (10^6 psi)	447 (65(b)
	268 to 461 (39 to 67)
Shear modulus, GPa (10^6 psi)	186 (27)(b)
	110 to 193 (16 to 28)
Bulk modulus, GPa (10^6 psi)	241 (35)(b)
	227 (33)
Poisson's ratio	0.19(b)
	0.18 to 0.19
Hardness:	
Knoop (50-g load)	2000 to 2750(b)
Knoop (100-g load)	2000 to 2400(b)
	1800 to 5900
Vickers (100-g load)	~3200
Rockwell A	93

(a) Tested at room temperature unless otherwise noted. (b) Single crystal. (c) At 1872 K (2910 °F). (d) At 2199 K (3990 °F). (e) At 1255 K (1800 °F). (f) At 1477 K (2200 °F)

Physical properties of vanadium, niobium, and tantalum carbides

Carbide	Crystal system	Theoretical density, g/cm^3	Melting point K	Melting point °F
VC	Cubic	5.48	2972 ± 50	4890 ± 90
V_2C	Hexagonal	5.75	2438 ± 25	3930 ± 45
NbC	Cubic	7.82	3772 ± 75	6330 ± 135
Nb_2C	Orthorhombic(a)	7.85	3360 ± 50	5590 ± 90
TaC	Cubic	14.50	4152 ± 28	7015 ± 50
Ta_2C	Hexagonal(b)	15.00	3602 ± 100(b)	6025 ± 180(b)

(a) Hexagonal above about 1470 K (2200 °F). (b) Ta_2C reportedly exists as α and β phases; the α phase (*C*6 structure) is stable up to about 2200 ± 28 K (3500 ± 50 °F), and the high-temperature β phase (*L*'3 structure) is stable between 2200 and about 3600 K (3500 and about 6000 °F)

Mechanical properties of vanadium, niobium, and tantalum carbides

	Reported values(a) for:		
Property	VC	NbC	TaC
Bend strength, MPa (ksi)	...	~289 (~42)	
	...	...	275 to 310 (40 to 45)(b)
	...	...	~378 (~55)(c)
Compressive strength, MPa (ksi)	607 (88)	~2374 (~345)	...
Tensile strength, MPa (ksi)	...	244 (36)	96 to 291 (14 to 42)
Modulus of elasticity, GPa (10^6 psi)	268 to 420 (39 to 61)	330 to 537 (48 to 78)	241 to 722 (35 to 105)
Shear modulus, GPa (10^6 psi)	157 (23)	197 to 245 (29 to 36)	215 to 227 (31 to 33)
Bulk modulus, GPa (10^6 psi)	389 (57)	296 to 378 (43 to 55)	248 to 343 (36 to 50)
Poisson's ratio	0.22	0.22	0.24
	...	0.25(d)	...
Hardness:			
Microhardness(e), kg/mm^2	2000 to 3000	1900 to 2600	1600 to 2400
Rockwell A	~91	~90	~88

(a) Tested at room temperature unless otherwise noted. (b) At 294 to 1273 K (70 to 1830 °F). (c) At 2033 K (3200 °F). (d) At 2200 K (3500 °F). (e) 30- to 100-g load

Applications of silicon carbide

Industries	Environmental conditions	Applications	Primary benefits
Oil and gas	High temperature, abrasive, high fluid pressures	Nozzles, bearings, seal faces, valve seats, choke inserts	Abrasion and wear resistance
Chemical processing	Strong acids (conc. HNO_3; H_2SO_4; HCl; HF) and strong bases (NaOH)	Seal faces, bearings, pump sleeves, pump components, heat exchangers	Wear and corrosion resistance, impermeability
	High-temperature oxidation	Gasifier tubes, thermocouple tubes	Corrosion resistance at high temperatures
Automotive and truck, aircraft, **aerospace**	Engine combustion	Combustion components, turbocharger rotors, gas-turbine vanes, blades, rocket nozzles	Low friction and high strength, low inertial loads thermal shock, resistance
Automotive and truck	Engine oils	Valve-train components	Low friction wear resistance
Sand blasting	High-velocity abrasive	Nozzles	Abrasion resistance
Paper	Pulp-black liquor (50% NaOH)	Seal faces, sleeves, bearings	Corrosion resistance
	Pulp/paper slurry	Suction-box covers foils, forming boards	Abrasion resistance, low friction
Heat treating, **furnacing steel**	High temperature	Thermocouple tubes, radiant tubes, recuperators burner components	Temperature and corrosion resistance impermeability
Mining and Mineral processing	Abrasive	Linings, pump components	Abrasion resistance
Nuclear	Boronated high-temperature water	Seal faces, bushings	Radiation resistance

Selected crystal structures of silicon carbide

Structure	Structure type	Pearson symbol	Space group	Lattice spacings, nm
Cubic (3*C*)	ZnS/Zinc blende	*cF*8	$F\bar{4}3m$	$a = 0.4358$
Hexagonal (2*H*)	ZnS/Wurtzite	*hP*4	$P6_3mc$	$a = 0.30763$, $c = 0.50480$
Hexagonal (6*H*)	SiC	*hP*12	$P6_3mc$	$a = 0.30806$, $c = 1.51173$
Rhombohedral (15*R*)	...	*hR*48	$R3m$	$a = 0.3082$, $c = 6.049$

Effect of crystal orientation on hardness of silicon carbide

Crystal structure	Indentation test plane	Orientation of Knoop indenter	Knoop hardness(a), kg/mm^2
6*H* polytype	(0001) face	Parallel (1010)	2917
	(0001) face	Parallel (1120)	2954
	(1010) face	Parallel *c*-axis	2129
	(1010) face	Perpendicular *c*-axis	2700
	(1120) face	Parallel *c*-axis	2391
	(1120) face	Perpendicular *c*-axis	2717
Cubic (β)	(100) face	Parallel edge	2525(b)
			2733(c)
	(100) face	Perpendicular edge	2732(b)
			2852(c)
	(111) face	Parallel edge(d)	2758(b)
			2878(c)
	(111) face	Perpendicular edge(d)	2772(b)
			2828(c)

(a) 100-g load. (b) Natural face. (c) Polished face. (d) Frequent cracking

Summary of silicon carbide properties

Property	Value
Thermal	
Expansion coefficient, 10^{-6}/°C, of:	
3.186 g/cm^3 (sublimed)	4.51 to 4.73 (RT to 1250 °C)
3.190 g/cm^3 (CVD at 25 °C)	4.78 (RT to 1250 °C)
3.10 g/cm^3 (sintered α)	4.02 (RT to 700 °C)
Thermal conductivity, W/m · K, of:	
High-purity SiC	490
Sintered α	125.6
3.186 g/cm^3 (sublimed)	110 (at 100 °C)
Sintered α	102.6 (at 200 °C)
3.186 g/cm^3 (sublimed)	57 (at 1000 °C)
Hot-pressed (2% alumina)	78 (at 100 °C)
Sintered α	77.5 (at 400 °C)
Hot-pressed (2% alumina)	36 (at 1000 °C)
CVD	13.7 (at 100 °C)
CVD	9.95 (at 1000 °C)
Specific heat, J/g · K, of sintered α	0.67
Mechanical	
Modulus of elasticity, GPa (10^6 psi), of:	
Sublimed (0% porosity)	475 (69)
Sublimed (3.186 g/cm^3)	459 to 476 (66.5 to 69)
CVD (3.190 g/cm^3)	441 to 469 (69 to 68)
Hot-pressed (2% alumina)	414 to 445 (60 to 64.5)
Sublimed (3.157 g/cm^3)	420 to 438 (61 to 63.5)
Sintered α (3.10 g/cm^3)	410 (59.5)
Reaction-bonded (3.12 g/cm^3)	382 to 394 (55.5 to 57)
Recrystallized (2.237 g/cm^3)	143 (20.7)
Shear modulus, GPa (10^6 psi) of:	
Sublimed (3.157 g/cm^3)	181 (26)
Reaction-bonded (3.12 g/cm^3)	162 (23.5)
Recrystallized	59.5 (8.6)
Poisson's ratio (sintered α)	0.14
Weibull modulus (sintered α)	10
Fracture toughness, MPa$\sqrt{m}$ (ksi$\sqrt{in.}$)	4.60 (42)
Flexural strength(a), MPa (ksi), of:	
Sublimed (3.186 g/cm^3)	228 to 261 (33 to 37.8)
Hot-pressed (2% alumina)	552 to 862 (80 to 125)
CVD (3.190 g/cm^3)	255 to 465 (37 to 67.4)
Compressive strength (sintered α), MPa (ksi)	4600 (667)
Physical	
Refractive index	2.65 to 2.7
Birefringence ($N_e - N_o$)	0.042 to 0.097
Dispersion (optical)	~0.085
General corrosion, mg/cm^2 · y, in:	1.8(b)
98% H_2SO_4 at 100 °C	55(c)
50% NaOH at 100 °C	2.5(b)
	≥1000(c)
53% HF at 25 °C	(b)
	7.9(c)
85% H_3PO_4 at 100 °C	(b)
	8.8(c)
70% HNO_3 at 100 °C	(b)
	0.5(c)
45% KOH at 100 °C	(b)
	1000(c)
25% HCl at 70 °C	(b)
	0.9(c)
10% HF + 57% HNO_3 at 25 °C	(b)
	1000(c)

(a) Three-point bend test of flexural strength with a specimen 3 × 3 × 25 mm (0.12 × 0.12 × 1 in.). (b) Sintered α structure. (c) Corrosion in a wear environment from pin-on-disk friction

Physical properties of sintered silicon carbide materials

Property	Reported value at: Room temperature	600 °C (1110 °F)	1000 °C (1830 °F)	1200 °C (2190 °F)
Reaction-sintered SiC				
Modulus of elasticity, GPa (10^6 psi)	365 (53)	344 (50)	338 (49)	331 (48)
Thermal conductivity, W/m · K (cal/cm · s · °C)	82.5 (0.197)	25 (0.060)	...	...
Thermal expansion, %	...	0.3	0.45	0.55
Sintered α SiC				
Modulus of elasticity, GPa (10^6 psi)	410 (59.5)	...	...	...
Thermal conductivity, W/m · K (cal/cm · s · °C)	90.4 (0.216)	50 (0.12)	...	...
Thermal expansion, %	...	0.28	0.44	0.54

Properties of sintered α-silicon carbide

Property	Reported values, at temperatures indicated: Room temp.	1000 °C (1830 °F)	1200 °C (2190 °F)	1400 °C (2550 °F)	1500 °C 2730 °F)	1650 °C (3000 °F)
Hardness (Knoop), kg/mm^2	2800	...	...	...	...	...
Wet abrasion (Riley-Stoker)	3.4	...	...	...	...	...
Density, g/cm^3	3.14 to 3.18	...	...	...	...	...
Modulus of elasticity, GPa (10^6 psi)	406 (58.9)	378 (54.9)	...	...	...	...
Shear modulus, GPa (10^6 psi)	178 (25.8)	169 (24.5)	...	...	...	...
Poisson's ratio	0.142	0.118	...	...	...	...
Flexural strength(a), MPa (ksi):						
In air	459 (66.6)	442 (64.1)	450 (65.3)	432 (62.7)	404 (58.6)	...
In argon	446 (64.8)	...	...	...	...	494 (71.6)
Weibull modulus(b)	12.3	...	...	...	...	...
Fracture toughness(c), MPa$\sqrt{m}$ (ksi$\sqrt{in.}$)	4.6 (4.2)	6.4 (5.8)	...	...	...	...
Oxidation	...	...	...	Not detectable		

(continued)

Properties of sintered α-silicon carbide (continued)

Property	Reported values, at temperatures indicated				
	Room temp.	200 ° (390 °F)	400 °C (750 °F)	600 °C (1110 °F)	1500 °C (2730 °F)
Thermal diffusivity(d), cm^2/s	0.413	0.230	0.185	0.140	...
Specific heat(e), J/kg · K (cal/g · °C)	670 (0.160)	921 (0.220)	1060 (0.252)	1120 (0.268)	1400 (0.334)
Thermal conductivity, W/m · K (cal/cm · s · °C)	87.1 (0.208)	67.0 (0.160)	61.5 (0.147)	49.4 (0.118)	...
				RT to 700 °C (RT to 1290 °F)	700 to 2000 °C (1290 to 3630 °F)
Coefficient of thermal expansion, 10^{-4}/°C (10^{-4}/°F)				4.02 (2.23)	5.32 (2.96)

(a) Four-point. (b) Two-parameter. (c) Double torsion and SENB. (d) Laser flash. (e) Drop calorimeter

Mechanical properties of silicon carbide materials

Property	Temperature, K	Temperature, °F	Reported value
CVD SiC			
Compressive strength, MPa (ksi)	293	68	>345 (>50)
Modulus of elasticity, GPa (10^6 psi)	1570	2370	>200 (>29)
	273	32	480 (70)
	297	75	420 (61)
	1210	1725	370 (49)
	1490	2220	340 (54)
	1670	2550	270 (39)
Microhardness (Knoop), kg/mm^2	297	75	3000(a)
	1070	1470	790(b)
	1670	2550	410(b)
Recrystallized SiC			
Compressive strength, MPa (ksi)	297	75	689(100)
Modulus of elasticity, GPa (10^6 psi)	297	75	210(30)
Hot-pressed SiC			
Compressive strength, MPa (ksi)	297	75	1380 to 3450 (200 to 500)
	810	1000	1720 to 2750 (250 to 400)
	1144	1600	2060 to 3100 (300 to 450)
	293	68	~3900 (~567)
	813	1000	~4100 (~596)
Tensile strength, MPa (ksi)	810	1000	280 (40)
	1255	1800	240 (35)
	1420	2100	190 (28)
	1530	2300	270 (39)
	1640	2500	190 (28)
Modulus of elasticity, GPa (10^6 psi)	297	75	430 to 448 (62 to 65)
	1640	2500	380 (55)
Poisson's ratio	293	68	0.17
Microhardness (Knoop), kg/mm^2	297	75	2500(a)
SiC whiskers			
Tensile strength, MPa (ksi)	297	75	690 to 11,380 (100 to 1650)
Modulus of elasticity, GPa (10^6 psi)	297	75	190 (28)
	297	75	90 to 855 (13 to 124)
	297	75	380 to 655 (55 to 95)
Sintered SiC			
Modulus of elasticity, GPa (10^6 psi)	295	72	406 (59)
	810	1000	392 (57)
	1255	1800	385 (56)
	1754	2700	372 (54)
Reaction-sintered SiC			
Modulus of elasticity, GPa (10^6 psi)	293	68	430 (62)
	293	68	370 (53)
	770	930	345 (50)
	1270	1830	340 (49)
	1370	2010	330 (48)
	1570	2370	320 (47)
Bulk modulus, GPa (10^6 psi)	297	75	96 (14)
Torsional modulus, GPa (10^6 psi)	293	68	155 (22.5)
Shear modulus, GPa (10^6 psi)	297	75	165 to 190 (24 to 27.5)
Poisson's ratio	293	68	0.13 to 0.24
Single-crystal SiC			
Microhardness (Knoop), kg/mm^2	297	75	2200 to 2950(a)

(a) 100-g load. (b) 950-g load

Properties of silicon-carbide-base high-temperature structural ceramic materials

Type of silicon carbide	Bulk density, g/cm³	Porosity, %	Flexural Strength, MPa, at: 20 °C	1000 °C	1400 °C	Modulus of elasticity, GPa, at 20 °C	Thermal expansion, 10^{-6}/K, at 20 to 1400 °C
Reaction-bonded	2.7	16	250	250	250	280	4.5
Silicon-impregnated	3.1	0	400	500	250	380	4.3
Recrystallized	2.6	20	100	100	100	200	4.5
Sintered	3.0	5	500	450	400	400	4.6
Hot-pressed	3.2	0	550	550	450	420	4.6

Properties of RBSC

Material	Bulk density, g/cm³	Porosity, vol%	Flexural strength four-point (RT) MPa	ksi	Modulus of elasticity GPa	10^6 psi
Norton NC-435	2.96	2.4	395	57	343	49.7
UKAEA Refel 1	3.10	0.9	309	45	363	52.6
Coors Si/SiC, SC-2	3.10	0	306	44	400	58

Shaping methods for silicon carbide ceramics

	Shaping methods						
Type of silicon carbide	Dry pressing Axial	Isostatic	Injection molding	Warm molding	Extruding	Slip casting	Hot pressing
Reaction-bonded	X	X	X	X	X	X	
Silicon-infiltrated	X	X	X	X	X	X	
Recrystallized	X	X				X	
Sintered	X	X				X	
Hot-pressed							X

Physical properties of thorium, uranium, and plutonium carbides

Carbide	Crystal system	Theoretical density, g/cm³	Melting point K	°F
ThC	Cubic	10.67	2772 ± 33(a)	4530 ± 60(a)
ThC_2	(b)	9.60	2927 ± 28(c)	4810 ± 50(c)
UC	Cubic	13.63	2769 ± 33(d)	4525 ± 60(d)
U_2C_3	Cubic	12.88	2002(e)	3145(e)
UC_2	Tetragonal-cubic(f)	11.68	2722 ± 28	4440 ± 50
PuC	Cubic	13.5 to 14.1	1927(g)	3010(g)
Pu_2C_3	Cubic	12.70	2322	3720
PuC_2	Tetragonal(h)	10.9	2022 to 2522	3180 to 4080

(a) Earlier tests indicated a value of 2898 K (4755 °F). (b) Transitions from monoclinic to tetragonal and from tetragonal to cubic occur at 1672 to 1700 K (2550 to 2600 °F) and 1750 to 1772 K (2690 to 2730 °F), respectively. (c) A value of 2823 K (4620 °F) has also been reported. (d) Several values ranging from 2548 to 2866 K (4127 to 4694 °F) have been reported. (e) U_2C_3 has a body-centered-cubic structure that retains its composition up to about 2002 K (3145 °F). (f) Exists in high-temperature phases above 1772 K (2730 °F). (g) Peritectic decomposition. (h) PuC_2 exists in a high-temperature phase above 1922 K (3000 °F) produced by the decomposition of body-centered-cubic Pu_2C_3. Some researchers report a face-centered-cubic structure for PuC_2

Room-temperature mechanical properties of uranium carbides

Property	Reported values for: UC	U_2C_3	UC_2
Bend strength, MPa (ksi)	55 to 83 (8 to 12)	89 to 110 (13 to 16)	55 to 69 (8 to 10)
	103 (15)	...	...
Compressive strength, MPa (ksi)	351 (51	454 (66)	...
	296 (43) (a)	...	...
	124 (18) (b)	...	...
Modulus of elasticity(c), GPa (10^6 psi)	179 to 220 (26 to 32)	179 to 220 (26 to 32)	...
Microhardness, Knoop, kg/mm²	500 to 800	650 to 800	~500
	935	...	620

(a) Stress applied parallel to pressing direction of specimen. (b) Stress applied perpendicular to pressing direction of specimen. (c) Static method

Nitrides

Properties of ceramic nitrides

Property	Silicon oxynitride (Si_2N_2O)	Aluminum nitride (AlN)	Hexagonal boron nitride: Parallel to platelets	Hexagonal boron nitride: Perpendicular to platelets	Cubic boron nitride
Theoretical density, g/cm^3	2.90	3.20	2.27	2.27	3.48
Coefficient of thermal expansion from 25 to 1000 °C (77 to 1830 °F), 10^{-6}/°C	4.3	4.4 to 5.7	2 to 6	1 to 2	...
Specific heat, J/kg · °C (Btu/lb · °F), at:					
25 °C (77 °F)	...	800 (0.19)	...	780 (0.18)	...
1000 °C (1830 °F)	...	1570 (0.37)	...	1950 (0.47)	...
Thermal conductivity, W/m · °C (Btu/ft · h · °F), at:					
25 °C (77 °F)	8 to 10 (4.6 to 5.8)	50 to 170 (29 to 98)	20 (11.5)	33 (19)	...
1000 °C (1830 °F)	4 (2.3)	20 to 60 (11.5 to 35)	13 (7.5)	27 (15.6)	...
Modulus of elasticity, GPa (10^6 psi)	275 to 280 (39 to 40)	260 to 350 (37 to 51)	100 (14.5)	20 (2.9)	150 (22)
Flexural strength, MPa (ksi)	450 to 480 (65 to 70)	235 to 370 (34 to 54)	Low	Low	High

Typical properties of aluminum nitride

Property	Value
Density, g/cm^3	3.25 to 3.30
Flexural strength, MPa (ksi)	207 to 310 (30 to 45)
Thermal expansion, 10^{-6}/°C, from 25 to 500 °C (77 to 930 °F)	4.3 to 4.6
Thermal conductivity at 25 °C (77 °F), W/m · °C	130 to 200
Dielectric constant at 1 MHz	8.6 to 9.0
Dielectric loss at 1 MHz	0.001

Physical properties of beryllium, magnesium, and calcium nitrides

Nitride	Crystal system	Theoretical density, g/cm^3	Melting or decomposition temperature
αBe_3N_2	Cubic	2.70	(a)
βBe_3N_2	Hexagonal	2.71	(b)
Mg_3N_2	Cubic	2.71	(c)
Ca_3N_2	Rhombohedral	2.18	(d)
αCa_3N_2	Cubic	2.61 to 2.64	(e)
βCa_3N_2	Pseudohexagonal	2.67	(f)
γCa_3N_2	Orthorhombic	2.73	(g)

(a) Transforms to βBe_3N_2 at about 1866 K (2900 °F). (b) This high-temperature phase vaporizes congruently above 1644 K (2500 °F). Sublimation reportedly occurs at 2255 K (3600 °F) in vacuum. (c) Reportedly exists in three allotropic forms: $\alpha \rightarrow \beta$ transition occurs at about 823 K (1022 °F), and $\beta \rightarrow \gamma$ transition at 1061 K (1450 °F). Vaporizes above 1310 K (1900 °F) by a complex mechanism. (d) Decomposes in air; stable under argon. (e) αCa_3N_2 is an intermediate phase which vaporizes congruently above 993 K (1330 °F). (f) A low-temperature phase which is metastable up to 593 K (608 °F) and converts to α phase at elevated temperatures (g) A high-temperature phase which forms above 1310 K (1900 °F) but is not stable

Mechanical properties of aluminum nitride

Property	Reported value(a)
Bend strength, MPa (ksi)	265 to 970 (385 to 140)
	186 (27)(b)
	125 (18.1)(c)
Compressive strength, MPa (ksi)	2070 (300)
Modulus of elasticity, GPa (10^6 psi)	274 to 344 (39.7 to 50)
	317 (46)(b)
	276 (40)(c)
Poisson's ratio	0.25
Hardness, kg/mm^2:	
Knoop	1225
Vickers	1400

(a) Tested at room temperature unless otherwise noted. (b) At 1273 K (1832 °F). (c) At 1673 K (2522 °F)

Typical properties of boron nitride

Property	ASTM test method	Temperature, °C	100% BN: Parallel to pressing direction	100% BN: Perpendicular to pressing direction	60% SiO_2 - 40% BN: Parallel to pressng direcition	60% SiO_2 - 40% BN: Perpendicular to pressing direction
Compressive strength, MPa	ACMA-1	25	310	234	317	289
Modulus of elasticity (sonic), GPa	...	25	65	41	94	106
Modulus of rupture, GPa	...	25	80.70	96.11	97.90	106.2
		1000	62.71	67.09		
		1350	36.09	36.20	77.22	77.91
Density, g/cm^3	C 20	...	2.08		2.12	
Hardness	...	...	385 (K100)		89.5 (15T)	
Water absorption, % of weight gain(a)	...	25	1.1		0.04	
Thermal expansion, 10^{-6}/°C	...	75 to 500	2.3	1.1	1.8	0.20
	...	75 to 1000	2.8	0.9	2.5	0.40
	...	75 to 1500	6.7	1.8	...	...
	...	75 to 2000	6.2	2.0	...	...
Thermal conductivity, W/m · °C	...	100	23.1	44.0	9.1	25.2
	...	350	18.1	36.3	7.2	19.8
	...	700	16.3	31.6	6.8	15.9

(continued)

Typical properties of boron nitride (continued)

Property	ASTM test method	Temperature, °C	100% BN Parallel to pressing direction	100% BN Perpendicular to pressing direction	60% SiO_2 - 40% BN Parallel to pressng direcition	60% SiO_2 - 40% BN Perpendicular to pressing direction
Maximum use temperature, °C;						
In inert or reducing atmosphere	...	...	2000		1400	
In oxidizing atmosphere	...	...	985		1400	
Dielectric constant (1 MHz)	D 150	...	4.08		3.7	
Dielectric strength, V/μm	...	...	37.4		38.8	
Dissipation factor (1 MHz)	D 150	...	0.00034		0.0015	
Loss factor (1 MHz)	D 150	...	0.00139		0.006	
MIL-I-10A grade	...	...	...		L-542	
Volume resistivity, 10^{12} Ω · cm	...	23	>200		250	
		150	>30		>32	
Surface resistivity, 10^{12} Ω	...	23	>250		>250	
		150	>250		>250	
Insulation resistance, 10^{12} Ω	...	23	>200		>250	
		150	0.25		0.25	
Oxidation rate in air (weight loss), mg/cm^2 · h	...	712	0.046		0.0059	
		1000	0.622		0.0110	
Typical chemical analysis, %:						
Total boron (B)	...	...	40		18	
Total nitride (N)	...	...	50		22	
Boric oxide (B_2O_3)	...	...	6		0.2	
Calcium (Ca)	...	...	0.2		0.03	
Silica (SiO_2)	...	...	0.2		59	
Other	...	...	3.6		0.77	

(a) In 168 h at 25 °C and 80 to 100% relative humidity

Mechanical properties of hot-pressed boron nitride

Property	Reported values(a) Parallel to pressing direction	Reported values(a) Perpendicular to pressing direction
Bend strength, MPa (ksi)	48 to 97 (7 to 14)	41 to 110 (6 to 16)
	41 to 62 (6 to 9)(b)	48 to 69 (7 to 10)(b)
	28 to 76 (4 to 11)(c)	28 to 76 (4 to 11)(c)
	...	83 (12)(d)
	...	117 (17)(e)
Tensile strength, MPa (ksi)	41 to 55 (6 to 8)	45 to 62 (6.5 to 9)
	4 (0.6)(b)	4 (0.6)(b)
	4 (0.6)(f)	4 (0.6)(f)
	10 (1.5)(e)(g)	~15 (~2.2)(e)(g)
	28 (4.0)(g)(h)	~28 (~4.0)(g)(h)
Compressive strength, MPa (ksi)	41 to 317 (6 to 46)	52 to 290 (7.5 to 42)
Modulus of elasticity, GPa (10^6 psi)	41 to 97 (6 o 14)	41 to 103 (6 to 15)
Hardness (Knoop, 100-g load), kg/mm^2	210 to 390	

(a) Tested at room temperature unless otherwise noted. (b) At 1273 K (1832 °F). (c) At 1623 K (2462 °F). (d) At 2073 K (3272 °F). (e) At 2273 K (3632 °F). (f) At 1772 K (2730 °F). (g) Strength increase might represent carbide formation from test environment. (h) At 2477 K (4000 °F)

Physical properties of chromium, molybdenum, and tungsten nitrides

Nitrides(a)	Crystal system	Theoretical density, g/cm^3	Melting point K	Melting point °F
CrN	Cubic	6.14 ± 0.02	~1772(b)	~2730(b)
Cr_2N	Hexagonal	6.51	...	...
MoN	Hexagonal	9.18	~1022(b)	~1380(b)
Mo_2N	Cubic	>8.04	~1172(b)	~1650(b)
Mo_3N	Tetragonal	...	(c)	(c)
WN	Hexagonal	12.1	~873(b)	~1112(b)
W_2N	Cubic	12.2	1072 to 1144(b)	1470 to 1600(b)

(a) These nitrides are considered to be relatively unstable compounds; their reactivity with nitrogen decreases in the order Cr→Mo→W. (b) Dissociation. (c) Stable only up to 873 K (1112 °F)

Thermal properties of chromium nitrides

Property	Temperature K	Temperature °F	Reported values(a) for: CrN	Reported values(a) for: Cr_2N
Specific heat, J/kg · K (Btu/lb. · °F)	294	70	711 (0.17)	586 (0.14)
	533	500	753 (0.18)	669 (0.16)
	811	1000	795 (0.19)	711 (0.17)
Thermal conductivity, W/m · K (Btu/ft · h · °F)	294	70	12 (7)	22 (13)
Coefficient of linear thermal expansion, 10^{-6}/K (10^{-6}/°F)	273 to 1000	32 to 1340	3.5(2)	...
	294 to 477	70 to 400	0.7 (0.37)	5.9 (3.3)
	477 to 700	400 to 800	1.5 (0.83)	8.1 (4.5)
	700 to 811	800 to 1000	3.1 (1.75)	8.6 (4.8)

(a) Values are approximate

Thermal properties of dimolybdenum nitride (Mo_2N)

Property	Temperature K	Temperature °F	Reported value(a)
Specific heat, J/kg · K (Btu/lb · °F)	294	70	293 (0.07)
	533	500	376 (0.09)
	811	1000	418 (0.10)
Thermal conductivity, W/m · K (Btu/ft · h · °F)	294	70	17 (10)
Coefficient of linear thermal expansion, 10^{-6}/K (10^{-6}/°F)	294 to 477	70 to 400	1.8 (1.0)
	477 to 700	400 to 800	4.1 (2.3)
	700 to 811	800 to 1000	5.4 (3.0)

(a) Values are approximate

Physical properties of mononitrides of rare-earth metals

Metal	Nitride	Crystal system	Theoretical density, g/cm^3	Melting or decomposition temperature(a) K	Melting or decomposition temperature(a) °F
Scandium	ScN	Cubic	4.21	4800	2923
Yttrium	YN	Cubic	5.90	4840	2493
Lanthanum	LaN	Cubic	6.85	...	...
Cerium	CeN	Cubic	8.09	4660	2843
Praseodymium	PrN	Cubic	7.49	...	...
Neodymium	NdN	Cubic	7.69	...	...
Samarium	SmN	Cubic	8.50	...	...
Europium	EuN	Cubic	8.77	...	...
Gadolinium	GdN	Cubic	9.10	...	...
Terbium	TbN	Cubic	9.57	...	...
Dysprosium	DyN	Cubic	9.93	...	...
Holmium	HoN	Cubic	10.26	...	...
Erbium	ErN	Cubic	10.35	...	...
Thulium	TmN	Cubic	10.84	...	...
Ytterbium	YbN	Cubic	11.33	...	...
Lutetium	LuN	Cubic	11.59	...	...

(a) Values are approximate

Thermal properties of silicon nitride ceramics

Property	RBSN	HPSN	SSN	SRBSN	HIP-SN	Sialon
Relative density, % theoretical	70 to 88	99 to 100	95 to 99	93 to 99	99 to 100	97 to 99
Coefficient of thermal expansion from 25 to 1000 °C (77 to 1830 °F), 10^{-6}/°C	3.0	3.2 to 3.3	2.8 to 3.5	3.0 to 3.5	3.0 to 3.5	3.0 to 3.7
Thermal conductivity at:						
25 °C (77 °F), W/m · °C (Btu/ft · h · °F)	7 to 14 (4 to 8)	30 to 43 (17 to 25)	15 to 31 (8 to 18)	...	22 (12.27)	15 to 22 (8 to 13)
1000 °C (1830 °F), W/m · °C (Btu/ft · h · °F)	1.4 to 3 (0.8 to 1.7)	5 to 10 (3 to 6)	4 to 5 (2.3 to 3)	...	...	2.5 (1.4)
Specific heat at:						
25 °C (77 °F), J/kg · °C (Btu/lb · °F)	700 to 1100 (0.17 to 0.26)	680 to 800 (0.16 to 0.19)	...	...	...	...
1000 °C (1830 °F), J/kg · °C (Btu/lb · °F)	1250 (0.30)	1200 (0.29)	...	...	...	...

Thermal shock resistance of silicon nitride ceramics

	Critical temperature change (ΔT_c) for:			
	Water quench		Oil quench	
Material type	Δ °C	Δ °F	Δ °C	Δ °F
RBSN	200 to 600	360 to 1080	750 to 1250	1350 to 2250
HPSN	400 to 800	720 to 1440	>1400	>2520
SSN	600 to 750	1080 to 1350	>1400	>2520
HIP-RBSN	800	1440	...	...
Sialon	300 to 550	540 to 990	...	...

Density and flexural strength of sintered silicon nitride

Composition	Fabrication method	Density, g/cm³	Room temperature flexural strength(a), MPa
Si_3N_4-$15Y_2O_3$-$3Al_2O_3$	Cold isopressed	3.141	432 (3-point)
Si_3N_4-$8Y_2O_3$-$4Al_2O_3$	Injection molded	3.04 to 3.09	380 to 402 (4-point)
Si_3N_4-$6Y_2O_3$-$2Al_2O_3$	Injection molded	3.13	385.8 (4-point)

(a) Unmatched

Properties of silicon-nitride-base high-temperature structural ceramic materials

Type of silicon nitride	Bulk density, g/cm³	Porosity, %	Flexural strength, MPa, at: 20 °C	1000 °C	1400 °C	Modulus of elasticity, GPa, at 20 °C	Thermal expansion 10^{-6}/K, at 20 to 1400 °C
Reaction bonded(a)	1.9 to 2.2	...	150	150	150	120	3.0
Reaction bonded(b)	2.4 to 2.6	...	250	250	300	160	3.0
Hot pressed	3.2	0	700	700	400	320	3.2

(a) Axial dry pressed, extruded. (b) Isopressed, slip cast, injection molded, warm molded

Mechanical properties of silicon nitride ceramics

	Material type							
Property	RBSN	HPSN	SSN	SRBSN	HIP-SN	HIP-RBSN	HIP-SSN	Sialon
Young's modulus, GPa (10^6 psi)	120 to 150 (17.4 to 36.2)	310 to 330 (44.9 to 47.8)	260 to 320 (37.7 to 46.4)	280 to 300 (40.6 to 43.5)	...	310 to 330 (44.9 to 47.8)	...	300 (43.5)
Poisson's ratio	0.20	0.27	0.25	0.23	0.23	0.27	...	0.23
Flexural strength, MPa (ksi) at:								
25 °C (77 °F)	150 to 350 (21.7 to 50.7)	450 to 1000 (65.2 to 145)	600 to 1200 (86.9 to 173.8)	500 to 800 (72.5 to 115.9)	600 to 1200 (86.9 to 173.8)	500 to 800 (72.5 to 115.9)	600 to 1200 (86.9 to 173.8)	750 to 950 (108.7 to 137.7)
1350 °C (2460 °F)	140 to 340 (20.2 to 49.3)	250 to 450 (36.2 to 65.2)	340 to 550 (49.3 to 79.7)	350 to 450 (50.7 to 65.2)	350 to 550 (50.7 to 79.7)	250 to 450 (36.2 to 65.2)	300 to 520 (43.5 to 75.3)	300 to 550 (43.5 to 79.7)
Weibull modulus	19 to 40	15 to 30	10 to 25	10 to 20	...	20 to 30	...	15
Fracture toughness, MPa$\sqrt{m}$ (ksi$\sqrt{in.}$)	1.5 to 2.8 (1.3 to 2.5)	4.2 to 7.0 (3.8 to 6.3)	5.0 to 8.5 (4.5 to 7.7)	5.0 to 5.5 (4.5 to 5.0)	4.2 to 7.0 (3.8 to 6.3)	2.0 to 5.8 (1.8 to 5.3)	4.0 to 8.0 (3.6 to 7.2)	6.0 to 8.0 (5.4 to 7.2)

Room-temperature mechanical properties of reaction-sintered silicon nitride

Property	Reported value
Bend strength, MPa (ksi)	117 to 241 (17 to 35)
Compressive strength, MPa (ksi)	345 to 690 (50 to 100)
Tensile strength, MPa (ksi)	69 to 172 (10 to 25)
Modulus of elasticity, GPa (10^6 psi)	96 to 220 (13.9 to 32)
Poisson's ratio	0.25 to 0.26
Microhardness, kg/mm²:	
100-g load	1700 to 2200
50-g load	2300 to 3000
25-g load	2670 to 3260

Properties of RBSN

Type of RBSN/form of silicon compact	Predominant phase	Density, g/cm^3	Modulus of elasticity, GPa	Modulus of rupture at 2.2 g/cm^3, MPa (ksi)	at 2.35 g/cm^3, MPa (ksi)	at 2.4 g/cm^3, MPa (ksi)	at 2.55 g/cm^3, MPa (ksi)	Fracture surface energy at 25 °C, W/m^2	at 2000 °C, W/m^2	Cryolite resistance (loss rate) at air/bath notch 2.29 g/cm^3, mm/yr (in./yr)	2.46 g/cm^3, mm/yr (in./yr)
High-purity/isostatically pressed high-purity silicon	β	2.2	55 (8)	96 (14)	...	...	...	23	20	...	...
Slip cast/slip cast technical-grade silicon	α	2.2 to 2.4	145 (21)	165 (24)	...	186 (27)	...	20	14	168 (6.6)	137 (5.4)

Room-temperature mechanical properties of hot-pressed silicon nitride (HS-130)

Property	Reported value
Bend strength, MPa (ksi)	707 (103)(a)
	680 (98.6)(b)
	860 (125)(c)
	898 (130)(c)
	763 (111)(d)
	790 to 913 (115 to 132)(e)
Compressive strength, MPa (ksi)	689 to 2760 (100 to 400)
Tensile strength, MPa (ksi)	397 to 434 (57.5 to 62.9)(f)
	360 (52.2)
Modulus of elasticity, GPa (10^6 psi)	290 to 307 (42 to 44.5)
Impact strength, J (in. · lbf)	0.23 to 0.40 (2 to 3.5)
Poisson's ratio	0.22

(a) Laboratory material containing 1% MgO, high-phase; tested in four-point loading. (b) Tested in four-point loading; maximum stress perpendicular to pressing direction. (c) Tested in three-point loading; maximum stress perpendicular to pressing direction. (d) Tested in three-point loading; maximum stress parallel to pressing direction. (e) Laboratory material containing 1% MgO, high-α phase; tested in three-point loading. (f) Expanding ring test; maximum stress perpendicular to pressing direction

Room-temperature values of bend strength, modulus of elasticity, and hardness for Si_2ON_2

Fabrication method	Density, g/cm^3	Bend strength(a) MPa	ksi	Modulus of elasticity GPa	10^6 psi	Hardness, Knoop, kg/mm^2
Cold pressed(b)	1.95	32.4	4.7	70.3	10.2	...
	2.09	37.9	5.5	75.8	11.0	...
	2.31	67.6	9.8	113.8	16.5	...
Hot pressed(c)	2.65	210.3	30.5	191.7	27.8	...
	2.70	...	...	...	...	1580

(a) Tested in three-point loading on a 200-mm (8-in.) span for cold pressed specimens, and on a 50-mm (2-in.) span for hot pressed specimens. (b) Si_2ON_2 content, 90% or higher. (c) Si_2ON_2 content, 80%

Properties of reaction-bonded Si_2ON_2

Material density	Density g/cm^3	Percent of theoretical(a)	Porosity Open, %	Closed, %	Modulus of rupture at 20 °C, MPa (ksi)	at 1000 °C, MPa (ksi)	at 1450 °C, MPa (ksi)	Modulus of elasticity, GPa (10^6 psi)	Oxidation in air(b) at 24 h, mg/cm^2	at 100 h, mg/cm^2	Cryolite resistance (loss rate) at air/bath notch, mm/yr (in./yr)
Normal	2.0 to 2.2	71 to 78	26 to 19	3	55 (8)	55 (8)	34 (5)	83 (12)	4.9	6.4	890 (35)
High	2.5 to 2.6	89 to 94	7 to 1	...	290 (42)	214 (31)	46 (6.7)	221 (31.7)	0.8	1.2	710 (28)

(a) Theoretical density = 2.8 g/cm^3. (b) At 1100 °C (2010 °F)

Compatibility of silicon nitride with various media

Compatible	Marginally compatible	Not compatible
Hot concentrated mineral acids Molten aluminum, lead, tin, zinc, light alloys, silver, gold, brass, and nickel/silver at their normal foundry temperatures	Soda-lime and borosilicate glasses at 1470 K (2190 °F) Molten metals and alloys up to 1370 K (2000 °F) except as indicated for specific metals	Hot hydrofluoric acid Hot concentrated caustic solutions and fused caustic salts Molten magnesium, copper, nichrome, and stainless steels

Oxide additives used in densifying silicon nitride powder and its surface layer of silica

	Temperature of liquid formation, °C	
Additive (M_xO_y)	Silicate (M_xO_y-SiO_2)	Oxynitride (M_xO_y-SiO_2-Si_3N_4)
Li_2O	1030	1030
MgO	1543	1390
Y_2O_3	1650	1480
CeO_2	1560	1460
ZrO_2	1640	1590
CaO	1435	1435
Al_2O_3	1595	1470

Shaping methods for silicon nitride ceramics

	Shaping methods						
	Dry pressing						
Type of silicon nitride	Axial	Isostatic	Injection molding	Warm molding	Extruding	Slip casting	Hot pressing
Reaction bonded	x	x	x	x	x	x	
Hot pressed							x

Room-temperature bend strength of silicon nitride-metal joints

Joining techniques and conditions	Metal	Layer thickness, μm	Layer thickness, mil	Strength, MPa	Strength, ksi
Active metal brazing at 850 °C (1560 °F) and 3.3 MPa (0.5 ksi) for 15 min. except as noted	Ag-33.5wt%Cu-1.5wt%Ti(a)	10 to 15	0.4 to 0.6	650	95
	Ag-26.7wt%Cu-4.5wt%Ti(b)	15	0.6	650	95
	Al	<10	<0.4	380(c)	55(c)
	AA6061	<10	<0.4	470(c)	68(c)
	AA2017	<10	<0.4	400(c)	58(c)
Solid-state bonding at 1400 °C (2550 °F) and 100 MPa (15 ksi) for 30 min. except as noted	Mo(d)	100	4	420(e)	61(e)
	Nb	100	4	290(e)	42(e)
	Ta	100	4	380(e)	55(e)
Eutectic joining at 1200 °C (2190 °F) and 10 MPa (1.5 ksi) for 30 min.	Fe	200	8	330(c)	48(c)
	Ni	200	8	250(c)	36(c)
	Type 410	200	8	320(c)	46(c)
	Type 304	200	8	220(c)	32(c)

(a) Joined at 950 °C (1740 °F) for 1 h in argon. (b) Joined at 875 °C (1605 °F) for 0.5 h in argon. (c) Specimen bar was 3 × 3 × 35 mm (0.12 × 0.12 × 0.12 × 1.4 in.) for four-point bend test; upper span, 10 mm (0.4 in.), lower span, 30 mm (1.2 in.); cross head speed, 0.5 mm/min. (0.02 in./min.). (d) Same conditions, but at 1500 °C (2730 °F). (e) Specimen bar was 2 × 2 × 15 mm (0.08 × 0.08 × 0.6 in.) for three-point bend test; span, 10 mm (0.4 in.), cross head speed, 0.5 mm/min. (0.02 in./min.)

Physical properties of thorium, uranium, and plutonium nitrides

Nitride	Crystal system	Theoretical density g/cm^3	Melting point K	Melting point °F
ThN	Cubic	11.60	3093(a)	5108(a)
Th_3N_4	Hexagonal(b)	10.50	(c)	(c)
UN	Cubic	14.32	3123(d)	5160(d)
αU_2N_3	Cubic(e)	11.24	(f)	(f)
βU_2N_3	Hexagonal(e)	12.24	(g)	(g)
UN_2	Cubic(h)	11.73	...	...
PuN	Cubic	14.25	(j)	(j)

(a) Congruent melting under nitrogen pressure of about 1 atm; other values of 2903 and 3063 K (4766 and 5040 °F) are also reported. (b) Rhombohedral structure. (c) Decomposes to ThN above 1923 K (3000 °F). (d) Congruent melting under nitrogen pressure greater than 2 atm; decomposition occurs at lower nitrogen pressures. (e) Body-centered-cubic (α) phase transforms slowly to the hexagonal (β) phase at temperatures above 1093 K (2000 °F). The change also is a function of composition and nitrogen pressure. (f) Decomposes at elevated temperatures but is stable to about 1644 K (2500 °F) at 1 atm nitrogen pressure. (g) Converts αU_2N_3 below 1373 K (2000 °F). (h) Exists only at very high nitrogen pressures and low temperatures. (j) Reported melting points ranged from 2870 to about 3030 K (4700 to about 5000 °F) at 1 atm nitrogen pressure. However, these values were probably decomposition temperatures

Physical properties of titanium, zirconium, and hafnium mononitrides

Nitride	Crystal system	Theoretical density g/cm^3	Melting point K	Melting point °F
TiN	Cubic	5.44	3223 ± 50	5340 ± 90
ZrN	Cubic	7.35	3253 ± 55	5400 ± 100
HfN	Cubic	13.94	3660 + 43	6128 ± 77

Hardness values for TiN, ZrN, and HfN

Nitride	Mohs scale	Microhardness, kg/mm^2 Knoop, 100-g load	Microhardness, kg/mm^2 Vickers, 50-g load
TiN	9 to 10	~1800	1800 to 2100
ZrN	7 to 8	~1500	1500 to 1850
HfN	7 to 8	...	1600 to 2150

Physical properties of vanadium, niobium, and tantalum nitrides

Nitride	Crystal system	Theoretical density g/cm^3	Melting point(a) K	Melting point(a) °F
VN	Cubic	6.08 ± 0.02	2450 ± 139	3950 ± 250
V_2N	Hexagonal	5.99 ± 0.01	...	...
NbN	Cubic	8.36	~2477	~4000
Nb_2N	Hexagonal	8.31	~2589	~4200
TaN	Hexagonal	14.36	3366 ± 44	5600 ± 80
Ta_2N	Hexagonal	15.86	3223	5342

(a) Rapid loss of nitrogen at elevated temperatures precludes reliability of melting-point values, particularly for the mononitride phases

Coefficients of linear thermal expansion for vanadium, niobium, and tantalum nitrides

Nitride	Temperature range K	Temperature range °F	Coefficient of linear thermal expansion(a), 10^{-6}/K	Coefficient of linear thermal expansion(a), 10^{-6}/°F
VN	294 to 1366	70 to 2000	8.1	4.5
Nb_2N	294 to 1272	70 to 1830	3.2	1.8
TaN	294 to 977	70 to 1300	3.6	2.0
Ta_2N	294 to 1270	70 to 1830	5.2	2.9
Ta_2N	294 to 1644	70 to 2500	4.7	2.6

(a) Values are approximate

Borides

Crystal structure, density, and melting point of selected borides

Boride	Crystal structure	Lattice parameters, Å				Theoretical density, g/cm³	Melting point	
		a	*b*	*c*	β, °		°C	°F
AlB_2	Hexagonal	3.005	...	3.253	...	3.17	1654	3009
AlB_{10}	Orthorhombic	8.88	9.10	5.69	...	2.54	2421	4390
αAlB_{12}	Tetragonal	10.16	...	14.28	...	2.58	2162	3924
βAlB_{12}	Orthogonal	12.34	12.63	10.16	...	2.60	2212	4014
BaB_6	Cubic	4.262	...	...	...	4.35	2270	4118
$\delta Be_{5-x}B$	Tetragonal	3.368	...	7.050	...	1.94	1160	2120
Be_2B	Cubic	4.670	...	...	...	1.89	~1500	~2730
n-Be_2B_3	Tetragonal	7.25	...	8.46	...	...	...	...
BeB_2	Hexagonal	9.79	...	9.55	...	2.42	>1970	>3506
BeB_6	Tetragonal	10.16	...	14.28	...	2.33	2020 to 2120	3668 to 3848
δBeB_{12}	Hexagonal	5.46	...	12.42	...	2.42	2300	4170
CaB_6	Cubic	4.154	...	...	...	2.43	2235	4055
CeB_4	Tetragonal	7.202	...	4.093	...	5.74	2380	4316
CeB_6	Cubic	4.141	...	...	...	4.79	2550	4620
Co_4B	Orthogonal	4.408	5.220	6.630	...	...	...	...
$Co_{23}B_6$	Cubic	11.05	...	...	...	6.99	...	...
Co_3B	Orthogonal	4.408	5.223	6.629	...	8.13	1110	2030
Co_2B	Tetragonal	5.015	...	4.220	...	8.05	1260	2246
CoB	Orthogonal	3.948	5.243	3.037	...	7.32	...	...
Cr_4B	Orthogonal	4.26	7.38	14.71	...	6.28	1650	3002
Cr_2B	Orthogonal	7.409	14.712	4.250	...	6.58	1870	3398
Cr_5B_3	Tetragonal	5.370	...	10.188	...	6.61	1900	3452
CrB	Tetragonal	2.94	...	15.72	...	6.14	...	...
δCrB	Orthogonal	2.967	7.867	2.932	...	6.10	2100	3812
Cr_2B_3	Orthogonal	3.026	18.115	2.954	...	5.60	...	...
Cr_3B_4	Orthogonal	2.986	13.02	2.952	...	5.76	2070	3758
CrB_2	Hexagonal	2.973	...	3.071	...	5.20	2200	3992
CrB_4	Orthogonal	4.744	5.477	2.866	...	4.25	1400 to 1600	2552 to 2912
CrB_{41}	Orthogonal	10.964	...	23.848	...	2.65	...	...
CuB_{24}	Orthogonal	10.980	...	23.925	...	2.82	...	...
DyB_2	Hexagonal	3.287	...	3.847	...	8.49	2100	3812
DyB_4	Tetragonal	7.097	...	4.017	...	6.76	2500	4532
DyB_6	Cubic	4.097	...	...	...	5.49	2200	3992
DyB_{12}	Cubic	7.500	...	...	...	4.60	2100	3812
DyB_{66}	Cubic	23.45	...	...	...	2.71	2150	3902
ErB_2	Hexagonal	3.28	...	3.79	...	8.88	2185	3965
ErB_4	Tetragonal	7.068	...	3.993	...	7.01	2500	4532
ErB_{12}	Cubic	7.484	...	...	...	4.708	2080	3776
ErB_{66}	Cubic	23.33	...	...	...	2.73	2150	3902
EuB_6	Cubic	4.17	...	...	...	4.99	≥2580	≥4676
$Fe_{23}B_6$	Cubic	10.67	...	...	...	7.38	...	...
$Fe_{3.5}B$	Tetragonal	8.62	...	4.28	...	...	...	...
Fe_3B	Tetragonal	8.674	...	4.313	...	7.30	...	...
Fe_2B	Tetragonal	5.132	...	8.532	...	7.34	1410	2570
FeB	Orthorhombic	4.059	5.503	2.947	...	6.73	1650	3002
FeB_{49}	Orthorhombic	10.951	...	23.861	...	2.35	...	...
GdB_2	Hexagonal	3.318	...	3.933	...	7.92	2050	3722
Gd_2B_5	Monoclinic	7.181	7.193	7.196	102.16	6.74	2100	3812
GdB_4	Tetragonal	7.133	...	4.047	...	6.47	2650	4802
GdB_6	Cubic	4.107	...	...	...	5.32	2510	4550
GdB_{12}	Cubic	7.524	...	...	...	4.475	...	...
GdB_{66}	Cubic	23.47	...	...	...	2.68	2150	3902
HfB	Orthogonal	6.517	3.218	4.919	...	12.19	2100 ± 20	3812 ± 36
HfB_2	Hexagonal	3.142	...	3.476	...	11.19	3380 ± 20	6116 ± 36
HfB_{12}	Cubic	7.377	...	...	...	...	...	...
HoB_2	Hexagonal	3.281	...	3.811	...	8.72	2200	3992
HoB_4	Tetragonal	7.087	...	4.008	...	6.87	2500	4532
HoB_6	Cubic	4.095	...	...	...	5.56	2180	3956
HoB_{12}	Cubic	7.492	...	...	...	4.67	2100	3812
HoB_{66}	Cubic	23.38	...	...	...	2.74	2025	3677
$IrB_{0.9}$	Hexagonal	2.81	...	2.81	...	17.39	...	...
$IrB_{1.15}$	Tetragonal	2.810	...	10.26	...	16.73	...	...
$IrB_{1.35}$	Monoclinic	10.525	2.910	6.099	91.07	17.25	...	...
KB_6	Cubic	4.233	...	...	...	2.28	...	...
LaB_3	Tetragonal	3.82	...	3.96	...	4.92	...	...
LaB_4	Tetragonal	7.323	...	4.181	...	5.40	...	...
LaB_6	Cubic	4.157	...	...	...	4.71	2715	4919
LuB_2	Hexagonal	3.246	...	3.704	...	9.76	2250	4082
LuB_4	Tetragonal	7.036	...	3.974	...	7.37	2550	4622
LuB_{12}	Cubic	7.464	...	...	...	4.87	2170	3938

(continued)

Crystal structure, density, and melting point of selected borides (continued)

Boride	Crystal structure	Lattice parameters, Å: a	b	c	β, °	Theoretical density, g/cm³	Melting point: °C	°F
LuB_{66}	Cubic	23.412	...	...	...	2.76	2100	3812
MgB_2	Hexagonal	3.086	...	3.522	...	2.63	...	...
MgB_4	Orthogonal	5.464	7.472	4.428	...	...	...	...
MgB_6	Tetragonal	7.07	...	6.45	...	...	...	...
Mn_4B	Orthogonal	14.53	7.293	4.209	...	6.87	1285	2345
Mn_2B	Tetragonal	5.149	...	4.209	...	7.18	1580	2876
MnB	Orthogonal	4.147	5.561	2.977	...	6.36	1890	3434
Mn_3B_4	Orthogonal	3.302	12.86	2.960	...	5.98	1750	3180
MnB_2	Hexagonal	3.009	...	3.037	...	5.34	1988	3610
MnB_4	Monoclinic	5.503	5.367	2.949	122.71	4.45	2160	3920
βMnB_4	Tetragonal	6.28	...	8.38	...	...	...	...
Mo_2B	Tetragonal	5.547	...	4.739	...	9.23	2280 ± 12	4136 ± 22
αMoB	Tetragonal	3.105	...	16.97	...	8.77	...	...
βMoB	Orthogonal	3.16	8.61	3.08	...	...	2600 ± 8	4712 ± 14
MoB_2	Hexagonal	3.04	...	3.07	...	7.99	2375 ± 15	4307 ± 27
Mo_2B_5	Orthogonal	3.012	...	20.937	...	7.45	2140 ± 15	3884 ± 27
$Mo_{0.8}B_3$	Hexagonal	5.203	...	6.349	...	4.87	...	...
Mo_3B_2	Tetragonal	6.00	...	3.15	...	9.07	2070	3758
MoB_4	Hexagonal	5.214	...	6.358	...	6.18	1800	3272
MoB_{12}	Hexagonal	3.004	...	3.174	...	...	2020	3668
Nb_3B_2	Tetragonal	6.185	...	3.280	...	7.95	2080 ± 40	3776 ± 72
Nb_3B_4	Orthogonal	3.312	14.11	3.143	...	7.28	2935 ± 12	5315 ± 22
δNbB	Orthogonal	3.297	8.723	3.166	...	7.57	2917 ± 10	5283 ± 18
εNbB_2	Hexagonal	3.089	...	3.303	...	7.00	3036 ± 15	5497 ± 27
NdB_4	Tetragonal	7.217	...	4.102	...	5.83	2350	4262
NdB_6	Cubic	4.126	...	...	...	4.95	2610	4730
NdB_{66}	Cubic	23.42	...	...	...	2.63	2150	3902
Ni_3B	Orthogonal	5.211	6.619	4.389	...	8.20	1175	2147
Ni_2B	Tetragonal	4.991	...	4.247	...	8.05	1225	2237
Ni_4B_3	Monoclinic	6.430	4.882	7.818	103.3	7.43	...	...
Ni_4B_3	Orthogonal	11.96	2.98	6.57	...	...	...	...
NiB	Orthogonal	2.936	7.38	2.968	...	7.17	1590	2894
NiB_{12}	Cubic	7.385	...	...	...	...	2320	4208
NpB_2	Hexagonal	3.165	...	3.975	...	12.47	...	...
$OsB_{1.1}$	Hexagonal	2.876	...	2.871	...	16.32	...	...
Os_2B_3	Hexagonal	2.910	...	12.910	...	14.48	...	...
OsB_2	Orthogonal	4.684	2.872	4.076	...	12.83	...	...
OsB_2	Hexagonal	2.876	...	2.871	...	17.10	...	...
Os_2B_5	Hexagonal	2.91	...	12.91	...	15.23	...	...
Pd_2B	Orthogonal	4.692	5.127	3.110	...	9.93	...	...
Pd_3B_2	Hexagonal	6.49	...	3.43	...	18.22	~1020	~1868
PmB_6	Cubic	4.128	...	...	...	...	...	...
PrB_4	Tetragonal	6.099	...	12.063	...	5.01	2350	4262
PrB_6	Cubic	4.132	...	...	...	4.84	2610	4730
PtB	Hexagonal	3.358	...	4.058	...	17.25	~920	~1688
PuB	Cubic	4.918	...	...	...	13.94	2050	3722
PuB_2	Hexagonal	3.19	...	3.90	...	12.67	1825	3317
PuB_4	Tetragonal	7.10	...	4.014	...	9.27	2050	3722
PuB_6	Cubic	4.120	...	...	...	7.57	2100	3812
PuB_{66}	Cubic	23.43	...	...	...	...	...	...
Re_3B	Orthogonal	2.890	9.313	7.258	...	19.36	~2150	~3902
Re_7B_3	Hexagonal	7.50	...	4.88	...	18.63	~2000	~3632
ReB_2	Hexagonal	2.900	...	7.478	...	12.68	~2400	~4352
Rh_7B_3	Hexagonal	7.471	...	4.777	...	...	...	...
Rh_5B_4	Hexagonal	3.306	...	20.394	...	9.60	...	...
RhB	Hexagonal	3.309	...	4.224	...	...	...	...
Ru_7B_3	Hexagonal	7.467	...	4.714	...	...	~1660	~3020
$RuB_{\sim 1.1}$	Hexagonal	2.852	...	2.855	...	...	1500	2730
$Ru_{11}B_8$	Orthogonal	11.60	11.34	2.83	...	10.69	...	...
$RuB_{1.1}$	Hexagonal	2.852	...	2.855	...	9.38	...	...
Ru_2B_3	Hexagonal	2.905	...	12.810	...	8.32	1550	2822
RuB_2	Hexagonal	2.852	...	2.855	...	10.14	1600	2912
Ru_2B_5	Hexagonal	2.89	...	12.81	...	9.19	...	...
ScB_2	Hexagonal	3.146	...	3.518	...	3.67	2250	4082
ScB_{12}	Tetragonal	5.22	...	7.35	...	2.89	2040	3704
$Sc_{11}B_{305}$	Orthogonal	10.965	...	24.087	...	2.51	...	...
SmB_2	Hexagonal	3.310	...	4.019	...	7.49	...	...
Sm_2B_5	Monoclinic	7.183	7.191	7.216	102.03	6.47	~1980	~3596
SmB_4	Tetragonal	7.167	...	4.070	...	6.15	2400	4352
SmB_6	Cubic	4.132	...	...	...	5.06	2580	4676

(continued)

Crystal structure, density, and melting point of selected borides (continued)

Boride	Crystal structure	Lattice parameters, Å a	b	c	β, °	Theoretical density, g/cm^3	Melting point °C	°F
SmB_{66}	Cubic	23.48	...	...	...	2.66	2150	3902
SrB_6	Cubic	4.193	...	...	...	3.44	2230	4046
Ta_2B	Tetragonal	5.783	...	4.866	...	15.21	2417 ± 15	4383 ± 27
Ta_3B_2	Tetragonal	6.184	...	3.286	...	14.92	2180 ± 20	3956 ± 36
TaB	Orthogonal	3.280	8.671	3.156	...	14.19	3090 ± 15	5594 ± 27
δTa_3B_4	Orthogonal	3.29	14.0	3.13	...	13.60	2990 ± 20	5414 ± 36
TaB_2	Hexagonal	3.098	...	3.227	...	12.54	3037 ± 20	5499 ± 36
TbB_2	Hexagonal	3.280	...	3.860	...	8.34	2100	3812
TbB_4	Tetragonal	7.119	...	4.029	...	6.58	2600	4712
TbB_6	Cubic	4.105	...	...	...	5.37	2340	4244
TbB_{12}	Cubic	7.504	...	...	...	4.54	2200	3992
TbB_{66}	Cubic	23.43	...	...	...	2.70	2100	3812
TcB_2	Hexagonal	2.900	...	7.475	...	7.30	...	...
ThB_4	Tetragonal	7.261	...	4.114	...	8.43	>2200	>3992
ThB_6	Cubic	4.112	...	...	...	7.09	2150	3902
ThB_{12}	Cubic	7.612	...	...	...	...	...	...
ThB_{66}	Cubic	23.46	...	...	...	2.92	...	...
TiB	Orthogonal	6.12	3.06	4.56	...	4.56	2190 ± 25	3974 ± 45
Ti_3B_4	Orthogonal	3.259	13.73	3.042	...	4.56	...	...
TiB_2	Hexagonal	3.030	...	3.230	...	4.52	3225 ± 20	5837 ± 36
Ti_2B_5	Hexagonal	2.98	...	13.98	...	4.63	...	...
$Ti_{1.87}B_{50}$	Tetragonal	8.830	...	5.072	...	2.65	...	...
TmB_2	Hexagonal	3.250	...	3.739	...	9.25	2250	4082
TmB_4	Tetragonal	7.0572	...	3.9882	...	7.10	2550	4622
TmB_{12}	Cubic	7.476	...	...	...	4.76	2180	3956
TmB_{66}	Cubic	23.433	...	...	...	2.73	2100	3812
UB	Cubic	4.88	...	...	...	14.22	...	...
UB_2	Hexagonal	3.131	...	3.987	...	12.69	2385	4325
UB_4	Tetragonal	7.075	...	3.979	...	9.38	2495	4523
UB_{12}	Cubic	7.473	...	...	...	5.85	2235	4055
V_3B_2	Tetragonal	5.746	...	3.032	...	5.75	1900 ± 12	3452 ± 22
VB	Orthogonal	3.060	8.048	2.972	...	5.60	2570 ± 15	4658 ± 27
V_5B_6	Orthogonal	3.058	21.25	2.974	...	...	...	...
V_3B_4	Orthogonal	3.030	13.18	2.986	...	5.43	2610 ± 12	4730 ± 22
V_2B_3	Orthogonal	3.061	18.40	2.984	...	...	...	...
VB_2	Hexagonal	2.998	...	3.056	...	5.07	2747 ± 15	4977 ± 27
$V_{1.54}B_{50}$	Tetragonal	8.824	...	5.072	...	2.60	...	...
W_2B	Tetragonal	5.568	...	4.744	...	17.09	2670 ± 16	4838 ± 29
βWB	Orthogonal	3.19	8.40	3.07	...	...	2665 ± 16	4829 ± 29
δWB	Tetragonal	3.117	...	16.910	...	15.74	...	...
W_2B_5	Orthogonal	2.982	...	20.715	...	13.17	2365 ± 15	4289 ± 27
WB_4	Hexagonal	5.202	...	6.333	...	8.40	2020 ± 30	3668 ± 54
WB_{12}	Hexagonal	3.994	...	3.174	...	...	...	...
YB_2	Hexagonal	3.78	...	4.40	...	3.37	2100	3812
YB_4	Tetragonal	7.086	...	4.012	...	4.36	2800	5072
YB_6	Cubic	4.100	...	...	...	3.72	2600	4712
YB_{12}	Cubic	7.502	...	...	...	3.44	2200	3992
YB_{66}	Cubic	23.34	...	...	...	2.48	2100	3812
YbB_2	Hexagonal	3.250	...	3.732	...	9.47	~1500	~2732
YbB_4	Tetragonal	7.04	...	4.00	...	7.24	~1850	~3362
YbB_6	Cubic	4.416	...	...	...	5.55	2370	4298
YbB_{12}	Cubic	7.469	...	...	...	...	2200	3992
YbB_{66}	Cubic	23.30	...	...	...	2.75	2150	3902
ZrB	Cubic	4.65	...	...	...	6.48	2800	5072
ZrB_2	Hexagonal	3.169	...	3.530	...	6.10	3245 ± 18	5873 ± 32
ZrB_{12}	Cubic	7.408	...	...	...	3.63	2250 ± 40	4082 ± 72
Zr_6B_{311}	Orthogonal	10.956	...	24.020	...	2.60	...	...

Thermodynamic data for selected borides

Boride	Heat capacity at constant pressure, C_p — At 298 K, J/mol · K	At 298 K, Btu/mol · °F	At 1000 K, J/mol · K	At 1000 K, Btu/mol · °F	At 2000 K, J/mol · K	At 2000 K, Btu/mol · °F	Enthalpy of formation, ΔH_f(a) — At 298 K, kJ/mol	At 298 K, Btu/mol	At 1000 K, kJ/mol	At 1000 K, Btu/mol	At 2000 K, kJ/mol	At 2000 K, Btu/mol
AlB_2	43.64	0.02298	78.21	0.04118	...	...	–66.94	–63.53	–81.04	–76.91	...	...
AlB_{12}	149.58	0.078769	317.82	0.16736	439.28	0.23132	–200.83	–190.58	–223.12	–211.74	–211.55	–200.76
CeB_6	103.31	0.054403	165.81	0.087316	198.65	0.10461	–351.46	–333.54	–360.62	–342.23	–389.18	–369.33
CoB	34.63	0.01824	56.45	0.02973	...	...	–94.14	–89.34	–96.14	–91.24	...	...
Co_2B	58.98	0.03106	89.28	0.04701	...	...	–125.52	–119.12	–128.68	–122.12	...	...
CrB	35.82	0.01886	57.35	0.03020	...	...	–75.31	–71.47	–74.86	–71.04	...	...
CrB_2	53.59	0.02822	85.07	0.04480	...	...	–94.14	–89.34	–94.02	–89.22	...	...
FeB	50.21	0.02644	54.72	0.02882	...	...	–71.13	–67.50	–72.96	–69.24	...	...
Fe_2B	75.33	0.03967	87.78	0.04622	...	...	–71.13	–67.50	–77.82	–73.85	...	...
HfB_2	49.45	0.02604	81.67	0.04301	...	...	–335.98	–318.85	–334.90	–321.62	...	...
MgB_2	47.82	0.02518	71.71	0.03776	...	...	–92.05	–87.36	–106.55	–101.12	...	...
MgB_4	70.17	0.03695	115.58	0.06086	...	...	–105.02	–99.66	–124.48	–118.13	...	...
MnB	35.82	0.01886	57.35	0.03020	...	...	–75.31	–71.47	–79.94	–75.86	...	...
MnB_2	53.62	0.02824	85.10	0.04481	...	...	–94.14	–89.34	–99.08	–94.03	...	...
NiB	34.63	0.01824	56.45	0.02973	...	...	–100.42	–95.30	–102.36	–97.14	...	...
Ni_4B_3	128.12	0.067468	201.32	0.10602	...	...	–311.71	–295.81	–318.79	–302.53	...	...
TaB_2	48.12	0.02534	76.77	0.04043	96.71	0.05093	–209.20	–198.53	–209.77	–199.07	–208.02	–197.41
TiB	29.67	0.01562	51.92	0.02734	53.40	0.02812	–160.25	–152.08	–161.85	–153.60	–187.00	–177.46
TiB_2	44.28	0.02332	76.89	0.04049	94.54	0.04978	–323.84	–307.32	–326.59	–309.93	–347.87	–330.13
UB_2	57.29	0.03017	87.75	0.04621	123.71	0.06516	–164.43	–156.04	–165.31	–156.88	–177.82	–168.75
UB_4	78.91	0.04155	137.49	0.07240	...	...	–245.60	–233.07	–248.34	–235.67	...	...
UB_{12}	169.58	0.089301	311.12	0.16384	...	...	–433.04	–410.95	–440.10	–417.65	...	...
V_3B_2	97.10	0.05113	142.09	0.074824	194.84	0.10260	–303.76	–288.27	–303.62	–288.14	–298.47	–283.25
VB	36.02	0.01897	55.74	0.02935	75.28	0.03964	–138.49	–131.43	–138.44	–131.38	–136.74	–129.77
V_5B_6	191.05	0.10061	303.80	0.15998	407.43	0.21455	–763.58	–724.64	–763.31	–724.38	–754.80	–716.31
V_3B_4	119.01	0.062671	192.32	0.10128	256.85	0.13526	–486.60	–461.76	–486.43	–461.62	–481.34	–456.79
V_2B_3	83.00	0.04371	136.59	0.071928	181.57	0.09561	–345.18	–327.58	–345.06	–327.46	–341.68	–324.25
VB_2	46.98	0.02474	80.85	0.04258	160.28	0.08440	–203.76	–193.37	–203.69	–193.30	–202.02	–191.72
ZrB_2	48.37	0.02547	71.99	0.03791	82.66	0.04353	–322.59	–306.14	–326.65	–309.99	–340.36	–323.00

Boride	Gibbs free energy of formation, ΔG_f(a) — At 298 K, kJ/mol	At 298 K, Btu/mol	At 1000 K, kJ/mol	At 1000 K, Btu/mol	At 2000 K, kJ/mol	At 2000 K, Btu/mol
AlB_2	–65.33	–62.00	–58.83	–55.83	...	...
AlB_{12}	–206.70	–196.16	–213.62	–202.73	–205.26	–194.79
CeB_6	–341.53	–324.11	–318.08	–301.86	–262.45	–249.06
CoB	–96.14	–91.24	–88.04	–83.55	...	...
Co_2B	–123.69	–117.38	–118.36	–112.32	...	...
CrB	–73.68	–69.92	–70.43	–66.84	...	...
CrB_2	–91.37	–86.71	–85.52	–81.16	...	...
FeB	–69.49	–65.95	–68.05	–64.58	...	...
Fe_2B	–71.75	–68.09	–68.19	–64.71	...	...
HfB_2	–332.20	–315.26	–324.49	–307.94	...	...
MgB_2	–89.52	–84.95	–80.40	–76.30	...	...
MgB_4	–103.72	–98.43	–94.66	–89.84	...	...
MnB	–73.68	–69.92	–68.68	–65.18	...	...
MnB_2	–91.37	–86.71	–83.79	–79.52	...	...
NiB	–98.73	–93.69	–93.25	–88.49	...	...
Ni_4B_3	–304.98	–289.43	–283.50	–269.04	...	...
TaB_2	–206.53	–195.00	–200.18	–189.97	–191.02	–181.28
TiB	–159.71	–151.56	–157.35	–149.33	–146.93	–139.44
TiB_2	–319.69	–303.39	–308.34	–292.61	–347.87	–330.13
UB_2	–162.34	–154.06	–161.46	–153.23	–149.39	–141.77
UB_4	–244.80	–232.32	–245.37	–232.86	...	...
UB_{12}	–438.63	–416.26	–455.88	–432.63	...	...
V_3B_2	–300.29	–284.98	–292.06	–277.16	–281.90	–267.52
VB	–136.84	–129.86	–132.94	–126.16	–127.93	–121.41
V_5B_6	–732.60	–695.24	–659.69	–626.05	–558.53	–530.04
V_3B_4	–480.12	–455.64	–464.89	–441.18	–444.84	–422.15
V_2B_3	–340.36	–323.00	–329.06	–312.28	–314.06	–298.04
VB_2	–200.60	–190.37	–193.19	–183.34	–183.20	–173.86
ZrB_2	–318.16	–301.93	–306.34	–290.72	–279.60	–265.34

(a) In standard state, pure phase, at 0.1 MPa (1 atm)

Enthalpies of formation for selected borides

Boride	Free enthalpy of formation(a) KJ/mol	BTu/mol	Boride	Free energy of formation(a) KJ/mol	BTu/mol
BaB_6	–333.0	–316	Mo_3B_2	–175.7	–166.7
Be_4B	–78.7	–74.7	MoB	–68.2	–64.7
Be_2B	–69.9	–66.3	MoB_2	–96.2	–91.3
BeB_2	–64.9	–61.6	Mo_2B_5	–209.2	–198.5
BeB_4	–87.0	–82.6	NbB_2	–246.9	–234.3
BeB_6	–108.8	–103.3	NdB_6	–430.1	–408.2
BeB_9	–161.5	–153.3	PmB_6	–442.2	–419.6
CaB_6	–119.7	–113.6	PrB_6	–416.7	–395.4
CeB_6	–338.9	–321.6	ScB_2	–264.8	–251.3
CrB_2	–125.5	–119.1	SmB_6	–454.4	–431.3
DyB_6	–502.1	–476.5	SrB_6	–210.9	–200.1
ErB_6	–527.2	–500.3	TaB_2	–192.5	–182.7
EuB_6	–469.4	–445.5	TbB_6	–492.9	–467.8
FeB	–38.5	–36.5	ThB_4	<–217	<–206
GdB_6	–479.9	–445.4	ThB_6	<–276	<–262
HfB	–196.6	–186.6	TiB_2	–292.9	–278
HfB_2	–358.1	–339.8	Ti_2B_5	<–439	<–417
HoB_6	–514.6	–488.4	TmB_6	–539.7	–512.2
LaB_6	–469.9	–445.9	VB	–129.7	–123.1
LuB_6	–560.7	–532.1	VB_2	–259.4	–246.2
Mg_4B	–70.3	–66.7	W_2B	–100.4	–95.3
Mg_2B	–59.0	–56.0	WB	–71.1	–67.5
MgB_2	–55.6	–52.8	W_2B_5	–146.4	–138.9
MgB_4	–73.6	–69.8	YB_6	–100.4	–95.3
MgB_6	–93.7	–88.9	YbB_6	–550.2	–522.0
MgB_{12}	–143.9	–136.6	ZrB	<–163	<–155
MnB_2	–79.5	–75.4	ZrB_2	–320.9	–304.5
Mo_2B	–106.7	–101.3	ZrB_{22}	<–502.1	<–476.5

(a) From the elements, in standard state, pure phase, at 0.1 MPa (1 atm) and 298 K

Thermal expansion and thermal conductivity of selected borides

Boride	Thermal expansion, 10^{-6}/K α	α_a	α_c	Temperature range °C	°F	Thermal conductivity W/m · K	Btu/ft · h · °F	Temperature °C	°F
BeB_6	6.8 ± 0.5	...	...	20 to 800	68 to 1470	46.8 ± 1.7	27 ± 1	20	68
	...	...	...	...	...	36.4 ± 1.7	21 ± 1	20	68
CaB_6	6.5 ± 0.5	...	...	20 to 800	68 to 1470	39.3 ± 1.7	22.7 ± 1	20	68
	...	...	...	...	...	23.0 ± 1.7	13.3 ± 1	20	68
CeB_6	7.3 ± 0.5	...	...	20 to 800	68 to 1470	33.9 ± 0.9	19.6 ± 0.5	20	68
Co_3B	...	...	...	...	...	17.0	9.8	...	...
Co_2B	...	...	...	...	...	14.0	8.1	...	...
CoB	...	...	...	...	...	17.0	9.8	...	...
Cr_4B	...	...	...	...	...	11.0 ± 0.4	6.4 ± 0.2	20	68
Cr_2B	14.2	...	...	27 to 1027	80 to 1880	10.9	6.3	20	68
	15.0	...	...	1027 to 2027	1880 to 3680	...	...	...	...
Cr_5B_3	13.7	...	...	27 to 1027	80 to 1880	15.8	9.1	20	68
	14.2	...	...	1027 to 2027	1880 to 3680	...	...	...	...
CrB	12.3	...	...	27 to 1027	80 to 1880	20.1	11.6	20	68
	12.6	...	...	1027 to 2027	1880 to 3680	...	...	...	...
Cr_3B_4	11.8	...	...	27 to 1027	80 to 1880	20.5	11.8	20	68
	12.1	...	...	1027 to 2027	1880 to 3680	...	...	...	...
CrB_2	10.5	...	...	27 to 1027	80 to 1880	31.8	18.4	20	68
	11.8	...	...	1027 to 2027	1880 to 3680	34.0	19.7	1027	1880
	11.0	...	...	0 to 1200	30 to 2190	55.0	31.8	2027	3680
Cr_2B_5	...	...	...	...	...	18.0 ± 0.8	10.4 ± 0.5	20	68
DyB_4	5.9	...	...	25 to 1027	77 to 1880	118	68.2	27	80
DyB_{12}	4.2	...	...	...	...	...	...	...	...
ErB_4	7.6	...	...	20 to 1000	68 to 1830	...	...	...	...
ErB_{12}	3.7	...	...	...	...	...	...	...	...
EuB_6	6.9 ± 0.5	...	...	20 to 800	68 to 1470	23.0 ± 0.9	13.2 ± 0.5	20	68
	...	...	...	...	...	18.0	10.4	27	80
Fe_2B	...	11.8 ± 0.3	8.9 ± 0.6	20 to 800	68 to 1470	30.1	17.4	20	68
FeB	~12	...	...	400 to 1000	750 to 1830	12.0	6.9	20	68
GdB_4	7.0	...	...	20 to 1000	68 to 1830	148.5	85.8	27	80
GdB_6	8.7 ± 0.5	...	...	20 to 800	68 to 1470	20.5 ± 1.3	11.8 ± 0.8	20	68
HfB_2	6.3	...	...	27 to 1027	80 to 1880	51.6	29.8	27	80
	6.8	...	...	1027 to 2027	1880 to 3680	60.0	34.7	1027	1880
	7.6	...	...	20 to 2205	68 to 4000	143.0	82.7	2027	3680
HoB_4	7.85	...	...	20 to 1000	68 to 1830	127.6	73.8	27	80

(continued)

Thermal expansion and thermal conductivity of selected borides (continued)

Boride	Thermal expansion, 10^{-6}/K α	α_a	α_c	Temperature range °C	°F	Thermal conductivity W/m · K	Btu/ft · h · °F	Temperature °C	°F
HoB_6	3.0	...	...	...	...	...	...	...	...
HoB_{12}	3.6	...	...	...	...	...	...	...	...
LaB_4	...	7.17 ± 1.16	8.36 ± 1.03	...	...	...	...	...	...
LaB_6	6.4 ± 0.5	...	...	20 to 800	68 to 1470	47.7 ± 4.2	27.6 ± 2.4	...	...
	...	...	...	...	...	45.0	26.0	...	...
LuB_{12}	3.4	...	...	...	...	...	...	...	...
Mn_4B	...	...	...	...	...	5.0	2.9	...	...
Mn_2B	...	...	...	...	...	6.6	3.8	...	...
MnB	...	...	...	...	...	7.7	4.5	...	...
Mn_3B_4	...	...	...	...	...	8.6	5.0	...	...
MnB_2	...	...	...	...	...	10.2	5.9	...	...
Mo_2B	5.0	...	...	25 to 500	77 to 930	...	...	...	...
MoB_2	7.7	7	10	300 to 900	570 to 1650	...	...	...	...
Mo_2B_5	8.6	...	...	27 to 1027	80 to 1880	~50	~28.9	20	68
	9.9	...	...	1027 to 2027	1880 to 3680	~27	~15.6	900	1650
MoB_4	6.5	...	...	...	...	...	...	...	...
Nb_3B_2	13.8	...	...	27 to 1027	80 to 1880	12.0	6.9	27	80
	13.9	...	...	1027 to 2027	1880 to 3680	...	...	...	...
NbB	12.9	...	...	27 to 1027	80 to 1880	15.6	9.0	27	80
	13.4	...	...	1027 to 2027	1880 to 3680	...	...	...	...
Nb_3B_4	9.9	...	...	27 to 1027	80 to 1880	20.5	11.8	27	80
	10.3	...	...	1027 to 2027	1880 to 3680	24.0	13.9	27	80
NbB_2	8.0	...	...	27 to 1027	80 to 1880	23.5	13.6	1027	1880
	8.5	...	...	1027 to 2027	1880 to 3680	40.3	23.3	2027	3680
	8.6	...	...	20 to 1650	68 to 3000	...	...	...	...
NdB_4	5.84	...	...	27 to 1027	80 to 1880	...	...	...	...
NdB_6	7.30 ± 1.0	...	...	20 to 800	68 to 1470	47.3 ± 3.3	27.3 ± 2	20	68
Ni_3B	...	...	...	...	...	41.8	24.2	...	...
Ni_2B	...	...	...	...	...	54.8	31.6	...	...
NiB	...	...	...	...	...	21.9	12.7	...	...
PrB_4	5.0	...	...	27 to 1027	80 to 1880	...	...	...	...
PrB_6	7.50 ± 0.5	...	...	20 to 800	68 to 1470	41.0	23.7	...	...
ScB_2	...	6.8 ± 0.5	7.6 ± 0.5	20 to 600	68 to 1110	...	...	...	...
	...	4.4	4.9	30 to 1100	86 to 2010	...	...	...	...
ScB_4	4.1	...	...	...	...	...	...	...	...
SmB_6	6.8 ± 0.5	...	...	20 to 800	68 to 1470	13.8 ± 1.7	8.0 ± 1	20	68
SrB_6	6.7 ± 0.5	...	...	20 to 800	68 to 1470	26.4 ± 2.1	15.3 ± 1.2	20	68
TaB_2	8.2	...	...	27 to 1027	80 to 1880	16.0	9.2	27	80
	8.8	...	...	1027 to 2027	1880 to 3680	16.1	9.3	1027	1880
	8.4	...	...	20 to 1650	68 to 3000	36.2	20.9	2027	3680
	...	5.85	7.14	30 to 1000	86 to 1830	...	...	...	...
	...	7.82	8.40	1000 to 2300	1830 to 4170	...	...	...	...
TbB_4	6.55	...	...	27 to 1027	80 to 1880	126.4	73.1	27	80
TbB_6	7.8 ± 1.0	...	...	20 to 800	68 to 1470	20.1 ± 1.3	11.6 ± 0.8	20	68
TbB_{12}	3.2	...	...	...	...	...	...	...	...
ThB_4	7.9	...	...	20 to 1770	68 to 3220	~25	~14.5	20	68
	...	...	...	...	...	~18	~16.2	730	1346
	...	...	...	...	...	~31	~17.9	1230	2246
	...	...	...	...	...	41	23.7	1730	3146
ThB_6	7.8 ± 0.5	...	...	20 to 800	68 to 1470	44.8 ± 5.0	25.9 ± 3	...	...
TiB_2	...	6.63	8.65	30 to 1000	86 to 1830	64.4	37.2	27	80
	...	8.00	11.20	1000 to 2300	1830 to 4170	69.9	40.4	1027	1880
	4.6	...	...	27 to 1027	80 to 1880	122.2	70.6	2027	3680
	5.2	...	...	1027 to 2027	1880 to 3680	98	56.6	25	77
	8.6	...	...	20 to 2205	68 to 4000	115.2	66.6	27	80
	...	7.30	10.27	300 to 1300	570 to 2370	37	21.4	27 to 400	80 to 750
TmB_4	6.5	...	...	27 to 1027	80 to 1880	158.2	91.4	27	80
TmB_{12}	3.85	...	...	...	...	...	...	...	...
UB_2	...	9	8	20 to 205	68 to 400	51.9	30.0	20	68
UB_4	7.0	...	...	20 to 1000	68 to 1830	4.0	2.3	20	68
UB_{12}	4.6	...	...	...	...	...	...	...	...
VB_2	7.6	...	...	27 to 1027	80 to 1880	42.3	24.4	27	80
	8.3	...	...	1027 to 2027	1880 to 3680	42.0	24.3	1027	1880
	...	...	...	...	...	67.4	39.0	2027	3680
W_2B	6.7	...	...	...	...	...	...	...	...
WB	~6.9	...	...	220 to 2205	68 to 4000	...	...	...	...
W_2B_5	7.8	...	...	27 to 1027	80 to 1880	~52	~30.1	20	68
	8.8	...	...	1027 to 2027	1880 to 3680	~26	~15.0	800	1470
WB_4	5.8	...	...	...	...	...	...	...	...
YB_4	...	7.6 ± 0.5	6.4 ± 0.6	...	...	29.2	16.9	27	80
YB_6	6.02 ± 0.6	...	...	...	...	29.3 ± 4.2	16.9 ± 2.3	20	68
YB_{12}	6.6 ± 0.6	...	...	...	...	...	...	...	...

(continued)

Thermal expansion and thermal conductivity of selected borides (continued)

Boride	Thermal expansion, 10^{-6}/K			Temperature range		Thermal conductivity		Temperature	
	α	α_a	α_c	°C	°F	W/m · K	Btu/ft · h · °F	°C	°F
YbB_6	5.8 ± 0.5	...	...	20 to 800	68 to 1470	25.1 ± 1.7	14.5 ± 1	20	68
YbB_{12}	2.0	...	...	–196 to 20	–320 to 68	...	...	...	...
	3.9	...	...	20 to 300	68 to 570	...	...	...	...
	5.8	...	...	300 to 1000	570 to 1830	...	...	...	...
ZrB_2	5.9	...	...	27 to 1027	80 to 1880	57.9	33.5	27	80
	6.5	...	...	1027 to 2027	1880 to 3680	64.4	37.2	1027	1880
	8.3	...	...	20 to 2205	68 to 4000	133.9	77.4	2027	3680
	...	...	...	...	...	~87	~50.3	20 to 2000	68 to 3680

Room-temperature hardnesses of selected borides

Boride	Hardness, GPa	Hardness, 10^6 psi	Load, N	Load, lbf
AlB_2	9.6 ± 1.0	1.4 ± 0.15	1	0.23
AlB_{10}	26.0 ± 1.2	3.8 ± 0.17	1	0.23
αAlB_{12}	23.5 ± 0.3	3.4 ± 0.04	1	0.23
βAlB_{12}	22.5 ± 1.0	3.3 ± 0.15	1	0.23
BaB_6	29.4 ± 2.8	4.3 ± 0.41	1	0.23
Be_5B	6.1	0.9	...	...
Be_2B	8.7	1.3	...	...
BeB_2	31.2	4.5	...	...
BeB_6	25.3	3.7	...	...
CaB_6	26.9 ± 2.2	3.9 ± 0.32	1	0.23
	16.2	2.3	1	0.23
CeB_6	30.8 ± 1.9	4.5 ± 0.3	0.3	0.07
Co_3B	11.3	1.6	0.5	0.11
Co_2B	11.3	1.6	0.5	0.11
CoB	11.3	1.6	0.5	0.11
	23.0	3.3	1	0.23
CoB_2	25.3	3.7	0.5	0.11
Cr_4B	12.2 ± 0.6	1.8 ± 0.09	0.5	0.11
Cr_2B	13.2 ± 1.0	1.9 ± 0.15	0.5	0.11
Cr_5B_3	13.9 to 14.9	2.0 to 2.2	0.5	0.11
CrB	11.8 to 12.7	1.7 to 1.8	1	0.23
Cr_3B_4	13.7 to 14.7	2.0 to 2.1	1	0.23
CrB_2	20.6 ± 0.8	3.0 ± 0.1	0.5	0.11
	10.9	1.6	1	0.23
DyB_4	18.6	2.7	...	...
DyB_6	23.5 ± 1.0	3.4 ± 0.15	1	0.23
ErB_4	18.6	2.7	...	...
ErB_6	27.5 ± 1.0	4.0 ± 0.15	1	0.23
EuB_6	26.1	3.8	1	0.23
	18.0	2.6	1	0.23
Fe_2B	13.1 ± 0.5	1.9 ± 0.07	...	...
	17.7	2.6	1	0.23
FeB	16.2 ± 0.5	2.3 ± 0.07	...	...
	18.6	2.7	1	0.23
GdB_4	18.6 ± 0.5	2.7 ± 0.07	...	...
GdB_6	18.1 ± 0.5	2.6 ± 0.07	...	...
HfB_2	28.4 ± 5.0	4.1 ± 0.7	0.3	0.07
	23.5	3.4	1.6	0.36
	21.2	3.1	1	0.23
HoB_4	16.5	2.4	...	...
HoB_{12}	26.5 ± 1.0	3.8 ± 0.15	1	0.23
$IrB_{1.15}$	16.2	2.3	...	...
LaB_6	19.7	2.9	...	...
LuB_{12}	28.4 ± 1.0	4.1 ± 0.15	1	0.23
Mn_4B	10.3 ± 0.5	1.5 ± 0.07	...	...
Mn_2B	17.7 ± 0.5	2.6 ± 0.07	...	...
MnB	20.1 ± 0.5	2.9 ± 0.07	...	...
Mn_3B_4	19.6 ± 0.5	2.8 ± 0.07	...	...
MnB_2	16.7 ± 0.5	2.4 ± 0.07	...	...
MnB_4	35.3 ± 1.0	5.1 ± 0.15	...	...
Mo_2B	24.5	3.6	0.5	0.11
αMoB	23.0	3.3	0.5	0.11
βMoB	24.5	3.6	0.5	0.11
MoB_2	11.8	1.7	0.5	0.11
Mo_2B_5	23.0	3.3	0.5	0.11
	29.5	4.3	1.0	0.23
	21.3	3.1	1.0	0.23
Nb_3B_2	22.5	3.3	0.5	0.11
NbB	21.5	3.1	0.5	0.11
Nb_3B_4	22.5	3.3	0.3	0.07
NbB_2	25.5	3.7	0.3	0.07
	20.9	3.0	1	0.23
NdB_4	19.1	2.8	...	...
NdB_6	24.9 ± 1.7	3.6 ± 0.25	0.7	0.16
Ni_3B	11.7	1.7	...	...
Ni_2B	14.0	2.0	...	...
Ni_4B_3	14.6	2.1	...	...
NiB	15.2	2.2	...	...
$OsB_{1.2}$	16.1	2.3	...	...
$OsB_{1.6}$	18.4	2.7	...	...
OsB_2	28.4	4.1	...	...
Pd_3B	4.6	0.7	...	...
Pd_5B_2	5.8	0.8	...	...
PrB_4	18.9	2.7	...	...
PrB_6	24.2 ± 1.0	3.5 ± 0.15	1	0.23
$PtB_{1.1}$	9.2	1.3	...	...
ReB_2	30.4	4.4	...	...
Ph_7B_3	7.6	1.1	0.5	0.11
$RhB_{\sim 1.1}$	11.9	1.7	0.5	0.11
Ru_7B_3	11.1	1.6	...	...
$Ru_{11}B_8$	12.8	1.8	...	...
$RuB_{\sim 1.1}$	13.8	2.0	...	...
Ru_2B_3	14.9	2.2	...	...
RuB_2	22.2	3.2	...	...
ScB_2	17.4 ± 0.3	2.5 ± 0.04	2	0.45
SmB_6	24.5 ± 3.0	3.5 ± 0.44	1	0.23
	13.6 ± 1.6	1.9 ± 0.2	...	...
SrB_6	28.6 ± 0.9	4.1 ± 0.1	0.3	0.07
Ta_3B_2	27.2	3.9	0.5	0.11
TaB	30.7	4.5	0.5	0.11
Ta_3B_4	32.9	4.8	0.5	0.11
TaB_2	24.5 ± 0.4	3.5 ± 0.06	0.3	0.07
	19.6	2.8	0.5	0.11
	24.5 ± 1	3.6 ± 0.15	...	...
TbB_4	18.6	2.7	...	...
TbB_6	22.6	3.3	1	0.23
TbB_{12}	25.5 ± 1.0	3.7 ± 0.15	1	0.23
ThB_4	...	...	...	...
ThB_6	17.1 ± 1.2	2.5 ± 0.17	0.2	0.05
ThB_{76}	22.7	3.3	...	...
TmB_4	25.5 ± 1.0	3.7 ± 0.15	1	0.23
TiB	22.7	3.3	...	...
TiB_2	33.0 ± 0.6	4.8 ± 0.09	0.3	0.07
	25.5	3.7	1	0.23
	...	...	...	...
	25.5	3.7	1	0.23
	20.8, 23.0	3.0, 3.3	5	1.13
UB_2	14.8	2.1	...	...
UB_4	11.1	1.6	1	0.23
UB_{12}	25.8	3.7	1	0.23

(continued)

Room-temperature hardnesses of selected borides (continued)

Boride	Hardness, GPa	Hardness, 10^6 psi	Load, N	Load, lbf	Boride	Hardness, GPa	Hardness, 10^6 psi	Load, N	Load, lbf
V_3B_2	22.4	3.2	0.5	0.11	YB_6	25.3 ± 1.0	3.7 ± 0.15	...	...
V_3B_4	23.0	3.3	0.5	0.11	YB_{12}	24.5 ± 1.5	3.6 ± 0.22	...	...
VB_2	27.5 ± 0.1	4.0 ± 0.01	0.3	0.07	YbB_6	26.1	3.8	1.0	0.23
	20.6	3.0	1	0.23		15.1 ± 0.3	2.2 ± 0.04	...	...
W_2B	23.7 ± 1.2	3.4 ± 0.17	1.5	0.34	YbB_{12}	32.4 ± 1.0	4.7 ± 0.15	0.3	0.07
WB	36.3	5.3	0.5	0.11	ZrB	34.3 to 35.3	5.0 to 5.1	0.3	0.07
WB_2	26.1 ± 0.1	3.8 ± 0.01	0.3	0.07	ZrB_2	22.1 ± 0.2	3.2 ± 0.03	0.3	0.07
W_2B_5	26.1 ± 0.1	3.8 ± 0.01	0.3	0.07		17.9	2.6	1	0.23
	24.5	3.6	1	0.23	ZrB_{12}	27.0 to 28.0	3.9 to 4.1	...	...
WB_4	39.2 ± 2.0	5.7 ± 0.3	0.5	0.11		25.3	3.7	1	0.23
YB_4	27.9 ± 1.5	4.0 ± 0.22	...	...					

Room-temperature mechanical properties of selected borides

Boride	Modulus of elasticity, GPa	Modulus of elasticity, 10^6 psi	Poisson's ratio	Flexural strength, MPa	Flexural strength, ksi	Compressive strength, GPa	Compressive strength, 10^6 psi	Fracture toughness, $MPa\sqrt{m}$	Fracture toughness, $ksi\sqrt{in.}$
BaB_6	385	56	...	...	...	...	...	...	...
CaB_6	451	65	...	138	20	...	...	...	...
CeB_6	379	55	...	...	...	...	...	...	...
CrB_2	211	31	...	607	88	1.3	0.19	...	...
EuB_6	...	...	...	183	27	...	...	...	...
Fe_2B	284	41	...	...	...	...	...	...	...
FeB	343	50	...	...	...	...	...	...	...
GdB_6	...	...	...	206(a)	30(a)	...	...	...	...
HfB_2	500	72.5	0.12	350 ± 70	51 ± 10	...	...	...	...
LaB_6	479	69.5	...	126	18	...	...	...	...
αMoB	...	...	...	345	50	...	...	...	...
Mo_2B_5	672	97.5	...	...	...	...	...	...	...
NbB_2	637	92.4	...	...	...	...	...	...	...
SmB_6	...	...	...	148	21	...	...	...	...
TaB_2	257	37	...	...	...	...	...	...	...
	248	36	...	...	...	...	...	...	...
TbB_4	...	...	...	137	20	...	...	...	...
ThB_4	148	21	...	137	20	...	...	...	...
TiB_2	551	80	0.11	240	35	...	...	6.7 ± 0.2	6.1 ± 0.2
	...	...	...	300 to 370	49 to 54	...	...	...	...
	545	79	0.093	...	...	...	...	4.9 ± 0.4	4.5 ± 0.4
	500	73	0.10	...	...	5.7 ± 0.17	0.8 ± 0.02	8.0 ± 0.3	7.3 ± 0.3
	530	77	...	...	...	1.3	0.19	6.0 ± 0.7	5.5 ± 0.6
	...	...	...	350 to 400	51 to 58	...	...	...	...
UB_4	440	64	...	413	60	...	...	...	...
VB_2	268	39	...	...	...	...	...	...	...
	262	38	...	...	...	...	...	...	...
W_2B_5	775	112	...	...	...	...	...	...	...
ZrB_2	343	50	...	305	44	...	...	...	...
	500	73	0.11	275	40	...	...	...	...
ZrB_{12}	...	...	...	...	...	1.6	0.23	...	...

(a) 8% porosity

Room-temperature electrical properties of selected borides

Boride	Electrical resistivity, $10^{-6}\ \Omega \cdot cm$	Work function, 10^{19} J	Work function, 10^{18} eV	Hall constant, 100 m^3/K	Boride	Electrical resistivity, $10^{-6}\ \Omega \cdot cm$	Work function, 10^{19} J	Work function, 10^{18} eV	Hall constant, 100 m^3/K
BaB_6	77	5.52	34.4	−57.5	Co_3B	28	...	...	...
Be_5B	30	...	...	...	Co_2B	33	...	...	...
Be_2B	10^3	...	...	...	CoB	76	...	...	−104
BeB_2	2×10^4	...	...	...	Cr_2B	107	4.13	25.8	−0.8
BeB_4	10^{12}	...	...	...	Cr_5B_3	49	4.39	27.4	−0.9
BeB_6	10^{13}	...	...	...	CrB	46	4.86	30.3	−1.0
BeB_9	6×10^{13}	...	...	...	Cr_3B_4	60	5.03	31.4	−1.0
CaB_6	222	4.58	28.6	−91.0	CrB_2	30	5.12	31.9	−1.2
CeB_6	29	4.69	29.3	−4.18	DyB_4	35	...	...	...

(continued)

Room–temperature electrical properties of selected borides (continued)

Boride	Electrical resistivity, $10^{-6}\,\Omega \cdot cm$	Work function 10^{19} J	Work function 10^{18} eV	Hall constant, 100 m^3/K
DyB_6	...	5.65	35.3	...
DyB_{12}	14	...	...	–4.6
ErB_4	50	...	...	...
ErB_6	...	5.39	33.6	...
ErB_{12}	16	...	...	–4.5
EuB_6	85	7.84	48.9	–50.2
Fe_2B	38	5.28	32.9	...
FeB	80	5.76	35.9	...
GdB_4	31	2.32	14.5	...
GdB_6	45	3.30	20.6	–4.39
HfB_2	11	5.76	35.9	–18.0
	46(a)	...	...	...
HoB_4	30	5.47	34.1	...
HoB_6	...	...	...	–5.5
HoB_{12}	15	...	...	...
LaB_4	24	...	...	...
LaB_6	15	4.29	26.8	–4.96
LuB_6	...	4.80	30.0	...
LuB_{12}	14	...	...	–4.8
Mn_4B	85	...	...	...
Mn_4B	40	...	...	...
MnB	57	...	...	...
Mn_3B_4	62	...	...	...
MnB_2	71	6.62	41.3	...
Mo_2B	40	...	...	...
αMoB	45	...	...	...
βMoB	25	...	...	...
MoB_2	45	...	...	...
Mo_2B_5	26	5.95	37.1	–6.6
Nb_3B_2	45	5.40	33.7	–0.46
NbB	40	5.78	36.1	–0.60
Nb_3B_4	34	6.24	38.9	–1.1
NbB_2	26	6.52	40.7	–1.5
	53(a)	...	...	...
NdB_4	39	...	...	...
NdB_6	20	6.35	39.6	–4.39
Ni_3B	21	...	...	0.0
Ni_2B	14	...	...	0.53
NiB	50	...	...	0.63
Pd_3B	10	...	...	...
Pd_5B_2	26	...	...	...
PrB_4	40	...	...	...
PrB_6	20	3.46	21.6	–4.33
Rh_7B_3	88	...	...	...
$RhB_{\sim1.1}$	880	...	...	...
ScB_2	7 to 15	4.64	28.9	...
SmB_6	207	7.04	43.9	1.54
SrB_6	111	4.27	26.6	–76.3
TaB	100	...	...	...
TaB_2	33	6.56	40.9	–2.1
	75(a)	...	...	...
TbB_4	32	...	...	...
TbB_6	37	5.22	32.6	–4.57
TbB_{12}	12	...	...	–4.2
ThB_6	15	4.67	29.1	–2.19
TiB	40	...	...	...
TiB_2	9	6.57	41.0	–19.6
TmB_4	35	...	...	...
TmB_6	...	4.50	28.1	...
TmB_{12}	17	...	...	–4.7
UB_2	...	5.29	33.0	...
UB_4	98	5.42	33.8	...
UB_{12}	23	4.74	29.6	–0.24
VB	35	...	...	...
VB_2	23	6.45	40.2	–0.82
	71(a)	...	...	...
W_2B_5	22	6.12	38.2	–6.9
YB_2	39	...	...	–3.05
YB_4	29	3.33	20.8	–21.3
YB_6	40	5.52	34.4	–4.56
YB_{12}	95	7.36	45.9	–4.9
YbB_6	47	5.01	31.3	–83.6
YbB_{12}	185	...	...	–8.4
ZrB_2	10	5.99	37.4	–19.0

(a) At 1027 °C (1880 °F)

Optical appearance of selected borides

Boride	Color
Be_2B	Gray with rose shade
BeB_2	Dark gray
BeB_6	Brick red
BaB_6	Black with violet shade
CaB_6	Black
CeB_4	Gray-brown
CeB_6	Blue-violet
CrB_2	Gray
DyB_4	Gray-brown
DyB_6	Blue
DyB_{100}	Black
ErB_4	Gray-brown
ErB_6	Blue
ErB_{100}	Black
EuB_6	Dark gray
Fe_2B	Gray
FeB	Gray
GdB_4	Gray-brown
GdB_6	Blue
GdB_{100}	Black
HfB_2	Gray
HoB_4	Gray-brown
HoB_6	Blue
HoB_{100}	Black
LaB_6	Purple-violet
LuB_6	Blue
LuB_{100}	Black
MgB_2	Dark brown
MgB_6	Dark brown
MgB_{12}	Dark brown
MnB	Red-brown
MnB_2	Red-brown
Mo_2B_5	Light gray
NbB_2	Gray
NdB_6	Blue
PrB_4	Gray-brown
PrB_6	Blue
ScB_2	Gray
SmB_4	Gray-brown
SmB_6	Blue
SmB_{100}	Black
SrB_6	Black with green shade
TaB_2	Gray
TbB_4	Gray-brown
TbB_6	Blue
TbB_{100}	Black
ThB_6	Red-violet
TiB_2	Gray

(continued)

Optical appearance of selected borides (continued)

Boride	Color	Boride	Color
TmB_4	Gray-brown	W_2B_5	Light gray
TmB_6	Blue	YB_4	Gray-brown
TmB_{100}	Black	YB_6	Blue-violet
UB_4	Gray-steel	YbB_6	Black
UB_{12}	Black	YbB_{100}	Black
VB_2	Gray	ZrB_2	Gray

Magnetic properties of selected borides

Boride	Molar magnetic susceptibility, $\times 10^6$	Effective magnetic moment (Bohr magneton, μB)	Curie temperature, K
BaB_6	1526	1.9	...
CaB_6	813	1.4	...
CeB_6	...	2.5	344
CoB	Paramagnetic	...	477
CrB	Paramagnetic	...	...
CrB_2	390	...	...
DyB_4	...	10.4	...
DyB_6	...	10.6	...
ErB_4	...	9.5	...
EuB_6	...	8.1	...
Fe_2B	...	1.9	1013
FeB	...	1.8	598
GdB_4	...	8.1	...
GdB_6	...	8.0	60
HfB_2	–4.0	...	...
HoB_4	...	10.8	...
LaB_6	60	0.0	...
Mn_2B	Paramagnetic	...	...
MnB	...	2.7	578
Mn_3B_4	Antiferromagnetic	...	...
MnB_2	...	2.3	140
Mo_2B_5	61.5	...	...
NbB_2	8.0	...	...
NdB_6	...	3.5	455
Ni_3B	264	...	...
Ni_2B	94	...	...
NiB	–5.5	...	...
PrB_6	...	3.6	~0
ScB_2	~100	...	...
ScB_{12}	–65	...	...
SmB_6	...	~1.5	...
SrB_6	0.0	0.0	...
TaB_6	–64.8	...	...
TbB_4	...	7.7	...
TiB_2	31.3	9.4	...
TmB_4	...	...	...
UB_2	550	...	...
UB_4	1390	...	...
UB_{12}	70	...	...
VB	Paramagnetic	...	...
VB_2	43.1	...	...
W_2B_5	506	...	...
YB_2	Diamagnetic	...	...
YbB_6	...	4.6	2
ZrB_2	–67.7	...	...

Chemical resistance of selected borides

Boride	Chemical resistance, % soluble					Oxidation resistance at 1000 °C (1830 °F), mg/cm^2
	H_2O	HNO_3	HCl	H_2SO_4	NaOH	
Be_5B	...	...	...	...	...	258
Be_2B	...	...	...	...	...	132
BeB_2	...	...	...	...	...	22
BeB_4	...	...	...	...	...	30
BeB_6	...	...	...	...	...	64
BaB_6	...	100	3.0	0	1.9	45%(a)

(continued)

Chemical resistance of selected borides (continued)

Boride	Chemical resistance, % soluble					Oxidation resistance at 1000 °C (1830 °F), mg/cm^2
	H_2O	HNO_3	HCl	H_2SO_4	NaOH	
CaB_6	...	100	0.5	0	2.6	37%(a)
CeB_6	...	100	5.0	2.5	1.1	...
Co_2B	...	100	100	...	...	...
CoB	...	...	0	...	...	...
Cr_4B	...	...	...	...	...	~3.5
Cr_3B_2	...	...	...	...	...	~15.5
CrB	...	...	...	...	...	~1.0
Cr_3B_4	...	...	...	...	...	~4.5(a)
CrB_2	...	59	97	97	12	~7.8
FeB		100	0	0	...	...
GdB_6	...	100	9	100	0.6	6.0
HfB_2	...	97	94	98	...	(b)
LaB_6	...	100	3.8	100	1.5	...
MgB_2	100	100	100	100	...	...
MgB_{12}	0	100	0	0	...	...
MnB	...	100	100	100	...	...
Mn_3B	0	100	100	100	...	...
Mn_4B	0	100	100	100	...	...
Mo_2B_5	...	97	27	93	...	~2.5
NbB_2	...	1	9	97	100	32.5
NdB_6	...	100	10.1	100	1.1	...
Ni_3B	100	...	...	...	...	...
PrB_6	...	100	9.2	100	0.7	...
ScB_2	...	84	75	67	...	...
SmB_6	...	100	22	100	0.6	...
SrB_6	0	100	1.5	0	1.9	43%(a)
TaB_2	...	0	2	97	100	2.52(a)
ThB_4	0	100	100	100	...	...
ThB_6	0	0	0	0	...	...
TiB_2	...	100	95	50	100	30(c)
UB_2	...	100	0	0	0	...
UB_4	...	100	100	100	...	...
UB_{12}	...	100	0	0	...	...
VB_2	...	99	97	93	39	...
W_2B_5	...	100	4	98	...	~4
YB_4	...	100	0	100	0	...
YB_6	...	100	0	100	0	...
YB_{12}	...	>75	0	0	0	...
ZrB_2	...	100	98	100	100	30(d)

(a) At 900 °C (1650 °F). (b) 0.50 cm (0.02 in.) thick oxide layer after 100 h at 1300 °C (2370 °F) (c) After 100 h. (d) After 150 h. Note: Chemical and oxidation resistance are very dependent on the nature of the attack. Above values represent worst possible scenarios

Glasses

Compositions and Applications

Materials used in glass manufacture

Material	Purpose
Antimony oxide (Sb_2O_3)	Decolorizing and fining agent
Aplite (K, Na, Ca, Mg, alumina silicate)	Source of alumina
Aragonite ($CaCO_3$)	Source of calcium oxide
Arsenic oxide (As_2O_3)	Fining and decolorizing agent
Barite/barytes ($BaSO_4$)	Flux and fining agent, source of barium
Barium carbonate ($BaCO_3$)	Source of barium for specialty glasses
Borate materials	Sources of B_2O_3
Sodium tetraborate ($Na_2O \cdot 2B_2O_3 \cdot 10H_2O$)	
5 mol borax ($Na_2O \cdot 2B_2O_3 \cdot 5H_2O$)	
Anhydrous borax ($Na_2O \cdot 2B_2O_3$)	
Boric acid ($B_2O_3 \cdot 3H_2O$)	
Calcium-aluminum-silicate/calumite slag	Reducing agent, fining aid
Caustic soda/sodium hydroxide (NaOH)	Solution (50%) is used for batch wetting
Cerium oxide (CeO)	Utroviolet absorber used in speciality glasses
Chromite ($FeO \cdot Cr_2O_3$)	Colorant for green bottles
Cobalt oxide ($Co_2O_3 \cdot CoO$)	Strong colorant (blue)
Colemanite ($Ca_2B_6O_{11} \cdot 5H_2O$)	Source of B_2O_3
Cryolite (Na_3AlF_6)	Flux and opacifier in opal glasses
Cullet	Crushed or powdered glass, may be internal or foreign
Dolomite/dolomitic limestone [$CaMg(CO_3)_2$]	Source of calcium and magnesium
Feldspars (Ca, Mg, Na, K, alumina silicate)	Sources of alumina
Albite ($Na_2O \cdot Al_2O_3 \cdot 6SiO_2$)	
Anorthite ($CaO \cdot Al_2O_3 \cdot 2SiO_2$)	
Microcline ($K_2O \cdot Al_2O_3 \cdot 6SiO_2$)	
Mixtures of these, plus Fe_2O_3 and free quartz; see also nepheline	
Fluorspar (CaF_2)	Used with feldspar as an opacifier in opal glasses
Gypsum ($CaSO_4 \cdot 2H_2O$)	Flux and fining agent
Iron oxides/rouge (FeO, $Fe_2O_3 \cdot Fe_3O_4$)	Colorants
Lead oxides	Source of PbO for lead glasses
Litharge (PbO)	
Red lead (Pb_3O_4)	
Lead silicates ($2PbO \cdot SiO_2$, $PbO \cdot SiO_2$, $4PbO \cdot SiO_2$)	
Limestone/calcite ($CaCO_3$)	Source of CaO
Lithia minerals	Source of LiO_2, flux, melting accelerator
Lepidolite ($LiF \cdot KF \cdot Al_2O_3 \cdot 3SiO_2$)	
Spodumene ($LiO_2 \cdot Al_2O_3 \cdot 4SiO_2$)	
Manganese dioxide/pyrolusite (MnO_2)	Colorant
Nepheline syenite	Source of alumina, made up of nepheline and feldspars (approximately 60% SiO_2, 23% Al_2O_3, 10% Na_2O, 5% K_2O, <1% CaO, MgO, Fe_2O_3)
Potash (K_2O) and potassium carbonate (K_2CO_3)	Sources of K_2O
Potassium dichromate ($K_2Cr_2O_7$)	Colorant in some artware
Pyrite (FeS_2)	Colorant in amber glass
Salt cake/anhydrous sodium sulfate (Na_2SO_4)	Melting and fining aid
Selenium (Se)	Aids decolorizing, and is used in colored glasses
Silica sand/quartz (SiO_2) Feldspathic sand (mix of feldspar and quartz, source of SiO_2 and Al_2O_3)	Glass former
Soda ash (Na_2CO_3)	Major flux used in all soda-lime-silica glass
Sodium antimonate ($2Na_2O \cdot 2Sb_2O_5 \cdot H_2O$)	Fining and decolorizing agent
Sodium nitrate ($NaNO_3$)	Oxidizing and fining agent
Tin oxides (SnO and SnO_2)	Colorants used in artware

Typical batches

Material	For 100 units glass	Based on 455 kg (1000 lb) sand(a)	For 100 units glass	Based on 455 kg (1000 lb) sand(a)
Sand	65.3	1000	73.0	1000
Limestone	18.6	284.8	5.6	76.7
Dolomite	...	...	18.6	254.8
Feldspar	10.9	166.9	...	...
Soda ash	22.3	341.5	22.6	309.6
Salt cake	0.6	9.2	1.3	17.8
Rouge	...	...	0.07	1.0
Carbon	0.04	0.6	0.08	1.1
Total	**117.74**	**1803.0**	**121.25**	**1661.0**
Glass yield, %	85		82.4	

(a) Calculating batches based on 455 kg (1000 lb) sand is a glass industry convention

Basic raw materials used in glass fiber production

Major constituents	Minor constituents
Silica sand	Salt cake
Calcined alumina	Niter
Feldspar	Carbon
Nepheline syenite	Gypsum
Limestone	Iron oxide
Dolomite	Fluorspar
Burned dolomite	Rutile
Magnesite	Zinc oxide
Soda ash	Manganese oxide
Borax	Potassium carbonate
Boric acid	
Colemanite	
Ulexite	
Kaolin clay	
Beneficiated blast furnace slag	
Cullet	

Main categories of silicate glasses

	Typical composition	Uses	Properties
Soda-lime	70 to 75% SiO_2 12 to 16% Na_2O 10 to 15% CaO	Bottles, glasses, windows	Optically clear, durable
Lead ("crystal") glasses	55 to 65% SiO_2 18 to 38% PbO 13 to 15% Na_2O or K_2O	Decorative items	High refractive index
Borosilicate	70 to 80% SiO_2 7 to 13% Ba_2O_3 4 to 8% Na_2O or K_2O 2 to 7% Al_2O_3	Chemical apparatus, lamp and tube envelopes	Chemical durability, low thermal expansion
Quartz glass	100% SiO_2	High-temperature uses	High softening temperature, low thermal expansion
Aluminosilicate glasses	52 to 58% SiO_2 15 to 25% Al_2O_3 4 to 18% CaO	High-temperature uses, thermometers, combustion tubes, cookware	High softening temperature, low thermal expansion

Glass compositions

	Composition, wt %						
Material	**SiO_2**	**Al_2O_3**	**Na_2O**	**CaO**	**MgO**	**B_2O_3**	**Other**
Container glass	71.0 to 73.0	1.0 to 3.0	13.0 to 14.0	10.0 to 11.0	0.1 to 1.0	...	0.04 to 0.06 Fe_2O_3; 0.1 to 1.0 K_2O; 0.2 SO_3
Float glass (U.S.)	72.8 to 73.2	0.1 to 0.2	13.65 to 13.85	8.55 to 8.85	3.85 to 4.0	...	0.10 to 0.14 Fe_2O_3; 0.01 to 0.04 K_2O; 0.25 to 0.30 SO_3
E-glass	52 to 56	12 to 16	...	15 to 25	0 to 6	8 to 13	...
S-glass	64 to 66	22 to 24	...	<0.01	10 to 12	<0.01	0.1 Fe_2O_3; <0.1 Zr_2O_3
D-glass	73 to 75	0 to 1	...	0 to 2	0 to 2	18 to 21	...
Quartz glass	99.97	...	...	...	...	...	...
Leachable alkali-borosilicate glass							
Quaternary	62.7	3.5	6.6	...	...	26.9	...
Ternary	65.0	...	9.0	...	...	26.0	...
Photochromic opthalmic glass	56.46	6.19	4.08	...	...	18.15	1.81 Li_2O; 5.72 K_2O; 4.99 ZrO_2; 2.07 TiO_2; 0.207 Ag; 0.166 Cl; 0.137 Br; 0.006 CuO
Photochromic flat glass	60.4	11.8	5.9	...	...	17.7	2.1 Li_2O; 1.6 K_2O; 0.28 PbO; 0.16 Ag; 0.48 Cl; 0.10 Br; 0.22 F; 0.007 CuO

Glass compositions and properties

	Body clear, R_b	Cladding clear, R_c	Body opal, C_b	Cladding clear, C_c
Composition, wt%				
SiO_2	58	58	64	58
Al_2O_3	20	15	6	15
B_2O_3	...	4	5	6
Na_2O	13	...	3	...
K_2O	4	...	3	...
MgO	2	7	1	6
CaO	3	10	15	15
BaO	...	6	...	...
F	...	...	3	...
Softening point, °C (°F)	863 (1585)	910 (1670)	...	890 (1635)
Annealing point, °C (°F)	633 (1171)	712 (1314)	610 (1130)	710 (1310)
Strain point, °C (°F)	588 (1090)	665 (1230)	563 (1045)	670 (1240)
Coefficient of thermal expansion from 0 to 300 °C (32 to 570 °F), 10^{-7}/°C	92 (198)	46 (115)	71 (160)	48 (118)
Density, g/cm^3	2.48	2.63	2.47	2.57
Liquidus temperature, °C (°F)	1058 (1935	1114 (2037)	...	1089 (1990)
Young's modulus, GPa (10^6 psi)	74.5 (10.8)	86.2 (12.5)	75.1 (10.9)	85.5 (12.4)
Poisson's ratio	0.22	0.24	0.22	0.25
Stress in cladding, MPa (ksi)(a)	344 (50)		207 (30)	
Stress in cladding, from tempering, MPa (ksi)(a)	414 (60)		241 (35)	

(a) Laminate property

Typical compositions of glass types used in lamps

Glass	Composition, wt%								
	SiO_2	Na_2O	K_2O	B_2O_3	Al_2O_3	MgO	CaO	BaO	PbO
Soda-lime glass	72	16	1	...	2	4	3	...	...
Lead glass	63	7	7	...	2	...	...	...	21
Borosilicate glass	78	5	...	15	2	...	...	...	...
Aluminosilicate glass	61	...	...	1	16	...	10	12	...
Vycor	96	...	...	4	...	...	...	...	...
Quartz glass	>99.9	...	...	...	...	...	...	...	...

Compositions of glass used for wool and textile products

Raw material	Wool			Textile		
	Containers, Type A	Thermal insulation T_1	Thermal insulation T_2	Electrical, Type E	Chemical resistant, Type C	Strength/stiffness, Type S
SiO_2	72 to 72.5	63	58.6	52 to 56	64 to 68	64
Al_2O_3	0 to 2.0	5	3.2	12 to 16	3.5	25
MgO	2.5 to 4.0	3	4.2	0 to 6	2 to 4	10
CaO	5.5 to 10.0	14	8.0	16 to 25	11 to 15	...
$Na_2O + K_2O$	10 to 16	10	15.1	0 to 2	7 to 10	...
B_2O_3	...	5	10.1	5 to 10	4.6	...
BaO	...	...	...	...	0 to 1	...
TiO_2	...	...	...	0 to 1.5	...	...
F_2	...	...	...	0 to 1	...	...
Fe_2O_3	...	...	...	...	0.8	...
FeO	...	...	...	0.8	...	...

Compositional ranges and properties for insulation-type glasses

Compound/property	Composition range, wt%			
	Rock wool made from basalt melted in a furnace	Rock wool made from basalt and other materials melted in a cupola	Slag wool made from slag melted in a cupola	Wool glass
Silicon dioxide	45 to 48	41 to 53	38 to 52	55 to 70
Aluminum oxide	12 to 13.5	6 to 14	5 to 15	0 to 7
Iron oxide(a)	5 to 6	3 to 8	0 to 2	0.1 to 0.5
Boric oxide	...	...	5 to 15	3 to 12
Sodium oxide	2.5 to 3.3	1.1 to 3.5	0 to 1	13 to 18
Potassium oxide	0.8 to 2	0.5 to 2	0.3 to 2	...
Magnesium oxide	8 to 10	6 to 16	4 to 14	0 to 5
Calcium oxide	10 to 12	10 to 25	20 to 43	5 to 13
Barium oxide	...	...	0 to 8	0 to 3
Titanium oxide	2.5 to 3	0.9 to 3.5	0.3 to 1	0 to 0.5
Sulfur(b)	0 to 0.2	0 to 0.2	0 to 2	0 to 0.5
Fluorine	...	...	...	0 to 1.5
Phosphorus pentoxide	...	...	0 to 0.5	...
Viscosity = 1000 P at temperature, °C (°F)	...	...	...	915 to 1085 (1680 to 1985)
Liquidus temperature, °C (°F)	>Viscosity	>Viscosity	>Viscosity	880 to 955 (1615 to 1755)

(a) In rock and slag wool produced from materials melted in a cupola with coke as fuel, all iron oxide is reduced to FeO. During the spinning process, a surface layer may form in which the iron is oxidized to Fe_2O_3. Typically, 8 to 15% of the iron is oxidized to Fe_2O_3. In an electric furnace melting basalt, up to 50% of the iron is in the form of Fe_2O_3 and is more evenly distributed throughout the entire volume. (b) In wool glass, sulfur is oxidized to sulfate

Composition and properties of commercially available porous glass

Composition on basis of ignited weight:	
SiO_2	96.3
B_2O_3	2.95
Na_2O	0.04
$R_2O_3 + RO_2$	0.72(a)
Appearance	Opalescent
Refractive index	1.33(b)
Apparent density (dry), g/cm^3 (lb/in.^{3})	1.5 (0.054)(c)
Internal pore volume, %	28
Average pore diameter, nm	5
Internal surface area, m^2/g	200
Water adsorption at saturation, %	25
Modulus of rupture, MPa (ksi)	42 (6.0)(d)
Elastic modulus at 22 °C (72 °F), GPa (10^6 psi)	17.6 (2.5)
Loss tangent at 22 °C (72 °F), 100 Hz	0.007(e)
Dielectric constant at 22 °C (72 °F), 100 Hz	3.1(e)

(a) Chiefly $Al_2O_3 + ZrO_2$. (b) Depends on amount of moisture in pores. (c) Depends on relative humidity. (d) Abraded 6.4 mm $\left(\frac{1}{4}\text{ in.}\right)$ rods at 22 °C (72 °F). (e) Appreciably affected by moisture; values are for specimens activated at 400 °C (750 °F), cooled in a desiccator, and then immediately tested

Composition and properties of selected glasses used for laboratory glassware applications

Material	Composition, wt% SiO_2	B_2O_3	Al_2O_3	Na_2O	K_2O	CaO	MgO	ZnO	TiO_2	Density g/cm^3	Density lb/in.^{3}	Modulus of elasticity GPa	Modulus of elasticity 10^6 psi
Borosilicate glass:													
Low-expansion	81	13	2	4	...	...	...	...	...	2.23	0.0806	63	9.1
Alumina	72	11	6	7	1	1	...	...	...	2.36	0.0853	...	...
Soda-lime glass	73	...	2	14	...	7	4	...	...	2.40	0.0867	...	...
Zinc-titania cover glass	65	9	2	7	7	...	...	7	3	2.57	0.0929	74	10.7
High-silica glass(a)	96.5	3	0.5	...	...	...	...	...	...	2.18	0.0788	73	10.6

Material	Coefficient of thermal expansion 0 to 300 °C (32 to 570 °F) 10^{-7}/°C	0 to 300 °C (32 to 570 °F) 10^{-7}/°F	25 °C (75 °F) to setting point 10^{-7}/°C	25 °C (75 °F) to setting point 10^{-7}/°F	Strain point °C	Strain point °F	Annealing point °C	Annealing point °F	Softening point °C	Softening point °F	Working point °C	Working point °F	Refractive index
Borosilicate glass:													
Low-expansion	32.5	18.1	35	19	510	950	570	1060	821	1510	1252	2286	1.474
Alumina	50	28	53	29	533	991	576	1070	795	1465	1189	2172	1.491
Soda-lime glass	89	49	...	...	511	952	545	1015	724	1335	...	...	1.515
Zinc-titania cover glass	74	41	84	47	508	946	550	1020	720	1330	1008	1846	1.52
High-silica glass(a)	7.5	4.2	...	...	890	1635	1020	1870	1530	2785	...	...	1.458

(a) Vycor

Compositions and structures of bioactive glasses and glass-ceramics(a)

Constituent	Material 45S5 Bioglass	45S5F Bioglass	45S5.4F Bioglass	40S5B5 Bioglass	52S4.6 Bioglass	55S4.3 Bioglass	KGC Ceravital	KGS Ceravital	KGy213 Ceravital	A-W-GC	MB-GC
SiO_2	45	45	45	40	52	55	46.2	46	38	34.2	19 to 52
P_2O_5	6	6	6	6	6	6	...	...	...	16.3	4 to 24
CaO	24.5	12.25	14.7	24.5	21	19.5	20.2	33	31	41.9	9 to 3
$Ca(PO_3)_2$	...	...	...	...	...	...	25.5	16	13.5	...	...
CaF_2	...	12.25	9.8	...	...	...	...	...	...	0.5	...
MgO	...	...	...	...	...	...	2.9	...	...	4.6	5 to 15
MgF_2	...	...	...	...	...	...	...	...	...	...	...
Na_2O	24.5	24.5	24.5	24.5	21	19.5	4.8	5	4	...	3 to 5
K_2O	...	...	...	...	...	...	0.4	...	...	...	3 to 5
Al_2O_3	...	...	...	...	...	...	...	...	7	...	12 to 33
B_2O_3	...	...	...	5	...	...	...	...	...	...	...
Ta_2O_5/TiO_2	...	...	...	...	...	...	...	...	6.5	...	...
Structure	Glass	Glass	Glass	Glass	Glass	...	Glass-ceramic	Glass-ceramic	...	Glass-ceramic	Glass-ceramic

(a) Compositions in wt%

Properties

Self-diffusion coefficients of alkali ions in alkali silicate glasses

Glass composition	Alkali oxide, mol%	D_o		E		D at 400 °C (750 °F)	
		cm²/s	in.²/s	MJ/kmol	kcal/mol	cm²/s	in.²/s
Na_2O-SiO_2	10	7.9×10^{-4}	1.22×10^{-4}	73.2	17.5	1.64×10^{-9}	0.25×10^{-9}
	20	1.3×10^{-3}	0.20×10^{-3}	73.5	17.4	2.91×10^{-9}	0.45×10^{-9}
	30	2.0×10^{-3}	0.31×10^{-3}	67.4	16.1	1.18×10^{-8}	0.18×10^{-8}
K_2O-SiO_2	10	2.0×10^{-3}	0.31×10^{-3}	81.2	19.4	1.00×10^{-10}	0.16×10^{-10}
	20	3.8×10^{-4}	0.59×10^{-4}	77.0	18.4	4.03×10^{-10}	0.62×10^{-10}
	33.3	9.33×10^{-4}	1.45×10^{-4}	66.9	16.0	5.94×10^{-9}	0.86×10^{-9}
Rb_2O-SiO_2	20	3.39×10^{-4}	0.53×10^{-4}	75.7	18.1	4.5×10^{-10}	0.70×10^{-10}
Cs_2O-SiO_2	16.7	8.7×10^{-3}	1.35×10^{-3}	103	24.5	9.64×10^{-11}	1.49×10^{-11}
	16.7	1.3×10^{-3}	0.20×10^{-3}	78.2	18.7	1.1×10^{-11}	0.17×10^{-11}

Comparison of heat transfer coefficients of borosilicate glass to that of selected metals

Material	Tube wall thickness		Thermal conductivity (Btu/h · ft² · °F)/ft	Coefficients: Outer film		Inner film		Wall heat transfer		Overall heat transfer	
	m	ft		W/m² · K	Btu/ft² · h · °F	W/m² · K	Btu/ft² · h · °F	W/m² · K	Btu/ft² · h · °F	W/m² · K	Btu/ft² · h · °F
Borosilicate glass	0.0012	0.0039	0.66	64.7	11.4	53	9.4	960	169	28.4	5.00
Type 316 stainless	0.00049	0.0016	9.4	64.7	11.4	53	9.4	3.339×10^4	5875	29.2	5.15
Aluminum	0.00049	0.0016	128	64.7	11.4	53	9.4	4.5×10^5	8.0×10^4	29.3	5.16
Copper	0.00049	0.0016	225	64.7	11.4	53	9.4	7.98469×10^5	1.40625×10^5	29.3	5.16

Nonoptical properties of single-component and multicomponent oxide glasses

Material	Physical properties: Density		Poisson's ratio		Mechanical properties: Elastic modulus				Knoop hardness	
			20 °C (70 °F)	900 °C (1650 °F)	20 °C (70 °F)		900 °C (1650 °F)			
	g/cm³	lb/in.³			GPa	10^6 psi	GPa	10^6 psi	MPa	ksi
Multicomponent oxide glass	2.3 to 6.3	0.083 to 0.23	0.19 to 0.31(b)		(b)		(b)		250 to 650	36 to 94
Single-component oxide glass (fused silica)(a)	2.201 to 2.203	0.07952 to 0.7959	0.17	0.19	70	10	81	12	900	130

Material	Thermal properties: Linear thermal expansion coefficient, ppm/°C		Transformation point, T_g		Softening point (viscosity of $10^{6.6}$ Pa · s or $10^{7.6}$ P)		Specific heat		Thermal conductivity		Strain point	
	−30 to 70 °C (−20 to 160 °F)	0 to 100 °C (30 to 212 °F)	°C	°F	°C	°F	J/kg · °C	Btu/lb · °F	W/m · K	Btu · in./ft² · h · °F	°C	°F
Multicomponent oxide glass	3.7 to 14.6	...	340 to 770	640 to 1420	471 to 825	880 to 1520	310 to 890	0.074 to 0.21	0.51 to 1.28	3.5 to 8.9	...	...
Single-component oxide glass (fused silica)(a)	...	0.51	...	...	1597 to 1727	2907 to 3141	750	0.18	1.38	9.57	1027 to 1077	1880 to 1971

(a) Ranges cover glass melted from quartz as well as synthetic material prepared by pyrolysis of silicon-tetrachloride or gaseous silicon-organic compounds. The low thermal expansion and the high thermal conductivity offer advantages for large mirror blanks. Caution: Prolonged exposure to about 1000 °C (1830 °F) can cause devitrification. (b) Elastic modulus for multicomponent oxide glass is 40 to 129 GPa (6 to 19 × 10^6 psi); no temperature range specified

Properties of glasses used in lamps

	Density, g/cm³	Strain point °C	Strain point °F	Annealing point °C	Annealing point °F	Softening point °C	Softening point °F	Working point °C	Working point °F	Coefficient of thermal expansion at 0 to 300 °C (32 to 570 °F), 10^{-7}/K	Elastic modulus at 20 °C (70 °F) GPa	Elastic modulus at 20 °C (70 °F) 10^6 psi	Electrical resistivity at 350 °C (660 °F), Ω · cm
Soda-lime glass	2.5	490	915	520	970	700	1290	1015	1860	94	72	10.4	10^5
Lead glass	2.8	410	770	445	835	635	1175	1000	1830	93	61	8.8	10^7
Borosilicate	2.3	520	970	570	1060	800	1470	1200	2190	40	64	9.3	10^7
Aluminosilicate	2.6	770	1420	810	1490	1025	1875	1250	2280	45	88	12.8	10^{11}
Vycor	2.2	890	1635	1020	1870	1530	2785	...	...	7.5	72	10.4	10^8
Quartz glass	2.2	1070	1960	1140	2085	1670	3040	...	...	5.5	73	10.6	10^{10}

Properties of glasses used in non-CRT applications

Application(a)	Glass code	Glass composition type(b)	Strain point °C	Strain point °F	Annealing point °C	Annealing point °F	Softening point °C	Softening point °F	Maximum use temperature °C	Maximum use temperature °F	Thermal expansion at 300 °C (570 °F) 10^{-7}/°C	Thermal expansion at 300 °C (570 °F) 10^{-7}/°F
L	0211	Alkali zinc borosilicate	508	946	550	1020	720	1330	483	901	74	41
L	1724	Alkaline-earth boroaluminosilicate	674	1245	726	1340	926	1699	649	1200	44	24
L	1733	Alkaline-earth boroaluminosilicate	640	1180	689	1270	928	1702	615	1140	37	21
L	1729	Alkaline-earth aluminosilicate	799	1470	855	1570	1107	2025	774	1425	35	19
E, L	7059	Baria aluminoborosilicate	593	1100	639	1180	844	1570	568	1055	46	26
F, P	7555	Lead borate V	326	619	370	700	415	779	301	574	88	49
F, P	7568	Lead zinc borate D	...	...	...	...	320	610	...	...	88	49
F, P	7570	Lead borosilicate V	358	676	376	708	447	837	333	631	83	46
F, P	7575	Lead zinc borosilicate D	...	...	...	...	380	720	425	800	89	49
F, P	7599	Lead zinc borate D	...	...	...	...	320	610	400	750	89	49
J, S	7740	Soda borosilicate	510	950	560	1040	821	1510	485	905	32.5	18
L	7913	96% silica	890	1635	1020	1870	1530	2785	865	1590	7.5	4.2
L	7940	100% silica	990	1815	1075	1965	1585	2885	965	1770	5.6	3.1
J	8603	Alkali zinc silicate (glass)	...	...	...	...	...	...	450	840	84	47
P	8603	Alkali zinc silicate (opal)	...	...	...	...	...	...	550	1020	89	49
J	8603	Alkali zinc silicate (ceram)	...	...	...	...	...	...	750	1380	103	57
F, P	AS	Soda-lime silicate	511	951	554	1030	740	1365	486	907	81(c)	45(c)
L	AX	Borosilicate	522	972	568	1055	789	1450	497	927	49(c)	27(c)
L	AN	Alkaline-earth aluminoborosilicate	616	1140	661	1220	862	1585	591	1095	45(c)	25(c)
L	AQ	Vitreous silica	1000	1830	1120	2050	1600	2910	975	1785	6(c)	3(c)
E, L	NA 35	Borosilicate	650	1200	700	1290	940	1725	625	1155	37(d)	21(d)
E, L	NA 40	Alkaline-earth aluminosilicate	656	1215	...	...	...	...	631	1170	43(d)	24(d)
E, L	NA 45	Baria aluminoborosilicate	610	1130	658	1215	859	1580	585	1085	46(d)	26(d)
L	BLC	Alkali borosilicate	535	995	575	1065	775	1425	510	950	51(e)	28(e)
E, L	OA 2	Alkaline-earth aluminoborosilicate	635	1175	685	1265	895	1645	610	1130	47(e)	26(e)
F, P	LS-7105	Lead borate devitrifiable sealing glass D	...	...	...	...	400	750	...	...	85(f)	47(f)
F, P	GA-8	Lead borosilicate	...	...	...	...	490	915	...	...	81(f)	45(f)
F, P	GA-9	Lead borosilicate	...	...	...	...	430	805	...	...	90(f)	50(f)
F, P	GA-12	Alkali borosilicate	...	...	...	...	560	1040	...	...	73(f)	41(f)
F, P	GA-21	Lead borosilicate	...	...	...	...	450	840	...	...	83(f)	46(f)
F, P	PLS-2401	(VFD glaze)	...	...	...	...	545	1015	...	...	77(f)	43(f)
F, P	PLS-3108A	(VFD glaze)	...	...	...	...	590	1095	...	...	65(f)	36(f)
P	PLS-3130	(PDP dielectric)	...	...	...	...	485	905	...	...	81(f)	45(f)
F, P	CV-455	Devitrifying solder glass (for soda lime) D	...	...	...	...	365	670	...	...	86	48
P	PP-200	Alkali free dielectric glass	...	...	...	...	340	645	...	...	125	70
F, P	SG-100	Vitreous solder glass V	...	...	...	...	362	684	...	...	77	43
E, L	AF45	Borosilicate	627	1160	663	1225	876	1610	602	1115	45	25
L	D263	Alkali-zinc borosilicate	...	...	557	1035	736	1355	...	...	73	41

(continued)

Properties of glasses used in non-CRT applications (continued)

Application(a)	Glass code	Glass composition type(b)	Strain point °C	Strain point °F	Annealing point °C	Annealing point °F	Softening point °C	Softening point °F	Maximum use temperature °C	Maximum use temperature °F	Thermal expansion at 300 °C (570 °F) 10^{-7}/°C	Thermal expansion at 300 °C (570 °F) 10^{-7}/°F
P	G 017-340	Lead borate (for float) V	...	...	...	...	360	680	...	...	70(g)	39(g)
F, P	8596	Devitrifying solder glass D	...	...	...	...	370	700	...	...	87(g)	48(g)
F, L, P		Soda-lime silicate (Fourcault process)	510	950	536	995	687	1270	485	905	91	51
F, L, P		Soda-lime silicate	490	915	...	...	720	1330	465	870	84	47
F, L, P		Soda-lime silicate (Float process)	523	973	545	1015	737	1360	498	930	77.5	43

Application(a)	Glass code	Glass composition type(b)	Density g/cm³	Density lb/in.³	Volume resistivity, Ω · cm 250 °C (480 °F)	Volume resistivity, Ω · cm 350 °C (660 °F)	Dielectric constant at 1 MHz	Loss tangent at 20 °C (70 °F), %	Elastic modulus GPa	Elastic modulus 10^6 psi	Index of refraction at 589.3 nm	Poisson's ratio
L	0211	Alkali zinc borosilicate	2.57	0.0929	8.3	6.7	6.7	0.46	74.5	10.8	1.523	0.22
L	1724	Alkaline-earth boroaluminosilicate	2.64	0.0954	13.8	11.6	6.6	0.001	82.7	12.0	1.540	...
L	1733	Alkaline-earth boroaluminosilicate	2.49	0.0890	13.4	11.4	5.3	0.09	66.9	9.7	1.516	0.235
L	1729	Alkaline-earth aluminosilicate	2.56	0.0925	13.1	11.0	5.9	...	80.6	11.7	1.52	0.216
E, L	7059	Baria aluminoborosilicate	2.76	0.0997	13.1	11.0	5.84	0.10	67.6	9.8	1.533	0.280
F, P	7555	Lead borate V	5.70	0.206	10.5	13.74(j)	...	...	...	...	...	...
F, P	7568	Lead zinc borate D	...	...	...	...	...	...	...	...	...	...
F, P	7570	Lead borosilicate V	5.46	0.197	10.6	8.7	15.0	0.22	53.5	7.76	1.860	...
F, P	7575	Lead zinc borosilicate D	3.80	0.137	8.6	7.05	20.4(j)	0.91	51.3	7.44	...	0.25
F, P	7599	Lead zinc borate D	5.78	0.209	9.3	7.7	17.7(j)	1.1	44.8	6.5	...	0.27
J, S	7740	Soda borosilicate	2.23	0.0806	8.1	6.6	4.6	0.4	62.7	9.1	1.473	0.20
L	7913	96% silica	2.18	0.0788	9.7	8.1	3.8	0.04	66.2	9.6	1.458	0.19
L	7940	100% silica	2.20	0.0795	12.3	10.7	3.8	0.00	72.4	10.5	1.458	0.16
J	8603	Alkali zinc silicate (glass)	2.365	0.0854	6.27	4.90	7.62(k)	0.80(k)	76.9	11.15	...	0.22
P	8603	Alkali zinc silicate (opal)	2.380	0.0860	8.81	7.23	5.73(k)	0.40(k)	82.7	12.0	...	0.21
J	8603	Alkali zinc silicate (ceram)	2.407	0.0870	8.76	7.07	5.63(k)	0.30(k)	86.9	12.62	...	0.19
F, P	AS	Soda-lime silicate	2.49	0.0900	7.7(h)	5.9(i)	7.5	0.9	71.7	10.4	1.52	0.21
L	AX	Borosilicate	2.42	0.0874	8.0(h)	6.6(i)	5.9	0.9	68.9	10.0	1.50	0.18
L	AN	Alkaline-earth aluminoborosilicate	2.72	0.0983	15.5(h)	12.0(i)	6.3	0.07	73.8	10.7	1.54	0.22
L	AQ	Vitreous silica	2.20	0.0795	...	...	...	...	...	...	...	...
E, L	NA 35	Borosilicate	2.50	0.0903	...	...	5.3	0.07	70.3	10.2	1.516	0.24
E, L	NA 40	Alkaline-earth aluminosilicate	2.87	0.104	...	...	6.7	...	92.4	13.4	1.574	0.26
E, L	NA 45	Baria aluminoborosilicate	2.78	0.100	...	...	5.6	...	68.9	10.0	1.533	0.24
L	BLC	Alkali borosilicate	2.36	0.0853	6.9	...	5.7	0.60	...	...	1.493	...
E, L	OA 2	Alkaline-earth aluminoborosilicate	2.76	0.0997	13.0	...	6.3	0.10	...	...	1.54	...
F, P	LS-7105	Lead borate devitrifiable sealing glass D	6.37	0.230	...	...	...	...	...	...	...	...
F, P	GA-8	Lead borosilicate	5.37	0.194	...	...	...	...	...	...	...	...
F, P	GA-9	Lead borosilicate	5.77	0.208	...	...	...	...	...	...	...	...
F, P	GA-12	Alkali borosilicate	2.95	0.107	...	...	...	...	...	...	...	...
F, P	GA-21	Lead borosilicate	5.74	0.207	...	...	...	...	...	...	...	...
F, P	PLS-2401	(VFD glaze)	...	...	12.2	...	12.2	0.266	...	...	...	...
F, P	PLS-3108A	(VFD glaze)	...	...	14.0	...	9.2	0.40	...	...	...	...
P	PLS-3130	(PDP dielectric)	...	...	12.2	...	11.7	0.264	...	...	...	...
F, P	CV-455	Devitrifying solder glass (for soda lime) D	5.915	0.214	...	8.3	18.3	0.68	...	...	...	...
P	PP-200	Alkali free dielectric glass	6.62	0.239	...	...	...	...	...	...	...	...
F, P	SG-100	Vitreous solder glass V	6.698	0.242	...	9.0	31.8	0.80	...	...	...	...
E, L	AF45	Borosilicate	2.72	0.0983	13.8	11.5	6.2	0.015	66.0	9.57	1.5276	0.235
L	D263	Alkali-zinc borosilicate	2.51	0.0907	...	...	...	...	...	...	1.5225	...
P	G 017-340	Lead borate (for float) V	4.80	0.173	...	...	...	...	...	...	...	...
F, P	8596	Devitrifying solder glass D	6.43	0.232	...	...	...	...	...	...	...	...
F, L, P		Soda-lime silicate (Fourcault process)	2.47	0.0892	...	...	8.3	...	71.0	10.3	1.513	0.22
F, L, P		Soda-lime silicate	2.483	0.0897	...	...	7.75	...	69.6	10.1	...	0.22
F, L, P		Soda-lime silicate (Float process)	2.498	0.0903	...	...	...	...	73.1	10.6	...	0.22

(a) E, TFEL or EL; F, VFD; J, ink-jet printer; L, LCD; M, dot matrix printer; P, PDP; S, LED. (b) V, vitreous; D, devitrifying. (c) 50 to 200 °C (120 to 390 °F). (d) 100 to 300 °C (210 to 570 °F). (e) 30 to 380 °C (85 to 715 °F). (f) 30 to 300 °C (85 to 570 °F). (g) 20 to 250 °C (70 to 480 °F). (h) At 200 °C (390 °F). (i) At 300 °C (570 °F). (j) kHz at 20 °C (68 °F). (k) 100 kHz at 21 °C (70 °F)

Properties of consolidated 96% silica glass

Property	Value
Refractive index	1.458
Average thermal expansion coefficient (0 to 300 °C, or 32 to 572 °F), 10^{-7}/°C (10^{-7}/°F)	7.5 (4.2)
Density, g/cm^3 (lb/in.^{3})	2.18 (0.0788)
Annealing point, °C (°F)	1020 (1870)(a)
Specific heat, cal/g · °C	0.18
Thermal diffusivity, cm^2/s	0.009
Thermal conductivity, W/m · K (cal/cm · s · °C)	1.38 (0.0033)
Total normal emissivity, 100 °C (212 °F)	0.87
Elastic modulus at 22 °C (72 °F), GPa (10^6 psi)	68 (9.8)
Shear modulus at 22 °C (72 °F), GPa (10^6 psi)	28 (4.0)
Poisson's ratio	0.19
Modulus of rupture at 22 °C (72 °F), abraded surface, MPa (ksi)	48 (7.0)
Knoop hardness, 100-g load, kgf/mm^2	487
Dielectric constant at 22 °C (72 °F):	
1 MHz	3.8
8.6 GHz	3.8
Electrical resistivity, Ω · cm:	
Log ρ at 250 °C (480 °F)	9.7(b)
Log ρ at 350 °C (660 °F)	8.1(c)
Chemical durability, weight loss, mg/cm^2	0.0005(d)(e), 0.07(f), 0.90(g)

(a) 1080 °C (1975 °F) for special processed ware. (b) 11.4 for special processed ware. (c) 9.7 for special processed ware. (d) 5% HCl at 100 °C (212 °F) for 24 h. (e) Reconstructed Pyrex glass is ten times as durable in 5% HCl as Pyrex No. 7740 glass (which itself has excellent acid resistance). (f) N/50 Na_2CO_3 at 100 °C (212 °F) for 6 h. (g) 5% NaOH at 100 °C (212 °F) for 6 h

Property data for glasses and thoria

Material	Density, g/cm^3	Softening point °C	Softening point °F	Coefficient of thermal expansion, 10^{-6}/K	Stress MPa	Stress ksi
Corning 7740 pyrex	2.23	820	1510	3.6	–267	–38.7
Borosilicate glass	2.46	765	1410	7.9	–45	–6.5
Soda-lime glass	2.49	715	1320	10.5	88	12.8
ThO_2 spheres	9.90	...	...	8.5(a)	...	...

(a) At 0 to 500 °C (30 to 930 °F)

Properties of chalcogenide glasses

Glass(a)	R	Density, g/cm^3	Molar volume, 10^{-6} m^3/mol	T_{12}(c) °C	T_{12}(c) °F	E_η(d) kJ/mol	E_η(d) 10^3 Btu/lb · mol
Se	2.0	4.25	18.6	30	85	...	...
$As_{20}Se_{80}$	2.2	4.46	17.5	82	180	235	101
$As_{40}Se_{60}$	2.4	4.60	16.8	165	330	280	120
$As_{40}S_{60}$	2.4	3.20	15.4	200	390	300	129
$Ge_{10}As_{10}Se_{80}$	2.3	4.36	17.8	...	...	...	...
$Ge_{10}As_{30}Se_{60}$	2.5	...	...	212	415	370	159
$Ge_{40}As_{20}Se_{40}$(b)	3.0	4.59	16.4	376	710	585	252

(a) Atomic basis. (b) Chalcogen deficient. (c) $T_{12} = 10^{12}$ P $= 10^{11}$ Pa · s isokom. (d) E_η, activation energy for viscous flow

Properties and use of five chemically strengthened glasses

Corning code	Thickness mm	Thickness mils	Abraded modulus of rupture MPa	Abraded modulus of rupture ksi	Center tension MPa	Center tension ksi	Depth of compression mm	Depth of compression mils	Applications
0313	1.270	50	310	45	55	8.0	0.1778	7	Cladding for aircraft windshields
	2.159	85	310	45	31	4.5	0.1778	7	Tape reels
0315	2.159	85	448	65	62	9.0	0.2286	9	Spacecraft windows
	5.080	200	517	75	24	3.6	0.2286	9	High-strength lens systems
0319	1.270	50	227	33	89	13.0	0.3048	12	Cladding for aircraft windshields
	2.159	85	276	40	55	8.0	0.3048	12	Cladding for aircraft windshields
8111	2.00	79	220	32(a)	2.8	0.42	NA		Photochromic eyeglass lenses
8361	2.00	79	380	55(a)	NA		NA		White crown eyeglass lenses

NA, Not available. (a) Unabraded modulus of rupture

Properties of materials for spacecraft windows

Property	Fused silica (Corning Code 7940)	Aluminosilicate glass (Corning Code 1723)
Mechanical		
Elastic modulus, GPa (10^6 psi)	73 (10.5)	85.5 (12.2)
Modulus of rupture, MPa (ksi)	60 (8.7)	170 (24.5) (tempered)
Poisson's ratio	0.17	0.26
Knoop hardness (100 g load), kg/mm^2	500	595
Optical		
Refractive index	1.459	1.547
10% UV cut-off(a), nm	165	350
Birefringence constant, nm/cm/kg/cm^2	3.45	2.40
Thermal		
Strain point, °C (°F)	990 (1815)	670 (1240)
Coefficient of thermal expansion from 0 to 200 °C, ppm/°C	0.57	4.6
Thermal conductivity, W/m · °C	1.38	1.29
Physical		
Density, g/cm^3 (lb/ft^3)	2.2 (137)	2.63 (164)

(a) Measured through 10 mm (0.4 in.) of glass

Comparison of spectral properties of commercial solar glass windshields with a standard tinted glass windshield

Property	Standard tinted	Solar Reflective	Solar Batch
Illuminant A transmission	77%	71%	71%
Visible reflection	7%	10%	6%
Solar transmission	54%	41%	43%
Solar reflectance	6%	31%	5%
UV transmission	20%	16%	33%(a)

(a) 17% for windshields due to PVB interlayer

Properties of materials for space-based mirrors

Property	Ultralow-expansion glass (Corning 7971)	Lithium-aluminosilicate glass-ceramic (Schott Zerodur)
Average coefficient of thermal expansion in ppm/°C for: 5 to 35 °C (40 to 95 °F)	0	–0.05
20 to 300 °C (70 to 570 °F)	0.03	0.05
Density, g/cm^3 (lb/ft^3)	2.20 (137)	2.53 (158)
Knoop hardness (200 g load), kg/mm^2	460	580
Modulus of rupture, MPa (ksi)	50 (7.25)	76.5 (11)
Elastic modulus, GPa (10^6 psi)	67.6 (9.8)	91 (13.2)
Stress-optical coefficient, nm/cm/kg/cm^2	4.15	2.94
Service temperature:		
Maximum operating temperature, °C (°F)	800 (1470)	150 (300)
Short-term operating temperature, °C (°F)	1050 (1920)	625 (1160)
Thermal conductivity, W/m · °C	1.31	1.64
Thermal diffusivity, 10^{-3} cm^2/s	7.9	7.9

Properties of infrared transmitting glasses

Property	Calcium aluminate (Barr & Stroud BS 37A)	Alkaline earth germanate (Corning, 9754)
Mechanical		
Elastic modulus, GPa (10^6 psi)	...	84 (12)
Poisson's ratio	...	0.29
Modulus of rupture, MPa (ksi)	83 (12)	45 (6.5) (abraded)
Hardness	6 on Moh scale	590 HK
Density, g/cm^3 (lb/ft^3)	...	3.581 (223.5)
Thermal		
Average thermal expansion (25 to 300 °C), ppm/°C	8.35	6.2
Service temperature, °C (°F)	700 (1300)(a)	650 (1200)
Optical		
10% IR cut-off(b), μm	5.45	5.7
Index of refraction at:		
3 μm	1.627	1.625
4 μm	1.607	1.606

(a) Moisture protective coating good to 500 °C (930 °F). (b) Wavelength for 10% cut-off in 2 mm (0.08 in.) thickness

Properties of a sodium-aluminosilicate for high-strength and frangible applications

Property	Property value
Mechanical	
Elastic modulus, GPa (10^6 psi)	71 (10.3)
Shear modulus, GPa (10^6 psi)	29.5 (4.3)
Poisson's ratio	0.21
Modulus of rupture, MPa (ksi)	275 (40)
Knoop hardness, (100 g load), kg/mm^2	480
Physical	
Density, g/cm^3 (lb/ft^3)	2.46 (153)
Softening point, °C (°F)	870 to 880 (1600 to 1615)
Annealing point, °C (°F)	622 to 631 (1151 to 1168)
Strain point, °C (°F)	574 to 583 (1065 to 1080)
Water absorption, %	0.00
Porosity, %	0.00
Permeability	Impermeable under vacuum conditions
Thermal	
Coefficient of thermal expansion (25 to 300 °C), ppm/°C	8.8
Thermal (cont'd)	
Thermal conductivity (25 °C), W/m · °C	1.03
Specific heat, J/kg · °C	803
Optical	
Refractive index (20 °C)	N_f = 1.5129 N_d = 1.5068 N_c = 1.5043
N_u value	58.9
Transmittance (visible), %	91.6(a)
Electrical	
Loss tangent, %	0.012(b)
Dielectric constant	7.38(b)
Loss factor, %	0.088(b)
Volume resistivity, Ω · cm, at:	
25 °C	14.55
250 °C	6.8
350 °C	5.4

Material is Corning Code 0313. (a) Transmittance of visible wavelength through 2 mm (0.08 in.). (b) At 25 °C (77 °F) and 1 MHz

Properties of foam glass insulation compared to other selected insulation materials

	Ideal insulation properties(a)(b)									
Insulation material	Noncombustible	Moisture resistant	Low K value(c)	Constant K value	Dimensionally stable	Compatible with bitumen	Strong, rigid, impact-resistant	Resists deterioration	Surfaces provide secure attachment	Compatible with membrane
Foam glass	Y	Y	I	Y	Y	Y	Y	Y	Y	Y
Perlite	N(c)	N	I	N	Y	Y	Y	N(e)	Y	Y
Glass fiber	N(c)	N	L	N	Y	Y	LR	N(e)	Y	Y
Polystyrene	N	Y	L	N	N	N	LS	Y	N	N
Polyisocyanurate	N	Y	VL	N	N	C(d)	LS	Y	Y	C(d)
Phenolic	N	Y	VL	Y	N	Y	Y	Y	Y(f)	Y

(a) I, intermediate; VL, very low; C, caution; LR, low rigidity; LS, low strength. (b) Organic burns per NRCA criteria. (c) Very low, <0.20 ; low, 0.20 to 0.30; intermediate, 0.30 to 0.36; high, >0.37. (d) May cause blistering under BUR membranes. (e) Organic portions deteriorate with moisture. (f) If covered with acceptable insulation layer

Hardness of glasses

Glass	Vickers hardness, GPa	Vickers hardness, 10^6 psi
SiO_2	6.2	0.90
	4.7(a)	0.68(a)
GeO_2	2.4(a)	0.35(a)
B_2O_3	2.0	0.29
Soda-lime-silica	4.5(a)	0.65(a)
$12.5Na_2O$-$17.5CaO$-$70SiO_2$	5.5(a)	0.80(a)
$37MgO$-$13Al_2O_3$-$50SiO_2$	6.6	0.95
$18Y_2O_3$-$24Al_2O_3$-$58SiO_2$	8.1	1.17
$37.5Y_2O_3$-$18.6Al_2O_3$-$37.5Si_3N_4$-$6.4SiO_2$	11.4(a)	1.65(a)
$30Na_2O$-$70B_2O_3$	4.7	0.68
Sodium borosilicate	4.1(a)	0.59(a)
$10BaF_2$-$30ZnF_2$-$30YF_3$-$30ThF_4$	3.0	0.44
$57ZrF_4$-$36BaF_2$-$3LaF_3$-$4AlF_3$	2.5	0.36
Se	0.3	0.04
As_2Se_3	1.3	0.19
$Ge_{25}Se_{75}$	1.9	0.28
$Ge_{40}As_{15}S_{45}$	2.6	0.38
$Ge_{30}Sb_{10}Se_{60}$	1.9	0.28

(a) Knoop

Observed strengths of silicate glasses

Glass	Diameter, mm	Diameter, in.	Surface	Temperature, °C	Temperature, °F	Atmosphere	Load rate, mm/s	Load rate, in./min.	Minimum, GPa	Minimum, 10^6 psi	Mean, GPa	Mean, 10^6 psi	Maximum, GPa	Maximum, 10^6 psi
SiO_2	1	0.04	Pristine	25	77	Air	1.7	4.0	0.55	0.08	4.48	0.65	5.52	0.80
	0.03	0.001	Pristine	25	77	Air	1.7	4.0	2.76	0.40	5.86	0.85	9.65	1.40
	0.03	0.001	Pristine	25	77	Vacuum	1.7	4.0	4.14	0.60	7.59	1.1	11.0	1.60
Soda-lime	0.5	0.02	Pristine	25	77	Air	...	...	0.89	0.13	3.93	0.57	5.38	0.78
	0.5	0.02	After 1 h at 125 °C (225 °F)	25	77	Air	...	...	0.42	0.06	1.17	0.17	1.72	0.25
7740	0.5	0.02	Pristine	25	77	Air	...	...	0.69	0.10	2.93	0.43	4.69	0.68
	4	0.16	Etched	25	77	Air	...	...	1.21	0.18	1.73	0.25	2.62	0.38
Soda-lime	0.25	0.01	Abraded	25	77	H_2O gas	0.02	0.05	0.1	0.014	0.15	0.02	0.21	0.03
	0.25	0.01	Abraded	–170	–275	H_2O gas	0.02	0.05	0.07	0.010	0.09	0.013	0.10	0.015
E-glass	0.0075	0.0003	Pristine	–170	–275	N_2	0.17	0.39	5.0	0.72	5.79	0.84	6.07	0.88
	0.0075	0.0003	Pristine	200	390	N_2	0.17	0.39	1.38	0.20	1.52	0.22	1.72	0.25
Soda-lime	1.3	0.05	Slight abrasion	25	77	Air	...	...	0.35	0.05	0.45	0.07	0.55	0.08
	0.7	0.03	Etched	25	77	Air	...	...	1.73	0.25	2.14	0.31	2.62	0.38
	Rect.	Rect.	Abraded	25	77	Air	(a)	(a)	0.04	0.007	0.05	0.008	0.06	0.008
	Rect.	Rect.	Abraded	25	77	Air	(b)	(b)	0.03	0.004	0.03	0.005	0.04	0.006
	Rect.	Rect.	Abraded	–170	–275	N_2	(a)	(a)	0.05	0.007	0.12	0.018	0.19	0.027

(a) 5.5 MPa/s (0.80 ksi/s) (b) 6.9 kPa/s (1 psi/s)

Mechanical properties of glasses

Glass	Modulus of elasticity GPa	Modulus of elasticity 10^6 psi	Poisson's ratio	Surface energy N/m	Surface energy lbf/ft	Fracture toughness MPa$\sqrt{m}$	Fracture toughness ksi$\sqrt{in.}$
SiO_2	73	10.6	0.17	4.4	0.30	0.79	0.72
GeO_2	43	6.2	0.21	...	...	...	...
B_2O_3	17	2.5	0.26	...	...	...	...
Se	10	1.5	0.33	...	...	...	...
As_2Se_3	17	2.5	0.29	...	...	...	...
$Ge_3As_4Se_3$	29	4.2	0.26	...	...	...	...
Ge-As-Se	18	2.6	...	...	...	0.25	0.23
PbO-4B2O3	60	8.7	0.26	...	...	...	...
Aluminosilicate	75	10.8	...	4.0	0.27	0.91	0.83
Borosilicate	60	8.7	...	4.6	0.31	0.77	0.70
Soda-lime-silica	66	9.6	...	3.9	0.26	0.75	0.68
$20La_2O_3$-$30Al_2O_3$-$50SiO_2$	100	14.5	...	...	...	...	...
$37.5Y_2O_3$-$8.6Al_2O_3$-$37.5Si_3N_4$-$6.4SiO_2$	183	26.5	0.28	...	...	...	...

Inclusion levels in commercially available glass products

Application	Defects: Size mm	Defects: Size in.	Defects: Quantity no./dm^3(a)	Defects: Quantity no./lb
Bubbles				
Best art lead crystal	>0.03	>0.001	0.3	0.04
Color TV panel	>0.25	>0.010	0.06	0.01
	>0.05	>0.002	30	5
Float automotive	>0.5	>0.020	0.06	0.01
Float architectural	>1.0	>0.040	0.02	0.003
	>0.1	>0.004	0.8	0.15
Soda-lime tableware	>0.03	>0.001	400	60
Glass-ceramic ovenware	>0.2	>0.008	2000	400
Containers	>0.05	>0.002	3000	500
	>0.8	>0.031	20	3
Wool fiberglass	>0.2	>0.008	10^4	2000
Textile fiberglass	>0.007	>0.0003	0 to 900	0 to 160
Solid inclusions				
Float architectural	...	...	0.001	0.0002
Containers	>0.03	>0.001	0.06	0.01
Glass-ceramic ovenware	>0.5	>0.020	1	0.2

(a) dm, decimeter

Tolerance specifications for selected photochemically sensitive glasses

Capability	Glass composition type(a)	Hole and slot size tolerance ±µm	Hole and slot size tolerance ±in.	Centerline tolerance ±µm/cm	Centerline tolerance ±in./in.	Edge to vertical angle of etch, degrees	Minimum feature size µm	Minimum feature size in.	Maximum ratio of feature size to thickness	Thickness control ±µm	Thickness control ±in.	Flatness ±µm/cm	Flatness ±in./in.	Minimum thickness µm	Minimum thickness in.
Standard process	Fotoform glass	1	0.001	10	0.001	2 to 3	63	0.0025	8:1	50	0.002	10	0.001	500	0.020
	Fotoform opal	1	0.001	15	0.0015	2 to 3	75	0.003	8:1	50	0.002	10	0.001	500	0.020
	Fotoceram	1	0.001	15	0.0015	2 to 3	75	0.003	8:1	50	0.002	10	0.001	500	0.020
Ultimate precision	Fotoform glass	8	0.0003	1	0.0001	1 to 2	25	0.001	40:1	25	0.001	1(b)	0.0001(b)	100(b)	0.004(b)
	Fotoform opal	10.4	0.0004	2	0.0002	1 to 2	38	0.0015	40:1	13	0.0005	1(b)	0.0001(b)	75(b)	0.003(b)
	Fotoceram	10.4	0.0004	2	0.0002	1 to 2	38	0.0015	40:1	13	0.0005	1(b)	0.0001(b)	75(b)	0.003(b)

(a) Corning Glass Works Fotoform process product. (b) Dependent on size

Chemical corrosion of selected substrate glasses at 95 °C (205 °F)

Glass composition type	Corrosive					
	5% HCl 24 h		5% NaOH 6 h		H_2O 24 h	
	Weight loss, mg/cm^2	Visual appearance(a)	Weight loss, mg/cm^2	Visual appearance(a)	Weight loss, mg/cm^2	Visual appearance(a)
Soda-lime silicate	0.01	1	0.7	3	0.01	1
Soda borosilicate	<0.01	1	1.5	3	<0.01	1
Barium aluminoborosilicate	11	4	3	2	0.03	1
Alkaline-earth boroaluminosilicate	4	1	2	3	0.02	1
Alkaline-earth aluminosilicate	<0.01	1	1	3	<0.01	1

(a) Visual appearance key: 1, no change; 2, slight surface darkening or iridescence; 3, slight or moderate surface frost; 4, heavy surface frost or surface crazing

Glass-Ceramics

Typical properties of various glass-ceramics

Property	Value
Thermal expansion, 10^{-6}/°C, from 25 to 500 °C (77 to 930 °F)	3.0 to 7.0
Thermal conductivity at 25 °C (77 °F), W/m · °C	1 to 5
Dielectric constant at 1 MHz	4 to 8
Dielectric loss at 1 MHz	>0.002
Flexural strength, MPa (ksi)	140 (20)

Thermal expansion coefficients of crystalline species that may be present in glass-ceramics

Species	Coefficient of thermal expansion, 10^{-7}/°C	Temperature range, RT to °C (°F)
βeucryptite ($Li_2O \cdot Al_2O_3 \cdot SiO_2$)	−86	700 (1290)
βspodumene ($Li_2O \cdot Al_2O_3 \cdot 4SiO_2$)	9	1000 (1830)
Lithium disilicate ($Li_2O \cdot 2SiO_2$)	110	600 (1110)
Quartz (SiO_2)	132	300 (570)
	237	600 (1110)
Cristobalite (SiO_2)	125	100 (212)
	500	300 (570)
Tridymite (SiO_2)	175	100 (212)
	250	200 (390)

Properties of glass-ceramic compared to those of marble and granite used as building materials

Material	Mechanical properties: Strength: Bending, kg/cm²	Strength: Compressive, ton/cm²	Charpy shock, kg · cm/cm²(a)	Elastic modulus, kg/cm²	Thermal properties: Specific heat (at 50 °C, or 120 °F), cal/°C	Coefficient of thermal expansion 30 to 380 °C (85 to 715 °F), 10^{-6}/°C	Coefficient of thermal expansion, 10^{-6}/°F	Thermal conductivity W/m · K	Thermal conductivity Kcal/m · h · °C	Water absorption rate, %	Physical properties: Hardness, Mohs scale	Density g/cm³	Density lb/in.²	Chemical properties: Acid resistance 1% H_2SO_4(b)	Alkali resistance 1% NaOH(b)	Freeze resistance(c)
Glass-ceramic	510	1.2 to 5.6	2.5	5.2	0.19	6.2	3.3	1.6	1.4	0.00	6.5	2.7	0.098	0.08	0.05	0.028
Marble	170	0.9 to 2.3	2.1	2.8 to 8.4	0.19	8.0 to 26.0	4.4 to 14.5	2.2 to 2.3	1.9 to 2.0	0.30	3 to 5	2.7	0.098	10.3	0.30	0.23
Granite	150	0.6 to 3.0	2.0	4.3 to 6.1	0.19	5.0 to 15.0	2.8 to 8.3	2.1 to 2.4		0.35	5.5	2.6	0.094	1.0	0.10	0.25

(a) Energy needed for rupture by instantaneous load. (b) Weight loss of test piece 15 × 15 × 10 mm (0.59 × 0.59 × 0.39 in.) after 650 h immersion of 25 °C (75 °F). (c) Weight loss of testpiece 15 × 15 × 10 mm (0.59 × 0.59 × 0.39 in.) after 25 cycles: immersion of testpiece in water of 25 °C (75 °F) for 2 days exposure for 4 h in a temperature of –20 °C (–5 °F)

Base compositions and applications of transparent glass-ceramics based on βquartz solid solutions

Material	Composition, wt%: SiO_2	Al_2O_3	MgO	Na_2O	K_2O	ZnO	Fe_2O_3	Li_2O	BaO	P_2O_5	F	TiO_2	ZrO_2	As_2O_3	Commercial applications
Vision(a)	68.8	19.2	1.8	0.2	0.1	1.0	0.1	2.7	0.8	…	…	2.7	1.8	0.8	Transparent cookware
Zerodur(b)	55.5	25.3	1.0	0.5	…	1.4	0.03	3.7	…	7.9	…	2.3	1.9	0.5	Telescope mirrors
Ceran(b)	63.4	22.7	(d)	0.7	(d)	1.3	(d)	3.3	2.2	(d)	(d)	2.7	1.5	(d)	Black infrared transmission cooktop
Narumi(c)	65.1	22.6	0.5	0.6	0.3	…	0.03	4.2	…	1.2	0.1	2.0	2.3	1.1	Rangetops; stove windows

(a) Corning. (b) Schott. (c) Nippon Electric. (d) No data available

Primary phase βquartz-based glass-ceramics

Product type	Manufacturer	Coefficient of thermal expansion, 10^{-6}/K	Modulus of rupture, MPa (ksi)	Composition, wt%: SiO_2	Al_2O_3	Li_2O	Na_2O/K_2O	MgO/CaO	BaO/ZnO	As_2O_5	P_2O_5	TiO_2/ZrO_2	Other
Commercial													
Vision	Corning	…	…	68.8	19.2	2.7	0.2/0.1	1.8/…	0.8/1.0	0.8	…	2.7/1.8	0.1 Fe_2O_3, 50 ppm CoO 50 ppm Cr_2O_3
Zerodur	Schott	0.03	…	56.5	25.3	3.7	0.5/…	1.0/…	…/1.4	0.5	7.0	2.3/1.9	0.03 Fe_2O_3
Narum 1	Nippon Electric and Glass	…	…	65.1	22.6	4.2	0.6/0.3	0.5/…	…	1.1	1.2	2.0/2.3	0.03 Fe_2O_3
9623 (T1)	Corning	0 to 0.5	…	65	23	3.8	…	1.8/…	…/1.5	0.9	…	2.0/2.0	…
Noncommercial													…
T2	…	1.55	62 (9)	74	19.5	2	…	4.5/…	…	…	…	…/4	…
T3	…	3.1	83 (12)	65	25	…	…	10/…	…/6	1.0	…	…/10	…
T4	…	1.5	69 (10)	74	16.5	…	…	3.5/…	…/6	…	…	…/4	…

Note: All products are metastable phase and convert to βspodumene if heated above 900 to 1000 °C (1650 to 1830 °F)

Compositions, properties, and applications of glass-ceramics based on βspodumene (keatite) solid solutions

Material	Composition, wt%											Mechanical properties					Thermal properties Coefficient of thermal expansion, 10^{-6}/K		Commercial applications
	SiO_2	Al_2O_3	MgO	Na_2O	K_2O	ZnO	Fe_2O_3	Li_2O	TiO_2	ZrO_2	As_2O_3	Modulus of rupture MPa	Modulus of rupture ksi	Elastic modulus GPa	Elastic modulus 10^6 psi	Hardness, HK_{100}	0 to 500 °C (32 to 930 °F)	0 to 1000 °C (32 to 1830 °F)	
Corningware(a) (code 9608)	69.7	17.8	2.6	0.4	0.2	1.0	0.1	2.8	4.7	0.1	0.6	100	15	81	12.8	660	1.2	...	Cookware; hot plate tops
Cercor(a) (code 9455)	72.5	22.5	...	...	...	...	...	5.0	...	...	...	...	...	...	...	...	...	0.5	Heat exchangers; regenerators

(a) Manufactured by Corning Glass Works

Primary phase βspodumene-based glass-ceramics

Commercial product type	Manufacturer	Density, g/cm³	Coefficient of thermal expansion, 10^{-6}/K	Composition, wt%									
				SiO_2	Al_2O_3	B_2O_3	Li_2O	Na_2O/K_2O	MgO/CaO	ZnO	As_2O_3/Sb_2O_3	Fe_2O_3	TiO_2/ZrO_2
C101 Cervit	Owens-Illinois	...	...	66.4	21.4	...	3.9	.../0.1	.../3.6	...	.../0.4	...	1.8/1.9
9608	Corning	2.5	0.4 to 2	69.7	17.1	...	2.5	0.4/0.1	2.8/...	...	.../0.2	...	4.8/0.1
9617	Corning	...	...	67.4	20.4	...	3.5	0.3/0.1	1.6/...	1.2	0.4/...	...	4.8/...(+0.2 F)
9455	Corning	...	...	71.8	22.9	...	5.1	...	...	...	...	...	0.1/...
0336	Corning	...	...	64.6	20.0	2.0	3.5	0.6/0.2	1.8/...	2.2	0.8/...	...	4.4/...
Corningware	Corning	...	...	69.7	17.8	...	2.8	0.4/0.2	2.6/...	1.0	0.6/...	0.1	4.7/0.1
Cercor	Corning	...	...	72.5	22.5	...	5.0	...	...	...	...	...	...

Noncommercial product type	Density, g/cm³	Coefficient of thermal expansion, 10^{-6}/K	Modulus of rupture MPa	Modulus of rupture ksi	Composition, %				
					SiO_2	Al_2O_3	Total alkali	Total alkaline earth	Nucleant
L1	2.44	5.2	255	37	(a)	10	14	...	P_2O_5
L2	2.47	4.5	230	33	(a)	10	12	...	P_2O_5
L3	2.51	4.5	160	23	(a)	10	10	2(b)	P_2O_5
L5	2.59	5.0	175	25	(a)	10	10	9	P_2O_5

(a) Balance at SiO_2 assumed. (b) BaO only

Composition, properties, and applications of a cordierite-based glass-ceramic

Composition, wt%(a):	
SiO_2	56.1
Al_2O_3	19.8
MgO	14.7
CaO	0.1
TiO_2	8.9
As_2O_3	0.3
Fe_2O_3	0.1
Properties:	
Crystalline phases	Cordierite Cristobalite Rutile Mg-dititanate
Coefficient of thermal expansion (0 to 700 °C, or 32 to 1290 °F), 10^{-6}/K	4.5
Modulus of rupture, MPa (ksi)	250 (36)
Fracture toughness, MPa$\sqrt{m}$ (ksi $\sqrt{in.}$)	2.2 (2.0)
Elastic modulus, GPa (10^6 psi)	120 (17.4)
Thermal conductivity, W/m · K (cal/cm · s · °C)	38 (0.09)
Hardness, HK_{100}	700
Dielectric constant at 8.6 GHz	5.5
Loss tangent at 8.6 GHz	0.0003
Softening temperature, °C (°F)	>1300 (>2370)
Commercial applications	Radomes

(a) Corning Glass Works, Code 9606

Compositions, properties, and applications, of sheet silicate (fluormica) glass-ceramics

Material	Composition, wt%								Mechanical properties					
									Modulus of rupture		Elastic modulus		Hardness,	Fracture energy,
	SiO_2	Al_2O_3	MgO	K_2O	F	ZrO_2	B_2O_3	CeO_2	MPa	ksi	GPa	10^6 psi	HK_{100}	$J \cdot m^2$
Macor(a) (9658)	47.2	16.7	14.5	9.5	6.3	...	8.5	...	100	15	65	9.4	250	8.2
Dicor(b)	56 to 64	0 to 2	15 to 20	12 to 18	4 to 9	0 to 5	...	0.05	152	22	...	...	362	...

Material	Thermal properties					Electrical properties	Optical properties	Commercial applications
	Coefficient of thermal expansion at 25 to 600 °C (77 to 1110 °F), 10^{-6}/K	Thermal conductivity		Maximum use temperature (unstressed)		Dielectric strength (dc, 25 °C, or 80 °F, 0.25 mm, or 0.010 in. thick), V/cm	Refractive index	
		W/m · K	cal/cm · s · °C	°C	°F			
Macor(a) (9658)	12.9	1.3 to 1.7	0.0030 to 0.0040	900	1650	1200	...	Machinable components
Dicor(b)	7.2	1.7	0.0040	...	...	...	1.52	Dental restorations

(a) Corning. (b) Dentsply

Primary phase lithium disilicate-based glass-ceramics

Product type	Manufac-turer	Density, g/cm³	Coefficient of thermal expansion, 10^{-6}/K	Modulus of rupture, MPa (ksi)	Composition, wt%											
					SiO_2	Al_2O_3	Li_2O	Na_2O/K_2O	Ag	Au	CeO_2	SnO_2	Sb_2O_3	Total alkali	Total alkaline earth	Nucleation
Commercial Fotoform	Corning	...	...	...	79.6	4.0	9.3	1.6/4.1	0.11	0.001	0.014	0.003	0.4	...	...	...
Noncommercial L4	...	2.40	9.3	380 (55)	(a)	5	...	...	...	...	...	...	...	15	5(BaO)	P_2O_5

(a) Balance SiO_2

Machinable glass-ceramics

Product type	Manu-facturer	Density, g/cm^3	Coefficient of thermal expansion, 10^{-6}/K	Modulus of rupture		Elastic modulus		Composition(a)									
				MPa	ksi	GPa	10^6 psi	SiO_2	Al_2O_3	$Al_2O_3 \cdot Y_2O_3$	CaO	B_2O_3	K_2O	MgO	CeO_2	F	ZrO_2
Commercial																	
Macor	Corning	2.52 to 2.63	6.3 to 9.7	60 to 102	8.7 to 14.8	~6.0	~8.7	47.2	16.7	...	...	8.5	9.5	14.5	...	6.3	...
Dicor	Dentsply	...	...	...	...	...	...	56 to 64	0 to 2	...	...	...	12 to 18	15 to 20	0.06	4 to 9	0 to 5
Noncommercial																	
CAYS-20	...	3.23	5.3	...	...	105	15.2	50	...	30	20	...	...	...	...	...	...
CAYS-25	...	3.14	5.7	...	...	103	14.9	50	...	25	25	...	...	...	...	...	...
CAYS-30	...	3.10	5.8	...	...	100	14.5	50	...	20	30	...	...	...	...	...	...

(a) Commercial compositions given in wt%, noncommercial compositions given in mol% and are approximate

Primary phase enstatite noncommercial glass-ceramics

Product type	Density, g/cm^3	Coefficient of thermal expansion, 10^{-6}/K	Modulus of rupture		Composition, wt%					Comments
			MPa	ksi	SiO_2	Al_2O_3	MgO	Li_2O	ZrO_2	
E1	...	6.8	193 ± 15	28 ± 2	58	5.4	21	0.9	10.7	Enstatite, βspodumene, and zirconia
E2	...	8.0	200 ± 15	29 ± 2	54	...	33	...	13.0	Enstatite, zircon, and zirconia
M5	3.18	8.2	750	109	(a)	22	22	...	(b)	(Clino)enstatite

(a) Balance SiO_2. (b) Nucleation

Primary phase willemite (Zn_2SiO_4) and other zinc-containing phases of noncommercial glass-ceramics

Product type	Density, g/cm^3	Coefficient of thermal expansion, 10^{-6}/K	Modulus of rupture		Composition, wt%						
			MPa	ksi	SiO_2	Al_2O_3	B_2O_3	BaO	ZnO	Nucleant	Other
Z1	3.46	5.0	125	18	(a)	5 to 30	...	15	30	TiO_2	...
Z2	3.79	6.0	...	...	(a)	5 to 30	...	0	57	P_2O_5, TiO_2	...
Z3	3.70	3.0	95	14	(a)	5 to 30	...	15	42	P_2O_5, TiO_2	...
Z4	3.72	3.1	...	...	(a)	5 to 30	...	20	37	P_2O_5, TiO_2	...
Z5	3.68	3.8	...	...	(a)	5 to 30	...	5	47	P_2O_5, TiO_2	...
ZB1	...	4.1	180	26	20	20	30	...	20	...	Zn aluminate, Zn borate
ZB2	...	4.3	130	19	40	20	10	...	30	...	Quartz
ZB3	...	4.8	150	22	30	17	15	...	30	...	5 CaO, 3 P_2O_5
ZB4	...	4.9	115	17	26	16	15	...	35	...	5 BaO, 3 P_2O_5
ZB5	...	3.7	75	11	30	17	15	...	36	...	4 ZrO_2

(a) Balance SiO_2

Fiber-reinforced (46% SiC fiber) noncommercial glass-ceramics

Product type	Primary phase	Major components	Minor components
LAS I	βspodumene	SiO_2-Al_2O_3-Li_2O-MgO	ZnO, ZrO_2, BaO
LAS II	βspodumene	LAS I + Nb_2O_5	ZnO, ZrO_2, BaO
LAS III	βspodumene	LAS I + Nb_2O_5	ZrO_2
MAS	Cordierite	SiO_2-Al_2O_3-MgO	BaO
BMAS	Barium osmullite	SiO_2-Al_2O_3-BaO-MgO	...
Ternary mullite	Mullite	SiO_2-Al_2O_3-BaO	...
Hexacelsian	Hexacelsian	SiO_2-Al_2O_3-BaO	...

Other glass-ceramics

Product type	Coefficient of thermal expansion, 10^{-6}/K	Composition, wt%												Comments
		SiO_2	Al_2O_3	BaO	ZrO_2	MgO	ZnO	Cs_2O	Y/Mg	S	Al	O	N	
Mullite														
T7	1.61	77	23	...	...	...	...	...	...	...	...	...	...	Transparent, heat treat 1000 °C (1830 °F)
T8	3.65	50	40	10	...	...	...	...	...	...	...	...	...	Transparent, heat treat 1000 °C (1830 °F)
Gahnite (ZrO_{2-t})														
T6	3.2	70	17	...	6	...	13	4	...	...	...	...	...	Transparent, heat treat 800 to 1000 °C (1470 to 1830 °F)
Spinel (ZrO_{2-t})														
T5	3.3	70	19	...	8	5	6	...	...	...	...	...	...	Transparent, heat treat 800 to 950 °C (1470 to 1740 °F)
Sialon														
YG2	...	...	...	...	...	...	...	...	15.2/...	14.6	8.7	54.6	6.9	(b)
Y3	...	...	...	...	...	...	...	...	15/...	15	10	45	15	(b)
E″	...	...	...	...	...	...	...	...	.../20	13.6	10	44.2	12.1	(c)
D	...	...	...	...	...	...	...	...	.../18.2	12	11.5	42.3	11.5	(c)

(a) Except for Sialon products, which are specified as at%. (b) 1050 °C (1920 °F) process temperature, Y_2SiAlO_5N phase. (c) 900 °C (1650 °F) process temperature, β″-Sialon

Glass-Metal Seals

Properties and compositions of glasses for glass-metal seals

Product type	Manufac-turer	Density, g/cm^3	Coefficient of thermal expansion, 10^{-6}/K	Modulus of rupture, MPa (ksi)	Composition, wt% SiO_2	Al_2O_3	Li_2O/ MgO	BaO/ ZnO	Na_2O/ K_2O	CaO	P_2O_5/ B_2O_3	Fe_2O_3	Comments
Commercial													
7583	Corning	6.0	8.4	...	...	...	...	...	...	...	...	...	...
CVIII	Owens-Illinois	5.9	7	41 to 98 (6 to 14)(a)	...	...	...	...	...	...	...	...	...
Noncommercial(b)													
Austenitic SS (4-5 Al)	...	...	...	...	(c)	...	...	...	...	...	...	...	Preoxidize metal
304L SS	...	...	~20	...	(c)	...	...	...	...	...	...	...	...
304L SS	...	...	~17.5	...	83	<0.9	5.3/...	...	.../4.9	...	2.9/2.6	...	Proprietary heat treat
304L SS	...	...	13.5 to 17.5	...	80	5.1	5.3/...	...	.../4.5	...	2.6/1.4	...	Coefficient of thermal expansion depends on heat treat
Molybdenum	...	...	...	...	47	9.75	...	4.75/32.5	3.75/...	...	2.2/...	...	...
Chromed steel	...	...	12.3 to 14.2	...	66	3.0	8.5/...	.../7.0	5.5	3.0	2.8/...	2.2	650 to 750 °C (1200 to 1380 °F) heat treat
Inconel 718	...	...	~14	...	67.1	2.8	23.7/...	...	.../2.8	...	1.0/2.6	...	$Li_2Si_2O_5$ type
Inconel 718	...	...	...	...	71.7	5.1	12.6/...	...	.../4.9	...	2.5/3.2	...	...

(a) Dependent on heat treatment, inversely proportional to crystallite size. (b) For sealing to metal. (c) Composition is LAS type. (d) Also recommended for Hastelloy C-276

Examples of glass-ceramic compositions for glass-metal seals

Compound/property	Sealing glass/composition, wt% S glass-ceramics(a)	GC1014(b)	GC1008(b)	SLM(c)
SiO_2	67.1	67.0	71.0	69.3
Li_2O	23.7	24.5	25.0	20.2
Al_2O_3	2.8	...	3.0	2.2
K_2O	2.8	4.8	...	...
MgO	...	...	...	6.8
B_2O_3	2.6	2.2	...	...
P_2O_5	1.0	1.0	1.0	1.5
Coefficient of thermal expansion, 10^{-7}/°C	100	177	185	192

(a) For sealing to Inconel 718, Hastelloy C276, and type 430 stainless steel. (b) For sealing to type 304 stainless steel. (c) For sealing to copper

Plastics

General

Test Methods and Standards

ASTM test methods

Number	Title
Mechanical testing	
D 638	Tensile Properties of Plastics
D 695	Compressive Properties of Rigid Plastics
D 2344	Apparent Horizontal Shear Strength of Reinforced Plastics by Short Beam Method
D 3039	Tensile Properties of Oriented Fiber Composites
D 3518	In-Plane Shear Stress-Strain Response of Unidirectional Reinforced Plastics
D 790	Flexural Properties of Plastics and Electrical Insulating Materials
D 3410	Test for Compressive Properties of Oriented Fiber Composites
Fatigue	
D 3479	Tension-Tension Fatigue of Oriented Fiber Resin Matrix Composites
D 671	Flexural Fatigue of Plastics by Constant-Amplitude-of-Force
Impact	
D 256	Impact Resistance of Plastics and Electrical Insulating Materials
D 1822	Tensile-Impact Energy to Break Plastics and Electrical Insulating Materials
D 3029	Impact Resistance of Rigid Plastic Sheeting or Parts by Means of Tup (Falling Weight)
Creep	
D 2990	Tensile, Compressive, and Flexural Creep and Creep-Rupture of Plastics
D 2991	Stress Relaxation of Plastics
Physical properties	
D 570	Water Absorption
D 792	Specific Gravity and Density of Plastics by Displacement
D 1505	Density of Plastics by the Density-Gradient Technique
D 2734	Void Content of Reinforced Plastics
D 3355	Fiber Content of Unidirectional Fiber/Polymer Composites
Thermal properties	
D 648	Deflection Temperature of Plastics under Flexural Load
D 696	Coefficient of Linear Thermal Expansion of Plastics
E 228	Linear Thermal Expansion of Rigid Solids with a Vitreous Silica Dilatometer
C 117	Steady-State Thermal Transmission Properties by Means of the Guarded Hot Plate
Electrical properties	
D 149	Dielectric Breakdown Voltage and Dielectric Strength of Electrical Insulating Materials at Commercial Power Frequencies
D 150	A-C Loss Characteristics and Permittivity (Dielectric Constant) of Solid Electrical Insulating Materials
Wear resistance	
D 673	Mar Resistance of Plastics
D 1242	Resistance of Plastic Materials to Abrasion
Chemical resistance	
C 581	Chemical Resistance of Thermosetting Resins used in Glass Fiber Reinforced Structures
D 543	Resistance of Plastics to Chemical Reagents

ASTM D 4000-type material specifications

Material	ASTM test standard
Phenolic	D 4617
Polyamide (nylon)	D 4066
Polycarbonate	D 3935
Polyoxymethylene (acetal)	D 4181
Polyphenylene sulfide	D 4067
Polypropylene	D 4101
Polystyrene	D 4549
Styrene-acrylonitrile	D 4203
Thermoplastic elastomer, ether-ester	D 4550
Thermoplastic polyester (general)	D 4507
Styrene-maleic anhydride	D 4634
Thermoplastic elastomer-styrenic	D 4774
Acrylonitrile-butadiene-styrene	D 4673

Typical tests for electrical properties of plastics

Test	Description
Insulating properties	
Volume resistivity	A measure of the tendency to pass leakage current through the bulk of a material
Surface resistivity	A measure of the tendency to pass leakage current across the plastic part surface
Dielectric properties	
Dielectric strength	A measure of resistance to electrical breakdown with exposure to high voltage
Dielectric constant	A measure of electrical energy stored in a dielectric material
Dielectric loss factor	A measure of energy lost in an insulating material when subjected to an alternating field
Flammability properties	
Hot-wire ignition	A measure of resistance to ignition with exposure to high temperatures induced by excessive current flow in a wire
High-current arc ignition	A measure of resistance to ignition with exposure to high-current low-voltage electrical arcing

UL standards for plastics

Number	Title
UL 94	Tests for Flammability of Plastic Materials for Parts in Devices and Applications
UL 746A	Polymeric Materials—Short Term Property Evaluations
UL 746B	Polymeric Materials—Long Term Property Evaluations
UL 746C	Polymeric Materials—Use in Electrical Equipment Evaluations
UL 746D	Polymeric Materials—Fabricated Parts

UL 94 flame classes

Type	Rating	Requirements: Thickness	Maximum burning rate	Maximum individual flame time, s	Maximum total flame time, s	Glowing time, s	Flaming drips allowed
Horizontal flame class	HB	<3.2 mm (1/8 in.)	75 mm/min.	...	...	...	...
		>3.2 mm (1/8 in.)	38 mm/min.	...	...	...	...
Vertical flame class							
Flame applied twice for 10 s	V-0	...	...	10	50	30	No
	V-1	...	...	30	250	60	No
	V-2	...	...	30	250	60	Yes
Flame applied for 5 s on, 5 s off, five times	5V	...	...	...	...	60(a)	No

(a) Maximum flame plus glowing time. For a 5V rating, if any shrinkage, melting, or elongation occurs, the tests must be repeated on 150 × 150 mm (6 × 6 in.) plaque samples

Military specifications for thermoplastic and thermoset resins

Material	Specification number
Thermoplastic	
Polysulfone	MIL-P-46120B
Polyamide-imide	MIL-P-46179A
Polyetheretherketone	MIL-P-46183
Polyether-imide	MIL-P-46184
Polyether sulfone	MIL-P-46185
Thermoset	
Resin, polyester, low-pressure laminating	MIL-R-7575C
Resin, phenolic, laminating	MIL-R-9299C
Resin, epoxy, low-pressure laminating	MIL-R-9300B
Resin solution, silicone, low-pressure laminating	MIL-R-25506C
Resin, polyimide, heat resistant, laminating	MIL-R-83330

Typical Aerospace Material Specifications for plastics

Number	Title
AMS 3628C	Plastic Extrusion and Moldings, Polycarbonate, General Purpose
AMS 3646B	Polychlorotrifluoroethylene (PCTFE) Sheet, Molded, Unplasticized
AMS 3656D	Polytetrafluoroethylene Extrusions, Normal Strength, as Sintered, Radiographically Inspected
AMS 3684A	Resin, Polyimide, Sealing-High Temperature Resistant, 315 °C, or 600 °F, Unfilled
AMS 3709A	Syntactic Foam Tiles
AMS 3756	Polytetrafluoroethylene Moldings, Glass Fiber Filled 75 PTFE Resin, 25 Glass, as Sintered

Chemistries

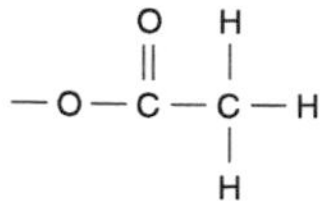

Acetate group

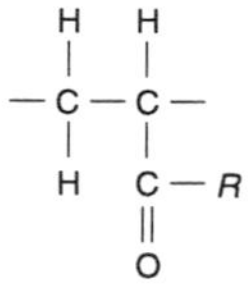

Acrylic group

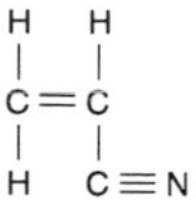

Acrylonitrile

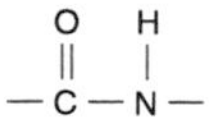

Amide group

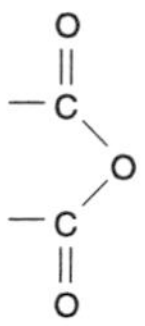

Anhydride group

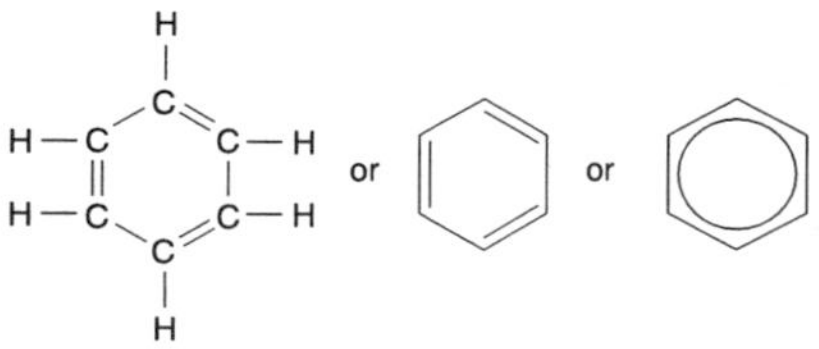

Benzene

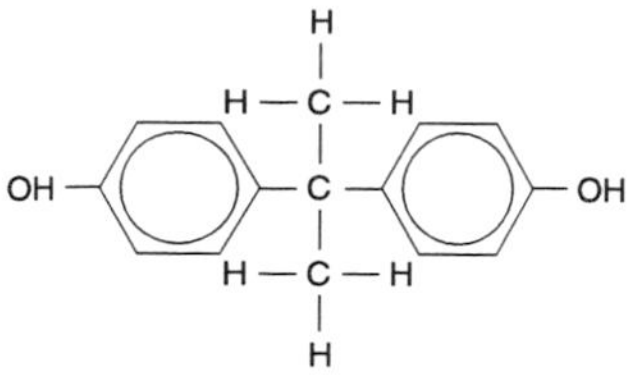

Bisphenol A

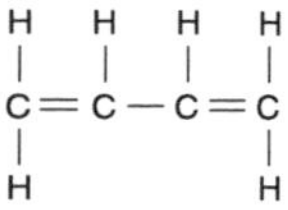

Butadiene

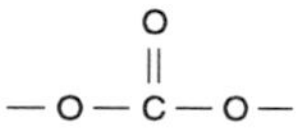

Carbonate group

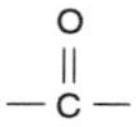

Carbonyl group

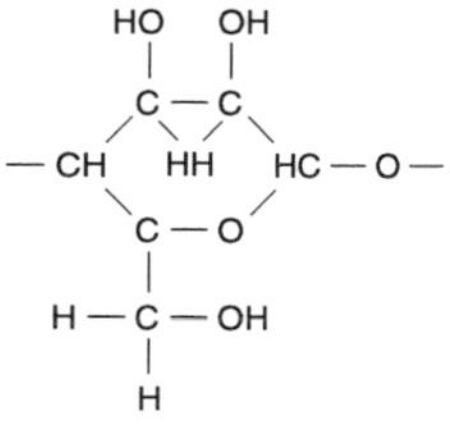

Cellulose

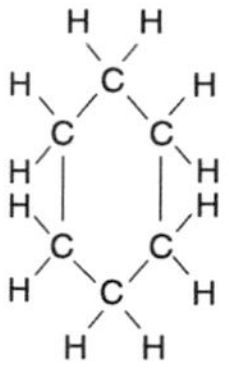

Cyclohexane

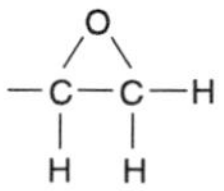

Epoxide or epoxy group

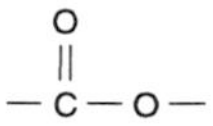

Ester group

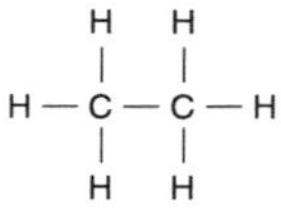

Ethane

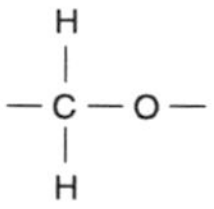

Ether group

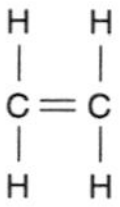

Ethylene

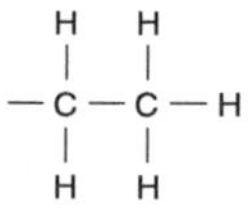

Ethyl group

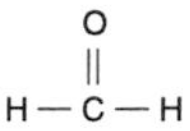

Formaldehyde

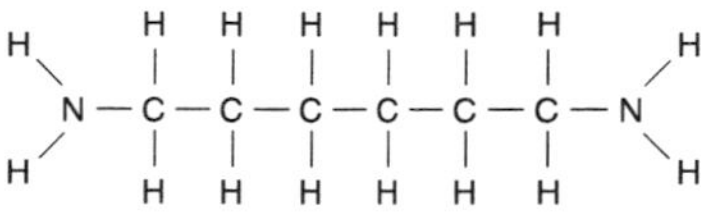

Hexamethylenediamine

—OH

Hydroxyl group

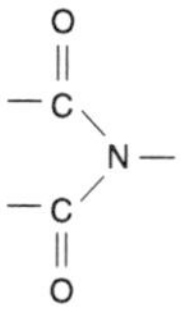

Imide group

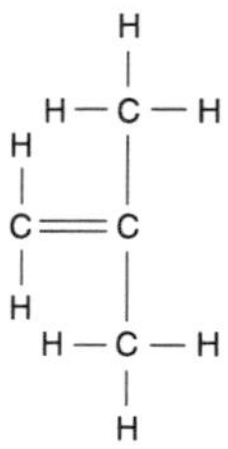

Isobutylene

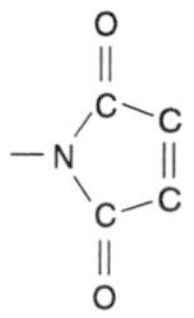

Maleimide group

Some chemical groups involved in the naming of polymers
Acetate group to methane

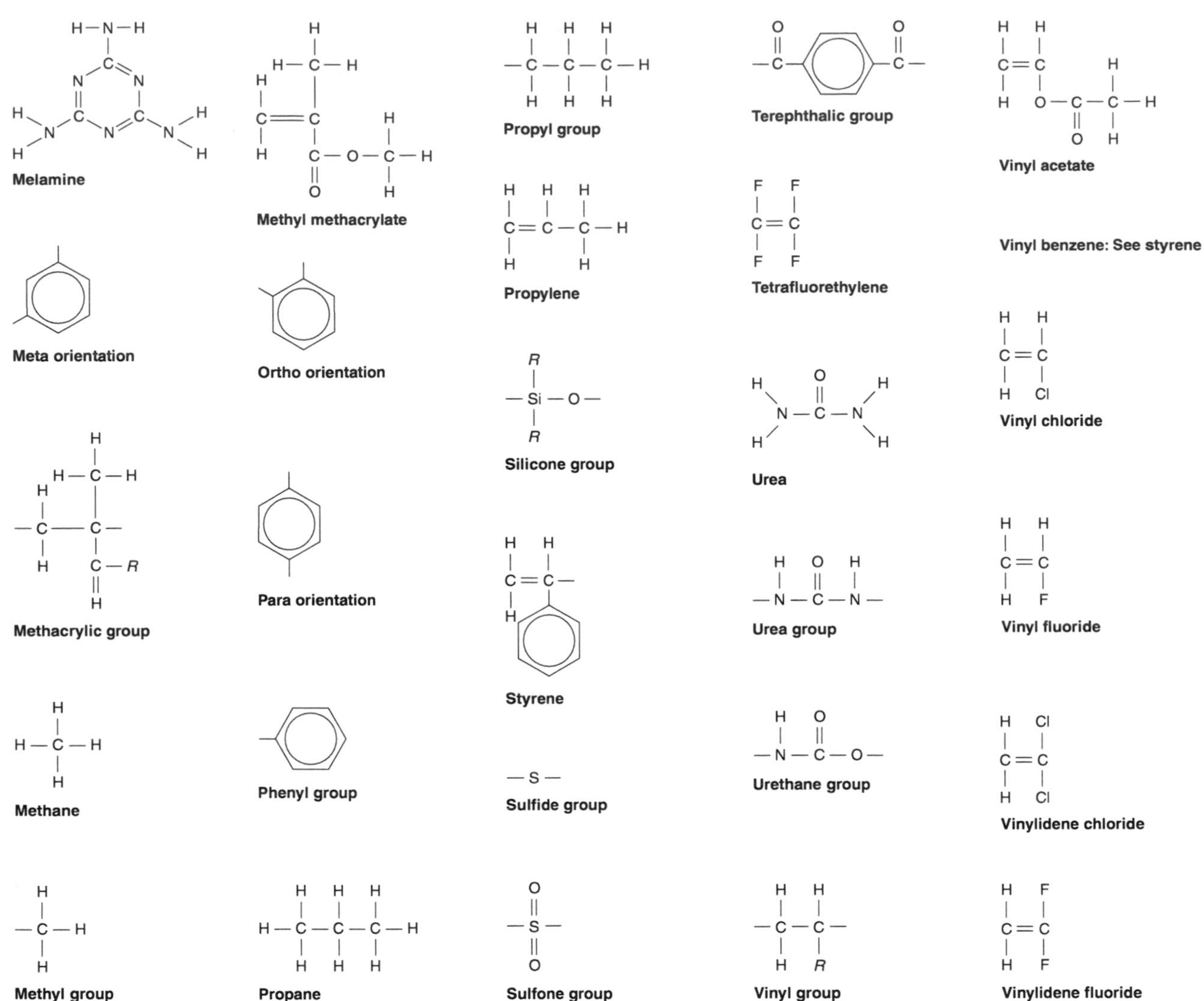

Some chemical groups involved in the naming of polymers (continued)
Acetate group to methane

Bond energies for common bonds in polymers

Bond	Bond energy kJ/mol	Bond energy kcal/g · mol
C – C	350	83
C – H	410	99
C – F	440	105
C – Cl	330	79
C – O	350	84
C – S	260	62
C – N	290	70
N – N	160	38
N – H	390	93
O – H	460	111
C = C	200	147
C = C	810	194
C = O	715	171
C = N	615	147

Applications

Common thermoplastic and thermoset molding compounds

Material	Principal applications	Potential subprocess problem areas(a): Feeding	Transporting	Injecting	Flowing
Thermoplastics					
ABS	Furniture, cabinets, containers, trim	B	H I	K L N O	Q
Acetal	Clock gears, miniature engineered parts	B	F H I J	O	P S
Acrylic	Automobile light lenses, plastic glazing	B	H I	K L N O	Q
Cellulose	Esters trim, moldings, screwdrivers	B	H I	K L N O	Q R
Polycarbonate	Auto bumpers, traffic lights, lenses	B C	F H I J	K L N O	Q
Polyester	Appliance parts, pump and electrical housings	B C	H I J	O	P S
Polyethylene	Houseware, food storage, dunnage	B	I	O	P
Fluoroplastics	Corrosion/solvent-resistant parts	B C	F I J	K L N O	Q
Polyimide	Aerospace items, electrical insulators	B C	H I J	K L N O	Q
Ionomer	Bumper rub strips, golf ball covers	B	H I	K L N O	P
Nylon	Auto parts, bearing retainers, appliances	B C	H I J	O	P S
Polyphenylene oxide and alloys	Auto instrument panels	B	H I (J)	K L N O	Q R
Polypropylene	Battery cases, auto parts, containers	B	H I	O	P S
Polystyrene	Toys, advertising displays, picture frames	B	H I	K L N O	Q
Polysulfone	Camera cases, aircraft parts, connectors	B C	H I J	K L N O	Q
Polyvinyl chloride	Soft steering wheels, trim items	B	F H I J	K L N O	Q R
Thermosets					
Alkyd	Switches, motor housings, pot/pan handles	A B	G H I J	K L M N O	Q R S T
Allyl	Electrical connectors, circuit boards	A B	G H I J	K L M N O	Q R S T
Epoxy	Electrical insulators, electronic cases	B E	G H I J	K L M N O	Q R S T
Polyester	Automotive structural parts	D E	G I J	K L M N O	P Q R S T
Polyimide	Aircraft components, aerospace parts	A D E	G H I J	K L M N O	P Q R S T
Melamine	Dinnerware, microwave cookware	B	G H I J	K L M N O	Q R S T
Phenolic	Distributor caps, plastic ash trays	A B	G H I J	K L M N O	Q R S T
Urethane	Automotive body panels, bumpers	A D E	G I	L M O	Q R T
Vinylester	Composite car/truck springs, wheels	B D E	G I J	K L M N O	P Q R S T

(a) Feeding:
- A–Moisture may chemically react to degrade the polymer base
- B–Drying is recommended to avoid splay in molded product
- C–Drying is essential to prevent molecular weight attrition
- D–Drying may volatilize monomers essential to the curing reaction
- E–Fiber reinforcement breakage may occur during force feeding

Transporting:
- F–Overheating may cause explosive depolymerization
- G–Overheating may cause premature curing of (thermoset) compound
- H–Venting is recommended to remove volatiles and reduce splay
- I–Fiber reinforcement lengths may be severely reduced
- J–Overheating may produce chemical changes in the base polymer

Injecting:
- K–Fast injection of the molding compound can lead to serious overheating
- L–Filled or reinforced compounds will exhibit much higher viscosity
- M–Open runners required as material can cure in closed channels
- N–Melt fracture may occur with high injection speeds
- O–Fiber-filler orientation will occur if molding compound is reinforced

Flowing:
- P–Major sink marks may develop if part sections are thick, not uniform
- Q–Weak knitlines may develop if compound packing pressure is low
- R–Large mold vents recommended to allow volatiles to escape
- S–Hot molds required to promote cure, crystalline growth
- T–Curing reaction may produce peak exotherm leading to degradation

Application areas generally associated with generic resin types, based on performance characteristics

Resin type	Applications	Typical neat resin properties
Polyester	Consumer products, tanks, pipes, pressure vessels, automotive structures	Tensile strength of 3.4 to 90 MPa (0.5 to 13 ksi); compressive strength of 90 to 210 MPa (13 to 30 ksi); up to 120 °C (250 °F) continuous use; low viscosity; fast reaction; can be catalyzed; high shrinkage
Vinyl ester	Consumer products, pipes, ducts, stacks, automotive structures, flooring, linings	Tensile strength of 60 to 90 MPa (9 to 13 ksi); elongation of 2 to 6%; up to 120 °C (250 °F) continuous use; low viscosity; fast reaction; can be catalyzed; intermediate shrinkage
Polybutadiene	Resin modifiers, coatings, adhesives, potting compounds	Good chemical resistance; up to 120 °C (250 °F) continuous use; high viscosity; fast reaction; can be catalyzed; low moisture pick-up
Epoxy	Adhesives, tooling, electronics, aerospace and automotive structures	Tensile strength of 55 to 130 MPa (8 to 19 ksi); excellent chemical resistance; up to 175 °C (350 °F) continuous use; high viscosity; can be catalyzed; intermediate reaction; low shrinkage
Polyimide	Primary and secondary aerospace structures in high-temperature areas, electronics	Tensile strength of 55 to 120 MPa (8 to 17 ksi); up to 315 °C (600 °F) continuous use; high viscosity; can be catalyzed; slow reaction; reaction by-products; microcracking
Bismaleimide	Similar to polyimide	Similar to polyimide, except that continuous use only up to 230 °C (450 °F); no reaction by-product
Low-performance thermoplastic	Automotive panels, appliance housings, gears, bearings, fixtures, consumer products	Amorphous or semicrystalline; high toughness; up to 120 °C (250 °F) continuous use; high processing temperatures and pressures; high viscosity

Selection of resins used with glass fiber reinforcements

Type of resin	Properties	Processes
Thermoset		
Polyesters(a)	Simplest, most versatile, most economical, and most widely used family of resin, having good electrical properties and good chemical resistance (especially to acids)	Compression molding; filament winding; hand lay-up; mat molding; pressure bag molding; continuous pultrusion; injection molding; spray-up; centrifugal casting; cold molding; comoform; encapsulation
Epoxies	Excellent mechanical properties, dimensional stability, and chemical resistance (especially to alkalis); low water absorption; self-extinguishing (when halogenated); low shrinkage; good abrasion resistance; very good adhesion properties	Compression molding; filament winding; hand lay-up; continuous pultrusion; encapsulation; centrifugal casting
Phenolics	Good acid resistance; good electrical properties (except arc resistance); high heat resistance	Compression molding; continuous laminating
Silicones	Highest heat resistance; low water absorption; excellent dielectric properties; high arc resistance	Compression molding; injection molding; encapsulation
Melamines	Good heat resistance; high impact strength	Compression molding
Diallyl phthalate	Good electrical insulation; low water absorption	Compression molding
Thermoplastic		
Polystyrene	Low cost; moderate heat distortion; good dimensional stability; good stiffness; good impact strength	Injection molding; continuous laminating
Nylon	High heat distortion; low water absorption; low elongation, good impact strength; good tensile and flexural strength	Injection molding; blow molding; rotational molding
Polycarbonate	Self-extinguishing; high dielectric strength; high mechanical properties	Injection molding
Styrene-acrylo-nitrile	Good solvent resistance; good long-term strength; good appearance	Injection molding
Acrylics	Good gloss, weather resistance, optical clarity, and color; excellent electrical properties	Injection molding; vacuum forming; compression molding; continuous laminating
Vinyls	Excellent weatherability; superior electrical properties; excellent moisture and chemical resistance; self-extinguishing	Injection molding; continuous laminating; rotational molding
Acetals	Very high tensile strength and stiffness; exceptional dimensional stability; high chemical and abrasion resistance; no known room-temperature solvent	Injection molding
Polyethylene	Good toughness; light weight; low cost; good flexibility; good chemical resistance; can be "welded"	Injection molding; rotational molding; blow molding
Fluorocarbons	Very high heat and chemical resistance; nonburning; lowest coefficient of friction; high dimensional stability	Injection molding; encapsulation; continuous pultrusion
Polyphenylene oxide modified	Very tough engineering plastic; superior dimensional stability; low moisture absorption; excellent chemical resistance	Injection molding
Polypropylene	Excellent resistance to stress or flex cracking; very light weight; hard, scratch-resistant surface; can be electroplated; good chemical and heat resistance; exceptional impact strength; good optical qualities	Injection molding; continuous laminating; rotational molding
Polysulfone	Good transparency; high mechanical properties, heat resistance, and electrical properties at high temperatures; can be electroplated	Injection molding

Advantages and disadvantages of thermoset and thermoplastic materials as housings for electronic assemblies

Housing or container material	Advantages	Disadvantages
Molded thermosets (epoxy, alkyd, phenolic, diallyl phthalate, etc.)	Many standard sizes available; good insulators; corrosion resistant; color or identification can be molded in; terminals can sometimes be molded in; cut off of resin-filled shell easier than for metal cans; same type of material can be used for shell and filling resin, resulting in good compatibility	Does not always adhere to resin, especially if silicone mold releases used to make shell; sealing of leakage joints can be difficult; physically weaker than steel, especially in thin sections; molding flash can cause fitting problems; cleaning of resin spillage can break shells
Molded thermoplastics (nylon, polyethylene, polystyrene, etc.)	Same as listed for thermosets except last two items; often less prone to cracking than thermosetting shells, although this depends on resiliency of material	Same as first three items listed for thermosets; adhesion can be poor due to excellent release characteristics of most thermoplastics; shell can distort from heat; cut off can be a problem, due to melting or softening of thermoplastics under mechanically generated heat

Advantages and disadvantages of various plastics as mold materials

Mold material	Advantages	Disadvantages
Cast epoxy	Good dimensional control; surface can be polished; can be made for inserts and multiple-part molds: long life and low maintenance	Dimensional control not quite as good as for machined-metal molds; requires mold release and cleaning; low thermal conductivity compared with metals
Cast plastisols	Parts easily removed from molds: molds are easy to make	Short useful life; poor dimensional control
Cast RTV silicone rubber	Same as for plastisols; better life than plastisols	Poor dimensional control, though better than plastisols
Machined TFE fluorocarbon	No mold release required; convenient to make for short runs and simple shapes; withstands high temperature cures	Poor dimensional control
Machined polyethylene and polypropylene	Same as listed for TFE fluorocarbon except high-temperature capability and lower cost	Poor dimensional control

General characteristics of amorphous and (partially) crystalline plastics

Material	Characteristics
Amorphous plastics	Transparent
	Low mold shrinkage
	Low and uniform coefficient of thermal expansion
	Low dependence of properties on temperature
Crystalline plastics	Resistance to organic solvents
	Good dynamic fatigue strength
	Load-bearing temperature range markedly increased by inorganic fiber reinforcement
	Enhancement of strength by orientation possible

Physical and Mechanical Properties

Short-time dielectric strength properties of selected thermoplastic and thermosetting plastics per ASTM D 149

Material	High-low V/cm	High-low V/mil	Material	High-low V/cm	High-low V/mil
PVC, chlorinated	590 to 480	1500 to 1220	PS-coacrylonitrile	200 to 160	510 to 410
PVC	590 to 160	1500 to 410	PMMA, molding grade	200 to 160	510 to 410
Ionomers	430 to 350	1100 to 890	Epoxy cast resins	200 to 160	510 to 410
PE, low density	390 to 180	990 to 460	PTFE	200 to 160	510 to 410
PE, medium density	390 to 180	990 to 460	Polyester, rigid cast	200 to 150	510 to 380
PE-copropylene (poly allomer)	370 to 320	940 to 810	ABS, high impact	200 to 140	510 to 360
PE-covinyl acetate	310 to 240	790 to 610	Ethyl cellulose	200 to 140	510 to 360
PS, general purpose	280 to 200	710 to 510	PC-ABS alloy	200 to 140	510 to 360
Polypropylene (PP)	260 to 200	660 to 510	ABS, medium impact	200 to 140	510 to 360
PVC, ABS modified	240	610	Polyvinyl formal	200	510
Polyphenylene sulfide (PPS)	240	610	Poly αmethylstyrene-comethyl methacrylate	180	460
Nylon 6/6	240	610	Nylon 12	180	460
Polychlorotrifluoroethylene	240 to 200	610 to 510	CP	180 to 120	460 to 300
PTFE-cohexafluoroethylene	240 to 200	610 to 510	PAE	170	430
PS, high heat	240 to 160	610 to 410	Nylon 11	170	430
PVDC	240 to 160	610 to 410	PS	170	430
PS, high impact	240 to 120	610 to 300	PMMA/PVC alloy	160	410
CN	240 to 120	610 to 300	Nylon 6	160	410
CA	240 to 100	610 to 250	PC	160	410
PI	220	560	PF resin	160 to 120	410 to 300
Silicone (SI), cast resin	220	560	CAB	160 to 100	410 to 250
PE-coethyl acrylate	220 to 180	560 to 460	POM	150	380
PPO	220 to 160	560 to 410	PVB	140	360
PS-cobutadiene	200 to 170	510 to 430	Melamine-phenol resin	130 to 90	330 to 230
POM copolymer	200	510	PVDF	100	250
PE, high density	200 to 190	510 to 480			

Volume resistivity of selected thermoplastic and thermosetting plastics per ASTM D 257

Material	High, Ω · m	Low, Ω · m	Material	High, Ω · m	Low, Ω · m
PTFE	10^{17}	...	PVC-copolypropylene	$>10^{13}$	...
PE	5.0×10^{16}	10^{15}	PE-copolypropylene (poly allomer)	$>10^{13}$	...
PTFE-co-hexafluoropropylene	$>2.0 \times 10^{16}$	...	Polyester, rigid cast	10^{13}	...
PCTFE	1.2×10^{16}	...	PE-covinyl acetate	10^{13}	...
PPO	10^{15}	...	POM	10^{13}	...
PI	$>10^{14}$(a)	...	Poly αmethylstyrene-comethyl methacrylate	10^{13}	...
PP	$>10^{14}$(a)	...	PVC, chlorinated	10^{13}	...
Ionomers	$>10^{14}$(a)	...	Nylon 6/6	10^{13}	10^{12}
PS, general purpose	$>10^{14}$(a)	...	SI, cast resin	10^{13}	10^{12}
PS, high impact	$>10^{14}$(a)	...	PS, high heat	10^{13}	10^{11}
PS-coacrylonitrile	$>10^{14}$(a)	...	Nylon 6	10^{13}	10^{11}
PVC	$>10^{14}$(a)	...	Epoxy cast resins	10^{13}	10^{10}
PSU	5.0×10^{14}	...	CAB	10^{13}	10^{8}
ABS, high impact	4.8×10^{14}	10^{14}	PMMA, molding grade	$>10^{12}$	...
PC-ABS alloy	4.0×10^{14}	2.2×10^{14}	PVDF	2.0×10^{12}	...
ABS, medium impact	2.7×10^{14}	...	POM, copolymer	10^{12}	...
PS-cobutadiene	2.5×10^{14}	5.0×10^{11}	Ethyl cellulose	10^{12}	10^{10}
PC	2.1×10^{14}	...	CA	10^{12}	10^{8}
PAE	1.5×10^{14}	...	Nylon 11	10^{11}	...
CN	1.5×10^{14}	...	Nylon 12	10^{11}	...
CP	10^{14}	10^{10}	Cycloaliphatic epoxy resin	10^{11}	...
PE-coethyl acrylate	10^{14}	...	Polyurethane (PUR), thermoplastic elastomer	1.1×10^{10}	2.0×10^{8}
Polymethyl-1-pentene	10^{14}	...	PF resin	10^{10}	10^{8}
PVDC	10^{14}	10^{12}	PVB	5.0×10^{7}	...

(a) Above sensitivity of instrument

Dissipation factor of selected thermoplastic and thermosetting plastics per ASTM D 150 at 60 Hz

Material	Low	High	Material	Low	High
Polymethyl-1-pentene	0.00007	...	Cycloaliphatic epoxy resin	0.005	...
PS, general purpose	0.001	0.0006	Ethyl cellulose	0.005	0.020
PTFE	0.0002	...	Poly α-methylstyrene-comethyl methacrylate	0.006	...
PTFE-cohexafluoropropylene	<0.0003(a)	...	PAE	0.006	...
PPO	0.00035	...	PS-coacrylonitrile	0.006	0.008
PS, high impact	0.0004	0.002	PVC	0.007	0.020
PP	<0.0005(a)	...	PVC-copolypropylene	0.008	0.010
PE, low density	<0.0005(a)	...	Nylon 6/6	0.01	...
PE, medium density	<0.0005(a)	...	PE-coethyl acrylate	0.01	0.02
PE, high density	<0.0005(a)	...	CP	0.01	0.04
PE-copropylene (poly allomer)	<0.0005(a)	...	CAB	0.01	0.04
PS, high heat	0.0005	0.003	CA	0.01	0.06
PSU	0.0008	...	Polyvinyl formal	0.013	...
PC	0.0009	...	PUR, thermoplastic elastomer	0.015	0.048
Ionomers	0.001	0.003	PVC, chlorinated	0.019	0.021
SI, cast resin	0.001	0.025	Nylon 11	0.03	...
PCTFE	0.0012	...	PVDC	0.03	0.045
PS-cobutadiene	0.002	0.003	Nylon 6	0.03	0.07
Epoxy cast resins	0.002	0.010	PMMA/PVC, alloy	0.04	...
Polyester, rigid cast resin	0.003	0.040	Nylon 12	0.04	...
PC-ABS alloy	0.003	0.007	PMMA, molding grade	0.04	0.06
ABS, high impact	0.003	0.008	MF resin	0.048	0.162
ABS, medium impact	0.003	0.008	PVDF	0.049	...
PE-covinyl acetate	0.003	0.02	PF resin	0.06	0.10
POM	0.0048	...	CN	0.09	0.12
POM, copolymer	0.005	...	PVB	0.115	...

(a) Below sensitivity of instrument

Electrical properties of selected thermoplastics and thermosets

Material	Dielectric strength, MV/m	Dielectric strength, V/mil	Dielectric constant at 60 Hz	Arc resistance, s	Critical tracking index, V
ABS	15.7	400	3.0	89	400+
ABS-PC	17.1	434	2.9	91	250+
DAP	15.4	390	4.0	150	600+
POM	19.7	500	3.7	220	600+
PMMA	19.7	500	3.7	...	...
PAR	15.2	385	3.1	78	200
LCP	20.1	510	4.6	192	175+
MF	11.2	285	8.9	180	600+
Nylon 6	16.5	420	3.8	60	600+
Nylon 6/6	23.6	600	4.0	60	600+
Nylon 12	16.1	410	3.5	120	600+
PAE	20.1	510	3.8	125	175+
PBT	15.7	400	3.3	184	600+
PC	15.0	380	3.2	10	100+
PBT-PC	17.7	450	3.4	99	260
PEEK	...	...	...	...	...
PEI	...	...	3.2	126	100+
PESV	16.1	410	...	20	150
PET	21.3	540	3.6	125	250+
PF	11.0	280	...	190	100+
PPO (modified)	15.7	400	3.9	34	400+
PPS	17.7	450	3.1	60	130
PSU	16.7	425	3.5	120	100+
SMA	18.9	480	...	...	600+
UP	14.8	375	2.7	...	600+

Dielectric constant of selected thermoplastic and thermosetting plastics per ASTM D 150 at 60 Hz

Material	Low	High	Material	Low	High
PTFE	2.0	...	PSU	3.14	...
Polytetrafluoropropylene-cohexafluoropropylene	2.1	...	Cycloaliphatic epoxy resin	3.2	...
Polymethyl-1-pentene	2.12	...	PVC	3.2	3.6
PP	2.2	...	PVC, chlorinated	3.3	3.8
PCTFE	2.24	...	PMMA, molding grade	3.3	3.9
PE, low density	2.25	2.35	PI	3.4	...

(continued)

Dielectric constant of selected thermoplastic and thermosetting plastics per ASTM D 150 at 60 Hz (continued)

Material	Low	High	Material	Low	High
PE, medium density	2.25	2.35	Poly α-methylstyrene-comethyl methacrylate	3.4	...
PE-copropylene (poly allomer)	2.3	...	Epoxy cast resins	3.5	5.0
PE, high density	2.30	2.35	CAB	3.5	6.4
Ionomers	2.4	2.5	CA	3.5	7.5
PC-ABS alloy	2.4	4.5	POM	3.7	...
ABS, high impact	2.4	5.0	POM, copolymer	3.7	...
ABS, medium impact	2.4	5.0	Nylon 11	3.7	...
PS, general purpose	2.45	3.1	Polyvinyl formal	3.7	...
PS, high heat	2.45	3.4	CP	3.7	4.3
PS, high impact	2.45	4.75	Nylon 6	3.6	...
PE-covinyl acetate	2.5	3.16	PMMA/PVC alloy	4.0	...
PS-cobutadiene	2.5	3.4	Nylon 6/6	4.0	...
PPO	2.58	...	Nylon 12	4.2	...
PS-coacrylonitrile	2.6	3.4	PVDC chloride	4.5	6.0
PE-coethyl acrylate	2.7	2.9	PF resin	5.0	6.5
SI, cast resin	2.75	4.2	PVB	5.60	...
PC	2.97	3.17	CN	7.0	7.5
Ethyl cellulose	3.0	4.2	Melamine-phenol resin	7.0	7.7
Polyester, rigid-cast resin	3.00	4.30	MF-resin	7.9	11.0
PVC-copolypropylene	3.1	3.7	PVDF	6.4	...
PAE	3.14	...			

Arc resistance of selected thermoplastic and thermosetting plastics per ASTM D 495

Material	High, s	Low, s	Material	High, s	Low, s
PCTFE	>360	...	Polyester, rigid cast resin	135	115
CA	310	50	PS, high heat	135	60
POM, copolymer	240	...	SI, cast resin	130	115
PE, medium density	235	200	POM	129	...
PI	230	...	PSU	122	75
PTFE	>200	...	PC-ABS alloy	120	70
CP	190	175	Epoxy cast resins	120	45
PP	185	136	PC	120	10
PAE	>180	...	Nylon 12	110	...
Melamine-phenol resin	130	130	PS-cobutadiene	95	...
PTFE-cohexafluoropropylene	>165	...	Ionomers	<90	...
PE, low density	160	135	ABS, high impact	85	50
Poly α-methylstyrene-comethyl methacrylate	150	...	ABS, medium impact	85	50
PS-coacrylonitrile	150	100	Ethyl cellulose	80	60
MF resin	145	100	PVC	80	60
Nylon 6/6	140	130	PPO	75	...
PS, general purpose	140	60	PVDF	70	30
PS, high impact	140	20	PMMA/PVC alloy	25	...

Thermal conductivity of PEs and other polymers

Material	Thermal conductivity at 20 °C (68 °F) $W/m^2 \cdot K$	Btu · in./s · ft^2 · °F
PE		
Low density	35	2.5
Medium density	35 to 42	2.5 to 2.9
High density	46 to 52	3.2 to 3.6
PVC	13 to 29	0.9 to 2.0
PS	10 to 14	0.7 to 1.0
Epoxy resin (Shell 828, diethanolamine), filled		
20 wt% mica	18	1.3
30 wt% mica	24	1.7
40 wt% mica	33	2.3
50 wt% mica	39	2.7
PUR		
20% closed cell	3.5	0.2
90% closed cell	1.7	0.1

Coefficients of linear thermal expansion for various polymers

Material	Coefficient of linear thermal expansion, 10^{-5}/K	Mold shrinkage, µm/m
PE		
Low density	10 to 20	20 to 40
High density	10 to 20	20 to 40
PP	2 to 20	10 to 30
Nylon 6/6	10	20
PS	6 to 8	2 to 6
PC	7	To 7
PBT		
Unfilled	6 to 10	9 to 20
Filled, glass fiber	3	2 to 8
Epoxy resin		
Unfilled	4 to 7	...
Filled, mica	2 to 6	2
Zinc	3.5	...
Copper	1.7	...
Silver	1.9	...

Polymer thermal and oxidative properties

Polymer	T_g (softens) °C	T_g (softens) °F	T_m (melts) °C	T_m (melts) °F	T_p (pyrolysis) °C	T_p (pyrolysis) °F	T_c (combustion) °C	T_c (combustion) °F	ΔH kJ/g	ΔH 10^3 Btu/lb	Limiting oxygen index
Nylon 6	50	120	215	420	431	810	450	840	39	16.8	20.8
Nylon 6/6	50	120	265	510	403	755	530	990	32	13.8	20.8
Polyester	85	185	255	490	433	810	480	900	24	10.3	20.5
Acrylic	100	212	>220	>430	290	555	>250	>480	32	13.8	18.2
PP	−20	-4	165	330	469	875	550	1020	44	18.9	18.6
Modacrylic	<80	<175	>240	>465	273	525	690	1275	...	...	29.5
PVC	<80	<175	>180	>355	>180	>355	450	840	21	9.0	38
Polyvinylidene chloride	−17	1	195	385	>220	>430	532	995	11	4.7	60
PTFE	126	260	>327	>620	400	750	560	1040	4	1.7	95
Aramid honeycomb core	275	525	375	705	410	770	>500	>930	30	12.9	29.4
Aramid	340	645	560	1040	>590	>1095	>550	>1020	...	...	29
Polybenzimidazole	>400	>750	...	...	>500	>930	>500	>930	...	...	41

Densities of typical engineering plastics

Material	Density, g/cm^3
Thermoplastic	
POM	1.42
PMMA	1.19
LCP	1.70
Nylon 6	1.13
Nylon 6/6	1.14
Nylon 12	1.06
PAE	1.52
PAR	1.19
PBT	1.31
PC	1.20
PEEK	1.32
PEI	1.27
PESV	1.37
PET	1.56
PPS	1.67
PSU	1.24
SMA	1.22
Thermoplastic alloys	
ABS	1.18
ABS-PC	1.12
PC-PBT	1.22
PPO (mod)	1.09
Thermoset	
DAP	1.79
MF	1.50
PF	1.36
UP	1.78

Properties comparison chart

Material	Amount of fiberglass, %	Specific gravity	Tensile strength, MPa	Tensile strength, ksi	Tensile modulus, GPa	Tensile modulus, 10^6 psi	Flexural strength, MPa	Flexural strength, ksi	Flexural modulus, GPa	Flexural modulus, 10^6 psi	Compressive strength, MPa	Compressive strength, ksi
Polyester SMC-R (low glass)	20	1.78	36.5	5.3	11.7	1.7	110	16.0	9.7	1.4	159	23.00
Polyester SMC-R (class A)	27	1.94	70	10.2	10.2	1.5	152	22.0	8.3	1.2	165	24.0
Polyester SMC-R (structural)	27	1.89	77.2	11.2	11.7	1.7	179	26.0	11.7	1.7	172	25
Acrylamate SMC-R (structural)	50	1.54	183	26.6	15.8	2.3	358	52.0	14.5	2.1	303	44.0
Acrylamate SMC-C (structural)	55	1.88	414	60.1 (0°)	29.6	4.3 (0°)	863	125.2 (0°)	27.6	4.0 (0°)	372	54.1 (0°)
			13.8	2.0 (90°)	6.9	1.0 (90°)	34.5	5.0 (90°)	4.1	0.6 (90°)	179	26 (90°)
Acrylamate SMC-C/R (structural)	C20/R30	1.81	286	41.5 (0°)	24.8	3.6 (0°)	450	65.2 (0°)	15.2	2.2 (0°)	337	49 (0°)
			105	15.2 (90°)	13.1	1.9 (90°)	314	45.5 (90°)	13.8	2.0 (90°)	248	36 (90°)
Epoxy SMC-R (structural)	63	1.85	269	39.0	23.4	3.4	476	69	19.3	2.8	228	33.0
Epoxy SMC-C (structural)	72	1.90	690	100 (0°)	44.7	6.5	1034	150 (0°)	41.4	6.0 (0°)	...	...
Polyester BMC compression	22.0	1.82	41.4	6.00	12.1	1.75	88.3	12.80	10.9	1.58	138	20.00
Polyester BMC injection (transfer)	22.0	1.82	33.5	4.86	10.5	1.53	87.2	12.65	10.9	1.44	...	...

(continued)

Properties comparison chart (continued)

Material	Amount of fiberglass, %	Specific gravity	Tensile strength, MPa	Tensile strength, ksi	Tensile modulus, GPa	Tensile modulus, 10^6 psi	Flexural strength, MPa	Flexural strength, ksi	Flexural modulus, GPa	Flexural modulus, 10^6 psi	Compressive strength, MPa	Compressive strength, ksi
Phenolic (RX 630)	...	1.78	82.7	12.0	16.6	2.4	185	26.8	22	3.2	276	40.0
Phenolic (RX 865)	...	1.86	55.2	8.0	17.2	2.5	131	19.0	22	3.2	241	35.0
Phenolic (HT 500)	...	1.75	58.6	8.5	...	...	93	13.5	14	2.0	172	25.0
Diallyl phthalate (RX1-520)	Short glass	1.83	62	9.0	...	...	124	18.0	14	2.0	172	25.0
Diallyl phthalate (RX3-2-530)	Long glass	1.76	58.6	8.5	...	...	124	18.0	11.7	1.7	193	28.0
Stampable polypropylene P-100	40	1.19	97	14.0	5.2	0.75	165	24.0	5.5	0.8	...	...
Stampable thermoplastic polyester	35	1.59	103	15.0	...	...	221	32.0	8.3	1.2	...	...

Material	Izod impact strength, notched, J/m	Izod impact strength, notched, ft · lbf/in.	Thermal coefficient of expansion, 10^{-6}/K	Water absorption, %, 24 h	Water absorption, %, 48 h	Mold shrinkage, mm/mm	Heat-deflection temperature at 1.82 MPa (0.264 ksi), °C	Heat-deflection temperature at 1.82 MPa (0.264 ksi), °F	Resin matrix heat-deflection temperature, °C	Resin matrix heat-deflection temperature, °F
Polyester SMC-R (low glass)	438	8.20	...	0.10	...	0.002	260	500	...	...
Polyester SMC-R (class A)	961	18.00	8.1	0.60	...	0.001(+)	232	450	...	...
Polyester SMC-R (structural)	1000	18.7	10.8	0.35	...	0	260	500	...	...
Acrylamate SMC-R (structural)	1549	29.0	18.5	0.15	...	0.006	...	...	135	275
Acrylamate SMC-C (structural)	1880	35.2	14.2 (0°)	0.15	...	0.003 (0°)	...	...	135	275
			26.4 (90°)	...	...	0.015 (90°)	...	...	...	...
Acrylamate SMC-C/R (structural)	1667	31.2	16.4 (0°)	0.15	...	0.005 (0°)	...	...	135	275
			20.3 (90°)	...	...	0.009 (90°)	...	...	...	...
Epoxy SMC-R (structural)	2350	44	12.6	0.08	...	0.002	302	575	...	...
Epoxy SMC-C (structural)	4359	85	8.0	0.07	...	0.002	302	575	...	...
Polyester BMC compression	227	4.26	11.8	0.20	...	0.001	260	500	...	...
Polyester BMC injection (transfer)	155	2.90	11.8	0.20	...	0.004	260	500	...	...
Phenolic (RX 630)	48.1	0.9	0.14	...	0.2	0.0015	193	380	...	...
Phenolic (RX 865)	69.4	1.3	0.13	0.05	0.2	0.002	274	525	...	...
Phenolic (HT 500)	53.4	1.0	...	...	0.10	0.002	288	550	...	...
Diallyl phthalate (RX1-520)	37.4	0.7	0.15	...	0.25	0.002	204	400	...	...
Diallyl phthalate (RX3-2-530)	267	5.0	0.12	...	0.35	0.002	288	550	...	...
Stampable polypropylene P-100	752	14.0	27.0	...	...	0.002	155	310	...	...
Stampable thermoplastic polyester	699	13.0	36.0	...	...	...	218	425	...	...

Properties of common plastics

	Tensile strength, MPa	Specific gravity	Maximum service temperature, °C	Heat distortion temperature, °C
Filament wound epoxy resin	680 to 1700	1.2 to 2.2	...	...
Fabric-reinforced epoxy resin	410 to 580	...	...	...
Non-reinforced epoxy resin	12 to 80	1.5 to 2.0	150 to 220	115 to 280
Fabric-reinforced polyester	17 to 38	...	...	...
Mat-reinforced polyester	70 to 165	...	...	...
Non-reinforced polyester	10 to 70	1.06 to 1.46	...	...
Glass-filled nylon	213	...	150 to 220	150 to 280
Glass-filled polystyrene	117	...	85 to 95	65 to 110
Nylon	8.6	1.09 to 1.14	110 to 150	70 to 80
Polystyrene	5.5	1.14 to 1.11	60 to 85	65 to 110
Polystyrene (high impact)	7.6	...	...	...
Acrylic resin	7	1.12 to 1.19	60 to 90	65 to 110
Poly(phenylene oxide)	7	1.06	...	200
Polysulphone	7	1.24	160	150
Acetal resin	6.8	1.41 to 1.43	80 to 110	125 to 140
Alkyd resin (molded)	6.8	2 to 2.24	150 to 185	150 to 220
Melamine-formaldehyde and phenolic resins	6.8	1.43 to 2.00	100 to 220	145 to 220
Urea-formaldehyde resin	6.8	1.47 to 1.52	...	...
Polycarbonate	6.8	1.20 to 1.51	120 to 135	150 to 160
Phenoxy resin	6.5	1.18 to 1.32	75	75
Phenolic resin (cast)	6.2	1.24 to 3.0	150 to 250	130 to 280
Poly(vinyl chloride)	6.2	1.16 to 1.55	65 to 75	60 to 75
Acrylonitrile-butadiene-styrene terpolymer (ABS)	5.8	1.01 to 1.07	80 to 110	100 to 130
Cellulose acetate	5.8	1.23 to 1.34	...	...
Poly(vinylidene chloride)	5.5	1.68 to 1.75	...	...
Cellulose propionate	5.1	1.18 to 1.24	...	...
Diallyl phthalate resin	4.8	1.31 to 1.70	150 to 250	130 to 280

(continued)

Properties of common plastics (continued)

	Tensile strength, MPa	Specific gravity	Maximum service temperature, °C	Heat distortion temperature, °C
Ethylcellulose	4.8	1.10 to 1.16	...	...
Cellulose acetate butyrate	4.7	1.15 to 1.25	...	...
Chlorinated polyether	4.1	1.4	...	...
Polychlorotrifluoroethylene	3.9	...	200	80 to 110
Polypropylene	3.4	0.9 to 0.91	...	...
Silicones	3.4	1.6 to 2.0	350	530
Polyethylene (high density)	3.0	...	...	...
Polytetrafluorethylene	2.4	2.1 to 2.4	280	...
Polyethylene (low to medium density)	7.16	0.91 to 0.97	...	...

Properties of polymers

Material	Linear coefficient of thermal expansion, 10^{-6}/K	Tensile modulus, GPa	Tensile modulus, 10^5 psi	Tensile strength, MPa	Tensile strength, ksi	Solidification temperature °C	Solidification temperature °F
Polymethyl methacrylate	70	2.76	4.00	65.6	9.5	97	207
Polyacrylonitrile	66	3.76	5.45	62.0	9.0	95	203
Polycarbonate	68	2.38	3.45	65.6	9.5	150	302
Polystyrene	63	2.77	4.02	42.8	6.2	102	216
Polyvinyl chloride	75	3.27	4.75	46.9	6.8	80	176
Cast epoxy	55	2.41	3.50	58.7	8.5	170	338

Values of fracture toughness and glass transition temperature for selected plastics

	Fracture energy		T_g	
Material	kJ/m^2	ft · lbf/in.^{2}	°C	°F
Thermoset				
Polyimide-1	0.20	0.095	350	662
Polyimide-2	0.12	0.057	360	680
Tetrafunctional epoxy	0.076	0.036	260	500
Thermoplastic				
Polyarylsulfone	5.5	2.6	...	...
Poly(amide-imide)	3.4	1.6	275	527
Polysulfone	3.1	1.5	174	345
Polyethersulfone	2.6	1.2	230	446
Polyimide-4	2.1	1.0	365	689
Polyimide-3	0.81	0.38	326	619
Polyphenylene sulfide	0.21	0.10	...	...

K_{Ic} values at 20 °C (68 °F)

	K_{Ic}	
Material	MPa$\sqrt{m}$	ksi$\sqrt{in.}$
PMMA	0.7 to 1.6	0.6 to 1.5
PS	0.7 to 1.1	0.6 to 1.0
PC	2.2	2.0
Polyether sulfone (PESV)	1.2	1.1
High-impact PS	1.0 to 2.0	0.91 to 1.8
Acrylonitrile-butadiene-styrene (ABS)	2.0	1.8
PVC	2.0 to 4.0	1.8 to 3.6
PP	3.0 to 4.5	2.7 to 4.1
PE	1.0 to 6.0	0.91 to 5.5
POM (acetal)	~4	~3.6
Nylon	2.5 to 3.0	2.3 to 2.7
Epoxy	0.6	0.55
Polyester	0.6	0.55
Polyethylene terephthalate (PET)	~5	~4.6

Notched Izod impact strength of rigid plastics at 24 °C (75 °F)

Plastic	Impact strength, notched J/m	ft · lbf/in.
PS	13.3 to 21.3	0.25 to 0.40
Cellulose acetate	53.3 to 298	1.0 to 5.6
Cellulose nitrate	267 to 373	5.0 to 7.0
Ethyl cellulose	187 to 320	3.5 to 6.0
Nylon 6/6	53.3 to 160	1.0 to 3.0
Nylon 6	53.3 to 160	1.0 to 3.0
Polyoxymethylene (POM)	107 to 160	2 to 3
Polyethylene (PE), low density	>853	>16
PE, high density	27 to 1070	0.5 to 20
Polypropylene (PP)	27 to 107	0.5 to 2
Polyvinyl formal (PVF)	53.3 to 1070	1 to 20
Phenol-formaldehyde (PF), general purpose	13.3 to 18.7	0.25 to 0.35
PF, cloth-filled	53.3 to 160	1 to 3
PF, glass fiber-filled	530 to 1600	10 to 30
Polytetrafluoroethylene (PTFE)	107 to 215	2.0 to 4.0
Nylon 6/12	53.3 to 74.6	1.0 to 1.4
Nylon 1	96	1.8
PPO, 25% glass fibers	74.6 to 80	1.4 to 1.5
Polyester, glass fiber-filled	2 to 20	1.8 to 18.2
Epoxy, glass fiber-filled	10 to 30	9.1 to 27.3
Polyimide (PI)	0.9	0.82

Environmental Effects

Effect of chemical environments on engineering plastics

Material	Aromatic hydrocarbons 25	Aromatic hydrocarbons 93	Aliphatic hydrocarbons 25	Aliphatic hydrocarbons 93	Halogenated solvents 25	Halogenated solvents 93	Esters and ketones 25	Esters and ketones 93	Alcohols 25	Alcohols 93	Amines 25	Amines 93	Weak bases and salts 25	Weak bases and salts 93	Strong bases 25	Strong bases 93	Strong acids 25	Strong acids 93	Strong oxidants 25	Strong oxidants 93	24 h water absorption
PA	1	1	1	1	1	2	1	1	1	2	1	2	1	2	2	3	5	5	5	5	0.20 to 1.90
Polycarbonate	5	5	1	1	5	5	5	5	...	2 to 5	...	5	1	5	5	5	1	1	1	1	0.15 to 0.35
Polyester	2	5	1	3 to 5	3	5	2	3 to 4	...	...	...	...	1	3 to 4	2	5	3	4 to 5	2	3 to 5	0.06 to 0.09
PTFE	1	1	1	1	1	1	1	1	1	1	1	1	1	1	1	1	1	1	1	1	0
Polyimide	1	1	1	1	1	1	1	1	...	...	...	...	2	3	4	5	3	4	2	5	0.30 to 0.40
Polyphenylene oxide	4	5	2	3	4	5	2	3	...	2 to 5	...	5	1	1	1	1	1	2	1	2	0.06 to 0.07
PPS	1	2	1	1	1	2	1	1	1	1	...	1 to 3	1	1	1	1	1	1	1	3	<0.05
Polysulfone	4	4	1	1	5	5	3	4	1	1	...	5	1	1	1	1	1	1	1	1	0.20 to 0.30
Diallyl phthalate	1 to 2	2 to 4	2	3	2	4	3 to 4	4 to 5	...	...	...	...	2	3	2	4	1 to 2	2 to 3	2	4	0.20 to 0.70
Phenolics	1	1	1	1	1	1	2	2	1	1	...	...	2	3	3	5	1	1	4	5	0.10 to 2.00

Chemical environment at 25 and 93 °C (77 and 200 °F)

Ratings: 1, no effect or inert; 2, slight effect; 3, mild effect; 4, softening or swelling; 5, severe degradation

Chemical and environmental resistance of various plastics

Plastic	Hydrocarbons: Aliphatic	Hydrocarbons: Aromatic	Hydrocarbons: Fully halogenated	Hydrocarbons: Partially halogenated	Alcohols: Monohydric	Alcohols: Polyhydric	Oils, fats, waxes	Phenols	Ketones	Esters	Ethers
ABS	3	4	4	4	3	2	2	4	4	4	4
Acrylic	2	4	4	4	4	2	2	4	4	4	4
Allyic esters	1	2	2	2	1	1	1	3 to 4	3	3	2
Cellulosic esters	3	4	4	4	4	1	3	4	4	4	3
Chlorinated polyalkylene ether	2	2	3 to 4	4	2	1	2	2	3 to 4	3	2
Epoxy	1	1 to 2	1	2	1	1	1	3	3	2	3
FEP	1	1	1	2	1	1	1	1	1	1	2
Melamine formaldehyde	2	1 to 2	1	1	1	1	2	2	2	2	1
Phenolic	1	1	1	1	2	1	1	1	2	2	1
Polyacetal	2	2	2	2 to 3	2	2	2	4	2	2	2
Polyalkylene ether	3	4	4	4	4	4	3 to 4	4	4	4	4
Polyamide (nylon)	2	3	3	4	2	2	2	4	2	3	2
Polyaromatic ether	3	4	4	4	4	2	2	1	4	4	4
Polybutadiene (1, 2)	2	3	...	3	1	1	...	2	2	2	2
Polycarbonate	3 to 4	4	2 to 3	4	2	2	3 to 4	4	4	4	4
Polyesters (TP)	2	3 to 4	2 to 4	4	1 to 2	1	...	4	4	3 to 4	4
Polyesters (unsaturated)	2	4	4	4	2	2	2	4	4	4	2
Polyethylene	3	3	4	4	3	2	1	1	2	2	2
Polyimide	1	1	1	1	1	1	1	2	1	1	1
Polyphenylene	2	2	2	2	1	1	1	2	2	2	1
Polyphenylene sulfide	1	2	3	4	1	1	...	2	2	1	2
Polypropylene	3	3	4	3 to 4	2	1	2	2	2	2	3
Polystyrene	3 to 4	4	4	4	2	2	4	4	4	4	4
Polysulfone	4	4	4	4	3 to 4	2	3	4	4	4	4
Polyurethane	3	4	4	4	4	2 to 3	2	4	4	4	4
Polyvinyl acetate	3 to 4	4	4	4	4	3	3	4	4	4	4
Polyvinyl chloride	2	4	4	4	2	2	2	4	4	4	4
Polyvinyl chloride vinylidene chloride	2	2 to 3	4	4	1	2	2	3	4	4	4
Polyvinyl fluoride	1	2	1	1	1	1	1	2	4	3	2
Polyvinylidene fluoride	1	2	2	2 to 3	1	1	1	3	3	3	2
Silicone	3	3 to 4	4	4	2	1	1	4	2	3	3
Styrene acrylonitrile	2	4	4	4	3	2	2	4	4	4	4
TFE	1	1	1	2	1	1	1	1	1	1	2
Urea-formaldehyde	2	2	1	1	1 to 2	1	2	2	2	2 to 3	1
Polybutylene	2	2	4	3	2	1	2	2	2	2	2
Polyxylylene	1	1	1	2	1	1	1	2	3	3	2
Vinyl esters	2	2	1	2	1	1	1	3	3	3	3

Plastic	Acids: Inorganic Concentrated	Acids: Inorganic Dilute	Acids: Organic Concentrated	Acids: Organic Dilute	Acids: Oxidizing Concentrated	Acids: Oxidizing Dilute	Bases: Concentrated	Bases: Dilute	Salts: Acid	Salts: Neutral	Salts: Base	Sunlight and weathering
ABS	2	2	4	3 to 4	2	2	2	2	2	1	2	2 to 3
Acrylic	4	2	4	3	4	3	3 to 4	2	2	2	2	2
Allyic esters	3	3	3	2	4	3	3	2	2	1	2	1
Cellulosic esters	4	3	4	4	4	4	4	3	2	2	3	2 to 3
Chlorinated polyalkylene ether	3	2	3	2	4	3	3	2	2	2	2	3
Epoxy	3	1	3	2	4	3	1	1	1	1	1	2
FEP	1	1	1 to 2	1	1	1	1	1	1	1	1	1
Melamine formaldehyde	4	3	2	2	4	3	4	2	1	1	2	2
Phenolic	3 to 4	2 to 3	2	3	4	4	4	4	1	1	3	2
Polyacetal	4	4	4	3	4	4	4	4	3	2	2	2 to 3
Polyalkylene ether	4	4	4	4	4	4	4	4	4	4	4	4
Polyamide (nylon)	4	2	4	3	4	4	2	1	4	2	3	2 to 3
Polyaromatic ether	2	1	2	1	4	2	1	1	1	1	1	1
Polybutadiene (1, 2)	3	1	2 to 3	1	3	2	2	2	1	1	2	...
Polycarbonate	3	2	3	3	4	2	4	4	2	1	3	2
Polyesters (TP)	2 to 3	2	3 to 4	2	4	3 to 4	4	3 to 4	2	2	3 to 4	2
Polyesters (unsaturated)	2	1	2	2	4	3	3	2	1	1	2	2
Polyethylene	1 to 2	1	1	1	4	2	2	2	1	1	1	4
Polyimide	3	2	2	1	4	2	4	3	2	2	4	1
Polyphenylene	2	1	2	1	2	1	2	1	1	1	1	1
Polyphenylene sulfide	2	2	2	2	4	3	2	1	2	2	2	2

(continued)

Chemical and environmental resistance of various plastics (continued)

Plastic	Acids: Inorganic Concentrated	Acids: Inorganic Dilute	Acids: Organic Concentrated	Acids: Organic Dilute	Acids: Oxidizing Concentrated	Acids: Oxidizing Dilute	Bases: Concentrated	Bases: Dilute	Salts: Acid	Salts: Neutral	Salts: Base	Sunlight and weathering
Polypropylene	1	1	2	1	4	2	1	1	1	1	1	4
Polystyrene	3	2	4	3	4	3	3	1	2	1	2	3
Polysulfone	2	2	3	2	4	2	2	1	2	1	2	1 to 2
Polyurethane	4	3	4	3	4	4	2	2	2	1	1	3
Polyvinyl acetate	4	4	4	4	4	4	4	4	4	4	4	2 to 3
Polyvinyl chloride	2	1	4	2	2	1	1	1	1	1	1	2
Polyvinyl chloride vinylidene chloride	2	2	4	2	3	2	3	1	2	2	2	2
Polyvinyl fluoride	3	1	3	1	2	1	1	1	1	1	1	1
Polyvinylidene fluoride	1	1	2	1	1 to 2	1	1	1	1	1	1	1
Silicone	2	2	3	2	4	1	2	1	1	1	1	1
Styrene acrylonitrile	2	2	3	2	4	2	2	1	2	1	2	3
TFE	1	1	1 to 2	1	1	1	1	1	1	1	1	1
Urea-formaldehyde	4	3	2	2	4	3	4	3	2	2	2	2
Polybutylene	1	1	2	1	4	2	1	1	1	1	1	4
Polyxylylene	1	1	1	1	4	2	1	1	1	1	1	2
Vinyl esters	2	1	2	2	4	3	3	2	1	1	2	2

Ratings: 1, excellent; 2, good; 3, fair; 4, poor

Water absorption properties of selected thermoplastics and thermosetting plastics per ASTM D 570

Material	Low, %	High, %	Material	Low, %	High, %
Polybutene	<0.1	...	PS-coacrylonitrile	0.20	0.30
Polytetrafluoroethylene (PTFE)-cohexafluoropropylene	0.01	...	PC/ABS alloy	0.20	0.35
PE, low density	<0.015	...	Polysulfone (PSU)	0.22	...
PVC, chlorinated	0.02	0.15	Polyoxymethylene (POM) copolymer	0.22	...
PS, general purpose	0.03	0.10	Polyaryl ether (PAE)	0.25	...
PVC	0.03	0.40	Nylon 12	0.25	...
PE-coethyl acrylate	0.04	...	POM	0.25	...
Polyvinylidene fluoride (PVDF)	0.04	...	ABS, medium impact	0.30	...
PE-covinyl acetate	0.05	0.13	PMMA, molding grade	0.30	0.40
PS, high impact	0.05	0.20	Melamine-formaldehyde (MF) resin	0.30	0.50
PS, high heat	0.05	0.40	Melamine-phenol resin	0.30	0.65
Polyphenylene oxide (PPO)	0.06	...	Polyimide (PI)	0.32	...
PVC-copropylene	0.07	0.40	Nylon 11	<0.40	...
Epoxy cast resins	0.08	0.15	Ethyl cellulose	0.8	1.8
Polyvinylidene chloride (PVDC)	0.10	...	Cellulose acetate butyrate (CAB)	0.9	2.0
Phenol-formaldehyde (PF) resin	0.10	0.20	Polyvinyl formal	1.0	1.3
Polymethyl methacrylate (PPMA)/PVC alloy	0.13	...	Cellulose nitrate (CN)	1.0	2.0
Polycarbonate (PC)	0.15	0.18	Polyvinyl butyral (PVB)	1.0	2.0
Polyester, rigid-cast resin	0.15	0.60	Nylon 6	1.3	1.9
PS-cobutadiene	0.19	0.39	Cellulose propionate (CP)	1.3	2.0
PVC, ABS modified	0.20	...	Ionomers	1.4	...
ABS, high impact	0.20	...	Nylon 6/6	1.5	...
Poly α-methylstyrene-comethyl methacrylate	0.20	...	Cellulose acetate (CA)	1.7	4.0

Thermoplastics

Structures and Properties

Chemical structures, glass transition temperatures, and melting temperatures of representative hydrocarbon thermoplastic polymers

Chemical name	T_s °C	T_s °F	T_m °C	T_m °F	Mer chemical structure
Polyethylene					H H —C—C— H H
HDPE	−90 or −20	−130 or −5	137	280	
LPDE	−110 or −20	−165 or −5	115	240	
Polypropylene					H CH_3 —C—C— H H
Atactic	−18	0	176	350	
Isotactic	−10	15	176	350	
Polyisobutylene	−70, −60	−95, −75	128	260	H CH_3 —C—C— H CH_3
Polyisoprene					H CH_3 H H —C—C=C—C— H H
Cis: natural rubber	−73	−100	28	80	
Trans: gutta percha	...	...	...	...	
Polymethylpentene (poly-4-methyl-1-pentene)	29	85	250	480	H H —C—C— H H—C—H C—H CH_3 CH_3
Polybutadiene (poly-1, 2-butadiene, butadiene rubber)					H H H H —C—C=C—C— H H H
Syndiotactic	−90	−130	154	310	
Isotactic	−90	−130	120	250	
Polystyrene					H H —C—C— H C_6H_5
Atactic	100, 105	212, 220	(a)		
Isotactic	100, 105	212, 220	240	465	

(a) Polymer is generally 95% or more noncrystalline. Any T_m given is for remaining crystalline portion or for crystalline version

Chemical structures, glass transition temperatures, and melting temperatures of representative nonhydrocarbon carbon-chain thermoplastic polymers

Chemical name	T_g		T_m		Mer chemical structure
	°C	°F	°C	°F	
Polyvinyl chloride (vinyl)	87	190	212	415	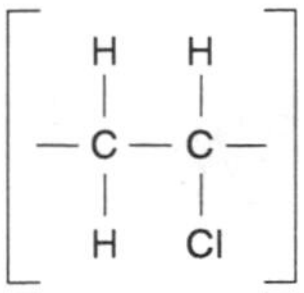 H H | | —C—C— | | H Cl
Polyvinyl fluoride	–20	–5	200	390	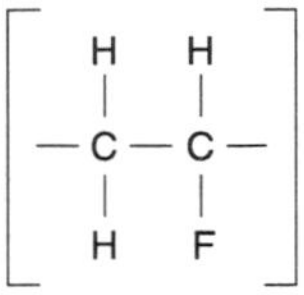 H H | | —C—C— | | H F
Polyvinylidene chloride	–17	1	198	390	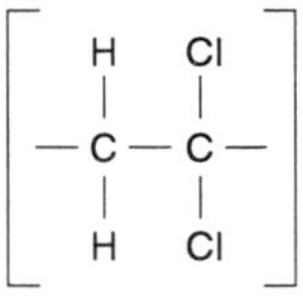 H Cl | | —C—C— | | H Cl
Polyvinylidene fluoride	–35	–30	...	...	 H F | | —C—C— | | H F
Polytetrafluoroethylene	–97, 126	–140, 260	327	620	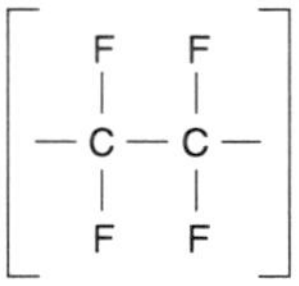 F F | | —C—C— | | F F
Polychlorotrifluoroethylene	45	115	220	430	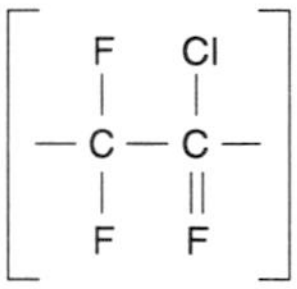 F Cl | | —C—C— | || F F
Polychloroprene (chloroprene rubber, or neoprene)	–50	–60	80	175	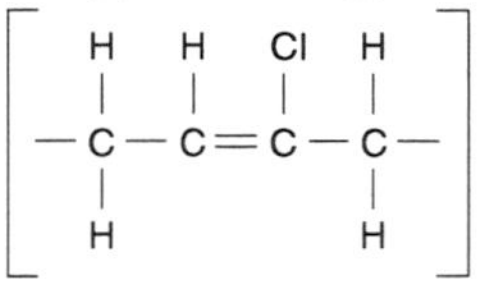 H H Cl H | | | | —C—C=C—C— | | H H
Polyacrylonitrile	104, 130	220, 265	317	600	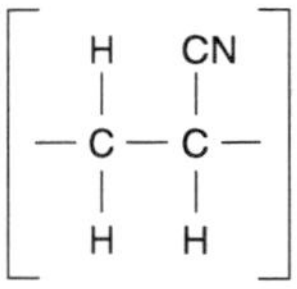 H CN | | —C—C— | | H H
Polyvinyl alcohol	85	185	258	495	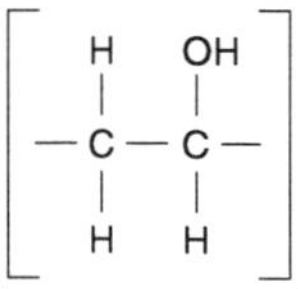 H OH | | —C—C— | | H H
Polyvinyl acetate	29	85	...	...	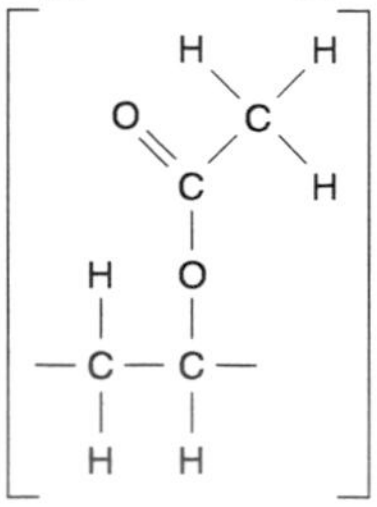 H H O C C H | H O | | —C—C— | | H H

(continued)

Chemical structures, glass transition temperatures, and melting temperatures of representative nonhydrocarbon carbon-chain thermoplastic polymers (continued)

Chemical name	T_g °C	T_g °F	T_m °C	T_m °F	Mer chemical structure
Polyvinyl carbazole	150, 208	300, 405	...	...	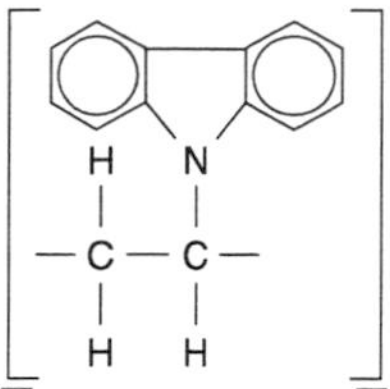 [—C(H)(H)—C(H)(N-carbazole)—]
Polymethyl methacrylate					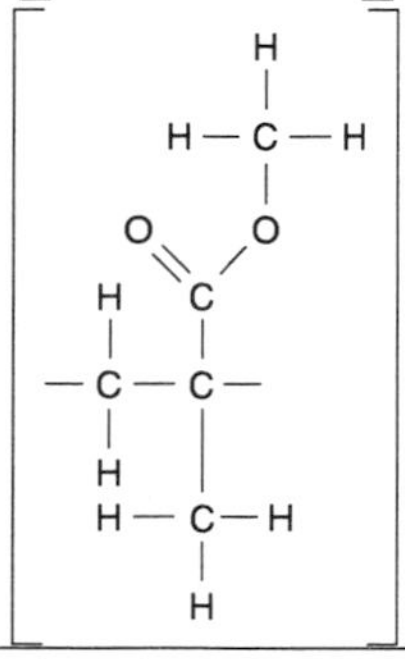 [—C(H)(H)—C(CH$_3$)(C(=O)—O—CH$_3$)—]
Syndiotactic	3	35	105, 120	220, 250	
Isotactic	3	35	45	115	

Chemical structures, glass transition temperatures, and melting temperatures of representative heterochain thermoplastic polymers

Chemical name	T_g °C	T_g °F	T_m °C	T_m °F	Mer chemical structure
Polyethylene oxide	–67 to –27	–90 to –15	62 to 72	145 to 160	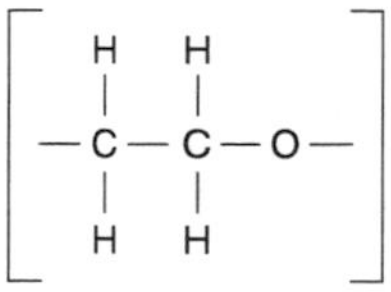 [—C(H)(H)—C(H)(H)—O—]
Polyoxymethylene	–85	–120	175	345	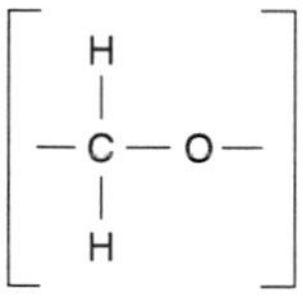 [—C(H)(H)—O—]
Polyamide					
Nylon 6	50	120	215	420	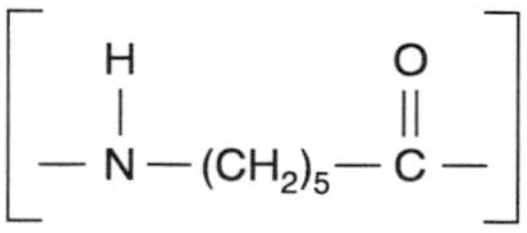 [—N(H)—(CH$_2$)$_5$—C(=O)—]
Nylon 6/10	40	105	227	440	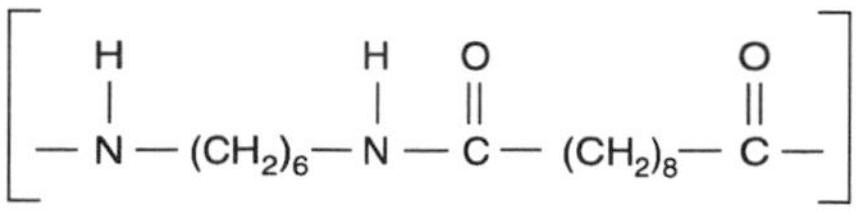 [—N(H)—(CH$_2$)$_6$—N(H)—C(=O)—(CH$_2$)$_8$—C(=O)—]
Polyethylene terephthalate	69	155	265	510	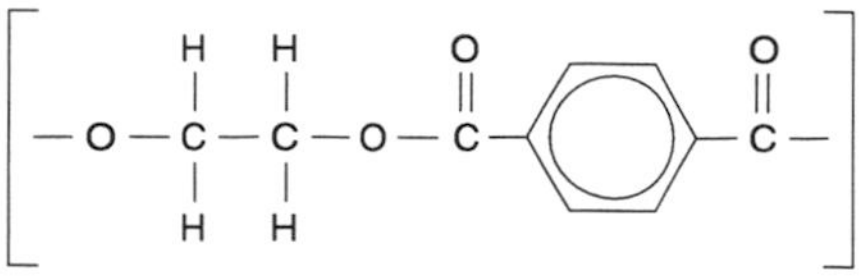 [—O—C(H)(H)—C(H)(H)—O—C(=O)—C$_6$H$_4$—C(=O)—]

(continued)

Chemical structures, glass transition temperatures, and melting temperatures of representative heterochain thermoplastic polymers

Chemical name	T_g °C	T_g °F	T_m °C	T_m °F	Mer chemical structure
Polycarbonate	150	300	265	510	[—O—C₆H₄—C(CH₃)₂—C₆H₄—O—C(=O)—]
Polydimethyl siloxane (silicone rubber)	–123	–190	–54	–65	[—Si(CH₃)₂—O—]

Chemical structures, glass transition temperatures, and melting temperatures of representative thermoplastic polymers for high-temperature service

Chemical name	T_g °C	T_g °F	T_m °C	T_m °F	Mer chemical structure
Poly *p*-phenylene terephthalamide (aromatic polyamide or aramid)	375	705	~640(a)	~1185	[—C₆H₄—N(H)—C(=O)—C₆H₄—C(=O)—N(H)—]
Polyaromatic ester	...	...	421	790	[—C₆H₄—O—C₆H₄—O—C₆H₄—C(=O)—]
Polyetheretherketone	143	290	334	635	[—C₆H₄—O—C₆H₄—O—C₆H₄—C(=O)—]
Polyphenylene sulfide	85	185	285	545	[—C₆H₄—S—]
Polyamide-imide	277 to 289	530 to 550	(b)		[—N(H)—C(=O)—C₆H₃(C(=O))₂N—R—]
Polyether sulfone	225	435	(b)		[—C₆H₄—S(=O)₂—C₆H₄—O—]

(continued)

Chemical structures, glass transition temperatures, and melting temperatures of representative thermoplastic polymers for high-temperature service

Chemical name	T_g °C	T_g °F	T_m °C	T_m °F	Mer chemical structure
Polyether-imide	215	420	(b)		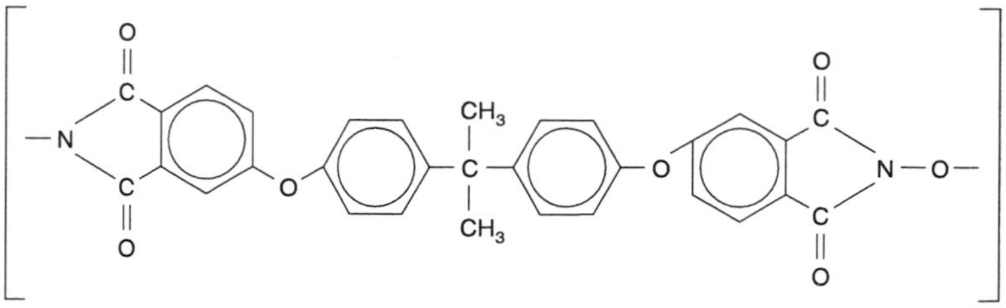
Polysulfone	193	380	(b)		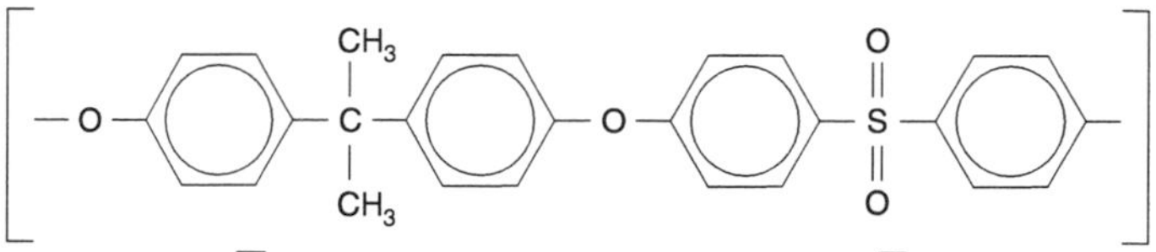
Polyimide (thermoplastic)	280 to 330	535 to 625	(b)		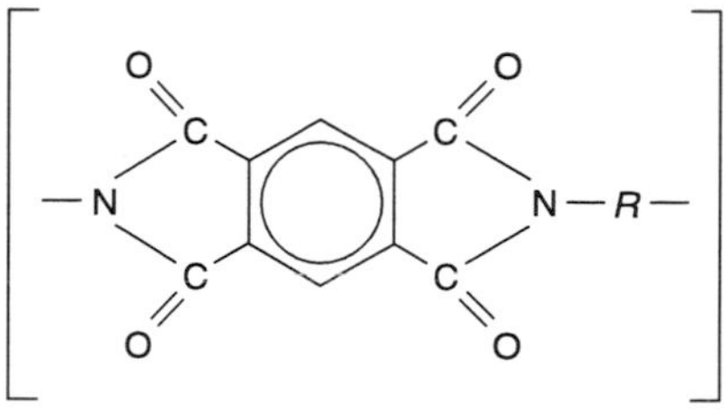

(a) T_d = 500 °C (930 °F). R contains at least one aromatic ring. (b) Polymer is generally 95% or more noncrystalline. Any T_m given is for remaining crystalline portion or for crystalline version

Electrical properties of thermoplastic materials

Property	ASTM method	Acetal	ABS	Acrylic	Cellulose acetate	Cellulose acetate butyrate	Cellulose propionate	Chlorinated polyether	Chlorotri-fluoro-ethylene	Nylon (polyamide)	Polycar-bonate
Arc resistance	D 495	129	90	No track	200	...	180	...	>360	140	120
Dielectric constant											
At 60 Hz	D 150	3.8	3.0	4.0	7.5	6.4	4.0	3.1	2.8	5.5	3.2
At 1 MHz	...	3.8	3.0	3.5	7.0	6.3	4.0	3.0	2.7	4.9	3.0
At 1 GHz	...	3.8	3.0	3.2	7.0	6.2	3.6	2.9	2.5	4.7	3.0
Dissipation factor											
At 60 Hz	D 150	0.004	0.003	0.04	0.01	0.02	0.01	0.01	0.001	0.01	0.0009
At 1 GHz	...	0.004	0.005	0.02	0.01	0.05	0.01	0.01	0.09	0.03	0.01
Dielectric strength, Step-by-step, MV/m (V/mil)	D 149	16 (400)	14 (350)	14 (350)	8 (2.00)	10 (250)	12 (300)	16 (400)	18 (450)	12.8 (320)	14.5 (364)
Volume resistivity, Ω · m	D 257	10^{12}	10^{14}	10^{12}	10^{11}	10^{12}	10^{13}	10^{13}	10^{16}	10^{13}	10^{10}

Property	ASTM method	Polyethylene, low-density	Polyethylene, med-density	Polyethylene, high-density	Poly-propylene	Poly-styrene	Poly-sulfone	Polyphenylene oxide	Phenoxy	Polyvinyl chloride	Styrene-acrylonitrile	Tetrafluoro-ethylene
Arc resistance	D 495	140	200	200	185	100	122	75	...	80	150	<200
Dielectric constant												
At 60 Hz	D 150	2.4	2.4	2.4	2.6	3.4	3.1	2.6	4.1	3.6	3.4	2.1
At 1 MHz	...	2.4	2.4	2.4	2.6	3.2	3.1	2.6	4.1	3.3	2.5	2.1
At 1 GHz	...	2.4	2.4	2.4	2.6	3.1	3.1	2.6	3.8	3.4	3.1	2.1
Dissipation factor												
At 60 Hz	D 150	<0.0005	<0.0005	<0.0005	<0.0005	0.0004	0.0008	0.0004	0.001	0.007	0.004	<0.0002
At 1 MHz	...	<0.0005	<0.0005	<0.0005	<0.0005	0.0004	0.001	...	0.002	0.009	0.007	<0.0002
At 1 GHz	...	<0.0005	<0.0005	<0.0005	<0.0005	0.0004	0.005	0.0009	0.03	0.006	0.007	<0.0002
Dielectric strength, Step-by-step, MV/m (V/mil)	D 149	168 (420)	20 (500)	22 (550)	18 (450)	12 (300)	16 (400)	16 (400)	16 (400)	15 (375)	12 (300)	17.2 (430)
Volume resistivity, Ω · m	D 257	10^{14}	10^{14}	10^{14}	10^{14}	10^{14}	10^{15}	10^{11}	10^{11}	10^{14}	10^{14}	10^{16}

Factors affecting the conductivities of thermoplastic compounds

Aspect ratio:	A continuous pathway of particles is essential to electrical conductivity. The greater the aspect ratio, the less additive needed to make the resin conductive
Loading level:	The amount of additive used in a compound. The higher the loading, the greater the conductivity. The lowest loading of an additive required to initiate conductivity is called "critical concentration"
Resin matrix:	Depending on the type of resin chosen, more or less additive may be required to meet conductivity requirements
Processing:	Improper processing can cause the fiber additives to break into shorter lengths and decrease their effectiveness. Conductivity is enhanced by proper tooling, gating, and processing
Conductivity:	The more conductive the additive is, the more conductive the finished product will be

Coefficients of thermal expansion for selected thermoplastics

Resin	Glass content, %	Coefficient of thermal expansion(a), 10^{-6}/°C	10^{-6}/°F
Polybutylene terephthalate (PBT)	30	25	14
Polyethylene terephthalate (PET)	30	29	16
	43	22	12
Polyphenylene sulfide (PPS)	40	20	11
Polyetherimide (PEI)	30	20	11
Polyamide (6/6 nylon)	25	23	13
Liquid crystal polymer (LCP)	30	5	3

(a) In direction of flow

Thermal properties of thermoplastics

Material	Heat-deflection temperature at 1.82 MPa (0.264 ksi) °C	°F	UL index °C	°F	Thermal conductivity W/m · K	Btu · in./h · ft^2 · °F	Coefficient of thermal expansion, 10^{-5}/K
Acrylonitrile-butadiene-styrene (ABS)	99	210	60	140	0.27	1.9	5.3
ABS-polycarbonate alloy (ABS-PC)	115	240	60	140	0.25	1.7	3.5
Diallyl phthalate (DAP)	285	545	130	265	0.36	2.5	2.7
Polyoxymethylene (POM)	136	275	85	185	0.37	2.6	3.7
Polymethyl methacrylate (PMMA)	92	200	90	195	0.19	1.3	3.4
Polyarylate (PAR)	155	310	...	...	0.22	1.5	3.1
Liquid crystal polymer (LCP)	311	590	220	430	...	...	0.5
Melamine-formaldehyde (MF)	183	360	130	265	0.42	2.9	2.2
Nylon 6	65	150	75	165	0.23	1.6	2.5
Nylon 6/6	90	195	75	165	0.25		4.0
Amorphous nylon 12	140	285	65	150	0.25		7.0
Polyarylether (PAE)	160	320	160	320	...	...	3.0
Polybutylene terephthalate (PBT)	...	...	120	250	...	...	4.5
PC	129	265	115	240	0.20		3.8
PBT-PC	129	265	105	220	...	...	2.8
PEEK	...	...	250	480	0.25	1.7	2.6
Polyether-imide (PEI)	210	410	170	340	0.22	1.5	3.1
Polyether sulfone (PESV)	203	395	170	340	...	...	5.5
PET	224	435	140	285	0.17		1.5
Phenol-formaldehyde (PF)	163	325	150	300	0.25	1.7	1.6
Unsaturated polyester (UP)	279	535	130	265	0.12	0.8	1.6
Modified polyphenylene oxide alloy (PPO) (mod)	100	212	80	175	...	...	3.8
Polyphenylene sulfide (PPS)	260	500	200	390	0.17		3.0
Polysulfone (PSU)	174	345	140	285	0.26	1.8	3.1
Styrene-maleic anhydride terpolymer (SMA)	103	215	80	175	...	...	...

Linear coefficients of thermal expansion of thermoplastics

Material	10^{-6}/K
Polymethyl methacrylate	50 to 90
Polyacrylonitrile	66
Cellulose acetate	100 to 150
Nylon 6	80 to 83
Nylon 11	100
Polycarbonate	68
Polyethylene terephthalate	65
Polyphenylene sulfide	49
Polyethylene, branched	100 to 220
Polypropylene	81 to 100
Polystyrene	50 to 83
Polyvinyl chloride	50 to 100

Deflection temperature under load at 1.82 MPa (0.264 ksi) for engineering thermoplastics

Engineering thermoplastic	Deflection temperature under load °C	°F
PAI	274	524
PESV	203	397
PEI	200	392
PSU	174	345
PEEK	166	330
Acetals	136	277
PC	135	275
PPS(a)	135	275
PPE	129	265
PPO	129	265
Nylon 6/6	90	194
ABS	86	187
PBT	54	130
PET(a)	38	100

(a) Estimated because values refer to neat (unreinforced, unmodified) resin, and engineering commercial grades of PPS and PET are always reinforced. Values for neat PPS and PET would not appear on suppliers' data sheets. Data generated using ASTM D 648

Plastics behavior as a function of temperature

Plastics	Temperature, °C (°F) −20 (−4)	−10 (14)	0 (32)	10 (50)	20 (68)	30 (85)	40 (105)	50 (120)
Polystyrene	A	A	A	A	A	A	A	A
Polymethyl methacrylate	A	A	A	A	A	A	A	A
Glass-filled nylon (dry)	A	A	A	A	A	A	A	B
Polypropylene	A	A	A	A	B	B	B	B
Polyethylene terephthalate	B	B	B	B	B	B	B	B
Acetal	B	B	B	B	B	B	B	B
Nylon (dry)	B	B	B	B	B	B	B	B
Polysulfone	B	B	B	B	B	B	B	B
High-density polyethylene	B	B	B	B	B	B	B	B
Rigid polyvinyl chloride	B	B	B	B	B	B	C	C
Polyphenylene oxide	B	B	B	B	B	B	C	C
Acrylonitrile-butadiene-styrene	B	B	B	B	B	B	C	C
Polycarbonate	B	B	B	B	C	C	C	C
Nylon (wet)	B	B	B	C	C	C	C	C
Polytetrafluoroethylene	B	C	C	C	C	C	C	C
Low-density polyethylene	C	C	C	C	C	C	C	C

A, brittle even when unnotched; B, brittle, in the presence of a notch; C, tough

Physical properties of principal optical plastics

Properties	ASTM method	Methyl methacrylate	Polystyrene	Polycarbonate	Methyl methacrylate styrene copolymer
Refractive index	D 542				
n_D		1.491 (7)	1.590 (3)	1.586 (0)	1.564
n_F		1.497 (8)	1.604 (1)	1.593 (4)	1.574 (4)
n_C		1.489 (2)	1.584 (9)	1.576	1.558 (3)
Abbe value	D 542	57.2	30.8	34.0	35
dn/dt, 10^{-5}/K	...	−12.5	−12.0	−14.3	−14.0
Haze, %	D 1003	<2	<3	<3.0	<3
Luminous transmittance, 3.2 mm (0.125 in.) thickness	D 1003	92	87 to 90	85 to 91	90
Critical angle, degrees	...	42.1	39.0	39.1	39.8
Deflection temperature at 2 °C/min. (3.6 °F/min.), °C (°F)	D 648				
At 1.8 MPa (0.264 ksi)		198	180	280	...
At 0.46 MPa (0.066 ksi)		214	230	270	212
Coefficient of linear thermal expansion, 10^{-5}/K	D 696	3.6	3.5	3.8	3.6
Recommended maximum continuous service temperature for normal parts, °C (°F)	...	80 (175)	80 (175)	115 (240)	80 (175)

(continued)

Physical properties of principal optical plastics (continued)

Properties	ASTM method	Methyl methacrylate	Polystyrene	Polycarbonate	Methyl methacrylate styrene copolymer
Water absorption, immersed 24 h at 23 °C (73 °F), %	D 570	0.3	0.2	0.15	0.15
Specific gravity	D 792	1.19	1.06	1.20	1.09
Hardness, 6.4 mm (0.25 in.) sample	D 785	M97	M 90	M 70	M 75
Impact strength, Izod notch, J/m (ft · lbf/in.)	D 256	0.3 to 0.5	0.35	12 to 17	...
Dielectric strength, MV/m (V/mil)	D 149	19.7 (500)	19.7 (500)	15.7 (400)	17.7 (450)
Dielectric constant	D 150				
At 60 Hz		3.7	2.6	2.90	3.40
At 1 MHz		2.2	2.45	2.88	2.90
Power factor	D 150				
At 60 Hz		0.05	0.0002	0.0007	0.006
At 1 MHz		0.03	0.0002 to 0.0004	0.0075	0.013
Volume resistivity, Ω · m	D 257	10^{16}	$>10^{14}$	8×10^{14}	10^{13}

Properties of principal optical plastics

Properties	PMMA	PS	NAS	SAN	PC	Methyl pentene	ABS	ADC	BK 7 glass
Refractive index									
n_D (589 nm)	1.491	1.590	1.533 to 1.567	1.567 to 1.571	1.586	1.467	1.538	1.504	1.517
n_C (656.3 nm)	1.488	1.585	1.558	1.563	1.581	1.464	...	1.501	1.514
n_F (486.1 nm)	1.496	1.604	1.575	1.578	1.598	1.473	...	1.510	1.522
Abbe value	61.4	31.1	32.4	37.9	34.5	51.9	...	56	64.6
Rate of change in index with temperature, dn/dt, 10^{-5}/°C (10^{-5}/°F)	−12.5 (−6.9)	−12.0 (−6.7)	−14.0 (−7.8)	...	−11.8 to −14.3 (−6.5 to −7.9)	...	...	−14.3 (−7.9)	0.3 (0.17)
Coefficient of linear expansion, 10^{-5}/K	6.74 at 70 °C	6.0 to 8.0	...	6.5 to 6.7	6.6 to 7.0	...	...	11.4 at 25 to 75 °C; 14.4 at 75 to 125 °C	0.71
Deflection temperature, 2 °C/min. (3.6 °F/min.)									
At 1.8 MPa (0.264 ksi), °C (°F)	92 (198)	82 (180)	...	99 to 104 (210 to 220)	142 (290)	...	90 (195)	...	...
At 445 kPa (66 psi), °C (°F)	101 (215)	110 (230)	100 (212)		146 (295)	...	84 (185)	...	...
Recommended maximum continuous service temperature, °C (°F)	92 (198)	82 (180)	93 (200)	79 to 88 (175 to 190)	124 (255)	...	...	100 (212)	...
Thermal conductivity, W/m · K (Btu · in./ · ft^2 · °F)	0.21 (1.45)	0.10 to 0.14 (0.69 to 0.97)	0.19 (1.32)	0.12 (0.83)	0.19 (1.32)	0.17 (1.18)	...	0.21 (1.45)	0.011 (0.076)
Haze, %	2	3	3	3	3	5	12	3	...
Luminous transmittance 3.175 mm (0.125 in.), %	92	88	90	88	89	90	79 to 90.6(a)	93	99.9
Water absorption, immersed 24 h at 23 °C (75 °F), %	0.3	0.2	0.15	0.2 to 0.35	0.15	...	...	0.2	...
Mold shrinkage, %	0.2 to 0.6	0.2 to 0.6	0.2	0.2 to 0.6	0.5 to 0.7	1.5 to 3.0	...	...	...

(a) 79% for 6.35 mm (0.25 in.) and 90.6% for 0.381 mm (0.015 in.)

Values of Poisson's ratio for selected unfilled thermoplastics

Polymer	Poisson's ratio
Acetal	0.35
Nylon 6/6	0.39
Modified PPO	0.38
Polycarbonate	0.36
Polystyrene	0.33
PVC	0.38
TFE (tetrafluorethylene)	0.46
FEP (fluorinated ethylene propylene)	0.48

Mechanical properties based on ASTM D 4000 or ISO 1043

Material	Tensile strength MPa	Tensile strength ksi	Tensile modulus GPa	Tensile modulus 10^6 psi	Flexural strength MPa	Flexural strength ksi	Impact strength J/m	Impact strength ft · lbf/in.	Hardness, Rockwell	Flame rating, UL 94
ABS	41	6.0	2.3	0.33	72.4	10.5	347	6.5	R103	HB
ABS-PC	59	8.5	2.6	0.38	89.6	13.0	560	10.5	R117	HB
DAP	48	7.0	10.3	1.50	117	17.0	37	0.7	E80	HB
POM	69.0	10.0	3.2	0.47	98.6	14.3	133	2.5	R120	HB
PMMA	72.4	10.5	3.0	0.43	110	16.0	21	0.4	M68	HB
PAR	68	9.9	2.1	0.30	82.7	12.0	288	5.4	R122	HB
LCP	110	16.0	11.0	1.60	124	18.0	101	1.9	R80	V0
MF	52	7.5	9.65	1.40	93.1	13.5	16	0.3	M120	V0
Nylon 6	81.4	11.8	2.76	0.40	113	16.4	59	1.1	R119	V2
Nylon 6/6	82.7	12.0	2.83	0.41	110	16.0	53	1.0	R121	V2
Nylon 12	81.4	11.8	2.3	0.34	113	16.4	64	1.2	R122	V2
PAE	121	17.6	8.96	1.30	138	20.0	64	1.2	M85	V0
PBT	52	7.5	2.3	0.34	82.7	12.0	53	1.0	R117	HB
PC	69.0	10.0	2.3	0.34	96.5	14.0	694	13.0	R118	V2
PBT-PC	55	8.0	2.2	0.32	86.2	12.5	800	15.0	R115	HB
PEEK	93.8	13.6	3.5	0.51	110	16.0	59	1.1	...	V0
PEI	105	15.2	3.0	0.43	152	22.0	53	1.0	M109	V0
PESV	84.1	12.2	2.6	0.38	129	18.7	75	1.4	M88	V0
PET	159	23.0	8.96	1.30	245	35.5	101	1.9	R120	HB
PF	41	6.0	5.9	0.85	62	9.0	21	0.4	M105	HB
PPO (mod)	54	7.8	2.5	0.36	88.3	12.8	267	5.0	R115	V0
PPS	138	20.0	11.7	1.70	179	26.0	69	1.3	R123	V0
PSU	73.8	10.7	2.5	0.36	106	15.4	64	1.2	M69	HB
SMA	31	4.5	1.9	0.27	55	8.0	133	2.5	R95	HB
UP	41	6.0	5.5	0.80	82.7	12.0	32	0.6	M88	HB

Mechanical, thermal, and physical properties of transparent structural materials

Material	Military specification	Mechanical properties: Tensile strength MPa	Tensile strength ksi	Compressive strength(a) MPa	Compressive strength(a) ksi	Modulus of elasticity(b) GPa	Modulus of elasticity(b) 10^6 psi	Design stress MPa	Design stress ksi
Polycarbonate	MIL-P-83310	68 to 72	9.8 to 10.5	86	12.5	2.1 to 2.4	0.3 to 0.35	28	4
Stretched acrylic	MIL-P-25690	79 to 82	11.5 to 11.9	120	17	3.1 to 3.4	0.45 to 0.49	21	3
Chemically tempered compound	MIL-G-25661-V	240 to 360	35 to 52	...	...	72	10.5	...	...

Material	Military specification	Thermal properties: Coefficient of thermal expansion(c) 10^{-6}/K	Coefficient of thermal expansion(c) 10^{-6}/°F	Specific heat(d) kJ/kg · K	Specific heat(d) Btu/lb · °F	Coefficient of thermal conductivity(e) W/m · K	Coefficient of thermal conductivity(e) Btu · in./ft² · h · °F	Physical properties: Density(e) g/cm³	Density(e) lb/in.³	Poisson's ratio	Specific gravity
Polycarbonate	MIL-P-83310	6.25	3.47	1.17	0.28	0.22	1.5	1.19	0.043	0.36(f)	1.2
Stretched acrylic	MIL-P-25690	...	...	1.47	0.35	0.17	1.15	1.19	0.043	0.35	1.19
Chemically tempered compound	MIL-G-25661-V	0.88	0.49	0.84	0.20	...	...	2.52	0.091	0.22	2.44

(a) ASTM D 694 and FTMS 406-1021. (b) ASTM D 747 and FTMS 406-1011. (c) ASTM D 696 and FTMS 406-2031. (d) ASTM C 351-54. (e) ASTM C 177. (f) AFFDL-TR-25-2 (Air Force Dynamics Laboratory)

Typical properties of thermoplastic resins

Thermoplastic resin	T_g, °C	T_m, °C	T_p, °C	Tensile strength, MPa	Tensile modulus, GPa	Elongation %
Acetal copolymer	...	175	232	69	3.2	40 to 75
Nylon 6	...	220	288	41 to 166	2.6	30 to 100
Nylon 6/6	...	265	327	95	1.6 to 3.8	15 to 80
Nylon 12	155	209	275	62	0.2 to 1.2	250 to 390
Polybutylene terephthalate	...	126	195	30	0.2	300 to 380
Polycarbonate	150	...	295	66	2.4	110 to 120
Polyetheretherketone	144	340	380	103	1.1	30 to 150
Polyethylene terephthalate	80	265	316	72	4.1	30 to 300
Polyphenylene sulfide	88	290	338	86	3.3	1 to 3
Polypropylene	-20	175	260	41	1.6	100 to 600
Polyurethane	...	137	265	62	2.1	60 to 180
Styrene maleic anhydride	114	...	265	48	3.3	1.8 to 30

High-performance thermoplastics

Material	Glass transition temperature °C	°F	Processing temperature °C	°F	Tensile properties at 25 °C (77 °F): Yield strength MPa	ksi	Modulus GPa	10^6 psi	Strain at break, %	Fracture toughness, resin J/m²	in. · lbf/in.²	Notched Izod J/m	ft · lbf/in.
PEEK	143	290	–400	–690	100	14.5	3.10	0.450	>40	>4025	>23	85	1.6
Polyarylene ketone (APC-HTX)	205	400	–420	–725	...	...	...	...	...	...	...	...	...
Polyarylene ketone (PXM 8505)	265	510	...	...	87.6	12.7	2.48	0.360	13	...	...	...	...
PPS (Ryton)	90	195	–343	–585	82.7	12.0	4.34	0.630	5	...	...	160	3.0
Polyarylene sulfide (PAS-2)	215	420	–329	–560	100	14.5	3.24	0.470(a)	7.3	...	...	42	0.8
Polyarylamide (J-polymer)	120	250	–343	–585	...	...	...	...	...	...	...	...	...
PAI (Torlon)	275	525	–400	–690	63.4	9.2	4.60	0.667	1.4	3400	19.4	144	2.7
PAI (Torlon AIX638)	243	470	–350	–600	90	13.0	2.75	0.400	30	3500	20	...	...
PEI (Ultem 1000)	217	425	–343	–585	105	15.2	2.96	0.430	60	3325	19	53	1.0
PI (Avimid K-III)	251	485	343 to 360	–585 to –615	102	14.8	3.76	0.546	14	1925	11	...	...
PI (LARC-TPI)	264	510	–343	–585	120	17.3	3.72	0.540	4.8	...	...	...	...
Polyimide sulfone ($PISO_2$)	273	525	–343	–585	62.7	9.1	4.96	0.719	1.3	1400	8	...	...
PSU (Udel P-1700)	190	375	–300	–510	70.3	10.2	2.48	0.360	>50	2450	14	64	1.2
Polyaryl sulfone (Radel A400)	220	430	–330	–560	71.7	10.4	2.14	0.310	60	3500	20	64	1.2
Polyarylether sulfone (Victrex 4100G)	230	450	–300	–510	84.1	12.2	2.62	0.380(a)	>40	1925	11	85	1.6
Polyester (Xydar SRT-300)	350	660	–400	–690	138	20.0	16.5	2.40	4.9	1210	6.9	128	2.4

(a) Flexural modulus

Properties of impact-modified plastics

Properties	Impact-resistant polystyrene (IPS): Semi-impact resistant	Impact resistant	High-impact resistant	Styrene-acrylonitrile copolymer (SAN): Not reinforced	Reinforced by 35 wt% glass fibers	Acrylic-styrene-acrylonitrile (ASA)
Density, g/cm³	1.04	1.04	1.04	1.08	1.36	1.07
Melt flow index, g/10 min.	1.5 to 20	3 to 20	3 to 12	...	...	...
Impact strength						
At 20 °C (68 °F), kJ/m² (ft · lbf/ft²)	40 to 60 (2750 to 4100)	60 to 80 (4100 to 5500)	No break	20 to 25 (1400 to 1700)	10 to 18 (690 to 1250)	No break
At –40 °C (–40 °F), kJ/m² (ft · lbf/ft²)	35 to 50 (2400 to 3450)	50 to 70 (3450 to 4800)	70 (4800) No break	...	...	20 to 50 (1400 to 3450)
Notched impact strength						
At 20 °C (68 °F), kJ/m² (ft · lbf/ft²)	5 to 6 (340 to 410)	5 to 8 (340 to 550)	8 to 14 (550 to 960)	2 to 3 (140 to 210)	4 to 5 (270 to 340)	7 to 14 (480 to 960)
At –40 °C (–40 °F), kJ/m² (ft · lbf/ft²)	3 to 5 (210 to 340)	4 to 6 (270 to 410)	6 to 12 (410 to 820)	...	...	...

Thermoplastic structural foam properties

Properties	Polyethylene	Flame retardant acrylonitrile-butadiene-styrene	Polyphenylene ether alloy: Standard V-0 grade	10% glass fiber	Polycarbonate: 7% glass fiber	30% glass fiber(a)	Polyphenylene ether/polyamide		Thermoplastic polyester	Flame-retardant polyester: 6.3 mm (0.25 in.) thick sample	9.5 mm (0.375 in.) thick sample
Physical											
Specific gravity	0.6	...	...	...	...	1.23	...	...	...	1.05	1.05
Density reduction, %	15	15	15	15	15	...	15	10	15	15	15
Mold shrinkage, %	1.8	0.5 to 0.8	0.5 to 0.7	0.4 to 0.6	0.4 to 0.7	0.2 to 0.4	1.2 to 1.6	0.2 to 0.4	0.5 to 0.7	0.5 to 0.7	0.5 to 0.7
Mechanical											
Tensile strength, MPa (ksi)	12.4 (1.8)	20.0 (2.9)	28.3 (4.1)	41.4 (6.0)	44.8 (6.5)	66.9 (9.7)	51.2 (7.42)	84.8 (12.3)	72.4 (10.5)	15.9 (2.3)	17.2 (2.5)
Tensile modulus, GPa (10^6 psi)	...	1.41 (0.205)	1.86 (0.270)	2.62 (0.380)	2.28 (0.330)	...	1.52 (0.220)	3.38 (0.490)	...	1.52 (0.220)	1.52 (0.220)
Elongation, %	...	...	...	...	4	3	...	...	4	...	...
Flexural strength, MPa (ksi)	18.6 (2.7)	41.4 (6.0)	60.0 (8.7)	55.8 (8.1)	86.2 (12.5)	120 (17.5)	64.8 (9.4)	84.8 (12.3)	114 (16.5)	39.3 (5.7)	4.20 (6.1)
Flexural modulus, GPa (10^6 psi)	0.827 (0.120)	1.69 (0.245)	2.28 (0.330)	3.03 (0.440)	2.79 (0.400)	6.72 (0.975)	1.62 (0.235)	4.31 (0.625)	5.17 (0.750)	1.90 (0.275)	1.62 (0.235)
Impact strength, falling dart test, J (in. · lbf)	...	4 (36)	8 (72)	3 (28)	20 (170)	11 (100)	33 (288)	27 (240)	4 (36)	3 (29)	9 (85)

(continued)

Thermoplastic structural foam properties (continued)

Properties	Polyethylene	Flame retardant acrylonitrile-butadiene-styrene	Polyphenylene ether alloy Standard V-0 grade	Polyphenylene ether alloy 10% glass fiber	Polycarbonate 7% glass fiber	Polycarbonate 30% glass fiber(a)	Polyphenylene ether/polyamide		Thermoplastic polyester	Flame-retardant polyester 6.3 mm (0.25 in.) thick sample	Flame-retardant polyester 9.5 mm (0.375 in.) thick sample
Thermal											
Deflection temperature											
At 1.8 MPa (0.264 ksi), °C (°F)	35 (94)	70 (162)	80 (180)	90 (190)	135 (273)	140 (280)	135 (275)	205 (398)	150 (300)	75 (168)	76 (169)
At 455 kPa (66 psi), °C (°F)	...	80 (177)	90 (194)	...	145 (290)	...	185 (365)	>205 (>400)	...	84 (183)	86 (186)
Flammability, UL 94	HB	V-0/5V	V-0/5V	V-0/5V	V-0/5V	V-0/5V	...	...	...	V-0/5V	V-0/5V

(a) Test values determined on 6.4 mm (0.250 in.) thick samples

Tensile strength at yield of engineering thermoplastics

Engineering thermoplastic	Tensile strength MPa	Tensile strength ksi
PAI	186.2	27.0
PEI	104.8	15.2
PEEK	100.0	14.5
PESV	84.1	12.2
Nylon 6/6	82.7	12.0
PSU	70.3	10.2
Acetals	68.9	10.0
PPO	66.2	9.6
PC	65.5	9.5
PPS (a)	65.5	9.5
PET(a)	62.1	9.0
PPE	53.8	7.8
PBT	51.7	7.5
ABS	48.3	7.0

(a) Estimated because values refer to neat (unreinforced, unmodified) resin, and engineering commercial grades of PPS and PET are always reinforced. Values for neat PPS and PET would not appear on suppliers' data sheets. Data generated using ASTM D 638

Flexural modulus of engineering thermoplastics

Engineering thermoplastic	Flexural modulus GPa	Flexural modulus 10^6 psi
PAI	4.1	0.66
PEEK	3.8	0.55
PPS(a)	3.8	0.55
PEI	3.3	0.48
Acetals	2.9	0.41
Nylon 6/6	2.9	0.41
PET(a)	2.8	0.40
ABS	2.7	0.39
PSU	2.7	0.39
PESV	2.6	0.375
PPE	2.5	0.36
PPO	2.5	0.36
PC	2.3	0.34
PBT	2.3	0.34

(a) Estimated because values refer to neat (unreinforced, unmodified) resin, and engineering commercial grades of PPS and PET are always reinforced. Values for neat PPS and PET would not appear on suppliers' data sheets. Data generated using ASTM D 790

Izod notched impact strength of engineering thermoplastics

Engineering thermoplastic	Izod impact strength J/m	Izod impact strength ft · lbf/in.
PC	747.6	14.0
PPE	267.0	5.0
PPO	267.0	5.0
ABS	213.6	4.0
PAI	133.5	2.5
PEEK	96.1	1.8
PESV	85.4	1.6
Acetals	74.8	1.4
PSU	69.4	1.3
Nylon 6/6	53.4	1.0
PBT	53.4	1.0
PEI	53.4	1.0
PET(a)	26.7	0.5
PPS(a)	26.7	0.5

(a) Estimated because values refer to neat (unreinforced, unmodified) resin, and engineering commercial grades of PPS and PET are always reinforced. Values for neat PPS and PET would not appear on suppliers' data sheets. Data generated using ASTM D 256

Interlaminar fracture toughness of thermoplastic resins

Material	J/m^2	ft · lbf/ft^2
Polyphenylene sulfide	720	50
Polyetherimide	950	65
Polyamideimide	1050	70
Polysulfone	1175	80
Polyether etherketone	1600	110

Shear and tensile creep moduli at 100s, 20 °C (68 °F), and small strains

Material	Creep moduli				
	Shear, G_c		Tensile, E_c		
	GPa	10^6 psi	GPa	10^6 psi	E_c/G_c
Polymethyl methacrylate	1.17	0.17	3.12	0.453	2.66
Polypropylene, density of 0.910 g/cm^3 from center of 38 mm (1.5 in.) thick block	0.552	0.08	1.51	0.219	2.73
Polypropylene, density of 0.907 g/cm^3, injection molding	0.526	0.07	1.59	0.230	3.02
Polyethylene, density of 0.955 g/cm^3, injection molding	0.393	0.05	1.05	0.152	2.68
Oxymethylene copolymer, injection moldings, dry	0.90	0.13	2.52	0.365	2.79
Nylon 6/6	3.45	0.50	9.93	1.44	2.88
Nylon 6/6 + 0.33 weight fraction of glass fiber, injection molding with 65% relative humidity	0.884	0.13	6.26	0.908	7.08
Oak, with specimen axis along the grain	0.848	0.12	9.07	1.32	10.7
Carbon fiber reinforced epoxy resin	3.88	0.56	176	25.5	45.3

Wear factors, coefficients of friction, and PV limits of reinforced and lubricated thermoplastic against steel

Material	PTFE content, wt %	Silicone content, wt %	Glass fiber content, wt %	Carbon fiber content, wt %	Aramid fiber content, wt %	Wear factor(a)	Coefficient of friction(b)	PV limit at 100 fpm psi-fpm
ABS(c)			Unmodified			3500	0.35	...
	15	...	...	...	...	300	0.16	4,000
	...	2	...	...	...	80	0.14	...
SAN(d)			Unmodified			3000	0.33	...
	15	...	...	...	...	200	0.14	5,000
	...	2	...	...	...	70	0.13	...
Polystyrene			Unmodified			3000	0.32	1,500
	...	2	...	...	...	37	0.08	9,000
	15	...	...	...	...	175	0.14	...
Polycarbonate			Unmodified			2500	0.38	500
	15	...	...	...	...	75	0.15	20,000
	13	2	...	...	...	40	0.09	23,000
	20	...	...	...	...	70	0.14	22,000
	13	2	30	...	...	27	0.19	30,000
	...	...	...	30	...	85	0.17	5,500
Polyetherimide			Unmodified			1000	0.17	...
	...	...	30	...	...	130	0.24	...
	...	...	...	30	...	75	0.23	...
	15	...	30	...	...	35	0.20	...
Polyethylene	20	...	...	...	...	45	0.13	...
Polysulfone			Unmodified			1500	0.37	5,000
	15	...	...	...	...	46	0.14	...
	...	...	30	...	...	160	0.22	...
	15	...	30	...	...	55	0.19	35,000
Polyether sulfone	15	...	30	...	...	60	0.20	30,000
	15	...	...	30	...	40	0.17	33,000
Polyimide			Unmodified			100	0.29	300,000
	10	...	...	15	...	28	0.12	1,000,000
Acetal			Unmodified			65	0.21	3,500
	15	...	...	...	...	20	0.16	...
	...	2	...	...	...	20	0.11	12,000
	18	2	...	...	...	7	0.10	18,000
	...	...	...	20	...	40	0.14	20,000
	20	...	...	...	...	13	0.12	16,000
	...	2	...	...	...	27	0.12	9,000
Polypropylene	20	...	...	...	...	33	0.11	5,000
	15	...	30	...	...	36	0.09	12,000
Polyphenylene sulfide			Unmodified			540	0.24	3,000
	20	...	...	...	...	55	0.10	...
	15	...	...	30	...	75	0.15	35,000
	...	...	...	30	...	160	0.20	20,000
	13	2	30	...	...	50	0.22	30,000
Nylon 6/6			Unmodified			200	0.28	2,500
	20	...	...	...	...	12	0.18	17,500
	...	2	...	...	...	40	0.09	6,000
	18	2	...	...	...	6	0.08	30,000
	...	...	...	30	...	20	0.20	27,000
	...	...	30	...	...	75	0.31	10,000
	13	2	...	30	...	6	0.11	43,000
	...	...	...	...	20	62	0.25	...
	10	...	...	...	15	23	0.25	...

(continued)

Wear factors, coefficients of friction, and PV limits of reinforced and lubricated thermoplastic against steel (continued)

Material	PTFE content, wt %	Silicone content, wt %	Glass fiber content, wt %	Carbon fiber content, wt %	Aramid fiber content, wt %	Wear factor(a)	Coefficient of friction(b)	PV limit at 100 fpm psi-fpm
Polyurethane			Unmodified			340	0.37	1,500
	...	...	...	60	0.32			
	...	2	...	...	...	55	0.31	...
	15	...	30	...	...	35	0.25	10,000
Polyether (PBT)			Unmodified			210	0.25	...
	20	...	...	...	...	15	0.17	15,500
	18	2	...	...	...	9	0.13	...
	...	...	...	30	...	24	0.15	22,000
	13	2	30	...	...	12	0.12	24,000
	...	...	...	...	20	95	0.21	...
Polyester elastomer			Unmodified			1000	0.59	...
	...	2	...	...	...	30	0.22	...
	15	...	...	...	...	40	0.25	...
	13	2	...	...	...	5	0.21	...
Modified PPO(e)			Unmodified			3000	0.39	500
	15	...	...	...	...	100	0.16	...
	15	...	30	...	...	45	0.22	22,000
Polyether ether keytone	...	...	...	30	...	60	0.13	...
	20	...	...	...	...	130	0.23	...
	15	...	...	15	...	60	0.20	40,000
Bronze (oil impregnated)	...	...	...	...	...	100	0.20	...
Phenolic:								
Wood flour and PTFE	...	...	...	...	...	30	0.26	...
Cellulose and PTFE	...	...	...	...	...	300	0.16	35,000
Teflon PTFE (milled glass)	...	...	...	...	...	8	0.16	...

(a) 10^{-10} in.^{3} · min/ft · lbf · h. (b) Dynamic, at 40 psi and 50 fpm. (c) Acrylonitrile butadiene styrene. (d) Styrene acrylonitrile. (e) Polyphenylene oxide

Friction coefficients for various materials

Material	Friction coefficient
PTFE	0.12 to 0.22 ...
PE, rigid	0.20 to 0.25 (X2.0)
PP	0.25 to 0.30 (X1.5)
POM	0.20 to 0.35 (X1.5)
PA	0.30 to 0.40 (X1.5)
PBT	0.35 to 0.40 ...
PS	0.40 to 0.50 (X1.2)
SAN	0.45 to 0.55 ...
PC	0.45 to 0.55 (X1.2)
PMMA	0.50 to 0.60 (X1.2)
ABS	0.50 to 0.65 (X1.2)
PE, flexible	0.55 to 0.60 (X1.2)
PVC	0.55 to 0.60 (X1.0)

PE, polyethylene; PP, polypropylene; POM, polyoxymethylene; PS, polystyrene; SAN, styrene-acrylonitrile; PMMA, polymethyl methacrylate; PVC, polyvinyl chloride

Properties of selected plastics

Plastic	Flammability(a)	Heat deflection (DTUL)(b) °C at 1.82 MPa	°F at 264 psi	Thermal conductivity(c) W/m · K	Btu · in/ft^2 · h · °F	Dielectric strength(d) V/µm	V/mil	Volume resistivity(e), Ω/cm	Relative permittivity(f), at 60 Hz	Arc resistance(g), s	Mold shrinkage(h), mm/mm (in. / in.)
Acetal	HB	128	230	0.225	1.56	19.7	500	10^{15}	3.70	129	0.020
Nylon 6	HB	86 to 103	155 to 185	0.173	1.20	12.0	305	4.5×10^{13}	4.00	...	0.005
Nylon 6/6	V-2	93	167	0.245	1.70	15.2	385	10^{15}	4.00	120	0.008
Polycarbonate	V-2	150	270	0.195	1.35	15.0	380	$>10^{16}$	2.96	120	0.005 to 0.007
Polypropylene	HB	64 to 78	115 to 140	0.175	1.21	23.6	600	$>10^{17}$	2.20	125	0.020
Polyphenylene sulfide	V-0	153	275	0.241	1.67	15.0	380	10^{16}	...	...	0.007
Acrylonitrile butadiene styrene (ABS)	HB	111 to 122	200 to 220	0.138	0.96	13.8 to 19.7	350 to 500	2.7×10^{16}	2.80	...	0.004 to 0.009
Polyphenylene oxide (PPO)	V-1	118	212	0.133	0.92	15.7	400	$>10^{16}$	2.65	75	0.005 to 0.007
Polystyrene acrylonitrile (SAN)	HB	122	220	0.101	0.70	20.3	515	4.4×10^{16}	3.50	65	...
Polyester (PBT)	HB	68 to 103	122 to 185	0.147 to 0.241	1.02 to 1.67	16.5 to 21.7	420 to 550	4×10^{16}	3.10	184	0.015 to 0.020
Polyester (PET)	HB	56 to 59	100 to 106	0.125	0.87	...	...	$>10^{16}$	...	...	0.02 to 0.025

(a) Test method UL-94. (b) ASTM test method D 648. (c) ASTM test method C 177. (d) ASTM test method D 149. (e) ASTM test method D 257. (f) ASTM test method D 150. (g) ASTM test method D 495. (h) ASTM test method D 955

Properties of thermoplastics

Material	Supplier	T_g °C	T_g °F	T_m °C	T_m °F	Tensile strength MPa	Tensile strength ksi	Tensile modulus MPa	Tensile modulus ksi	Fracture toughness, G_{Ic} J/m²	Fracture toughness, G_{Ic} ft · lbf/ft²
Udel P-1700	Union Carbide	190	375	None	…	76	11	2200	320	3200	220
Radel	Union Carbide	220	430	None	…	…	…	…	…	5500	380
Vitrx PES 200P	ICI	220	430	None	…	83	12	2410	350	2600	180
Ultem	General Electric	220	430	None	…	110	16	3300	480	3700	250
Torlon	Amoco	275	530	None	…	193	28	4800	700	3400	230
Ryton	Phillips	85	185	285	545	65	9.5(a)	3800	550(a)	210(b)	15
PEEK	ICI	143	290	343	650	…	…	…	…	…	…
Avimid K-11	Du Pont	277	530	None	…	110	16	2850	415	14000	960
CM-X	Ausimont U.S.A.	…	…	327	620	37	5.5(b)	3800	550	(b)	…

(a) 15% crystallinity. (b) High crystallinity

Impact resistance of selected engineering thermoplastics

Thermoplastic	Notched Izod 23 °C (73 °F) J/cm	Notched Izod 23 °C (73 °F) ft · lbf/in.	Notched Izod –29 °C (–20 °F) J/cm	Notched Izod –29 °C (–20 °F) ft · lbf/in.	Notched Izod –40 °C (–40 °F) J/cm	Notched Izod –40 °C (–40 °F) ft · in./in.	Falling dart(a) 23 °C (73 °F) J	Falling dart(a) 23 °C (73 °F) in. · lbf	Falling dart(a) –29 °C (–20 °F) J	Falling dart(a) –29 °C (–20 °F) in. · lbf	Falling dart(a) –40 °C (–40 °F) J	Falling dart(a) –40 °C (–40 °F) in. · lbf
Acetal	0.53	1.0	…	…	…	…	…	…	…	…	…	…
Nylon, amorphous	5.9 to 11	11 to 21	2.1 to 4.8	4 to 9	…	…	…	…	…	…	…	…
Nylon 6(b)	0.53 to 0.80	1.0 to 1.5	……	…	…	…	…	…	…	…	…	…
Impact-modified nylon 6(b)	9.6	18	2.9	5.5	…	…	65.1	576	57.0	504	…	…
Nylon 6/6(b)	0.3 to 0.6	0.5 to 1.2	…	…	…	…	…	…	…	…	…	…
Toughened nylon 6/6(b)	9.1 to 12	17 to 22	…	…	…	…	…	…	…	…	…	…
33% glass-toughened nylon 6/6(b)	2.2	4.1	…	…	…	…	…	…	…	…	…	…
Nylon/ABS	9.98	18.7	1.0	1.8	…	…	73.2	648	35.3	312	…	…
Nylon/PPO	2.1	4.0	1.3	2.5	…	…	50.8	450	39.6	350	…	…
Polyarylate	2.9	5.5	…	…	2.1	4.0	…	…	…	…	…	…
Polycarbonate	6.4 to 8.5	12 to 16	…	…	1.3 to 5.98	2.5 to 11.2	…	…	…	…	…	…
Polycarbonate/ABS	4.5 to 6.4	8.5 to 12	2.7 to 6.14	5 to 11.5	…	…	38.0 to 61.0	336 to 540	29.8 to 46.1	264 to 408	…	…
Polycarbonate/PBT	6.9 to 8.5	13 to 16	…	…	1.6 to 6.4	3 to 12	55.6	492	48.8	432	…	…
Polycarbonate/PET	9.1	17	6.9	13	…	…	57.0	504	50.2	444	…	…
Polyester, aromatic (LCP)	0.53 to 5.3	1 to 10	…	…	…	…	…	…	…	…	…	…
Thermoplastic polyester, PBT	0.4 to 0.53	0.8 to 1.0	…	…	…	…	…	…	…	…	…	…
Impact-modified PBT	8.0 to 8.5	15 to 16	7.5 to 8.0	14 to 15	…	…	43.4 to 54.2	384 to 480	43.4	384	…	…
Impact-modified PBT/PET	8.0 to 8.5	15 to 16	…	…	…	…	54.2	480	…	…	…	…
30% glass-reinforced PET, thermoplastic polyester	1.0	1.9	…	…	1.0	1.8	…	…	…	…	…	…
Toughened 30% glass PET	1.4 to 2.3	2.6 to 4.4	…	…	1.2 to 1.6	2.3 to 3.0	…	…	…	…	…	…
Polyether ether ketone	0.59 to 0.80	1.1 to 1.5	…	…	…	…	…	…	…	…	…	…
Polyetherimide	0.3 to 0.53	0.6 to 1.0	…	…	…	…	33.9 to 36.2	300 to 320	…	…	…	…
Polyether sulfone	0.53 to 0.80	1.0 to 1.5	…	…	…	…	…	…	…	…	…	…
Polyphenylene oxide, modified	2.7 to 5.3	5 to 10	…	…	1.3 to 1.9	2.5 to 3.5	14.7 to 33.9	130 to 300	…	…	3.4 to 11.3	30 to 100
Polyphenylene sulfide	0.3 to 0.75	0.5 to 1.4	…	…	…	…	…	…	…	…	…	…
Polyphthalate carbonate	5.3	10	…	…	…	…	…	…	…	…	…	…
Polysulfone	0.69	1.3	…	…	0.64	1.2	…	…	…	…	…	…
Polyurethane, engineering thermoplastic	0.5 to 2.1	1.4	…	…	…	…	…	…	…	…	…	…

(a) Gardner or falling dart. (b) Dry, as molded

Fatigue endurance of reinforced engineering thermoplastics

Material	Glass fiber content, %	Carbon fiber content, %	Cyclic failure stress(a), At 10^4 cycles MPa	Cyclic failure stress(a), At 10^4 cycles ksi	Cyclic failure stress(a), At 10^7 cycles MPa	Cyclic failure stress(a), At 10^7 cycles ksi
Acetal copolymer	30	…	62	9.0	48	7.0
Nylon 6(b)	30	…	48	7.0	40	5.75
Nylon 6/6	…	…	45	6.5	36	5.2
	40	…	72	10.5	63	9.1
Nylon 6/6(b)	…	…	23	3.4	21	3.1
	30	…	55	8.0	41	5.9
	40	…	62	9.0	48	7.0
	…	30	90	13.0	55	8.0
	…	40	103	15.0	59	8.5

(continued)

Fatigue endurance of reinforced engineering thermoplastics (continued)

Material	Glass fiber content, %	Carbon fiber content, %	Cyclic failure stress(a), At 10^4 cycles, MPa	ksi	At 10^7 cycles, MPa	ksi
Nylon 6/10(b)	30	...	47	6.8	40	5.5
	40	...	55	8.0	48	7.0
Polycarbonate	20	...	62	9.0	34	5.0
	40	...	100	14.5	41	6.0
Polyester, PBT	30	...	76	11.0	35	5.1
	...	30	90	13.0	45	6.5
Polyetheretherketone	...	30	124	18.0	121	17.5
Polyether sulfone	30	...	110	16.0	34	5.0
	40	...	131	19.0	43	6.2
	...	30	152	22.0	46	6.7
Modified polyphenylene oxide	30	...	50	7.2	33	4.75
Polyphenylene sulfide	...	30	90	13.0	66	9.5
Polysulfone	30	...	97	14.0	31	4.5
	40	...	110	16.0	38	5.5

(a) Specimens test in accordance with ASTM D 671 at 1800 cycles per minute. (b) Moisture conditioned, 50% relative humidity

Melt viscosities of thermoplastic resins

Material	Temperature, °C	°F	Viscosity, Pa · s	P
Polyamideimide	350	662	10^6	10^7
Polysulfone	300	570	10^4	10^5
Polyetherimide	305	580	10^5	10^6
Polyphenylene sulfide	313	595	10^4	10^5
Polyetheretherketone	400	750	10^3	10^4

Effect of flame retardant on mechanical properties

Plastic	Flame retardant type (a)	Amount of additive, % (b)	Change in tensile strength At yield, %	At break, %	Elongation at break, %	Flexural yield strength, %	Flexural modulus, %	Heat deflection temperature at 455 kPa, %	Impact strength, Izod, notched, %
Polyethylene	NP	30	10 to 15	...	...	...	...	50 to 75	−15 to −20
	SP	25 to 35	10 to 15	−20 to −20	−5 to −10	...	...	50 to 60	−15 to −20
Polypropylene	NP	40	−25 to −35	−5 to −10	−60 to 70	7 to −12	50 to 70	...	−15 to −75
	SP	20	...	−9 to −11	...	...	...	...	...
Polystyrene	NP	20 to 30	...	−15 to −55	−25 to −40	−25 to −45	0	0	−10 to −70
	SP	17 to 20	...	−2 to −8	9 to 12	...	...	...	...
Impact polystyrene	NP	20	0	0	−5 to −75	...	2 to 15	−3 to −10	−20 to −40
	SP	15 to 20	...	...	...	...	...	...	...
ABS	NP	15 to 20	0	0 to −10	...	0	0	0 to −10	−3 to −70
	SP	15 to 20	...	...	...	...	...	...	...

(a) NP, nonplasticizing; SP, semiplasticizing. (b) Blend of two to three parts additive per part antimony trioxide. (c) At yield

Solubility parameters of plastics and polymers

Polymer	Solubility parameter, $\sqrt{cal/cm3}$	Polymer	Solubility parameter, $\sqrt{cal/cm3}$
Polytetrafluoroethylene	6.2	Polyvinyl acetate	9.4
Polydimethyl siloxane	7.3	Polymethyl methacrylate	9.0 to 9.5
Polyethylene	7.9 to 8.1	Polyvinyl chloride	9.38 to 9.7
Polypropylene	7.9	Bisphenol A polycarbonate	9.5
Polyisobutylene	7.7 to 8.05	Polyvinylidene chloride	9.8
Styrene-butadiene rubber (25% styrene)	8.09 to 8.6	Polyethylene terephthalate	10.7
Polyisoprene (natural rubber)	7.9 to 8.35	Cellulose dinitrate	10.48 to 10.56
Polybutadiene	8.4 to 8.6	Cellulose diacetate	11.35
Polysulfide rubbers	9.0 to 9.4	Epoxide resins	11.0
Polystyrene	8.5 to 9.7	Acetal resin (homopolymer)	11.1
Polychloroprene	9.18 to 9.25	Polyamide 6/6	13.6
Nitrile rubbers (25% acrylonitrile)	9.38 to 9.5	Polyacrylonitrile	15.4

Retention of tensile strength by selected thermoplastics after exposure to various chemical media

Chemical medium	Retention of tensile strength(a), %, by: PPS(b)	Nylon 6/6	PC(c)	PSO(d)	Modified PPO(e)
Acids					
Hydrochloric, 37%	100	0	0	100	100
Sulfuric, 30%	100	0	100	100	100
Acetic, glacial	98	0	67	91	78
Bases					
Ammonium hydroxide, 28%	100	85	0	100	100
Sodium hydroxide, 30%	100	89	7	100	100
N-butylamine	49	91	0	0	0
Aniline	96	85	0	0	0
Hydrocarbons					
Cyclohexane	100	90	75	99	0
Toluene	98	76	0	0	0
Diesel fuel	100	87	100	100	36
Gasoline	100	80	99	100	0
Organic solvents					
Chloroform	87	57	0	0	0
Chlorobenzene	100	73	0	0	0
Ethylene chloride	72	65	0	0	0
Butyl alcohol	100	87	94	100	84
Cyclohexanol	100	84	74	95	27
Phenol	100	0	0	0	0
Methyl ethyl ketone	100	87	0	0	0
Ethyl acetate	100	89	0	0	0

(a) After exposure for 24 h at 93 °C (200 °F). (b) Polyphenylene sulfide. (c) Polycarbonate. (d) Polysulfone. (e) Polyphenylene oxide

Acetal

$$\left[-\overset{H}{\underset{H}{\overset{|}{\underset{|}{C}}}}-O- \right]_n$$

Molecular structure of polyacetal resin

Thermal, physical, and chemical properties of acetal polymers (POM)

Property	Homopolymer	Copolymer	25% glass-reinforced copolymer
Heat-deflection temperature at 1.82 MPa, °C	125	110	160
Maximum resistance to continuous heat, °C	100	110	125
Coefficient of linear thermal expansion, 10^{-5}/°C	10.0	8.5	5.0
Compressive strength, MPa	106	110	117
Flexural strength, MPa	97	90	193
Impact strength, Izod, cm · N/cm of notch	80.1	69.4	96.1
Tensile strength, MPa	69	62	129
Elongation, %	30	50	3
Hardness, Rockwell M	94	78	79
Specific gravity	1.142	1.41	1.61
Dielectric constant	3.2	3.7	4.0
Water absorption, %	0.25	0.25	0.3
Resistance to chemicals at 25 °C(a):			
Nonoxidizing acids (20% H_2SO_4)	U	U	U
Oxidizing acids (10% HNO_3)	U	U	U
Aqueous salt solutions (NaCl)	S	S	S
Aqueous alkalis (C_2NOH)	S	S	S
Polar solvents (C_2H_5OH)	S	S	S
Nonpolar solvents (C_6H_6)	Q	Q	Q
Water	S	S	S

(a) S, satisfactory; Q, questionable; U, unsatisfactory

Selected properties of acetal (injection molded or extruded)

Property	ASTM test method	Reported value
Mechanical		
Yield strength, MPa (ksi)	D 638	62 to 69 (9 to 10)
Tensile modulus, GPa (10^6 psi)	D 638	2.8 to 3.6 (0.41 to 0.52)
Elongation (break), %	D 638	25 to 40
Compressive strength, MPa (ksi)	D 695	36 (5.2)
Flexural yield strength, MPa (ksi)	D 790	90 to 97 (13 to 14)
Flexural modulus, GPa (10^6 psi)	D 790	2.8 (0.40)
Impact strength, Izod(a), J/cm (ft · lbf/in.)	D 256	0.53 to 0.80 (1.0 to 1.5)
Hardness, Rockwell		M80 to 90
Thermal		
Coefficient of linear thermal expansion, 10^{-5}/°C (10^{-5}/°F)	D 696	15.3 to 18.0 (8.5 to 10.0)
Specific heat, kJ/kg · K (Btu/lbf · °F)	C 351	1.7 (0.4)
Continuous service temperature, °C (°F)		91 (195)
Electrical		
Volume resistivity, Ω · cm	D 257	10^{15}
Dielectric strength, V/10^{-3} mm (V/10^{-3} in.)	D 149	20 (500)
Dielectric constant, (50 to 100 Hz)	D 150	3.7
Dissipation factor (50 to 100 Hz)	D 150	0.005
Processing barameters		
Processing temperature, °C (°F)	...	170 to 230 (340 to 450)
Density, g/cm^3	D 792	1.42
Specific volume, cm^3/kg (in.^{3}/lb)	D 792	685 (19)
Linear mold shrinkage, mm/mm (in./in.)	D 955	0.02 to 0.025
Water absorption, % in 24 h	D 570	0.3

(a) Notched, at room temperature

Tabulated tensile creep properties of polyoxymethylene

Supplier	Trade name and grade designation	Description	Test temperature °C	Test temperature °F	Initial applied stress MPa	Initial applied stress ksi	Creep (apparent) modulus(a) At 1 H MPa	At 1 H ksi	At 10 h MPa	At 10 h ksi	At 30 h MPa	At 30 h ksi	At 100 h MPa	At 100 h ksi	At 300 h MPa	At 300 h ksi	At 10^3 h MPa	At 10^3 h ksi	At latest test point MPa	At latest test point ksi	Time at latest test point 10^3 h
DuPont	Delrin 570	20% glass fiber-reinforced homopolymer, general-purpose, injection molding	23	73	3.5	0.5	8410	1220	7580	1100	6340	920	5500	800	4830	700	4410	640	2620	380	20
					7.0	1.0	6480	940	5720	830	5170	750	4620	670	4200	610	3720	540	2070	300	20
					10.5	1.5	6070	880	5170	750	4960	720	4270	620	4070	590	3450	500	2620	380	10
					14.0	2.0	6000	870	5380	780	4830	700	4340	630	3860	560	3380	490	2200	320	20
			60	140	3.5	0.5	4690	680	4070	590	3450	500	2830	410	2410	350	2070	300	1450	210	20
					7.0	1.0	3650	530	3170	460	2760	400	2410	350	2140	310	1930	280	1380	200	20
					14.0	2.0	2900	420	2210	320	2000	290	1720	250	1520	220	1310	190	760	110	10
			85	185	3.5	0.5	2830	410	2340	340	2000	290	1790	260	1650	240	1310	190	760	110	10
					7.0	1.0	2620	380	2070	300	1650	240	1520	220	1380	200	1240	180	620	90	10
					10.5	1.5	2280	330	2070	300	1930	280	1650	240	1520	220	1380	200	1100	160	10
					17.5	2.5	2070	300	1590	230	1450	210	1240	180	...	...	...	...	...	...	...
			90	195	14.0	2.0	2280	330	1720	250	1590	230	1310	190	1240	180	1100	160	...	...	...

(a) Calculated from total creep strain or deflection (before rupture and onset of yielding)

Recommended processing factors for injection-moldable acetal resins

Grade	Melt temperature range °C	Melt temperature range °F	Melt viscosity(a), Pa · s	Injection speed	Mold temperature °C	Mold temperature °F	Maximum, regrind, %	Gating diameter(b) mm	Gating diameter(b) in.	Shrinkage range, mm/mm (in./in.)	Minimum cycle time(b), s
Homopolymers											
Unmodified base resin(c)	205 to 225	400 to 440	240 to 1300	Low to medium	80 to 105	180 to 220	30(d)	1.6 to 1.9	0.063 to 0.075	0.020	20
Impact modified	195 to 215	380 to 420	330 to 1300	Low to medium	10 to 70	50 to 160	35	1.6 to 1.9	0.063 to 0.075	0.020	20
Copolymers											
Unmodified base resin:											
Low viscosity(e)	175 to 200	350 to 390	180	Medium to high	65 to 95	150 to 200	25	2.03 to 3.81	0.080 to 0.150	0.020 to 0.022	20
Medium viscosity(f)	180 to 200	360 to 390	250	Medium	65 to 95	150 to 200	25	2.54 to 3.81	0.100 to 0.150	0.020 to 0.023	30
High viscosity(g)	190 to 205	370 to 400	400	Medium	65 to 95	150 to 200	25	3.18 to 5.08	0.125 to 0.200	0.018 to 0.022	35
Glass fiber reinforced:											
12.5%	190 to 205	370 to 400	325	Medium	65 to 95	150 to 200	25	2.54 to 3.81	0.100 to 0.150	0.006 to 0.020	30
25%	195 to 210	380 to 410	400	Medium	65 to 95	150 to 200	25	3.18 to 5.08	0.125 to 0.200	0.004 to 0.018	30
Mineral filled:											
Low filler, high flow	180 to 200	360 to 390	200	Low	65 to 95	150 to 200	25	2.54 to 3.81	0.100 to 0.150	0.016 to 0.020	25
High filler, medium flow	190 to 205	370 to 400	280	Low	65 to 95	150 to 200	25	3.18 to 5.08	0.125 to 0.200	0.014 to 0.018	30
Impact modified:											
Medium	180 to 200	360 to 390	...	Medium	65 to 95	150 to 200	25	2.54 to 3.81	0.100 to 0.150	0.018 to 0.020	30
High	190 to 205	370 to 400	...	Medium	65 to 95	150 to 200	25	3.18 to 5.08	0.125 to 0.200	0.017 to 0.020	30

(a) Melt viscosity measured at 1000/s at the median temperature of the recommended melt-temperature range. (b) Data based on a wall thickness of 3.18 mm (1.8 in.). (c) Includes glass filled, PTFE filled, all viscosities. (d) 100% regrind can be used if clean and free from deterioration. (e) 45 melt index. (f) 9.0 melt index. (g) 2.5 melt index

Acrylics

$$\left[-\overset{\displaystyle H}{\underset{\displaystyle H}{\overset{|}{\underset{|}{C}}}} - \overset{\displaystyle H}{\underset{\displaystyle CN}{\overset{|}{\underset{|}{C}}}} - \right]_n$$

Molecular structure of polyacrylonitrile

Nonimpact-modified acrylic grade applications

Market	Application: Injection-molded	Extruded
Automotive	Signal light devices for traffic, aircraft, marine, and bus use; emergency flashers; nameplates and emblems; automotive instrument panels; automotive glazing	...
Medical	Blood cuvettes; medical spikes; urine meters	...
Industrial	Display shelving; video discs; molded letters for signs; lighting diffusers and louvers; HID refractors; instrument panel covers	Lighting; sings; tubing; display shelving and enclosures; mill shapes
Consumer	Drinking tumblers; faucet knobs; household and bathroom accessories; stationery accessories; camera, projection and viewer lenses; lighting	Furniture components; manometers
Miscellaneous	Christmas decorations; drinking tumblers and food service sets	...

Impact-modified acrylic grade applications

Market	Application: Injection-molded	Extruded
Industrial	Lighting; display shelving; meter covers	Signs; lighting louvers; glazing; display shelving; profiles
Medical	Bone marrow mixer; medical spikes; chest drainage units; surgical instruments	...
Consumer appliances	Refrigerator trays; shelving; microwave oven doors	...
Consumer miscellaneous	Bird feeders; picture frames; cassettes	...
Consumer recreation	...	Sanitaryware (rigidized and coextrusion); swimming pool panels; spas (rigidized and coextrusion); recreational vehicle components (rigidized)
Automotive and marine	Signal light devices; boat windows	...

Physical properties of various grades of acrylic-base impact-modified acrylics

Property	ASTM test method	Low	Medium	High
Refractive index number	D 542	1.49	1.49	1.49
Specific gravity	D 792	1.18	1.17	1.15
Light transmission, %	D 1003	92	92	90
Haze, %	D 638	3	3	4
Tensile strength, MPa (ksi)	D 638	59 (8.5)	47 (6.8)	38 (5.5)
Flexural modulus, GPa (10^6 psi)	D 790	2.622 (0.380)	2.484 (0.360)	1.863 (0.270)
Impact strength,				
J/m (ft · lbf/in.)	D 256, notched, Izod; falling dart, 3.2 mm (0.13 in.) thick, 6 mm (0.24 in.) radius	21.3 (0.4)	32.0 (0.6)	53.4 (1.0)
J (ft · lbf)		213 (4)	374 (7)	530 (10)
Heat deflection temperature under load, at 1.82 MPa (0.264 ksi), °C (°F)	D 648	91 (196)	74 (165)	71 (160)
Rockwell hardness, M scale	D 785	81	65 to 68	38 to 45
Water absorption, %	D 570 wt gain, 24 h	0.3	0.3	0.4
Melt flow rate, g/10 min.	D 1238 condition I	6	3.2 to 9.0	1.0 to 3.5
Continuous-use temperature, °C (°F)	No load	74 to 85 (165 to 185)	74 to 85 (165 to 185)	71 to 83 (160 to 180)
Flammability	UL 94(a)	94 HB	94 HB	94 HB

(a) Underwriters Laboratories test method

Physical properties of styrene-butadiene-base impact-modified acrylics

Property	ASTM test method	Flow grade: Stiff	Flow grade: Medium	Flow grade: Easy
Specific gravity	D 792	1.11	1.12	1.11
Light transmission, %	D 1003	87	87	88
Haze, %	D 1003	9	8	6
Tensile strength, MPa (ksi)	D 638	48.3 (7.00)	55.2 (8.00)	56.7 (6.30)
Flexural modulus, GPa (10^6 psi)	D 790	2.415 (0.350)	2.760 (0.400)	2.070 (0.300)
Impact strength, J/m (ft · lbf/in.)	D 256, notched, Izod	107 (2.0)	54 (1.0)	96 (1.8)
Heat deflection temperature at 1.82 MPa (0.264 ksi), °C (°F)	D 648	85 (186)	87 (188)	86 (186)
Rockwell hardness, M scale	D 785	45	68	34
Water absorption, %	D 570 wt gain, 24 h	0.3	0.3	0.3
Melt flow rate, g/10 min.	D 1238 condition I	1.2	3.0	9.0
Continuous-use temperature, °C (°F)	No load	74 to 85 (165 to 185)	74 to 85 (165 to 185)	71 to 83 (160 to 180)
Flammability	UL 94(a)	HB	HB	HB

(a) Underwriters Laboratories test method

Thermal, physical, and chemical properties of typical acrylics

Property	Cast acrylics	Acrylic-PVC alloy
Heat-deflection temperature at 1.82 MPa, °C	95	70
Maximum resistance to continuous heat, °C	75	60
Coefficient of linear thermal expansion, 10^{-5}/°C	7.0	6.0
Compressive strength, MPa	103	58
Flexural strength, MPa	97	72
Impact strength, Izod, cm · N/cm of notch	21.4	8.0
Tensile strength, MPa	66	45
Elongation, %	4	100
Hardness, Rockwell	M80	R100
Specific gravity	1.18	1.25
Dielectric constant	3.0	3.5
Resistance of chemicals at 25 °C(a):		
Nonoxidizing acids (20% H_2SO_4)	S	S
Oxidizing acids (10% HNO_3)	U	Q
Aqueous salt solutions (NaCl)	S	S
Aqueous alkalis (NaOH)	S	S
Polar solvents (C_2H_5OH)	S	S
Nonpolar solvents (C_6H_6)	Q	Q
Water	S	S

(a) S, satisfactory; Q, questionable; U, unsatisfactory

Selected properties of acrylics

Property	ASTM test method	Reported value for: Cast acrylic Molded acrylic	Molded high-sheet and rod	Molded high-impact acrylic
Mechanical				
Yield strength, MPa (ksi)	D 638	48 to 69 (7.0 to 10.0)	55 to 76 (8.0 to 11.0)	38 to 62 (5.5 to 9.0)
Tensile modulus, GPa (10^6 psi)	D 638	2.6 to 3.1 (0.38 to 0.45)	2.4 to 3.1 (0.35 to 0.45)	1.4 to 2.4 (0.20 to 0.35)
Elongation (break), %	D 638	2 to 9	2 to 7	8 to 13
Compressive strength, MPa (ksi)	D 695	83 to 125 (12 to 18)	76 to 130 (11 to 19)	41 to 97 (6 to 14)
Flexural yield strength, MPa (ksi)	D 790	90 to 115 (13 to 17)	83 to 115 (12 to 17)	55 to 90 (8 to 13)
Flexural modulus, GPa (10^6 psi)	D 790	2.8 to 3.4 (0.040 to 0.50)	2.8 to 3.4 (0.40 to 0.50)	2.1 to 2.8 (0.30 to 0.40)
Impact strength, Izod(a), J/cm (ft · lbf/in.)	D 256	0.2 to 0.3 (0.3 to 0.5)	0.2 to 0.3 (0.3 to 0.6)	0.4 to 1.3 (0.8 to 2.4)
Hardness, Rockwell	...	M85 to 105	M80 to 100	R105 to 120
Thermal				
Thermal conductivity, W/m · K (Btu · in./ft^2 · h · °F)	C 177	0.19 to 0.29 (1.3 to 2.0)	0.19 to 0.29 (1.3 to 2.0)	0.17 to 0.26 (1.2 to 1.8)
Coefficient of linear thermal expansion, 10^{-5}/°C (10^{-5}/°F)	D 696	9.0 to 16 (5.0 to 9.0)	9.0 to 16 (5.0 to 9.0)	9.0 to 14 (5.0 to 8.0)
Continuous service temperature, °C (°F)		68 to 88 (155 to 190)	60 to 71 (140 to 160)	71 to 82 (160 to 180)
Electrical				
Volume resistivity, Ω · cm	D 257	$>10^{15}$	$>10^{15}$	2.0×10^{16}
Dielectric strength, V/10^{-3} mm (V/10^{-3} in.)	D 149	18 to 21 (450 to 530)	17 (430)	16 to 20 (400 to 500)
Dielectric constant (50-100 Hz)	D 150	3.5 to 4.5	3.5 to 4.5	3.5 to 3.9
Dissipation factor (50-100 Hz)	D 150	0.050 to 0.060	0.050 to 0.060	0.030 to 0.040

(continued)

Selected properties of acrylics (continued)

Property	ASTM test method	Reported value for: Cast acrylic Molded acrylic	Molded high- sheet and rod	Molded high- impact acrylic
Processing parameters				
Processing temperature, °C (°F)	...	165 to 230 (325 to 450)	...	150 to 205 (300 to 400)
Density, g/cm^3	D 792	1.17 to 1.19	1.17 to 1.20	1.11 to 1.18
Specific volume, cm^3/kg (in.^{3}/lb)	D 792	865 to 830 (24 to 23)	865 to 795 (24 to 22)	940 to 865 (26 to 24)
Linear mold shrinkage, mm/mm (in./in.)	D 955	0.001 to 0.004	...	0.004 to 0.008
Water absorption, % in 24 h	D 570	0.3 to 0.4	0.3 to 0.4	0.2 to 0.3

(a) Notched, at room temperature

Typical molding conditions for impact-modified acrylics

Conditions	Flow grade: Grade 5	Grade 6	Grade 8
Barrel temperatures, °C (°F)			
Rear	232 (450)	227 (440)	238 (460)
Center	238 (460)	232 (450)	243 (470)
Front	243 (470)	238 (460)	249 (480)
Nozzle	243 (470)	238 (460)	249 (480)
Melt temperature, °C (°F)	254 (480)	249 (480)	260 (500)
Mold temperature, °C (°F)	38 to 71 (100 to 160)	38 to 71 (100 to 160)	38 to 71 (100 to 160)
Screw speed, rpm (rad/s)	60 to 100 (6.3 to 10.5)	60 to 100 (6.3 to 10.5)	60 to 100 (6.3 to 10.5)
Injection fill speed, s	2 to 8	2 to 8	2 to 8
Cycle time, s	40 to 50	40 to 50	40 to 50

Part thickness, 3.2 mm (0.13 in.)

Typical pultrusion formulation for an acrylic IPN

Material	Content, wt %
Interpol 047-1000	83.5
t-butylperbenzoate	0.5
t-butylperoctoate	1.5
Mold release	1.0
Stypol 040-7263	15.5

Typical prepreg formulation for an acrylic IPN

Material	Content, wt %
Interpol 047-1050	82.0
USP-245 (peroxide catalyst)	1.0
Stypol (040-7263)	17.0

Viscosity build

Time, h	CP
0	140
0.5	175
1.0	250
2.0	500
20.0	48,000
72.0	900,000

Acrylonitrile-Butadiene-Styrene (ABS)

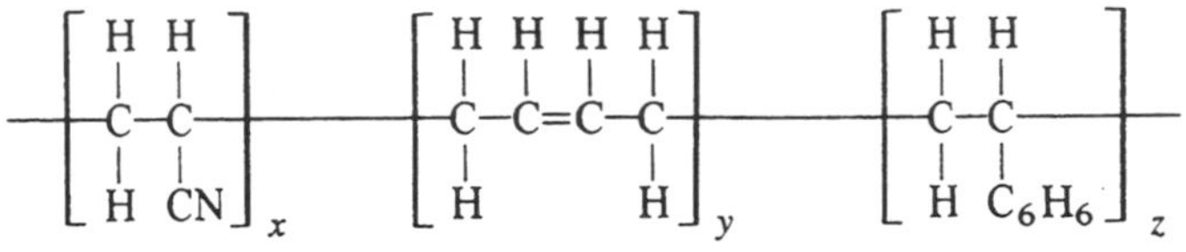

$$-\left[\begin{array}{cc} H & H \\ | & | \\ C & - C \\ | & | \\ H & CN \end{array}\right]_x - \left[\begin{array}{cccc} H & H & H & H \\ | & | & | & | \\ C & - C & = C & - C \\ | & & & | \\ H & & & H \end{array}\right]_y - \left[\begin{array}{cc} H & H \\ | & | \\ C & - C \\ | & | \\ H & C_6H_6 \end{array}\right]_z -$$

Molecular structure of acrylonitrile-butadiene-styrene copolymer

Electrical properties of ABS

Property	ASTM test method	Typical value
DC properties		
Volume resistivity, $10^{13}\ \Omega \cdot m$	D 257	2 to 4
Surface resistivity, $10^{14}\ \Omega$	D 257	2.8 to 13
AC properties		
Dissipation factor		
At 1 kHz	D 150	0.0069 to 0.015
At 1 MHz	D 150	0.0083 to 0.031
Dielectric constant		
At 1 kHz	D 150	2.89 to 3.5
At 1 MHz	D 150	2.87 to 3.2

Thermal, physical, and chemical properties of typical ABS plastics

Property	Extrusion-grade ABS	20% glass reinforced ABS
Heat deflection temperature at 1.82 MPa, °C	90	100
Maximum resistance to continuous heat, °C	80	90
Coefficient of linear thermal expansion, 10^{-5}/°C	9.5	2.0
Compressive strength, MPa	48	97
Flexural strength, MPa	62	103
Impact strength, Izod, cm · N/cm of notch	320.3	53.4
Tensile strength, MPa	34	76
Elongation, %	60	5
Hardness, Rockwell	R 60	M85
Specific gravity	1.03	1.2
Dielectric constant	0.25	0.4
Resistance to chemicals at 25 °C(a):		
Nonoxidizing acids (20% H_2SO_4)	S	S
Oxidizing acids (10% HNO_3)	U	U
Aqueous salt solutions (NaCl)	S	S
Aqueous alkalis (NaOH)	S	S
Polar solvents (C_2H_5OH)	Q	Q
Nonpolar solvents (C_6H_6)	Q	Q

(a) S, satisfactory; Q, questionable; U, unsatisfactory

Selected properties of extruded and molded ABS thermoplastics

Property	ASTM test method	Reported value for: Extruded ABS	Molded high-impact ABS	Molded medium-impact ABS	Molded heat-resistant ABS	Molded flame-resistant ABS	Molded transparent ABS
Mechanical							
Yield strength, MPa (ksi)	D 638	22 to 54 (3.2 to 7.8)	32 to 43 (4.7 to 6.3)	41 to 55 (6.0 to 8.0)	48 to 55 (7.0 to 8.0)	35 to 48 (5.1 to 7.0)	39 to 43 (5.6 to 6.2)
Tensile modulus, GPa (10^6 psi)	D 638	0.97 to 2.5 (0.14 to 0.36)	1.4 to 2.2 (0.20 to 0.32)	1.5 to 2.8 (0.22 to 0.40)	2.4 to 2.9 (0.35 to 0.42)	2.4 to 2.8 (0.35 to 0.40)	2.0 to 2.3 (0.29 to 0.34)
Elongation (break), %	D 638	30 to 90	10 to 50	10 to 25	10 to 20	15 to 20	40 to 60
Compressive strength, MPa (ksi)	D 695	6.9 to 28 (1 to 4)	6.9 to 21 (1 to 3)	55 to 83 (8 to 12)	62 to 76 (9 to 11)	48 to 55 (7 to 8)	48 to 62 (7 to 9)
Flexural yield strength, MPa (ksi)	D 790	34 to 97 (5 to 14)	41 to 69 (6 to 10)	69 to 83 (10 to 12)	76 to 90 (11 to 13)	62 to 90 (9 to 13)	69 to 76 (10 to 11)
Flexural modulus, GPa (10^6 psi)	D 790	0.69 to 2.8 (0.10 to 0.40)	1.4 to 2.1 (0.20 to 0.30)	2.1 to 2.8 (0.30 to 0.40)	2.1 to 2.8 (0.30 to 0.40)	2.1 to 2.8 (0.30 to 0.40)	2.1 to 2.8 (0.30 to 0.40)
Impact strength, Izod(a), J/cm (ft · lbf/in.)	D 256	1.4 to 6.09 (2.6 to 11.4)	3.2 to 3.7 (6.0 to 7.0)	1.7 to 2.4 (3.1 to 4.5)	1.3 to 2.1 (2.5 to 4.0)	1.9 to 2.9 (3.5 to 5.5)	2.1 to 2.7 (4.0 to 5.0)
Hardness, Rockwell	...	R75 to 110	R85 to 100	R108 to 115	R105 to 115	R100 to 120	R98 to 105
Thermal							
Thermal conductivity, W/m · K (Btu · in./ft^2 · h · °F)	C 177	0.19 to 0.33 (1.3 to 2.3)	0.14 to 0.33 (1.0 to 2.3)	0.14 to 0.30 (0.98 to 2.1)	0.22 to 0.36 (1.5 to 2.5)	...	0.19 to 0.33 (1.3 to 2.3)
Coefficient of thermal expansion, 10^{-5}/°C (10^{-5}/°F)	D 696	10.8 to 23.0 (6.0 to 12.8)	11.0 to 16 (6.1 to 9.1)	8.1 to 11.3 (4.5 to 6.3)	7.7 to 10.8 (4.3 to 6.0)	...	...
Specific heat, kJ/kg · K (Btu/lbf · °F)	C 351	1.3 to 1.7 (0.3 to 0.4)	1.3 to 1.7 (0.3 to 0.4)	1.3 to 1.7 (0.3 to 0.4)	1.3 to 2.1 (0.3 to 0.5)	...	...
Continuous service temperature, °C (°F)	...	85 (185)	85 (185)	85 (185)	...	...	...
Electrical							
Volume resistivity, Ω · cm	D 257	1.0 to 5.0 × 10^{16}	1.0 to 5.0 × 10^{16}	1.0 to 5.0 × 10^{16}	...	1.5 × 10^{14}	2.5 × 10^{15}
Dielectric strength, V/10^{-3} mm (V/10^{-3} in.)	D 149	14 to 18 (350 to 450)	14 to 16 (350 to 400)	15 to 18 (375 to 450)	14 to 16 (360 to 400)	15 to 16 (380 to 400)	16 (400)
Dielectric constant (50 to 100 Hz)	D 150	2.7 to 3.1	2.7 to 3.1	2.7 to 3.1	2.7 to 3.1	...	3.7
Dissipation factor (50 to 100 Hz)	D 150	...	0.005 to 0.010	0.003 to 0.005	0.030 to 0.040	...	0.015
Processing parameters							
Processing temperature, °C (°F)	...	165 to 230 (325 to 450)	165 to 260 (325 to 500)	165 to 260 (325 to 500)	175 to 290 (350 to 550)	160 to 260 (320 to 500)	160 to 245 (320 to 475)
Density, g/cm^3	D 792	1.02 to 1.06	1.01 to 1.05	1.03 to 1.06	1.06 to 1.08	1.16 to 1.22	1.07
Specific volume, cm^3/kg (in.^{3}/lb)	D 792	975 to 940 (27 to 26)	975 to 940 (27 to 26)	975 to 940 (27 to 26)	940 to 905 (26 to 25)	865 to 795 (24 to 22)	940 (26)
Linear mold shrinkage, mm/mm (in./in.)	D 955	0.005 to 0.009	0.005 to 0.009	0.005 to 0.009	0.005 to 0.009	0.005 to 0.008	0.005 to 0.008
Water absorption, % in 24 h	D 570	0.2 to 0.4	0.2 to 0.4	0.2 to 0.4	0.2 to 0.4	...	0.4

(a) Notched, at room temperature

Range of mechanical properties of commercial ABS

Grade	Tensile strength MPa	Tensile strength ksi	Elongation at break, %	Notched impact strength J/m	Notched impact strength ft · lbf/in.
Injection molding	20 to 76	2.9 to 11	1 to 46	50 to 100	1 to 2
Injection molding and extrusion	29 to 49	4.2 to 7.1	2 to 46	100 to 400	2 to 8
Extrusion	2 to 59	0.3 to 8.5	2 to 40	100 to 700	2 to 14
Glass fiber filled	66 to 110	9.6 to 16	1 to 4	50 to >100	1 to 2

Typical mechanical and thermal properties of standard ABS grades

Property	ASTM test method	Medium impact	High impact	Very high impact
Mechanical				
Tensile strength at yield, MPa (ksi)	D 638	45 (6.5)	39 (5.6)	32 (4.7)
Tensile modulus, GPa (10^6 psi)	D 638	2.5 (0.36)	2.2 (0.32)	1.8 (0.26)
Flexural strength at yield, MPa (ksi)	D 790	76 (11)	66 (9.5)	54 (7.8)
Flexural modulus, GPa (10^6 psi)	D 790	2.8 (0.4)	2.2 (0.32)	1.8 (0.26)
Izod impact strength, 3.2 mm (0.13 in.) bar J/m (ft · lbf/in.)				
At 23 °C (73 °F)	D 256A	160 (3.0)	270 (5.0)	400 (7.5)
At -40 °C (-40 °F)	D 256A	59 (1.1)	75 (1.4)	120 (2.2)
Hardness, Rockwell R	D 785A	108 to 118	102 to 113	90 to 100
Specific gravity	D 792	1.03 to 1.07	1.01 to 1.05	1.01 to 1.04
Thermal				
Deflection temperature under load, °C (°F)				
At 0.46 MPa (0.066 ksi)	D 648	99 (210)	100 (212)	93 (200)
At 1.8 MPa (0.26 ksi)	D 648	121 (220)	106 (222)	89 (192)
Coefficient of linear thermal expansion, 10^{-5}/K	D 696	9	10	11
UL temperature index, °C (°F)	UL 746C(a)	60 to 75 (140 to 167)	60 to 75 (140 to 167)	60 (140)

(a) Underwriters Laboratories test method

Typical mechanical and thermal properties of ABS specialty grades

Property	ASTM test method	High heat	Plating	Flame retardant	Clear
Mechanical					
Tensile strength at yield, MPa (ksi)	D 638	46 (6.6)	43 (6.2)	46 (6.6)	43 (6.3)
Tensile modulus per, GPa (10^6 psi)	D 638	2.1 (0.30)	2.6 (0.38)	2.4 (0.35)	2.3 (0.33)
Flexural strength at yield, MPa (ksi)	D 790	77 (11.2)	77 (11.2)	76 (11)	72 (10.5)
Flexural modulus, GPa (10^6 psi)	D 790	2.3 (0.34)	2.6 (0.38)	2.5 (0.36)	2.4 (0.35)
Izod impact strength, 3.2 mm (0.13 in.) bar, J/m (ft · lbf/in.)	D 256A	220 (4.2)	320 (6.0)	210 (4.0)	130 (2.5)
Hardness, Rockwell R	D 785A	108 to 111	103 to 111	97 to 102	103
Specific gravity	D 792	1.05	1.06	1.21	1.07
Thermal					
Deflection temperature under load, °C (°F)					
At 0.46 MPa (0.066 ksi)	D 648	104 (220)	99 (210)	96 (205)	85 (185)
At 1.8 MPa (0.26 ksi)	D 648	102 (215)	93 (200)	88 (190)	77 (170)
Coefficient of linear thermal expansion, 10^{-5}/K	D 696	7.7	7.0		8.6
UL temperature index, °C (°F)	UL 746C(a)	60 to 75 (140 to 167)	60 (140)	60 to 75 (140 to 167)	60 (140)

(a) Underwriters Laboratories test method

Typical conditions for injection molding of ABS

Process parameter	Setting
Drying temperature, °C (°F)	82 to 88 (180 to 190)
Drying time, h	1 to 2
Mold temperature, °C (°F)	49 to 66 (120 to 150)
Melt temperature, °C (°F)	220 to 260 (425 to 500)
Barrel sets, °C (°F):	
Rear	195 (380)
Middle	210 to 215 (410 to 420)
Front	220 to 230 (430 to 450)
Fill speed	low-medium
Screw speed, rpm	50 to 60
Back pressure, kPa (psi)	690 (100)
Injection pressure, MPa (ksi)	69 to 83 (10 to 12)

Typical conditions for extrusion of ABS

Process parameter	Setting
Drying temperature, °C (°F)	82 (180)
Drying time, h	3
Melt temperature, °C (°F)	205 to 245 (400 to 475)
Barrel sets, °C (°F):	
Rear	190 to 220 (375 to 425)
Middle	205 to 230 (400 to 450)
Front	205 to 245 (400 to 475)
Die temperature, °C (°F)	220 to 250 (425 to 480)
Roll temperatures,(a) °C (°F):	
Top	79 to 115 (175 to 225)
Middle	66 to 79 (150 to 175)
Bottom	79 to 105 (175 to 225)

(a) Down stack

Typical conditions for thermoforming of ABS

Process parameter	Setting
Sheet temperature(a), °C (°F):	
Minimum any layer	110 (230)
Maximum surface	220 (425)
Normal	165 (325)
Heater temperature range, °C (°F)	260 to 815 (500 to 1500)
Heater normal temperature, °C (°F)	540 (1000)
Plug temperature for female mold, °C (°F)	95 to 150 (200 to 300)
Mold temperature, °C (°F):	
Female mold	49 to 115 (120 to 240)
Male mold	49 to 115 (120 to 240)
Plug speed, mm/s (in./s)	125 to 255 (5 to 10)
Plug size for female mold, % of mold	80 to 90
Heater distance from sheet, mm (in.):	
Top of sheet	150 to 305 (6 to 12)
Bottom of sheet	305 to 455 (12 to 18)
Maximum sheet heating rate, (s/mil) of sheet thickness:	
One-side heaters	0.02 (0.5)
Two-side heaters	0.03 (0.75)
Billow height, % of draw depth:	
Female mold	50
Male mold	75
Maximum sheet cooling rate, s/μm (s/mil) of sheet thickness:	
Forced air	0.01 (0.25)
Water fog	0.005 (0.125)

(a) As determined by optical pyrometer measurements. For sheet thicknesses greater than 4.44 mm (0.175 in.), minimum temperature must be higher than that shown. ABS/PVC resins should be processed at temperatures 11 to 17 °C (20 to 30 °F) lower. ABS/SMA resins should be processed at temperatures 8 to 17 °C (15 to 30 °F) higher

Cellulosics

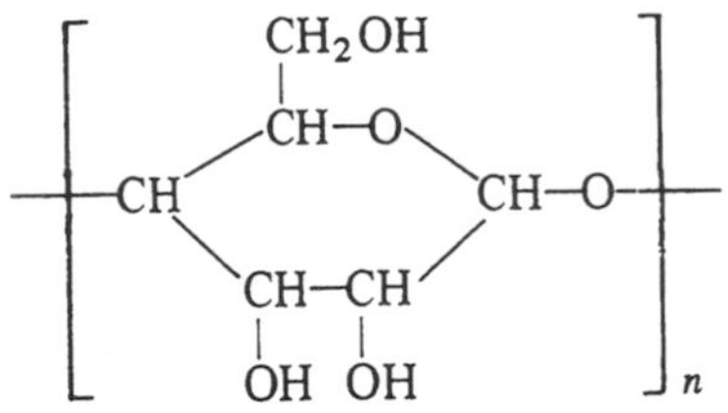

```
              CH2OH
               |
             CH—O
            /     \
 ——[——CH              CH—O——]——
            \     /
             CH—CH
             |   |
             OH  OH             n
```

Molecular structure of cellulose

Thermal, physical, and chemical properties of typical cellulosics

Property	Ethylcellulose	Cellulose triacetate	Cellulose acetate butyrate
Heat-deflection temperature at 1.82 MPa, °C	65	65	65
Maximum resistance to continuous heat, °C	60	60	60
Coefficient of linear thermal expansion, 10^{-5}/°C	15.0	12.5	14.0
Compressive strength, MPa	120	55	34.5
Flexural strength, MPa	41	55	41
Impact strength, Izod, cm · N/cm of notch	21.35	106.7	160
Tensile strength, MPa	34	42	34
Elongation, %	10	25	50
Hardness, Rockwell	R60	R80	R75
Specific gravity	1.1	1.3	1.2
Dielectric constant	3	4	4
Resistance of chemicals at 25 °C(a):			
Nonoxidizing acids (20% H_2SO_4)	Q	U	U
Oxidizing acids (10% HNO_3)	U	U	U
Aqueous salt solutions (NaCl)	S	S	S
Aqueous alkalis (NaOH)	S	S	S
Polar solvents (C_2H_5OH)	Q	Q	Q
Nonpolar solvents (C_6H_6)	U	U	U
Water	S	S	S

(a) S, satisfactory; Q, questionable; U, unsatisfactory

Properties of cellulose derivatives

Property	EC	CA	CAB	CAP	CN
T_m, °C (°F)	135 (275)	230 (445)	140 (60)	190 (90)	...
Molding temperature, °C (°F)					
Compression	120 to 200 (250 to 390)	125 to 215 (260 to 420)	125 to 200 (265 to 390)	130 to 205 (265 to 400)	85 to 95 (185 to 200)
Injection	175 to 260 (350 to 500)	170 to 255 (335 to 490)	170 to 250 (335 to 480)	170 to 270 (335 to 515)	...
Coefficient of thermal expansion, molded, 10^{-6}/K	100 to 200	80 to 180	110 to 170	110 to 170	80 to 120
Heat-deflection temperature, at 1.82 MPa (0.264 ksi), molded, °C (°F)	45 to 88 (115 to 190)	45 to 90 (111 to 195)	45 to 95 (113 to 202)	45 to 110 (111 to 228)	60 to 70 (140 to 160)
Water absorption, 24 h at 3.2 mm (1/8 in.) thick, %	0.08 to 1.8	1.7 to 6.5	0.9 to 2.2	1.2 to 2.8	1.0 to 2.0

Selected properties of molded cellulosics

Property	ASTM test method	Reported value for: Molded ethyl cellulose	Molded cellulose acetate	Molded cellulose acetate/butyrate	Molded cellulose acetate propionate	Molded cellulose nitrate
Mechanical						
Yield strength, MPa (ksi)	D 638	14 to 55 (2.0 to 8.0)	21 to 48 (3.0 to 7.0)	17 to 55 (2.5 to 8.0)	28 to 45 (4.0 to 6.5)	48 to 55 (7.0 to 8.0)
Tensile modulus, GPa (10^6 psi)	D 638	0.69 to 2.1 (0.10 to 0.30	0.6 to 2.8 (0.08 to 0.40)	0.3 to 1.4 (0.05 to 0.20)	...	1.3 to 1.6 (0.19 to 0.23)
Elongation (break), %	D 638	...	6 to 60	40 to 45	...	40 to 45
Compressive strength, MPa (ksi)	D 695	69 to 240 (10 to 35)	14 to 62 (2 to 9)	14 to 140 (2 to 20)	...	150 to 240 (22 to 35)
Flexural yield strength, MPa (ksi)	D 790	28 to 83 (4 to 12)	28 to 83 (4 to 12)	21 to 62 (3 to 9)	41 to 55 (6 to 8)	62 to 76 (9 to 11)
Flexural modulus, GPa (10^6 psi)	D 790	...	1.4 (0.20)	0.69 to 1.4 (0.10 to 0.20)	0.69 to 1.4 (0.10 to 0.20)	...
Impact strength, Izod(a), J/cm (ft · lbf/in.)	D 256	1.1 to 4.5 (2.0 to 8.5)	1.6 to 3.7 (3.0 to 7.0)	0.5 to 2.7 (0.9 to 5.0)	...	2.7 to 3.7 (5.0 to 7.0)
Hardness, Rockwell	...	R50 to 115	R50 to 115	R30 to 100	...	R95 to 110
Thermal						
Thermal conductivity, W/m · K (Btu · in./ft^2 · h · °F)	C 177	0.16 to 0.29 (1.1 to 2.0)	0.17 to 0.32 (1.2 to 2.2)	0.16 to 0.32 (1.1 to 2.2)	0.17 to 0.33 (1.2 to 2.3)	0.19 (1.3)
Coefficient of linear expansion, 10^{-5}/°C (10^{-5}/°F)	D 696	18.0 to 36.0 (10.0 to 20.0)	14.4 to 32.4 (8.0 to 18.0)	19.8 to 30.6 (11.0 to 17.0)	...	14.4 to 21.6 (8.0 to 12.0)
Specific heat, kJ/kg · K (Btu/lbf · °F)	C 351	...	1.3 to 1.7 (0.3 to 0.4)	1.3 to 1.7 (0.3 to 0.4)	1.3 to 1.7 (0.3 to 0.4)	1.3 to 1.7 (0.3 to 0.4)
Electrical						
Volume resistivity, Ω · cm	D 257	10^{12} to 10^{14}	10^{10} to 10^{14}	10^{11} to 10^{15}	10^{11} to 10^{14}	10^{11}
Dielectric strength, V/10^{-3} mm (V/10^{-3} in.)	D 149	14 to 20 (350 to 500)	8 to 20 (200 to 500)	10 to 16 (250 to 400)	10 to 16 (250 to 400)	12 to 22 (300 to 550)
Dielectric constant (50 to 100 Hz)	D 150	3.0 to 4.2	3.5 to 7.5	3.5 to 6.2	3.5 to 6.2	7.0 to 7.5
Dissipation factor (50 to 100 Hz)	D 150	0.005 to 0.020	0.010 to 1.00	0.020 to 0.050	0.020 to 0.050	0.090 to 0.120
Processing parameters						
Processing temperature, °C (°F)	...	...	125 to 215 (260 to 420)	140 to 200 (280 to 390)	...	85 to 120 (185 to 250)
Density, g/cm^3	D 792	1.09 to 1.17	1.22 to 1.34	1.15 to 1.22	1.19 to 1.21	1.35 to 1.40
Specific volume, cm^3/kg (in.^{3}/lb)	D 792	905 to 865 (25 to 24)	830 to 725 (23 to 20)	865 to 830 (24 to 23)	865 to 830 (24 to 23)	760 to 725 (21 to 20)
Linear mold shrinkage, mm/mm (in./in.)	D 955	0.005 to 0.009	0.003 to 0.009	...	0.003 to 0.009	...

(a) Notched, at room temperature

Fluoroplastics

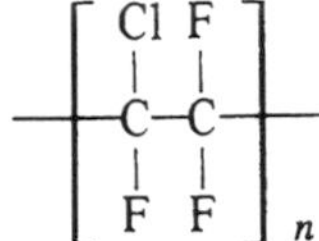

Polyvinyl fluoride

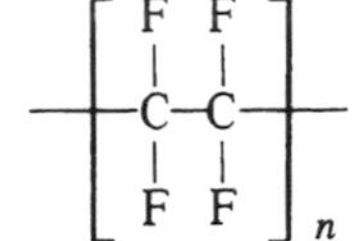

Polyvinylidene fluoride

$$-\left[CF_2 - CF_2 \right]_x - \left[CF_2 - CF(CF_3) \right]_y -$$

TFE HFP

Molecular structure of TFE-HFP copolymer

$$-\left[CClF - CF_2 \right]_n -$$

Chlorotrifluoroethylene

$$-\left[CF_2 - CF_2 \right]_n -$$

Polytetrafluoroethylene

Molecular structure of fluoroplastics

Properties for PTFE, FEP, PFA, and ETFE

Properties	ASTM test method	Polytetrafluoro-ethylene	Fluorinated perfluoroethylene propylene	Perfluoro alkoxy alkane	Ethylene tetrafluoroethylene
General					
Specific gravity	D 792	2.13 to 2.19	2.12 to 2.17	2.13 to 2.16	1.70
Linear mold shrinkage, mm/mm	D 955	0.030 to 0.060	0.030 to 0.060	0.040	0.030 to 0.040
Water absorption, %	D 570	None	<0.01	0.03	<0.03
Flammability	UL 94(h)	V-0	V-0	V-0	V-0
Thermal					
Heat deflection temperature under load					
at 455 kPa, °C (°F)	D 648	120 (250)	(a)	(a)	105 (220)
at 1.8 MPa, °C (°F)	D 648	50 (120)	(a)	(a)	70 (160)
Melting point, °C (°F)		325 (621)	250 to 280 (487 to 540)	300 to 310 (575 to 590)	270 (520)
Mechanical					
Unnotched Izod impact strength, J/mm (ft · lbf/in.)	D 256	0.11 (2.0)(b) 0.16 (3.0)(c)	0.15 (2.9), no break	(a)	(a)
Notched Izod impact strength, J/mm (ft · lbf/in.)	D 256	(a)	(a)	(a)	>1.1 (20)(d), no break(c)
Tensile strength, MPa (ksi)	D 638	28 to 45 (4.0 to 6.5)	21 to 28 (3.0 to 4.0)	28 to 30 (4.0 to 4.3)	45 (6.5)
Flexural strength, MPa (ksi)	D 790	No break	21 (3.0)	(a)	38 (5.5)
Flexural modulus, GPa (10^6 psi)	D 790	0.48 to 0.76 (0.070 to 0.110)	0.66 to 0.72 (0.095 to 0.105)	0.66 (0.095)	1.4 (0.200)
Hardness					
Durometer D	D 2240	50 to 65	55	60	75
Shore D	D 785	(a)	57	(a)	(a)
Rockwell R	D 785	45	45	(a)	(a)
Electrical					
Dielectric strength, MV/m (V/mil)	D 149	40 to 80 (1000 to 2000)(e)	83 (2100)(e)	80 (2000)(f)	16 (400)(g) <80 (2000)(e)
Dielectric constant at 1 MHz	D 150	2.1	2.1	2.1	2.6

(a) Not reported. (b) At 20 °C (70 °F). (c) At 23 °C (73 °F). (d) At -54 °C (-65 °F). (e) At 0.25 mm (0.010 in.). (f) Thickness not reported. (g) At -40 °C (-40 °F). (h) Underwriters Laboratories test method

Properties for CTFE, PVDF, and ECTFE

Properties	ASTM test method	Polychloro-trifluoroethylene	Polyvinylidene fluoride	Ethylene chlorotrifluoroethylene
General				
Specific gravity	D 792	2.10	1.75 to 1.78	1.68
Linear mold shrinkage, mm/mm or in./in.	D 955	0.010 to 0.015	0.030	0.020 to 0.026
Water absorption, %	D 570	None	0.04 to 0.06	<0.01
Flammability	UL 94(j)	V-0	V-0	V-0
Thermal				
Heat deflection temperature under load				
at 455 kPa, °C (°F)	D 648	130 (265)	150 (300)	115 (240)
at 1.8 MPa, °C (°F)	D 648	75 (167)	90 (195)	77 (170)
Melting point, °C (°F)		210 to 215 (412 to 420)	155 to 170 (312 to 334)	240 (464)
Mechanical				
Unnotched Izod impact strength, J/mm (ft · lbf/in.)	D 256	(a)	1.7 to 3.1 (32 to 58)(b)	(a)
Notched Izod impact strength, J/mm (ft · lbf/in.)	D 256	0.065 (1.2)(b)	0.16 to 0.55 (3.0 to 10.3)(b)	No break(c); 0.11 to 0.16 (2 to 3)(d)
Tensile strength, MPa (ksi)	D 638	32 to 39 (4.6 to 5.7)	35 to 52 (5.2 to 7.5)	31 to 48 (4.5 to 7.0)
Flexural strength, MPa (ksi)	D 790	51 to 64 (7.4 to 9.3)	59 to 75 (8.6 to 10.8);	48 (7.0)
Flexural modulus, GPa (10^6 psi)	D 790	1.3 (0.185)	1.2 to 1.4 (0.175 to 0.200)	1.7 (0.240)
Hardness				
Durometer D	D 2240	69	(a)	(a)
Shore D	D 785	74 to 79	70 to 80	75
Rockwell R	D 785	(a)	109	93
Electrical				
Dielectric strength, MV/m (V/mil)	D 149	25 (635)(e)	10.2 (260)(f); 50.4 (1280)(g)	78.7 (2000)(h); 19.3 (490)(g)
Dielectric constant at 1 MHz	D 150	3.0	6.1	2.5

(a) Not reported. (b) At 25 °C (77 °F). (c) At 23 °C (73 °F). (d) At -40 °C (-40 °F). (e) At 1.59 mm (0.0625 in.). (f) At 3.18 mm (0.125 in.). (g) At 0.20 mm (0.008 in.). (h) At 0.025 mm (0.001 in.). (j) Underwriters Laboratories test method

Selected properties of fluoroplastics

Property	ASTM test method	Reported value for: Molded fluorocarbon PCTFE	Molded fluorocarbon PTFE	Extruded or molded fluorocarbon FEP	Extruded or molded fluoro-carcarbon PVF_2	Extruded or molded fluorocarbon ETFE/ECTFE	Extruded or molded fluorocarbon PFA
Mechanical							
Yield strength, MPa (ksi)	D 638	32 to 39 (4.6 to 5.7)	17 to 45 (2.5 to 6.5)	19 to 21 (2.7 to 3.1)	50 to 59 (7.2 to 8.5)	...	29 (4.2)
Tensile modulus, GPa (10^6 psi)	D 638	1.0 to 2.1 (0.15 to 0.30)	0.28 to 0.48 (0.04 to 0.07)	0.34 (0.05)	1.2 to 1.4 (0.17 to 0.20)	2.8 to 3.1 (0.40 to 0.45)	3.0 (0.43)
Elongation (break), %	D 638	120 to 300	250 to 375	250 to 330	200 to 300	150 to 200	300
Compressive strength, MPa (ksi)	D 695	7 to 14 (1 to 2)	7 to 14 (1 to 2)	14 (2)	62 to 97 (9 to 14)	14 (2)	...
Flexural yield strength, MPa (ksi)	D 790	55 to 62 (8 to 9)	...	No break	14 (2)	17 (2.5)	...
Flexural modulus, GPa (10^6 psi)	D 790	0.69 (0.10)	0.69 (0.10)	0.69 (0.10)	1.4 (0.20)	...	0.69 (0.10)
Impact strength, Izod(a), J/cm (ft · lbf/in.)	D 256	1.3 to 1.4 (2.4 to 2.7)	1.6 (3.0)	No break	1.9 to 2.1 (3.6 to 4.0)	No break	No break
Hardness, Rockwell	...	R90 to 110	D50 to 55	D58	...	...	...
Thermal							
Thermal conductivity, W/m · K (Btu · in./ft^2 · h · °F)	C 177	0.20 to 0.26 (1.4 to 1.8)	0.27 (1.9)	0.20 (1.4)	0.17 (1.2)	0.22 (1.5)	0.26 (1.8)
Coefficient of linear thermal expansion, 10^{-5}/°C (10^{-5}/°F)	D 696	8.1 to 9.4 (4.5 to 5.2)	14.4 to 16.2 (8.0 to 9.0)	10.8 to 18.9 (6.0 to 10.5)	15.3 (8.5)	9.0 to 13.5 (5.0 to 7.5)	14.4 to 21.6 (8.0 to 12.0)
Specific heat, kJ/kg · K (Btu/lb · °F)	C 351	0.8 (0.2)	1.3 (0.3)	1.3 (0.3)	1.3 (0.3)	...	25 (6.0)
Continuous service temperature, °C (°F)	...	195 (380)	290 (550)	...	170 (340)	150 to 180 (300 to 355)	260 (500)
Electrical							
Volume resistivity, Ω · cm	D 257	10^{18}	10^{18}	10^{18}	5.0×10^{14}	10^{16}	10^{18}
Dielectric strength, V/10^{-3} mm (V/10^{-3} in.)	D 149	18 to 22 (450 to 550)	16 to 20 (400 to 500)	20 to 24 (500 to 600)	10 (260)	19 (490)	79 (2000)
Dielectric constant (50-100 Hz)	D 150	2.3 to 2.8	2.1	2.1	10	2.6	2.1
Dissipation factor (50-100 Hz)	D 150	0.002	<0.0002	<0.0003	0.050	0.0007	0.0002
Processing parameters							
Processing temperature, °C (°F)	...	240 to 290 (460 to 550)	...	315 to 400 (600 to 750)	205 to 290 (400 to 550)	230 to 260 (450 to 500)	345 to 400 (650 to 750)
Density, g/cm^3	D 792	2.10 to 2.20	2.14 to 2.30	2.13 to 2.17	1.77	1.70	2.10 to 2.20
Specific volume, cm^3/kg (in.^{3}/lb)	D 792	470 (13)	470 (13)	470 (13)	540 (15)	580 (16)	470 (13)
Linear mold shrinkage, mm/mm (in./in.)	D 955	0.010 to 0.015	0.010 to 0.015	0.030 to 0.060	0.030	0.020	0.040
Water absorption, % in 24 h	D 570	0.0	0.0	...	0.1	...	0.1

(a) Notched, at room temperature

Thermal properties of typical thermoplastic fluoropolymers

Property	PVDF	PTFE	PC-TFE	1/1 PE-TFE	1/1 PE-CTFE
Density, g/cm^3	1.78	2.2	2.13	1.70	1.68
T_g, °C (°F)	–45 (–50)	127 (260)	45 (115)	...	...
T_m, °C (°F)	170 (340)	327 (620)	218 (425)	270 (520)	245 (475)
Thermal conductivity, at 20 to 30 °C (68 to 95 °F) W/m^2 · K (Btu · in./s · ft^2 · °F)	...	0.18 (2.5)	0.18 (2.5)	...	...
Specific heat, at 40 °C (105 °F), J/kg · K (cal/g · °C)	...	960 (0.23)	...	...	...
Coefficient of thermal expansion, 10^{-5}/K	...	5.5	7	...	...

Thermal, physical, and chemical properties of typical polyfluorocarbon plastics

Property	PTFE	PCTFE	PVDF	PVF	PE-CTFE	PE-TFE
Heat-deflection temperature at 1.82 MPa, °C	100	100	80	90	115	120
Maximum resistance to continuous heat, °C	250	200	150	125	100	160
Coefficient of linear thermal expansion, 10^{-5}/°C	10	14	8.5	10	8	7
Compressive strength, MPa	28	38	...	...	41	48
Flexural strength, MPa	...	60	...	...	48	38
Impact strength, Izod cm · N/cm of notch	160	133.5	...	...	...	...
Tensile strength, MPa	24	34	55	...	48	48
Elongation, %	200	100	200	...	200	250
Hardness, Rockwell	D52	R80	R110	D64	R95	R50
Specific gravity	2.16	2.1	1.76	1.4	1.7	1.7
Dielectric constant	2	2.5	8	8	5	2.7
Water absorption, %	0	0	0	0	0	0
Resistance to chemicals at 25 °C(a):						
Nonoxidizing acids (20% H_2SO_4)	S	S	S	S	S	S
Oxidizing acids (10% HNO_3)	S	S	S	Q	Q	Q
Aqueous salt solutions (NaCl)	S	S	S	S	S	S
Aqueous alkalis (NaOH)	S	S	S	S	S	S
Polar solvents (C_2H_5OH)	S	S	S	S	S	S
Nonpolar solvents (C_6H_6)	S	S	S	S	Q	Q
Water	S	S	S	S	S	S

(a) S, satisfactory; Q, questionable; U, unsatisfactory

Yield strength of fluoropolymers at various temperatures

Temperature °C	Temperature °F	Yield strength MPa	Yield strength ksi
Tetrafluoroethylene			
	–420	130	19
	–320	110	16
	–200	80	11.5
	–100	55	7.7
	–68	25	3.8
	32	13	1.8
	73	9	1.3
	158	5.5	0.8
	250	3.5	0.5
Fluorinated perfluoroethylene-propylene			
	–420	165	24
	–320	130	19
	–200	95	14
	–100	60	9
	–68	30	4
	32	14	2
	73	13	1.8
	158	7	1.0
	250	3.5	0.5

Polyamides (Nylons)

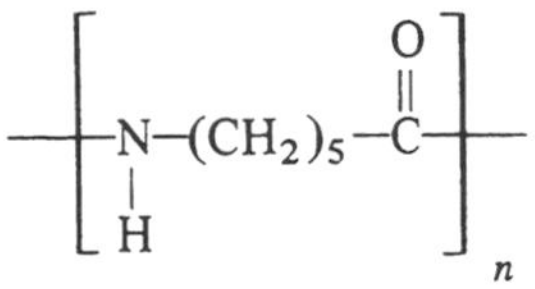

$$-\!\!\left[N(H)-(CH_2)_5-C(=O) \right]_n\!\!-$$

Nylon 6

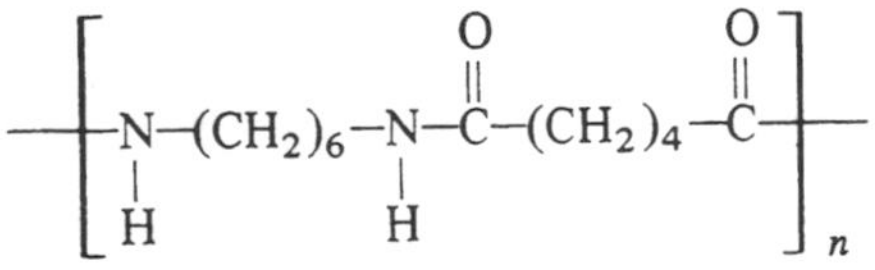

$$-\!\!\left[N(H)-(CH_2)_6-N(H)-C(=O)-(CH_2)_4-C(=O) \right]_n\!\!-$$

Nylon 6/6

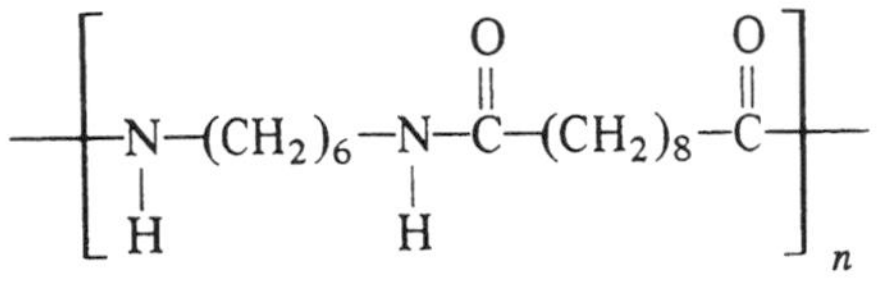

$$-\!\!\left[N(H)-(CH_2)_6-N(H)-C(=O)-(CH_2)_8-C(=O) \right]_n\!\!-$$

Nylon 6/10

$$-\!\!\left[N(H)-(CH_2)_{10}-C(=O) \right]_n\!\!-$$

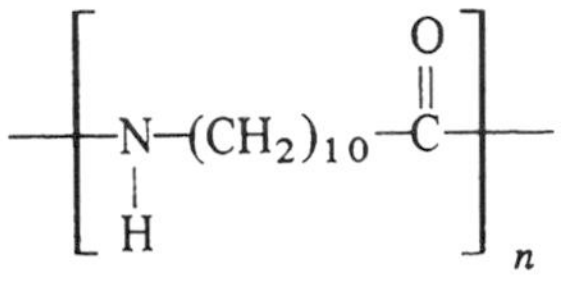

Nylon 11

Molecular structure of nylons

Thermal, physical, and chemical properties of typical polyamides

Property	Nylon 6/6	Nylon 6	Nylon 11	Nylon 6/10
Heat deflection temperature at 1.82 MPa, °C	75	80	55	80
Maximum resistance to continuous heat, °C	65	70	60	75
Melting point, °C	265	215	185	220
Coefficient of linear thermal expansion, 10^{-5}/°C	8.0	8.0	10	10
Tensile strength, MPa	83	62	50	55
Elongation, %	60	100	120	100
Specific gravity	1.14	1.13	1.05	1.09
Dielectric constant	4.0	4.0	3.5	4.5
Resistance to chemicals at 25 °C(a):				
Nonoxidizing acids (20% H_2SO_4)	U	U	U	U
Oxidizing acids (10% HNO_3)	U	U	U	U
Aqueous salt solutions (NaCl)	S	S	S	S
Aqueous alkalis (NaOH)	S	S	S	S
Polar solvents (C_2H_5OH)	Q	Q	Q	Q
Nonpolar solvents (C_6H_6)	S	S	S	S
Water	S	S	S	S

(a) S, satisfactory; Q, questionable; U, unsatisfactory

Thermal properties of representative polyamides

Property	Nylon 6	Nylon 12	Nylon 6/6	Nylon 6/10	Nylon 6/L	Nylon MXD/6
T_g, °C (°F)	50 to 70 (120 to 160)(a)	46 (115)	57 to 80 (135 to 175)(a)	50 (120)	142 (290)	115 (240)
T_m, °C (°F)	225 (440)	180 (360)	265 (510)	219 (425)	210 (410)	238 (460)
Melt-processing temperature, °C (°F)	225 to 290 (440 to 550)	180 to 270 (360 to 525)	270 to 325 (520 to 620)	230 to 290 (450 to 550)	...	...
Specific heat, J/kg · K (cal/g · K)	1670 (0.4)	1260 (0.3)	1670 (0.4)	1670 (0.4)	...	...
Coefficient of thermal expansion, 10^{-5}/K	8.3	10.0	8.0	9.0	...	...
Heat-deflection temperature, at 455 kPa (66 psi), °C (°F)	185 (365)	145 (293)	190 (374)	165 (330)	...	...
Water absorption, 24 h, 3.2 mm (1/8 in.), %	1.3 to 1.9	0.25	1.5	0.4	...	...

(a) Observed range is attributed to variable sample water content; T_g increases with dryness

Selected properties of nylons

Property	ASTM test method	Reported value for: Extruded or molded, aromatic or transparent nylon	Extruded or molded nylon 6	Extruded or molded elastomer copolymer nylon 6	Cast nylon 6	Extruded or molded nylon 6/6	Extruded or molded high-impact modified nylon 6/6
Mechanical							
Yield strength, MPa (ksi)	D 638	68 to 84 (9.8 to 12.2)	62 to 90 (9.0 to 13.0)	52 to 69 (7.5 to 10.0)	76 to 90 (11.0 to 13.0)	76 to 83 (11.0 to 12.0)	14 to 55 (2.0 to 8.0)
Tensile modulus, GPa (10^6 psi)	D 638	2.8 (0.40)	1.4 to 2.8 (0.20 to 0.40)	...	2.4 to 3.7 (0.35 to 0.54)	2.6 to 3.2 (0.38 to 0.47)	...
Elongation (break), %	D 638	70 to 150	100 to 320	150 to 275	20 to 60	60 to 300	40 to 240
Compressive strength, MPa (ksi)	D 695	97 (14)	69 to 90 (10 to 13)	...	62 to 105 (9 to 15)	105 (15)	34 (5)
Flexural yield strength, MPa (ksi)	D 790	90 (13)	34 to 97 (5 to 14)	34 to 83 (5 to 12)	34 to 110 (5 to 16)	41 to 110 (6 to 16)	...
Flexural modulus, GPa (10^6 psi)	D 790	2.8 (0.40)	2.8 (0.40)	0.69 to 2.1 (0.10 to 0.30)	1.4 to 3.4 (0.20 to 0.50)	1.4 to 4.1 (0.20 to 0.60)	1.4 to 2.1 (0.20 to 0.30)
Impact strength, Izod(a), J/cm (ft · lbf/in.)	D 256	...	0.53 to 1.6 (1.0 to 3.0)	No break	0.53 to 1.6 (1.0 to 3.0)	0.53 to 1.1 (1.0 to 2.0)	8.0 to 13.3 (15.0 to 25.0)
Hardness, Rockwell	...	M95	R90 to 120	R90 to 110	R95 to 120	R120	R112
Thermal							
Thermal conductivity, W/m · K (Btu · in./ft^2 · h · °F)	C 177	0.19 (1.3)	0.17 (1.2)	...	0.17 (1.2)	0.26 (1.8)	0.26 (1.8)
Coefficient of linear thermal expansion 10^{-5}/°C (10^{-5}/°F)	D 696	5.0 (2.8)	9.0 to 14 (5.0 to 8.0)	...	9.0 to 16 (5.0 to 9.0)	14 (8.0)	...
Specific heat, kJ/kg · K (Btu/lb · °F)	C 351	1.7 (0.4)	1.7 (0.4)	...	1.7 (0.4)	1.7 (0.4)	1.7 (0.4)
Continuous service temperature, °C (°F)	...	...	79 to 120 (175 to 250)	...	79 to 120 (175 to 250)	79 to 150 (175 to 300)	79 to 150 (175 to 300)
Electrical							
Volume resistivity, Ω · cm	D 257	10^{15}	10^{11} to 10^{14}	...	10^{14}	10^{11} to 10^{14}	10^{11} to 10^{14}
Dielectric strength, V/10^{-3} mm (V/10^{-3} in.)	D 149	14 (350)	16 (400)	18 (450)	14 (350)	24 (600)	20 (500)
Dielectric constant (50 to 100 Hz)	D 150	3.7	3.8 to 5.0	3.0 to 4.0	4.0	4.2	3.2
Dissipation factor (50 to 100 Hz)	D 150	0.02	0.010 to 0.100	0.010 to 0.080	0.020	0.020	0.010 to 0.015
Processing parameters							
Processing temperature, °C (°F)	...	255 to 315 (490 to 600)	225 to 270 (440 to 520)	230 to 300 (450 to 575)	...	270 to 325 (520 to 620)	270 to 305 (520 to 580)
Density, g/cm^3	D 792	1.12	1.12 to 1.14	1.08 to 1.14	1.15	1.13 to 1.15	1.08 to 1.11
Specific volume, cm^3/kg ($in.^3$/lb)	D 792	865 (24)	905 to 865 (25 to 24)	905 to 865 (25 to 24)	865 (24)	865 (24)	905 (25)
Linear mold shrinkage, mm/mm (in./in.)	D 955	0.005	0.002 to 0.005	0.008 to 0.020	...	0.005 to 0.015	0.012 to 0.019
Water absorption, % in 24 h	D 570	0.4	1.7 to 1.8	0.8 to 1.4	0.6	1.5	1.5

Property	ASTM test method	Reported value for: Extruded or molded copolymer nylon 6/6-6	Extruded or molded nylon 6/9	Extruded or molded nylon 6/10	Extruded or molded nylon 6/12	Extruded or molded nylon 11	Extruded or molded nylon 12
Mechanical							
Yield strength, MPa (ksi)	D 638	52 to 82.7 (7.5 to 12.0)	59 (8.5)	49 to 59 (7.1 to 8.5)	55 to 68.9 (8.0 to 10.0)	55 (8.0)	55 to 66 (8.0 to 9.5)
Tensile modulus, GPa (10^6 psi)	D 638	1.0 to 2.8 (0.15 to 0.40)	1.9 (0.28)	1.2 to 1.9 (0.17 to 0.28)	1.2 to 2.1 (0.18 to 0.30)	1.0 to 1.4 (0.15 to 0.20)	1.2 (0.18)
Elongation (break), %	D 638	100 to 285	1100	100 to 300	150 to 325	300	300
Compressive strength, MPa (ksi)	D 695	...	...	...	...	...	...
Flexural yield strength, MPa (ksi)	D 790	69 to 103 (10 to 15)	69 to 103 (10 to 15)	69 to 103 (10 to 15)	83 to 90 (12 to 13)	...	...
Flexural modulus, GPa (10^6 psi)	D 790	1.4 to 2.8 (0.20 to 0.40)	1.4 to 2.1 (0.20 to 0.30)	1.4 to 2.1 (0.20 to 0.30)	1.4 to 2.8 (0.20 to 0.40)	2.1 (0.30)	1.4 (0.20)
Impact strength, Izod(a), J/cm (ft · lbf/in.)	D 256	0.53 (1.0)	0.59 to 1.1 (1.1 to 2.1)	0.5 to 1.2 (0.9 to 2.3)	0.53 to 0.80 (1.0 to 1.5)	0.96 (1.8)	1.1 to 2.9 (2.0 to 5.5)
Hardness, Rockwell	...	R90 to 115	R111	R111	R115	R90 to 110	R105 to 110

(continued)

Selected properties of nylons (continued)

Property	ASTM test method	Reported value for: Extruded or molded, copolymer nylon 6/6-6	Extruded or molded nylon 6/9	Extruded or molded nylon 6/10	Extruded or molded nylon 6/12	Extruded or molded nylon 11	Extruded or molded nylon 12
Thermal							
Thermal conductivity, W/m · K (Btu · in./ft^2 · h · °F)	C 177	0.25 (1.7)	...	0.20 (1.4)	0.20 (1.4)	0.23 (1.6)	0.20 (1.4)
Coefficient of linear thermal expansion 10^{-5}/°C (10^{-5}/°F)	D 696	14 (8.0)	14 to 16 (8.0 to 9.0)	16 (9.0)	9.0 to (5.0)	18.0 (10.0)	18.0 (10.0)
Specific heat, kJ/kg · K (Btu/lb · °F)	C 351	1.7 (0.4)	1.7 (0.4)	1.7 (0.40)	1.3 to 1.7 (0.3 to 0.4)	2.1 (0.5)	1.3 (0.3)
Continuous service temperature, °C (°F)	...	79 to 135 (175 to 275)	...	79 to 120 (175 to 250)	79 to 150 (175 to 300)	79 to 150 (175 to 300)	
Electrical							
Volume resistivity, Ω · cm	D 257	10^{10}	10^{13}	10^{12}	10^{13}	10^{13}	10^{13}
Dielectric strength, V/10^{-3} mm (V/10^{-3} in.)	D 149	16 to 24 (400 to 600)	24 (600)	16 (400)	16 (400)	17 (425)	18 (450)
Dielectric constant (50 to 100 Hz)	D 150	...	3.7	3.9	3.9	3.7	4.2
Dissipation factor (50 to 100 Hz)	D 150	...	0.020	0.040	0.020	...	0.04
Processing parameters							
Processing temperature, °C (°F)	...	175 to 205 (350 to 400)	230 to 290 (450 to 550)	230 to 290 (450 to 550)	230 to 290 (450 to 550)	200 to 260 (390 to 500)	180 to 260 (360 to 500)
Density, g/cm^3	D 792	1.08 to 1.14	1.10	1.07 to 1.09	1.07	1.04	1.01 to 1.02
Specific volume, cm^3/kg (in.^{3}/lb)	D 792	905 to 865 (25 to 24)	905 (25)	905 (25)	905 (25)	940 (26)	905 (25)
Linear mold shrinkage, mm/mm (in./in.)	D 955	0.006 to 0.010	0.010 to 0.015	0.012	0.010	0.010	0.003 to 0.010
Water absorption, % in 24 h	D 570	1.5 to 2.0	0.4	0.2	0.4	0.3	0.3

(a) Notched, at room temperature

Property values for nylons 6 and 6/6

Dry, as-molded, approximately 0.2% moisture content

Property	Nylon 6		Nylon 6/6			
					Toughened	
	Molding and extrusion compound	Glass fiber reinforced, 30 to 35%	Molding compound	Glass fiber reinforced, 30 to 33%	Unreinforced	Glass fiber reinforced, 33%
Mechanical						
Tensile strength at break, MPa (ksi)	...	165 (24)	94.5 (13.7)	193 (28)	50 (7.0)	140 (20.3)
Elongation at break, %	30 to 100	2.2 to 3.6	15 to 60	2.5 to 34	125	4 to 6
Tensile yield strength, MPa (ksi)	80.7 (11.7)	...	55 (8.00)	170 (25)	...	...
Compressive strength, rupture or yield, MPa (ksi)	90 to 110 (13 to 16)	131 to 165 (19 to 24)	86.2 to 103 (12.5 to 15.0)(a)	165 to 276 (24 to 40)	...	103 (15)
Flexural strength, rupture, or yield, MPa (ksi)	108 (15.7)	240 (35)	114 to 117 (16.5 to 17.0)	283 (41)	59 (8.5)	206 (29.9)
Tensile modulus, GPa (10^6 psi)	2.6 (0.38)	8.62 to 10.0 (1.25 to 1.45)	1.59 to 3.79 (0.23 to 0.55)	9.0 (1.3)	...	...
Flexural modulus at 23 °C (73 °F), GPa (10^6 psi)	2.7 (0.39)	9.65 (1.40)	2.8 to 3.1 (0.41 to 0.45)	8.96 to 10.0 (1.30 to 1.45)	1.65 (0.240)	7.58 (1.10)
Izod impact, 3.2 mm (1/8 in.) thick specimen, notched, J/m (ft · lbf/in.)	32 to 53 (0.6 to 1.0)	117 to 181 (2.2 to 3.4)	29 to 53 (0.55 to 1.0)	85 to 240 (1.6 to 4.5)	907 (17.0)	240 (4.5)
Rockwell hardness	R119	M93 to 96	R120	R101 to 119	R100	R107
Thermal						
Coefficient of linear thermal expansion, 10^{-6}/K	80 to 83	16 to 80	80	15 to 54		
Deflection temperature under flexural load, °C (°F)						
At 1.82 MPa (0.264 ksi)	68 to 85 (155 to 185)	200 to 215 (392 to 420)	75 to 88 (167 to 190)	120 to 250 (252 to 490)	65 (150)	245 (470)
At 0.46 MPa (0.066 ksi)	185 to 190 (365 to 375)	215 to 220 (420 to 430)	230 to 246 (450 to 474)	125 to 260 (260 to 500)	...	260 (495)
Thermal conductivity, W/m · K (Btu · in./h · ft^2 · °F)	0.243 (1.69)	0.243 to 0.472 (1.69 to 3.27)	0.243 (1.69)	0.211 to 0.483 (1.46 to 3.35)	...	...
Physical						
Specific gravity	1.12 to 1.14	1.35 to 1.42	1.13 to 1.15	1.15 to 1.40	1.08	1.34
Water absorption, 3.2-mm (1/8-in.) thick specimen, 24 h, %	1.3 to 1.9	0.90 to 1.2	1.0 to 2.8	0.7 to 1.1	1.0	0.7
Saturation, %	8.5 to 10.0	6.4 to 7.0	8.5	5.5 to 6.5	...	...
Electrical						
Dielectric strength, 3.2-mm (1/8-in.) thick specimen, short time, MV/m (V/mil)	0.0157 (400)	15.7 to 17.7 (400 to 450)	0.0236 (600)	14.2 to 19.7 (360 to 500)	...	...

(a) Yield

Property values for nylons 6 and 6/6
Conditioned to 50% relative humidity

	Nylon 6		Nylon 6/6	
Property	**Molding and extrusion compound**	**Glass fiber reinforced, 30 to 35%**	**Molding compound**	**Glass fiber reinforced, 30 to 33%**
Tensile strength at break, MPa (ksi)	...	110 (16)	75 (11)	152 (22)
Elongation at break, %	300	...	150 to 300	5 to 7
Tensile yield strength, MPa (ksi)	51 (7.4)	...	45 (6.5)	...
Flexural strength, rupture or yield, MPa (ksi)	40 (5.8)	145 (21.0)	42 (6.1)	170 (25)
Tensile modulus, GPa (10^6 psi)	0.690 (0.100)	5.52 (0.800)	3.45 (0.500)	
Compressive modulus, GPa (10^6 psi)	1.70 (0.250)	...	...	...
Flexural modulus at 23 °C (73 °F), GPa (10^6 psi)	0.965 (0.140)	5.52 to 6.55 (0.800 to 0.950)	1.28 (0.185)	5.52 (0.800)
Izod impact, 3.2-mm (1/8-in.) thick specimen, notched, J/m (ft · lbf/in.)	160 (3.0)	197 to 292 (3.7 to 5.5)	45 to 112 (0.85 to 2.1)	138 to 159 (2.6 to 3.0)
Rockwell hardness	...	M78	...	...

Recommended processing factors for injection-moldable nylons

Grade	Melt-temperature range		Melt viscosity(a), Pa · s	Injection speed	Mold temperature		Maximum regrind, %	Gating diameter(b)		Shrinkage range, mm/mm (in./in.)	Minimum cycle time(b), s
	°C	°F			°C	°F		mm	in.		
Nylon 6											
Unmodified base resin:											
Low viscosity	240 to 270	460 to 520	120	Low to high	10 to 95	50 to 200	50	1.0 to 1.8	0.040 to 0.070	0.009 to 0.015	12 to 15
Medium viscosity	250 to 280	480 to 540	...	Low to high	10 to 95	50 to 200	50	1.5 to 2.3	0.060 to 0.090	0.009 to 0.015	12 to 15
High viscosity	240 to 270	460 to 520	120	Low to high	80 to 95	180 to 200	50	1.0 to 1.8	0.040 to 0.070	0.007 to 0.013	9 to 12
Glass reinforced:											
15%	245 to 275	470 to 530	...	High	80 to 95	180 to 200	25	1.5 to 2.3	0.060 to 0.090	0.004 to 0.008	12 to 18
33%	260 to 295	500 to 560	200	High	80 to 95	180 to 200	25	1.5 to 3.18	0.060 to 0.125	0.003 to 0.006	12 to 18
44%	265 to 300	510 to 570	...	High	80 to 95	180 to 200	25	1.5 to 3.18	0.060 to 0.125	0.002 to 0.005	12 to 18
Mineral reinforced, 35 to 40%	270 to 300	520 to 570	185	High	80 to 105	180 to 220	25	1.5 to 3.18	0.060 to 0.125	0.008 to 0.012	12 to 18
Mineral/glass reinforced, 35 to 40%	270 to 300	520 to 570	200	High	80 to 105	180 to 220	25	1.5 to 3.18	0.060 to 0.125	0.003 to 0.006	12 to 18
Impact modified:											
Low	250 to 280	480 to 540	160	Low to high	10 to 95	50 to 200	25	1.0 to 2.0	0.040 to 0.080	0.009 to 0.080	15 to 18
High	250 to 280	480 to 540	...	Low to high	10 to 95	50 to 200	25	1.5 to 2.3	0.060 to 0.090	0.010 to 0.016	15 to 18
Flame retarded:											
Unreinforced	240 to 260	460 to 500	...	High	80 to 95	180 to 200	25	1.5 to 2.3	0.060 to 0.090	0.009 to 0.015	12 to 15
25% glass reinforced	260 to 295	500 to 560	...	High	80 to 95	180 to 200	25	1.5 to 3.18	0.060 to 0.125	0.003 to 0.006	12 to 15
Nylon 6/6											
Unmodified base resin, low viscosity	270 to 295	520 to 560	1000 poise	High	40 to 95	100 to 200	50	1.0 to 1.8	0.040 to 0.070	0.015 to 0.020	12 to 15
Glass reinforced, 13, 33, and 43%	290 to 300	550 to 575	...	High	65 to 95	150 to 200	25	1.5 to 3.18	0.060 to 0.125	0.003 to 0.008	12 to 18
Mineral reinforced, 35 to 40%	280 to 300	540 to 570	...	High	80 to 105	108 to 200	25	1.5 to 3.18	0.060 to 0.125	0.012 to 0.017	12 to 18
Mineral/glass reinforced, 35 to 40%	280 to 300	540 to 570	...	High	80 to 105	180 to 220	25	1.5 to 3.18	0.060 to 0.125	0.010 to 0.015	12 to 18
Impact modified	290 to 295	550 to 560	...	High	40 to 95	100 to 200	25	1.5 to 3.18	0.060 to 0.125	0.015 to 0.020	12 to 15
Flame retarded:											
Unreinforced	250 to 270	480 to 520	...	High	40 to 95	100 to 200	25	1.5 to 2.3	0.060 to 0.090	0.015 to 0.020	12 to 15
25% glass reinforced	250 to 270	480 to 520	...	High	40 to 95	100 to 200	25	1.5 to 2.3	0.060 to 0.090	0.003 to 0.005	12 to 18
Supertough	290 to 295	550 to 560	...	High	40 to 95	100 to 200	25	1.5 to 2.3	0.060 to 0.090	0.015 to 0.020	12 to 15

(a) Melt viscosity measured at 1000/s at the median temperature of the recommended melt-temperature range. (b) Data based on a wall thickness of 3.18 mm (1/8 in.)

Polyamideimides (PAI)

Molecular structure of polyamideimide

Thermal, physical, and chemical properties of a typical polyamideimide (PAI)

Property	Value
Heat-deflection temperature at 1.82 MPa, °C	275
Maximum resistance to continuous heat, °C	225
Coefficient of linear thermal expansion, 10^{-5}/°C	3.6
Compressive strength, MPa	220
Flexural strength, MPa	210
Impact strength, Izod, cm · N/cm of notch	133.5
Tensile strength, MPa	186
Elongation, %	12
Hardness, Rockwell M	119
Specific gravity	1.4
Dielectric constant	4.0
Water absorption	0.2
Resistance to chemicals at 25% °C(a)	
Nonoxidizing acids (20% H_2SO_4)	S
Oxidizing acids (10% HNO_3)	U
Aqueous salt solutions (NaCl)	S
Aqueous alkalis (NaOH)	U
Polar solvents (C_2H_5OH)	S
Nonpolar solvents (C_6H_6)	S
Water	S

(a) S, satisfactory; Q, questionable; U, unsatisfactory

Selected properties of molded polyamideimide

Property	ASTM test method	Reported value
Mechanical		
Yield strength, MPa (ksi)	D 638	93.1 (13.5)
Elongation (break), %	D 638	3
Compressive strength, MPa (ksi)	D 695	240 (35)
Flexural yield strength, MPa (ksi)	D 790	160 to 165 (23 to 24)
Flexural modulus, GPa (10^6 psi)	D 790	4.8 (0.70)
Impact strength, Izod(a), J/cm (ft · lb/in.)	D 256	0.53 (1.0)
Hardness, Rockwell	...	R104
Thermal		
Coefficient of linear thermal expansion, 10^{-5}/°C (10^{-5}/°F)	D 696	6.3 (3.5)
Electrical		
Volume resistivity, Ω · cm	D 257	10^{14}
Dissipation factor (50 to 100 Hz)	D 150	0.005 to 0.007
Processing parameters		
Processing temperature, °C (°F)	...	345 to 385 (650 to 725)
Density, g/cm^3	D 792	1.40
Specific volume, cm^3/kg (in.^{3}/lb)	D 792	725 (20)
Linear mold shrinkage, mm/mm (in./in.)	D 955	0.006
Water absorption, % in 24 h	D 570	0.3

(a) Notched, at room temperature

PAI mechanical, thermal, electrical, and general properties

Property	ASTM test method	3% TiO_2, ½% PTFE	12% graphite, 3% PTFE	20% graphite, 3% PTFE	12% graphite, 8% PTFE	30% glass fiber	30% graphite fiber	40% glass fiber
Mechanical								
Tensile strength, MPa (ksi)	D 1708							
−196 °C (-321 °F)		218 (31.5)	...	130 (18.8)	...	204 (29.5)	158 (22.8)	...
23 °C (73 °F)		192 (27.8)	164 (23.7)	152 (22.0)	123 (17.8)	205 (29.7)	203 (29.4)	220 (31.8)
135 °C (275 °F)		117 (16.9)	113 (16.3)	113 (16.3)	104 (15.1)	160 (23.1)	158 (22.8)	172 (25.0)
232 °C (450 °F)		66 (9.5)	73 (10.6)	56 (8.1)	54 (7.8)	113 (16.3)	108 (15.7)	137 (19.9)
Tensile elongation, %	D 1708							
−196 °C (-321 °F)		6		3		4	3	...
23 °C (73 °F)		15	7	7	9	7	6	7
135 °C (275 °F)		21	20	15	21	15	14	6
232 °C (450 °F)		22	17	17	15	12	11	8
Tensile modulus, at 23 °C (73 °F), GPa (10^6 psi)	D 1708	4.9 (0.7)	6.6 (0.95)	7.8 (1.13)	6.0 (0.87)	10.8 (1.56)	22.3 (3.22)	...
Flexural strength, MPa (ksi)	D 790							
−196 °C (-321 °F)		287 (41.0)	...	203 (29.0)	...	381 (54.4)	315 (45.0)	...
23 °C (73 °F)		244 (34.9)	219 (31.2)	212 (30.2)	189 (27.0)	338 (48.3)	355 (50.7)	364 (52.0)
135 °C (275 °F)		174 (24.8)	165 (23.5)	157 (22.4)	144 (20.5)	251 (35.9)	263 (37.6)	298 (43.6)
232 °C (450 °F)		120 (17.1)	113 (16.2)	111 (15.8)	100 (14.3)	184 (26.2)	177 (25.2)	279 (38.9)
Flexural modulus, GPa (10^6 psi)	D 790							
−196 °C (-321 °F)		7.9 (1.14)	...	9.6 (1.39)	...	14.1 (2.04)	24.6 (3.57)	...
23 °C (73 °F)		5.0 (0.73)	6.9 (1.0)	7.3 (1.06)	6.3 (0.91)	11.7 (1.70)	19.9 (2.88)	14.5 (2.10)
135 °C (275 °F)		3.9 (0.56)	5.5 (0.79)	5.6 (0.81)	4.4 (0.64)	10.7 (1.55)	18.8 (2.72)	13.8 (2.0)
232 °C (450 °F)		3.6 (0.52)	4.5 (0.72)	5.1 (0.74)	4.3 (0.62)	9.9 (1.43)	15.7 (2.28)	13.8 (2.0)
Compressive strength, MPa (ksi)	D 695	220 (32.1)	170 (24.1)	120 (17.8)	130 (18.3)	260 (38.3)	250 (36.9)	322 (46.7)
Compressive modulus, GPa (10^6 psi)	D 695	...	4.0 (0.77)	5.3 (0.58)	...	7.9 (1.15)	9.9 (1.43)	...
Shear strength, at 23 °C (73 °F), MPa (ksi)	D 732	128 (18.5)	112 (16.1)	77 (11.1)	80 (11.5)	140 (20.1)	120 (17.3)	150 (22.7)
Izod impact strength (3.2 mm, or ⅛ in.), J/m (ft · lbf/in.)	D 256							
Notched		142 (2.7)	63 (1.2)	84 (1.6)	69 (1.3)	79 (1.5)	47 (0.9)	79 (1.5)
Unnotched		1062 (20.0)	404 (7.6)	250 (4.7)	...	504 (9.5)	340 (6.4)	430 (8.1)
Poisson's ratio		0.45	0.39	0.39	...	0.43	0.39	...
Thermal								
Deflection temperature at 1.82 MPa (0.264 ksi), °C (°F)	D 648	278 (532)	279 (534)	280 (536)	278 (532)	282 (539)	282 (540)	280 (536)
Coefficient of linear thermal expansion, 10^{-6}/K	D 696	30.6	25.2	25.2	27.0	16.2	9.0	12.6
Thermal conductivity, W/m · K (Btu · in/h · ft² · °F)	C 177	0.26 (1.8)	0.54 (3.7)	...	...	0.37 (2.5)	0.53 (3.6)	...
Flammability, UL		94V-0	94V-0	94V-0	94V-0	94V-0	94V-0	94V-0
Limiting oxygen index, %	D 2863	45	44	45	46	51	52	47
Electrical								
Dielectric constant	D 150							
10^3 Hz		4.2	6.0	7.3	6.8	4.4	...	4.3
10^6 Hz		3.9	5.4	6.6	6.0	4.2	...	4.6
Dissipation factor	D 150							
10^3 Hz		0.026	0.037	0.059	0.037	0.022	...	0.040
10^6 Hz		0.031	0.042	0.063	0.071	0.050	...	0.044
Volume resistivity, Ω · m	D 257	2×10^{15}	8×10^{13}	8×10^{13}	8×10^{13}	2×10^{15}	...	5×10^{14}
Surface resistivity, Ω	D 257	5×10^{18}	8×10^{17}	4×10^{17}	1×10^{18}	1×10^{18}	...	9×10^{17}
Dielectric strength (1.02 mm, or 0.040 in.), kV/mm (V/mil)	D 149	23.6 (580)	...	...	...	32.6 (840)	...	19.5 (490)
General								
Density, g/cm³	D 792	1.42	1.46	1.51	1.50	1.61	1.48	1.68
Hardness, Rockwell E	D 785	86	72	70	66	94	94	107

Properties of polyamideimide engineering resins with various reinforcements and additives

Nominal composition	Properties/characteristics	Applications
High-strength composites		
3% TiO_2 ½% fluorocarbon	Best impact resistance, most elongation, and good mold-release and electrical properties	Connectors, switches, relays, thrust washers, spline liners, valve seats, poppets, mechanical linkages, bushings, wear rings, insulators, cams, picker fingers, ball bearings, rollers, thermal insulators
30% glass fiber 1% fluorocarbon	High stiffness, good retention of stiffness at elevated temperatures, very low creep, and high strength	Burn-in sockets, gears, valve plates, fairings, tube clamps, impellers, rotors, housings, back-up rings, terminal strips, insulators, brackets
30% glass fiber	Can be molded to greater thickness with some sacrifice in mechanical properties	Same as above, but for parts requiring thicker cross sections
30% glass fiber 4% TiO_2 1% fluorocarbon	High stiffness, good retention of stiffness at elevated temperatures, very low creep, and high strength	Structural, electrical, valve plates, metal replacement

(continued)

Properties of polyamideimide engineering resins with various reinforcements and additives (continued)

Nominal composition	Properties/characteristics	Applications
High-strength composites (cont'd)		
30% graphite fiber 1% fluorocarbon	Best retention of stiffness at high temperatures, best fatigue resistance; electrically conductive	Metal replacement, housings, mechanical linkages, gears, fasteners, spline liners, cargo rollers, brackets, valves, labyrinth seals, fairings, tube clamps, standoffs, impellers, shrouds, potential use for EMI shielding
33% carbon fiber 1% fluorocarbon	Can be molded to greater thickness with some sacrifice in mechanical properties; electrically conductive	Injection molding of normally troublesome thick cross sections
Proprietary blend of carbon fibers and fluorocarbons	High stiffness and lubricity	Service requiring high stiffness and some lubricity, especially sliding vanes; potential use for EMI shielding
Wear-resistant composites		
12% graphite powder 8% fluorocarbon	Good for reciprocating motion or bearings subject to high loads at low speed; best wear resistance	Bearings, thrust washers, wear pads, strips, piston rings, seals
12% graphite powder 8% fluorocarbon	Designed for bearing use; good wear resistance, low coefficient of friction, and high compressive strength	Bearings, thrust washers, wear pads, strips, piston rings, seals vanes, valve seats
20% graphite powder 3% fluorocarbon	Better wear resistance at high speeds	Bearings, thrust washers, wear pads, strips, piston rings, seals vanes, valve seats
High performance composite		
40% glass fiber 1% fluorocarbon	Best cost-to-performance ratio	Switches, relays, terminal strips, wear bands, back-up rings, housings, impellers, brackets, thermal insulators

Wear characteristics of PAI bearing grades

	PV values		
Parameter	**10 000**	**45 000**	**50 000**
Pressure, kPa (psi)	345 (50)	345 (50)	6900 (1000)
Velocity, m/s (ft/min.)	1 (200)	4.5 (900)	0.25 (50)

	PAI grade		
Wear characteristics	**12% graphite, 3% PTFE**	**20% graphite, 3% PTFE**	**12% graphite, 8% PTFE**
Wear factor, K, 10^{-10} cm^3 · min/J · h (10^{-10} in.^{3} · min/ft · lbf · h)			
PV = 10 000	205 (17)	95 (8)	75 (6)
PV = 45 000	640 (53)	485 (40)	510 (42)
PV = 50 000	495 (41)	375 (31)	290 (24)
Coefficient of friction, static			
PV = 10 000	0.06	0.02	0.02
PV = 45 000	0.13	0.07	0.08
PV = 50 000	0.11	0.14	0.08
Coefficient of friction, kinetic			
PV = 10 000	0.27	0.19	0.19
PV = 45 000	0.14	0.15	0.13
PV = 50 000	0.12	0.11	0.11

Chemical resistance of PAI with 3% TiO_2 and ½% PTFE

Chemical	Retained tensile strength(a), %	Chemical	Retained tensile strength(a), %
Acids		**Aqueous solutions (10%, unless otherwise noted)**	
Acetic (10%)	100	Aluminum sulfate	100
Glacial acetic	100	Ammonium chloride	100
Acetic anhydride	100	Ammonium nitrate	98
Lactic	100	Ammonium sulfate	100
Benzene sulfonic	28	Barium chloride	100
Chromic (10%)	100	Bromine (saturated solution, 50 °C or 120 °F)	100
Formic (88%)	66	Calcium chloride	100
Hydrochloric (10%)	100	Calcium nitrate	96
Hydrochloric (37%)	95	Ferric chloride	99
Phosphoric (35%)	100	Magnesium chloride	100
Sulfuric (30%)	100	Potassium permanganate	100
Bases		Sodium bicarbonate	100
Ammonium hydroxide (28%)	81	Silver chloride	100
Sodium hydroxide (15%)	43	Sodium carbonate	100
Sodium hydroxide (30%)	7	Sodium chloride	100

(continued)

Chemical resistance of PAI with 3% TiO_2 and ½% PTFE (continued)

Chemical	Retained tensile strength(a), %
Sodium chromate	100
Sodium hypochlorite	100
Sodium sulfate	100
Sodium sulfide	84
Sodium sulfite	100
Water	100
Alcohols	
2-aminoethanol	9
Amyl ethanol	100
Butyl ethanol	100
Cyclohexanol	100
Ethylene glycol	100
Amines	
Aniline	97
n-butylamine	100
Dimethylaniline	100
Ethylenediamine	7
Morpholine	100
Pyridine	43
Aldehydes and ketones	
Acetophenone	100
Benzaldehyde	100
Cyclohexanone	100
Formaldehyde (37%)	100
Furfural	84
Methyl ethyl ketone	100
Chlorinated organics	
Acetyl chloride, 50 °C (120 °F)	100
Benzyl chloride, 50 °C (120 °F)	100
Carbon tetrachloride	100
Chlorobenzene	100
2-chloroethanol	100
Chloroform, 50 °C (120 °F)	100
Epichlorohydrin	100
Ethylene chloride	100
Esters	
Amyl acetate	100
Butyl acetate	100
Butyl phthalate	100
Ethyl acetate	100
Ethers	
Butyl ether	100
Cellosolve	100
p-dioxane, 50 °C (120 °F)	100
Tetrahydrofuran	100
Hydrocarbons	
Cyclohexane	100
Diesel fuel	99
Gasoline, 50 °C (120 °F)	100
Heptane	100
Mineral oil	100
Motor oil	100
Stoddard solvent	100
Toluene	100
Xylene	100
Nitriles	
Acetonitrile	100
Benzonitrile	100
Nitro compounds	
Nitrobenzene	100
Nitromethane	100
Miscellaneous	
Cresyldiphenyl phosphate	100
Sulfolane	100
Triphenylphosphite	100

(a) After 24h exposure at 93 °C (200 °F), unless otherwise noted

Summary of PAI flammability data

Oxygen index, ASTM D 2863

	Oxygen index, %
PAI with 3% TiO_2, ½% PTFE	45
PAI with 12% graphite, 3% PTFE	44
PAI with 20% graphite, 3% PTFE	45
PAI with 12% graphite, 8% PTFE	46
PAI with 30% glass fiber	51
PAI with 30% graphite fiber	52
PAI with 40% glass fiber	47

FAA(a) smoke density, National Bureau of Standards, NFPA 258. Specimen thickness, 1.3-1.5 mm (0.05-0.06 in.)

	3% TiO_2, ½% PTFE		30% glass fiber		30% graphite fiber	
	Smoldering	Flaming	Smoldering	Flaming	Smoldering	Flaming
Minimum light transmittance, %	92	6	96	56	95	28
Maximum specific optical density (D_m)	5	170	2	35	3	75
Time to 90% D_m, minutes	18.5	18.6	10.7	15.7	17.0	16.0

FAA toxic gas emission test, National Bureau of Standards, NFPA 258. Specimen thickness, 1.3-1.5 mm (0.05-0.06 in.)

	30% glass fiber		30% graphite fiber	
	Smoldering, ppm	Flaming, ppm	Smoldering, ppm	Flaming, ppm
Hydrochloric acid	0	<1	0	<1
Hydrofluoric acid	0	0	0	0
Carbon monoxide	<10	120	<10	100
Nitrogen oxides	<2	19	0	14
Hydrocyanic acid	0	4	0	5
Sulfur dioxide	0	0	0	4

(continued)

Summary of PAI flammability data (continued)
Ignition properties of PAI with 3% TiO_2, ½% PTFE. STM D 1929

	°C	°F
Flash ignition temperature	570	1058
Self-ignition temperature	620	1148

Vertical flammability class by Underwriters' Laboratories, UL 94

Thickness mm	in.	94V-0 ratings for various grades
0.20	0.008	3% TiO_2, ½% PTFE
0.51	0.020	3% TiO_2, ½% PTFE
1.17	0.046	3% TiO_2, ½% PTFE; 12% graphite, 3% PTFE; 30% glass fiber
1.47	0.058	3% TiO_2, ½% PTFE; 12% graphite, 3% PTFE; 30% glass fiber
2.44	0.096	3% TiO_2, ½% PTFE; 12% graphite, 3% PTFE
3.05	0.120	3% TiO_2, ½% PTFE; 12% graphite, 3% PTFE
3.18	0.125	All grades

Note: The test methods used to obtain these data measure response to heat and flame under controlled laboratory conditions detailed in the test method specified and may not provide an accurate measure of fire hazard under actual fire conditions. (a) FAA, Federal Aviation Administration

Polyarylates

Thermal, physical, and chemical properties of typical crystalline polyarylates (Ekonol)

Heat-deflection temperature at 1.82 MPa, °C	300
Maximum resistance to continuous heat, °C	250
Coefficient of linear thermal expansion, 10^{-5}/°C	71
Compressive strength, MPa	265
Flexural strength, MPa	63
Impact strength, Izod, cm · N/cm of notch	25
Tensile strength, MPa	60
Elongation, %	15
Hardness, Rockwell	R130
Specific gravity	1.44
Dielectric constant	3
Water absorption	0.02
Resistance to chemicals at 25% °C(a):	
Nonoxidizing acids (20% H_2SO_4)	S
Oxidizing acids (10% HNO_3)	Q
Aqueous salt solutions (NaCl)	S
Aqueous alkalis (NaOH)	S
Polar solvents (C_2H_5OH)	S
Nonpolar solvents (C_6H_6)	S
Water	S

(a) S, satisfactory; Q, questionable; U, unsatisfactory

Thermal, physical, and chemical properties of typical polyarylates

Heat-deflection temperature at 1.82 MPa, °C	175
Maximum resistance to continuous heat, °C	150
Coefficient of linear thermal expansion, 10^{-5}/°C	6.5
Compressive strength, MPa	93
Flexural strength, MPa	80
Impact strength, Izod, cm · N/cm of notch	215
Tensile strength, MPa	68
Elongation, %	50
Hardness, Rockwell	R125
Specific gravity	1.2
Dielectric constant	0.7
Water absorption	0.26
Resistance to chemicals at 25% °C(a):	
Nonoxidizing acids (20% H_2SO_4)	S
Oxidizing acids (10% HNO_3)	Q
Aqueous salt solutions (NaCl)	S
Aqueous alkalis (NaOH)	S
Polar solvents (C_2H_5OH)	S
Nonpolar solvents (C_6H_6)	S
Water	S

(a) S, satisfactory; Q, questionable; U, unsatisfactory

Properties of commercial amorphous PARs

Property	ASTM test method	Amoco, Ardel D-100	Hoechst Celanese, Durel 400	DuPont, LP-101
Physical				
Specific gravity, g/cm^3	D 792	1.21	1.21	1.21
Mold shrinkage, mm/mm	D 955	0.009	0.006	0.008
Light transmission, 3 mm (0.12 in.) specimen, %	D 1003	75	84 to 88	...
Haze, %	D 1003	7.0	1.5	...
Refractive index		1.61	1.61	...
Thermal				
Heat-deflection temperature at 1.82 MPa (0.264 ksi), °C (°F)	D 648	174 (345)	171 (340)	160 (320)
Coefficient of thermal expansion, 10^{-5}/K	D 696	7.20	6.3	5.04
Thermal conductivity, W/m · K (Btu · in./h · ft^2 · °F)	...	0.178 (1.25)	0.177 (1.23)	...
Flammability, UL 94, 3.18 mm (0.13 in.) specimen	...	V-0(a)	V-0	V-0
Oxygen index, %	D 2863	34	38	37
Mechanical				
Tensile strength, yield, MPa (ksi)	D 638	68.9 (10.0)	68.9 (10.0)	75.8 (11.0)
Tensile elongation, rupture, %	D 638	50	50	50
Flexural strength, MPa (ksi)	D 790	75.8 (11.0)	100 (14.5)	75.8 (11.0)
Flexural modulus, GPa (10^6 psi)	D 790	2.14 (0.310)	2.28 (0.331)	2.34 (0.339)
Notched Izod impact at 23 °C (73 °F), J/m (ft · lbf/ft)	D 256	224 (50)	294 (66)	117 (26)
Direct impact, 3 mm (0.12 in.) plaque, 18 mm (0.070 in.) dart, J (ft · lbf)	...	>90 (66)	>54 (40)	...
Electrical				
Dielectric strength, 3 mm (0.12 in.) specimen, kV/mm (V/mil)	D 149	15.75 (400)	18.31 (465)	13.39 (340)
Arc resistance, s	D 495	125	124	91
Dielectric constant				
At 60 Hz	D 150	3.34	3.08	3.11
At 1 MHz	D 150	3.30	2.93	3.12
Dissipation factor				
At 60 Hz	D 150	0.002	0.001	0.004
At 1 MHz	D 150	0.020	0.022	0.006
Volume resistivity, 10^{12} Ω · m	D 257	3.0	2.0	...

Injection-molded test specimens. (a) 1.59 mm (0.063 in.) specimen

Polyarylsulfones (PAS)

Molecular structure of polyarylsulfone

Inherent properties of PAS

Property	ASTM test method	Value
Melt flow at 400 °C (750 °F)	D 1238	10 to 30 g/10 min. (0.35 to 1.05 oz/10 min.)
Density, g/cm^3	D 1505	1.37
Molding shrinkage, %	D 955	0.6
Rockwell hardness	D 785	M85
Water absorption, 24 h, %	D 570	0.4
Equilibrium, %	D 570	1.85
Refractive index, %	...	1.651

Selected properties of polyarylsulfone (molded or extruded)

Property	ASTM test method	Reported value
Mechanical		
Yield strength, MPa (ksi)	D 638	89.6 (13.0)
Tensile modulus, GPa (10^6 psi)	D 638	2.6 (0.37)
Elongation (break), %	D 638	13 to 20
Compressive strength, MPa (ksi)	D 695	125 (18)
Flexural yield strength, MPa (ksi)	D 790	115 (17)
Flexural modulus, GPa (10^6 psi)	D 790	2.8 (0.40)
Impact strength, Izod(a), J/cm (ft · lbf/in.)	D 256	0.53 to 1.1 (1.0 to 2.0)
Hardness, Rockwell	...	M110
Thermal		
Thermal conductivity, W/m · K (Btu · in./ft^2 h · °F)	C 177	0.19 (1.3)
Coefficient of linear thermal expansion, 10^{-5}/°C (10^{-5}/°F)	D 696	8.5 (4.7)
Continuous service temperature, °C (°F)	...	190 (375)
Electrical		
Volume resistivity, Ω · cm	D 257	10^{16}
Dielectric strength, V/10^{-3} mm (V/10^{-3} in.)	D 149	14 (350)
Dielectric constant (50 to 100 Hz)	D 150	3.5
Dissipation factor (50 to 100 Hz)	D 150	0.003
Processing parameters	...	
Processing temperature, °C (°F)		370 to 425 (700 to 800)
Density, g/cm^3	D 792	1.36
Specific volume, cm^3/kg (in.^{3}/lb)	D 792	725 (20)
Linear mold shrinkage, mm/mm (in./in.)	D 955	0.007 to 0.009
Water absorption, % in 24 h	D 570	1.1

(a) Notched, at room temperature

Maximum working stress levels of PAS

	Maximum working stress	
Temperature	MPa	ksi
23 °C (73 °F)	27.6	4.00
60 °C (140 °F)	23.4	3.40
100 °C (212 °F)	18.6	2.70
125 °C (260 °F)	15.5	2.25
149 °C (300 °F)	13.1	1.90

Property retention of PAS at various levels of regrind usage

Property	Regrind level, %			
	0	25	50	100
Flexural strength, MPa (ksi)	119.3 (17.3)	121 (17.5)	118.6 (17.2)	108 (15.6)
Flexural modulus, GPa (10^6 psi)	2.96 (0.430)	2.97 (0.431)	2.96 (0.430)	2.72 (0.394)
Notched Izod impact strength, J/m (ft · lbf/in.)	69.9 (1.31)	72.6 (1.36)	74.2 (1.39)	73.1 (1.37)
Tensile impact strength, J/mm^2 (ft · lbf/in.^{2})	0.305 (145)	0.295 (140)	0.275 (130)	0.295 (140)

Combustion characteristics of PAS

Property	Test method	Ratings(a)
Flammability of 0.75 mm (0.030 in.) specimen	UL 94(b)	V-0
Oxygen	ASTM D 2863	33
Auto-ignition temperature, °C (°F)	ASTM D 1929	502 (935)
Smoke density (D_s) of 1.6 mm (0.063 in.) specimen		
D_s at 1.5 min	NBS flaming smoke chamber	0
D_s at 4.0 min	NBS flaming smoke chamber	1
D_m (D_s maximum)	NBS flaming smoke chamber	5 to 15
Gas toxicity, ppm		
SO_2 at 1.5 to 4.0 min.	...	15
SO_2 maximum	...	75
CO at 1.5 to 4.0 min.	...	50
CO, maximum	...	150

(a) These numerical flame spread ratings or flammability ratings are not intended to reflect hazards presented by these or any other materials under actual fire conditions. (b) Per test run in Amoco Performance Products Laboratory using this procedure

Polycarbonates (PC)

$$\left[-\!\!\!\bigcirc\!\!-\overset{\overset{CH_3}{|}}{\underset{\underset{CH_3}{|}}{C}}-\!\!\bigcirc\!\!-O-\overset{\overset{O}{\|}}{C}-O- \right]_n$$

Molecular structure of polycarbonate

Selected properties of polycarbonate (extruded or molded)

Property	ASTM test method	Reported value
Mechanical		
Yield strength, MPa (ksi)	D 638	55 to 66 (8.0 to 9.5)
Tensile modulus, GPa (10^6 psi)	D 638	2.1 to 2.4 (0.30 to 0.35)
Elongation (break), %	D 638	100 to 125
Compressive strength, MPa (ksi)	D 695	83 to 90 (12 to 13)
Flexural yield strength, MPa (ksi)	D 790	90 to 97 (13 to 14)
Flexural modulus, GPa (10^6 psi)	D 790	2.1 (0.30)
Impact strength, Izod(a), J/cm (ft · lbf/in.)	D 256	6.41 to 9.61 (12.0 to 18.0)
Hardness, Rockwell	...	M70 to 82
Thermal		
Thermal conductivity, W/m · K (Btu · in./ft^2 h · °F)	C 177	0.19 (1.3)
Coefficient of linear thermal expansion, 10^{-5}/°C (10^{-5}/°F)	D 696	11 to 13 (6.0 to 7.0)
Specific heat, kJ/kg · K (Btu/lb · °F)	C 351	1.3 (0.3)
Continuous service temperature, °C (°F)	...	120 to 135 (250 to 275)
Electrical		
Volume resistivity, Ω · cm	D 257	2.0×10^{16}
Dielectric strength, V/10^{-3} mm (V/10^{-3} in.)	D 149	15 (380)
Dielectric constant (50 to 100 Hz)	D 150	3.0
Dissipation factor (50 to 100 Hz)	D 150	0.0007
Processing parameters	...	
Processing temperature, °C (°F)		250 to 345 (480 to 650)
Density, g/cm^3	D 792	1.20
Specific volume, cm^3/kg (in.^{3}/lb)	D 792	830 (23)
Linear mold shrinkage, mm/mm (in./in.)	D 955	0.005 to 0.007
Water absorption, % in 24 h	D 570	0.2

(a) Notched, at room temperature

Thermal, physical, and chemical properties of typical polycarbonates

Property	Unfilled polycarbonate	20% glass-filled polycarbonate
Heat-deflection temperature, °C	130	145
Maximum resistance to continuous heat, °C	115	130
Coefficient of linear thermal expansion, 10^{-5}/°C	6.8	2.2
Compressive strength, MPa	86	124
Flexural strength, MPa	93	158
Impact strength, Izod, cm · N/cm of notch	534	106
Tensile strength, MPa	72	131
Elongation, %	110	4
Hardness, Rockwell	M70	M92
Specific gravity	1.2	1.4
Dielectric constant	3	4
Water absorption, %	0.15	0.25
Resistance to chemicals at 25% °C(a):		
Nonoxidizing acids (20% H_2SO_4)	Q	Q
Oxidizing acids (10% HNO_3)	U	U
Aqueous salt solutions (NaCl)	S	S
Aqueous alkalis (NaOH)	U	U
Polar solvents (C_2H_5OH)	S	S
Nonpolar solvents (C_6H_6)	U	U
Water	S	S

(a) S, satisfactory; Q, questionable; U, unsatisfactory

Recommended processing factors for injection-moldable polycarbonate

Grade	Melt-temperature range °C	Melt-temperature range °F	Mold temperature °C	Mold temperature °F	Shrinkage range, mm/mm (in./in.)	Minimum cycle time(a), s
Unmodified base resin:						
Fast cycling	270 to 315	520 to 600	65 to 95	150 to 200	0.005 to 0.007	15 to 20
Low viscosity	275 to 315	530 to 600	65 to 95	150 to 200	...	15 to 20
Medium viscosity	290 to 315	550 to 600	65 to 95	150 to 200	...	15 to 20
High viscosity	315 to 345	600 to 650	65 to 95	150 to 200	0.005 to 0.007	15 to 20
High heat (polythalate carbonate)	330 to 360	630 to 680	65 to 95	150 to 200	0.007 to 0.010	20 to 25
Glass reinforced:						
10%	300 to 325	570 to 620	80 to 115	180 to 240	0.002 to 0.004	20 to 25
20%	315 to 340	600 to 640	80 to 115	180 to 240	0.003	20 to 25
30%	315 to 345	600 to 650	80 to 115	180 to 240	0.0025	20 to 25
40%	315 to 350	600 to 660	95 to 125	200 to 260	0.002	20 to 25
Impact modified	290 to 315	550 to 600	70 to 95	160 to 200	0.005 to 0.007	20 to 25
Flame retarded:						
Fast cycling	270 to 305	520 to 580	65 to 95	150 to 200	0.005 to 0.007	15 to 20
Low viscosity	275 to 295	530 to 560	70 to 95	160 to 200	...	20 to 25
Medium viscosity	290 to 315	550 to 600	70 to 95	160 to 200	...	20 to 25
High viscosity	315 to 345	600 to 650	80 to 115	180 to 240	0.005 to 0.007	20 to 25

(a) Data based on a wall thickness of 3.18 mm (1/8 in.). Low to medium injection speed. Maximum regrind, 20 to 25%

Polybutylene Terephthalate (PBT)

Selected properties of polybutylene (extruded or molded)

Property	ASTM test method	Reported value
Mechanical		
Yield strength, MPa (ksi)	D 638	21 to 28 (3.0 to 4.0)
Tensile modulus, GPa (10^6 psi)	D 638	0.21 to 0.34 (0.03 to 0.05)
Elongation (break), %	D 638	350 to 450
Flexural modulus, GPa (10^6 psi)	D 790	0.69 (0.10)
Impact strength, Izod(a), J/cm (ft · lbf/in.)	D 256	No break
Hardness, Rockwell	...	D55 to 65
Thermal		
Thermal conductivity, W/m · K (Btu · in./ft^2 h · °F)	C 177	0.22 (1.5)
Coefficient of linear thermal expansion, 10^{-5}/°C (10^{-5}/°F)	D 696	16 (9.0)
Specific heat, kJ/kg · K (Btu/lb · °F)	C 351	1.7 to 2.1 (0.4 to 0.5)
Continuous service temperature, °C (°F)	...	105 (225)
Electrical		
Dielectric constant, (50 to 100 Hz)	D 150	2.2
Dissipation factor (50 to 100 Hz)	D 150	0.005
Processing parameters		
Processing temperature, °C (°F)	...	93 to 195 (199 to 380)
Density, g/cm^3	D 792	0.91
Specific volume, cm^3/kg (in.^{3}/lb)	D 792	1080 (30)
Linear mold shrinkage, mm/mm (in./in.)	D 955	0.003
Water absorption, % in 24 h	D 570	0.01

(a) Notched, at room temperature

Thermal, physical, and chemical properties of typical polybutylene terephthalate (PBT)

Property	Unfilled PBT	PBT with 30% fibrous glass
Heat-deflection temperature at 1.82 MPa , °C	65	200
Maximum resistance to continuous heat, °C	60	150
Coefficient of linear thermal expansion, 10^{-5}/°C	7.0	2.5
Compressive strength, MPa	75	120
Flexural strength, MPa	96	110
Impact strength, Izod, cm · N/cm of notch	53.4	50
Tensile strength, MPa	55	117
Elongation, %	100	4
Hardness, Rockwell	M70	M90
Specific gravity	1.35	1.5
Dielectric constant	4.0	4.0
Water absorption, %	0.05	0.05
Resistance to chemicals at 25% °C(a):		
Nonoxidizing acids (20% H_2SO_4)	S	S
Oxidizing acids (10% HNO_3)	Q	Q
Aqueous salt solutions (NaCl)	S	S
Aqueous alkalis (NaOH)	S	S
Polar solvents (C_2H_5OH)	S	S
Nonpolar solvents (C_6H_6)	S	S
Water	S	S

(a) S, satisfactory; Q, questionable

Thermoplastic Polyesters

$$HO-\overset{O}{\overset{\|}{C}}-R-\overset{O}{\overset{\|}{C}}-O-\overset{CH_3}{\overset{|}{CH}}-CH_2-O\overset{O}{\overset{\|}{C}}-CH{=}CH-\overset{O}{\overset{\|}{C}}-O-\overset{CH_3}{\overset{|}{CH}}-CH_2-OH$$

Molecular structure of a typical vinyl ester

Selected properties of thermoplastic polyester (extruded or molded)

Property	ASTM test method	Reported value
Mechanical		
Yield strength, MPa (ksi)	D 638	55 to 59 (8.0 to 8.5)
Tensile modulus, GPa (10^6 psi)	D 638	1.9 (0.28)
Elongation (break), %	D 638	100 to 300
Compressive strength, MPa (ksi)	D 695	55 to 105 (8 to 15)
Flexural yield strength, MPa (ksi)	D 790	83 to 110 (12 to 16)
Flexural modulus, GPa (10^6 psi)	D 790	2.1 to 2.8 (0.30 to 0.40)
Impact strength, Izod(a), J/cm (ft · lbf/in.)	D 256	0.4 to 0.53 (0.8 to 1.0)
Hardness, Rockwell	...	M80 to 95
Thermal		
Thermal conductivity, W/m · K (Btu · in./ft^2 · h · °F)	C 177	0.19 to 0.29 (1.3 to 2.0)
Coefficient of linear thermal expansion, 10^{-5}/°C (10^{-5}/°F)	D 696	11 to 16 (6.0 to 9.0)
Specific heat, kJ/kg · K (Btu/lb · °F)	C 351	1.3 to 2.1 (0.3 to 0.5)
Continuous service temperature, °C (°F)	...	93 to 120 (200 to 250)
Electrical		
Volume resistivity, Ω · cm	D 257	10^{15}
Dielectric strength, V/10^{-3} mm (V/10^{-3} in.)	D 149	23 (580)
Dielectric constant (50 to 100 Hz)	D 150	3.2
Dissipation factor (50 to 100 Hz)	D 150	0.002
Processing parameters		
Processing temperature, °C (°F)	...	205 to 275 (400 to 525)
Density, g/cm^3	D 792	1.30 to 1.40
Specific volume, cm^3/kg (in.^{3}/lb)	D 792	940 to 865 (26 to 24)
Linear mold shrinkage, mm/mm (in./in.)	D 955	0.015 to 0.020
Water absorption, % in 24 h	D 570	0.1

(a) Notched, at room temperature

Thermal, physical, and chemical properties of a typical polyester thermoplastic elastomer

Property	Value
Tensile strength, MPa	39
Elongation, %	350
Hardness, Shore D	72
Specific gravity	1.25
Izod notched impact strength, J/cm	2.1
Heat-deflection temperature at 500 kPa, °C	166
Softening point, °C	203
Coefficient of linear thermal expansion, 10^{-5}/°C	21
Water absorption, %	0.3
Resistance to chemicals at 25% °C(a):	
Nonoxidizing acids (20% H_2SO_4)	S
Oxidizing acids (10% HNO_3)	U
Aqueous salt solutions (NaCl)	S
Aqueous alkalis (NaOH)	Q
Polar solvents (C_2H_5OH)	S
Nonpolar solvents (C_6H_6)	Q

(a) S, satisfactory; Q, questionable; U, unsatisfactory

Thermal, physical, and chemical properties of a typical polyester

Property	Value
Heat-deflection temperature at 1.82 MPa, °C	65
Maximum resistance to continuous heat, °C	60
Melting point °C	240
Coefficient of linear thermal expansion, 10^{-5}/°C	7
Tensile strength, MPa	60
Elongation, %	50
Specific gravity	1.35
Dielectric constant	3.0
Resistance to chemicals at 25% °C(a):	
Nonoxidizing acids (20% H_2SO_4)	Q
Oxidizing acids (10% HNO_3)	Q
Aqueous salt solutions (NaCl)	S
Aqueous alkalis (NaOH)	Q
Polar solvents (C_2H_5OH)	Q
Nonpolar solvents (C_6H_6)	U
Water	S

(a) S, satisfactory; Q, questionable; U, unsatisfactory

Thermal and related properties of polyester films

Property	PCL	PBT	PET
Density, g/cm^3	...	1.31 to 1.38	1.38 to 1.41
T_g, amorphous, °C (°F)	40 (105)	60 to 70 (140 to 160)	70 to 80 (170 to 175)
T_m, °C (°F)	64 to 70 (150 to 160)	225 to 235 (440 to 455)	260 to 265 (500 to 510)
T_c, °C (°F)	...	...	125 to 180 (260 to 355)
Heat-deflection temperature, at 345 kPa (50 psi), °C (°F)	...	...	158 (315)
Water absorption, at 25 °C (77 °F), 24 h immersion, %	...	...	0.55

Recommended processing factors for injection-moldable thermoplastic polyesters

Grade	Melt-temperature range °C	Melt-temperature range °F	Melt viscosity(a) Pa · s	Injection speed	Mold temperature °C	Mold temperature °F	Maximum regrind, %	Gating diameter(b) mm	Gating diameter(b) in.	Shrinkage range(c), mm/mm (in. / in.)	Minimum cycle time(b), s
Polybutylene terephthalate (PBT)											
Unmodified base resin	245 to 255	470 to 490	...	Medium to high	91 to 121	195 to 250	25	1.3	0.050	0.018 to 0.022	35 to 50
Glass reinforced, 20 to 30%	250 to 265	480 to 510	...	High	79 to 91	175 to 195	25	1.5	0.060	0.004 to 0.007(F); 0.007 to 0.012(T)	30 to 40
Mineral/glass reinforced	250 to 260	480 to 500	...	High	79 to 91	175 to 195	25	1.5	0.060	0.007 to 0.010(F); 0.017 to 0.022(T)	30 to 45
Impact modified, unfilled	230 to 255	450 to 490	...	High	88 to 91	190 to 195	0	2.4	0.095	0.017 to 0.022	50 to 100
Flame retarded, 20 to 30%	250 to 265	480 to 510	...	High	79 to 91	175 to 195	25	1.5	0.060	0.004 to 0.007(F); 0.007 to 0.012(T)	30 to 45
Polyethylene terephthalate (PET)											
Glass reinforced:											
15%	270 to 295	520 to 560	...	Medium to high	82 to 121	180 to 250	25	0.51 to 2.54	0.020 to 0.100	0.004 to 0.008	15 to 18
30%	280 to 310	540 to 590	120 to 350	Medium to high	82 to 121	180 to 250	25	0.51 to 2.54	0.020 to 0.100	0.003 to 0.006	15 to 18
45%	280 to 310	540 to 590	110 to 350	Medium to high	82 to 121	180 to 250	25	0.51 to 2.54	0.020 to 0.100	0.002 to 0.004	15 to 18
55%	280 to 310	540 to 590	110 to 400	Medium to high	82 to 121	180 to 250	25	0.51 to 2.54	0.020 to 0.100	0.002 to 0.003	15 to 18
Mineral/glass reinforced, 35 to 40%	280 to 310	540 to 590	105 to 500	Medium to high	82 to 121	180 to 250	25	0.51 to 2.54	0.020 to 0.100	0.003 to 0.006	15 to 18
Impact modified:											
30%	275 to 300	530 to 575	600	High	93 to 121	200 to 250	25	0.51 to 2.54	0.020 to 0.100	0.002 to 0.009	15 to 18
35%	270 to 300	520 to 570	450	High	85 to 121	185 to 250	25	0.51 to 2.54	0.020 to 0.100	0.002 to 0.009	15 to 18
Flame retarded:											
15% glass reinforced	270 to 295	520 to 560	...	High	82 to 121	180 to 250	25	0.51 to 2.54	0.020 to 0.100	0.004 to 0.008	15 to 18
30% glass reinforced	280 to 300	540 to 570	120 to 300	High	82 to 121	180 to 250	25	0.51 to 2.54	0.020 to 0.100	0.015 to 0.009	15 to 18
35% mineral filled	280 to 300	540 to 570	110 to 300	High	82 to 121	180 to 250	25	0.51 to 2.54	0.020 to 0.100	0.015 to 0.009	15 to 18

(a) Melt viscosity measured at 1000/s at the median temperature of the recommended melt-temperature range. (b) Data based on a wall thickness of 3.18 mm (1/8 in.). (c) F, flow direction; T, transverse direction

Polyetheretherketones (PEEK)

Molecular structure of polyetheretherketone (PEEK)

PEEK key properties

Mechanical	Tough, ductile, abrasion resistant; excellent fatigue characteristics; load-bearing at high temperature
Thermal	High melting point; high continuous service temperature
Flammability	Low flammability, fire, smoke properties
Chemical	Essentially inert to organics; high degree of acid and alkali resistance; particularly resistant to high-temperature water/steam
Hard radiation	No significant degradation below 1100 Mrad
Processing	Easily processed on conventional equipment

Thermal, physical, and chemical properties of a typical polyetheretherketone (PEEK)

Property	Unfilled PEEK	40% glass-filled PEEK
Heat-deflection temperature at 1.82 MPa, °C	150	300
Maximum resistance to continuous heat, °C	125	225
Coefficient of linear thermal expansion, 10^{-5}/°C	5.5	2.2
Compressive strength, MPa	90	125
Flexural strength, MPa	110	250
Impact strength, Izod, cm · N/cm of notch	50	75
Tensile strength, MPa	70	107
Elongation, %	50	2
Hardness, Rockwell	R123	R123
Specific gravity	1.3	1.5
Dielectric constant	3.2	3.5
Water absorption, %	0.15	0.12
Resistance to chemicals at 25 °C(a):		
Nonoxidizing acids (20% H_2SO_4)	S	S
Oxidizing acids (10% HNO_3)	S	S
Aqueous salt solutions (NaCl)	S	S
Aqueous alkalis (NaOH)	S	S
Polar solvents (C_2H_4OH)	S	S
Nonpolar solvents (C_6H_6)	S	S
Water	S	S

(a) S, satisfactory

Mechanical properties of PEEK

Property	ASTM test method	Neat	30% glass reinforced	30% carbon reinforced
Tensile strength, MPa (ksi)				
At 23 °C (73 °F)	D 638	100 (14.5)	160 (23.5)	215 (31.2)
At 100 °C (212 °F)	D 638	66 (9.6)	130 (18.7)	185 (26.8)
At 150 °C (300 °F)	D 638	35 (5.1)	75 (10.9)	100 (15.5)
Elongation to break, %	D 638	>40	...	...
Flexural modulus, GPa (10^6 psi)				
At 23 °C (73 °F)	D 790	3.9 (0.565)	8.69 (1.26)	15.5 (2.25)
At 100 °C (212 °F)	D 790	3.0 (0.435)	...	12.2 (1.77)
At 150 °C (300 °F)	D 790	2.0 (0.290)	...	10.0 (1.45)
Izod impact strength, N/mm (ft · lbf/in.)				
Notched	D 256	0.09 (1.6)	0.12 (2.2)	0.06 (1.2)
Unnotched	D 256	No break	...	0.64 (12.0)
Heat-deflection temperature, °C (°F)	...	152 (305)	>300 (>575)	>300 (>575)
Coefficient of linear thermal expansion, 10^{-5}/K	...	5.5	2.1	1.5

PEEK processing characteristics

Property	Value
Melting point, °C (°F)	334 (633)
Glass transition temperature, °C (°F)	143 (289)
Temperature of maximum crystallization, °C (°F)	
From melt	256 (493)
From solid	185 (365)
Maximum crystallinity, %	48
Specific gravity	
Amorphous	1.265
Fully crystalline	1.320
Water absorption after 24 h at 40% RH, %	0.15
Bulk density of granules, kg/m^3 (lb/ft^3)	770 (48)
Melt viscosity range, 400 °C (752 °F), 1/1000 s, Pa · s	400 to 500
Processing temperature range, °C (°F)	371 to 399 (700 to 750)
Melt thermal stability at 400 °C (752 °F), h	>1
Mold shrinkage, %	1.1

Recommended processing factors for injection-moldable polyetheretherketone (PEEK) and polyethersulfone (PES)(a)

Material	Melt-temperature °C	Melt-temperature °F	Melt viscosity(a) Pa · s	Shrinkage range(c), mm/mm (in./in.)	Minimum cycle time(b), s
PEEK					
Unmodified base resin:					
General purpose	350 to 380	660 to 720	330	0.010	10 to 15
General purpose	350 to 380	660 to 720	250	0.010	10 to 15
Glass reinforced:					
20%	360 to 385	680 to 725	450	0.007	10 to 15
30%	360 to 385	680 to 725	480	0.005	10 to 15
Carbon fiber reinforced	370 to 395	700 to 740	550	0.005	10 to 15
PES					
Unmodified base resin:					
Low viscosity	340 to 360	645 to 680	180	0.006	10 to 15
General purpose	340 to 360	645 to 680	220	0.006	10 to 15
Medium viscosity	350 to 370	660 to 700	400	0.006	10 to 15
High viscosity	350 to 375	660 to 710	550	0.006	10 to 15
Glass reinforced:					
Easy flow, 20% glass	335 to 355	635 to 670	200	0.003	10 to 15
Easy flow, 30% glass	335 to 355	635 to 670	250	0.002	10 to 15
General purpose, 20% glass	350 to 370	660 to 700	250	0.003	10 to 15
General purpose, 30% glass	350 to 375	660 to 710	300	0.002	10 to 15
High viscosity, 20% glass	350 to 375	660 to 710	600	0.003	10 to 15

(a) Medium to high injection speed. Mold temperature, 160 to 170 °C (320 to 340 °F). Maximum regrind, 30%. Grating diameter, 1.0 to 2.0 mm (0.004 to 0.008 in.). (b) Melt viscosity measured at 1000/s at the median temperature of the recommended melt-temperature range. (c) Data based on a wall thickness of 3.18 mm (1/8 in.)

Polyetherimides (PEI)

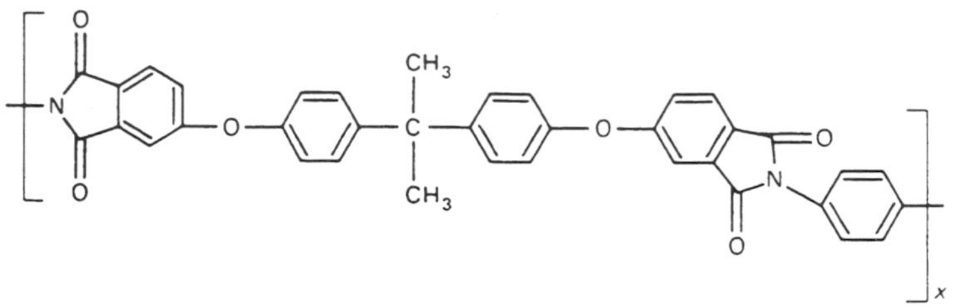

Molecular structure of polyetherimide (PEI)

Thermal, physical, and chemical properties of a typical polyetherimide (PEI)

Property	Unmodified PEI	10% glass-reinforced PEI	20% glass-reinforced PEI	30% glass-reinforced PEI
Heat-deflection temperature at 1.82 MPa, °C	190	200	205	210
Maximum resistance to continuous heat, °C	170	175	180	185
Coefficient of linear thermal expansion, 10^{-5}/°C	5.6	4.4	3.2	2.0
Compressive strength, MPa	140	155	169	176
Flexural strength, MPa	145	195	205	225
Impact strength, Izod, cm · N/cm of notch	133.5	146	213	267
Tensile strength, MPa	104	114	138	169
Elongation, %	6.0	6.0	3.0	3.0
Hardness, Rockwell M	109	115	120	125
Specific gravity	1.27	1.35	1.45	1.6
Dielectric constant	3.1	3.3	3.5	3.7
Water absorption, %	0.06	0.1	0.15	0.2
Resistance to chemicals at 25 °C(a):				
Nonoxidizing acids (20% H_2SO_4)	S	S	S	S
Oxidizing acids (10% HNO_3)	U	U	U	U
Aqueous salt solutions (NaCl)	S	S	S	S
Aqueous alkalis (NaOH)	Q	Q	Q	Q
Polar solvents (C_2H_5OH)	S	S	S	S
Nonpolar solvents (C_6H_6)	S	S	S	S
Water	S	S	S	S

(a) S, satisfactory; Q, questionable; U, unsatisfactory

Recommended processing factors for injection-moldable polyetherimide(a)

Material	Melt-temperature range °C	Melt-temperature range °F	Melt viscosity(b), Pa · s	Mold temperature °C	Mold temperature °F	Maximum regrind, %
Unmodified base resin	345 to 425	650 to 800	450	65 to 150	150 to 300	50
Glass reinforced:						
10% glass	355 to 425	675 to 800	400	65 to 195	150 to 380	40
20% glass	355 to 425	675 to 800	450	65 to 195	150 to 380	40
30% glass	355 to 425	675 to 800	500	65 to 195	150 to 380	40
Mineral reinforced	355 to 425	675 to 800	...	65 to 195	150 to 380	40
Mineral/glass reinforced	355 to 425	675 to 800	...	65 to 195	150 to 380	40
Impact modified	290 to 310	550 to 590	...	65 to 120	150 to 250	...
Flame retarded	290 to 310	550 to 590	...	-9 to 175	15 to 350	...

(a) Data based on a wall thickness of 3.18 mm (0.125 in.). Maximum injection speed, 262 cm^3/s (16 $in.^3/s$). (b) Melt viscosity measured at 1000/s at the median temperature of the recommended melt-temperature range

Polyether Sulfones (PES)

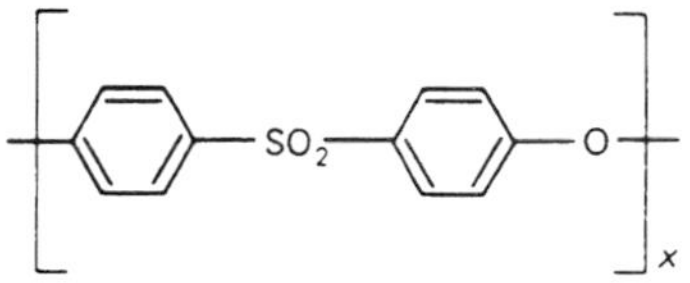

Molecular structure of polyether sulfone (PES)

Selected properties of polyether sulfone (extruded or molded)

Property	ASTM test method	Reported value
Mechanical		
Yield strength, MPa (ksi)	D 638	85.5 (12.4)
Tensile modulus, GPa (10^6 psi)	D 638	2.4 (0.35)
Elongation (break), %	D 638	50 to 100
Compressive strength, MPa (ksi)	D 695	97 (14)
Flexural yield strength, MPa (ksi)	D 790	125 to 130 (18 to 19)
Flexural modulus, GPa (10^6 psi)	D 790	2.8 (0.40)
Impact strength, Izod(a), J/cm (ft · lb/in.)	D 256	0.85 (1.6)
Hardness, Rockwell M	...	88
Thermal		
Thermal conductivity, W/m · K (Btu · in./ft^2 · h · °F)	C 177	0.14 to 0.22 (1.0 to 1.5)
Coefficient of linear thermal expansion, 10^{-5}/°C (10^{-5}/°F)	D 696	7.2 (4.0)
Specific heat, kJ/kg · K (Btu/lb · °F)	C 351	1.3 (0.3)
Continuous service temperature, °C (°F)	...	170 (340)
Electrical		
Volume resistivity, Ω · cm	D 257	10^{17}
Dielectric strength, V/10^{-3} mm (V/10^{-3} in.)	D 149	16 (400)
Dielectric constant (50 to 100 Hz)	D 150	3.1
Dissipation factor (50 to 100 Hz)	D 150	0.001
Processing parameters		
Processing temperature, °C (°F)	...	300 to 355 (575 to 675)
Density, g/cm^3	D 792	1.40
Specific volume, cm^3/kg (in.^{3}/lb)	D 792	725 (20)
Linear mold shrinkage, mm/mm (in./in.)	D 955	0.007
Water absorption, % in 24 h	D 570	0.4

(a) Notched, at room temperature

Polyethylenes (PE)

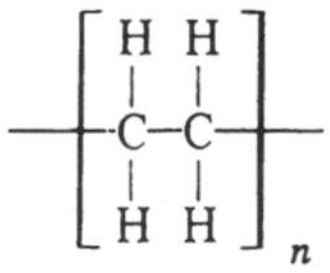

$$\left[\begin{array}{cc} H & H \\ | & | \\ -C & -C- \\ | & | \\ H & H \end{array} \right]_n$$

Molecular structure of polyethylene

Thermal, physical, and chemical properties of polyethylene

Property	LDPE	HDPE	UHMWPE
Heat-deflection temperature at 1.82 MPa, °C	40	85	85
Maximum resistance to continuous heat, °C	40	80	80
Coefficient of linear thermal expansion, 10^{-5}/°C	10.0	12.0	12.0
Compressive strength, MPa	...	21	...
Impact strength, Izod, cm · N/cm of notch	No break	106.7	No break
Tensile strength, MPa	6	28	38
Elongation, %	100	30	400
Hardness, Rockwell	D40	D40	R50
Specific gravity	0.91	0.95	0.94
Dielectric constant	2.3	2.3	2.3
Resistance to chemicals at 25 °C(a):			
Nonoxidizing acids (20% H_2SO_4)	S	S	S
Oxidizing acids (10% HNO_3)	Q	Q	Q
Aqueous salt solutions (NaCl)	S	S	S
Polar solvents (C_2H_5OH)	S	S	S
Nonpolar solvents (C_6H_6)	Q	Q	Q
Water	S	S	S

(a) S, satisfactory; Q, questionable

Selected properties of molded polyethylenes

Property	ASTM test method	Reported value for: Low-density polyethylene	Medium density polyethylene	High-density polyethylene	Polyallomer polyethylene	Cross-linkable polyethylene	UHMW polyethylene
Mechanical							
Yield strength, MPa (ksi)	D 638	4.1 to 16 (0.6 to 2.3)	8.3 to 24 (1.2 to 3.5)	21 to 38 (3.1 to 5.5)	21 to 28 (3.0 to 4.1)	11 to 32 (1.6 to 4.6)	17 to 24 (2.5 to 3.5)
Tensile modulus, GPa (10^6 psi)	D 638	0.07 to 0.28 (0.01 to 0.04)	0.14 to 0.41 (0.02 to 0.06)	0.41 to 1.2 (0.06 to 0.018)	...	0.34 to 3.4 (0.05 to 0.50)	0.14 to 0.76 (0.02 to 0.11)
Elongation (break), %	D 638	100 to 800	50 to 600	30 to 1300	400 to 500	100 to 450	300 to 500
Compressive strength, MPa (ksi)	D 695	...	...	21 to 28 (3 to 4)	...	18 (2.6)	...
Flexural yield strength, MPa (ksi)	D 790	...	34 to 48 (5 to 7)	...	...	14 to 48 (2 to 7)	...
Flexural modulus, GPa (10^6 psi)	D 790	0.07 (0.01)	0.69 (0.10)	0.69 to 2.1 (0.10 to 0.30)	0.69 (0.10)	0.69 to 2.8 (0.10 to 0.40)	0.69 (0.10)
Impact strength, Izod(a), J/cm (ft · lbf/in.)	D 256	No break	0.3 to 8.0 (0.5 to 15.0)	0.3 to 10.7 (0.5 to 20.0)	No break	No break	No break
Hardness, Rockwell	...	D40 to 50	D50 to 60	D60 to 70	R50 to 85	D55 to 80	D60 to 70
Thermal							
Thermal conductivity, W/m · K (Btu · in./ft^2 · h · °F)	C 177	0.30 (2.1)	0.30 to 0.42 (2.1 to 2.9)	0.43 to 0.52 (3.0 to 3.6)	0.087 to 0.17 (0.6 to 1.2)	...	...
Coefficient of linear thermal expansion, 10^{-5}/°C (10^{-5}/°F)	D 696	18.0 to 36.0 (10.0 to 20.0)	25.2 to 28.8 (14.0 to 16.0)	21.6 to 23.4 (12.0 to 13.0)	14.4 to 21.6 (8.0 to 12.0)	18.0 to 63.0 (10.0 to 35.0)	12.6 (7.0)
Specific heat, kJ/kg · K (Btu/lb · °F)	C 351	2.1 to 2.5 (0.5 to 0.6)	2.1 to 2.5 (0.5 to 0.6)	2.1 to 2.5 (0.5 to 0.6)	2.1 (0.5)	...	...
Continuous service temperature, °C (°F)	...	60 to 90 (140 to 190)	60 to 90 (140 to 190)	60 to 90 (140 to 190)	...	...	...
Electrical							
Volume resistivity, Ω · cm	D 257	10^{16}	10^{16}	10^{16}	10^{15}	...	10^{16}
Dielectric strength, V/10^{-3} mm (V/10^{-3} in.)	D 149	18 to 20 (450 to 500)	18 to 20 (450 to 500)	18 to 20 (450 to 500)	0.08 (2)	...	...
Dielectric constant, (50 to 100 Hz)	D 150	2.2 to 2.4	2.2 to 2.4	2.2 to 2.4	...	...	...
Dissipation factor (50 to 100 Hz)	D 150	<0.0005	<0.0005	<0.0005	0.005	...	0.0002
Processing parameters							
Processing temperature, °C (°F)	...	150 to 315 (300 to 600)	150 to 315 (300 to 600)	150 to 315 (300 to 600)	...	120 to 400 (250 to 750)	260 to 650 (500 to 1200)
Density, g/cm^3	D 792	0.900 to 0.925	0.926 to 0.940	0.941 to 0.965	0.89	0.95 to 1.45	0.94
Specific volume, cm^3/kg (in.^{3}/lb)	D 792	1080 (30)	1050 (29)	1010 (28)	...	...	1080 (30)
Linear mold shrinkage, mm/mm (in./in.)	D 955	0.015 to 0.050	0.015 to 0.050	0.020 to 0.050	0.010 to 0.020	0.007	...
Water absorption, % in 24 h	D 570	1.5	1.5	1.5	<0.1	<0.1	<0.1

(a) Notched, at room temperature

UHMWPE properties

Property	ASTM test method	Typical values
Mechanical		
Density, g/cm^3	D 792	0.926 to 0.934
Tensile strength at yield, MPa (ksi)	D 638	21 (3.1)
Tensile strength at break, MPa (ksi)	D 638	48 (7.0)
Elongation at break, %	D 638	350
Elastic modulus, GPa (10^6 psi)		
At 23 °C (73 °F)	D 638	0.69 (0.10)
At –269 °C (–450 °F)	D 638	2.97 (0.43)
Izod impact strength, kJ/m (ft Ã lbf/in.) notch		
At 23 °C (73 °F)	D 256(a)	1.6 (30)
At –40 °C (–40 °F)	D 256(a)	1.1 (21)
Hardness, Shore D	D2240	62 to 66
Abrasion resistance		100
Water absorption, %	D 570	Nil
Relative solution viscosity, dL/g	D 4020	2.3 to 3.5
Thermal		
Crystalline melting range, powder, °C (°F)	Polarizing microscope	138 to 142 (280 to 289)
Coefficient of linear expansion, 10^{-4}/K		
At 20 to 100 °C (68 to 212 °F)	D 696	2
At –200 to –100 °C (–330 to –150 °F)	D 696	0.5
Electrical		
Volume resistivity, Ω · m	D 257	5×10^{14}
Dielectric strength, kV/cm (V/mil)	D 149	900 (2300)
Dielectric constant	D 150	2.30
Dissipation factor, $\times 10^{-4}$		
At 50 Hz	D 150	1.9
At 1 kHz		0.5
At 0.1 MHz		2.5
Surface resistivity, wt% carbon black, Ω		
0.2% for color	D 257	10^{14}
2.5% for UV protection	D 257	10^{13}
6.5% for antistatic applications	D 257	10^5
16.7% for conductive applications	D 257	10^3

(a) Samples had two notches (15° ± ½°) on opposite sides to a depth of 5 mm (0.20 in.)

Polyethylene Terephthalate (PET)

Thermal, physical, and chemical properties of typical polyethylene terephthalate (PET)

Property	PET	PET with 30% fibrous glass
Heat-deflection temperature at 1.82 MPa, °C	100	226
Maximum resistance to continuous heat, °C	100	160
Coefficient of linear thermal expansion, 10^{-5}/°C	6.5	2.9
Compressive strength, MPa	86	172
Flexural strength, MPa	112	234
Impact strength, Izod, cm · N/cm of notch	26.7	50
Tensile strength, MPa	62	158
Elongation, %	100	2.5
Hardness, Rockwell M	96	100
Specific gravity	1.35	1.56
Dielectric constant	3.6	4.0
Water absorption, %	0.2	0.05
Resistance to chemicals at 25 °C(a):		
Nonoxidizing acids (20% H_2SO_4)	S	S
Oxidizing acids (10% HNO_3)	Q	Q
Aqueous salt solutions (NaCl)	S	S
Aqueous alkalis (NaOH)	S	S
Polar solvents (C_2H_5OH)	S	S
Nonpolar solvents (C_6H_6)	S	S
Water	S	S

(a) S, satisfactory; Q, questionable

Polymethyl Methacrylate (PMMA)

Physical properties of various unmodified polymethyl methacrylates

Property	ASTM test method	Flow grade(a)		
		Grade 5	Grade 6	Grade 8
Heat deflection temperature under load, annealed, at 1.82 MPa (0.264 ksi), °C (°F)	D 648	74 (165)	80 (176)	90 to 102 (194 to 216)
Continuous-use temperature, °C (°F)	No load	60 to 74 (140 to 165)	63 to 80 (145 to 176)	71 to 102 (160 to 216)
Melt flow rate, g/10 min.	D 1238, condition I	24	17	2 to 8
Refractive index number	D 542	1.49	1.49	1.49
Specific gravity	D 792	1.18	1.18	1.19
Light transmission, %	D 1003	92	92	92
Tensile strength, MPa (ksi)	D 638	61 (8.8)	66 (9.6)	70 (10.2)
Flexural modulus, GPa (10^6 psi)	D 790	3.036 (0.440)	3.036 (0.440)	3.105 (0.450)
Impact strength, J/m (ft · lbf/in.)	D 256, notched, Izod	12.3 (0.23)	12.3 (0.23)	12.3 (0.23)
Hardness, Rockwell M	D 785	84	89	97
Water absorption, %	D 570, wt gain, 24 h	0.3	0.3	0.3

(a) Flow grade per ASTM D 788

Thermoplastic Polyimides

Class of monomeric compound	Chemical name	Chemical structure
Aromatic diamines	*meta*- and *para*-phenylene diamines (MPDA, PPDA)	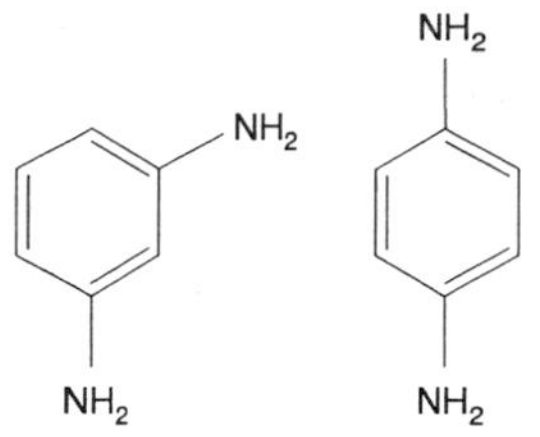 MPDA PPDA
	2,2-bis[4-(4-amino-phenoxy)phenyl] hexafluoropropane (4-BDAF)	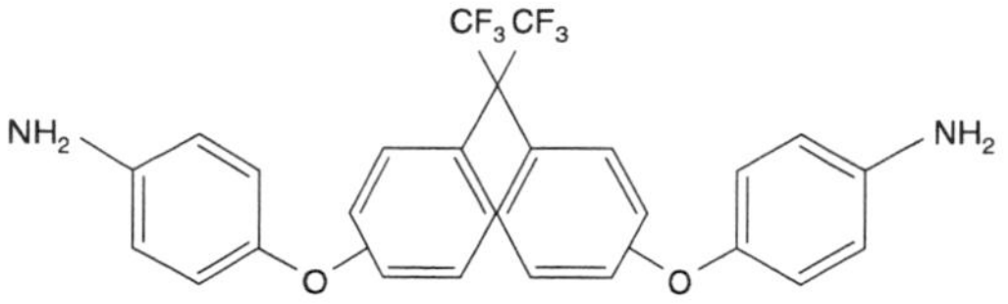
	Di(methylthio)toluene-diamine (ETHACURE 300)	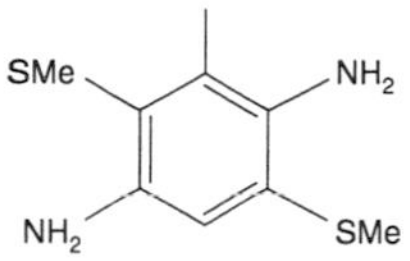
Aromatic dianhydrides	2,2-bis(3,4-dicarboxyphenyl) hexafluoropropane dianhydride (6-FDA)	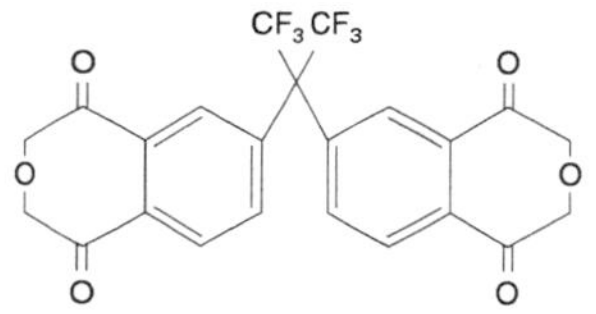
	1,2,4,5-Benzene-tetracarboxylic acid dianhydride (PMDA)	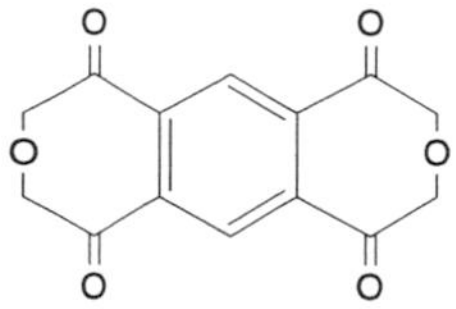
	3,3′,4,4′-Benzophenone-tetracarboxylic acid dianhydride (BTDA)	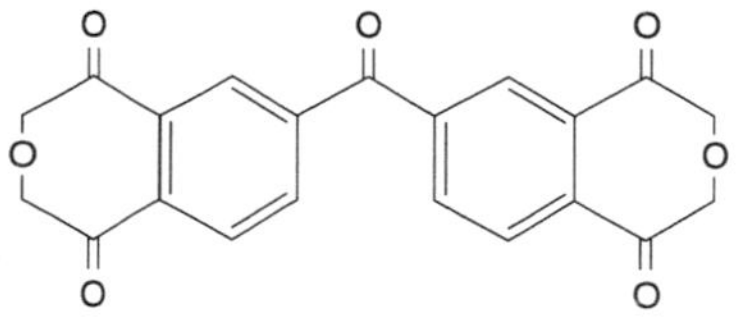

Monomeric constituents used to prepare key polyimides

Thermal, physical, and chemical properties of typical polyimides (PI)

Property	Thermoplastic	Glass-filled thermoset (50%)
Heat-deflection temperature at 1.82 MPa, °C	315	350
Maximum resistance to continuous heat, °C	300	325
Coefficient of linear thermal expansion, 10^{-5}/°C	5.0	1.3
Compressive strength, MPa	241	234
Flexural strength, MPa	172	145
Impact strength, Izod, cm · N/cm of notch	80	294
Tensile strength, MPa	96.5	44
Elongation, %	8	0.5
Hardness, Rockwell	E60	M118
Specific gravity	1.4	1.6

(continued)

Thermal, physical, and chemical properties of typical polyimides (PI) (continued)

Property	Thermoplastic	Glass-filled thermoset (50%)
Dielectric constant	3.4	3.5
Water absorption, %	0	0.2
Resistance to chemicals at 25 °C(a):		
Nonoxidizing acids (20% H_2SO_4)	Q	Q
Oxidizing acids (10% HNO_3)	Q	Q
Aqueous salt solutions (NaCl)	S	S
Aqueous alkalis (NaOH)	U	U
Polar solvents (C_2H_5OH)	S	S
Nonpolar solvents (C_6H_6)	S	S
Water	S	S

(a) S, satisfactory; Q, questionable; U, unsatisfactory

Comparison of thermal characterization results obtained on postcured film and compression-molded samples of polyimide polymers

Polymer sample physical state	Specific polymer tested	First significant endotherm or T_g obtained using DSC		First significant weight loss temperature in air obtained using TGA	
		°C	°F	°C	°F
Postcured, cast film	EYMYD L-30N	432	809	440	824
	Avimid N	400	752	450	842
Postcured, compression molded specimen	EYMYD L-30N	434	813	440	824
	Avimid N	400	752	450	842

All measurements made on a DuPont 993 thermal analyzer equipped with appropriate DSC and TGA modules. Analyses performed in air at a 5 °C/min. (9 °F/min.) heat-up rate

Thermal characterization results obtained on commercially available PIs and PBIs

Polymer	First significant endotherm or T_g obtained on postcured film using DSC(a)		First significant weight loss temperature in air using TGA: As-cast		Postcured(a)	
	°C	°F	°C	°F	°C	°F
Avimid N	400	752	440	824	450	842
PBI	360	680	400	752	430	806
EYMYD L-30N	410	770	430	806	440	824

All measurements made on a DuPont 993 thermal analyzer equipped with appropriate DSC and TGA modules. Analyses performed in air at a 5 °C/min. (9 °F/min.) heat-up rate. (a) The Avimid N, PBI, and EYMYD L-30N film samples were postcured for 2 h in an air-circulating oven at 370 °C (700 °F)

Liquid Crystal Polymers (LCP)

Thermal, physical, and chemical properties of typical liquid crystal polymers (LCP)

Property	Unfilled LCP	50% talc-filled LCP
Heat-deflection temperature at 1.82 MPa, °C	350	325
Maximum resistance to continuous heat, °C	250	250
Compressive strength, MPa	42	42
Flexural strength, MPa	125	110
Impact strength, Izod, cm · N/cm of notch	135	70
Tensile strength, MPa	135	70
Elongation, %	4.0	3.0
Hardness, Rockwell R	60	76
Specific gravity	1.35	1.84
Dielectric constant	3	3.5
Water absorption	0	0
Resistance to chemicals at 25 °C(a):		
Nonoxidizing acids (20% H_2SO_4)	S	S
Oxidizing acids (10% HNO_3)	S	S
Aqueous salt solutions (NaCl)	S	S
Aqueous alkalis (NaOH)	S	S
Polar solvents (C_2H_5OH)	S	S
Nonpolar solvents (C_6H_6)	S	S
Water	S	S

(a) S, satisfactory

Typical LCP properties
30% glass reinforced

Properties	ASTM test method	Vectra A130	Xydar RC210
Physical			
Specific gravity, g/cm^3	D 792	1.6	1.6
Water absorption, %	D 570	<0.04	<0.1
Mechanical			
Tensile strength, MPa (ksi)	D 638	180 (26)	140 (20)
Tensile modulus, GPa (10^6 psi)	D 638	15 (2.2)	16 (2.3)
Tensile elongation, %	D 638	2.3	1.7
Notched Izod impact strength, J/m (ft · lbf/in.)	D 256	110 (2.0)	110 (2.0)
Thermal			
Heat-deflection temperature, at 1.8 MPa (0.264 ksi), °C (°F)	D 648	230 (445)	346 (655)
Electric, °C (°F)	UL 746C(a)	220 (430)	240 (465)
Electrical			
Dielectric strength, kV/mm (V/mil)	D 149	43 (1100)	25 (640)
Arc resistance, s	D 495	137	188
Flammability			
Burn	UL 94(a)	V-0 on 0.4 mm (1/64 in.) part	V-0 on 0.8 mm (1/32 in.) part
Oxygen index, %	D 2863	37	46

(a) Underwriters Laboratories test method

Polyphenylene Oxide (PPO)

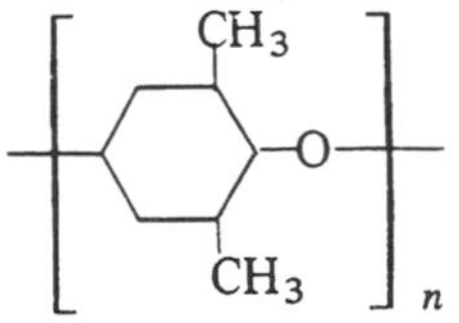

Molecular structure of polyphenylene oxide

Thermal, physical, and chemical properties of a polyphenylene oxide (PPO)

Property	PPO	Glass-filled PPO
Heat-deflection temperature at 1.82 MPa, °C	100	145
Maximum resistance to continuous heat, °C	80	130
Coefficient of linear thermal expansion, 10^{-5}/°C	5.0	2.0
Compressive strength, MPa	96	123
Flexural strength, MPa	89	144
Impact strength, Izod, cm · N/cm of notch	270	107
Tensile strength, MPa	55	120
Elongation, %	50	4
Hardness, Rockwell R	115	115
Specific gravity	1.1	1.1
Dielectric constant	2.8	3.0
Resistance to chemicals at 25 °C(a):		
Nonoxidizing acids (20% H_2SO_4)	S	S
Oxidizing acids (10% HNO_3)	Q	Q
Aqueous salt solutions (NaCl)	S	S
Aqueous alkalis (NaOH)	S	S
Polar solvents (C_2H_5OH)	S	S
Nonpolar solvents (C_6H_6)	U	U
Water	S	S

(a) S, satisfactory; Q, questionable; U, unsatisfactory

Selected properties of phenylene oxides

Property	ASTM test method	Reported value for: Extruded phenylene oxide	Molded phenylene oxide
Mechanical			
Yield strength, MPa (ksi)	D 638	46 to 66 (6.6 to 9.5)	54 to 79.3 (7.8 to 11.5)
Tensile modulus, GPa (10^6 psi)	D 638	2.5 to 2.6 (0.36 (0.38)	2.5 to 2.6 (0.36 to 0.38)
Elongation (break), %	D 638	50 to 60	50 to 60
Compressive strength, MPa (ksi)	D 695	110 to 115 (16 to 17)	110 to 115 (16 to 17)
Flexural yield strength, MPa (ksi)	D 790	90 to 97 (13 to 14)	90 to 97 (13 to 14)
Flexural modulus, GPa (10^6 psi)	D 790	2.8 (0.40)	2.8 (0.40)
Impact strength, Izod(a), J/cm (ft · lbf/in.)	D 256	2.7 to 3.7 (5.0 to 7.0)	2.7 (5.0)
Hardness, Rockwell R	...	114	115 to 119
Thermal			
Thermal conductivity, W/m · K (Btu · in./ft^2 · h · °F)	C 177	0.22 (1.5)	0.22 (1.5)
Coefficient of linear thermal expansion, 10^{-5}/°C (10^{-5}/°F)	D 696	5.4 to 7.2 (3.0 to 4.0)	9.4 (5.2)
Specific heat, kJ/kg · K (Btu/lb · °F)	C 351	1.3 (0.3)	1.3 (0.3)
Continuous service temperature, °C (°F)	...	80 to 105 (175 to 220)	80 to 105 (175 to 220)
Electrical			
Volume resistivity, Ω · cm	D 257	10^{16} to 10^{17}	10^{16} to 10^{17}
Dielectric strength, V/10^{-3} mm (V/10^{-3} in.)	D 149	16 to 24 (400 to 600)	16 to 22 (400 to 500)
Dielectric constant, (50 to 100 Hz)	D 150	2.6	2.6
Dissipation factor (50 to 100 Hz)	D 150	0.0004	0.004

(continued)

Selected properties of phenylene oxides (continued)

Property	ASTM test method	Reported value for: Extruded phenylene oxide	Molded phenylene oxide
Processing parameters			
Processing temperature, °C (°F)	...	205 to 40 (400 to 460)	205 to 315 (400 to 600)
Density, g/cm^3	D 792	1.06 to 1.10	1.06 to 1.10
Specific volume, cm^3/kg (in.^{3}/lb)	D 792	940 to 905 (26 to 25)	940 to 905 (26 to 25)
Linear mold shrinkage, mm/mm (in./in.)	D 955	0.005 to 0.007	...
Water absorption, % in 24 h	D 570	0.1	0.6

(a) Notched, at room temperature

Recommended processing factors for injection-moldable polyphenylene oxide (PPO)

Grade	Melt-temperature range °C	°F	Injection speed	Mold temperature °C	°F	Maximum regrind, %	Gating diameter(a) mm	in.	Shrinkage range, mm/mm (in./in.)	Minimum cycle time(b), s
Unmodified base resin(c)	260 to 315	500 to 600	Medium to high	65 to 120	150 to 250	25	1.52 to 3.18 (T); 3.18 to 6.35 (W)	0.060 to 0.125 (T); 0.125 to 0.250 (W)	0.005 to 0.007	30 to 60
Glass reinforced	290 to 325	550 to 620	Low to high	90 to 105	190 to 220	25	1.52 to 3.18 (T); 3.18 to 6.35 (W)	0.060 to 0.125 (T); 0.125 to 0.250 (W)	0.002 to 0.005	30 to 60
Flame retarded	230 to 315	450 to 600	Medium to high	65 to 120	150 to 250	25	1.52 to 3.18 (T); 3.18 to 6.35 (W)	0.060 to 0.125 (T); 0.125 to 0.250 (W)	0.005 to 0.007	30 to 60
Automotive	275 to 310	530 to 590	Medium to high	75 to 105	170 to 220	25	Gates vary	Gates vary		50 to 90

(a) T, thickness; W, width. (b) Data based on a wall thickness of 3.18 mm (1/8 in.). (c) Multiple grades

Polyphenylene Sulfide (PPS)

Molecular structure of polyphenylene sulfide

Thermal, physical, and chemical properties of a typical polyphenylene sulfide (PPS)

Property	Unfilled PPS	40% glass-filled PPS
Heat-deflection temperature at 1.82 MPa, °C	135	250
Maximum resistance to continuous heat, °C	110	200
Coefficient of linear thermal expansion, 10^{-5}/°C	5.0	2.2
Compressive strength, MPa	110	145
Flexural strength, MPa	96	207
Impact strength, Izod, cm · N/cm of notch	21	75
Tensile strength, MPa	74	141
Elongation, %	1.1	1
Hardness, Rockwell R	123	123
Specific gravity	1.3	1.6
Dielectric constant	3.8	4.6
Water absorption, %	0.02	0.03
Resistance to chemicals at 25 °C(a):		
Nonoxidizing acids (20% H_2SO_4)	S	S
Oxidizing acids (10% HNO_3)	S	S
Aqueous salt solutions (NaCl)	S	S
Aqueous alkalis (NaOH)	S	S
Polar solvents (C_2H_5OH)	S	S
Nonpolar solvents (C_6H_6)	S	S
Water	S	S

(a) S, satisfactory

Selected properties of molded phenylene sulfide

Property	ASTM test method	Reported value
Mechanical		
Yield strength, MPa (ksi)	D 638	68.9 (10.0)
Tensile modulus, GPa (10^6 psi)	D 638	3.4 (0.50)
Elongation (break), %	D 638	3
Flexural yield strength, MPa (ksi)	D 790	140 (20)
Flexural modulus, GPa (10^6 psi)	D 790	4.1 (0.60)
Impact strength, Izod(a), J/cm (ft · lbf/in.)	D 256	0.16 (0.30)
Hardness, Rockwell R	...	124
Thermal		
Thermal conductivity, W/m · K (Btu · in./ft^2 · h · °F)	C 177	0.29 (2.0)
Coefficient of linear thermal expansion, 10^{-5}/°C (10^{-5}/°F)	D 696	9.9 (5.5)
Continuous service temperature, °C (°F)	...	260 (500)
Electrical		
Volume resistivity, Ω · cm	D 257	10^{16}
Dielectric strength, V/10^{-3} mm (V/10^{-3} in.)	D 149	24 (600)
Dielectric constant (50 to 100 Hz)	D 150	3.1
Dissipation factor (50 to 100 Hz)	D 150	0.0004
Processing parameters		
Processing temperature, °C (°F)	...	330 to 370 (625 to 700)
Density, g/cm^3	D 792	1.34
Specific volume, cm^3/kg (in.^{3}/lb)	D 792	760 (21)
Linear mold shrinkage, mm/mm (in./in.)	D 955	0.010
Water absorption, % in 24 h	D 570	0.2

(a) Notched, at room temperature

Properties of PPS biaxially oriented/heat set film

Property	Value
Tensile yield strength, MPa (ksi)	90 to 110 (13 to 16)
Tensile strength, MPa (ksi)	125 to 190 (20 to 30)
Elongation, %	40 to 60
Elmendorf tear, N/mm (gf/mil)	1.5 to 3.0 (4 to 8)
Haze, %	2 to 10
Shrinkage, %	2.5
Coefficient of linear thermal expansion, 10^{-5}/K	2.2
Coefficient of hygroscopic expansion, 10^{-5}/% RH	0.18
Dielectric constant at 1 kHz	3.0
Dissipation factor at 1.1 kHz	0.0005
Volume resistivity, $\Omega \cdot$ m	1.3×10^{15}

Typical conditions for injection molding of polyphenylene sulfide molding compounds

Machine	Reciprocating screw machine preferred
Predrying of resin	Drying for 2 h at 150 to 165 °C (300 to 325 °F) recommended
Clamp tonnage	0.35 to 0.56 t / cm^2 (2.5 to 4 tons/in.^{2}) of projected area
Cylinder temperature	315 to 345 °C (600 to 650 °F) preferred
Injection pressure	Maximum allowable without flashing
Injection rate	High for good surface; moderate for minimum warpage
Cycle time	15 to 50 s
Screw speed	Medium range
Back pressure	345 kPa (50 psi) or less–just enough to yield consistent shot-to-shot weight
Mold temperature	95 to 150 °C (200 to 300 °F) preferred
Purge	Purge before and after with a low-melt-flow polyethylene
Mold release	Coat mold cavities with a high-temperature mold release, particularly in ribs and bosses. High-temperature silicones, fluorocarbon release sprays, and stearate dusts have been found to be effective. Use of release should be continued after each shot until the press is on cycle. Use of release may then be discontinued or, for difficult-to-eject parts, only periodic application may be required

Recommended processing factors for injection-moldable polyphenylene sulfide(a)

Grade	Melt temperature range °C	°F	Melt viscosity(b), Pa · s	Gating diameter(c) mm	in.
Glass reinforced:					
30%	300 to 355	575 to 675	180	0.76	0.030
40 to 45%	300 to 355	575 to 675	200	0.76	0.030
40% glass filled, modified	300 to 355	575 to 675	470	0.76	0.030
Glass/mineral reinforced:					
30% glass/30% mineral	310 to 355	590 to 675	210	1.3	0.050
30% glass/35% mineral	310 to 355	590 to 675	240	1.3	0.050

(a) High injection speed, mold temperature, 135 to 150 °C (275 to 300 °F); maximum regrind, 30 to 35%. (b) Melt viscosity measured at 1000/s; data taken at 300 °C (572 °F). (c) Data based on a wall thickness of 3.18 mm (0.125 in.)

Polypropylene

$$\left[\begin{array}{cc} H & H \\ | & | \\ -C & -C- \\ | & | \\ H & CH_3 \end{array} \right]_n$$

Molecular structure of polypropylene

Thermal, physical, and chemical properties of typical polypropylene and polymethyl pentene

Property	PP	TPX
Heat-deflection temperature at 1.82 MPa, °C	80	55
Maximum resistance to continuous heat, °C	70	50
Coefficient of linear thermal expansion, 10^{-5}/°C	9.0	11.7
Compressive strength, MPa	45	38
Flexural strength, MPa	48	34.5
Impact strength, Izod, cm · N/cm of notch	27	27
Tensile strength, MPa	34.5	24
Elongation, %	100	15
Hardness, Rockwell	R80	L70
Specific gravity	0.90	0.83
Dielectric constant	2.3	2.1
Resistance to chemicals at 25 °C(a):		
Nonoxidizing acids (20% H_2SO_4)	S	S
Oxidizing acids (10% HNO_3)	Q	Q
Aqueous salt solutions (NaCl)	S	S
Aqueous alkalis (NaOH)	S	S
Polar solvents (C_2H_5OH)	S	S
Nonpolar solvents (C_6H_6)	Q	Q
Water	S	S

(a) S, satisfactory; Q, questionable

Selected properties of polypropylenes

Property	ASTM test method	Reported value for: Extruded or molded polypropylene	Reported value for: Extruded or molded high-impact polypropylene
Mechanical			
Yield strength, MPa (ksi)	D 638	30 to 38 (4.3 to 5.5)	19 to 31 (2.8 to 4.5)
Tensile modulus, GPa (10^6 psi)	D 638	1.1 to 1.6 (0.16 to 0.23)	0.69 to 1.2 (0.10 to 0.18)
Elongation (break), %	D 638	200 to 700	350 to 500
Compressive strength, MPa (ksi)	D 695	41 to 55 (6 to 8)	28 to 48 (4 to 7)
Flexural yield strength, MPa (ksi)	D 790	41 to 55 (6 to 8)	...
Flexural modulus, GPa (10^6 psi)	D 790	1.4 to 2.1 (0.20 to 0.30)	0.69 to 1.4 (0.10 to 0.20)
Impact strength, Izod(a), J/cm (ft · lbf/in.)	D 256	0.27 to 1.2 (0.5 to 2.2)	0.53 to 6.41 (1.0 to 12.0)
Hardness, Rockwell R	...	80 to 110	50 to 85
Thermal			
Thermal conductivity, W/m · K (Btu · in./ft^2 · h · °F)	C 177	0.10 (0.7)	0.12 to 0.17 (0.8 to 1.2)
Coefficient of linear thermal expansion, 10^{-5}/°C (10^{-5}/°F)	D 696	11 to 18.0 (6.0 to 10.0)	11 to 16.2 (6.0 to 9.0)
Specific heat, kJ/kg · K (Btu/lb · °F)	C 351	2.1 (0.5)	2.1 (0.5)
Continuous service temperature, °C (°F)	...	110 (230)	...
Electrical			
Volume resistivity, Ω · cm	D 257	10^{16}	10^{16}
Dielectric strength, V/10^{-3} mm (V/10^{-3} in.)	D 149	20 to 24 (500 to 600)	20 to 26 (500 to 650)
Dielectric constant (50 to 100 Hz)	D 150	2.2 to 2.6	2.3
Dissipation factor (50 to 100 Hz)	D 150	>0.0005	>0.0003

(continued)

Selected properties of polypropylenes (continued)

Property	ASTM test method	Reported value for: Extruded or molded polypropylene	Extruded or molded high-impact polypropylene
Processing parameters			
Processing temperature, °C (°F)	...	175 to 290 (350 to 550)	175 to 290 (350 to 550)
Density, g/cm^3	D 792	0.90 to 0.91	0.89 to 0.90
Specific volume, cm^3/kg (in.^{3}/lb)	D 792	1120 (31)	1120 (31)
Linear mold shrinkage, mm/mm (in./in.)	D 955	0.010 to 0.020	0.010 to 0.020
Water absorption, % in 24 h	D 570	0.01	0.01

(a) Notched, at room temperature

Physical properties of polypropylene

Property	ASTM test method	Homopolymer	Copolymer
Tensile strength, MPa (ksi)	D 638	31 to 41 (4.5 to 6.0)	21.4 (3.1)
Elongation, %	D 638	100 to 600	300
Flexural modulus, GPa (10^6 psi)	D 790	1.2 to 1.7 (0.170 to 0.250)	0.9 (0.130)
Notched Izod impact strength, J/m (ft · lbf/in.)	D 256	20 to 53 (0.4 to 1.0)	763 (14.0)
Deflection temperature under load, °C (°F)	D 648		
At 1.82 MPa (0.264 ksi)		50 to 60 (120 to 140)	43 (110)
At 0.45 MPa (0.066 ksi)		110 to 120 (225 to 250)	85 (185)

Selected properties of stamped polypropylene

Property	Directionalized: Primary direction	Directionalized: Transverse direction	Isotropic
Flexural strength, MPa (ksi)	240 (35)	160 (32)	165 (24)
Flexural modulus, GPa (10^6 psi)	8.62 (1.25)	4.5 (0.65)	5.5 (0.80)
Tensile strength, MPa (ksi)	160 (23)	97 (14)	97 (14)
Tensile modulus, GPa (10^6 psi)	8.62 (1.25)	4.5 (0.65)	5.2 (0.75)
Deflection temperature at 1.82 MPa (264 psi), °C (°F)	155 (310)	155 (310)	155 (310)
Specific gravity	1.19	1.19	1.19
Glass content, %	40	40	40

Comparison of PP homopolymer and HIPS

Comparison factor	HIPS	PP
Thermal properties	...	Better
Impact strength, including low-temperature impact	Better	...
Dimensional stability	Better	...
Processibility	Better	...
Water vapor transmission	...	Better
Chemical resistance	...	Better
Stiffness	Better	...
Printability/paintability	Better	...

Polystyrene

$$\left[\begin{array}{cc} H & H \\ | & | \\ -C- & -C- \\ | & | \\ H & C_6H_6 \end{array}\right]_n$$

Molecular structure of polystyrene

Thermal, physical, and chemical properties of typical polystyrenes

Property	Unfilled PS	Impact PS	30% glass-filled PS	SAN
Heat-deflection temperature at 1.82 MPa	90	90	105	100
Maximum resistance to continuous heat, °C	75	70	95	85
Coefficient of linear thermal expansion, 10^{-5}/°C	7.5	8.0	4.0	6.0
Compressive strength, MPa	90	45	103	90
Flexural strength, MPa	83	50	117	100
Impact strength, Izod, cm · N/cm of notch	21	80	80	30
Tensile strength, MPa	41	41	83	60
Elongation, %	1.5	3	1	1.5
Hardness, Rockwell M	60	35	60	80
Specific gravity	1.04	1.04	1.2	1.07
Dielectric constant	2.5	3.0	3.0	3.5
Resistance to chemicals at 25 °C(a):				
Nonoxidizing acids (20% H_2SO_4)	S	S	S	S
Oxidizing acids (10% HNO_3)	Q	Q	Q	Q
Aqueous salt solutions (NaCl)	S	S	S	S
Aqueous alkalis (NaOH)	S	S	Q	S
Polar solvents (C_2H_5OH)	S	S	S	S
Nonpolar solvents (C_6H_6)	U	U	U	U
Water	S	S	S	S

(a) S, satisfactory; Q, questionable; U, unsatisfactory

Selected properties of molded polystyrenes

Property	ASTM test method	Reported value for: Molded polystyrene	Heat- and impact-resistant molded polystyrene	Impact- and flame resistant molded polystyrene
Mechanical				
Yield strength, MPa (ksi)	D 638	34 to 82.7 (5.0 to 12.0)	10 to 48 (1.5 to 7.0)	26 to 34 (3.8 to 5.0)
Tensile modulus, GPa (10^6 psi)	D 638	2.8 to 3.4 (0.40 to 0.50)	1.0 to 3.4 (0.15 to 0.50)	1.9 to 2.2 (0.28 to 0.32)
Elongation (break), %	D 638	1.0 to 2.5	10 to 90	13 to 25
Compressive strength, MPa (ksi)	D 695	76 to 110 (11 to 16)	28 to 62 (4 to 9)	...
Flexural yield strength, MPa (ksi)	D 790	55 to 97 (8 to 14)	21 to 83 (3 to 12)	34 to 48 (5 to 7)
Flexural modulus, GPa (10^6 psi)	D 790	2.8 to 3.4 (0.40 to 0.50)	0.69 to 3.4 (0.10 to 0.50)	2.1 (0.30)
Impact strength, Izod(a), J/cm (ft · lbf/in.)	D 256	0.1 to 0.2 (0.2 to 0.4)	0.3 to 0.4 (0.5 to 0.8)	0.53 to 0.75 (1.0 to 1.4)
Hardness, Rockwell M	...	65 to 80	10 to 50	10 to 15
Thermal				
Thermal conductivity, W/m · K (Btu · in./ft^2 · h · °F)	C 177	0.10 to 0.14 (0.7 to 1.0)	0.04 to 0.12 (0.3 to 0.8)	...
Coefficient of linear thermal expansion, 10^{-5}/°C (10^{-5}/°F)	D 696	11 to 14 (6.0 to 8.0)	9.0 to 37.8 (5.0 to 21.0)	...
Specific heat, kJ/kg · K (Btu/lb · °F)	C 351	1.3 (0.3)	1.3 (0.3)	...
Continuous service temperature, °C (°F)	...	70 to 95 (160 to 205)	60 to 70 (140 to 160)	...
Electrical				
Volume resistivity, Ω · cm	D 257	10^{16}	10^{16}	10^{15}
Dielectric strength, V/10^{-3} mm (V/10^{-3} in.)	D 149	20 to 28 (500 to 700)	12 to 24 (300 to 600)	20 (500)
Dielectric constant, (50 to 100 Hz)	D 150	2.5 to 3.1	2.5 to 4.8	3.2
Dissipation factor (50 to 100 Hz)	D 150	0.0001 to 0.0006	0.0004 to 0.0020	0.0006

(continued)

Selected properties of molded polystyrenes (continued)

Property	ASTM test method	Reported value for: Molded polystyrene	Heat- and impact-resistant molded polystyrene	Impact- and flame resistant molded polystyrene
Processing parameters				
Processing temperature, °C (°F)	...	150 to 260 (300 to 500)	175 to 315 (350 to 600)	150 to 260 (300 to 500)
Density, g/cm^3	D 792	1.04 to 1.09	1.04 to 1.10	1.10 to 1.20
Specific volume, cm^3/kg (in.^{3}/lb)	D 792	940 (26)	940 (26)	905 (25)
Linear mold shrinkage, mm/mm (in./in.)	D 955	0.001 to 0.006	0.002 to 0.006	0.002 to 0.006
Water absorption, % in 24 h	D 570	0.0 to 0.1	0.1 to 0.6	...

(a) Notched, at room temperature

Properties of various grades of impact polystyrene

Property	Crystal PS	Medium-impact PS	Extrusion-grade HIPS	Injection molding grade HIPS
Mechanical				
Melt flow rate, condition G, g/10 min.	2.5	14.0	3.0	9.0
Tensile strength at yield, MPa (ksi)	...	25.5 (3.7)	19.3 (2.8)	24.8 (3.6)
Tensile strength at break, MPa (ksi)	52 (7.6)	28 (4.0)	21 (3.1)	26 (3.7)
Tensile modulus, GPa (10^6 psi)	2.48 (0.360)	1.79 (0.260)	1.59 (0.230)	1.65 (0.240)
Elongation at break, %	3	35	45	50
Flexural strength, MPa (ksi)	90 (13)	41 (6)	37.9 (5.5)	39.3 (5.7)
Flexural modulus, GPa (10^6 psi)	3.17 (0.460)	2.07 (0.300)	1.72 (0.250)	2.07 (0.300)
Izod impact strength, notched, 23 °C (73 °F), J/m (ft · lbf/in.)	16.0 (0.3)	64.1 (1.2)	123 (2.3)	133 (2.5)
Hardness, Rockwell L	76(a)	70	50	55
Thermal				
Vicat softening temperature, °C (°F)	109 (230)	98 (210)	101 (215)	98 (210)
Deflection temperature under load, 1.82 MPa (0.264 ksi), °C (°F)	90 (195)	82 (180)	87 (190)	82 (180)

(a) Hardness, Rockwell M

Important properties and constants of HIPS

Property	Value
Coefficient of thermal expansion, 10^{-5}/K	9
Linear mold shrinkage, mm/mm or in./in.	0.004 to 0.007
Thermal conductivity at 23 °C (73 °F), W/m · K (Btu · in./h · ft^2 · °F)	0.124 (0.860)
Thermal diffusivity at 23 °C (73 °F), $10^{-4}\times$ cm^2/s ($10^{-4}\times$ in.^{2}/s)	9.51 (1.47)
Heat capacity at 23 °C (73 °F), kJ/kg · K (Btu/lb · K)	1.25 (0.537)
Density, g/cm^3	1.04
Volume resistivity, Ω · m	$>1\times10^{14}$
Dielectric strength, kV/cm (V/mil)	177 (450)
Dielectric constant at 1 MHz	2.5
Dissipation factor at 1 MHz	4×10^{-4}
Surface resistivity, Ω	$>10^{13}$
Water absorption, %	<0.1
Oxygen permeability, cm^3-mil/100 in.^{2}, 24 h, atm	350
Water vapor transmission, g-mil/100 in.^{2}, 24 h	8
Poisson's ratio	0.34
Glass transition temperature, °C (°F)	100 (212)

Typical properties of specialty HIPS

Property	Ignition-resistant HIPS	Very-high-impact PS	High-gloss, high-impact PS	Glass-filled HIPS (10% type E glass)
Mechanical				
Melt flow rate, condition G, g/10 min.	5.0	2.5	4.5	2.2
Tensile strength at yield, MPa (ksi)	26.9 (3.9)	19.3 (2.8)	23.4 (3.4)	39.3 (5.7)
Tensile strength at break, MPa (ksi)	25.5 (3.7)	20.7 (3.0)	24.8 (3.6)	39.3 (5.7)
Tensile modulus, GPa (10^6 psi)	1.79 (0.260)	1.65 (0.240)	1.72 (0.250)	2.14 (0.310)
Elongation at break, %	40	50	35	4.0
Flexural strength, MPa (ksi)	40.0 (5.8)	33.1 (4.8)	42.7 (6.2)	57.9 (8.4)
Flexural modulus, GPa (10^6 psi)	2.1 (0.300)	1.86 (0.270)	1.93 (0.280)	3.19 (0.463)
Izod impact strength, notched, 23 °C (73 °F), J/m (ft · lbf/in.)	107 (2.0)	214 (4.0)	133 (2.5)	133 (2.5)

(continued)

Typical properties of specialty HIPS (continued)

Property	Ignition-resistant HIPS	Very-high-impact PS	High-gloss, high-impact PS	Glass-filled HIPS (10% type E glass)
Thermal				
Vicat softening temperature, °C (°F)	100 (212)	101 (215)	101 (215)	111 (230)
Heat-deflection temperature under load, at 1.82 MPa (0.264 ksi), °C (°F)	87 (190)	87 (190)	85 (185)	92 (200)
Density at 23 °C (73 °F), g/cm^3	1.15	1.0	...	1.13
Flammability, UL 94 rating	V-0	...	...	...
Gardner gloss at 20 °C (68 °F)	...	...	85	...

Effect of chemicals on HIPS

Chemical class	Rating at room temperature(a)
Nonoxidizing mineral acids	S
Organic acids	S
Oxidizing acids	NR
Aqueous alkalies	S
Aqueous salt solutions	S
Alcohols	F
Aliphatic amines	S
Aromatic amines	NR
Esters	NR
Ethers	NR
Polyglycols	S
Polyglycol ethers	NR
Hydrocarbons	NR
Ketones or aldehydes	NR
Vegetable oils	F
Essential oils	NR

(a) S, satisfactory; F, fair, possible effect; NR, not recommended

Polysulfones

Molecular structure of polysulfone

PSU physical properties

Property	ASTM test method	Value
Color	...	Light yellow
Clarity	...	Transparent
Refractive index	...	1.63
Density, g/cm^3	D 1505	1.24
Glass transition temperature, °C (°F)	...	185 (365)
Water absorption, at 24 h, %	D 570	0.22
Melt flow, at 343 °C (650 °F), g/10 min.	(a)	8
Mold shrinkage, mm/mm	...	0.007
Hardness, Rockwell M	D 785	69

(a) Similar to melt index test (ASTM D 1238) for lower-melting thermoplastics. Number indicates flow rate from 2.10 mm (0.08 in.) diam orifice under 298 kPa (43 psi)

PSU thermal properties

Property	ASTM test method	Value
Heat-deflection temperature at 1.8 MPa (0.264 ksi), °C (°F)	D 648	174 (345)
Coefficient of linear thermal expansion, 10^{-5}/K	D 696	5.1
UL temperature index, °C (°F)	UL 746(a)	160 (320)
Thermal conductivity, W/m · K (Btu · in./ft^2 · h · °F)	C 117	0.26 (1.8)
Specific heat at 23 °C (75 °F), J/g · K (Btu/lb · °F)	...	1.00 (0.239)

(a) Underwriters' Laboratories test method

PSU electrical properties

Property	ASTM test method	Value
Volume resistivity, Ω · cm	D 257	5×10^{16}
Surface resistivity, Ω	D 257	3×10^{16}
Dielectric strength, 3.2 mm (1/8 in.) thickness, MV/m (MV/in.)	D 149	16.6 (0.42)
Dielectric constant		
At 60 kHz	D 150	3.15
At 1 kHz	D 150	3.14
At 1 MHz	D 150	3.10
At 1 GHz	D 150	3.00
Loss tangent		
At 60 Hz	D 150	0.0011
At 1 kHz	D 150	0.0013
At 1 MHz	D 150	0.0050
At 1 GHz	D 150	0.0040

Properties of aromatic sulfone polymers

Properties	PSU	PESV	PPSU
T_g, °C (°F)	190 (375)	220 to 230 (430 to 445)	...
Heat-deflection temperature, at 1.82 MPa (0.264 ksi), °C (°F)	175 (345)	200 (397)	205 (400)
Izod impact strength, notched, J/m (ft · lbf/in.)	65 (1.2)	75 (1.4)	...

Thermal, physical, and chemical properties of typical polysulfones

Property	Polysulfone	Polyether sulfone
Heat deflection temperature at 1.82 MPa, °C	175	205
Maximum resistance to continuous heat, °C	150	165
Coefficient of linear thermal expansion, 10^{-5}/°C	5.4	5.5
Compressive strength, MPa	96	96
Flexural strength, MPa	107	127
Impact strength, Izod, cm · N/cm of notch	80	80
Tensile strength, MPa	82	82
Elongation, %	25	25
Hardness, Rockwell M	69	88
Specific gravity	1.24	1.37
Dielectric constant	3.1	3.1
Water absorption, %	0.3	0.4
Resistance to chemicals at 25 °C(a):		
Nonoxidizing acids (20% H_2SO_4)	S	S
Oxidizing acids (10% HNO_3)	U	U
Aqueous salt solutions (NaCl)	S	S
Aqueous alkalis (NaOH)	S	S
Polar solvents (C_2H_5OH)	S	S
Nonpolar solvents (C_6H_6)	Q	Q
Water	S	S

(a) S, satisfactory; Q, questionable; U, unsatisfactory

Selected properties of polysulfone (extruded or molded)

Property	ASTM test method	Reported value
Mechanical		
Yield strength, MPa (ksi)	D 638	68.9 to 75.8 (10.0 to 11.0)
Tensile modulus, GPa (10^6 psi)	D 638	2.5 (0.36)
Elongation (break), %	D 638	50 to 100
Compressive strength, MPa (ksi)	D 695	90 to 97 (13 to 14)
Flexural yield strength, MPa (ksi)	D 790	97 to 105 (14 to 15)
Flexural modulus, GPa (10^6 psi)	D 790	2.8 (0.40)
Impact strength, Izod(a), J/cm (ft · lb/in.)	D 256	0.7 (1.3)
Hardness, Rockwell M	...	69
Thermal		
Thermal conductivity, W/m · K (Btu · in./ft^2 · h · °F)	C 177	0.1 (0.8)
Coefficient of linear thermal expansion, 10^{-5}/°C (10^{-5}/°F)	D 696	9.4 to 10 (5.2 to 5.6)
Specific heat, kJ/kg · K (Btu/lb · °F)	C 351	1.3 (0.3)
Continuous service temperature, °C (°F)	...	175 to 190 (350 to 375)
Electrical		
Volume resistivity, Ω · cm	D 257	5.0×10^{16}
Dielectric strength, V/10^{-3} mm (V/10^{-3} in.)	D 149	17 (425)
Dielectric constant, (50 to 100 Hz)	D 150	3.1
Dissipation factor (50 to 100 Hz)	D 150	0.0008
Processing parameters		
Processing temperature, °C (°F)	...	290 to 400 (550 to 750)
Density, g/cm^3	D 792	1.24
Specific volume, cm^3/kg (in.^{3}/lb)	D 792	795 (22)
Linear mold shrinkage, mm/mm (in./in.)	D 955	0.007
Water absorption, % in 24 h	D 570	0.2

(a) Notched, at room temperature

PSU mechanical properties

Property	ASTM test method	Value
Tensile strength at yield, MPa (ksi)	D 638	70.3 (10.2)
Tensile modulus, GPa (10^6 psi)	D 638	248 (0.360)
Tensile elongation at break, %	D 638	75
Flexural strength, MPa (ksi)	D 790	106 (15.4)
Flexural modulus, GPa (10^6 psi)	D 790	2.69 (0.390)
Compressive strength at yield, MPa (ksi)	D 695	96 (14)
Compressive modulus, GPa (10^6 psi)	D 695	2.58 (0.374)
Shear strength at yield, MPa (ksi)	D 732	41.4 (6.00)

(continued)

PSU mechanical properties (continued)

Property	ASTM test method	Value
Tensile impact, kJ/m^2 ($ft \cdot lbf/in.^2$)	D 1822	420 (200)
Poisson's ratio at 0.5% strain	...	0.37
Notched Izod at 6.2 mm (0.25 in.) thickness, J/m (ft · lbf/in.)	D 256	64 (1.2)
Notched Izod at 3.1 mm (0.125 in.) thickness, J/m (ft · lbf/in.)	D 256	69 (1.3)
Unnotched Izod at 3.1 mm (0.125 in.) thickness, J/m (ft · lbf/in.)	D 256	No break
Notched Izod at 3.1 mm (0.125 in.) thickness and –40 °C (–40 °F), J/m (ft · lbf/in.)	D 256	64 (1.2)

Working stress levels for polysulfone in water at various temperatures

Water temperature		Steady load		Intermittent load	
°C	°F	MPa	ksi	MPa	ksi
22	72	20.7	3.0	24.1	3.5
60	140	10.3	1.5	13.8	2.0
82	180	3.4	0.5	6.9	1.0
99	210	0.34	0.05	...	...

Permeability of polysulfone to various gases (ASTM D 1434)

Gas	Permeability, 10^{-13}cm/s(a) at 25 °C
Ammonia (NH_3)	6,400
Carbon dioxide (CO_2)	5,700
Helium (He)	11,700
Hydrogen (H_2)	10,800
Methane (CH_4)	220
Nitrogen (N_2)	240
Oxygen (O_2)	1,380
Sulfur hexafluoride (SF_6)	10.8
Dicholorofluoromethane (CCl_2F_2)	3.5
Dicholorotetrafluoroethane ($C_2Cl_2F_4$)	1.5

(a) $\frac{cm^3 \text{ at STP} \cdot cm}{cm^2 \cdot cm\,Hg \cdot s}$

PSU flammability and burning behavior

Property	Test method	Value
Flammability rating, 6.1 mm (0.25 in.) thickness	UL 94	V-0
Limiting oxygen index	ASTM D 286	30
Smoke emission (specific optical density, D_m, flaming condition)	NBS smoke chamber	90

Processing factors for injection-moldable polysulfone

Grade	Melt temperature range		Melt viscosity(a),	Injection speed		Mold temperature	
	°C	°F	Pa · s	g/s	oz/s	°C	°F
Unmodified base resin	330 to 400	625 to 750	800	15 to 20	0.5 to 0.7	95 to 150	200 to 300
Glass reinforced, 10 to 30%	355 to 400	670 to 750	850 to 950	18 to 24	0.6 to 0.8	120 to 150	250 to 300
Mineral reinforced	370 to 400	700 to 750	900 to 1000	10 to 15	0.4 to 0.5	120 to 150	250 to 300
Impact modified	260 to 315	500 to 600	600 to 750	15 to 25	0.5 to 0.9	70 to 120	160 to 250
Flame retarded	330 to 400	625 to 750	800	15 to 20	0.5 to 0.7	95 to 150	200 to 300
Other	270 to 295	520 to 560	700	18 to 24	0.6 to 0.8	120 to 250	250 to 300

(a) Melt viscosity measured at 1000/s at the median temperature of the recommended melt-temperature range. Maximum regrind, 25%

Polyurethanes

$$-\left[O-(CH_2)_x-O-\overset{O}{\overset{\|}{C}}-\underset{H}{\underset{|}{N}}-(CH_2)_y-\underset{H}{\underset{|}{N}}-\overset{O}{\overset{\|}{C}}\right]_n-$$

Molecular structure of polyurethanes

Thermal, physical, and chemical properties of a typical thermoplastic polyurethane elastomer (TPU)

Heat-deflection temperature at 1.82 MPa, °C	70
Coefficient of linear thermal expansion, 10^{-5}/°C	15
Tensile strength, MPa	20
Elongation, %	600
Specific gravity	1.25
Hardness, Shore A	80
Resistance to chemicals at 25 °C(a):	
Nonoxidizing acids (20% H_2SO_4)	Q
Oxidizing acids (10% HNO_3)	U
Aqueous salt solutions (NaCl)	S
Aqueous alkalis (NaOH)	Q
Polar solvents (C_2H_5OH)	U
Nonpolar solvents (C_6H_6)	Q
Water	S

(a) S, satisfactory; Q, questionable; U, unsatisfactory

Thermal and related properties of polyurethane resins

Thermal and related properties	Polyurethane resin (cast)	Urethane elastomer	Urethane rigid foam
Cure process parameters			
Mold pressure, MPa (ksi)	...	5.2 to 13.8 (0.75 to 2.0)	...
Mold temperature, °C (°F)(a)	...	145 to 205 (293 to 400)	...
Mold shrinkage, mm/mm	...	0.009 to 0.030	...
T_g, °C (°F)	135 (275)	...	...
Cured material properties			
Water absorption, 24 h, 3.2 mm (1/8 in.) thick, %	0.20 to 0.60	0.70 to 0.90	<1%
Specific gravity (b)	1.10 to 1.50	1.11 to 1.25	0.56 to 0.64
Continuous service temperature, °C (°F)(b)	90 to 120 (190 to 250)	90 (190)	160 (325)
Heat-deflection temperature, at 1.82 MPa (0.264 ksi), °C (°F)	50 to 205 (120 to 400)	...	...
Specific heat, kJ/kg · K (Btu/lb · °F)(b)	1.3 to 2.3 (0.30 to 0.55)	1.7 to 1.9 (0.40 to 0.45)	1.4 (0.33)
Coefficient of thermal expansion, 10^{-6}/K (10^{-6}/°F)(b)	70 to 100 (39 to 56)	100 to 200 (56 to 111)	80 (45)
Thermal conductivity, W/m · K (Btu/ft · h · °F)(b)	0.17 to 0.21 (0.100 to 0.121)	0.07 to 0.30 (0.041 to 0.178)	0.06 to 0.12 (0.033 to 0.067)

(a) At 5.2 to 14 MPa (0.750 to 2 ksi). (b) At room temperature

Properties of rigid block PUR foams, at varying densities, and different flammability classifications

Property	Rigid-block PUR foam		
Density (DIN 53420), g/cm^3	0.032	0.050	0.090
Compression strength (DIN 53421), MPa (ksi)			
Parallel to foam flow direction	0.20 (0.029)	0.35 (0.051)	0.70 (0.102)
Perpendicular to foam flow direction	0.11 (0.015)	0.20 (0.029)	0.60 (0.087)

(continued)

Properties of rigid block PUR foams, at varying densities, and different flammability classifications (continued)

Property	Rigid-block PUR foam		
Tensile strength (DIN 53430), MPa (ksi)	0.20 (0.029)	0.27 (0.039)	0.90 (0.131)
Elongation at break, %	4.0	5.2	5.0
Elastic modulus, MPa (ksi)	6.2 (0.899)	6.5 (0.943)	29.2 (4.23)
Dimensional stability (DIN 53431), %			
3 h at -30 °C (-20 °F)	–0.3	–0.1	–.1
5 h at 130 °C (265 °F)	+2	+0.5	+0.2
Open cells (ASTM D 2856), %	9	12	8
Thermal conductivity (DIN 52616), W/m · K (Btu · in./h · ft^2 · °F)	0.021 (0.1458)	0.021 (0.1458)	0.027 (0.1875)
Flammability tests			
Chimney test (DIN 4102)	B1/B2	...	...
Small burner test (DIN 4102)	...	B2	B2
Epiradiateur test	M1/M2	M3	...
Swiss fire prevention test	V	III	III

Values will be obtained only for continuous rigid block production under optimal processing conditions

Effect of postcure on physical properties of a polyester TPUR

Properties	Values, based on IRHD hardness 85	95
No postcure		
Tensile strength, MPa (ksi)	30 (4.4)	27 (3.9)
Elongation at break, %	660	450
Compression set, 22 h at 70 °C, %	50	51
With postcure, 24 h at 110 °C		
Tensile strength, MPa (ksi)	33 (4.8)	31 (4.5)
Elongation at break, %	600	400
Compression set, 22 h at 70 °C, %	28	35

Chemical base, polyethylene adipate/MDI/1, 4-butanediol

Range of physical properties of available TPUR elastomers

Property	Values
Hardness, Shore	70A to 80D
Tensile strength, MPa (ksi)	20 to 55 (3 to 8)
Elongation, %	100 to 800
Tensile stress at 100% elongation, MPa (ksi)	2 to 35 (0.3 to 5)
Flexural modulus, GPa (10^6 psi)	<0.02 to 1.035 (<0.003 to 0.250)
Tear strength, N/mm (lb/in.)(a)	25 to 260 (138 to 1400)
Bayshore rebound, %	20 to 60
Glass transition temperature, low, °C (°F)	–70 to 15 (–19 to 30)

(a) Linear inch

Glass transition temperature (T_g) of various polyols used to produce TPUR elastomers

Polyol type(a)	T_g (b) °C	°F
Polypropylene glycol	–73	–100
Polytetramethylene glycol	–100	–150
Poly (1,4-butanediol adipate)	–71	–95
Poly (ε-caprolactone)	–72	–98
Poly (1,6-hexanediol) carbonate	–62	–80

(a) All polyols tested had a molecular weight of 2000. (b) Tested under DIN 53504

Comparative hydrolysis resistance of 85 Shore A TPUR

	Tensile strength over time, days															
	0		2		4		6		8		10		12		15	
Polyol	MPa	ksi	MPa	ksi	MPa	ksi	MPa	ksi	MPa	ksi	MPa	ksi	MPa	ksi	MPa	ksi
Short-chain diol adipate	42	6.090	24	3.480	0	0	0	0	0	0	0	0	0	0	0	0
Short-chain diol adipate + carbodiimide stabilizer	40	5.800	34	4.930	24	3.480	18	2.610	4	0.580	2	0.290	1	0.145	0	0
Long-chain diol adipate	44	6.380	38	5.510	16	2.320	4	0.580	2	0.290	1	0.145	0	...	0	0
Long-chain diol adipate + carbodiimide stabilizer	45	6.526	41	5.946	40	5.800	37	5.370	36	5.220	30	4.350	26	3.770	3	0.435
Poly (ε-caprolactone)	42	6.090	35	5.080	16	2.320	4	0.580	3	0.435	1	0.145	0	0	0	0
Poly (ε-caprolactone) + carbodiimide stabilizer	40	5.800	38	5.510	36	5.220	35	5.080	33	4.790	30	4.350	24	3.480	4	0.580
Polypropylene glycol containing ether groups	35	5.080	22	3.190	20	2.900	16	2.320	15	2.108	12	1.740	9	1.300	5	0.725
Tetramethylene oxide	35	5.080	18	2.610	15	2.180	13	1.890	12	1.740	11	1.600	10	1.450	8	1.160

Aged in water at 100 °C (212 °F). Chemical basis is MDI/1, 4-butanediol/polyol

Selected properties of polyurethanes

Property	ASTM test method	Reported value for: Cast liquid polyurethane	Thermoplastic elastomer polyurethane
Mechanical			
Yield strength, MPa (ksi)	D 638	14 to 69 (2.0 to 10.0)	14 to 55 (2.0 to 8.0)
Tensile modulus, GPa (10^6 psi)	D 638	0.69 to 6.9 (0.10 to 1.0)	0.069 to 2.1 (0.01 to 0.30)
Elongation (break), %	D 638	200 to 1000	100 to 500
Compressive strength, MPa (ksi)	D 695	140 (20)	140 (20)
Flexural yield strength, MPa (ksi)	D 790	6.9 to 34 (1 to 5)	6.9 to 62 (1 to 9)
Flexural modulus, GPa (10^6 psi)	D 790	0.69 (0.10)	2.1 (0.30)
Impact strength, Izod(a), J/cm (ft · lb/in.)	D 256	>13.3 (>25.0)	No break
Hardness, Rockwell	...	D90	M29
Thermal			
Thermal conductivity, W/m · K (Btu · in./ft^2 · h · °F)	C 177	0.20 (1.4)	0.07 to 0.29 (0.5 to 2.0)
Coefficient of linear thermal expansion, 10^{-5}/°C (10^{-5}/°F)	D 696	18.0 to 36.0 (10.0 to 20.0)	18.0 to 36.0 (10.0 to 20.0)
Specific heat, kJ/kg · K (Btu/lb · °F)	C 351	1.7 (0.4)	1.7 to 19 (0.4 to 4.5)
Electrical			
Volume resistivity, Ω · cm	D 257	10^{12} to 10^{15}	10^{11} to 10^{13}
Dielectric strength, V/10^{-3} mm (V/10^{-3} in.)	D 149	16 to 20 (400 to 500)	21 (525)
Dissipation factor (50 to 100 Hz)	D 150	0.015 to 0.017	0.015 to 0.048
Processing parameters			
Processing temperature, °C (°F)	...	85 to 120 (185 to 250)	150 to 230 (300 to 450)
Density, g/cm^3	D 792	1.10 to 1.50	1.05 to 1.25
Specific volume, cm^3/kg (in.^{3}/lb)	D 792	975 to 795 (27 to 22)	940 to 795 (26 to 22)
Linear mold shrinkage, mm/mm (in./in.)	D 955	0.020	0.001 to 0.030
Water absorption, % in 24 h	D 570	0.0 to 1.5	0.7 to 0.9

(a) Notched, at room temperature

Styrene Acrylonitrile (SAN)

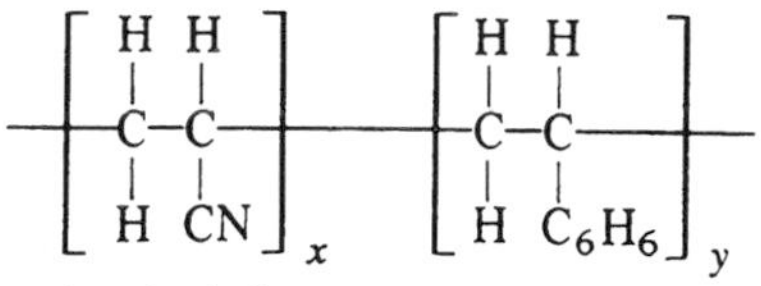

$$\left[\begin{array}{c}H\ \ H\\ |\ \ \ |\\ -C-C-\\ |\ \ \ |\\ H\ \ CN\end{array}\right]_x\left[\begin{array}{c}H\ \ H\\ |\ \ \ |\\ -C-C-\\ |\ \ \ |\\ H\ \ C_6H_6\end{array}\right]_y$$

Acrylonitrile   Styrene

Molecular structure of styrene acrylonitrile copolymer

Selected properties of molded styrene acrylonitrile

Property	ASTM test method	Reported value
Mechanical		
Yield strength, MPa (ksi)	D 638	62 to 83 (9.0 to 12.0)
Tensile modulus, GPa (10^6 psi)	D 638	2.8 to 3.9 (0.40 to 0.56)
Elongation (break), %	D 638	1.5 to 4.0
Compressive strength, MPa (ksi)	D 695	97 to 115 (14 to 17)
Flexural yield strength, MPa (ksi)	D 790	97 to 130 (14 to 19)
Flexural modulus, GPa (10^6 psi)	D 790	3.4 to 4.1 (0.50 to 0.60)
Impact strength, Izod(a), J/cm (ft · lbf/in.)	D 256	0.2 to 0.3 (0.3 to 0.5)
Hardness, Rockwell M	...	80 to 90
Thermal		
Thermal conductivity, W/m · K (Btu · in./ft^2 · h · °F)	C 177	0.1 (0.8)
Coefficient of linear thermal expansion, 10^{-5}/°C (10^{-5}/°F)	D 696	6.5 to 6.8 (3.6 to 3.8)
Specific heat, kJ/kg · K (Btu/lb · °F)	C 351	1.3 (0.3)
Continuous service temperature, °C (°F)	...	80 to 90 (175 to 190)
Electrical		
Volume resistivity, Ω · cm	D 257	10^{16}
Dielectric strength, V/10^{-3} mm (V/10^{-3} in.)	D 149	16 to 20 (400 to 500)
Dielectric constant (50 to 100 Hz)	D 150	2.6 to 3.4
Dissipation factor (50 to 100 Hz)	D 150	0.006 to 0.008
Processing parameters		
Processing temperature, °C (°F)	...	175 to 300 (350 to 575)
Density, g/cm^3	D 792	1.20 to 1.33
Specific volume, cm^3/kg (in.^{3}/lb)	D 792	940 to 905 (26 to 25)
Linear mold shrinkage, mm/mm (in./in.)	D 955	0.002 to 0.007
Water absorption, % in 24 h	D 570	0.2 to 0.3

(a) Notched, at room temperature

SAN, OSA, and ASA properties

Property	ASTM test method	SAN, 30%	SAN, glass filled	OSA	ASA
Mechanical					
Density, g/cm^3	D 792	1.07 to 1.08	1.22	1.02	1.05 to 1.06
Tensile strength, MPa (ksi)	D 638	57 to 75 (8.3 to 10.9)	139 (20)	39 (5.7)	36 to 39 (5.2 to 5.7)
Tensile modulus, GPa (10^6 psi)	D 638	3.4 to 3.6 (0.49 to 0.52)	11 (1.6)	2.1 (0.30)	...
Flexural modulus, GPa (10^6 psi)	D 790	3.1 to 3.6 (0.45 to 0.52)	10.3 (1.49)	...	1.6 to 1.7 (0.23 to 0.25)
Elongation at yield, %	D 638	2.0 to 4.5	1.6	30	...
Notched Izod impact strength, J/m (ft · lbf/ft)	D 256	13 to 24 (2.9 to 5.4)	60 (13.5)	750 (170)	500 to 550 (110 to 125)
Hardness, Rockwell	D 785	M78 to 85	M123	R102	R96
Thermal					
Heat-deflection temperature, °C (°F)					
At 1.82 MPa (0.264 ksi)	D 648	99 to 109 (210 to 230)	100 (212)	99 (210)	88 (190)
At 0.455 MPa (0.066 ksi)	D 648	103 to 115 (220 to 240)	108 (225)	...	96 (205)
Vicat softening point, °C (°F)	D 1525	108 to 113 (225 to 235)	...	106 (225)	...
Temperature index, °C (°F)	UL 746B(a)	60 (140)	74 (165)	...	...
Flammability	UL 94(a)	94HB	94HB	94HB	94HB
Linear coefficient of thermal expansion, 10^{-3}/K	D 696	6.7 to 6.8	1.9	8.0	5.9

(continued)

SAN, OSA, and ASA properties (continued)

Property	ASTM test method	SAN, 30%	SAN, glass filled	OSA	ASA
Electrical					
Dielectric strength, kV/mm (V/mil)	D 149	0.0076 to 0.010 (0.20 to 0.25)	0.012 (0.30)	...	...
Dielectric constant at 1 MHz	D 150	2.8 to 3.0	3.6	...	...
Power factor at 1 MHz	D 150	0.008 to 0.010	0.008	...	...

(a) Underwriters' Laboratories test method

Tabulated tensile creep properties of SAN

Supplier	Trade name and grade designation	Description	Test temperature °C	Test temperature °F	Initial applied stress MPa	Initial applied stress ksi	Creep (apparent) modulus(a) At 1 h MPa	At 1 h ksi	At 10 h MPa	At 10 h ksi	At 30 h MPa	At 30 h ksi	At 100 h MPa	At 100 h ksi	At 300 h MPa	At 300 h ksi	At 10^3 h MPa	At 10^3 h ksi	At latest test point MPa	At latest test point ksi	Time at latest test point, h	Time at rupture at initial applied stress in air, h
Dow	Tyril 867	General-purpose	23	73	30.3	4.40	3450	500	3280	475	3140	455	2930	425	2650	385	2240	325	1720	250	3500	4390
					33.9	4.92	3450	500	3280	475	3140	455	2930	425	2650	385	2170	315	2000	290	1705	1410
					37.9	5.50	3380	490	3030	440	2830	410	2450	355	1970	285	...	...	1970	285	309	284
					41.6	6.03	3280	475	3000	435	2760	400	2280	330	...	...	...	...	1900	275	190	152
					44.5	6.46	3240	470	2960	430	2720	395	...	...	...	...	...	...	2550	370	47	55
					47.4	6.87	3170	460	...	...	...	...	...	...	...	...	...	...	3030	440	6	4.7

(a) Calculated from total creep strain or deflection (before rupture and onset of yielding)

SAN environmental properties and glass transition temperature

Property	ASTM test method	SAN	SAN, 30% glass filled
Environmental			
Clarity, %	D 1003	87 to 89	40
Water absorption, %	D 510	0.15 to 0.40	0.15
FDA approvable	...	No	No
Affected by			
Weak acids	D 543	Not attacked	Not attacked
Strong acids	D 543	Minimal attack	Minimal attack
Weak alkalies	D 543	Not attacked	Not attacked
Strong alkalies	D 543	Minimal attack	Minimal attack
Low molecular weight organic			
Solvents	D 543	Minimal attack	Minimal attack
Alcohols	D 543	Minimal attack	Minimal attack
Other			
Glass transition temperature, °C (°F)	D 3418	109 to 116 (230 to 240)	...

Thermoplastic Alloys and Blends

Applications of alloys and blends

Polymer-polymer combination	Blend type	Typical application
Polyester(a)-PC	Crystalline-amorphous	Automobile bumpers
PA-PPO	Crystalline-amorphous	Automobile fenders
Polyester(a)-PPO	Crystalline-amorphous	Auto body panels
PA-PAR	Crystalline-amorphous	Auto body panels
Polyester(a)-PAR	Crystalline-amorphous	Auto body panels
PA-ABS	Crystalline-amorphous	Power tool housings
Polyester(a)-PSU	Crystalline-amorphous	Electric connectors
Polyetheretherketone-PESV	Crystalline-amorphous	Microcircuitry packaging
PPO-PS	Amorphous-amorphous	Business machine housings
PC-acrylonitrile-styreneacrylate	Amorphous-amorphous	Exterior building components
ABS-PC	Amorphous-amorphous	Auto instrument panels
SMA-PC	Amorphous-amorphous	Auto instrument panels
PSU-ABS	Amorphous-amorphous	Food service trays
PBT-PET	Crystalline/crystalline	Appliance housings
PP-EPDM	Crystalline/elastomer	Automobile bumper covers
PA-polyolefin	Crystalline/crystalline	Solvent barrier containers
Polyacetal-elastomer	Crystalline/elastomer	Golf shoe soles
PA-EPDM	Crystalline/elastomer	Sporting goods
Polyester(a)-COPE	Crystalline/elastomer	Winch housings
PC-TPUR	Amorphous/elastomer	Automobile bumpers

(a) PBT or PET

Examples of crystalline, amorphous, and elastomer blend components

Crystalline polymers	Amorphous polymers	Elastomers
Polyethylene (PE)	PC	EPDM
Polypropylene (PP)	PAR	Butadiene rubbers
Polyacetal	PSU	Copolyester elastomer (COPE)
PA 6	Polyether sulfone (PESV)	Copolyamide elastomer (COPA)
PA 6/6	Polyether-imide (PEI)	Styrene block copolymers (SBC)
PBT	Styrenics	Thermoplastic polyurethane (TPUR)
	Acrylics	

Effect of incompatible polymer blends on volume resistivity on compression-molded plaques ($\Omega \cdot m$)

Polymer	Carbon black(a), %						
	3	4	5	6	7	8	10
PVC	...	$>10^6$	...	...	...	...	2
PVC/EPDM(b)	...	...	...	...	...	...	0.70
PVC/EPDM(c)	...	5.1×10^2	...	...	...	...	...
GPPS	$>10^6$	...	...	...	3.2	0.65	...
PS/SBS(d)	10	2	0.9	0.5	0.3	...	...

(a) Regular Ketjenblack EC. (b) Weight ratio 60/40. (c) Weight ratio 83/17. (d) Weight ratio 70/30. GP, general purpose

Properties of amorphous-crystalline blends

Property	Components		
	PC	PBT	Blend PC-PBT
Specific gravity	1.2	1.31	1.24
Mold shrinkage, %	0.5	1.5	0.9
Coefficient of linear thermal expansion, 10^{-5}/K	12.2	14.0	12.8
Heat deflection temperature at 1.8 MPa (0.264 ksi), °C (°F)	132 (270)	66 (150)	80 (175)
Tensile strength, MPa (ksi)	66 (9.6)	57 (8.3)	58 (8.4)
Elongation, %	110	50	120
Flexural modulus, GPa (10^6 psi)	2.38 (0.345)	2.59 (0.376)	2.10 (0.305)
Notched Izod impact strength, J/m (ft · lbf/ft)	850 (190)	50 (10)	745 (170)
Chemical resistance	Fair	Excellent	Good

Properties of ABS and PC blends versus properties of their components

Property	Components		Theoretical blend	Commercial blends	
	ABS(a)	PC	50/50 ABS-PC	PC-ABS	ABS-PC
Specific gravity	1.07	1.2	1.14	1.12	1.10
Water absorption, 24 h, %	0.32	0.15	0.24	0.21	...
Coefficient of thermal expansion 10^{-5}/K	10.8	12.2	11.5	6.3	9.2
Heat deflection temperature at 1.8 MPa (0.264 ksi), °C (°F)	100 (212)	132 (270)	116 (240)	116 (240)	88 (190)
Tensile strength, MPa (ksi)	43 (6.2)	66 (9.6)	55 (8.0)	59 (8.6)	43 (6.2)
Flexural strength, MPa (ksi)	76 (11)	93 (13)	85 (12)	90 (13)	72 (10)
Flexural modulus, GPa (10^6 psi)	2.4 (0.35)	2.35 (0.341)	2.375 (0.344)	2.415 (0.350)	2.275 (0.330)
Notched Izod impact, J/m (ft · lbf/ft)	230 (52)	850 (190)	540 (120)	560 (130)	450 (100)
Hardness, Rockwell R	112	70(b)	...	117	105

(a) Heat resistant. (b) Hardness, Rockwell M

Typical mechanical and thermal properties of ABS alloys

Property	ASTM test method	ABS-PVC	ABS-PC	ABS-SMA	ABS-PA
Mechanical					
Tensile strength at yield, MPa (ksi)	D 638	41 (6.0)	52 (7.6)	36 (5.2)	36 (5.2)
Tensile modulus, GPa (10^6 psi)	D 638	2.4 (0.35)	2.5 (0.36)	2.2 (0.32)	1.3 (0.18)
Flexural strength at yield, MPa (ksi)	D 790	66 (9.5)	86 (12.5)	59 (8.5)	
Flexural modulus, GPa (10^6 psi)	D 790	2.4 (0.35)	2.3 (0.34)	2.2 (0.32)	1.0 (0.15)
Izod impact strength, 3.2 mm (0.13 in.) bar J/m (ft · lbf/in.)	D 256A	430 (8.0)	450 (8.5)	160 (3.0)	960 (18)
Hardness, Rockwell R	D 785A	96 to 110	111 to 120	95 to 109	75
Specific gravity	D 792	1.03 to 1.07	1.01 to 1.05	1.01 to 1.04	1.06
Thermal					
Deflection temperature under load, °C (°F)					
At 0.46 MPa (0.066 ksi)	D 648	77 (170)	113 (235)	...	92 (197)
At 1.8 MPa (0.26 ksi)	D 648	71 (160)	110 (230)	113 (230)	...
Coefficient of linear thermal expansion, 10^{-5}/K	D 696	8.3	6.7	9.0	10
UL temperature index, °C (°F)	UL 746C(a)	50 to 80 (122 to 176)	95 (203)	50 (122)	60 (140)

(a) Underwriters Laboratories test method

Mechanical properties of PC-polyester alloys

Property	Blend composition	
	50% PET, 50% PC	50% PBT, 50% PC
Tensile strength, MPa (ksi)	57.9 (8.4)	55 (8)
Tensile elongation, %	155	...
Flexural modulus, GPa (10^6 psi)	2.28 (0.330)	2.21 (0.320)
Heat-deflection temperature at 1.82 MPa (0.264 ksi), °C (°F)	143 (290)	98 (210)
Notched Izod impact, J/m (ft · lbf/in.)	91 (1.7)	96 (1.8)
Unnotched Izod impact, J/m (ft · lbf/in.)	590 (11)	...

Mechanical properties of nylon PPO blends

Properties	Product	
	Noryl GTX-910	Noryl GTX-820
Glass fibers, %	0	20
Tensile strength, MPa (ksi)	59.3 (8.6)	93.1 (13.5)
Ultimate elongation, %	60	10
Flexural modulus, GPa (10^6 psi)	2.14 (0.310)	3.86 (0.560)
Heat-deflection temperature at 1.82 MPa (0.264 ksi), °C (°F)	144 (290)	188 (370)
Notched Izod impact, J/m (ft · lbf/in.)	210 (4.0)	80 (1.5)

Mechanical properties of various grades of PPO-PS blends

Property	General grade	Intermediate heat resistant	Flame retardant, intermediate heat resistant	High heat	Glass-reinforced
Estimated polystyrene content, %	80	50	50	25	30
T_g, °C (°F)	115 (240)	145 (290)	145 (295)	165 (330)	160 (320)
Heat-deflection temperature at 1.82 MPa (0.264 ksi), °C (°F)	100 (212)	129 (265)	129 (265)	149 (300)	149 (300)
Notched Izod impact, J/m (ft · lbf/in.)	270 (5)	270 (5)	270 (5)	530 (10)	120 (2.3)
Flexural modulus, GPa (10^6 psi)	2.48 (0.360)	2.48 (0.360)	2.48 (0.360)	2.41 (0.350)	7.58 (1.10)
Flexural strength, MPa (ksi)	88 (12.8)	93 (13.5)	93 (13.5)	104 (15.1)	138 (20.0)
Flammability (UL 94)	V-1 at 1.52 mm (0.060 in.)	Horizontal burn	V-1 at 1.5 mm (0.058 in.)	V-0 at 1.55 mm (0.061 in.)	Horizontal burn
UL continuous-use temperature, °C (°F)	80 (175)	90 (195)	105 (220)	105 (220)	90 (195)

Properties of a commercial S/MA-polycarbonate alloy

Property	ASTM test method	Value
Mechanical		
Tensile strength, MPa (ksi)	D 638	48.2 (7.0)
Elongation, %	D 638	85
Flexural strength, MPa (ksi)	D 790	93.0 (13.5)
Flexural modulus, GPa (10^6 psi)	D 790	2.41 (0.350)
Izod impact, J/m (ft · lbf/in.)	D 256	800 (15)
Mold shrinkage, cm/cm or in./in.	D 955	0.007
Gardner, J (in. · lbf)	...	>51 (>450)
Thermal		
Vicat softening point, °C (°F)	D 1525	143 (290)
Deflection temperature under load, 1.82 MPa (0.264 ksi), 3.2 mm (⅛ in.) specimen, °C (°F)	D 648	113 (236)
Other		
Specific gravity, g/cm^3	D 792	1.13

Thermosets

General

Representative chemical structures and starting materials for common thermosets

Chemical family name	Starting materials	Representative chemical structure
Phenol-formaldehyde (phenolic)	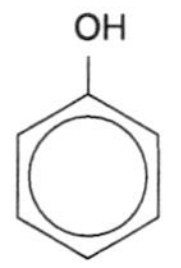 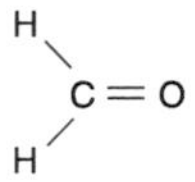 Phenol + Formaldehyde	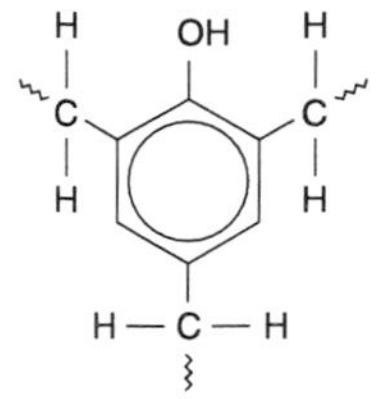
Melamine-formaldehyde (melamine)	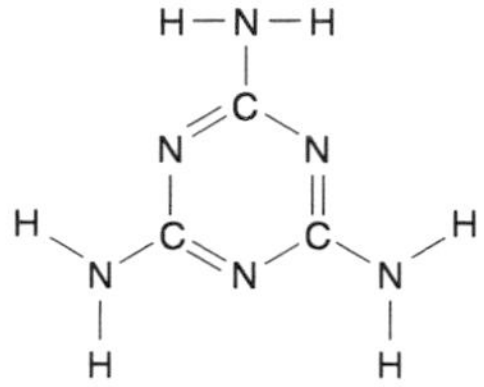 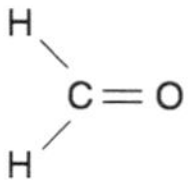 Melamine + Formaldehyde	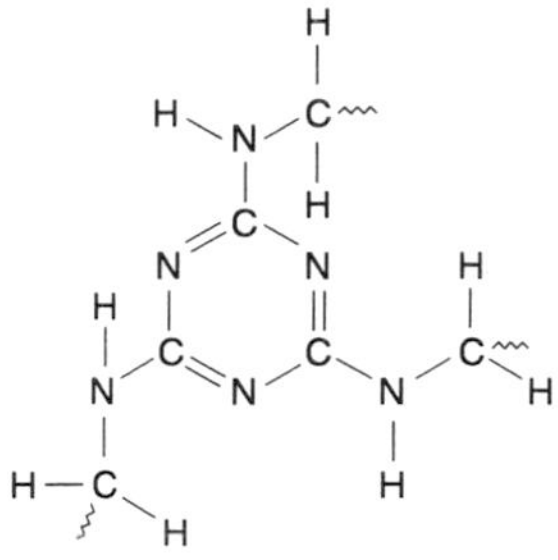
Epoxy	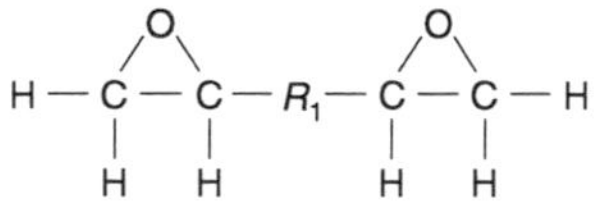 Epoxide resin (one type) + Curing agent (one type)	
Polyimide (bismaleimide type)	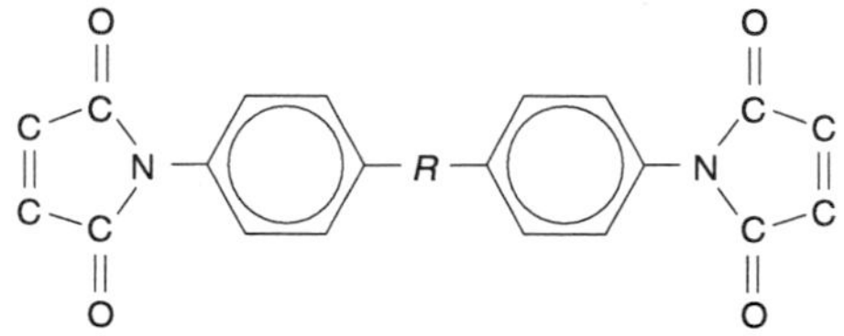 Bismaleimide resin 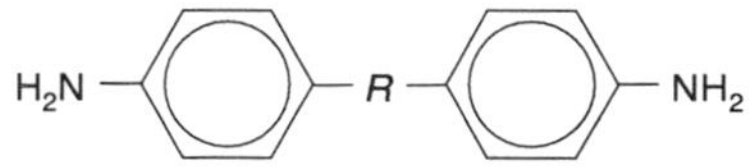 + Curing agent (one type)	Cross linked through maleimide end caps

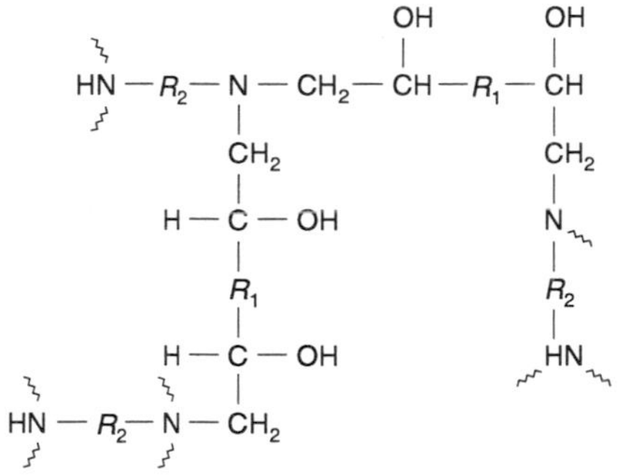

Characteristics of thermosetting resin families

	Major thermosets			
	Phenolic	Unsaturated polyester	Amino	Epoxy
General characteristics				
Chemical	Excellent resistance to humidity, acids, and oils; not resistant to strong oxidizers	Low moisture uptake	Excellent resistance to fats, oils, and waxes; poor resistance to strong alkalies	Good chemical resistance
Electrical	Excellent electrical properties	Excellent electrical properties	Excellent electrical properties	Good to excellent electrical properties
Flammability	Low flammability and low smoke generation	Limited high-temperature performance	Low flammability	
Processing	Requires heat and pressure for cure	Versatile processing; no volatiles during cure	Requires heat and pressure for cure	Versatile processing; no volatiles during cure; excellent adhesion to filler and fibers
Common reinforcements	Wood flour, mica, glass flakes, fibers	Glass fiber, calcium carbonate	Alpha-cellulose, glass fiber	Glass or carbon fibers, silica
Long-term heat resistance, Underwriters' Laboratories temperature index(a), °C (°F)	150 (300)	105 to 130 (220 to 265)	100 (212)(b) 130 to 150 (265 to 300)(c)	130 (265)
Fabrication methods				
Compression molding	Yes	Yes	Yes	Yes
Injection molding	Yes	Yes	Yes	Yes
Transfer molding	Yes	Yes	Yes	Yes
Autoclave molding	Yes	Yes	...	Yes
Filament winding	...	Yes	...	Yes
Spray-up	...	Yes	...	...
Pultrusion	...	Yes	...	Yes
Wet lay-up	...	Yes	...	Yes

(a) In UL 746B. (b) UF. (c) MF

Thermoset engineering plastic applications

Resin family	Applications
Phenolic	Molded articles, friction materials, laminates
Amino	
UF	Electrical components, closures for jars and bottles
MF	Dinnerware, electrical components
Unsaturated polyester	Construction components (tanks, flat and corrugated sheet)
	Transportation equipment (automotive body panels)
	Boats and marine equipment
	Cast products (bathroom fixtures)
Epoxy	Laminates (structural and electrical)
	Tooling, encapsulation (electronics packaging), molded items

Matrix properties associated with thermoset resin families

Property	Epoxy(a)	Cyanate	Toughened bismaleimide
Residual chlorine, ppm	200 to 2000	<10	<10
Specific gravity	1.2 to 1.25	1.1 to 1.35	1.2 to 1.3
Tensile strength, MPa (ksi)	48 to 90 (7 to 13)	69 to 90 (10 to 13)	35 to 90 (5 to 13)
Tensile modulus, GPa (10^6 psi)	3.1 to 3.8 (0.45 to 0.55)	3.1 to 3.4 (0.45 to 0.50)	3.4 to 4.1 (0.50 to 0.60)
Tensile strain-at-break, %	1.5 to 8	2 to 5	1.5 to 3
Fracture toughness(b), J/m^2 (in. · lbf/in.^{2})	70 to 210 (0.4 to 1.2)	105 to 210 (0.6 to 1.2)	70 to 105 (0.4 to 0.6)
Water absorbed, saturated at 100 °C (212 °F), %	2 to 6	1.3 to 2.5	4.0 to 4.5
Heat deflection temperature, °C (°F)			
Dry	150 to 240 (300 to 460)	230 to 260 (450 to 500)	>250 (>480)
Water-saturated at 100 °C (212 °F)	100 to 150 (212 to 300)	150 to 200 (212 to 390)	200 to 250 (390 to 480)
Coefficient of thermal expansion, 10^{-6}/K	60 to 70	60 to 70	60 to 65
Thermogravimetric analysis onset, °C (°F)	260 to 340 (500 to 640)	400 to 420 (750 to 790)	360 to 400 (680 to 750)
Flammability, UL 94			
Unmodified rating	Burns	V-0 to burns	V-0 to burns
Bromine for V-0, %	18 to 20	0 to 12	0 to 12
Dielectric constant at 1 MHz	3.8 to 4.5	2.7 to 3.2	3.4 to 3.7
Dissipation factor at 1 MHz	2 to 5×10^{-2}	1 to 5×10^{-3}	3 to 9×10^{-3}
Cure temperature, °C (°F)	150 to 220 (300 to 430)	177 to 250 (350 to 480)	220 to 300 (430 to 570)
Mold shrinkage(c), mm/mm	0.006	0.004	0.007

(a) Diepoxides and tetraepoxides cured with aromatic diamines. (b) Double torsion method. (c) Transfer molded parts filled 65 wt% with fused silica; cured at 177 °C (350 °F) and measured at 25 °C (75 °F)

Typical resin properties

Resin system	Dielectric constant at 1 MHz(a)	Glass transition temperature(a) °C	°F
FR-4 epoxy	3.50 to 3.60	125 to 135	255 to 275
Polyfunctional FR-4	3.50 to 3.60	140 to 150	285 to 300
High-temperature, one-component epoxy system	3.90 to 4.00	170 to 180	340 to 355
Bismaleimide triazine epoxy	3.20 to 3.30	180 to 190	355 to 375
Polyimide epoxy	3.50 to 3.60	250 to 260	480 to 500
Cyanate esters	2.80 to 3.00	240 to 250	465 to 480
Polyimide	3.30 to 3.40	>260	>500
PTFE (melting point)	2.03 to 2.09	327	620

(a) Numbers for cast resin samples

Specific gravity of thermoset molding compounds

Thermoset compounds	Specific gravity
Nylon-filled phenolic	1.24 to 1.28
Cellulose-filled phenolic	1.35 to 1.38
Cellulose-filled melamine	1.47 to 1.52
Aramid-filled epoxy	1.30 to 1.32
Graphite-filled phenolic	1.44 to 1.55
Graphite-filled epoxy	1.48 to 1.54
Glass-filled phenolic	1.78 to 1.95
Glass-filled epoxy	1.70 to 1.93
Glass-filled melamine	1.79 to 2.00
Glass-filled polyester	1.82 to 1.98
Glass-filled vinyl ester	1.84 to 1.90
Mineral-filled phenolic	1.85 to 2.10
Mineral-filled epoxy	1.85 to 1.93
Mineral-filled melamine	1.70 to 1.75

Advantages and limitations of various adhesive types

Adhesive type	Advantages	Limitations
Phenolics	Very high bond strength	Used mostly for structural (and possible corrosive) applications; difficult to process at low temperatures
Polyurethanes	Easy to rework	Not suitable for temperatures above 120 °C (250 °F), relatively high outgassing, some decomposition
Polyamides	Easy to rework	High moisture absorption; high outgassing; variations in electrical-insulation properties, especially when exposed to high humidity
Polyimides	Very high temperature stability	High cure temperatures; solvents required as vehicles
Silicones	High-temperature stability; easy to rework; high purity; low outgassing	Moderate to poor bond strength; high coefficient of thermal expansion
Epoxies	Some are easy to rework (by thermomechanical means); some are low outgassers; easy to process; can be filled to 60 to 70% with a variety of conductive or nonconductive fillers	Depending on type of curing agent used and degrees of cure: outgassing, catalyst leaching, corrosivity
Cyanoacrylates	Very rapid setting (≤10 s); very high initial bond strengths	Bond strengths, often degrade under moist or elevated-temperature (≥150 °C, or 300 °F) conditions

Typical characteristics of thermoset adhesives

Type	Form	Cure temperature, °C (°F)	Maximum use temperature, °C (°F)	Advantages	Disadvantages
Epoxy	Two-part paste	Room or accelerated at 93 to 178 (200 to 350)	Generally below 82 (180)	Ease of storage at room temperature; ease of mixing and use; long shelf life; gap filling when filled	No generally as strong or environmentally resistant as typical heat-cured epoxies
	One-part film	121 (250)	To 82 (180)	Covers large areas; bondline thickness control; wide variety of formulas; higher-temperature curing materials; better environmental properties	Store at -18 °C (0 °F); short shelf life; high-temperature cure; brittle and low peel strength
		149 (300) 178 (350)	149 to 177 (300 to 350)		

(continued)

Typical characteristics of thermoset adhesives (continued)

Type	Form	Cure temperature, °C (°F)	Maximum use temperature, °C (°F)	Advantages	Disadvantages
Acrylic	Two-part liquid or pastes	Room to 100 (212)	105 (221)	Fast setting; easy to mix and use; good moisture resistance; tolerant of surface contamination	Strong, objectionable odor; limited pot life
Polyurethane	One or two parts	Room or heat cure		Good peel; good for cryogenic use	Moisture sensitive before and after cure
Silicone	One- and two-part pastes	Room to 260 (500)	To 260 (500)	High peel and impact resistance; easy to use; good heat and moisture resistance	High cost; low strength
Hot melt	One-part	Melt at 190 to 232 (375 to 450)	18 to 171 (120 to 340)	Rapid application; fast setting; low cost; indefinite shelf life; nontoxic; no mixing	Poor heat resistance; special equipment required; poor creep resistance; low strength; high melt temperature
Bismaleimide (BMI)	One-part paste or film	>178 (350) and 246 (475) postcure	232 (450)	Structural bonds with bismaleimide composites; higher temperature than epoxies; no volatiles; good shelf life	Brittle and low peel; limited formulas available
Polyimide	Thermoplastic liquids; one- and two-part pastes	260 (500) and postcure	204 to 260 (400 to 500)	High-temperature resistance; structural strength	High cost; low peel strength; high cure and postcure temperatures; volatiles for some forms
Phenolic-based	One-part films	163 to 177 (325 to 350)	To 177 (350)	High-temperature use	Low peel strength

Use-temperature guide to structural adhesives

	Use temperature, °C (°F)								
Adhesive	−253 (−423)	−196 (−320)	−73 (−100)	−54 (−65)	Room	82 (180)	149 (300)	216 (420)	260 (500)
Epoxy-nitrile modified	L/V	L/V	L/E	L/E	H/E	M/V	L/Mod	...	...
Epoxy-nylon	L/E	L/E	L/E	L/E	H/V	L/G	L/Mod	...	...
Epoxy-phenolic	...	...	L/V	L/V	L/V	L/V	L/G	G	G
Vinyl-phenolic	...	...	...	L/V	M/E	H/E	L/Mod	...	...
Nitrile-phenolic	Mod	E	E	L/E	H/V	M/G	L/Mod	...	...
Bismaleimides	...	...	...	Mod	L/G	L/G	L/G	L/V	...
Polyimides	...	...	...	L/V	L/G	L/G	L/G	L/G	L/G
Polyurethanes	H/V	H/V	H/V	H/G	H/G	H/Mod	H/P	...	...
Acrylics	...	...	...	L/P	H/E	M/G	L/P	...	...

Peel: L, low; M, medium; H, high.Lap shear: P, poor; Mod, moderate; G, good; V, very good; E, excellent. Peel is indicated first, followed by lap shear: peel/lap shear

Comparative characteristics of thermoset adhesives

Resin base	Disadvantages	Advantages	Major uses
Polyesters	Considerable shrinkage, brittle on impact, poor hot strength	Fair strength, low viscosity, low-temperature cure, good electrically	Repairs, compatible with explosives and radomes
Epoxy	Generally rigid, poor hot strength, somewhat toxic	High strength, low shrinkage low-temperature cure, fair electrically	FRP-to-metal joints radomes, and aircraft structural parts
Phenolics	Requires solvents and high-temperature cures, bad electrically, may be corrosive	Good hot strength nontoxic, inexpensive	High-temperature ceramic-to-FRP joints
Rubber phenolics	Require solvents and high-temperature cures, bad electrically	Moderate strength and peel resistance	Metallic joints, shock-absorbing parts
Epoxy phenolics	Rigid; requires heat cures, bad electrically	High heat resistance, fair strength, good cryogenic strength	Aircraft structural parts where high temperature or extremely low temperature is required
Silicones	Low strength, requires solvents	Extremely high heat resistance, good arc resistance	High temperature, silicone adherends and in copolymer adhesives
Polyimides	Rigid, requires heat cure, may be corrosive	Extremely high heat resistance, good electrically	High-temperature, long-age metal-to-metal aircraft parts

Clear-casting mechanical properties

Property	Orthophthalic	Isophthalic	BPA fumarate	Chlorendic	Dicyclopentadiene
Barcol hardness	...	40	34	40	...
Tensile strength, MPa (ksi)	55 (8)	75 (11)	40 (6)	20 (3)	39 (5.7)
Tensile modulus, GPa (10^6 psi)	3.45 (0.50)	3.38 (0.49)	2.83 (0.41)	3.38 (0.49)	3.31 (0.48)
Tensile elongation, %	2.1	3.3	1.40	...	1.27
Flexural strength, MPa (ksi)	85 (12)	131 (19)	110 (16)	117 (17)	59 (8.5)
Flexural modulus, GPa (10^6 psi)	3.45 (0.50)	3.59 (0.52)	3.38 (0.49)	3.93 (0.57)	3.45 (0.50)
Compressive strength, MPa (ksi)	...	117 (17)	103 (15)	103 (15)	...
Heat-deflection temperature, °C (°F)	80 (175)	90 (195)	130 (265)	140 (285)	110 (230)

Recommended wall thicknesses for molded thermoset plastic parts

	Small parts		General parts		Large parts	
Material	mm	in.	mm	in.	mm	in.
Phenolic	1.575 to 3.175	0.062 to 0.125	2.36 to 4.75	0.093 to 0.187	4.75 to 25.4	0.187 to 1.00
Urea	1.575 to 3.175	0.062 to 0.125	2.36 to 4.75	0.093 to 0.187	4.75 to 9.52	0.187 to 0.375
Melamine	1.575 to 3.175	0.062 to 0.125	2.36 to 4.75	0.093 to 0.187	4.75 to 9.52	0.187 to 0.375
DAP	1.143 to 2.362	0.045 to 0.093	1.98 to 3.96	0.078 to 0.156	3.175 to 9.52	0.125 to 0.375
Alkyd	1.98 to 3.175	0.078 to 0.125	2.54 to 4.75	0.100 to 0.187	4.75 to 12.7	0.187 to 0.500
Polyester:						
Granular	1.98 to 3.175	0.078 to 0.125	2.54 to 4.75	0.100 to 0.187	4.75 to 12.7	0.187 to 0.500
Bulk molding compound	1.143 to 2.632	0.045 to 0.093	1.98 to 3.96	0.100 to 0.187	3.175 to 9.52	0.125 to 0.375
Sheet molding compound	1.575 to 2.632	0.062 to 0.093	2.36 to 4.75	0.093 to 0.187	4.75 to 9.52	0.187 to 0.375

DAP

Thermal, physical, and chemical properties of a typical fibrous-glass-filled allylic plastic (DAP)

Property	Value
Heat-deflection temperature at 1.82 MPa, °C	200
Coefficient of linear thermal expansion, 10^{-5}/°C	150
Compressive strength, MPa	186
Flexural strength, MPa	131
Impact strength, Izod, cm · N/cm of notch	106
Tensile strength, MPa	58
Elongation, %	4
Hardness, Rockwell	E80
Specific gravity	1.7
Water absorption	0.14
Dielectric constant	4
Resistance to chemicals at 25 °C(a):	
Nonoxidizing acids (20% H_2SO_4)	S
Oxidizing acids (10% HNO_3)	U
Aqueous salt solutions (NaCl)	S
Aqueous alkalis (NaOH)	Q
Polar solvents (C_2H_5OH)	S
Nonpolar solvents (C_6H_6)	U
Water	S

(a) S, satisfactory; Q, questionable; U, unsatisfactory

Thermal and related properties of allyl resins

Thermal and related properties	Allyl diglycol carbonate neat resin	Diallyl phthalate (DAP) Glass fiber filled	Diallyl phthalate (DAP) Mineral filled
Cure process parameters			
Mold pressure, MPa (ksi)	...	3.4 to 27.6 (0.5 to 4.0)	3.4 to 27.6 (0.5 to 4.0)
Compression mold temperature, °C (°F)	130 to 160 (270 to 320)	145 to 195 (290 to 380)(a)	130 to 165 (270 to 330)(a)
Mold shrinkage, mm/mm	...	0.001 to 0.005	0.005 to 0.007
Cured material properties			
Water absorption, 24 h, 3.2 mm (1/8 in.) thick, %	0.20	0.12 to 0.35	0.20 to 0.50
Specific gravity(b)	1.30 to 1.40	1.61 to 1.85	1.65 to 1.68
Continuous service temperature, °C (°F)	100 (212)	150 to 205 (300 to 400)	150 to 205 (300 to 400)
Heat-deflection temperature, at 1.82 MPa (0.264 ksi), °C (°F)	60 to 90 (140 to 190)	165 to 260 (325 to 500)	165 to 260 (325 to 500)
Burning rate, mm/mm (in./min)	8.9 (0.35) to self-extinguishing	Self-extinguishing to nonburning	Self-extinguishing to nonburning
Specific heat, kJ/kg · K (BTu/lbf · °F)(b)	2.3 (0.55)	1.26 to 1.33 (0.30 to 0.32)	1.26 (0.30)
Coefficient of thermal expansion, 10^{-6}/K (10^{-6}/°F)	80 to 140 (45 to 79)	10 to 35 (5.5 to 20)	10 to 42 (5.5 to 2.3)
Thermal conductivity, W/m · K (Btu/ft · h · °F)(b)	0.199 to 0.210 (0.115 to 0.120)	0.20 to 0.60 (0.12 to 0.36)	0.29 to 1.02 (0.168 to 0.600)

(a) At 3.5 to 28 MPa (0.5 to 4 ksi). (b) At room temperature

Amino

Thermal, physical, and chemical properties of typical amino plastics

Property	Cellulose-filled MF	Cellulose-filled UF
Heat-deflection temperature at 1.82 MPa, °C	150	130
Maximum resistance to continuous heat, °C	100	75
Coefficient of linear thermal expansion, 10^{-5}/°C	4	3
Compressive strength, MPa	276	221
Flexural strength, MPa	86	96.5
Impact strength, Izod, cm · N/cm of notch	16	16
Tensile strength, MPa	69	55
Elongation, %	0.7	0.7
Hardness, Rockwell	M115	M110
Specific gravity	1.5	1.5
Resistance to chemicals at 25 °C(a):		
Nonoxidizing acids(20% H_2SO_4)	S	S
Oxidizing acids (10% HNO_3)	U	U
Aqueous salt solutions (NaCl)	S	S
Aqueous alkalis (NaOH)	S	S
Polar solvents (C_2H_5OH)	S	S
Nonpolar solvents (C_6H_6)	Q	Q
Water	S	S

(a) S, satisfactory; Q, questionable; U, unsatisfactory

Applications of amino resins

Type	Applications
Molding compounds	
Cellulose-filled ureas	Wiring devices (circuit breakers, wall plates, receptacles), closures, toothpaste tube inserts, buttons, toilet seats, knobs, handles
Cellulose-filled melamines	Molded dinnerware, ashtrays, buttons, utensil handles, appliance components, knobs, handles
Wood flour filled melamines	Industrial electrical parts, military specifications
Glass- and mineral-filled melamines	Military specifications
Laminating melamine resins	Industrial laminates, decorative laminates for counter- and tabletops, wall paneling, and furniture sufacing
Adhesives	Particle boards, plywood, boat hulls, agricultural waste composites, flooring, furniture assemblies
Coating resins	Improving properties of paper, treatment of textiles for crease proofing and other properties, cross-linking water and solvent-soluble coatings (for automotive appliance and coil coatings)

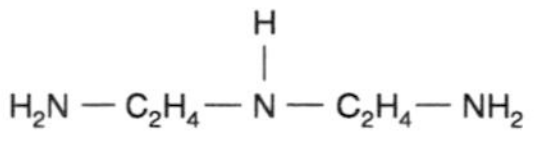

$H_2N-C_2H_4-N(H)-C_2H_4-NH_2$

Diethylenetriamine (DTA)

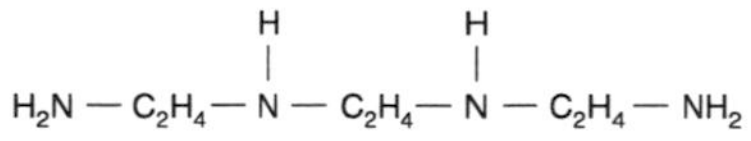

$H_2N-C_2H_4-N(H)-C_2H_4-N(H)-C_2H_4-NH_2$

Triethylenetetramine (TETA)

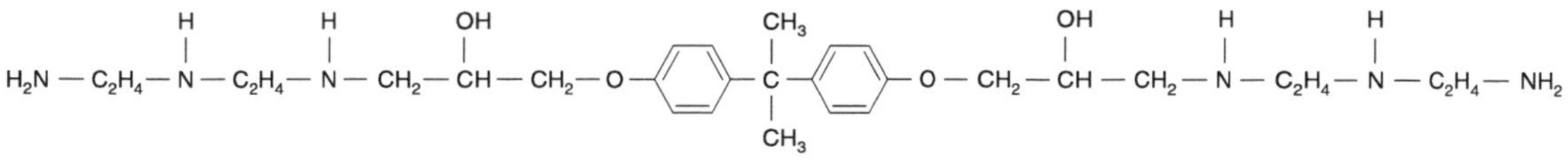

$H_2N-C_2H_4-N(H)-C_2H_4-N(H)-CH_2-CH(OH)-CH_2-O-C_6H_4-C(CH_3)_2-C_6H_4-O-CH_2-CH(OH)-CH_2-N(H)-C_2H_4-N(H)-C_2H_4-NH_2$

Typical amine adduct (DTA + DGEBA)

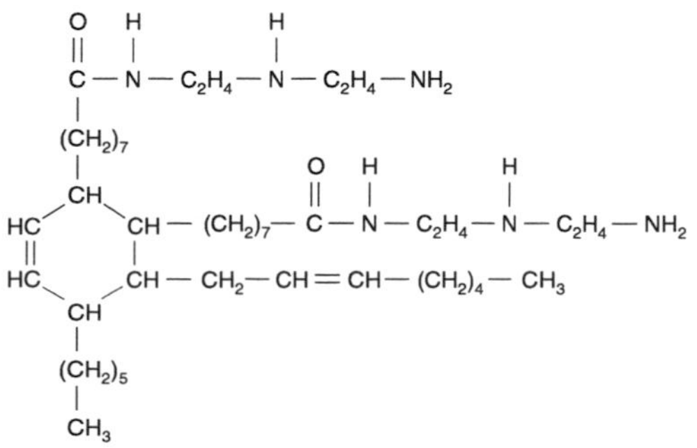

$C(=O)-N(H)-C_2H_4-N(H)-C_2H_4-NH_2$ | $(CH_2)_7$ | CH (ring: HC=HC, CH, CH) — $CH-(CH_2)_7-C(=O)-N(H)-C_2H_4-N(H)-C_2H_4-NH_2$; $CH-CH_2-CH=CH-(CH_2)_4-CH_3$; $CH-(CH_2)_5-CH_3$

Typical C_{36} polyamide (linoleic acid dimer + DTA)

Aliphatic amine curing agents and adducts

Thermal and related properties of amino resins

Thermal and related properties	Melamine-formaldehyde No filler	Melamine-formaldehyde Cellulose filler	Urea-formaldehyde, alpha cellulose filler
Cure process parameters			
Mold pressure, MPa (ksi)	13.8 to 34.5 (2.0 to 5.0)	10.3 to 41.4 (1.5 to 6.0)	13.8 to 55.2 (2.0 to 8.0)
Mold temperature, °C (°F)	150 to 165 (300 to 330)(a)	145 to 180 (290 to 360)(b)	150 to 260 (300 to 500)(c)
Mold shrinkage, mm/mm	0.011 to 0.012	0.006 to 0.008	0.007
T_g, °C (°F)	None(d)	...	...
Cured material properties			
Water absorption, 24 h, 3.2 mm (1/8 in.) thick, %	0.30 to 0.50	0.34 to 0.80	0.60
Specific gravity(e)	1.48	1.45 to 1.52	1.48 to 1.50
Continuous service temperature, °C (°F)	100 (210)	120 (250)	75 (170)
Heat-deflection temperature, at 1.82 MPa (0.264 ksi), °C (°F)	150 (298)	130 (266)	130 to 135 (266 to 275)
Flammability rating	...	...	94V-0
Burning rate	Self-extinguishing	Self-extinguishing	Self-extinguishing
Specific heat, kJ/kg · K (Btu/lbf · °F)(e)	...	...	1.68 (0.40)
Coefficient of thermal expansion, 10^{-6}/K (10^{-6}/°F)(e)	...	45 (25.0)	27 to 29 (14.9 to 16.0)
Thermal conductivity, W/m · K (Btu/ft · h · °F)(e)	...	0.265 to 0.314 (0.156 to 0.185)	0.285 to 0.409 (0.168 to 0.241)

(a) At 21 to 35 MPa (3 to 5 psi). (b) At 10 to 42 MPa (1.5 to 6 ksi). (c) At 14 to 55 MPa (2 to 8 ksi). (d) Based on private communication. American Cyanamid Company. (e) At room temperature

Epoxies

$$CH_2=CH-CH_3 \xrightarrow{Cl_2} CH_2=CH-CH_2Cl \xrightarrow{HOCl} ClCH_2-CH(OH)-CH_2Cl \xrightarrow{NaOH} H_2C\overset{O}{-}CH-CH_2Cl$$

$$C_6H_6 \xrightarrow{CH_2=CH-CH_3} C_6H_5-CH(CH_3)_2 \xrightarrow{[O]} C_6H_5-C(CH_3)_2-COOH \xrightarrow{H^+}$$

$$C_6H_5OH \xrightarrow{(CH_3)_2C{=}O} HO-C_6H_4-C(CH_3)_2-C_6H_4-OH$$

$$HO-C_6H_4-C(CH_3)_2-C_6H_4-OH + ClCH_2CH\overset{O}{-}CH_2 \xrightarrow{NaOH}$$

$$CH_2\overset{O}{-}CHCH_2-[O-C_6H_4-C(CH_3)_2-C_6H_4-O-CH_2-CHCH_2]_n-O-C_6H_4-C(CH_3)_2-C_6H_4-O-CH_2CH\overset{O}{-}CH_2$$

Epoxy resin synthesis

Epoxy properties and tests

Ingredient	Property	Test method
Epoxy resins	Epoxide content	Titration
	Viscosity/softening point	Viscometer/Duran or rheometer
	Residual chlorides	Titration
	Moisture content	Titration/Karl Fisher
	Molecular weight distribution	GPC
	Characterization	HPLC/infrared spectroscopy
Hardener (amine)	Amine content	Titration
	Purity	Melting point refractive index, HPLC
Catalyst	Purity	Melting point
	Cation	Atomic absorption
Modifier (inorganic)	Particle size	Sedigraph/particle sizer
	Moisture	Moisture analyzer/Karl Fisher
Modifier (organic)	Viscosity	Rheometer
	Reactivity	Titration

GPC, gel permeation chromatography; HPLC, High-performance liquid chromatography

Selected epoxy resins

Resin	Supplier	Formula
DER 332	Dow Chemical Co.	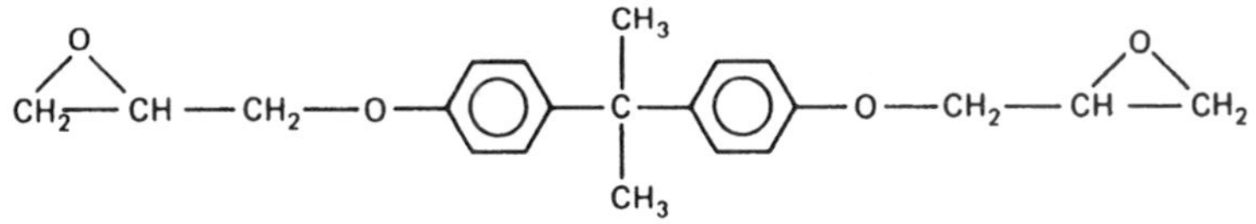
EPON 826	Shell Chemical Co.	
EPI-REZ 509	Interez Inc.	
Araldite GY 6008		
(Diglycidyl ether of bisphenol A)	Ciba-Geigy Corp.	
EPN 1139	Ciba-Geigy Corp.	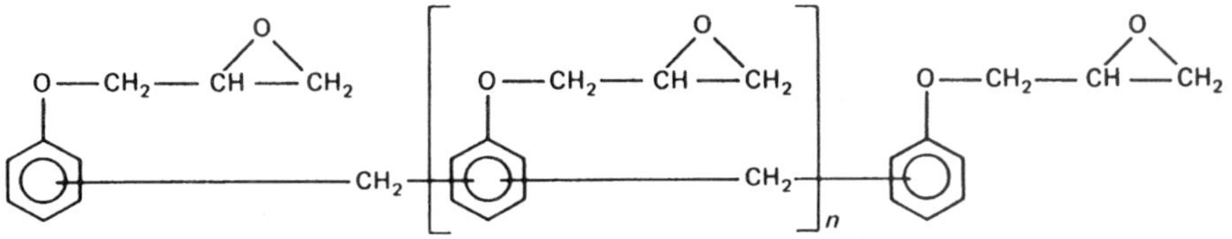
DEN 431		
(Polyglycidyl ether of phenol-formaldehyde novolac)	Dow Chemical Co.	
EPI-REZ 5022	Interez Inc.	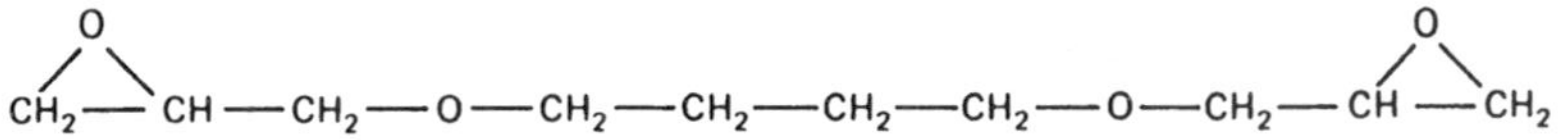
RD-2		
Diglycidyl ether of butanediol)	Ciba-Geigy Corp.	
Tonox 60/40		
40% *m*-phenylenediamine		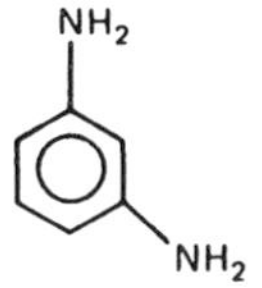
60% 4,4′-methylenedianiline	UniRoyal	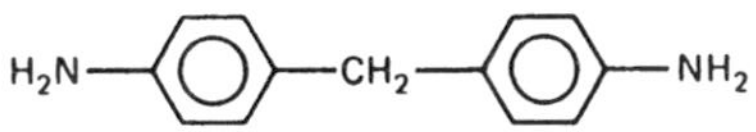

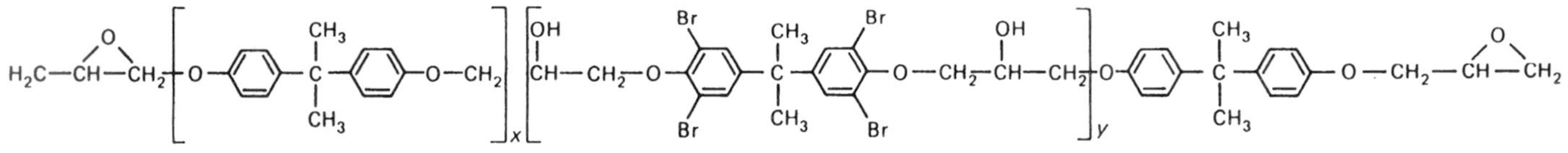

Molecular structure of brominated epoxy resins

Butylene glycol diglycidyl ether (BGDGE)

Diglycidyl ether of bisphenol F (DGEBPF)

3, 4-epoxycyclohexyl methyl-
3′, 4′-epoxycyclohexane carboxylate

Vinyl cyclohexene diepoxide (VCDO)

Triglycidyl ether of triphenyl methane (TGETPM)

Resorcinol diglycidyl ether (RDE)

Tetraglycidyl-4, 4′ (4-aminophenyl)-*p*-diisopropyl benzene

Tetraglycidyl-4, 4′ (4-amino-3, 5-dimethylphenyl)-*p*-diisopropylbenzene

Molecular structure of some commercial epoxy resins

Epoxy resins used in aerospace prepregs

Name	Supplier	Formula
Araldite MY 0510 (Triglycidyl) *p*-aminophenol	Ciba-Geigy Corp.	
Araldite MY 720 (*n*, *n*, *n′*, *n′*-tetraglycidyl-4′,4′-methylenebisbenzenamine)	Ciba-Geigy Corp.	
EPON 826 (Diglycidyl ether of bisphenol A)	Shell Chemical Co.	
DER 330 (Diglycidyl ether of bisphenol A)	Dow Chemical Co.	
Epiclon 830 (Diglycidyl ether of bisphenol F)	Dinippon	
Araldite ECN 1235 (Epoxy novolac)	Ciba-Geigy Corp.	

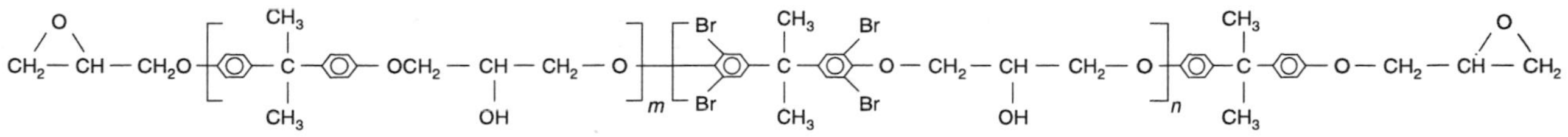

(a)

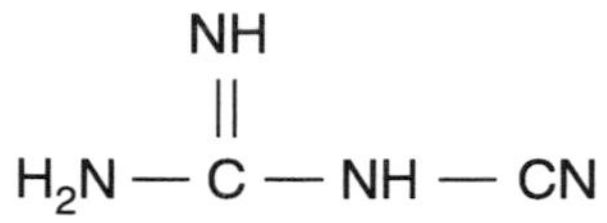

$$H_2N-\underset{}{\overset{NH}{\overset{\|}{C}}}-NH-CN$$

(b)

Chemical structure of (a) a brominated FR-4 type epoxy and (b) curing agent (dicyandiamide)

Thermal and related properties of epoxy resins

Thermal and related properties	Neat resin	Short glass fiber molding compound
Cure process parameters		
Mold pressure, MPa (ksi)	...	2.07 to 2.28 (0.30 to 0.33)
Mold temperature, °C (°F)	...	150 to 165 (300 to 330)(a)
Mold shrinkage, mm/mm	0.001 to 0.004	0.001 to 0.005
T_g, °C (°F)	60 to 175 (140 to 347)	125 (259)
Cured material properties		
Water absorption, 24 h, 3.2 mm (1/8 in.) thick, %	0.080 to 0.150	0.05 to 0.20
Specific gravity(b)	1.11 to 1.40	0.60 to 2.00
Continuous service temperature, °C (°F)	120 to 290 (250 to 550)	150 to 260 (300 to 500)
Heat-deflection temperature, at 1.82 MPa (0.264 ksi), °C (°F)	45 to 290 (115 to 550)	150 to 275 (300 to 525)
Flammability rating	...	94V-0
Specific heat, kJ/kg · K (Btu/lbf · °F)(b)	1.05 (0.25)	0.80 (0.19)
Coefficient of thermal expansion, 10^{-6}/K (10^{-6}/°F)	45 to 65 (25 to 36)	11 to 35 (6 to 19.5)
Thermal conductivity, W/m · K (Btu/ft · h · °F)(b)	0.17 to 0.20 (0.10 to 0.12)	0.17 to 0.40 (0.10 to 0.24)

(a) At 2.1 to 35 MPa (0.3 to 5 ksi). (b) At room temperature

Thermal, physical, and chemical properties of a typical epoxy resin coating

Property	Value
Heat-deflection temperature at 1.82 MPa, °C	140
Maximum resistance to continuous heat, °C	130
Coefficient of linear thermal expansion, 10^{-5}/°C	5
Tensile strength, MPa	50
Elongation, %	5
Hardness, Shore D	85
Specific gravity	112
Water absorption, %	1
Dielectric constant	4
Resistance to chemicals at 25 °C(a):	
Nonoxidizing acids (20% H_2SO_4)	S
Oxidizing acids (10% HNO_3)	U
Aqueous salt solutions (NaCl)	S
Aqueous alkalis (NaOH)	S
Polar solvents (C_2H_5OH)	S
Nonpolar solvents (C_6H_6)	S
Water	S

(a) S, satisfactory; U, unsatisfactory

Thermal, physical, and chemical properties of typical molded epoxy plastics (EP)

Property	Epoxy plastic	Glass-filled EP	Glass-sphere-filled EP
Heat-deflection temperature at 1.82 MPa, °C	140	150	115
Maximum resistance to continuous heat, °C	120	135	110
Coefficient of linear thermal expansion, 10^{-5}/°C	2.5	2.0	2.5
Compressive strength, MPa	120	207	83
Flexural strength, MPa	124	103	41
Impact strength, Izod, cm · N/cm of notch	53.4	53.4	10.6
Tensile strength, MPa	52	83	41
Elongation, %	5	4	1
Hardness, Rockwell	M90	M105	...
Specific gravity	1.2	1.8	0.8
Dielectric constant	4	4	4
Water absorption	0.2	0.1	0.1
Resistance to chemicals at 25 °C(a):			
Nonoxidizing acids(20% H_2SO_4)	S	S	S
Oxidizing acids (10% HNO_3)	U	U	U
Aqueous salt solutions (NaCl)	S	S	S
Aqueous alkalis (NaOH)	S	S	S
Polar solvents (C_2H_5OH)	S	S	S
Nonpolar solvents (C_6H_6)	S	S	S
Water	S	S	S

(a) S, satisfactory; U, unsatisfactory

Selected properties of epoxies

Property	ASTM test method	Cast epoxy	Cast flexible epoxy	Molded epoxy novolac	Molded epoxy cycloaliphatic
Mechanical					
Yield strength, MPa (ksi)	D 638	34 to 82.7 (5.6 to 12.0)	17 to 68.9 (2.5 to 10.0)	48 to 82.7 (7.0 to 12.0)	62 to 82.7 (9.0 to 12.0)
Tensile modulus, GPa (10^6 psi)	D 638	2.4 to 2.8 (0.35 to 0.40)	...	2.8 to 3.4 (0.40 to 0.50)	3.3 to 3.4 (0.48 to 0.50)
Elongation (break), %	D 638	3 to 6	25.0 to 70.0	2 to 6	2 to 8
Compressive strength, MPa (ksi)	D 695	105 to 170 (15 to 25)	14 to 90 (2 to 13)	140 to 170 (20 to 25)	205 to 345 (12 to 13)
Flexural yield strength, MPa (ksi)	D 790	90 to 145 (13 to 21)	14 to 83 (2 to 12)	69 to 83 (10 to 12)	83 to 90 (12 to 13)
Flexural modulus, GPa (10^6 psi)	D 790	1.4 to 3.4 (0.20 to 0.50)	...	...	2.8 to 3.4 (0.40 to 0.50)
Impact strength, Izod(a), J/cm (ft · lbf/in.)	D 256	0.1 to 0.53 (0.2 to 1.0)	1.9 to 2.7 (3.5 to 5.0)	...	...
Hardness, Rockwell	...	M80 to 110	...	...	...
Thermal					
Thermal conductivity, W/m · K (Btu · in./ft^2 · h · °F)	C 177	0.17 to 0.23 (1.2 to 1.6)	...	...	...
Coefficient of thermal expansion, 10^{-5}/°C (10^{-5}/°F)	D 696	8.1 to 11 (4.5 to 6.2)	3.6 to 14 (2.0 to 8.0)	3.1 to 4.0 (1.7 to 2.2)	2.9 to 5.4 (1.6 to 3.0)
Specific heat, kJ/kg · K (Btu/lbf · °F)	C 351	0.8 to 1.3 (0.2 to 0.3)	...	...	...
Continuous service temperature, °C (°F)	...	80 to 90 (175 to 190)	40 to 50 (100 to 125)	230 to 260 (450 to 500)	230 to 260 (450 to 500)
Electrical					
Volume resistivity, Ω · cm	D 257	10^{12} to 10^{17}	10^{14}	2.0 to 5.0 × 10^{14}	>10^{16}
Dielectric strength, V/10^{-3} mm (V/10^{-3} in.)	D 149	12 to 20 (300 to 500)	10 to 16 (250 to 400)	11 to 16 (280 to 400)	16 (400)
Dielectric constant (50-100 Hz)	D 150	3.5 to 5.0	3.0 to 5.0	4.5 to 5.5	3.4
Dissipation factor (50-100 Hz)	D 150	0.002 to 0.010	0.010 to 0.040	0.003	0.006
Processing parameters					
Processing temperature, °C (°F)	...	...	...	120 to 165 (250 to 330)	...
Density, g/cm^3	D 792	1.10 to 1.40	1.05 to 1.30	1.20 to 1.70	1.22
Specific volume, cm^3/kg (in.^{3}/lb)	D 792	760 to 725 (21 to 20)	760 to 725 (21 to 20)	795 to 540 (22 to 15)	795 (22)
Linear mold shrinkage, mm/mm (in./in.)	D 955	0.001 to 0.010	0.001 to 0.010	...	...
Water absorption, % in 24 h	D 570	0.1	...	0.1 to 0.2	0.1 to 0.7

(a) Notched, at room temperature

Properties of commercial grades of BPA epoxy resins

Average molar weight	Average wpe(a)	Approximate value of *n*	Viscosity at 25 °C (80 °F)	Softening point(b)	
				°C	°F
350	182	0	80	...	...
380	188	0.15	140	...	...
600	310	0.9	Semisolid	40	105
900	475	2.0	Solid	70	160
1400	900	3.7	Solid	100	212
2900	1750	9.0	Solid	130	265
3750	3200	11.9	Solid	150	300

(a) Weight per epoxide, that is, grams of resins needed to provide 1 molar equivalent of epoxide. Also referred to as EEW (epoxide equivalent weight) and EMM (epoxy molar mass). All three items are interchangeable. (b) By Durran's mercury method

Physical properties of one-component epoxy adhesives

Property	Epoxy-film tape	Thermosetting hot-melt	Epoxy paste
Appearance	Black film on carrier	Semirigid green solid	Red-brown paste
Viscosity	...	Extrude at 60 to 80 °C (140 to 175 °F)	300 Pa · s
Specific gravity	1.14	1.35	1.44
Elongation at break, %	...	1 to 2	7
Sag, mm (in.)	0 (2 mm, or 0.8 in., layer/200 °C, or 390 °F)	0 (6 mm, or 0.2 in., bead/171 °C, or 340 °F)	<3 (0.12)(6 mm, or 0.2 in., bead/171 °C, or 340 °F)
Cure schedule	24 h/25 °C (77 °F) or 30 min./100 °C (212 °F)	4 h/25 °C (77 °F)	5 days/25 °C (77 °F) or 2 h/88 °C (190 °F)
Shelf life	3 mo/25 °C (77 °F)	3 mo/25 °C (77 °F)	3 mo/25 °C (77 °F)

Mechanical properties of two-component epoxy pastes

Property	General purpose	Fast setting	High performance
Aluminum lap-shear strength, MPa (ksi)(a)			
At -60 °C (-75 °F)	20 (2.9)	10 (1.5)	29 (4.2)
At 25 °C (77 °F)	18 (2.6)	20 (2.9)	31 (4.5)
At 82 °C (180 °F)	<2 (0.3)	<8 (1.2)	18 (2.6)
At 121 °C (250 °F)	...	...	6.9 (1.0)
T-peel strength, N/mm (lbf/in.)	...	...	2 (11)
Thermal aging, MPa (ksi)	15 (2.2) (30 days/60 °C, or 140 °F)	21 (3.0) (30 days/60 °C, or 140 °F)	26 (3.8) (14 days/121 °C, or 250 °F)
Humidity aging, MPa (ksi)	12 (1.7) (40 °C, or 105 °F/92% RH)	7 (1.0) (54 °C, or 130 °F/95% RH)	20 (2.9) (54 °C, or 130 °F/95% RH)
Chemical resistance, MPa (ksi)			
Gasoline (90 days)	17 (2.4)	...	...
JP-4 (7 days)	...	...	34 (4.9)
Other substrate lap-shear strength, MPa (ksi), at 25 °C (77 °F)	23 (3.3) (copper)	...	10 (1.5) (polyether-imide)

RH, relative humidity. (a) Aluminum lap-shear specimens tested according to ASTM D 1002, using chromic acid etched 2024-T3 aluminum

Mechanical properties of one-component epoxy adhesives

Property	Epoxy film tape	Thermosetting hot melt	Epoxy paste
CRS lap-shear strength, MPa (ksi)(a)			
At -30 °C (-22 °F)	21 (3.0)	23 (3.3)	23 (3.3)
At 25 °C (77 °F)	16 (2.3)	20 (2.9)	22 (3.2)
At 82 °C (180 °F)	15 (2.2)	17 (2.5)	19 (2.8)
Salt spray endurance (500 h), MPa (ksi)	15 (2.2	17 (2.5)	18 (2.6)

(a) CRS, cold-rolled steel (oily) lap-shear specimens tested according to ASTM D 1002

Typical properties and performance of epoxy molding compound cured with novolac resin or dianhydride, and silicone molding compound

	Type of encapsulant		
	Epoxy cured with		
Property/performance	Novolac resin	Dianhydride	Silicon
Glass transition temperature, °C (°F)	140 to 170 (285 to 340)	160 to 220 (320 to 430)	90 to 120 (195 to 248)
Electric/dielectric at high temperature	Acceptable	Good	Excellent
Adhesion/lead seal	Acceptable	Excellent	Poor
Reliability under moisture stress	Excellent	Test dependent	Test dependent
Sensitivity of moldability to humidity	Moderate	Profound	Moderate

Mechanical properties of a rubber-toughened epoxy (CTBN-toughened DGEBA)

CTBN concentration, %	Fracture energy, kJ/m^2	Tensile strength, MPa	Modulus of elasticity, GPa
0	0.12	72	3.3
4.5	1.05	70	2.3
10	2.72	56	2.2
15	3.43	45	2.0
20	3.59	20	1.0
30	2.00	17	0.1

Epoxy resin curing agents for electrical applications

		General curing conditions, 25 °C (80 °F)									
		Typical cure				Postcure(b)				Heat deflection temperature(c)	
				Temperature			Temperature				
Curing agent type and typical products	Concentration, phr(a)	Time, h	Days	°C	°F	Time, h	°C	°F	Pot life at 25 °C (80 °F), h	°C	°F
Aliphatic amines and derivatives, room-temperature cure:											
Diethylenetriamine (DTA)	12	...	7	25	80	1	200	390	¼ to ½	124	255
Triethylenetetramine (TETA)	14	...	7	25	80	1	200	390	¼ to ½	123	255
Polyamides (Versamides)(d)	30 to 50	...	7	25	80	None	...	...	2 to 3	55	130
Amine adducts(e)	26	...	4	25	80	1	150	300	¼ to ½	120	250
Aliphatic/cycloaliphatic amines, moderate-temperature cure:											
Diethylaminopropylamine (DEAPA)	8	0.5	...	115	240	None	...	...	3 to 4	100	212
Bis (*p*-aminocyclohexyl) methane (PACM-20)	29	1.0	...	150	300	3	150	300	1.5	149	300
Isophorone diamine (IPD)	23	1.0	...	100	212	3	150	300	1.0	146	295
N-aminoethylpiperazine (AEP)	20	1.0	...	150	300	3	150	300	¼ to ½	110	230
2-ethyl-4-methyl imidazole (EMI-24)	10	8.0	...	60	140	None	...	...	4 to 6	110	230
Same	4	4.0	...	60	140	2	150	300	20+	160	320
Aromatic amines, elevated-temperature cure:											
Metaphenylenediamine (MPDA)	14	2.0 + 2.0	...	80(175)/150	300	2	150	300	5 to 6	150	300
4, 4′-methylenedianiline (MDA)	28	2.0 + 2.0	...	80 (175)/150	300	2	150	300	5 to 6	160	320
4, 4′-diaminodiphenylsulfone (DDS) + 1 phr BF_3–MEA	30(f)	2.0 + 2.0	...	125(257)/200	390	2	200	390	(g)	175	350
Aromatic amine eutectics(h)	20	2.0 + 2.0	...	80(175)/150	300	2	150	300	6 to 8	145	290
Carboxylic acid anhydrides, elevated-temperature cure:											
Hexahydrophthalic anhydride (HHPA) + phr BDMA(i)	78	3.0 + 1.0	...	90(195)/150	300	3	200	390	24	132	270
Nadic methyl anhydride (NMA) - 1 phr BDMA	90	3.0 + 1.0	...	120(250/150	300	3	200	390	60 to 80	144	290
Chlorendic anhydride (CA)(j)	117	Gel + 4.0	...	25(80)/150	300	3	200	390	(j)	197	385
Dodecenylsuccinic anhydride (DDSA) + 1 phr BDMA	134	4.0 + 1.0	...	90(195)/150	300	3	200	390	120	74	165
Methyltetrahydrophthalic anhydride (MTHPA) + 1 phr BDMA	80	1.0 + 1.0	...	100(212)/150	300	4	150	300	24	130	265
Catalytic Lewis acids, elevated-temperature cure:											
Boron trifluoride monoethylamine (BF_3-MEA)	3	3.0 + 4.0	...	120(250/200	390	4	200	390	>250	174	345
Latent curing agents, elevated-temperature cure:											
Dicyanamide (DICY)	6	1.0	...	175	350	1.0	175	350	∞	135	275(k)

(a) Parts per 100 parts resin by weight for the DGEBA-WPE = 189. (b) Used to obtain heat deflection temperature. Postcure should be made at or above HDT. (c) By ASTM D 648. (d) Based on polyamide (Versamide) resin; other grades are available. (e) Based on curing agent U, Shell Chemical Company. Numerous proprietary types are available. (f) Less than stoichiometric (33 phr) because BF_3-MEA also assists curing. (g) Mixture is too viscous for this measurement. Prepreg useful life is 10 to 20 days. (h) Data on curing agent Z, Shell Chemical Company. Many others available. (i) BDMA, benzyldimethylamine. (j) Chlorendic anhydride seldom used alone or at stoichiometry (203 phr) due to fast gel time. (k) T_g as measured by thermal mechanical analysis on a laminate

Mechanical and electrical properties of selected epoxy resin castings

Curing agent type	Anhydride	Aliphatic amine	Aromatic amine	Catalytic	High-temperature anhydride	Epoxy novolac anhydride	Dianhydride
Typical curing agent	HHPA	DETA	MPDA	BF_3-MEA	NMA	NMA	PMDA
Parts, phr	78	12	14	3	90	87.5	55
Resin	DGEBA	DGEBA	DGEBA	DGEBA	DGEBA	Novolac	DGEBA
Catalyst, type/phr	BDMA/1	...	...	...	BDMA/1	BDMA/1.5	...
Cure cycle, h/°C (°F)	4/150 (300)	24/25 (77) + 2/200 (390)	2/80 (175) + 2/150 (300)	2/120 (250) + 2/200 (390)	4/150 (300) + 3/200 (390)	2/90 (195) + 4/165 (330) + 16/200 (390)	4/150 (300) + 14/200 (390)
Heat deflection temperature, °C (°F)	130 (265)	125 (255)	150 (300)	174 (345)	170 (340)	195 (385)	280 (535)
Tensile strength at 25 °C (77 °F), MPa (ksi)	72 (10.5)	75 (10.9)	85 (12.4)	43 (6.2)	75 (10.5)	66 (9.6)	22 (3.2)
Tensile modulus at 25 °C (77 °F), GPa (10^6 psi)	2.80 (0.400)	2.87 (0.410)	3.30 (0.480)	2.70 (0.390)	3.40 (0.500)	2.94 (0.428)	2.70 (0.390)
Elongation, %	5.6	6.3	5.1	3.0	2.7	3.2	...
Tensile strength at 100 °C (212 °F), MPa (ksi)	37 (5.3)	32 (4.6)	45 (6.6)	29 (9.2)	46 (6.7)	...	14 (2.0)(a)
Tensile modulus at 100 °C (212 °F), GPa (10^6 psi)	2.10 (0.300)	1.80 (0.260)	2.20 (0.320)	1.90 (0.271)	1.40 (0.200)	...	...
Elongation, %	11.1	9.0	7.2	9.1	7.2	...	...
Flexural strength at 25 °C (77 °F), MPa (ksi)	126 (18.3)	103 (15.0)	131 (19.0)	112 (16.3)	112 (17.0)	147 (21.4)	59 (8.6)
Flexural modulus at 25 °C (77 °F), GPa (10^6 psi)	3.22 (0.470)	2.48 (0.360)	2.80 (0.400)	...	4.80 (0.700)	3.69 (0.537)	3.61 (0.380)
Compressive strength at 25 °C (77 °F), MPa (ksi)	111 (16.2)	224 (32.6)	234 (34.2)	...	116 (16.9)	159 (23.1)	254 (37.0)
Compressive modulus at 25 °C (77 °F), GPa (10^6 psi)	5.09 (0.740)	1.86 (0.271)	2.13 (0.310)	...	0.73 (0.106)	2.22 (0.303)	2.41 (0.350)
Volume resistivity, Ω · cm							
At 25 °C (77 °F)	4×10^{14}	2×10^{16}	10^{16}	...	2×10^{14}	10^{16}	...
At 100 °C (212 °F)	2×10^{14}	5×10^{12}	...	...	3×10^{12}	...	...
At 150 °C (300 °F)	10^{10}	...	...	...	3×10^{11}	...	...
At 200 °C (390 °F)	5×10^{7}	...	...	...	10^{9}	...	7×10^{11}

(continued)

Mechanical and electrical properties of selected epoxy resin castings (continued)

Curing agent type	Anhydride	Aliphatic amine	Aromatic amine	Catalytic	High-temperature anhydride	Epoxy novolac anhydride	Dianhydride
Dielectric constant, 60 Hz							
At 25 °C (77 °F)	3.3	3.49	4.60	3.53	3.36	3.43	3.57(b)
At 100 °C (212 °F)	3.3	4.55	4.65	3.51	3.39	...	3.70(c)
At 150 °C (300 °F)	...	5.8	5.40	3.69	3.72	...	3.732
Dielectric constant, 1 kHz							
At 25 °C (77 °F)	3.3	3.26	4.50	3.56	3.26	3.45	3.52(b)
At 100 °C (212 °F)	3.4	4.65	4.60	3.37	3.01	...	3.65(c)
At 150 °C (300 °F)	4.0	4.83	4.90	3.59	3.29	...	3.70
Dielectric constant, 1 MHz							
At 25 °C (77 °F)	3.2	3.33	3.85	3.20	2.99	3.20	3.34(b)
At 100 °C (212 °F)	3.4	4.36	4.30	3.25	3.90	...	3.52(c)
At 150 °C (300 °F)	3.6	4.23	4.60	3.30	3.29	...	3.61
Dissipation factor, 60 Hz							
At 25 °C (77 °F)	0.005	0.005	0.008	0.008	0.008	0.0066	0.007(b)
At 100 °C (212 °F)	0.003	0.002	0.008	0.008	0.009	...	0.005(c)
At 150 °C (300 °F)	0.003	0.003	...	...	0.009	...	0.008
Dissipation factor, 1 kHz							
At 25 °C (77 °F)	0.007	0.006	0.017	0.009	0.006	0.0058	0.008(b)
At 100 °C (212 °F)	0.004	0.056	0.006	0.011	0.003	...	0.004(c)
At 150 °C (300 °F)	0.07	0.048	0.035	0.046	0.039	...	0.004
Dissipation factor, 1 MHz							
At 25 °C (77 °F)	0.013	0.034	0.038	0.024	0.021	0.016	0.022(b)
At 100 °C (212 °F)	0.015	0.048	0.020	0.015	0.013	...	0.019(c)
At 150 °C (300 °F)	0.02	0.033	0.015	0.012	0.013	...	0.013

HHPA, hexahydrophthalic anhydride; DETA, diethylenetriamine; MPDA, metaphenylenediamine; BF_3-MEA, boron trifluoride monoethylamine complex; NMA, nadic methyl anhydride; PMDA, pyromellitic dianhydride; BDMA, benzyldimethylamine. (a) Measured at 150 °C (300 °F). (b) Measured at 23 °C (73 °F). (c) Measured at 93 °C (200 °F)

Phenolics

Phenolic properties and test

Ingredients	Property	Test method
Phenolic resin	Phenol	Titration
	Molecular weight	GPC
	Characterization	HPLC/infrared spectroscopy
	Solids	Evaporation
Modifier (organic)	Viscosity	Rheometer

Typical properties of phenolic resin

Property	Value
Specific gravity	1.08 to 1.09
Solids content, %	60 to 62
Viscosity at 25 °C (77 °F), Pa · s (cP)	0.12 to 0.20 (120 to 200)
Refractive index	1.518 to 1.525
Cure time at 165 °C, (329 °F), s	85 to 105
Free formaldehyde, %	11.5 to 13.5
Free phenol, %	0 to 0.5
Trace elements and sodium	<5 ppm each
Potassium, lithium, iron	< 10 ppm total

Phenolic matrix properties and test methods

Material	Property	Test method(s)
Uncured resin	Composition	HPLC; infrared spectroscopy; GPC
	Processibility	Solids; gel; volatile content
	Chemical activity	DSC
Cured neat resin	Completeness of cure	Glass transition; solvent extraction
Uncured impregnated system	Characterization	HPLC; infrared spectroscopy; DSC
	Processibility	Resin content; volatile content; flow; gel; tack/drape; fiber weight
Cured impregnated system	Completeness of cure	DSC
	Thermal properties	TGA; flammability
	Electrical properties	Dielectric

Classification of phenolic molding compounds

Type	Description and applications
General purpose	Wood flour filled (specific gravity of 1.36 to 1.43 and impact strength up to 18 J/m, or 0.34 ft · lbf/in.); used in wiring and small-appliance parts
Impact	Cellulose, mineral, and glass fiber filled to give impact strengths up to 910 J/m (17 ft · lbf/in.)
Nonbleeding	Compounds specifically for the closure industry are wood flour filled, single-stage resins. These compounds are formulated with carbon black rather than dye to eliminate bleeding upon contact with alcohol
Electrical	Mineral-filled with very low water absorption and therefore improved insulation resistance
Heat-resistant	Generally mineral-filled with a wide range of densities and impact strengths. They will stand short-term exposure of more than 260 °C (500 °F)
Special-purpose	Compounds developed for chemical resistance, special color, combinations of properties

Selected properties of molded phenolics

Property	ASTM test method	Reported value
Mechanical		
Yield strength, MPa (ksi)	D 638	48 to 55 (7.0 to 8.0)
Tensile modulus, GPa (10^6 psi)	D 638	5.2 to 6.9 (0.75 to 1.0)
Elongation (break), %	D 638	1 to 2
Compressive strength, MPa (ksi)	D 695	83 to 195 (12 to 28)
Flexural yield strength, MPa (ksi)	D 790	83 to 105 (12 to 15)
Impact strength, Izod(a), J/cm (ft · lbf/in.)	D 256	0.1 to 0.2 (0.2 to 0.4)
Hardness, Rockwell	...	M124 to 128
Thermal		
Thermal conductivity, W/m · K (Btu · in./ft^2 · h · °F)	C 177	0.13 to 0.22 (0.9 to 1.5)
Coefficient of linear thermal expansion, 10^{-5}/°C (10^{-5}/°F)	D 696	4.5 to 11 (2.5 to 6.0)
Specific heat, kJ/kg · K (Btu/lbf · °F)	C 351	1.7 (0.4)
Continuous service temperature, °C (°F)	...	150 to 175 (300 to 350)
Electrical		
Volume resistivity, Ω · cm	D 257	10^{11} to 10^{12}
Dielectric strength, V/10^{-3} mm (V/10^{-3} in.)	D 149	12 to 16 (300 to 400)
Dielectric constant (50-100 Hz)	D 150	5.0 to 5.6
Dissipation factor (50-100 Hz)	D 150	0.06 to 0.10
Processing parameters		
Processing temperature, °C (°F)	...	130 to 160 (270 to 320)
Density, g/cm^3	D 792	1.25 to 1.30
Specific volume, cm^3/kg (in.^{3}/lb)	D 792	795 to 760 (22 to 21)
Linear mold shrinkage, mm/mm (in./in.)	D 955	0.010 to 0.012
Water absorption, % in 24 h	D 570	0.1 to 0.2

(a) Notched, at room temperature

Thermal, physical, and chemical properties of typical phenolic plastics

Property	Wood flour-filled	Mineral-filled	Glass-reinforced
Heat-deflection temperature at 1.82 MPa, °C	165	200	250
Maximum resistance to continuous heat, °C	160	175	175
Coefficient of linear thermal expansion, 10^{-5}/°C	3.0	2.0	1.5
Compressive strength, MPa	172	172	120
Impact strength, Izod, cm · N/cm of notch	21.5	21.5	75
Tensile strength, MPa	48	41	60
Elongation, %	0.5	0.5	0.2
Hardness, Rockwell	M100	M110	E70
Specific gravity	1.4	1.5	1.85
Water absorption, %	0.4	0.03	0.5
Dielectric constant	6	8	5
Resistance to chemicals at 25 °C(a):			
Nonoxidizing acids(20% H_2SO_4)	S	S	S
Oxidizing acids (10% HNO_3)	Q	Q	Q
Aqueous salt solutions (NaCl)	S	S	S
Aqueous alkalis (NaOH)	Q	Q	Q
Polar solvents (C_2H_5OH)	S	S	S
Nonpolar solvents (C_6H_6)	S	S	S
Water	S	S	S

(a) S, satisfactory; Q, questionable

Thermal and related properties of phenolic resins

Thermal and related properties	Neat resin	Chopped glass fiber molding compound
Cure process parameters		
Mold pressure, MPa (ksi)	17 to 26 (0.25 to 4.0)	1.9 to 27.6 (0.28 to 4.0)
Mold temperature, °C (°F)	130 to 160 (270 to 320)(a)	140 to 175 (280 to 350)(a)
Mold shrinkage, mm/mm	0.010 to 0.012	0.0001 to 0.0040
Tg, °C (°F)	300 (572)	...
Cured material properties		
Water absorption, 24 h, 3.2 mm (1/8 in.) thick, %	0.010 to 0.20	0.03 to 1.20
Specific gravity(b)	1.23 to 1.32	1.65 to 1.95
Continuous service temperature, °C (°F)	250 to 350	350 to 550
Heat-deflection temperature, at 1.82 MPa (0.264 ksi), °C (°F)	120 to 175 (250 to 350)	150 to 315 (300 to 600)
Flammability rating	94V-1	94V-0
Coefficient of thermal expansion, 10^{-6}/K (10^{-6}/ °F)	25 to 60 (13.8 to 33.3)	8 to 20 (4.4 to 11.4)
Specific heat, kJ/kg · K (Btu/lbf · °F)(b)	1.4 to 1.8 (0.34 to 0.42)	0.85 to 1.25 (0.2 to 0.3)
Thermal conductivity, W/m · K (Btu/ft · h · °F)(b)	0.12 to 0.24 (0.072 to 0.144)	0.32 to 0.60 (0.19 to 0.35)

(a) At 1.7 to 28 MPa (0.25 to 4 ksi). (b) At room temperature

Typical properties of various structural phenolic adhesives

Property	Neoprene phenolic	Nitrile phenolic	Vinyl phenolic	Epoxy phenolic
Forms	Tapes	Tapes, pastes	Emulsions, tapes, separate components(a)	Tapes, pastes
Typical cure conditions	40 to 60 min. at 150 to 230 °C (300 to 450 °F) with 0.17 to 3.5 MPa (25 to 500 psi) contact pressure	From 1 to 2 h at 150 °C (300 °F) to 8 min. at 230 °C (425 °F) with 0.07 to 1.4 MPa (10 to 200 psi) contact pressure	Emulsions: 2 to 6 min. at 60 to 80 °C (140 to 180 °F); contact pressure to 18 kg/linear cm (100 lb/linear in.) Nonemulsions: 15 to 30 min. at 150 °C (300 °F); contact pressure to 0.70 MPa (100 psi)	40 to 60 min. at 160 to 175 °C (325 to 350 °F) with 0.70 MPa (100 psi) contact pressure
Tensile lap-shear strength, MPa (ksi)(b)				
At -55 °C (-70 °F)	34.5 to 41.4 (5 to 6)	22.1 to 32.7 (3.2 to 4.6)	15.2 to 20.5 (2.2 to 3.0)	17.2 to 20.5 (2.5 to 3.0)
At 25 °C (75 °F)	17.2 to 24.1 (2.5 to 3.5)	22.1 to 32.7 (3.2 to 4.6)	15.2 to 20.5 (2.2 to 3.0)	17.2 to 20.5 (2.5 to 3.0)
At 80 °C (180 °F)	...	...	6.7 to 10.3 (1.0 to 1.5)	...
At 150 °C (300 °F)	...	10.3 to 17.2 (1.5 to 2.5	...	10.3 to 12.4 (1.5 to 1.8)
Heat resistance, °C (°F)	80 to 95 (180 to 200)	260 (500)	80 (180)	>260 (500)
Applications	Metal-to-metal bonding; bonding of rubbers, ceramics, glasses, thermosets	Bonding of metals, plastics, rubbers, and frictional materials; bonding of honeycomb	Bonding of honeycomb to paper, metal, or fibrous glass; bonding foams or porous materials to metals	Metal-to-metal bonding; bonding of honeycomb

(a) Liquid phenolic plus powder polyvinyl formal. (b) Aluminum to aluminum

Thermoset Polyesters

Selected properties of cast thermoset polyesters

Property	ASTM test method	Rigid TS polyester	Flexible TS polyester
Mechanical			
Yield strength, MPa (ksi)	D 638	41 to 89.6 (6.0 to 13.0)	3.4 to 21 (0.5 to 3.0)
Tensile modulus, GPa (10^6 psi)	D 638	2.1 to 4.1 (0.30 to 0.60)	...
Elongation (break), %	D 638	5	40 to 300
Compressive strength, MPa (ksi)	D 695	90 to 205 (13 to 30)	14 to 105 (2 to 15)
Flexural yield strength, MPa (ksi)	D 790	105 to 140 (15 to 20)	34 to 105 (5 to 15)
Flexural modulus, GPa (10^6 psi)	D 790	1.4 to 5.5 (0.20 to 0.80)	0.69 (0.10)
Impact strength, Izod(a), J/cm (ft · lbf/in.)	D 256	0.1 to 0.2 (0.2 to 0.4)	3.7 (7.0)
Hardness	...	Rockwell M70 to 115	Shore D85 to 95
Thermal			
Thermal conductivity, W/m · K (Btu · in./ft^2 · h · °F)	C 177	0.17 (1.2)	...
Coefficient of linear thermal expansion, 10^{-5}/°C (10^{-5}/°F)	D 696	9.0 to 18.0 (5.0 to 10.0)	...
Specific heat, kJ/kg · K (Btu/lbf · °F)	C 351	1.3 to 2.1 (0.3 to 0.5)	...
Continuous service temperature, °C (°F)	...	120 to 150 (250 to 300)	65 to 120 (150 to 250)
Electrical			
Volume resistivity, Ω · cm	D 257	10^{13}	10^{12}
Dielectric strength, V/10^{-3} mm (V/10^{-3} in.)	D 149	16 to 20 (400 to 500)	9.8 to 16 (250 to 400)
Dielectric constant (50 to 100 Hz)	D 150	3.0 to 4.4	4.5 to 8.0
Dissipation factor (50 to 100 Hz)	D 150	0.003 to 0.030	0.030 to 0.300
Processing parameters			
Density, g/cm^3	D 792	1.10 to 1.50	1.00 to 1.20
Specific volume, cm^3/kg (in.^{3}/lb)	D 792	905 to 685 (25 to 19)	975 to 830 (27 to 23)
Water absorption, % in 24 h	D 570	0.1 to 0.6	0.5 to 2.5

(a) Notched, at room temperature

Polyester matrix properties and test methods

Material	Property	Test method
Uncured resin	Composition	Infrared spectroscopy
		HPLC
	Processibility	RDS viscosity
		Gel
		Volatile content
Cured neat resin	Completeness of cure	DSC
Uncured impregnated system	Characterization	HPLC
	Processibility	Resin content
		Flow
		Gel
		Tack/drape
		Fiber weight
Cured impregnated system	Completeness of cure	DSC
	Laminate properties	Ply thickness
		Fiber volume
		Hardness

Thermal and related properties of polyester resins

Thermal and related properties	Neat resin	Resin and 10 to 40 wt% chopped glass fiber
Cure process parameters		
Mold pressure, MPa (ksi)	...	3.4 to 13.8 (0.5 to 2.0)
Mold temperature, °C (°F)	...	140 to 175 (280 to 350)(a)
Mold shrinkage, mm/mm	...	0.001 to 0.012
T_g, °C (°F)	110 (230)	...
Cured material properties		
Water absorption, 24 h, 3.2 mm (⅛ in.) thick, %	...	0.06 to 0.28
Specific gravity(b)	1.10 to 1.46	1.6 to 2.1
Continuous service temperature, °C (°F)	120 to 150 (250 to 300)	150 to 175 (300 to 350)
Heat-deflection temperature, at 1.82 MPa (0.264 ksi), °C (°F)	50 to 205 (120 to 400)	190 to 205 (375 to 400)
Flammability rating	...	94V-0
Specific heat, kJ/kg · K (Btu/lbf · °F)(b)	1.3 to 2.3 (0.30 to 0.55)	1.0 (0.25)
Coefficient of thermal expansion, 10^{-6}/K (10^{-6}/°F)	55 to 100 (31 to 55)	20 to 33 (11 to 18)
Thermal conductivity, W/m · K (Btu/ft · h · °F)(b)	0.17 to 0.22 (0.10 to 0.13)	0.4 to 0.6 (0.24 to 0.38)

(a) At 3.5 to 14 MPa (0.5 to 2 ksi). (b) At room temperature

Types of polyester resin

Type	Anhydride	Glycol
General purpose	Orthophthalic, maleic	Ethylene, diethylene, or propylene
Corrosion resistant	Isophthalic, maleic, bisphenol A, chlorendic	Propylene
Flame resistant	Brominated tetra hydropththalic, tetrabromopththalic, chlorendic anhydride	Dibromoneopentyl glycol

Components of polyester resins

Anhydride(a)	Diluent	Use
Orthophthalic	Styrene	General low cost
Isophthalic	Styrene	Better mechanical
Isophthalic	Vinyl toluene	Better mechanical, less volatile
Orthophthalic	Diallylphthalate	Improve electrical, less volatile
Isophthalic	Methyl methacrylate	Outdoor exposure
Tetrabromophthalic	Styrene	Fire retardant
Isophthalic % bisphenol A	Styrene	Corrosion resistance

(a) In addition to maleic or fumaric anhydride

Polyester properties and tests

Ingredient	Property	Test method
Polyester	Reactivity	Titration of peroxide
	Molecular weight	GPC
	Purity	HPLC
		H_2O determinations

Typical polyester formulations

% MEKP (0.5% cobalt naphthenate)	Gel time at 30 °C (86 °F), min.
2.0	9.0
1.0	18.5
0.75	25.0
0.5	38.5

% BPO (0.2% dimethyl aniline)	Gel time at 30 °C (86 °F), min.
2.0	4.5
1.0	7.5
0.5	12.0
0.25	21.0

Thermal, physical, and chemical properties of glass-reinforced unsaturated polyesters

Property	Value
Heat-deflection temperature at 1.82 MPa, °C	200
Maximum resistance to continuous heat, °C	160
Coefficient of linear thermal expansion, 10^{-5}/°C	2.5
Compressive strength, MPa	172
Flexural strength, MPa	83
Impact strength, Izod, cm · N/cm of notch	160
Tensile strength, MPa	69
Elongation, %	1.5
Hardness, Rockwell	M50
Specific gravity	2
Dielectric constant	5
Resistance to chemicals at 25 °C(a):	
Nonoxidizing acids (20% H_2SO_4)	S
Oxidizing acids (10% HNO_3)	Q
Aqueous salt solutions (NaCl)	S
Aqueous alkalis (NaOH)	Q
Polar solvents (C_2H_5OH)	S
Nonpolar solvents (C_6H_6)	S
Water	S

(a) S, satisfactory; Q, questionable

Catalyst-promotor-inhibitor systems for room-temperature-cure polyester resins

Application or end use	System, %	Gel time starting at room temperature, min	Approximate time at 21 to 24 °C (70 to 75 °F) for development of 35 Barcol hardness
Gel coats	MEKP-1.5(a) Cobalt naphthenate-0.4(b) (assessory promoters usually omitted because of tendency to discolor)	30 (high filler content)	6 to 8 (can proceed with lay-up over gel coat in 30 to 45 min.)
For normal lay-up resins	MEKP-1.0 Cobalt naphthenate-0.4	32	6 to 8
For fast-cure resins	MEKP-1.0 Cobalt naphthenate-0.4 Dimethyl aniline-0.1	16	2 to 2.5
For fast-cure resins	MEKP-1.0 Cobalt naphthenate-0.4 Quaternary ammonium salt-0.1	15	2 to 2.5
Alternate room-temperature cure	Cyclohexanone peroxide(c)-1.0 Cobalt naphthenate-0.4	30	~6 to 8
Alternate room-temperature cure	Bis-I-hydroxy cyclohexyl peroxide(c)-1.0	30	~6 to 8
Alternate room-temperature cure	Benzoyl peroxide-1.0 Dimethyl aniline-0.1	20	2
Effect of inhibitor	MEKP-1.0 Cobalt naphthanate-0.4 Hydroquinone-0.1	∞	∞

(a) Percentages based on 100 parts polyester resin. (b) Concentration of cobalt metal, 6%. (c) Peroxides costlier than MEKP

Bismaleimides

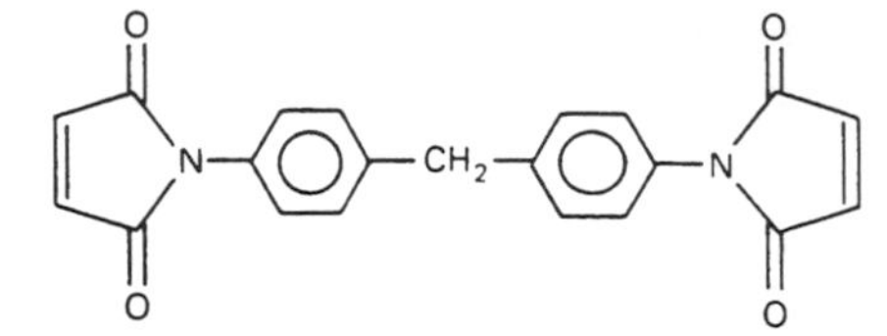

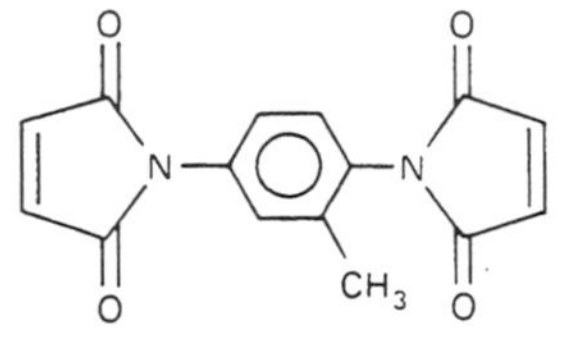

Kerimid 601

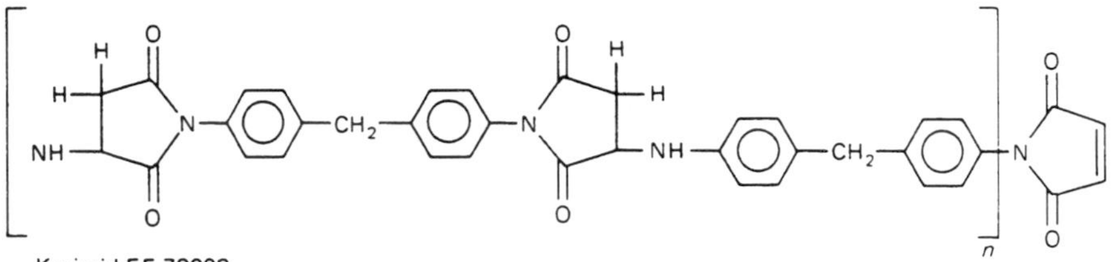

Kerimid FE 70003

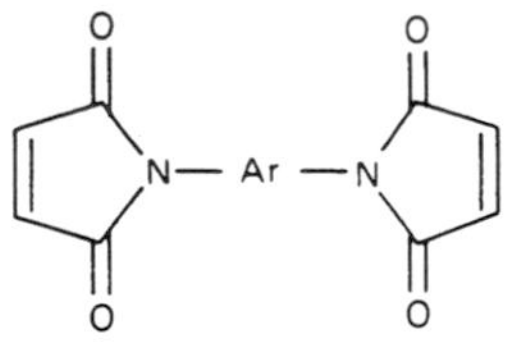

Completely aromatic

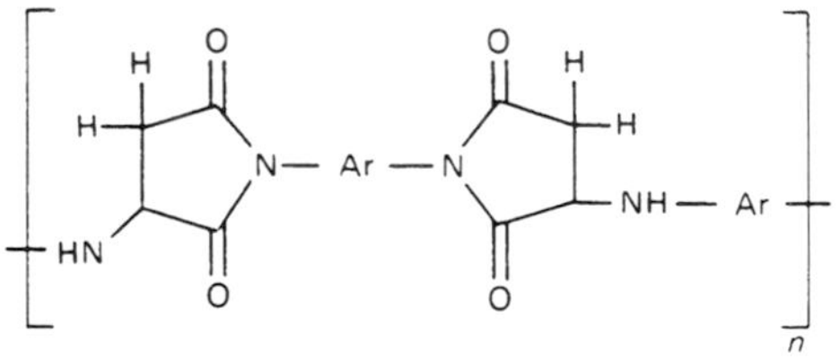

Rhone-Poulenc
Kerimid 353

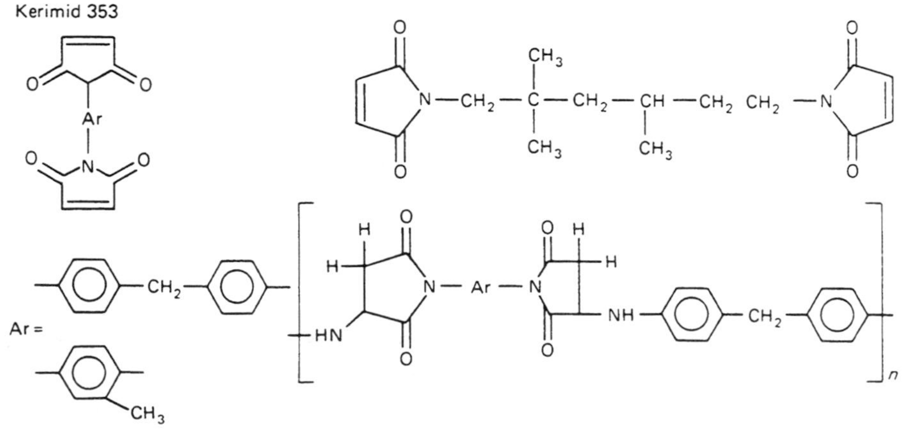

Ar =

Commercial bismaleimides

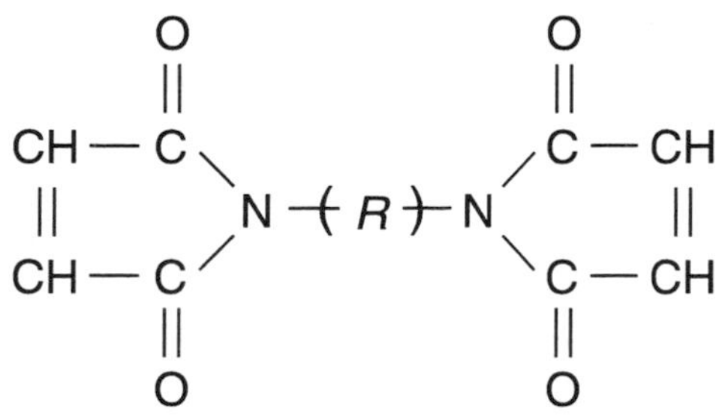

Bismaleimide (polyimide precursor)

Selected properties of 5245C modified bismaleimide resin

Property	Reported value	Property	Reported value
Density, g/cm^3	1.25	Flexural modulus (dry), GPa (10^6 psi):	
T_g, °C (°F)	220 (428)	At RT	3.3 (0.49)
Tensile strength (dry) at RT, MPa (ksi)	84 (12.2)	At 93 °C (200 °F)	3.1 (0.46)
Tensile modulus (dry) at RT, GPa (10^6 psi)	33 (0.48)	At 130 °C (270 °F)	2.7 (0.40)
Tensile strain-to-failure (dry) at RT, %	2.9	Flexural modulus (wet)(a), GPa (10^6 psi):	
Flexural strength (dry), at MPa (ksi):		At 93 °C (200 °F)	2.9 (0.43)
At RT	145 (21.0)	At 130 °C (270 °F)	2.7 (0.39)
At 93 °C (200 °F)	115 (16.7)	Izod impact strength, unnotched, J/m (ft · lbf/in.)	410 (7.7)
At 132 °C (270 °F)	107 (15.5)	Fracture toughness, J/m^2 (in. · lbf/in.^{2})	67 (0.38)
Flexural strength (wet)(a), MPa (ksi):		Moisture absorption (72-h water boil), %	1.7
At 93 °C (200 °F)	96 (14.0)	Coefficient of thermal expansion, 10^{-6}/K	72
At 130 °C (270 °F)	83 (12.1)		

(a) Wet condition, 40-h water boil, specimen held 5 min. at test temperature before loading

Thermal and related properties of bismaleimide resins

Thermal and related properties	Neat resin	68.3 vol% T300 carbon fiber and resin	57.7 vol% E-glass fiber and resin
Cure process parameters			
Mold pressure, MPa (ksi)	0.59 to 0.69 (0.085 to 0.10)	...	...
Mold temperature, °C (°F)	175 to 190 (350 to 375)(a)	210 (410)(b)	210 (410)(b)
	230 to 245 (450 to 475)	210 (450)(b)(c)	210 (450)(b)(c)
T_g, °C (°F)	230 to 345 (450 to 650)(d)	265 (510)	255 (490)
Cured material properties			
Specific gravity	1.23 to 1.29	1.60	2.00
Continuous service temperature, °C (°F)			
Short term	315 (600)	315 (600)	315 (600)
Long term	230 (450)	230 (450)	230 (450)
Flammability rating	Low	...	...
Coefficient of thermal expansion, 10^{-6}/K (10^{-6}/°F)	30 to 50 (17 to 27)	...	...

(a) At 490 to 700 kPa (85 to 100 psi). (b) For 5 h. (c) Postcure. (d) Dry

Selected properties of compimide bismaleimide resins

Property	C795	C800	C183	C796	C353/TM-122(a)	C796/TM-122(b)
Density, g/cm^3	1.32	...	1.31	...	...	...
T_g (dry), °C (°F)	290 (554)	290 (554)	250 (482)	...	...	...
Morphology	Amorphous	Amorphous	Amorphous	Amorphous	Amorphous	Amorphous
Flexural strength (dry), MPa (ksi):						
At RT	110 (15.9)	92 (13.2)	106 (15.4)	76 (11.0)	110 (15.9)	114 (16.5)
At RT after isothermal aging for 500 h at 200 °C (392 °F) in circulating air	100 (14.5)	...	...	...	...	...
At RT after isothermal aging for 500 h at 250 °C (478 °F) in circulating air	108 (15.6)	...	...	...	...	...
At RT after isothermal aging for 500 h at 250 °C (478 °F) in circulating air and then exposed to 200 °C (392 °F)	65 (9.4)	...	58 (8.4)	...	...	...
At 200 °C (392 °F) after isothermal aging for 500 h at 200 °C (392 °F) in circulating air	71 (10.2)	...	...	...	...	...

(continued)

Selected properties of compimide bismaleimide resins (continued)

Property	C795	C800	C183	C796	C353/TM-122(a)	C796/TM-122(b)
Flexural strength (dry), MPa (ksi):(cont'd)						
At 200 °C (392 °F) after isothermal aging for 500 h at 250 °C (478 °F) in circulating air	62 (8.9)	46 (6.6)	...	31 (4.5)	64 (9.3)	73 (10.5)
At 250 °C (478 °F) after isothermal aging for 500 h at 250 °C (478 °F) in circulating air	41 (5.9)	...	...	...	...	...
Flexural modulus (dry), GPa (10^6 psi):						
At RT	5.5 (0.79)	3.9 (0.56)	4.1 (0.59)	4.6 (0.66)	3.7 (0.53)	3.9 (0.56)
At RT after isothermal aging for 500 h at 200 °C (392 °F) in circulating air	5.5 (0.79)	...	...	...	...	...
At RT after isothermal aging for 500 h at 250 °C (478 °F) in circulating air	5.3 (0.77)	...	...	...	...	...
At 200 °C (392 °F) after isothermal aging for 500 h at 200 °C (392 °F) in circulating air	4.7 (0.68)	...	3.2 (0.47)	...	...	...
At 200 °C (392 °F) after isothermal aging for 500 h at 250 °C (478 °F)	3.4 (0.49)	2.1 (0.30)	...	3.0 (0.43)	2.5 (0.36)	2.62 (0.38)
At 250 °C (478 °F) after isothermal aging for 500 h at 250 °C (478 °F) in circulating air	4.5 (0.65)	...	...	...	...	...
Flexural elongation (dry), %:						
At RT	2.4	...	2.6	1.7	2.7	3.0
At 200 °C (392 °F)	...	...	1.8	...	...	...
At 250 °C (478 °F)	2.2	...	...	1.0	2.6	3.0
Tensile strength (dry) at RT, MPa (ksi)	89 (13)	...	...	...	...	...
Tensile strain-to-failure, %:						
Dry, at RT	2.2	...	...	...	...	...
Wet, at 250 °C (478 °F)	2.0	...	...	...	...	...
Fracture toughness, J/m^2 (in. · lbf/in.^{2})	40 (0.23)	160 (0.80)	180 (1.0)	63 (0.36)	389 (2.2)	230 (1.27)
Gel time at 170 °C (338 °F), min.	25	50	45	>30	...	...
Complex viscosity at 110 °C (230 °F), Pa · s (cP)	0.4 to 2.8 (400 to 2800)	0.2 to 2.5 (220 to 2500)	2.0 to 8.0 (2000 to 8000)	0.4 to 3.5 (400 to 3500)	...	...
Heat of polymerization, J/g (Btu/lb)	265 (0.10)	260 (0.10)	260 (0.10)	>200 (>0.85)	...	...
Melting range, °C (°F)	110 to 120 (230 to 248)	...	100 to 140 (212 to 284)	...	...	...
Coefficient of thermal expansion to 250 °C (478 °F), 10^{-6}/K	73.4	...	66	...	...	...
Moisture absorption, wt %	4.85	...	...	...	...	...

(a) Weight ratio for C353/TM-122, 76/24; TM-122 is 4, 4′-bis (2-propenylphenoxy) diphenylsulfone. (b) Weight ratio for C796/TM-122, 70/30; TM-122 is 4, 4′-bis (2-propenylphenoxy) diphenylsulfone

Bismaleimide properties and tests

Ingredient	Property	Test method
Bismaleimide resin	Viscosity	Rheometer
	Composition	HPLC/infrared spectroscopy
Modifier (organic)	Viscosity	Rheometer
	Molecular weight	GPC

Bismaleimide matrix properties and test methods

Material	Property	Test method
Uncured resin	Composition	HPLC
		Infrared spectroscopy
		GPC
		DSC
	Processibility	RDS viscosity
		Gel time
		Volatile content
Cured neat resin	Completeness of cure	Glass transition
	H_2O/solvent resistance	Glass transition, wet
		Moisture weight gain
		Solvent weight gain
	Resin toughness	Cleavage (G_{Ic})
Uncured impregnated system	Characterization	HPLC
		Infrared spectroscopy
		DSC
	Processibility	Resin content
		Fiber content
		Flow
		Gel
		RDS viscosity
		Volatiles
		Tack/drape
Cured impregnated system	Completeness of cure	Glass transition
		DSC
	Moisture resistance	Weight gain
	Thermal properties	Thermal conductivity
	Laminate properties	Ply thickness
		Fiber volume

Selected properties of bismaleimide resins

Material	Density, g/cm³	Uncured melting or softening temperature, °C	Uncured melting or softening temperature, °F	Tensile strength (dry) at RT, MPa	Tensile strength (dry) at RT, ksi	Tensile strength (dry) at 232 °C (450 °F), MPa	Tensile strength (dry) at 232 °C (450 °F), ksi
Hexcel F178	...	...	...	56	8.14	...	...
Narmco 5250-2	1.24	...	...	62	9.0	...	...
U.S. Polymeric V378A	1.26	...	...	78	11.3	44	6.4
Univ. Dayton BPA-BMI	1.26	70	155	48	7.0	...	...
Kerimid 70003	1.25	...	...	...	...	...	...

Material	Flexural modulus (dry) at RT, GPa	Flexural modulus (dry) at RT, 10^6 psi	Fracture toughness at RT, J/m²	Fracture toughness at RT, in. · lbf/in.²	T_g, (dry), °C	T_g, (dry), °F
Hexcel F178	...	...	29.4	0.17	260	500
Narmco 5250-2	2.9	0.43	100	0.56	321	610
U.S. Polymeric V378A	...	...	...	...	...	...
Univ. Dayton BPA-BMI	...	...	...	...	280	536
Kerimid 70003	...	...	82	0.46	320	620

Material	Tensile strain-to-failure (dry) at RT, %	Tensile modulus (dry) at RT, GPa	Tensile modulus (dry) at RT, 10^6 psi	Flexural strength (dry) at RT, MPa	Flexural strength (dry) at RT, ksi
Hexcel F178	1.3	...	...	...	...
Narmco 5250-2	1.7	2.7	0.39	138	20
U.S. Polymeric V378A	6.6	...	...	...	...
Univ. Dayton BPA-BMI	1.5	3.4	0.50	...	...
Kerimid 70003	2.2	...	...	...	...

Material	T_g, (wet), °C	T_g, (wet), °F	Coefficient of thermal expansion, 10^{-6}/K	Moisture absorption, wt %
Hexcel F178	140	284	...	3.7
Narmco 5250-2	...	...	...	3.3 (steam 96 h)
U.S. Polymeric V378A	...	...	...	...
Univ. Dayton BPA-BMI	...	...	...	1.7
Kerimid 70003	...	...	21	1.7 (100 h BW)

Selected properties of Matrimid 5292 bismaleimide

Property	Reported value
Density, g/cm³	1.23
Melting point	
Viscosity Component A	150 to 160 °C (302 to 320 °F)
Viscosity Component B, amber liquid, at 25 °C (77 °F), mPa · s	12000 to 20000
T_g, (by TMA), °C (°F)	273 (523)
T_g, (by DMA), °C (°F):	
Dry	295 (563)
Wet	305 (581)
Morphology	Amorphous, cross-linked
Tensile strength (dry), MPa (ksi):	
At RT	82 (11.9)
At 150 °C (300 °F)	51 (7.4)
At 204 °C (400 °F)	40 (5.8)
Tensile modulus (dry), GPa (10^6 psi):	
At RT	4.3 (0.62)
At 150 °C (300 °F)	2.4 (0.35)
At 204 °C (400 °F)	2.0 (0.29)

Property	Reported value
Tensile strain-to-failure, %:	
At RT	2.3
At 150 °C (300 °F)	2.6
At 204 °C (400 °F)	2.3
Tensile strength (wet), MPa (ksi):	
At RT	66 (9.6)
At 150 °C (300 °F)	30 (4.3)
Tensile modulus (wet), GPa (10^6 psi):	
At RT	3.8 (0.55)
At 150 °C (300 °F)	1.9 (0.27)
Flexural strength (dry) at RT, MPa (ksi)	167 (24.2)
Flexural modulus (dry) at RT, GPa (10^6 psi)	4.0 (0.59)
Moisture absorption, wt %	1.40
Fracture toughness (compact tension), J/m² (in. · lbf/in.²)	170 (0.97); 216 (1.22)

Selected properties of Keramid 601 and 353 bismaleimides

Property	601	353
Density, g/cm^3	1.30	...
Melting point, °C (°F)	80 (177)	70 to 125 (158 to 298)
T_g, °C (°F)	290 (554)	285 (545)
Morphology	Amorphous	Amorphous
Gel time at 170 °C (338 °F), min.	...	30
Melt viscosity, Pa · s (cP)	...	0.115 to 0.130 (115 to 130)
Tensile strength (dry), MPa (ksi):		
At RT	63.4 (9.2)	...
At 200 °C (392 °F)	42.0 (6.1)	...
Tensile strain-to-failure, %:		
At RT	3.1	...
At 200 °C (392 °F)	4.9	...
Flexural strength (dry), MPa (ksi):		
At RT	150 (21.7)	60 (8.7)
At 250 °C (482 °F)	...	50 (7.3)
After 210 h at 250 °C (482 °F)	103 (15.0)	...
After 1650 h at 250 °C (482 °F)	82 (11.9)	...
Flexural modulus (dry), GPa (10^6 psi):		
At RT	...	5.6 (0.81)
At 250 °C (478 °F)	...	3.4 (0.49)
Flexural elongation (dry), %:		
At RT	...	1.8
At 250 °C (478 °F)	...	1.2
Fracture toughness at RT:		
K_{Ic}, MPa$\sqrt{m}$ (ksi$\sqrt{in.}$)	382 (348)	...
G_{Ic}, J/m^2 (in. · lbf/in.^{2})	34 (0.19)	30 (0.17)
Thermogravimetric weight loss under nitrogen, wt %	1, up to 400 °C (750 °F)	...
Moisture absorption, wt %:		
At RT in 24 h	0.3	...
At 100 °C in 2 h	1.0	...
Coefficient of thermal expansion, 10^{-6}/K	61	...

Constituent properties of bismaleimide resins

Property	Fiberite X-86	Cycom 3100	Hysol EA9102	Hysol EA9655	Hexcel F650	Fiberite 987A
Density, g/cm^3	1.22	...	...	...	...	1.27
Uncured melting or softening temperature, °C (°F)	70 to 125 (159 to 257)	...	...	...	...	...
Tensile strength (dry), MPa (ksi):						
At RT	...	...	58.5 (8.5)	52 (7.5)	49.3 (7.15)	...
At 177 °C (350 °F)	...	...	43 (6.3)	27 (3.9)	...	...
Tensile strength (wet), at RT, MPa (ksi)	...	...	...	39 (5.7)	...	...
Tensile strain-to-failure (dry), %:						
At RT	...	2.1	2.2	1.6	...	1.2
At 177 °C (350 °F)	...	2.2	3.3	2.2	...	...
Tensile strain-to-failure (wet), at 177 °C (350 °F), %	...	3.0	...	...	...	...
Tensile modulus (dry), at RT, GPa (10^6 psi)	...	...	...	...	...	4.2 (0.61)
Flexural strength (dry), MPa (ksi):						
At RT	130 (18.9)	117 (17)	...	...	...	121 (17.6)
At 177 °C (350 °F)	85 (12.3)	93 (13.5)	...	...	...	53.7 (7.8)
At 232 °C (450 °F)	72 (10.5)	...	...	...	...	41 (6.0)
Flexural strength (wet), at 177 °C (350 °F), MPa (ksi)	46 (6.7)	58 (8.5)	...	...	...	38 (5.5)
Flexural modulus (dry), GPa (10^6 psi):						
At RT	4.6 (0.66)	...	...	...	...	43 (0.62)
At 177 °C (350 °F)	3.0 (0.44)	...	...	...	...	2.2 (0.32)
At 232 °C (450 °F)	2.6 (0.38)	...	...	...	...	1.8 (0.26)
Flexural modulus (wet), at 177 °C (350 °F), GPa (10^6 psi)	1.7 (0.25)	...	...	...	...	1.6 (0.23)
Fracture toughness at RT, J/m^2 (in. · lbf/in.^{2})	...	...	...	...	67 (0.38)	...
T_g (dry), °C (°F)	290 (554)	300 (527)	298 (568)	253 (489)	>316 (>600)	320 (608)
T_g (wet), °C (°F)	...	...	210 (410)	118 (244)	...	...
Moisture absorption, wt%	4.4(a)	...	...	...	4.3	2.97(a)

(a) Equilibrium water boil

Polyimides

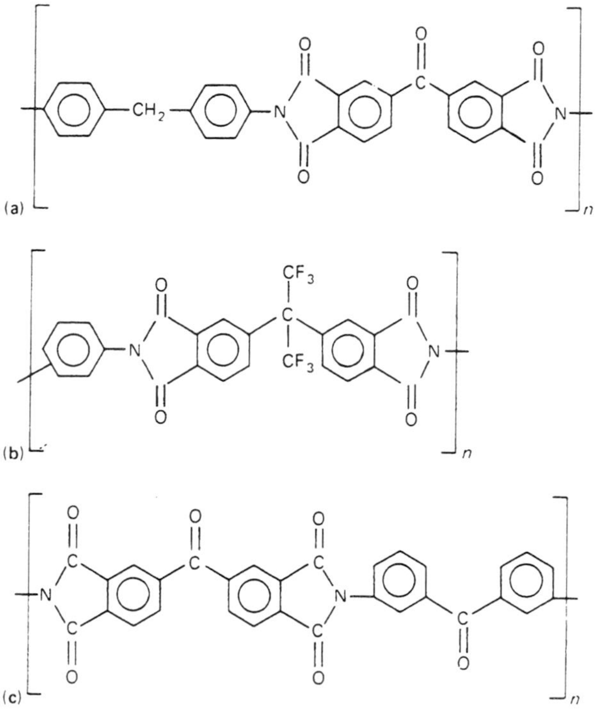

Condensation polyimide resins

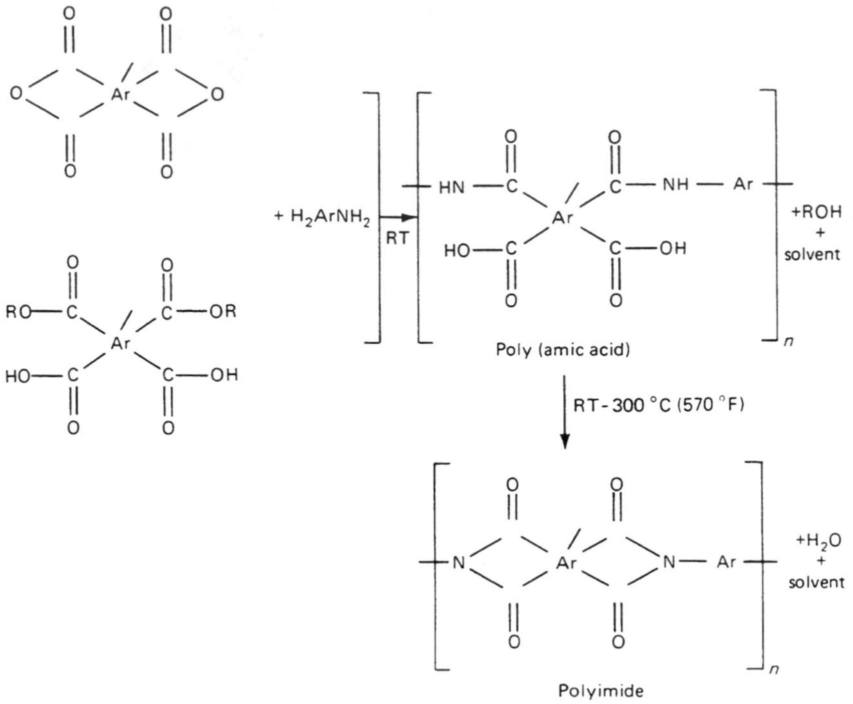

$$\text{Ar(CO)}_2\text{O(CO)}_2\text{O} \ \text{or} \ \text{Ar(COOR)}_2\text{(COOH)}_2 + H_2ArNH_2 \xrightarrow{RT} \left[-HN-\overset{O}{\overset{\|}{C}}-Ar(COOH)_2-\overset{O}{\overset{\|}{C}}-NH-Ar- \right]_n + ROH + \text{solvent}$$

Poly (amic acid)

RT - 300 °C (570 °F)

$$\left[-N(CO)_2Ar(CO)_2N-Ar- \right]_n + H_2O + \text{solvent}$$

Polyimide

Condensation polyimides

4, 4' carbonyl-bis-1, 2-benzendedicarboxylic acid, -diethyl ester (BTDE)

1, 3-bis (3-aminophenoxy) benzene (APB)

3-aminophenyl acetylene (APA)

+ E1OH

+ H_2O

Molecular structure of PMR acetylene end-capped Theramid AL-600

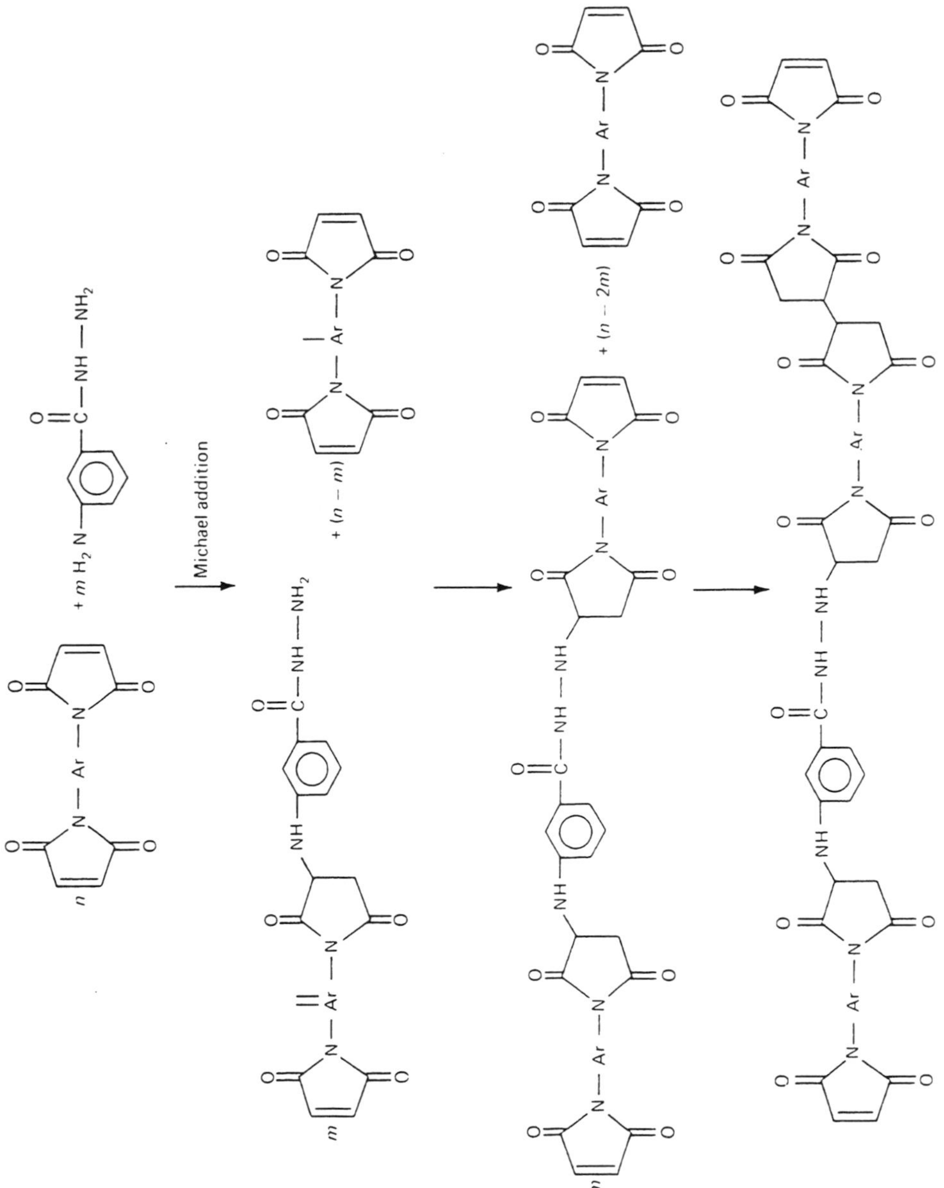
Michael addition
+ (n − m)
+ (n − 2m)

Polyimide properties and tests

Ingredient	Property	Test method
Polyimide resin	Ingredient ratio	HPLC/infrared titration
	Purity	HPLC
	Functional groups	Titration

Thermal and related properties of polyimide resins

Thermal and related properties	Neat resin	50% glass fiber and resin (molding compound)
Cure process parameters		
Mold pressure, MPa (ksi)	1.4 to 17.2 (0.2 to 2.5)	1.4 to 6.9 (0.2 to 10)
Mold temperature, °C (°F)	290 to 315 (500 to 600)(a) 315 (600)(c)	175 to 250 (350 to 480)(b)
Mold shrinkage, mm/mm	0.0126	0.0005 to 0.0040
T_g, °C (°F)	315 to 370 (600 to 698)(d)	...
Cured material properties		
Water absorption, 24 h, 3.2 mm (⅛ in.) thick, %	0.24 to 0.40	0.20
Specific gravity(e)	1.19 to 1.43	1.60 to 1.95
Continuous service temperature, °C (°F)	500 to 600	480 to 500
Heat-deflection temperature, at 1.82 MPa (0.264 ksi), °C (°F)	305 to 360 (582 to 680)	290 to 350 (0.15 to 0.27)
Burning rate	Nonburning	...
Specific heat, kJ/kg · K (Btu/lbf · °F)(e)	1.05 to 1.5 (0.25 to 0.35)	0.63 to 1.13 (550 to 660)
Coefficient of thermal expansion, 10^{-6}/K (10^{-6}/°F)	25 to 80 (12.7 to 45)	10 to 27 (5.5 to 15.1)
Thermal conductivity, W/m · K (Btu/ft · h · °F)(e)	0.10 to 0.34 (0.058 to 0.20)	0.34 to 0.49 (0.20 to 0.29)

(a) At 1.4 to 17 MPa (0.2 to 2.5 ksi). (b) At 1.4 to 70 MPa (0.2 to 10 ksi). (c) Postcure. (d) Dry. (e) At room temperature

Physical and mechanical properties of polyimides

Material	Density, g/cm³	Tensile strength		Modulus of elasticity	
		MPa	ksi	GPa	10^6 psi
Avimid K-III	1.31	102	15.0	3.8	0.55
Skybond 701	1.35	69	10.0	4.1	0.60
PMR-15	1.32	38.6	5.6	3.9	0.57
NR-150B2	1.40	110	16.0	4.1	0.60
Thermid MC-600	1.34	83	12.0	4.1	0.60
UpJohn 2080	1.40	120	17.1	1.3	0.19
BMIs	1.22 to 1.30	41 to 82	6 to 12	4.1 to 4.8	0.60 to 0.70
Ultem 1000	1.27	104	15.2	3.0	0.43
Torlon 4203	1.38	186	27.0	4.4	0.64

Material	Izod impact strength, notched		Strain-to failure, %
	J/m	ft · lbf/in.	
Avimid K-III	...	...	14
Skybond 701	53.4	1.0	1.00
PMR-15	53.4	1.0	1.5
NR-150B2	42.7	0.8	6.0
Thermid MC-600	48	0.9	1.5
UpJohn 2080	37.4	0.7	10.0
BMIs	...	...	1.3 to 2.3
Ultem 1000	53.4	1.0	60
Torlon 4203	133.5	2.5	20

Material	Flexural strength		Flexural modulus	
	MPa	ksi	GPa	10^6 psi
Avimid K-III	...	...	...	...
Skybond 701	...	...	...	...
PMR-15	176	25.5	4.0	0.58
NR-150B2	...	...	...	...
Thermid MC-600	145	21.0	4.5	0.66
UpJohn 2080	117	17.0	3.4	0.48
BMIs	76 to 145	11 to 21	3.4 to 4.8	0.50 to 0.70
Ultem 1000	145	21	3.4	0.48
Torlon 4203	211	30.7	4.5	0.66

(continued)

Physical and mechanical properties of polyimides (continued)

Material	Glass transition temperature, T_g		Fracture toughness	
	°C	°F	J/m²	in. · lbf/in.²
Avimid K-III	250	482	1900	11.0
Skybond 701	330	626	...	...
PMR-15	340	644	280	1.57
NR-150B2	340	644	2400	13.4
Thermid MC-600	320	608	...	...
UpJohn 2080	280	536	...	...
BMIs	230 to 290	446	34 to 260	0.19 to 1.45
Ultem 1000	210	426	...	...
Torlon 4203	267	512	3900	21.9

Selected properties of polyimides

Property	ASTM test	Molded polyimide	Encapsulated polyimide
Mechanical			
Yield strength, MPa (ksi)	D 638	118 (17.1)	18 (2.6)
Tensile modulus, GPa (10^6 psi)	D 638	1.3 (0.19)	...
Elongation (break), %	D 638	10	1
Compressive strength, MPa (ksi)	D 695	205 (30)	69 (10)
Flexural yield strength, MPa (ksi)	D 790	200 (29)	69 (10)
Flexural modulus, GPa (10^6 psi)	D 790	3.4 (0.50)	4.1 (0.60)
Impact strength, Izod(a), J/cm (ft · lbf/in.)	D 256	0.37 (0.7)	0.37 to 0.43 (0.7 to 0.8)
Hardness, Rockwell	...	E99	D50
Thermal			
Thermal conductivity, W/m · K (Btu · in./ft² · h · °F)	C 177	0.1 (0.7)	0.3 (1.9)
Coefficient of linear thermal expansion, 10^{-5}/°C (10^{-5}/°F)	D 696	9.0 (5.0)	8.1 (4.5)
Specific heat, kJ/kg · K (Btu/lbf · °F)	C 351	1.3 (0.3)	...
Continuous service temperature, °C (°F)	...	260 to 425 (500 to 800)	...
Electrical			
Volume resistivity, Ω · cm	D 257	10^{16}	10^{15}
Dielectric strength, V/10^{-3} mm (V/10^{-3} in.)	D 149	22 (560)	29 (725)
Dielectric constant (50 to 100 Hz)	D 150	3.4	...
Dissipation factor (50 to 100 Hz)	D 150	0.0005	...
Processing parameters			
Processing temperature, °C (°F)	...	315 (600)	...
Density, g/cm³	D 792	1.43	1.55
Specific volume, cm³/kg (in.³/lb)	D 792	685 (19)	650 (18)
Linear mold shrinkage, mm/mm (in./in.)	D 955	...	0.003
Water absorption, % in 24 h	D 570	0.3	0.1

(a) Notched, at room temperature

Selected properties of PMR-15 polyimide

Property	PMR-15 polyimide
Density, g/cm³	1.32
T_g after 316 °C (600 °F), 16-h postcure, °C (°F)	340 (662)
Morphology	Amorphous, cross-linked
Tensile strength (dry) at RT, MPa (ksi):	38.6 (5.6)
Tensile modulus (dry), at RT, GPa (10^6 psi):	39 (0.57)
Tensile strain-to-failure, %:	1.1
Flexural strength (dry), MPa (ksi):	
At RT	176 (25.5)
At 288 °C (550 °F)	73 (10.7)
At 316 °C (600 °F)	72 (10.4)
At 343 °C (650 °F)	52 (7.6)
Flexural modulus (dry), GPa (10^6 psi):	
At RT	4.0 (0.58)
At 288 °C (550 °F)	2.3 (0.34)
At 316 °C (600 °F)	1.9 (0.27)
At 343 °C (650 °F)	1.8 (0.26)
Fracture toughness:	
K_{Ic}, MPa$\sqrt{m}$ (ksi$\sqrt{in.}$)	1100 (1010); 648 (590)
G_{Ic}, J/m² (ft · lbf/in.²)	280 (1.6); 94 (0.52)
Izod impact strength, notched, J/m (ft · lbf/in.)	53.37 (1.0)
Moisture absorption, equilibrium moisture absorption (95% RH, 71 °C, or 160 °F), wt %	4.2
T_g, °C (°F):	
After 316 °C (1-h) cure	320 (608)
After 316 °C (16-h) cure	340 (662)
Weight loss after 1000 h at 288 °C (550 °F), %:	
In flowing air (100 cm³/min)	0.3
After 2000 h	0.8
After 3000 h	2.0
Coefficient of thermal expansion, 10^{-6}/K	14

Polyimide matrix properties and test methods

Material	Property	Test method
Uncured resin	Composition	HPLC
		Infrared spectroscopy
		GPC
	Processibility	RDS viscosity
		Gel time
		Volatile content
Cured neat resin	Completeness of cure	Glass transition
		DSC
	H_2O/solvent resistance	Glass transition, wet
		Solvent weight gain
Uncured impregnated system	Characterization	HPLC
		Infrared spectroscopy
		DSC
	Processibility	Resin content
		Fiber weight
		Flow
		Gel
		RDS viscosity
		Volatiles
		Tack/drape
Cured impregnated system	Completeness of cure	Glass transition
		DSC
	Moisture resistance	Weight gain
	Thermal properties	TGA
		Thermal conductivity
	Laminate properties	Ply thickness
		Fiber volume

Selected properties of Thermid resins

Property	Thermid MC-600	Thermid IP-600
Density, g/cm^3	1.37	1.34
Moisture absorption (24-h water boil), wt %	1.24	1.18
Uncured melting range, °C (°F)	190 to 210 (374 to 410)	149 to 171 (300 to 340)
T_g, °C (°F),		
After 371 °C (700 °F), 8-h postcure	320 (608)	300 (572)
After 371 °C (700 °F), 16-h postcure	350 (662)	350 (662)
Morphology	Amorphous, cross-linked	Amorphous, cross-linked
Heat of reaction, J/g (Btu/lb)	335 (0.15)	335 (0.15)
Tensile strength (dry) MPa (ksi):		
At RT	82.7 (12.0)	58 (8.5)
Tensile modulus (dry), GPa (10^6 psi):		
At RT	4.1 (0.60)	5.0 (0.73)
At 316 °C (600 °F)	...	28 (4.01)
Tensile strain-to-failure, %:		
At RT	1.5	1.2
At 316 °C (600 °F)	...	4.2
Compressive strength (dry):		
At RT, MPa (ksi)	172 (25)	...
Flexural strength (dry), MPa (ksi):		
At RT	146.6 (21)	106 (15.3)
After 1000 h	92 (13.4)	...
Flexural strength (dry), MPa (ksi):		
At 316 °C (600 °F)	29 (4.2)	44 (6.4)
After 1000 h	18 (2.6)	...
Flexural modulus (dry), GPa (10^6 psi):	4.6 (0.66)	...
Izod impact, notched, J/m (ft · lbf/in.)	80 (1.5)	...
Coefficient of thermal expansion, 10^{-6}/K	35 to 50	...
Weight loss, %		
After 500 h at 316 °C (600 °F)	2.89	...
After 1000 h at 316 °C (600 °F)	4.04	...
Dielectric constant (1 MHz)	3.496	...
Dissipation factor (1 MHz)	0.0096	...

Polyurethanes

Thermal, physical, and chemical properties of a typical PUR coating deposited from a two-package system

Property	Value
Heat-deflection temperature at 1.82 MPa, °C	75
Maximum resistance to continuous heat, °C	70
Coefficient of linear thermal expansion, 10^{-5}/°C	15.0
Tensile strength, MPa	6.89
Elongation, %	200
Hardness, Shore	D50
Specific gravity	1.2
Dielectric constant	6
Resistance to chemicals at 25 °C(a):	
Nonoxidizing acids (20% H_2SO_4)	Q
Oxidizing acids (10% HNO_3)	U
Aqueous salt solutions (NaCl)	S
Aqueous alkalis (NaOH)	Q
Polar solvents (C_2H_5OH)	U
Nonpolar solvents (C_6H_6)	Q
Water	S

(a) S, satisfactory; Q, questionable; U, unsatisfactory

Selected properties of liquid cast polyurethane

Property	ASTM test	Reported value
Mechanical		
Yield strength, MPa (ksi)	D 638	14 to 69 (2.0 to 10.0)
Tensile modulus, GPa (10^6 psi)	D 638	0.69 to 6.9 (0.10 to 1.0)
Elongation (break), %	D 638	200 to 1000
Compressive strength, MPa (ksi)	D 695	140 (20)
Flexural yield strength, MPa (ksi)	D 790	6.9 to 34 (1 to 5)
Flexural modulus, GPa (10^6 psi)	D 790	0.69 (0.10)
Impact strength, Izod(a), J/cm (ft · lbf/in.)	D 256	>13.3 (>25.0)
Hardness, Rockwell	...	D90
Thermal		
Thermal conductivity, W/m · K (Btu · in./ft^2 · h · °F)	C 177	0.20 (1.4)
Coefficient of linear thermal expansion, 10^{-5}/°C (10^{-5}/°F)	D 696	18.0 to 36.0 (10.0 to 20.0)
Specific heat, kJ/kg · K (Btu/lbf · °F)	C 351	1.7 (0.4)
Electrical		
Volume resistivity, Ω · cm	D 257	10^{12} to 10^{15}
Dielectric strength, V/10^{-3} mm (V/10^{-3} in.)	D 149	16 to 20 (400 to 500)
Dissipation factor (50 to 100 Hz)	D 150	0.015 to 0.017
Processing parameters		
Processing temperature, °C (°F)	...	85 to 120 (185 to 250)
Density, g/cm^3	D 792	1.10 to 1.50
Specific volume, cm^3/kg (in.^{3}/lb)	D 792	975 to 795 (27 to 22)
Linear mold shrinkage, mm/mm (in./in.)	D 955	0.020
Water absorption, % in 24 h	D 570	0.0 to 1.5

(a) Notched, at room temperature

Silicones

Thermal, physical, and chemical properties of typical silicone coatings

Property	Value
Maximum resistance to continuous heat, °C	250
Coefficient of linear thermal expansion, 10^{-5}/°C	40
Tensile strength, MPa	5
Elongation, %	200
Hardness, Shore A	65
Specific gravity	1.2
Dielectric constant	3
Water absorption, %	0.2
Resistance to chemicals at 25 °C(a):	
Nonoxidizing acids (20% H_2SO_4)	Q
Oxidizing acids (10% HNO_3)	U
Aqueous salt solutions (NaCl)	S
Aqueous alkalis (NaOH)	S
Polar solvents (C_2H_5OH)	S
Nonpolar solvents (C_6H_6)	Q
Water	S

(a) S, satisfactory; Q, questionable; U, unsatisfactory

Selected properties of flexible cast silicone

Property	ASTM test	Reported value
Mechanical		
Yield strength, MPa (ksi)	D 638	2.1 to 6.9 (0.3 to 1.0)
Tensile modulus, GPa (10^6 psi)	D 638	620 (90)
Elongation (break), %	D 638	100 to 700
Compressive strength, MPa (ksi)	D 695	0 to 6.9 (0 to 1)
Hardness, Rockwell	...	A15 to 60
Thermal		
Thermal conductivity, W/m · K (Btu · in./ft^2 · h · °F)	C 177	0.14 to 0.29 (1.0 to 2.0)
Coefficient of linear thermal expansion, 10^{-5}/°C (10^{-5}/°F)	D 696	14 to 54 (8.0 to 30.0)
Density, g/cm^3	D 792	0.99 to 1.50
Water absorption, % in 24 h	D 570	0.1
Electrical		
Volume resistivity, Ω · cm	D 257	10^{14} to 10^{15}
Dielectric strength, V/10^{-3} mm (V/10^{-3} in.)	D 149	22 (550)
Dissipation factor (50 to 100 Hz)	D 150	0.001 to 0.025

Typical property ranges of silicone conformal coatings

Property, as supplied	Value
Viscosity, Pa · s	0.120 to 70
Flash point, closed cup, °C (°F)	5 to 100 (40 to 212)
Nonvolatile content, %	25 to 100
Shelf life, months	6 to 12
Pot life, days	0.5 to 30
Cure rate, based on:	
m/min. (ft/min.)	1.5 to 3.0 (5 to 10)
kW/m (W/in.)	8 (200)
kJ/m^2 (ft · lbf/ft^2)	15 (1030)
Useful temperature range, °C (°F)	–55 to 150 (–67 to 300)
Skin-over time, min.	5 to 60
Tack-free time, h	0.5 to 4
Cure time, 0.64 mm (25 mils) thick, h	24
Cure time, 3.2 mm (0.125 in.) thick, h	72
Full cure, days	7
Electrical	
Dielectric strength, kV/mm (V/mil)	20 to 50 (500 to 1200)
Volume resistivity, Ω · cm	5×10^{14} to 5×10^{16}
Dielectric constant at 25 °C (77 °F)	
At 100 Hz	2.5 to 3.0
At 100 kHz	2.6 to 3.1
Dissipation factor at 25 °C (77 °F)	
At 100 Hz	0.0016 to 0.0018
At 100 kHz	0.0004 to 0.002
Physical	
Durometer hardness, Shore A	20 to 70
Tensile strength, MPa (ksi)	1.7 to 6.2 (0.250 to 0.900)
Elongation, %	50 to 1200
Tear strength, die B, N/mm (lbf/in.)	3.5 to 10.5 (20 to 60)
Volume expansion at 25 to 100 °C (77 to 212 °F), 10^{-4}/K	2 to 10

Thermal and related properties of silicone resins

Thermal and related properties	Neat resin	Silica-filled heat-vulcanized molding compound	Silica-filled, carbon fiber reinforced, two-part RTV
Cure process parameters			
Mold temperature, °C (°F)	...	165 (330)(a)	Room temperature(b)
Mold shrinkage, mm/mm	0 to 0.006	0.0030 to 0.0067	0.002 to 0.006
T_g, °C (°F)	−125 (−193)	...	...
Cured material properties			
Water absorption, 24 h, 3.2 mm (⅛ in.) thick, %	0.12(c)	0.10	...
Specific gravity(d)	0.99 to 1.50	1.86 to 1.88	1.46
Continuous service temperature, °C (°F)	260 (500)	175 to 260 (350 to 500)	205 (400)
Heat-deflection temperature, at 1.82 MPa (0.264 ksi), °C (°F)	...	225 to 345 (435 to 650)	...
Flammability rating	Self-extinguishing	...	...
Specific heat, kJ/kg · K (Btu/lbf · °F)(d)	...	0.80 to 0.84 (0.19 to 0.20)	1.1 (0.27)
Coefficient of thermal expansion, 10^{-6}/K (10^{-6}/°F)	80 to 300 (44 to 166)	55 to 30 (13 to 18)	250 (140)
Thermal conductivity, W/m · K (Btu/ft · h · °F)(d)	0.22 (0.13)	0.37 to 0.49 (0.22 to 0.29)	0.34 to 0.49 (0.20 to 0.29)

(a) In compression mold. (b) For 24 h. (c) For 7 days at 25 °C (77 °F). (d) At room temperature

Vinyls

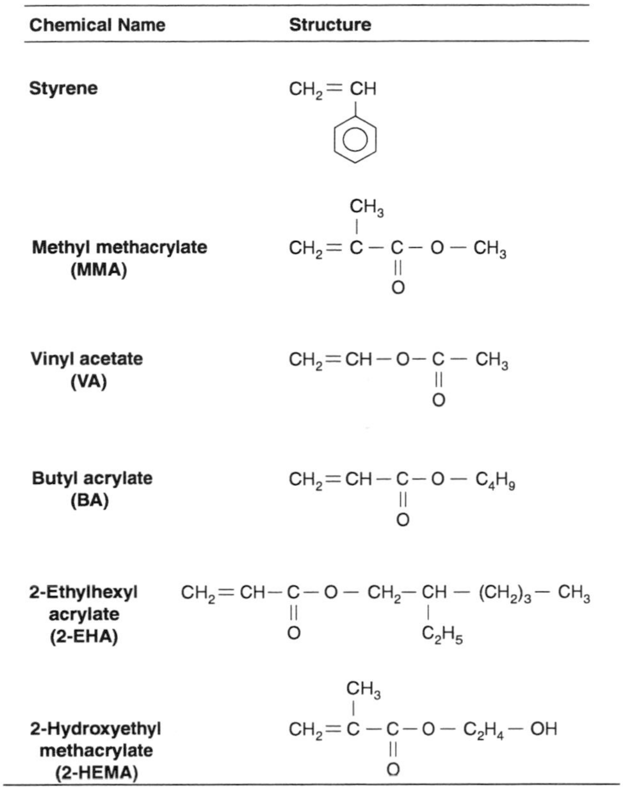

Chemical Name	Structure
Styrene	$CH_2{=}CH{-}C_6H_5$
Methyl methacrylate (MMA)	$CH_2{=}C(CH_3){-}C({=}O){-}O{-}CH_3$
Vinyl acetate (VA)	$CH_2{=}CH{-}O{-}C({=}O){-}CH_3$
Butyl acrylate (BA)	$CH_2{=}CH{-}C({=}O){-}O{-}C_4H_9$
2-Ethylhexyl acrylate (2-EHA)	$CH_2{=}CH{-}C({=}O){-}O{-}CH_2{-}CH(C_2H_5){-}(CH_2)_3{-}CH_3$
2-Hydroxyethyl methacrylate (2-HEMA)	$CH_2{=}C(CH_3){-}C({=}O){-}O{-}C_2H_4{-}OH$

Vinyl monomers

$$H_2C{=}C(CH_3){-}C({=}O){-}O{-}H_2C{-}CH(OH){-}CH_2{-}R{-}CH_2{-}CH(OH){-}CH_2{-}O{-}C({=}O){-}C(CH_3){=}CH_2$$

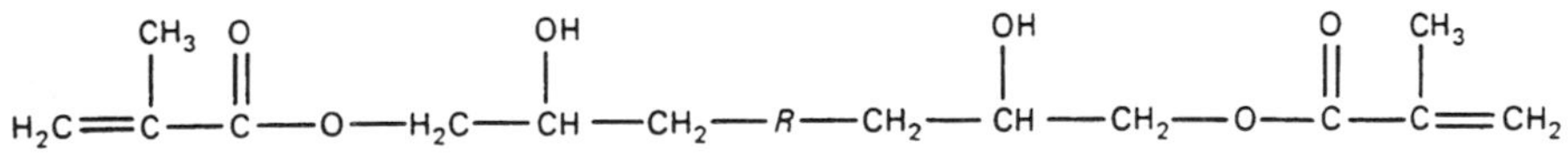

Typical vinyl ester

Gel times of a vinyl ester resin with various initiators

Initiator	Gel time, min.(a) at 25 °C (80 °F)	80 °C (175 °F)	120 °C (250 °F)
2% benzoylperoxide (BPO) 50%) + 0.1% dimethylaniline (DMA)	10	...	...
1.5% methy ethyl ketone peroxide (MEKP) (50%) + 3% Co-octoate (1%) +0.1% DMA	14	...	...
1.5% MEKP + 3% Co-octoate (1%) + 0.015% DMA	34	...	...
2% BPO (50%) + 0.01% DMA	117	...	...
2% MEKP (50%)	700	6	2
2% MEKP (50%)	...	25	15
1% BPO (50%)	...	25	15
1% cumylhydroperoxide	...	32	10
1% *t*-butylperbenzoate	...	120	6
1% *t*-butylcumylperoxide	...	360	9

(a) 5g (140 oz.), isothermally

Typical vinyl ester formulation, by parts

(Percentages based on 100 parts vinyl ester resin)

Vinyl ester	100%
Promoter (6% cobalt naphthenate)	0.2 to 0.5%
Activator (100% dimethyl aniline)	0.0 to 0.2%
Catalyst (9% MEKP)	0.9 to 2.5%

Typical vinyl ester formulation, by weight

Compound	Parts by weight
Vinyl ester resin	40
Styrene	30
Polyvinyl acetate	10
Calcium carbonate	150
AS P400 clay	17
Zinc stearate	4
1-butyl perbenzoate	0.5

Thermal and related properties of PVC and other vinyl polymers

	PVC						
Property	Rigid	Plasticized	30% glass-filled	Chlorinated PVC	PVDC	PVFM	PVB
T_g, °C (°F)	75 to 105 (170 to 220)	(a)	75 to 105(b) (170 to 220)	110 (230)	...	105 (220)	49 (120)
T_m, °C (°F)	(c)	...	...	(c)	210 (410)	(c)	(c)
Molding temperature, °C (°F)							
Compression	140 to 205 (285 to 400)	140 to 195 (285 to 385)	...	170 to 205 (350 to 400)	104 to 175 (220 to 350)	150 to 175 (300 to 350)	140 to 160 (280 to 320)
Injection	150 to 215 (300 to 415)	160 to 195 (320 to 385)	130 to 210 (270 to 405)	160 to 225 (325 to 440)	150 to 205 (300 to 400)	150 to 205 (300 to 400)	120 to 170 (250 to 340)
Heat-deflection temperature, at 1.82 MPa (0.264 ksi), °C (°F)	140 to 170 (285 to 340)	...	155 (310)	202 to 234 (395 to 450)	130 to 150 (265 to 300)	150 to 170 (300 to 340)	...
Water absorption, 24 h at 3.2 mm (1/8 in.) thick, %	0.04 to 4.0	0.15 to 0.75	0.008	0.02 to 0.15	0.1	0.5 to 3.0	1.0 to 2.0

(a) Variable; can be lower than 75 to 105 °C (165 to 220 °F) depending on type and concentration of plasticizer. (b) Irrespective of the filler. (c) Amorphous

Mechanical properties of unreinforced vinyl ester resins

Property	BPA vinyl ester	Novolac vinyl ester	Resilient vinyl ester
Tensile strength, MPa (ksi)	80 (11.6)	80 (11.6)	78 (170)
Tensile modulus, GPa (10^6 psi)	3.1 (0.45)	3.2 (0.46)	2.9 (0.42)
Tensile elongation, at break, %	5.5	3.5	10
Flexural strength, MPa (ksi)	124 (18.0)	120 (17.4)	120 (17.4)
Flexural modulus, GPa (10^6 psi)	3.2 (0.46)	3.3 (0.48)	2.8 (0.41)
Heat-deflection temperature, °C (°F)	100 (212)	150 (300)	75 (167)

Gel time variation for Derakane 411-45 resin

Material			
MEKP, wt %	Cobalt naphthanate, wt %	2, 4 pentandione, wt %	Gel time, min.
10	0.25	0.0	21
1.0	0.25	0.05	23
1.0	0.25	0.1	60
1.0	0.25	0.2	180
1.0	0.25	0.3	265

Delayed gel times for Derakane 411-45 resin

	Material				Typical exotherm	
MEKP, wt %	Cobalt naphthenate, wt %	2,4 pentanedione, wt %	Typical gel time, min.	Typical peak time, min.	°C	°F
1.0	0.25	0.0	21	37	40	108
1.0	0.25	0.05	23	39	60	138
1.0	0.25	0.1	60	74	55	132
1.0	0.25	0.2	180	191	70	153
1.0	0.25	0.3	265	280	60	147

Derakane vinyl ester resins

		Heat distortion temperature			
Product name	Type	°C	°F	Resin/styrene ratio	Applications
Derakane 411-45	Bisphenol A epoxy	100	215	55/45	General
Derakane 510N	Brominated bisphenol A epoxy	120	250	64/36	Flame retardant
Derakane 8084	Flexibilized bisphenol A epoxy	80	180	60/40	Toughened
Derakane 470-36	Epoxy novolac	140 to 150	290 to 300	64/36	High temperatures

Selected properties of molded vinyls

Property	ASTM test	Vinyl PVA-PVC(a)	Vinyl PVC-PVC$_2$	Vinyl Cl-PVC	Vinyl PVC-PVB
Mechanical					
Yield strength, MPa (ksi)	D 638	41 to 52 (6.0 to 7.5)	21 to 34 (3.0 to 5.0)	52 to 62 (7.5 to 9.0)	3.4 to 21 (0.5 to 3.0)
Tensile modulus, GPa (10^6 psi)	D 638	...	0.34 to 0.55 (0.05 to 0.08)	2.5 to 3.3 (0.36 to 0.48)	...
Elongation (break), %	D 638	40 to 80	100 to 250	5 to 65	150 to 450
Compressive strength MPa (ksi)	D 695	55 to 90 (8 to 13)	14 to 21 (2 to 3)	62 to 150 (9 to 22)	...
Flexural yield strength, MPa (ksi)	D 790	69 to 110 (10 to 16)	28 to 41 (4 to 6)	97 to 115 (14 to 17)	...
Flexural modulus, GPa (10^6 psi)	D 790	...	...	2.8 to 31 (0.4 to 4.5)	...
Impact strength, Izod(a), J/cm (ft · lbf/in.)	D 256	0.21 to 11 (0.4 to 20)	0.16 to 0.53 (0.3 to 1.0)	0.53 to 3.0 (1.0 to 5.6)	...
Hardness	...	Rockwell D65 to 85	Rockwell M50 to 65	Rockwell 118 to 122	Shore A10 to 100
Thermal					
Thermal conductivity, W/m · K (Btu · in./ft^2 · h · °F)	C 177	0.14 to 0.22 (1.0 to 1.5)	0.07 (0.5)	0.13 (0.9)	...
Coefficient of linear thermal expansion, 10^{-5}/ °C (10^{-5}/°F)	D 696	9.0 to 18.0 (5.0 to 10.0)	15 (8.5)	6.8 (3.8)	...
Specific heat, kJ/kg · K (Btu/lbf · °F)	C 351	0.8 to 1.7 (0.2 to 0.4)	1.3 (0.3)	1.3 (0.3)	...
Continuous service temperature, °C (°F)	...	65 to 75 (150 to 165)	75 to 100 (170 to 212)	110 (230)	...
Electrical					
Volume resistivity, Ω · cm	D 257	10^{16}	10^{14} to 10^{16}	10^{15}	5.0×10^{10}
Dielectric strength, V/10^{-3} mm (V/10^{-3} in.)	D 149	14 to 20 (350 to 500)	16 to 24 (400 to 600)	47 to 59 (1200 to 1500)	14 (350)
Dielectric constant (50 to 100 Hz)	D 150	3.2 to 4.0	4.5 to 6.0	3.1	5.6
Dissipation factor (50 to 100 Hz)	D 150	0.007 to 0.020	0.030 to 0.045	0.020	0.115
Processing parameters					
Processing temperature, °C (°F)	...	150 to 215 (300 to 415)	120 to 205 (250 to 400)	175 to 225 (350 to 440)	140 to 170 (280 to 340)
Density, g/cm^3	D 792	1.30 to 1.58	1.65 to 1.72	1.50 to 1.58	1.05
Specific volume, cm^3/kg (in.^{3}/lb)	D 792	760 to 685 (21 to 19)	615 to 580 (17 to 16)	650 (18)	940 (26)
Linear mold shrinkage, mm/mm (in./in.)	D 955	0.001 to 0.005	0.005 to 0.025	0.003 to 0.007	...
Water absorption, % in 24 h	D 570	0.0 to 0.4	0.1	0.5	1.0 to 2.0

(a) Rigid material. (b) Notched, at room temperature

Characteristics of PVC and related polymers

Material	Characteristics
PVC	
flexible	Low to moderate heat resistance; tensile strength, 1 to 3.5 ksi; elongation, 200 to 450%; 100% modulus, 1.2 to 2.8 ksi Note: Flexible PVC contains plasticizers, and various amounts of other ingredients. Plasticizers are added to enhance the workability and flexibility of the material
rigid	Hard and tough; resistance to abrasion; excellent dielectric characteristics; excellent resistance to moisture and good environmental stability; heat resistance is low to moderate
PVC copolymers (for example, vinyl chloride-vinylacetate)	Less rigid than rigid PVC; creep and tensile strength properties are inferior to rigid PVC
Chlorinated PVC (CPVC)	Very good heat resistance (withstands temperature about 33 °C, or 60 °F, higher than other types of vinyls); good resistance to creep; dimensionally stable; resistant to chemical degradation (includes acids, alkalies, most organic solvents, oil and grease)
Polyvinylidene chloride (PVDC) (Saran)	Best resistance to acids and most alkalies; resistant to degradation by hydrocarbons (aliphatic and aromatic), alcohols, and esters; best performance in the presence of organic or aqueous vapors

Production and Machining

Thermoplastic and thermoset processing comparison

Process	Process pressure MPa	Process pressure ksi	Maximum equipment pressure MN	Maximum equipment pressure tonf	Maximun size m^2	Maximun size ft^2	Pressure limited	Ribs	Bosses	Vertical walls	Spherical shape	Box sections	Slides/ cores	Weld-able	Good finish, both sides	Varying cross section
Thermoplastics																
Injection	15 to 45	2 to 7	30	3370	0.75	8.0	y	y	y	y	n	n	y	y	y	y
Injection compression	20	2.9	30	3370	1.5	16	y	y	y	n	n	n	y	y	y	y
Hollow injection	15	2.2	30	3370	2.0	20	y	y	y	y	n	y	y	y	y	y
Foam injection	5	0.7	15	1690	3.0	30	y	y	y	...	n	y	y	y	y	y
Sandwich molding	20	2.9	30	3370	1.5	16	y	y	y	y	n	n	y	y	y	y
Compression	20	2.9	30	3370	1.5	16	y	y	y	y	n	n	y	y	y	y
Stamping	20	2.9	30	3370	1.5	16	y	n	n	n	n	n	n	y	y	n
Extrusion	n/a	n/a	n/a	n/a	n/a	n/a	n/a	y	n	n/a	n	y	n	y	y	y
Blow molding	1	0.15	10	1120	2.0	20	n	n	n	y	y	y	y	y	n	n
Twin-sheet forming	1	0.15	10	1120	6.0	65	n	n	n	y	y	y	n	y	n	n
Twin-sheet stamping	1	0.15	30	3370	6.0	65	n	n	n	n	n	y	n	y	y	n
Thermoforming	0.1	0.015	n/a	n/a	...	...	n	n	n	y	n	n	y	y	n	n
Filament winding	0	0	n/a	n/a	...	...	n/a	y	n	y	y	y	n	y	n	y
Rotational casting	0.1	0.015	n/a	n/a	...	...	n	n	n	y	y	n	n	y	n	n
Thermosets																
Compression																
Powder	60	8.7	30	3370	0.5	5	y	y	y	y	n	n	y	n	y	y
Sheet molding compound	6 to 20	0.85 to 3	30	3370	4 to 5	45 to 55	y	y	y	y	n	n	y	n	y	y
Cold-press molding	1	0.15	30	3370	...	...	n	n	y	y	n	n	n	n	y	y
Hot-press molding	5	0.75	30	3370	6.0	65	y	n	y	y	n	n	n	n	y	y
High-strength sheet molding compound	4 to 10	0.60 to 1.5	30	3370	3.0	30	y	y	y	y	n	n	n	n	y	y
Prepreg	0.5 to 5	0.07 to 0.75	30	3370	6.0	65	y	n	n	y	n	n	n	n	y	y
Vacuum bag	0.1	0.015	n/a	n/a	...	...	n	n	y	y	n	y	n	n	n	y
Hand lay-up	0	0	n/a	n/a	...	...	n	n	y	y	n	y	n	n	n	y
Injection																
Powder	100	14.5	10	1120	0.1	1.1	y	y	y	y	n	n	y	n	y	y
Bulk molding compound	30	4.5	30	3370	1.0	11	y	y	y	y	n	n	y	n	y	y
ZMC	30	4.5	30	3370	1.0	11	y	y	y	y	n	n	y	n	y	y
Stamping	3	0.45	30	3370	6.0	65	y	n	n	y	n	n	n	n	y	n
Reaction injection molding	1	0.15	10	1120	...	...	y	y	y	n	n	y	n	y	y	y
Resin transfer molding, or resinject	0.1	0.015	10	1120	...	...	n	y	n	y	n	y	n	n	y	y
High-speed resin transfer molding, or fast resinject	2	0.3	30	3370	...	...	n	y	n	y	n	y	n	n	y	y
Foam polyurethane	0.5	0.07	n/a	...	...	...	n	y	y	y	y	y	n	n	y	y
Reinforced foam	1	0.15	30	3370	3.0	30	y	y	y	y	n	y	n	n	y	y
Filament winding	n/a	n/a	n/a	n/a	...	...	n/a	y	n	y	y	y	n	n	(a)	y
Pultrusion	n/a	n/a	n/a	n/a	n/a	n/a	n/a	y	n	n/a	n	y	n	n	y	y

Note: y, yes; n, no; n/a, not applicable. (a) one side of filament-wound article will exhibit a strong fiber pattern

Subjective assessment of thermoplastic and thermoset processing

Aspect	Advantage	Reasons
Prepreg formulation		
Viscosity	TS	Lower
Solvents	TS	Greater choice
Hand	TS	More flexible
Tack	TS	Prepolymer variable
Storage	TP	Not reactive
Quality control	TP	Fewer variables
Composite fabrication		
Lay-up	TS	Ease of handling
Degassing	TP	Fewer volatiles
Temperature changes	TP	Fewer
Maximum temperature	TS	Lower
Pressure changes	TP	Fewer
Maximum pressure	TS	Lower
Cycle time	TP	Lower
Postcure cycle	TP	Not required
Repair	TP	Remelt
Post forming	TP	Remelt

TP, thermoplastic; TS, thermoset

Molding and Forming Processes

Common molding processes

Process	Speed	Cost	Size limitation
Injection molding	Fast	Low	Small to medium large
Blow molding	Fast to medium	Low	Small to medium large
Transfer molding	Fast	Low	Small to medium
Extrusion	Fast	Low	Small to medium
Reaction injection molding	Fast	Low	Medium to large
Pultrusion	Fast	Low	Small to medium
Compression molding	Fast to medium	Low to medium	Small to very large
Filament winding	Medium	Low to medium	Medium to large
Rotocast	Medium	Medium	Medium to large
Vacuum form	Medium	Medium	Medium to large
Resin transfer molding	Medium to slow	Medium to high	Medium to large
Autoclave	Slow	Medium to high	Medium to large
Hand lay-up or spray-up	Slow	High	Medium to very large

Molding process advantages and disadvantages

Process	Advantages	Disadvantages
Compression	Highly filled materials can be molded; most amenable for large-area projection at low thickness; isotropic (non-directional) properties; easily controlled	Rather labor intensive with more material handling; high material losses; secondary operations are required; poor control of thickness; cannot mold with delicate inserts
Transfer	More machine intensive; shorter cures; good control of thickness; suitable for incorporation of delicate inserts	Requires better process control; complex and expensive mold; selective in processing material; some anisotropy (directional) properties; some material loss in central feed system
Injection	Process amenable to automation; rapid curing cycle; good control of dimension and flash thickness; excellent reproducibility of molded parts; thin sections with delicate inserts can be molded	Greater skill required in processing; machines and molds are complex and expensive; process not adaptable to interruption; injection molding grade materials in closer tolerances are required; care required to minimize effect of weld lines; material lost in sprues and runner

Premold and postmold operations

Process	Preprocess	Postprocess	Remarks
Injection molding	Pellets to dryer	Degate	Deflash as required
Blow molding	Pellet to dryer	Trim pinch-off	
Transfer molding	Make preforms	Degate	Deflash as required
Extrusion	Pellets to dryer	Trim	Deflash as required
Reaction injection molding	Add liquid chemicals to reservoirs	Degate	Deflash and/or trim as required
Pultrusion	Place fibers in head, add liquid resin	Cut off trim	Compression molding may follow
Compression molding	Cut material blanks; preheat as required	Deflash and trim	Surface fill may be required
Filament winding	Place winding mandrel on machine	Cut off and trim	Compression molding may follow
Rotocast	Place shapes or charge inside mold	Varies	
Vacuum form	Cut material blanks to size	Cut off clamping flange	
Resin transfer molding	Cut, fabricate, and place preform	Degate and trim	
Autoclave	Hand lay-up from preform; vacuum bag	Demold and trim	
Hand lay-up and spray-up	None	Varies	

Injection molding temperature guidelines

	Temperature	
Process component	°C	°F
Resin drying(a)		
for 4.5 h	135	275
for 4.0 h	150	300
for 2.5 h	175	350
Barrel processing		
Rear	335 to 350	635 to 660
Middle	340 to 360	645 to 680
Front	350 to 380	660 to 715
Nozzle	350 to 380	660 to 715
Melt	340 to 400	680 to 750
Mold cavity	160 to 220	320 to 430

(a) Do not dry below 135 °C (275 °F)

Injection molding tolerances

Parameter	Tolerance limits: Low cost	Commercial	Precision
Focal length	±3%/5%	±2%/3%	±0.5%/1%
Radius of curvature	±3%/5%	±2%/3%	±0.5%/1%
Figured, spherical	10 to 30 f(a)	5 to 10 f(a)	1 to 5 f(a)
Figured, flat	...	...	...
Irregularity	3 to 7 f(a)	1 to 3 f(a)	1 to ½ f(a)
Surface quality	80/50	60/50	40/20
Concentricity	±3 min	±2 min	±1 min
Vertex thickness, mm (mil)	±0.40 to 0.60 (±0.010 to 0.015)	±0.28 to 0.16 (±0.007 to 0.004)	±0.12 to 0.08 (±0.003 to 0.002)
Diameter or length, mm (mil)	±0.40 to 0.60 (±0.010 to 0.015)	±0.28 to 0.16 (±0.007 to 0.004)	±0.12 to 0.08 (±0.003 to 0.002)
Repeatability tolerances lens-to-lens	1 to 2%	0.5 to 1%	≤0.3 to 0.5%

(a) f, optical fringe

Pultrusion product characteristics

Size	Forming guide system and equipment pulling capacity influence size limitation
Shape	Straight, constant cross sections; some curved sections possible
Length	No limit
Reinforcements	Fiberglass, aramid fiber, carbon fiber, and thermoplastic
Resin systems	Polyester, vinyl ester, and epoxy
Reinforcement, wt %	All roving, 40 to 80%; mat and roving, 25 to 50%; 55% woven roving or biaxial materials and mat, 40 to 70%
Mechanical strength	Medium to high, primarily unidirectional, approaching isotropic
Labor intensity	Low to medium
Mold cost	Low to medium
Production rate	Shape and thickness related

Design guidelines for use in resin transfer molding (RTM) materials selection

General question	Specific areas of concern
Mold	
How many parts are required from the process in a given time period?	Primarily, tooling and pumping/dispensing; secondarily, release, cleaners, resin
Is design life a consideration?	Tooling, pumping/dispensing, cleaners, release agents
What are the dimensional requirements?	Tooling, cleaners, release agents
What are the strength requirements?	Tooling, cleaners, release agents
What are the surface finish requirements?	Tooling, cleaning, release agents, resin system, reinforcement
Part	
What are the performance requirements of the part?	Reinforcement, resin system
Production	
What are the shop environmental requirements?	Resin system, cleaners, release agents, tooling, pumping/dispensing
What are the capabilities of personnel?	All
What are the cost objectives?	All

Typical parts currently manufactured using RTM

Use	Part
Industrial	Solar collectors, electrostatic precipitator plates, fan blades, business-machine cabinetry, water tanks
Recreational	Canoe paddles, television antennae, snowmobiles
Construction	Seating, bathtubs, roof sections
Aerospace	Airplane wing ribs, cockpit hatch covers, airplane escape doors
Automobile	Crash members, leaf springs, car bodies, bus shelters

Advantages and disadvantages of various RTM reinforcement forms

Reinforcement form	Advantages/disadvantages
Continuous strand mat	Good formability, wash resistance, high bulk factor, high part fill-out, uses glass fibers
Woven roving/fabric	High strength (biaxial), good formability
Unidirectional roving/fabric	High strength (unidirectional), high stiffness, good formability
Chopped-strand mat	Low formability, low wash resistance, low cost, high bulk factor, uses glass fibers
Preforms	Highly complex forms possible, little forming/handling necessary, high initial cost
Veils/surfacing mats	Good surface quality, wear resistance

Breakdown of costs of RTM versus hand lay-up

Item	Resin transfer molding	Hand lay-up
Product weight, kg (lb)	30 (65)	33 (75)
Production rate, pieces/month	1000	1000
Direct laborers	14	30
Materials cost:		
Resin	28.4%	27.4%
Glass fibers	27.7%(a)	26.4%
Others	0.3%	2.7%
Subtotal	56.4%	56.5%
Depreciation cost:		
Mold	9.0%(b)	3.0%(c)
Equipment	1.8%	0.7%
Subtotal	10.8%	3.7%
Scrap	1.6%	0.0%
Manufacturing cost	14.4%	39.6%
Fuel cost	0.0%	0.2%
Subtotal	16.0%	39.8%
Total	83.2%	100.0%

(a) Including the costs of auxiliary materials, fuel, manufacturing, depreciation of preformer, and preforming screen. (b) Life of the mold is assumed to be 2000 pieces. (c) Life of the mold is assumed to be 200 pieces

Breakdown of costs of RTM versus other molding techniques

Item	Resin transfer molding	Sheet molding compounds	Injection molding
Process operation			
Production volume	5000 to 10,000/press	25,000/press	30,000/press
Fixed assets	Moderate	High	High
Labor	High	Moderate	Moderate
Skill dependency	Considerable	Very low	Lowest
Operation	Movements/intersections	Flowing, neat	Flowing, neat
Inspection/control			
Raw materials	Yes	Compounds for degradation	Compounds for degradation
Products	Visual with attention	Visual, easy	Visual, easy
Finishing	Trim flash, and so on	Very little	Very little
Products			
Complexity	Preform limit	Yes	Best
Size	Big parts for low investment	Big parts if flat	Not very big parts
Tolerance	Good	Very good	Very good
Surface appearance	Gel-coated	Very good	Very good
Voids/wrinkles	Occasional	Extremely rare	Least
Reproducibility	Skill dependent	Yes	Yes
Cores/inserts	Possible	Not easy	Possible
Strength	Moderate	Best	Very good
Material usage			
Raw-material cost	Lowest	Highest	High
Handling/applying	Skill dependent	Easy	Automatic feed
Inventory	Raw materials	Dependent on number of types	Dependent on number of types
Precision	Skill dependent	Very good	Automatic feed
Waste	<3%	Very low	Attention runner
Scrap	Skill dependent	Cuts reusable	Low
Reinforcement flexibility	Yes	No	No
Mold			
Initial cost	Moderate	Very high	Very high
Cycle life	3000 to 4000 parts	Years	Years
Handling	With care	Metal	Metal
Preparation	In-factory	Special shop	Special shop
Maintenance	In-factory	Special machine shop/equipment	Special machine shop/equipment

Processing Parameters for Specific Materials

Typical molding conditions for unmodified acrylics

Conditions	Flow grade		
	Grade 5	Grade 6	Grade 8
Barrel temperatures, °C (°F)			
Rear	171 (340)	182 (360)	216 (420)
Center	182 (360)	193 (380)	227 (440)
Front	193 (380)	204 (400)	238 (460)
Nozzle	193 (380)	204 (400)	238 (460)
Melt temperature, °C (°F)	104 (400)	216 (420)	249 (480)
Mold temperature, °C (°F)	71 (160)	71 (160)	82 (180)
Screw speed, rpm (rad/s)	100 (10.5)	100 (10.5)	100 (10.5)
Injection fill speed, s	2 to 8	2 to 8	2 to 8
Cycle time, s	40 to 50	40 to 50	40 to 50

Part thickness, 3.2 mm (0.13 in.)

Production methods for cellular polymers

Type of polymer	Extrusion	Expandable formulation	Spraying	Froth foaming	Compression molding	Injection molding	Sintering	Leaching
Cellulose acetate	X							
Epoxy resin		X	X	X				X
Phenolic resin		X						
Polyethylene	X	X			X	X	X	X
Polystyrene	X	X				X	X	
Silicones		X						
Urea-formaldehyde resin				X				
Urethane polymers		X	X	X		X		
Latex foam rubber				X				
Natural rubber	X	X			X			
Synthetic elastomers	X	X			X			
Polyvinyl chloride	X	X		X	X	X		X
Ebonite					X			
Polytetrafluoroethylene							X	

Recommended LCP drying

	Vectra A130	Xydar RC210
Inlet air temperature, °C (°F)	150 (300)	150 (300)
Inlet air dew point, °C (°F)	0 (32)	-30 (-20)
Residence time, h	4	8

Dessicant bed hopper or tray driers are recommended

Typical molding conditions for HIPS

Condition	
Barrel temperature	
Rear zone, °C (°F)	160 to 210 (320 to 410)
Middle zone, °C (°F)	175 to 240 (350 to 465)
Front zone, °C (°F)	190 to 270 (375 to 520)
Melt temperature, °C (°F)	180 to 280 (355 to 535)
Mold temperature, °C (°F)	25 to 85 (80 to 185)
Back pressure, kPa (bar)	70 to 3100 (0.7 to 31)
Injection pressure, MPa (bar)	27.5 to 170 (275 to 1700)

Typical thermoforming conditions for HIPS

Temperature	°C	°F
Heater	500 to 800	930 to 1470
Sheet	160	320
Plug	110 to 150	230 to 300
Mold	90	195

Typical extrusion conditions for HIPS

General temperature profile	°C	°F
Barrel		
Zone 1 (feed)	150 to 205	300 to 400
Zone 2	160 to 220	320 to 430
Zone 3	175 to 230	350 to 445
Zone 4 (exit)	190 to 245	375 to 475
Die	190 to 245	375 to 475
Melt	175 to 260	350 to 500
Polishing rolls, down stack		
Top	65 to 100	150 to 212
Middle	38 to 90	100 to 195
Bottom	43 to 100	110 to 212

Typical acrylamate SRIM processing conditions

Parameter	Value
Injection flow	10 to 100 kg/min. (22 to 220 lb/min.), depending on part geometry, reinforcement-fiber loading, and filler loading in the resin stream
Shot size	As required within limits of the liquid time interval and the metering system
Reactant temperature	Ambient
Mold temperature	100 to 120 °C (212 to 250 °F)
Liquid time	6 to 17 s
Mold release	2006 wax (Chemtrend Inc., Park Chemical Company) 2044 wax-base coat (Chemtrend Inc.) (Test patch on other than steel or aluminum tools)
Machine ratio	Approximately 2 parts resin to 1 part isocyanate, with dry air blanket on resin tank and dry nitrogen blanket on isocyanate tank
Resin stream	7 to 14 MPa (1 to 2 ksi)
Isocyanate stream	7 to 14 MPa (1 to 2 ksi)

Secondary Operations, Including Machining

Secondary operations, material remain

Operation	Purpose	Relative cost
Deflash	Clean up mold parting line separation	Low, if tumble or wheel abrade; medium to high, otherwise
Degate	Remove sprue from injection, transfer, reaction injection, or resin transfer molding	Low
Drill	Produce small hole in a part	Low
Ream	Precision enlargement of existing hole	Low
Bore	Produce large, precision-sized holes	Medium
Turn	Create surface of revolution on part	Low to medium
Tap	Create internally threaded hole	Low
Mill	Create flat surface on part	Medium
Router	Remove material for irregular shape	Medium to high
Grind	Achieve precision dimension on flat or round surface; remove stocks on very hard material	Low to medium
Surface cleanup	Usually, hand sand	Medium to high
Surface contour	Contour cutting, contour grinding, three-dimensional numerical control machining	Very high

Secondary operations, additions

Operation	Purpose	Relative cost
Paint	Appearance, environmental protection	Low to medium
Plate	Appearance	Low, vacuum, metallize sputter
Bulk cost	Noise and vibration control protection	Medium, tank dip
Film decoration	Appearance, electrical conductivity	Medium
Patch and repair	Appearance, function	High

Secondary operations, assembly

Operation	Relative cost
Adhesive bond	Low to medium
Weld	Medium
Mechanical threaded fasteners	Medium
Rivets	Low to medium
Clips	Low
Snap-together	Low

Miscellaneous secondary operations

Operation	Remarks
Heat treat	Convert thermoplastic to cross-linked thermoset Force end reaction of trapped volatiles and liquids
Inspect	Visual, dimensional, and/or structural
Nondestructive testing	Critical-structure parts require this on either a 100% or statistical basis
Packing and packaging	Loose in a box, egg-crated, 100% wrapped, and so on; may be reusable or throw-away

Thread milling of plastics

Materials	Hardness, Rockwell	Condition(a)	Speed m/min.	Speed fpm	Feed mm/tooth/rev	Feed in./tooth/rev	HSS tool material ISO	HSS tool material AISI
Thermoplastics								
Acrylic, acetal, polycarbonate, polysulfone, polystyrene	M60 to 120	C, M, or E	90	300	0.050	0.002	S4, S2	M2, M7
Acrylonitrile-butadiene-styrene (ABS), polyarylether, polypropylene, polyethylene, cellulose acetate	R50 to 120	C, M, or E	105	340	0.050	0.002	S4, S2	M2, M7
Fluorocarbons: tetrafluoroethylene (TFE), chlorotrifluoroethylene (CTFE)	R74 to 95	M or E	90	300	0.050	0.002	S4, S2	M2, M7
Polyamides (nylons): unfilled–types 6, 6/6, 6/12, 11, 12	R78 to 120	M or E	115	375	0.050	0.002	S4, S2	M2, M7

(continued)

Thread milling of plastics (continued)

Materials	Hardness, Rockwell	Condition(a)	Speed m/min.	Speed fpm	Feed mm/tooth/rev	Feed in./tooth/rev	HSS tool material ISO	HSS tool material AISI
Thermosets								
Epoxy, melamine, phenolic	M100 to 128	C or M	115	375	0.050	0.002	S4, S2	M2, M7
Furan, polybutadiene	R40 to 100	C	58	190	0.038	0.0015	S9, S11	T15, M42
Silicone	Shore A15 to 65	C or M	46	150	0.038	0.0015	S9, S11	T15, M42
Polyimide	E40 to 50	M or E	115	375	0.050	0.002	S4, S2	M2, M7
Polyurethane	Shore A65 to 95	C	58	190	0.038	0.0015	S9, S11	T15, M42
	Shore D55 to 75	C	69	225	0.050	0.002	S4, S2	M2, M7
Allyl (DAP)	M95 to 100	C	90	300	0.050	0.002	S4, S2	M2, M7
Allyl, fiber-filled	M108 to 115	F and M	58	190	0.038	0.0015	S9, S11	T15, M42

(a) C, cast; M, molded; E, extruded; F, filled

Face milling of plastics

Materials	Hardness, Rockwell	Condition(a)	Depth of cut		Speed		High speed steel tool: Feed tooth		High speed steel tool: Tool material	
			mm	in.	m/min.	fpm	mm	in.	ISO	AISI
Thermoplastics										
Acrylic, acetal, polycarbonate, polysulfone, polystyrene	M60 to 120	C, M, E	1	0.040	120	400	0.13	0.005	S4, S2	M2, M3
			4	0.150	105	350	0.20	0.008	S4, S2	M2, M3
			8	0.300	90	300	0.25	0.010	S4, S2	M2, M3
Acrylonitrile-butadiene-styrene (ABS), polyarylether, polypropylene, polyethylene, cellulose acetate	R50 to 120	C, M, or E	1	0.040	135	450	0.13	0.005	S4, S2	M2, M3
			4	0.150	120	400	0.20	0.008	S4, S2	M2, M3
			8	0.300	105	350	0.25	0.010	S4, S2	M2, M3
Fluorocarbons: tetrafluoroethylene (TFE), chlorotrifluoroethylene (CTFE)	R74 to 95	M or E	1	0.040	120	400	0.13	0.005	S4, S2	M2, M3
			4	0.150	105	350	0.20	0.008	S4, S2	M2, M3
			8	0.300	90	300	0.25	0.010	S4, S2	M2, M3
Polyamides (nylons): unfilled–types 6, 6/6, 6/12, 11, 12	R78 to 120	M or E	1	0.040	150	500	0.15	0.006	S4, S2	M2, M3
			4	0.150	135	450	0.25	0.010	S4, S2	M2, M3
			8	0.300	120	400	0.36	0.014	S4, S2	M2, M3
Polyamides (nylons): 35% glass-reinforced–types 6, 6/6	R78 to 120	F and M	1	0.040	...	...	...	...	...	...
			4	0.150	...	...	...	...	...	...
			8	0.300	...	...	...	...	...	...
Polyamides (nylons): 35% glass-reinforced–types 6/10, 6/12	E40 to 50	F and M	1	0.040	...	...	...	...	...	...
			4	0.150	...	...	...	...	...	...
			8	0.300	...	...	...	...	...	...
Thermosets										
Epoxy, melamine, phenolic	M100 to 128	C or M	1	0.040	150	500	0.13	0.005	S4, S2	M2, M7
			4	0.150	135	450	0.20	0.008	S4, S2	M2, M7
			8	0.300	105	350	0.25	0.010	S4, S2	M2, M7
Furan, polybutadiene	R40 to 100	C	1	0.040	76	250	0.13	0.005	S9, S11	T15, M42
			4	0.150	60	200	0.20	0.008	S9, S11	T15, M42
			8	0.300	46	150	0.25	0.010	S9, S11	T15, M42
Silicone	Shore A15 to 65	C or M	1	0.040	60	200	0.13	0.005	S9, S11	T15, M42
			4	0.150	53	175	0.20	0.008	S9, S11	T15, M42
			8	0.300	38	125	0.25	0.010	S9, S11	T15, M42
Silicone, glass–filled	M80 to 90	F and M	1	0.040	...	...	...	...	...	...
			4	0.150	...	...	...	...	...	...
			8	0.300	...	...	...	...	...	...
Polyimide	E40 to 50	M or E	1	0.040	150	500	0.13	0.005	S4, S2	M2, M7
			4	0.150	135	450	0.20	0.008	S4, S2	M2, M7
			8	0.300	150	350	0.25	0.010	S4, S2	M2, M7
Polyimide, glass-filled	M109 to 115	F and M	1	0.040	...	...	...	...	...	...
			4	0.150	...	...	...	...	...	...
			8	0.300	...	...	...	...	...	...
Polyurethane	Shore A65 to 95	C	1	0.040	76	250	0.13	0.005	S9, S11	T15, M42
			4	0.150	60	200	0.20	0.008	S9, S11	T15, M42
			8	0.300	46	150	0.25	0.010	S9, S11	T15, M42
	Shore D55 to 75	C	1	0.040	90	300	0.13	0.005	S4, S2	M2, M7
			4	0.150	76	250	0.20	0.008	S4, S2	M2, M7
			8	0.300	60	200	0.25	0.010	S4, S2	M2, M7
Allyl (DAP)	M95 to 100	C	1	0.040	120	400	0.13	0.005	S4, S2	M2, M3
			4	0.150	105	350	0.20	0.008	S4, S2	M2, M3
			8	0.300	90	300	0.25	0.010	S4, S2	M2, M3

(continued)

Face milling of plastics (continued)

Materials	Hardness, Rockwell	Condition(a)	Depth of cut mm	Depth of cut in.	Speed m/min.	Speed fpm	High speed steel tool: Feed tooth mm	Feed tooth in.	Tool material ISO	Tool material AISI
Allyl, glass-filled	E80 to 87	F and M	1	0.040	...	...	...	...	...	...
			4	0.150	...	...	...	...	...	...
			8	0.300	...	...	...	...	...	...
Allyl, fiber-filled	M108 to 115	F and M	1	0.040	76	250	0.13	0.005	S9, S11	T15, M42
			4	0.150	60	200	0.20	0.008	S9, S11	T15, M42
			8	0.300	53	175	0.25	0.010	S9, S11	T15, M42

Materials	Hardness, Rockwell	Condition(a)	Depth of cut mm	Depth of cut in.	Speed: Brazed m/min.	Brazed fpm	Indexable m/min.	Indexable fpm	Carbide tool (uncoated): Feed per tooth mm	Feed per tooth in.	Tool material grade ISO	Tool material grade AISI
Thermoplastics												
Acrylic, acetal, polycarbonate, polysulfone, polystyrene	M60 to 120	C, M, or E	1	0.040	200	650	200	650	0.13	0.005	K20, M20	C-2
			4	0.150	185	600	185	600	0.18	0.007	K20, M20	C-2
			8	0.300	170	550	170	550	0.23	0.009	K20, M20	C-2
Acrylonitrile-butadiene-styrene (ABS), polyarylether, polypropylene, polyethylene, cellulose acetate	R50 to 120	C, M, or E	1	0.040	230	750	230	750	0.102	0.004	K20, M20	C-2
			4	0.150	215	700	215	700	0.18	0.007	K20, M20	C-2
			8	0.300	200	650	200	650	0.23	0.009	K20, M20	C-2
Fluorocarbons: tetrafluoroethylene (TFE), chlorotrifluoroethylene (CTFE)	R74 to 95	M or E	1	0.040	200	650	200	650	0.102	0.004	K20, M20	C-2
			4	0.150	185	600	185	600	0.18	0.007	K20, M20	C-2
			8	0.300	170	550	170	550	0.23	0.009	K20, M20	C-2
Polyamides (nylons): unfilled–types 6, 6/6, 6/12, 11, 12	R78 to 120	M or E	1	0.040	260	850	260	850	0.15	0.006	K20, M20	C-2
			4	0.150	230	750	230	750	0.20	0.008	K20, M20	C-2
			8	0.300	215	700	215	700	0.25	0.010	K20, M20	C-2
Polyamides (nylons): 35% glass-reinforced–types 6, 6/6	R78 to 120	F and M	1	0.040	200	650	200	650	0.102	0.004	K20, M20	C-2
			4	0.150	185	600	185	600	0.15	0.006	K20, M20	C-2
			8	0.300	170	550	170	550	0.20	0.008	K20, M20	C-2
Polyamides (nylons): 35% glass-reinforced–types 6/10, 6/12	E40 to 50	F and M	1	0.040	170	550	170	550	0.102	0.004	K20, M20	C-2
			4	0.150	150	500	150	500	0.15	0.006	K20, M20	C-2
			8	0.300	135	450	135	450	0.20	0.008	K20, M20	C-2
Thermosets												
Epoxy, melamine, phenolic	M100 to 128	C or M	1	0.040	260	850	260	850	0.102	0.004	K20, M20	C-2
			4	0.150	230	750	230	750	0.15	0.006	K20, M20	C-2
			8	0.300	215	700	315	700	0.20	0.008	K20, M20	C-2
Furan, polybutadiene	R40 to 100	C	1	0.040	145	475	145	475	0.102	0.004	K20, M20	C-2
			4	0.150	130	420	130	425	0.15	0.006	K20, M20	C-2
			8	0.300	115	375	115	375	0.20	0.008	K20, M20	C-2
Silicone	Shore A15 to 65	C or M	1	0.040	145	475	145	475	0.102	0.004	K20, M20	C-2
			4	0.150	130	425	130	425	0.15	0.006	K20, M20	C-2
			8	0.300	115	375	115	375	0.20	0.008	K20, M20	C-2
Silicone, glass-filled	M80 to 90	F and M	1	0.040	130	425	130	425	0.102	0.004	K20, M20	C-2
			4	0.150	115	375	115	375	0.15	0.006	K20, M20	C-2
			8	0.300	100	325	100	325	0.20	0.008	K20, M20	C-2
Polyimide	E40 to 50	M or E	1	0.040	260	850	260	850	0.13	0.005	K20, M20	C-2
			4	0.150	230	750	230	750	0.20	0.008	K20, M20	C-2
			8	0.300	200	650	200	650	0.25	0.010	K20, M20	C-2
Polyimide, glass-filled	M109 to 115	F and M	1	0.040	160	525	160	525	0.13	0.005	K20, M20	C-2
			4	0.150	145	475	145	475	0.20	0.008	K20, M20	C-2
			8	0.300	130	425	130	425	0.25	0.010	K20, M20	C-2

(continued)

Face milling of plastics (continued)

Materials	Hardness, Rockwell	Condition(a)	Depth of cut mm	Depth of cut in.	Speed: Brazed m/min.	Speed: Brazed fpm	Speed: Indexable m/min.	Speed: Indexable fpm	Carbide tool (uncoated): Feed per tooth mm	Feed per tooth in.	Tool material grade ISO	Tool material grade AISI
Polyurethane	Shore A65 to 95	C	1	0.040	145	475	145	475	0.102	0.004	K20, M20	C-2
			4	0.150	130	425	130	425	0.15	0.006	K20, M20	C-2
			8	0.300	115	375	115	375	0.20	0.008	K20, M20	C-2
	Shore D55 to 75	C	1	0.040	160	525	160	525	0.102	0.004	K20, M20	C-2
			4	0.150	145	475	145	475	0.15	0.006	K20, M20	C-2
			8	0.300	130	425	130	425	0.20	0.008	K20, M20	C-2
Allyl (DAP)	M95 to 100	C	1	0.040	200	650	200	650	0.15	0.006	K20, M20	C-2
			4	0.150	185	600	185	600	0.20	0.008	K20, M20	C-2
			8	0.300	170	550	170	550	0.25	0.010	K20, M20	C-2
Allyl, glass-filled	E80 to 87	F and M	1	0.040	130	425	130	425	0.15	0.006	K20, M20	C-2
			4	0.150	115	375	115	375	0.20	0.008	K20, M20	C-2
			8	0.300	100	325	100	325	0.25	0.010	K20, M20	C-2
Allyl, fiber-filled	M108 to 115	F and M	1	0.040	160	525	160	525	0.15	0.006	K20, M20	C-2
			4	0.150	145	475	145	475	0.20	0.008	K20, M20	C-2
			8	0.300	130	425	130	425	0.25	0.010	K20, M20	C-2

(a) C, cast; M, molded; E, extruded; F, filled

Side and slot milling of plastics (with arbor-mounted cutters)

Materials	Hardness, Rockwell	Condition(a)	Depth of cut mm	Depth of cut in.	High speed steel tool: Speed m/min.	Speed fpm	Feed tooth mm	Feed tooth in.	Tool material ISO	Tool material AISI
Thermoplastics										
Acrylic, acetal, polycarbonate, polysulfone, polystyrene	M60 to 120	C, M, or	1	0.040	90	300	0.102	0.004	S4, S2	M2, M3
			4	0.150	79	260	0.13	0.005	S2, S5	M2, M3
			8	0.300	67	220	0.18	0.007	S4, S2	M2, M3
Acrylonitrile-butadiene-styrene (ABS), polyarylether, polypropylene, polyethylene, cellulose acetate	R50 to 120	C, M, E	1	0.040	105	340	0.102	0.004	S4, S2	M2, M3
			4	0.150	90	300	0.13	0.005	S4,S2	M2, M3
			8	0.300	79	260	0.18	0.007	S4, S2	M2, M3
Fluorocarbons: tetrafluoroethylene (TFE), Chlorotrifluoroethylene (CTFE)	R74 to 95	M or E	1	0.040	90	300	0.102	0.004	S4, S2	M2, M3
			4	0.150	79	260	0.13	0.005	S4, S2	M2, M3
			8	0.300	67	220	0.18	0.007	S4, S2	M2, M3
Polyamides (nylons): unfilled–types 6, 6/6, 6/12, 11, 12	R78 to 120	M or E	1	0.040	115	380	0.102	0.004	S4, S2	M2, M3
			4	0.130	105	340	0.15	0.006	S4, S2	M2, M3
			8	0.300	90	300	0.20	0.008	S4, S2	M2, M3
Polyamides (nylons): 35% glass reinforced–types 6, 6/6	R78 to 120	F and M	1	0.040	...	...	...	...	...	...
			4	0.150	...	...	...	...	...	...
			8	0.300	...	...	...	...	...	...
Polyamides (nylons): 35% glass reinforced–types 6/10, 6/12	E40 to 50	F and M	1	0.040	...	...	...	...	...	...
			4	0.150	...	...	...	...	...	...
			8	0.300	...	...	...	...	...	...
Thermosets										
Epoxy, melamine, phenolic	M100 to 128	C or M	1	0.040	115	380	0.102	0.004	S4, S2	M2, M7
			4	0.150	105	340	0.13	0.005	S4, S2	M2, M7
			8	0.300	79	260	0.18	0.007	S4, S2	M2, M7

(continued)

Side and slot milling of plastics (with arbor-mounted cutters) (continued)

Materials	Hardness, Rockwell	Condition(a)	Depth of cut mm	Depth of cut in.	Speed m/min.	Speed fpm	High speed steel tool: Feed tooth mm	Feed tooth in.	Tool material ISO	Tool material AISI
Furan, polybutadiene	R40 to 100	C	1	0.040	60	200	0.102	0.004	S9, S11	T15, M42
			4	0.150	46	150	0.13	0.005	S9, S11	T15, M42
			8	0.300	35	115	0.18	0.007	S9, S11	T15, M42
Silicone	Shore A15 to 65	C or M	1	0.040	46	150	0.102	0.004	S9, S11	T15, M42
			4	0.150	41	135	0.13	0.005	S9, S11	T15, M42
			8	0.300	29	95	0.18	0.007	S9, S11	T15, M42
Silicone, glass-filled	M80 to 90	F and M	1	0.040	...	...	...	...	...	...
			4	0.150	...	...	...	...	...	...
			8	0.300	...	...	...	...	...	...
Polyimide	E40 to 50	M or E	1	0.040	115	380	0.102	0.004	S4, S2	M2, M7
			4	0.150	105	340	0.13	0.005	S4, S2	M2, M7
			8	0.300	79	260	0.18	0.007	S4, S2	M2, M7
Polyimide, glass-filled	M109 to 115	F and M	1	0.040	...	...	...	...	...	...
			4	0.150	...	...	...	...	...	...
			8	0.300	...	...	...	...	...	...
Polyurethane	Shore A65 to 95	C	1	0.040	60	200	0.102	0.004	S9, S11	T15, M42
			4	0.150	46	150	0.13	0.004	S9, S11	T15, M42
			8	0.300	35	115	0.18	0.007	S9, S11	T15, M42
	Shore D55 to 75	C	1	0.040	67	220	0.102	0.004	S4, S2	M2, M7
			4	0.150	60	200	0.13	0.005	S4, S2	M2, M7
			8	0.300	46	150	0.18	0.007	S4, S2	M2, M7
Allyl (DAP)	M95 to 100	C	1	0.040	90	300	0.102	0.004	S4, S2	M2, M3
			4	0.150	79	260	0.13	0.005	S4, S2	M2, M3
			8	0.300	67	220	0.18	0.007	S4, S3	M2, M3
Allyl, glass-filled	E80 to 87	F and M	1	0.040	...	...	...	...	...	...
			4	0.150	...	...	...	...	...	...
			8	0.300	...	...	...	...	...	...
Allyl, fiber-filled	M108 to 115	F and M	1	0.040	60	200	0.102	0.004	S9, S11	T15, M42
			4	0.150	46	150	0.13	0.005	S9, S11	T15, M42
			8	0.300	41	135	0.18	0.007	S9, S11	T15, M42

Materials	Hardness, Rockwell	Condition(a)	Depth of cut mm	Depth of cut in.	Speed: Brazed m/min.	Brazed fpm	Indexable m/min.	Indexable fpm	Carbide tool (uncoated): Feed per tooth mm	Feed per tooth in.	Tool material grade ISO	Tool material grade AISI
Thermoplastics												
Acrylic, acetal, polycarbonate, polysulfone, polystyrene	M60 to 120	C, M, or E	1	0.040	145	475	145	475	0.102	0.004	K20, M20	C-2
			4	0.150	135	450	135	450	0.13	0.005	K20, M20	C-2
			8	0.300	120	400	120	400	0.15	0.006	K20, M20	C-2
Acrylonitrile-butadiene-styrene (ABS), polyarylether, polypropylene, polyethylene, cellulose acetate	R50 to 120	C, M, or E	1	0.040	170	550	170	550	0.102	0.004	K20, M20	C-2
			4	0.150	160	525	160	525	0.13	0.005	K20, M20	C-2
			8	0.300	145	475	145	475	0.15	0.006	K20, M20	C-2
Fluorocarbons: tetrafluoroethylene (TFE), chlorotrifluoroethylene (CTFE)	R74 to 95	M or E	1	0.040	145	475	145	475	0.075	0.003	K20, M20	C-2
			4	0.150	135	450	135	450	0.102	0.004	K20, M20	C-2
			8	0.300	120	400	120	400	0.13	0.005	K20, M20	C-2
Polyamides (nylons): unfilled–types 6, 6/6, 6/12, 11, 12	R78 to 120	M or E	1	0.040	200	650	200	650	0.102	0.004	K20, M20	C-2
			4	0.150	170	550	170	550	0.13	0.005	K20, M20	C-2
			8	0.300	160	525	160	525	0.18	0.007	K20, M20	C-2

(continued)

Side and slot milling of plastics (with arbor-mounted cutters) (continued)

Materials	Hardness, Rockwell	Condition(a)	Depth of cut		Speed				Carbide tool (uncoated)			
					Brazed		Indexable		Feed per tooth		Tool material grade	
			mm	in.	m/min.	fpm	m/min.	fpm	mm	in.	ISO	AISI
Polyamides (nylons): 35% glass-reinforced–types 6, 6/6	R78 to 120	F and M	1	0.040	145	475	145	475	0.075	0.003	K20, M20	C-2
			4	0.150	135	450	135	450	0.102	0.004	K20, M20	C-2
			8	0.300	120	400	120	400	0.13	0.005	K20, M20	C-2
Polyamides (nylons): 35% glass-reinforced–types 6/10, 6/12	E40 to 50	F and M	1	0.040	120	400	120	400	0.075	0.003	K20, M20	C-2
			4	0.150	115	380	115	380	0.102	0.004	K20, M20	C-2
			8	0.300	105	340	105	340	0.13	0.005	K20, M20	C-2
Thermosets												
Epoxy, melamine, phenolic	M100 to 128	C or M	1	0.040	200	650	200	650	0.075	0.003	K20, M20	C-2
			4	0.150	170	550	170	550	0.102	0.004	K20, M20	C-2
			8	0.300	160	525	160	525	0.13	0.005	K20, M20	C-2
Furan, polybutadiene	R40 to 100	C	1	0.040	110	360	110	360	0.075	0.003	K20, M20	C-2
			4	0.150	100	320	100	320	0.102	0.004	K20, M20	C-2
			8	0.300	85	280	85	280	0.13	0.005	K20, M20	C-2
Silicone	Shore A15 to 65	C or M	1	0.040	110	360	110	360	0.075	0.003	K20, M20	C-2
			4	0.150	100	320	100	320	0.102	0.004	K20, M20	C-2
			8	0.300	85	280	85	280	0.13	0.005	K20, M20	C-2
Silicone, glass-filled	M80 to 90	F and M	1	0.040	100	320	100	320	0.075	0.003	K20, M20	C-2
			4	0.150	85	280	85	280	0.102	0.004	K20, M20	C-2
			8	0.300	73	240	73	240	0.13	0.005	K20, M20	C-2
Polyimide	E40 to 50	M or E	1	0.040	200	650	200	650	0.102	0.004	K20, M20	C-2
			4	0.150	170	550	170	550	0.13	0.005	K20, M20	C-2
			8	0.300	145	475	145	475	0.18	0.007	K20, M20	C-2
Polyimide, glass-filled	M109 to 115	F and M	1	0.040	120	390	120	390	0.102	0.004	K20, M20	C-2
			4	0.150	110	360	110	360	0.13	0.005	K20, M20	C-2
			8	0.300	100	320	100	320	0.18	0.007	K20, M20	C-2
Polyurethane	Shore A65 to 95	C	1	0.040	110	360	110	360	0.075	0.003	K20, M20	C-2
			4	0.150	100	320	100	320	0.102	0.004	K20, M20	C-2
			8	0.300	85	280	85	280	0.13	0.005	K20, M20	C-2
	Shore D55 to 75	C	1	0.040	120	390	120	390	0.075	0.003	K20, M20	C-2
			4	0.150	110	360	110	360	0.102	0.004	K20, M20	C-2
			8	0.300	100	320	100	320	0.13	0.005	K20, M20	C-2
Allyl (DAP)	M95 to 100	C	1	0.040	145	475	145	475	0.102	0.004	K20, M20	C-2
			4	0.150	135	450	135	450	0.13	0.005	K20, M20	C-2
			8	0.300	120	400	120	400	0.18	0.007	K20, M20	C-2
Allyl, glass-filled	E80 to 87	F and M	1	0.040	100	320	100	320	0.102	0.004	K20, M20	C-2
			4	0.150	85	280	85	280	0.13	0.005	K20, M20	C-2
			8	0.300	73	240	73	240	0.18	0.007	K20, M20	C-2
Allyl, fiber-filled	M108 to 115	F and M	1	0.040	120	390	120	390	0.102	0.004	K20, M20	C-2
			4	0.150	110	360	110	360	0.13	0.005	K20, M20	C-2
			8	0.300	100	320	100	320	0.18	0.007	K20, M20	C-2

(a) C, cast; M, molded; E, extruded; F, filled

Drilling of plastics

Materials	Hardness, Rockwell	Condition(a)	Speed m/min.	Speed fpm	Feed, for reamer diameter of: 1.5 mm (1/16 in.) mm/rev	1.5 mm (1/16 in.) ipr	3 mm (1/8 in.) mm/rev	3 mm (1/8 in.) ipr	6 mm (1/4 in.) mm/rev	6 mm (1/4 in.) ipr	12 mm (1/2 in.) mm/rev	12 mm (1/2 in.) ipr
Thermoplastics												
Acrylic, acetal, polycarbonate, polysulfone, polystyrene	M60 to 120	C, M, or E	30	100	0.025	0.001						
			60	200	...	...	0.050	0.002	0.102	0.004	0.13	0.005
Acrylonitrile-butadiene-styrene (ABS), polyarylether, polypropylene, polyethylene, cellulose acetate	R50 to 120	C, M, or E	46	150	0.025	0.001						
			76	250	...	...	0.050	0.002	0.102	0.004	0.13	0.005
Fluorocarbons: tetrafluoroethylene (TFE), chlorotrifluoroethylene (CTFE)	R74 to 95	M or E	30	100	0.025	0.001						
			60	200	...	...	0.050	0.002	0.102	0.004	0.13	0.005
Polyamides (nylons): unfilled–types 6, 6/6, 6/12, 11, 12	R78 to 120	M or E	46	100	0.025	0.001						
			76	250	...	...	0.050	0.002	0.102	0.004	0.13	0.005
Polyamides (nylons): 35% glass-reinforced–types 6, 6/6	R78 to 120	F and M	23	75	0.025	0.001						
			46	100	...	...	0.050	0.002	0.102	0.004	0.13	0.005
Polyamides (nylons): 35% glass-reinforced–types 6/10, 6/12	E40 to 50	F and M	20	65	0.025	0.001						
			38	125	...	...	0.050	0.002	0.102	0.004	0.13	0.005
Thermosets												
Epoxy, melamine, phenolic	M100 to 128	C or M	30	100	0.025	0.001						
			60	200	...	...	0.050	0.002	0.075	0.003	0.102	0.004
Furan, polybutadiene	R40 to 100	C	30	100	0.025	0.001						
			60	200	...	...	0.050	0.002	0.075	0.003	0.102	0.004
Silicone	Shore A15 to 65	C or M	20	65	0.025	0.001						
			38	125	...	...	0.050	0.002	0.075	0.003	0.102	0.004
Silicone, glass-filled	M80 to 90	F and M	20	65	0.025	0.001						
			38	125	...	...	0.025	0.001	0.050	0.002	0.075	0.003
Polyimide	E40 to 50	M or E	30	100	0.025	0.001						
			60	200	...	...	0.050	0.002	0.102	0.004	0.13	0.005
Polymide, glass-filled	M109 to 115	F and M	20	65	0.025	0.001						
			38	125	...	...	0.025	0.001	0.102	0.002	0.075	0.003
Polyurethane	Shore A65 to 95	C	30	100	0.025	0.001						
			60	200	...	...	0.050	0.002	0.075	0.003	0.102	0.004
	Shore D55 to 75	C	46	150	0.025	0.001						
			76	250	...	...	0.050	0.002	0.102	0.004	0.13	0.005
Allyl (DAP)	M95 to 100	C	30	100	0.025	0.001						
			60	200	...	...	0.050	0.002	0.102	0.004	0.13	0.005
Allyl, glass-filled	E80 to 87	F and M	20	65	0.025	0.001						
			30	100	...	...	0.050	0.002	0.102	0.004	0.13	0.005
Allyl, fiber-filled	M108 to 115	F and M	23	75	0.025	0.001						
			46	150	...	...	0.050	0.002	0.102	0.004	0.15	0.006

Materials	Hardness, Rockwell	Condition(a)	Feed, for reamer diameter of: 18 mm (3/4 in.) m/rev	18 mm (3/4 in.) ipr	25 mm (1 in.) mm/rev	25 mm (1 in.) ipr	35 mm (1-1/2 in.) mm/rev	35 mm (1-1/2 in.) ipr	50 mm (2 in.) mm/rev	50 mm (2 in.) ipr	Tool material grade ISO	Tool material grade AISI or C
Thermoplastics												
Acrylic, acetal, polycarbonate, polysulfone, polystyrene	M60 to 120	C, M, or E	0.15	0.006	0.20	0.008	0.25	0.010	0.30	0.012	S2, S3	M10, M7, M1
Acrylonitrile-butadiene-styrene (ABS), polyarylether, polypropylene, polyethylene, cellulose acetate	R50 to 120	C, M, or E	0.15	0.006	0.20	0.008	0.25	0.010	0.30	0.012	S2, S3	M10, M7, M1

(continued)

Drilling of plastics (continued)

Materials	Hardness, Rockwell	Condition(a)	Feed, for reamer diameter of: 18 mm (3/4 in.) m/rev	ipr	25 mm (1 in.) mm/rev	ipr	35 mm (1-1/2 in.) mm/rev	ipr	50 mm (2 in.) mm/rev	ipr	Tool material grade ISO	AISI or C
Fluorocarbons: tetrafluoroethylene (TFE), chlorotrifluoroethylene (CTFE)	R74 to 95	M or E	0.15	0.006	0.20	0.008	0.25	0.010	0.30	0.012	S2, S3	M10, M7, M1
Polyamides (nylons): unfilled–types 6, 6/6, 6/12, 11, 12	R78 to 120	M or E	0.15	0.006	0.20	0.008	0.25	0.010	0.30	0.012	S2, S3	M10, M7, M1
Polyamides (nylons): 35% glass-reinforced–types 6, 6/6	R78 to 120	F and M	0.15	0.006	0.20	0.008	0.25	0.010	0.30	0.012	S9, S11	T15, M42
Polyamides (nylons): 35% glass-reinforced–types 6/10, 6/12	E40 to 50	F and M	0.15	0.006	0.20	0.008	0.25	0.010	0.30	0.012	S9, S11	T15, M42
Thermosets												
Epoxy, melamine, phenolic	M100 to 128	C or M	0.13	0.005	0.15	0.006	0.20	0.008	0.25	0.010	S2, S3	M10, M7, M1
Furan, polybutadiene	R40 to 100	C	0.13	0.005	0.15	0.006	0.20	0.008	0.25	0.010	S2, S3	M10, M7, M1
Silicone	Shore A15 to 65	C or M	0.15	0.006	0.20	0.008	0.20	0.008	0.25	0.010	S2, S3	M10, M7, M1
Silicone, glass-filled	M80 to 90	F and M	0.102	0.004	0.15	0.006	0.20	0.008	0.25	0.010	S9, S11	T15, M42
Polyimide	E40 to 50	M or E	0.15	0.006	0.20	0.008	0.25	0.010	0.30	0.012	S2, S3	M10, M7, M1
Polymide, glass-filled	M109 to 115	F and M	0.102	0.004	0.15	0.006	0.20	0.008	0.25	0.010	S9, S11	T15, M42
Polyurethane	Shore A65 to 95	C	0.13	0.005	0.15	0.006	0.20	0.008	0.25	0.010	S2, S3	M10, M7, M1
	Shore D 55 to 75	C	0.15	0.006	0.20	0.008	0.25	0.010	0.30	0.012	S2, S3	M10, M7, M1
Allyl (DAP)	M95 to 100	C	0.15	0.006	0.20	0.008	0.25	0.010	0.30	0.012	S2, S3	M10, M7, M1
Allyl, glass-filled	E80 to 87	F and M	0.15	0.006	0.20	0.008	0.25	0.010	0.30	0.012	S9, S11	T15, M42
Allyl, fiber-filled	M108 to 115	F and M	0.20	0.008	0.25	0.010	0.30	0.012	0.40	0.015	S9, S11	T15, M42

(a) C, cast; M, molded; E, extruded; F, filled

Reaming of plastics (roughing or finishing)

Materials	Hardness, Rockwell	Condition(a)	Speed m/min.	fpm	Feed, for reamer diameter of: 3 mm (1/8 in.) mm/rev	ipr	6 mm (1/4 in.) mm/rev	ipr	12 mm (1/2 in.) mm/rev	ipr
Thermoplastics										
Acrylic, acetal, polycarbonate, polysulfone, polystyrene	M60 to 120	C, M, or E	20 40	65 130	0.050	0.002	0.050	0.002	0.102	0.004
Acrylonitrile-butadiene-styrene (ABS), polyarylether, polypropylene, polyethylene, cellulose acetate	R50 to 120	C, M, or E	30 50	100 165	0.050	0.002	0.050	0.002	0.102	0.004
Fluorocarbons: tetrafluoroethylene (TFE), chlorotrifluoroethylene (CTFE)	R74 to 95	M or E	20 40	65 130	0.050	0.002	0.050	0.002	0.102	0.004
Polyamides (nylons): unfilled–types 6, 6/6, 6/12, 11, 12	R78 to 120	M or E	20 50	65 165	0.050	0.002	0.050	0.002	0.102	0.004
Polyamides (nylons): 35% glass-reinforced–types 6, 6/6	R78 to 120	F and M	15 30	50 100	0.050	0.002	0.050	0.002	0.075	0.003
Polyamides (nylons): 35% glass-reinforced–types 6/10, 6/12	E40 to 50	F and M	14 26	45 85	0.050	0.002	0.050	0.002	0.075	0.003
Thermosets										
Epoxy, melamine, phenolic	M100 to 128	C or M	20 40	65 130	0.050	0.002	0.050	0.002	0.102	0.004
Furan, polybutadiene	R40 to 100	C	20 40	65 130	0.050	0.002	0.050	0.002	0.102	0.004

(continued)

Reaming of plastics (roughing or finishing) (continued)

Materials	Hardness, Rockwell	Condition(a)	Speed m/min.	Speed fpm	Feed, for reamer diameter of: 3 mm (1/8 in.) mm/rev	3 mm (1/8 in.) ipr	6 mm (1/4 in.) mm/rev	6 mm (1/4 in.) ipr	12 mm (1/2 in.) mm/rev	12 mm (1/2 in.) ipr
Silicone	Shore A15 to 65	C or M	14 26	45 85	0.050	0.002	0.050	0.002	0.102	0.004
Silicone, glass-filled	M80 to 90	F and M	14 26	45 85	0.050	0.002	0.050	0.002	0.075	0.003
Polyimide	E40 to 50	M or E	20 40	65 130	0.050	0.002	0.050	0.002	0.102	0.004
Polymide, glass-filled	M109 to 115	F and M	14 26	45 85	0.050	0.002	0.050	0.002	0.075	0.003
Polyurethane	Shore A65 to 95	C	20 40	65 130	0.050	0.002	0.050	0.002	0.075	0.003
	Shore D 55 to 75	C	30 50	100 165	0.050	0.002	0.050	0.002	0.102	0.004
Allyl (DAP)	M95 to 100	C	20 40	65 130	0.050	0.002	0.050	0.002	0.102	0.004
Allyl, glass-filled	E80 to 87	F and M	14 26	45 85	0.050	0.002	0.050	0.002	0.075	0.003
Allyl, fiber-filled	M108 to 115	F and M	14 26	45 85	0.050	0.002	0.050	0.002	0.075	0.003

Materials	Hardness, Rockwell	Condition(a)	Feed, for reamer diameter of: 25 mm (1 in.) mm/rev	25 mm (1 in.) ipr	35 mm (1-1/2 in.) mm/rev	35 mm (1-1/2 in.) ipr	50 mm (2 in.) mm/rev	50 mm (2 in.) ipr	Tool material grade ISO	Tool material grade AISI or C
Thermoplastics										
Acrylic, acetal, polycarbonate, polysulfone, polystyrene	M60 to 120	C, M, or E	0.15	0.006	0.20	0.008	0.25	0.010	S3, S4, S2, K10	M1, M2, M7, C-2
Acrylonitrile-butadiene-styrene (ABS), polyarylether, polypropylene, polyethylene, cellulose acetate	R50 to 120	C, M, or E	0.15	0.006	0.20	0.008	0.25	0.010	S3, S4, S2,K10	M1, M2, M7, C-2
Fluorocarbons: tetrafluoroethylene (TFE), chlorotrifluoroethylene (CTFE)	R74 to 95	M or E	0.15	0.006	0.20	0.008	0.25	0.010	S3, S4, S2, K10	M1, M2, M7, C-2
Polyamides (nylons): unfilled–types 6, 6/6, 6/12, 11, 12	R78 to 120	M or E	0.15	0.006	0.20	0.008	0.25	0.010	S3, S4, S2, K10	M1, M2, M7, C-2
Polyamides (nylons): 35% glass-reinforced–types 6, 6/6	R78 to 120	F and M	0.075	0.003	0.102	0.004	0.13	0.005	S9, S11, K10	T15, M42, C-2
Polyamides (nylons): 35% glass-reinforced–types 6/10, 6/12	E40 to 50	F and M	0.075	0.003	0.102	0.004	0.13	0.005	S9, S11, K10	T15, M42, C-2
Thermosets										
Epoxy, melamine, phenolic	M100 to 128	C or M	0.15	0.006	0.20	0.008	0.25	0.010	S3, S4, S2, K10	M1, M2, M7, C-2
Furan, polybutadiene	R40 to 100	C	0.15	0.006	0.20	0.008	0.25	0.010	S3, S4, S2, K10	M1, M2, M7, C-2
Silicone	Shore A15 to 65	C or M	0.15	0.006	0.20	0.008	0.25	0.010	S9, S11, K10	T15, M42, C-2
Silicone, glass-filled	M80 to 90	F and M	0.075	0.003	0.102	0.004	0.13	0.005	S9, S11, K10	T15, M42, C-2
Polyimide	E40 to 50	M or E	0.15	0.006	0.20	0.008	0.25	0.010	S9, S11, K10	M1, M2, M7, C-2
Polymide, glass-filled	M109 to 115	F and M	0.075	0.003	0.102	0.004	0.13	0.005	S9, S11, K10	T15, M42, C-2
Polyurethane	Shore A65 to 95	C	0.075	0.003	0.102	0.004	0.13	0.005	S3, S4, S2, K10	M1, M2, M7, C-2
	Shore D55 to 75	C	0.15	0.006	0.20	0.008	0.25	0.010	S3, S4, S2, K10	M1, M2, M7, C-2
Allyl (DAP)	M95 to 100	C	0.15	0.006	0.20	0.008	0.25	0.010	S3, S4, S2, K10	M1, M2, M7, C-2
Allyl, glass-filled	E80 to 87	F and M	0.075	0.003	0.102	0.004	0.13	0.005	S9, S11, K10	T15, M42, C-2
Allyl, fiber-filled	M108 to 115	F and M	0.075	0.003	0.102	0.004	0.13	0.005	S9, S11, K10	T15, M42, C-2

(a) C, cast; M, molded; E, extruded; F, filled

Boring of plastics

Materials	Hardness, Rockwell	Condition(a)	Depth of cut mm	Depth of cut in.	Speed m/min.	Speed fpm	Feed tooth mm	Feed tooth in.	High speed steel tool material ISO	High speed steel tool material AISI
Thermoplastics										
Acrylic, acetal, polycarbonate, polysulfone, polystyrene	M60 to 120	C, M, or E	0.25	0.010	120	400	0.050	0.002	S4, S5	M2, M3
			1.25	0.050	100	325	0.102	0.004	S4, S5	M2, M3
			2.5	0.100	84	275	0.15	0.006	S4, S5	M2, M3
Acrylonitrile-butadiene-styrene (ABS), Polyarylether, polypropylene, polyethylene, cellulose acetate	R50 to 120	C, M, or E	0.25	0.010	135	435	0.050	0.002	S4, S5	M2, M3
			1.25	0.050	105	350	0.102	0.004	S4, S5	M2, M3
			2.5	0.100	100	325	0.15	0.006	S4, S5	M2, M3
Fluorocarbons: tetrafluoroethylene (TFE), chlorotrifluoroethylene (CTFE)	R74 to 95	M or E	0.25	0.010	125	405	0.050	0.002	S4, S5	M2, M3
			1.25	0.050	100	325	0.102	0.004	S4, S5	M2, M3
			2.5	0.100	84	275	0.15	0.006	S4, S5	M2, M3
Polyamides (nylons): unfilled–types 6, 6/6, 6/12, 11, 12	R78 to 120	M or E	0.25	0.010	150	500	0.050	0.002	S4, S5	M2, M3
			1.25	0.050	120	400	0.102	0.004	S4, S5	M2, M3
			2.5	0.100	105	350	0.20	0.008	S4, S5	M2, M3
Polyamides (nylons): 35% glass-reinforced–types 6, 6/6	R78 to 120	F and M	0.25	0.010	...	...	...	...	...	...
			1.25	0.050	...	...	...	...	...	...
			2.5	0.100	...	...	...	...	...	...
Polyamides (nylons): 35% glass-reinforced–types 6/10, 6/12	E40 to 50	F and M	0.25	0.010	...	...	...	...	...	...
			1.25	0.050	...	...	...	...	...	...
			2.5	0.100	...	...	...	...	...	...
Thermosets										
Epoxy, melamine, phenolic	M100 to 128	C or M	0.25	0.010	150	500	0.050	0.002	S4, S5	M2, M3
			1.25	0.050	120	400	0.102	0.004	S4, S5	M2, M3
			2.5	0.100	110	360	0.20	0.008	S4, S5	M2, M3
Furan, polybutadiene	R40 to 100	C	0.25	0.010	76	250	0.050	0.002	S9, S11	T15, M42
			1.25	0.050	60	200	0.102	0.004	S9, S11	T15, M42
			2.5	0.100	49	160	0.20	0.008	S9, S11	T15, M42
Silicone	Shore A15 to 65	C or M	0.25	0.010	60	200	0.050	0.002	S9, S11	T15, M42
			1.25	0.050	49	160	0.102	0.004	S9, S11	T15, M42
			2.5	0.100	43	140	0.20	0.008	S9, S11	T15, M42
Silicone, glass-filled	M80 to 90	F and M	0.25	0.010	...	...	...	...	...	...
			1.25	0.050	...	...	...	...	...	...
			2.5	0.100	...	...	...	...	...	...
Polyimide	E40 to 50	M or E	0.25	0.010	150	500	0.050	0.002	S4, S5	M2, M3
			1.25	0.050	120	400	0.102	0.004	S4, S5	M2, M3
			2.5	0.100	110	360	0.20	0.008	S4, S5	M2, M3
Polyimide, glass-filled	M109 to 115	F and M	0.25	0.010	...	...	...	...	...	...
			1.25	0.050	...	...	...	...	...	...
			2.5	0.100	...	...	...	...	...	...
Polyurethane	Shore A65 to 95	C	0.25	0.020	76	250	0.050	0.002	S9, S11	T15, M42
			1.25	0.050	60	200	0.102	0.004	S9, S11	T15, M42
			2.5	0.100	49	160	0.20	0.008	S9, S11	T15, M42
	Shore D55 to 75	C	0.25	0.010	90	300	0.050	0.002	S4, S5	M2, M3
			1.25	0.050	73	240	0.102	0.004	S4, S5	M2, M3
			2.5	0.100	60	200	0.20	0.008	S4, S5	M2, M3
Allyl (DAP)	M95 to 100	C	0.25	0.010	120	400	0.050	0.002	S4, S5	M2, M3
			1.25	0.050	100	325	0.102	0.004	S4, S5	M2, M3
			2.5	0.300	84	275	0.15	0.006	S4, S5	M2, M3
Allyl, glass-filled	E80 to 87	F and M	0.25	0.010	...	...	...	...	...	...
			1.25	0.050	...	...	...	...	...	...
			2.5	0.100	...	...	...	...	...	...
Allyl, fiber-filled	M108 to 115	F and M	0.25	0.010	76	250	0.050	0.002	S9, S11	T15, M42
			1.25	0.050	60	200	0.102	0.004	S9, S11	T15, M42
			2.5	0.100	49	160	0.20	0.008	S9, S11	T15, M42

(continued)

Materials	Hardness, Rockwell	Condition(a)	Depth of cut mm	Depth of cut in.	Speed, Brazed m/min.	Speed, Brazed fpm	Speed, Indexable m/min.	Speed, Indexable fpm	Carbide tool (uncoated) Feed per tooth mm	Feed per tooth in.	Tool material grade ISO	Tool material grade AISI
Thermoplastics												
Acrylic, acetal, polycarbonate, polysulfone, polystyrene	M60 to 120	C, M, or E	0.25	0.010	135	450	160	530	0.050	0.002	K20, M20	C-2
			1.25	0.050	110	360	130	425	0.102	0.004	K20, M20	C-2
			2.5	0.100	100	325	115	385	0.20	0.008	K20, M20	C-2
Acrylonitrile-butadiene-styrene (ABS), polyarylether, polypropylene, polyethylene, cellulose acetate	R50 to 120	C, M, or E	0.25	0.010	160	530	190	625	0.050	0.002	K20, M20	C-2
			1.25	0.050	130	425	150	500	0.102	0.004	K20, M20	C-2
			2.5	0.100	115	385	140	455	0.20	0.008	K20, M20	C-2
Fluorocarbons: tetrafluoroethylene (TFE), chlorotrifluoroethylene (CTFE)	R74 to 95	M or E	0.25	0.010	135	450	160	530	0.050	0.002	K20, M20	C-2
			1.25	0.050	110	360	130	425	0.102	0.004	K20, M20	C-2
			2.5	0.100	100	325	115	385	0.20	0.008	K20, M20	C-2
Polyamides (nylons): unfilled–types 6, 6/6, 6/12, 11, 12	R78 to 120	M or E	0.25	0.010	180	595	215	700	0.050	0.002	K20, M20	C-2
			1.25	0.050	145	475	170	560	0.102	0.004	K20, M20	C-2
			2.5	0.100	125	415	150	490	0.20	0.008	K20, M20	C-2
Polyamides (nylons): 35% glass-reinforced–types 6, 6/6	R78 to 120	F and M	0.25	0.010	135	450	160	530	0.050	0.002	K20, M20	C-2
			1.25	0.050	110	360	130	425	0.102	0.004	K20, M20	C-2
			2.5	0.100	100	325	115	385	0.15	0.006	K20, M20	C-2
Polyamides (nylons): 35% glass-reinforced–types 6/10, 6/12	E40 to 50	F and M	0.25	0.010	115	370	135	435	0.050	0.002	K20, M20	C-2
			1.25	0.050	90	300	105	350	0.102	0.004	K20, M20	C-2
			2.5	0.100	84	275	95	315	0.15	0.006	K20, M20	C-2
Thermosets												
Epoxy, melamine, phenolic	M100 to 128	C or M	0.25	0.010	185	600	215	700	0.050	0.002	K20, M20	C-2
			1.25	0.050	145	475	170	560	0.102	0.004	K20, M20	C-2
			2.5	0.100	125	415	150	490	0.20	0.008	K20, M20	C-2
Furan, polybutadiene	R40 to 100	C	0.25	0.010	105	350	120	400	0.050	0.002	K20, M20	C-2
			1.25	0.050	84	275	95	315	0.102	0.004	K20, M20	C-2
			2.5	0.100	72	235	84	275	0.20	0.006	K20, M20	C-2
Silicone	Shore A15 to 65	C or M	0.25	0.010	105	340	120	400	0.050	0.002	K20, M20	C-2
			1.25	0.050	84	275	95	315	0.102	0.004	K20, M20	C-2
			2.5	0.100	72	235	84	275	0.15	0.006	K20, M20	C-2
Silicone, glass-filled	M80 to 90	F and M	0.25	0.010	90	300	105	345	0.050	0.002	K20, M20	C-2
			1.25	0.050	72	235	84	275	0.102	0.004	K20, M20	C-2
			2.5	0.100	64	210	75	245	0.15	0.006	K20, M20	C-2
Polyimide	E40 to 50	M or E	0.25	0.010	180	585	210	690	0.050	0.002	K20, M20	C-2
			1.25	0.050	140	465	170	550	0.102	0.004	K20, M20	C-2
			2.5	0.100	130	425	150	500	0.20	0.008	K20, M20	C-2
Polyimide, glass-filled	M109 to 115	F and M	0.25	0.010	115	375	135	440	0.050	0.002	K20, M20	C-2
			1.25	0.050	90	300	105	350	0.102	0.004	K20, M20	C-2
			2.5	0.100	84	275	95	315	0.20	0.008	K20, M20	C-2
Polyurethane	Shore A65 to 95	C	0.25	0.020	105	350	120	400	0.050	0.002	K20, M20	C-2
			1.25	0.050	84	275	95	315	0.102	0.004	K20, M20	C-2
			2.5	0.100	72	235	84	275	0.15	0.006	K20, M20	C-2
	Shore D55 to 75	C	0.25	0.010	115	375	135	440	0.050	0.002	K20, M20	C-2
			1.25	0.050	90	300	105	350	0.102	0.004	K20, M20	C-2
			2.5	0.100	84	275	95	315	0.20	0.008	K20, M20	C-2
Allyl (DAP)	M95 to 100	C	0.25	0.010	135	450	160	530	0.050	0.002	K20, M20	C-2
			1.25	0.050	110	360	130	425	0.102	0.004	K20, M20	C-2
			2.5	0.300	100	320	115	375	0.20	0.008	K20, M20	C-2
Allyl, glass-filled	E80 to 87	F and M	0.25	0.010	90	295	105	345	0.050	0.002	K20, M20	C-2
			1.25	0.050	72	235	84	275	0.102	0.004	K20, M20	C-2
			2.5	0.100	64	210	75	245	0.20	0.008	K20, M20	C-2
Allyl, fiber-filled	M108 to 115	F and M	0.25	0.010	115	370	135	435	0.050	0.002	K20, M20	C-2
			1.25	0.050	90	295	105	350	0.102	0.004	K20, M20	C-2
			2.5	0.100	81	265	95	315	0.20	0.008	K20, M20	C-2

(a) C, cast; M, molded; E, extruded; F, filled

End milling-slotting of plastics

Materials	Hardness, Rockwell	Condition(a)	Axial depth of cut mm	Axial depth of cut in.	m/min.	fpm	Feed per tooth, 10 mm (3/8 in.) mm	Feed per tooth, 10 mm (3/8 in.) in.	Feed per tooth, 12 mm (1/2 in.) mm	Feed per tooth, 12 mm (1/2 in.) in.
Thermoplastics										
Acrylic, acetal, polycarbonate, polysulfone, polystyrene	M60 to 120	C, M, or E	0.75	0.030	120	400	0.075	0.003	0.102	0.004
			3	0.125	105	350	0.075	0.003	0.075	0.003
			Diam/2	Diam/2	90	300	0.050	0.002	0.050	0.002
			Diam/1	Diam/1	76	250	0.025	0.001	0.025	0.001
Acrylonitrile-butadiene-styrene (ABS), Polyarylether, polypropylene, polyethylene, cellulose acetate	R50 to 120	C, M, or E	0.75	0.030	150	500	0.075	0.003	0.102	0.004
			3	0.125	135	450	0.075	0.003	0.075	0.003
			Diam/2	Diam/2	120	400	0.050	0.002	0.050	0.002
			Diam/1	Diam/1	105	350	0.025	0.001	0.025	0.001
Fluorocarbons: tetrafluoroethylene (TFE), chlorotrifluoroethylene (CTFE)	R74 to 95	M or E	0.75	0.030	120	400	0.075	0.003	0.102	0.004
			3	0.125	105	350	0.075	0.003	0.075	0.003
			Diam/2	Diam/2	90	300	0.050	0.002	0.050	0.002
			Diam/1	Diam/1	76	250	0.025	0.001	0.025	0.001
Polyamides (nylons): unfilled–types 6, 6/6, 6/12, 11, 12	R78 to 120	M or E	0.75	0.030	150	500	0.075	0.003	0.102	0.004
			3	0.125	135	450	0.075	0.003	0.075	0.003
			Diam/2	Diam/2	120	400	0.050	0.002	0.050	0.002
			Diam/1	Diam/1	105	350	0.025	0.001	0.025	0.001
Polyamides (nylons): 35% glass-reinforced–types 6, 6/6	R78 to 120	F and M	0.75	0.030	120	400	0.075	0.003	0.075	0.003
			3	0.125	105	350	0.075	0.003	0.075	0.003
			Diam/2	Diam/2	90	300	0.050	0.002	0.050	0.002
			Diam/1	Diam/1	76	250	0.025	0.001	0.025	0.001
Polyamides (nylons): 35% glass-reinforced–types 6/10, 6/12	E40 to 50	F and M	0.75	0.030	105	350	0.075	0.003	0.075	0.003
			3	0.125	90	300	0.075	0.003	0.075	0.003
			Diam/2	Diam/2	76	250	0.050	0.002	0.050	0.002
			Diam/1	Diam/1	60	200	0.025	0.001	0.025	0.001
Thermosets										
Epoxy, melamine, phenolic	M100 to 128	C or M	0.75	0.030	150	500	0.075	0.003	0.102	0.004
			3	0.125	135	450	0.075	0.003	0.102	0.004
			Diam/2	Diam/2	120	400	0.050	0.002	0.075	0.003
			Diam/1	Diam/1	105	350	0.025	0.001	0.050	0.002
Furan, polybutadiene	R40 to 100	C	0.75	0.030	90	300	0.075	0.003	0.102	0.004
			3	0.125	84	275	0.075	0.003	0.102	0.004
			Diam/2	Diam/2	76	250	0.050	0.002	0.075	0.003
			Diam/1	Diam/1	60	300	0.025	0.001	0.050	0.002
Silicone	Shore A15 to 65	C or M	0.75	0.030	90	300	0.075	0.003	0.102	0.004
			3	0.125	84	275	0.075	0.003	0.102	0.004
			Diam/2	Diam/2	76	250	0.050	0.002	0.075	0.003
			Diam/1	Diam/1	60	200	0.025	0.001	0.050	0.002
Silicone, glass-filled	M80 to 90	F and M	0.75	0.030	105	350	0.075	0.003	0.102	0.004
			3	0.125	90	300	0.075	0.003	0.102	0.004
			Diam/2	Diam/2	84	275	0.050	0.002	0.075	0.003
			Diam/1	Diam/1	76	250	0.025	0.001	0.050	0.002
Polyimide	E40 to 50	M or E	0.75	0.030	150	500	0.075	0.003	0.102	0.004
			3	0.125	135	450	0.075	0.003	0.102	0.004
			Diam/2	Diam/2	120	400	0.050	0.002	0.075	0.003
			Diam/1	Diam/1	105	350	0.025	0.001	0.050	0.002
Polyimide, glass-filled	M109 to 115	F and M	0.75	0.030	135	450	0.075	0.003	0.102	0.004
			3	0.125	120	600	0.075	0.003	0.102	0.004
			Diam/2	Diam/2	103	350	0.050	0.002	0.075	0.003
			Diam/1	Diam/1	90	300	0.025	0.001	0.050	0.002

(continued)

End milling-slotting of plastics (continued)

Materials	Hardness, Rockwell	Condition(a)	Axial depth of cut mm	Axial depth of cut in.	m/min.	fpm	Feed per tooth, for slot width of: 10 mm (3/8 in.) mm	10 mm (3/8 in.) in.	12 mm (1/2 in.) mm	12 mm (1/2 in.) in.
Polyurethane	Shore A65 to 95	C	0.75	0.030	90	300	0.075	0.003	0.102	0.004
			3	0.125	84	275	0.075	0.003	0.102	0.004
			Diam/2	Diam/2	76	250	0.050	0.002	0.075	0.003
			Diam/1	Diam/1	60	200	0.025	0.001	0.050	0.002
	Shore D55 to 75	C	0.75	0.030	105	350	0.075	0.003	0.102	0.004
			3	0.125	90	300	0.075	0.003	0.102	0.004
			Diam/2	Diam/2	84	275	0.050	0.002	0.075	0.003
			Diam/1	Diam/1	76	250	0.025	0.001	0.050	0.002
Allyl (DAP)	M95 to 100	C	0.75	0.030	105	350	0.075	0.003	0.102	0.004
			3	0.125	90	300	0.075	0.003	0.102	0.004
			Diam/2	Diam/2	84	275	0.050	0.002	0.075	0.003
			Diam/1	Diam/1	76	250	0.025	0.001	0.050	0.002
Allyl, glass-filled	E80 to 87	F and M	0.75	0.030	120	400	0.075	0.003	0.102	0.004
			3	0.125	105	350	0.075	0.003	0.102	0.004
			Diam/2	Diam/2	90	300	0.050	0.002	0.075	0.003
			Diam/1	Diam/1	76	250	0.025	0.001	0.050	0.002
Allyl, fiber-filled	M108 to 115	F and M	0.75	0.030	135	450	0.075	0.003	0.102	0.004
			3	0.125	130	425	0.075	0.003	0.102	0.004
			Diam/2	Diam/2	120	400	0.050	0.002	0.075	0.003
			Diam/1	Diam/1	105	350	0.025	0.001	0.050	0.002

Materials	Hardness, Rockwell	Condition(a)	Axial depth of cut mm	Axial depth of cut in.	Feet per tooth, for slot width of: 18 mm (3/4 in.) m/min.	18 mm (3/4 in.) fpm	25-50 mm (1-2 in.) mm	25-50 mm (1-2 in.) in.	HSS tool material (except as noted) ISO	AISI
Thermoplastics										
Acrylic, acetal, polycarbonate, polysulfone, polystyrene	M60 to 120	C, M, or E	0.75	0.030	0.13	0.005	0.15	0.006	S4, S5, S2	M2, M3, M7
			3	0.125	0.102	0.004	0.13	0.005		
			Diam/2	Diam/2	0.075	0.003	0.102	0.004		
			Diam/1	Diam/1	0.050	0.002	0.075	0.003		
Acrylonitrile-butadiene-styrene (ABS), Polyarylether, polypropylene, polyethylene, cellulose acetate	R50 to 120	C, M, or E	0.75	0.030	0.13	0.005	0.15	0.006	S4, S5, S2	M2, M3, M7
			3	0.125	0.102	0.004	0.13	0.005		
			Diam/2	Diam/2	0.075	0.003	0.102	0.004		
			Diam/1	Diam/1	0.050	0.002	0.075	0.003		
Fluorocarbons: tetrafluoroethylene (TFE), chlorotrifluoroethylene (CTFE)	R74 to 95	M or E	0.75	0.030	0.13	0.005	0.15	0.006	S4, S5, S2	M2, M3, M7
			3	0.125	0.102	0.004	0.13	0.005		
			Diam/2	Diam/2	0.075	0.003	0.102	0.004		
			Diam/1	Diam/1	0.050	0.002	0.075	0.003		
Polyamides (nylons): unfilled–types 6, 6/6, 6/12, 11, 12	R78 to 120	M or E	0.75	0.030	0.13	0.005	0.15	0.006	S4, S5, S2	M2, M3, M7
			3	0.125	0.102	0.004	0.13	0.005		
			Diam/2	Diam/2	0.075	0.003	0.102	0.004		
			Diam/1	Diam/1	0.050	0.002	0.075	0.003		
Polyamides (nylons): 35% glass-reinforced–types 6, 6/6	R78 to 120	F and M	0.75	0.030	0.102	0.004	0.13	0.005	K10	C-2
			3	0.125	0.075	0.003	0.13	0.005		
			Diam/2	Diam/2	0.075	0.003	0.102	0.004		
			Diam/1	Diam/1	0.050	0.002	0.075	0.003		
Polyamides (nylons): 35% glass-reinforced–types 6/10, 6/12	E40 to 50	F and M	0.75	0.030	0.102	0.004	0.13	0.005	K10	C-2
			3	0.125	0.075	0.003	0.13	0.005		
			Diam/2	Diam/2	0.075	0.003	0.102	0.004		
			Diam/1	Diam/1	0.050	0.002	0.075	0.003		

(continued)

End milling-slotting of plastics (continued)

Materials	Hardness, Rockwell	Condition(a)	Axial depth of cut		Feet per tooth, for slot width of: 18 mm (3/4 in.)		25-50 mm (1-2 in.)		HSS tool material (except as noted)	
			mm	in.	m/min.	fpm	mm	in.	ISO	AISI
Thermosets										
Epoxy, melamine, phenolic	M100 to 128	C or M	0.75	0.030	0.20	0.008	0.25	0.010	S4, S5, S2	M2, M3, M7
			3	0.125	0.20	0.008	0.25	0.010		
			Diam/2	Diam/2	0.15	0.006	0.20	0.008		
			Diam/1	Diam/1	0.102	0.004	0.15	0.006		
Furan, polybutadiene	R40 to 100	C	0.75	0.030	0.13	0.005	0.15	0.006	S9, S11	T15, M42
			3	0.125	0.13	0.005	0.15	0.006		
			Diam/2	Diam/2	0.102	0.004	0.13	0.005		
			Diam/1	Diam/1	0.075	0.003	0.102	0.004		
Silicone	Shore A15 to 65	C or M	0.75	0.030	0.13	0.005	0.15	0.006	S9, S11	T15, M42
			3	0.125	0.13	0.005	0.15	0.006		
			Diam/2	Diam/2	0.102	0.004	0.13	0.005		
			Diam/1	Diam/1	0.075	0.003	0.102	0.004		
Silicone, glass-filled	M80 to 90	F and M	0.75	0.030	0.13	0.005	0.15	0.006	K10	C-2
			3	0.125	0.13	0.005	0.15	0.006		
			Diam/2	Diam/2	0.120	0.004	0.13	0.005		
			Diam/1	Diam/1	0.075	0.003	0.102	0.004		
Polyimide	E40 to 50	M or E	0.75	0.030	0.20	0.008	0.25	0.010	S4, S5, S2	M2, M3, M7
			3	0.125	0.20	0.008	0.25	0.010		
			Diam/2	Diam/2	0.15	0.006	0.20	0.008		
			Diam/1	Diam/1	0.102	0.004	0.15	0.006		
Polyimide, glass-filled	M109 to 115	F and M	0.75	0.030	0.13	0.005	0.15	0.006	K10	C-2
			3	0.125	0.13	0.005	0.15	0.008		
			Diam/2	Diam/2	0.102	0.004	0.13	0.005		
			Diam/1	Diam/1	0.075	0.003	0.102	0.004		
Polyurethane	Shore A65 to 95	C	0.75	0.030	0.13	0.005	0.15	0.006	S9, S11	T15, M42
			3	0.125	0.13	0.005	0.15	0.006		
			Diam/2	Diam/2	0.102	0.004	0.13	0.005		
			Diam/1	Diam/1	0.075	0.003	0.102	0.004		
	Shore D55 to 75	C	0.75	0.030	0.15	0.006	0.20	0.008	S4, S5, S2	M2, M3, M7
			3	0.125	0.15	0.006	0.20	0.008		
			Diam/2	Diam/2	0.13	0.005	0.18	0.007		
			Diam/1	Diam/1	0.102	0.004	0.15	0.006		
Allyl (DAP)	M95 to 100	C	0.75	0.030	0.15	0.006	0.20	0.008	S4, S5, S2	M2, M3, M7
			3	0.125	0.15	0.006	0.20	0.008		
			Diam/2	Diam/2	0.13	0.005	0.15	0.006		
			Diam/1	Diam/1	0.102	0.004	0.13	0.005		
Allyl, glass-filled	E80 to 87	F and M	0.75	0.030	0.13	0.005	0.15	0.006	K10	C-2
			3	0.125	0.13	0.005	0.15	0.006		
			Diam/2	Diam/2	0.102	0.004	0.13	0.005		
			Diam/1	Diam/1	0.075	0.003	0.13	0.005		
Allyl, fiber-filled	M108 to 115	F and M	0.75	0.030	0.13	0.005	0.15	0.006	K10	C-2
			3	0.125	0.13	0.005	0.15	0.006		
			Diam/2	Diam/2	0.102	0.004	0.13	0.005		
			Diam/1	Diam/1	0.075	0.003	0.13	0.005		

(a) C, cast; M, molded; E, extruded; F, filled

Circular sawing of plastics (with HSS blade)

Materials	Hardness, Rockwell	Condition(a)	Solid stock diameter or thickness mm	in.	Pitch mm/ tooth	Pitch in./ tooth	Cutting speed m/min.	fpm	Feed mm/ tooth	Feed in./ tooth	HSS tool material ISO	AISI
Thermoplastics												
Acrylic, acetal, polycarbonate, polysulfone, polystyrene	M60 to 120	C, M, or E	6 to 80	¼ to 3	3 to 12	0.12 to 0.50	90	300	0.102	0.004		
			80 to 160	3 to 6	10 to 18	0.40 to 0.70	76	250	0.102	0.004	S4, S2	M2, M7
			160 to 250	6 to 9	15 to 20	0.60 to 0.80	60	200	0.15	0.006		
			250 to 400	9 to 15	18 to 25	0.70 to 1.00	46	150	0.15	0.006		
Acrylonitrile-butadiene-styrene (ABS), Polyarylether, polypropylene, polyethylene, cellulose acetate	R50 to 120	C, M, or E	6 to 80	¼ to 3	3 to 12	0.12 to 0.50	105	350	0.102	0.004		
			80 to 160	3 to 6	10 to 18	0.40 to 0.70	90	300	0.102	0.004	S4, S2	M2, M7
			160 to 250	6 to 9	15 to 20	0.60 to 0.80	67	220	0.15	0.006		
			250 to 400	9 to 15	18 to 25	0.70 to 1.00	52	170	0.15	0.006		
Fluorocarbons: tetrafluoroethylene (TFE), chlorotrifluoroethylene (CTFE)	R74 to 95	M or E	6 to 80	¼ to 3	3 to 12	0.12 to 0.50	90	300	0.102	0.004		
			80 to 160	3 to 6	10 to 18	0.40 to 0.70	76	250	0.102	0.004	S4, S2	M2, M7
			160 to 250	6 to 9	15 to 20	0.60 to 0.80	60	200	0.15	0.006		
			250 to 400	9 to 15	18 to 25	0.70 to 1.00	46	150	0.15	0.006		
Polyamide (nylons): unfilled–types 6, 6/6, 6/12, 11, 12	R78 to 120	M or E	6 to 80	¼ to 3	3 to 12	0.12 to 0.50	105	350	0.102	0.004		
			80 to 160	3 to 6	10 to 18	0.40 to 0.70	90	300	0.102	0.004	S4, S2	M2, M7
			160 to 250	6 to 9	15 to 20	0.60 to 0.80	67	220	0.15	0.006		
			250 to 400	9 to 15	18 to 25	0.70 to 1.00	52	170	0.15	0.006		
Thermosets												
Epoxy, melamine, phenolic	M100 to 128	C or M	6 to 80	¼ to 3	3 to 12	0.12 to 0.50	105	350	0.102	0.004		
			80 to 160	3 to 6	10 to 18	0.40 to 0.70	90	300	0.102	0.004	S4, S2	M2, M7
			160 to 250	6 to 9	15 to 20	0.60 to 0.80	67	220	0.15	0.006		
			250 to 400	9 to 15	18 to 25	0.70 to 1.00	52	170	0.15	0.006		
Furan, polybutadiene	R40 to 100	C	6 to 80	¼ to 3	3 to 12	0.12 to 0.50	46	150	0.075	0.003		
			80 to 160	3 to 6	10 to 18	0.40 to 0.70	37	120	0.075	0.003	S4, S	M2, M7
			160 to 250	6 to 9	15 to 20	0.60 to 0.80	30	100	0.13	0.005		
			250 to 400	9 to 15	18 to 25	0.70 to 1.00	24	80	0.13	0.005		
Silicone	Shore A15 to 65	C or M	6 to 80	¼ to 3	3 to 12	0.12 to 0.50	30	100	0.075	0.003		
			80 to 160	3 to 6	10 to 18	0.40 to 0.70	24	80	0.075	0.003	S4, S2	M2, M7
			160 to 250	6 to 9	15 to 20	0.60 to 0.80	15	50	0.13	0.005		
			250 to 400	9 to 15	18 to 25	0.70 to 1.00	15	50	0.13	0.005		
Polyimide	E40 to 50	M or E	6 to 80	¼ to 3	3 to 12	0.12 to 0.50	105	350	0.102	0.004		
			80 to 160	3 to 6	10 to 18	0.40 to 0.70	90	300	0.102	0.004	S4, S2	M2, M7
			160 to 250	6 to 9	15 to 20	0.60 to 0.80	67	220	0.15	0.006		
			250 to 400	9 to 15	18 to 25	0.70 to 1.00	52	170	0.15	0.006		
Polyurethane	Shore A65 to 95	C	6 to 80	¼ to 3	3 to 12	0.12 to 0.50	46	150	0.075	0.003		
			80 to 160	3 to 6	10 to 18	0.40 to 0.70	37	120	0.075	0.003	S4, S2	M2, M7
			160 to 250	6 to 9	15 to 20	0.60 to 0.80	30	100	0.102	0.004		
			250 to 400	9 to 15	18 to 25	0.70 to 1.00	24	80	0.102	0.004		
	Shore D55 to 75	C	6 to 80	¼ to 3	3 to 12	0.12 to 0.50	60	200	0.102	0.004		
			80 to 160	3 to 6	10 to 18	0.40 to 0.70	46	150	0.102	0.004	S4, S2	M2, M7
			160 to 250	6 to 9	15 to 20	0.60 to 0.80	37	120	0.15	0.006		
			250 to 400	9 to 15	18 to 25	0.70 to 1.00	30	100	0.15	0.006		
Allyl (DAP)	M95 to 100	C	6 to 80	¼ to 3	3 to 12	0.12 to 0.50	76	250	0.102	0.004		
			80 to 160	3 to 6	10 to 18	0.40 to 0.70	60	200	0.102	0.004	S4, S2	M2, M7
			160 to 250	6 to 9	15 to 20	0.60 to 0.80	46	150	0.15	0.006		
			250 to 400	9 to 15	18 to 25	0.70 to 1.00	37	120	0.15	0.006		
Allyl, fiber-filled	M108 to 115	F and M	6 to 80	¼ to 3	3 to 12	0.12 to 0.50	46	150	0.075	0.003		
			80 to 160	3 to 6	10 to 18	0.40 to 0.70	37	120	0.075	0.003	S4, S2	M2, M7
			160 to 250	6 to 9	15 to 20	0.60 to 0.80	30	100	0.13	0.005		
			250 to 400	9 to 15	18 to 25	0.70 to 1.00	24	80	0.13	0.005		

(a) C, cast; M, molded; E, extruded; F, filled

Electronic Materials

General Information

General Information

Selected constants

Constant	Symbol	cgs unit	SI units
Gases			
Avogadro's number	N_A	6.02×10^{23} mole^{-1}	6.02×10^{23} mole^{-1}
Boltzmann constant	k_B	1.38×10^{-16} erg/K	1.38×10^{-23} J/K
Gas constant	R	1.987 cal(mole · K)$^{-1}$	8.31 J(mole · K)$^{-1}$
Atomic			
Planck's constant	h	6.63×10^{-27} erg · s	6.63×10^{-34} J · s
Electron volt	eV	1.60×10^{-12} erg	1.60×10^{-19} J
		23.05 kcal · mole^{-1}	5.50 kJ · mole^{-1}
Rydberg	Ry	13.60 eV	1.097×10^{7} m^{-1}
Bohr radius	r_0	0.529×10^{-8} cm	0.529×10^{-10} m
Particles			
Electron rest mass	m	9.11×10^{-28} g	9.11×10^{-31} kg
Electron charge	e	4.80×10^{-10} esu	1.60×10^{-19} C
Temperature			
International Practical Scale	t	0 °C	+ 273.15 K
Thermodynamic scale	T	−273.15 °C	0 K
Electrical			
Resistivity	ρ	Ω · cm	Ω · meter
Conductivity	σ	(Ω · cm)$^{-1}$	(Ω · meter)$^{-1}$
Other constants			
Angstrom	Å	10^{-8} cm	10^{-10} m
Micron	μ	10^{-4} cm	10^{-6} m
Speed of light in vacuum	c	2.998×10^{10} cm · s^{-1}	2.998×10^{8} m · s^{-1}
Dielectric constant of vacuum	ε_0	1 (dimensionless)	8.85×10^{-12} Fm^{-1}

Designation of electron states

n	Maximum l	Maximum number of states	Designation of states
1	0	2	*s*
2	1	6	*p*
3	2	10	*d*
4	3	14	*f*

Calculated Fermi energies and temperatures for free electrons(a)

Element	E_F, (eV)(a)	T_F, (K · 10^{-4})
Li	4.7	5.5
Na	3.1	3.7
K	2.1	2.4
Rb	1.8	2.1
Cs	1.5	1.8
Cu	7.0	8.2
Ag	5.5	6.4
Au	5.5	6.4

(a) 1 eV = 23,050 cal/mole

Allowed combinations of quantum numbers, up to $n = 4$

n	l	Designation	m_l	m_s	Number	Total
1	0	1*s*	0	± ½	2	2
2	0	2*s*	0	± ½	2	8
2	1	2*p*	−1, 0, 1	± ½	6	
3	0	3*s*	0	± ½	2	
3	1	3*p*	−1, 0, 1	± ½	6	18
3	2	3*d*	−2, −1, 0, 1, 2	± ½	10	
4	0	4*s*	0	± ½	2	
4	1	4*p*	−1, 0, 1	± ½	6	32
4	2	4*d*	−2, −1, 0, 1, 2	± ½	10	
4	3	4*f*	−3, −2, −1, 0, 1, 2, 3	± ½	14	

Comparison of enthalpies of atomization, fusion, and formation of ionic, metallic, and covalent solids(a)

Crystal type	Example	ΔH^{at}	ΔH^{F}	$\Delta H^{F}/\Delta H^{at}$	ΔH_f^0	$\Delta H^{F}/\Delta H_f^0$
Ionic	NaCl	77	3.4	0.044	49	0.070
	KF	87	3.4	0.044	68	0.050
Metallic	Sn	71	1.72	0.024	...	...
	Ga	62	1.34	0.022	...	...
	Al	73	2.55	0.035	...	...
Covalent	Si	106	7.2	0.068	...	...
	Ge	80	6.7	0.084	...	...
III-V	AlSb	83	9.77	0.117	11.5	0.848
	GaAs	73	12.6	0.173	10	0.795
	GaSb	69	7.78	0.113	5.52	0.675
	InAs	65	8.79	0.135	6.5	0.739
	InSb	64	5.70	0.089	4.0	0.703

(a) All energies are in Kcal $(g \cdot atom)^{-1}$

The equivalent conductances of the separate ions

Ion	0°	18°	25°	50°	75°	100°	128°	156°	Ion	0°	18°	25°	50°	75°	100°	128°	156°
K	40.4	64.6	74.5	115	159	206	263	317									
Na	26	43.5	50.9	82	116	155	203	249	$\frac{1}{2}SO_4$	41	68	79	125	177	234	303	370
NH_4	40.2	64.5	74.5	115	159	207	264	319									
Ag	32.9	54.3	63.5	101	143	188	245	299	$\frac{1}{2}C_2O_4$	39	63	73	115	163	213	27.5	336
$\frac{1}{2}$Ba	33	55	65	104	149	200	262	322									
$\frac{1}{2}$Ca	30	51	60	98	142	191	252	312	$\frac{1}{2}C_4H_5O_7$	36	60	70	113	161	214	...	...
$\frac{1}{2}$La	35	61	72	119	173	235	312	388	$\frac{1}{4}Fe(CN)_4$	58	95	111	173	244	321	...	...
Cl	41.1	65.5	75.5	116	160	207	264	318	H	240	314	350	465	565	644	722	777
NO_2	40.4	61.7	70.6	104	140	178	222	263	OH	105	172	192	284	360	439	525	592
$C_2H_2O_2$	20.3	34.6	40.8	67	96	130	171	211									

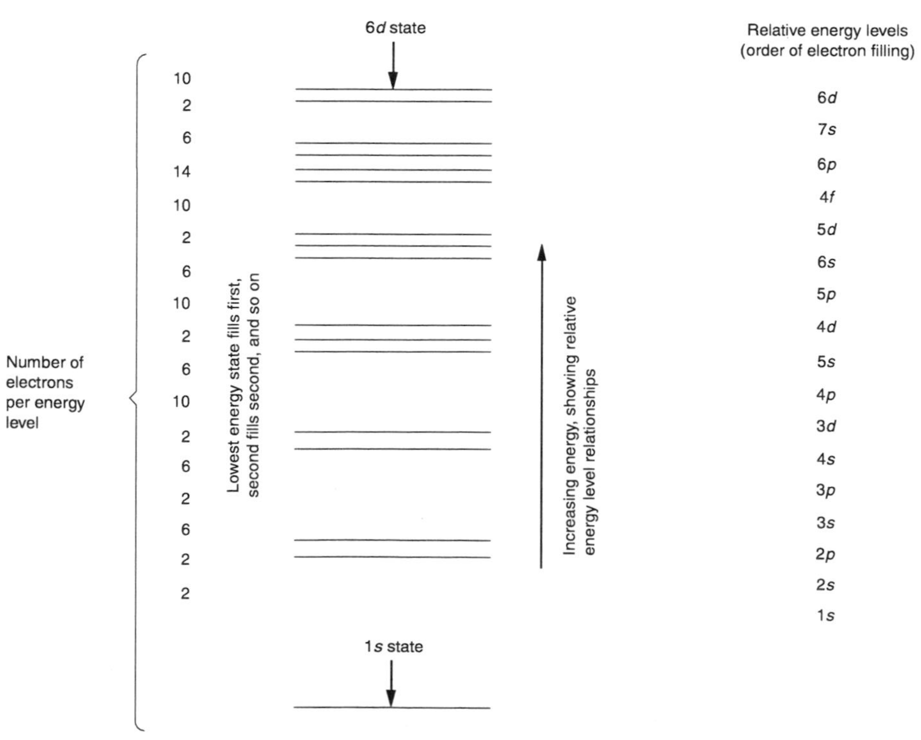

Scientific energy level diagram for electrons in a typical atom
Energy levels increase from bottom to top. Note number of electrons (when filled) per level and order of level filling

Ground state electron configuration of atoms

Z(a)	Element	Outer configuration	Z(a)	Element	Outer configuration	Z(a)	Element	Outer configuration	Z(a)	Element	Outer configuration
1	H	$1s$	29	Cu	$3d^{10} 4s$	55	Cs	$6s$	87	Fr	$7s$
2	He	$1s^2$	30	Zn	$4s^2$	56	Ba	$6s^2$	88	Ra	$7s^2$
3	Li	$2s$	31	Ga	$4s^2 4p$	57	La	$5d\, 6s^2$	89	Ac	$6d\, 7s^2$
4	Be	$2s^2$	32	Ge	$4s^2 4p^2$	58	Ce	$4f^2 6s^2$	90	Th	$6d^2 7s^2$
5	B	$2s^2 2p$	33	As	$4s^2 4p^3$	59	Pr	$4f^3 6s^2$	91	Pa	$6d^3 7s^2$
6	C	$2s^2 2p^2$	34	Se	$4s^2 4p^4$	60	Nd	$4f^4 6s^2$	92	U	$6d^4 7s^2$
7	N	$2s^2 2p^3$	35	Br	$4s^2 4p^5$	61	Pm	$4f^5 6s^2$	93	Np	$5f^5 7s^2$
8	O	$2s^2 2p^4$	36	Kr	$4s^2 4p^6$	62	Sm	$4f^6 6s^2$	94	Pu	$5f^5 6d\, 7s^2$
9	F	$2s^2 2p^5$	37	Rb	$5s$	63	Eu	$4f^7 6s^2$	95	Am	$5f^6 6d\, 7s^2$
10	Ne	$2s^2 2p^6$	38	Sr	$5s^2$	64	Gd	$4f^7 5d\, 6s^2$	96	Cm	$5f^7 6d\, 7s^2$
11	Na	$3s$	39	Y	$4d\, 5s^2$	65	Tb	$4f^8 5d\, 6s^2$	97	Bk	$5f^8 6d\, 7s^2$
12	Mg	$3s^2$	40	Zr	$4d^2 5s^2$	66	Dy	$4f^9 5d\, 6s^2$	98	Cf	$5f^9 6d\, 7s^2$
13	Al	$3s^2 3p$	41	Nb	$4d^4 5s$	67	Ho	$4f^{10} 5d\, 6s^2$	99	Es	$5f^{10} 6d\, 7s^2$
14	Si	$3s^2 3p^2$	42	Mo	$4d^5 5s$	68	Er	$4f^{11} 5d\, 6s^2$	100	Fm	$5f^{11} 6d\, 7s^2$
15	P	$3s^2 3p^3$	43	Tc	$4d^6 5s$	69	Tm	$4f^{12} 5d\, 6s^2$			
16	S	$3s^2 3p^4$	44	Ru	$4d^7 5s$	70	Yb	$4f^{13} 5d\, 6s^2$			
17	Cl	$3s^2 3p^5$	45	Rh	$4d^8 5s$	71	Lu	$4f^{14} 5d\, 6s^2$			
18	Ar	$3s^2 3p^6$	46	Pd	$4d^{10}$	72	Hf	$5d^2 6s^2$			
19	K	$4s$	47	Ag	$4d^{10} 5s$	73	Ta	$5d^3 6s^2$			
20	Ca	$4s^2$	48	Cd	$5s^2$	74	W	$5d^4 6s^2$			
21	Sc	$3d\, 4s^2$	49	In	$5s^2 5p$	75	Re	$5d^5 6s^2$			
22	Ti	$3d^2 4s^2$	50	Sn	$5s^2 5p^2$	76	Os	$5d^6 6s^2$			
23	V	$3d^3 4s^2$	51	Sb	$5s^2 5p^3$	77	Ir	$5d^9$			
24	Cr	$3d^5 4s$	52	Te	$5s^2 5p^4$	78	Pt	$5d^9 6s$			
25	Mn	$3d^5 4s^2$	53	I	$5s^2 5p^5$	79	Au	$5d^{10} 6s$			
26	Fe	$3d^6 4s^2$	54	Xe	$5s^2 5p^6$	80	Hg	$6s^2$			
27	Co	$3d^7 4s^2$				81	Tl	$6s^2 6p$			
28	Ni	$3d^8 4s^2$				82	Pb	$6s^2 6p^2$			
						83	Bi	$6s^2 6p^3$			
						84	Po	$6s^2 6p^4$			
						85	At	$6s^2 6p^5$			
						86	Rn	$6s^2 6p^6$			

(a) Atomic number

Electrical resistivities and temperature coefficients of some elements near room temperature (20 °C)

Element	ρ, $10^6\ \Omega \cdot$ cm	α, $10^3\ \Omega/\Omega \cdot$ deg	Element	ρ, $10^6\ \Omega \cdot$ cm	α, $10^3\ \Omega/\Omega \cdot$ deg
Aluminum	2.6548	4.29	Molybdenum	5.2 (0 °C)	5.3
Antimony	39.0 (0 °C)	3.6	Nickel	6.84	6.92
Beryllium	4	25	Palladium	10.8	3.77
Bismuth	106.5 (0 °C)	5.6	Platinum	9.85	3.927
Cadmium	6.83 (0 °C)	4.2	Plutonium	141.4 (107 °C)	–2.08 (107 °C)
Calcium	3.91 (0 °C)	4.02 (0 °C)	Rhenium	19.3	3.95
Carbon (graphite)	13.75 (0 °C)	...	Rhodium	4.51	4.3
Chromium	12.9 (0 °C)	3	Silicon (impure)	10 (0 °C)	...
Cobalt	6.24	5.3	Silver	1.59	4.1
Copper	1.6730	4.3	Sodium	4.69	...
Germanium (impure)	46	...	Sulfur (yellow)	2×10^{23}	...
Gold	2.35	3.5	Tantalum	13.5	3.83
Indium	8.0 (0 °C)	5	Tellurium	4.35×10^5 (23 °C)	...
Iodine	1.3×10^{15}	...	Thallium	15 (0 °C)	...
Iridium	5.3	3.93	Thorium	15.7 (25 °C)	3.8
Iron	9.71	6.51	Tin	11 (0 °C)	3.64
Lead	20.648	3.68	Titanium	42	3.5
Lithium	9.35	5	Tungsten	5.3 (27 °C)	4.5
Magnesium	4.45	3.7	Uranium	11 (0 °C)	2.1 (27 °C)
Mercury	98.4 (50 °C)	0.97	Zinc	5.916	4.19

Changes in resistivity per atomic percent of alloying element in Au, Ag, and Cu

Alloying element	Change in resistivity in: Au	Ag	Cu
Cu	0.485(a)	0.068(a)	...
Ni	1.00	...	1.25(a)
Co	6.1	...	6.4
Fe	7.66	...	9.3
Mn	2.41	...	2.83
Cr	4.25	...	...
Ti	14.4	...	...
Ag	0.38	...	0.14
Pd	0.407	0.436	0.89
Rh	4.2	...	4.40
Au	...	0.38	0.55
Pt	1.02	1.59	2.51
Ir	...	...	6.1
Zn	0.96	0.62	0.335
Ga	2.2	2.28	1.40
Ge	5.2	5.52	3.75
As	...	8.46	6.8
Cd	0.64	0.382	0.21
In	1.41	1.78	1.10
Sn	3.63	4.32	2.85
Sb	...	7.26	5.45
Hg	0.41	0.79	1.00
Tl	...	2.27	...
Pb	...	4.64	...
Bi	...	7.3	...

(a) in μΩ · cm per atomic percent at 18 °C (65 °F)

Corrosion of electronic materials

Material	Device or component	Corrosion
Gold	Integrated circuits	Uniform films
Ni-plated Al	Electrical connectors	Galvanic
Aluminum	Bond pads	Galvanic
Aluminum	Metallized components	Pitting
Al-Mg alloy	Wire	Intergranular
Palladium	Electrical connectors	Fretting
Tin and tin/lead	Electrical connectors	Uniform
Stainless steel	Electrical connectors	Passive films
Gallium arsenide	Electrical connectors	Complex
Ni undercoat Au	Electrical connectors	Pitting
Tin/tin-lead alloy	Connector	Uniform

Solubility parameters, molar volumes, and electronegatives of selected elements

Element	δ	V	χ
Al	86	10.0	1.5
Ga	74	11.8	1.6
In	67	15.7	1.6
P	75	17.8	2.0
As	66	13.1	2.0
Sb	59	18.2	1.9
Si	88	11.7	1.8
Ge	76	13.7	1.8
Sn	65	16.3	1.8
Pb	51	18.3	1.8
S	60.6	15.5	2.5
Se	54.8	16.5	2.4
Te	47.4	20.4	2.1
Zn	58	9.2	1.5
Cd	45	13	1.5
Fe	117	7.1	1.8
Ni	124	6.6	1.7
Co	126	6.6	1.8

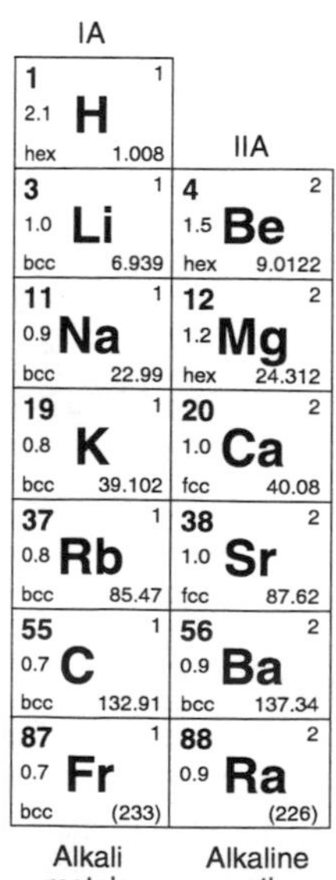

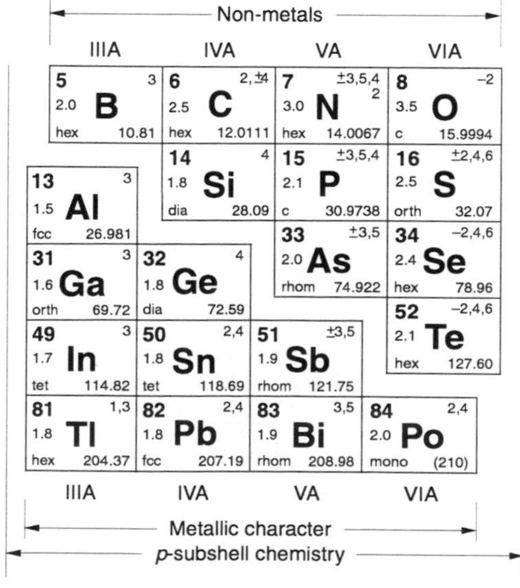

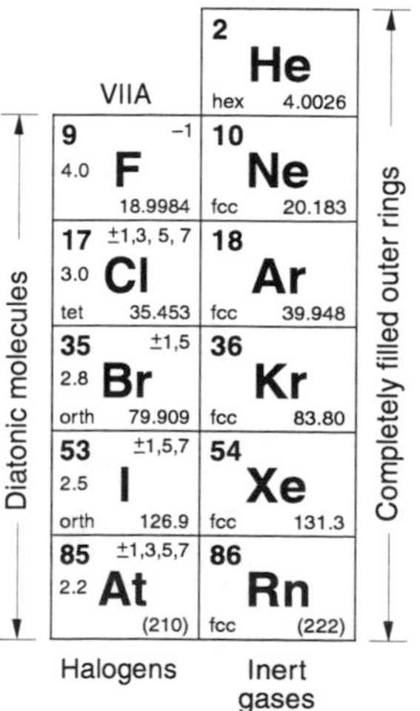

(a)

IIIB	IVB	VB	VIB	VIIB	VIII	VIII	VIII	IB	IIB
21 3 1.3 **Sc** hex 44.956	22 3.4 1.5 **Ti** hex 47.90	23 2,3,4,5 1.6 **V** bcc 50.942	24 2,3,6 1.6 **Cr** bcc 51.966	25 3,4,6,7 1.5 **Mn** c 54.938	26 2,3 1.8 **Fe** bcc 55.847	27 2,3 1.8 **Co** hex 58.933	28 2,3 1.8 **Ni** fcc 58.71	29 2,1 1.9 **Cu** fcc 63.54	30 2 1.6 **Zn** hex 65.37
39 3 1.3 **Y** hex 89.905	40 4 1.6 **Zr** hex 91.22	41 3,5 1.6 **Nb** bcc 92.91	42 3,4,5,6 1.8 **Mo** bcc 95.94	43 7 1.9 **T** hex (98)	44 2,3,4,6,8 2.2 **Ru** hex 101.07	45 2,3,4 2.2 **Rh** fcc 102.905	46 2,4 2.2 **Pd** fcc 106.4	47 1 1.9 **Ag** fcc 107.870	48 2 1.7 **Cd** hex 112.40
57 3 1.1 **La** hex 138.91	72 4 1.3 **Hf** hex 178.49	73 5 1.5 **Ta** bcc 180.95	74 2,3,4,5,6 1.7 **W** bcc 183.85	75 −1,2,4,6,7 1.9 **Re** hex 186.2	76 2,3,4,6,8 2.2 **Os** hex 190.2	77 2,3,4,6 2.2 **Ir** fcc 192.2	78 2,4 2.2 **Pt** fcc 195.09	79 1,3 2.4 **Au** fcc 196.967	80 1,2 1.9 **Hg** rhom 200.59
89 3 **Ac** (227)	104								

(b)

Lanthenides

58 3,4 1.1 **Ce** fcc 140.12	59 3,4 1.1 **Pr** hex 140.907	60 3 1.2 **Nd** hex 144.24	61 3 **Pm** hex (147)	62 2,3 1.2 **Sm** rhom 150.35	63 2,3 **Eu** bcc 151.96	64 3 1.1 **Gd** hex 157.25	65 3,4 1.2 **Tb** hex 158.924	66 3 **Dy** hex 162.50	67 3 1.2 **Ho** hex 164.93	68 3 1.2 **Er** hex 167.26	69 2,3 1.2 **Tm** hex 168.934	70 2,3 1.1 **Yb** fcc 173.04	71 3 1.2 **Lu** hex 174.97
90 4 1.3 **Th** fcc 232.03	91 4,5 1.5 **Pa** (231)	92 3,4,5,6 1.7 **U** orth 236.03	93 3,4,5,6 1.3 **Np** orth (237)	94 3,4,5,6 1.3 **Pu** (242)	95 3,4,5,6 1.3 **Am** (243)	96 3 **Cm** (247)	97 3,4 **Bk** (247)	98 3 **Cf** (249)	99 **E** (254)	100 **Fm** (253)	101 **Md** (256)	102 **No** (254)	103 **Lw** (257)

Actinides

Percent ionic character/single chemical bond

Difference in electronegativity	0.1	0.2	0.3	0.4	0.5	0.6	0.7	0.8	0.9	1.0	1.1
Ionic character, %	0.5	1	2	4	6	9	12	15	19	22	26
	1.2	1.3	1.4	1.5	1.6	1.7	1.8	1.9	2.0	2.1	2.2
	30	34	39	43	47	51	55	59	63	67	70
	2.3	2.4	2.5	2.6	2.7	2.8	2.9	3.0	3.1	3.2	3.3
	74	76	79	82	84	86	88	89	91	92	94

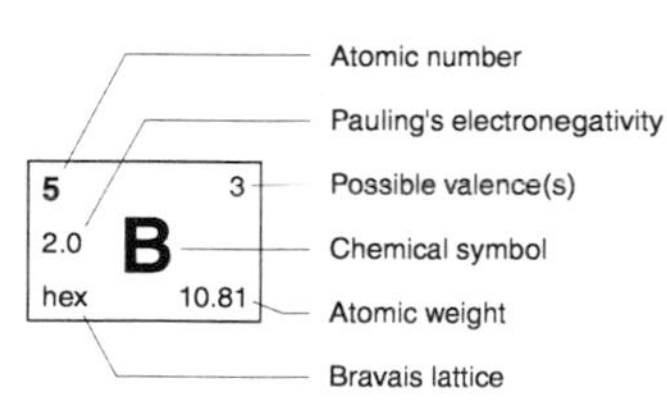

(c)

Microelectronics periodic table of the elements
(a) Semiconductors. (b) Metals (chemistry of the *d*-subshell). (c) Rare earths (chemistry of the *f*-subshell)

Property Data

General

Typical electrical properties reported on data sheet

Property	ASTM test method	Test description	Usefulness for design purposes
Dielectric strength	D 149	The voltage at which the insulating capabilities of a material break down. The value is normalized by the specimen wall thickness	Dielectric strength can be of value in an insulating application. Because the value is affected by time, temperature, moisture, and mechanical stress, it is not often used as a design factor
Volume resistivity	D 257	The ability of a material to resist current leakage through its thickness	Volume resistivity as reported in the data sheet has very limited design value. Temperature, moisture, and time must also be taken into consideration
Dielectric constant	D 150	A ratio of the capacitance of the material compared to the capacitance of a dimensionally equivalent vacuum	Dielectric constant is dependent on moisture, frequency, voltage, and temperature. Thus, it is more important for material selection than for design
Dissipation factor	D 150	The ratio of the amount of electrical current dissipated as heat to the amount of current transmitted through a material	The dissipation factor is dependent on moisture, frequency, voltage, and temperature. This property is important to maintaining efficiency in electrical components and to the prevention of breakdown caused by heat in the system. However, this property is more important to material selection than to design

Dielectric constants for various materials

Material	Dielectric constant, κ
Ceramics and glasses	
Silica glass	3.8
Boron nitride	4.1
αquartz	4.2
Cordierite	4.5
Glass-ceramic	4.8
Steatite	5.5
Silicon nitride	6 to 7
Forsterite	6.5
Mullite	6.5
Beryllia	6.7
Magnesium oxide	8.2 to 20
Aluminum nitride	8.8 to 8.9
Alumina	8.9 to 10.0
Barium tetratitanate	40
Silicon carbide	42.0
Titanium dioxide	100
Calcium titanate	160
Strontium titanate	320
Barium titanate	1000 to 5000
Barium zirconium titanate	20 000
Lead magnesium niobate	20 000
Nonpolar resins	
Polytetrafluoroethylene	2.0 to 2.2
Polypropylene	2.2
Polyethylene	2.3
Polystyrene	2.5 to 2.6
Polar resins	
Polycarbonate	2.9 to 3.0
Epoxy resins (unfilled)	3.0 to 4.5
Silicones (glass-filled)	3.1 to 4.5
Polyvinyl acetate	3.2
Polyvinyl chloride (rigid)	3.2 to 3.6
Polyethylene terephthalate	3.25
Nylon	3.5 to 4.6
Methylmethacrylate	3.6
Polyvinyl acetate	3.7 to 3.8
Phenolics (cellulose-filled)	4 to 15
Cellulose triacetate	4.7
Phenolics (mica-filled)	4.7 to 7.5
Phenolics (glass-filled)	5 to 7
Cellulose cotton fiber (dry)	5.4
Cellulose kraft fiber (dry)	5.9
Cellulose cellophane (dry)	6.6
Polyvinyl fluoride	8.4 to 8.5
Tricyanoethyl cellulose	15.2

Typical properties of dielectric materials

Material	Electrical resistivity, Ω · cm	Dielectric strength, 10^5 V/cm
Ceramics		
Alumina	10^{11} to 10^{14}	0.16 to 0.64
Cordierite	10^{12} to 10^{14}	0.16 to 0.99
Forsterite	10^{14}	0.94
Porcelains:		
Dry process	10^{12} to 10^{14}	0.16 to 0.95
Wet process	10^{12} to 10^{14}	0.35 to 1.6
Zirconia	10^{13} to 10^{15}	0.99 to 1.6
Steatite	10^{13} to 10^{15}	0.79 to 1.6
Titanates (Ba, Sr, Ca, Mg, Pb)	10^{8} to 10^{15}	0.02 to 1.2
Titanium dioxide	10^{12} to 10^{18}	0.39 to 0.83
Plastic resins		
Allyl, cast	10^{13} to 10^{14}	1.5
Aniline formaldehyde	10^{16} to 10^{17}	2.5
Epoxy, cast	10^{16} to 10^{17}	1.6
Melamine formaldehyde	10^{11} to 10^{14}	1.5
Methyl methacrylate	10^{14} to 10^{15}	1.8
Nylons	10^{13} to 10^{15}	1.2 to 2.4
Phenol formaldehyde	10^{12} to 10^{13}	1.4
Polyethylene	10^{12} to 10^{13}	1.8
Polystyrene	10^{17} to 10^{19}	2.0
Rubber, hard	10^{15}	1.9
Shellac	10^{9}	0.8
Vinyl chloride	10^{14} to 10^{16}	2.4
Fiber-reinforced resins		
Polyester	10^{13} to 10^{14}	1.4 to 2.0
Phenolic	10^{11} to 10^{12}	0.6 to 1.5
Epoxy	10^{14} to 10^{15}	1.4
Melamine	10^{10} to 10^{11}	0.7 to 1.2
Polyurethane	10^{11} to 10^{14}	1.3 to 3.5
Rubbers and elastomers		
Polyisoprene	10^{15}	2.4
Butadiene	10^{15}	...
Styrene-butadiene	10^{14}	2.4
Acrylonite butadiene	10^{10}	1.9
Polychloroprene	10^{11}	2.0
Isobutylene-isoprene	10^{17}	3.0
Polysulfide	10^{8}	1.3
Polymethane	10^{11}	2.0

Superconductivity

Properties of some superconducting elements

Element	ρ(a)	T_c, K	H_0, Oe	Element	ρ(a)	T_c, K	H_0, Oe
W	5.3	0.012	1070	αTh	15.7	1.37	162
Be	4	0.026	...	Pa	10.8	1.4	...
Ir	5.3	0.14	19	Re	19.3	1.7	193
αHf	35.1	0.165	...	Tl	15	2.4	171
αTi	42	0.39	56	In	8	3.4	293
Ru	7.6	0.49	66	βSn	11	3.7	309
Cd	6.8	0.52	30	Hg	98	4.15	412
Os	9.5	0.65	65	Ta	13.5	4.48	830
αU	30	0.68(?)	...	V	24.8	5.3	1020
αZr	45	0.55	47	βLa	61.5	5.9	1600
Zn	5.9	0.85	52	Pb	20.6	7.2	803
Mo	5.2	0.92	98	Tc	18.5	8.2	1410
Ga	15	1.09	59	Nb	15	9.2 to 9.4	1950
Al	2.6	1.19	99				

(a) In units of μΩ · cm near room temperature

Superconductivities of selected compounds

Compound	T_c, K	Compound	T_c, K
Nb_3Sn	18.05	V_3Ga	16.5
Nb_3Ge	23.2	V_3Si	17.1
Nb_3Al	17.5	$Pb_1Mo_{5.1}S_6$	14.4
NbN	16.0	Ti_3Co	3.44
$(SN)_x$ polymer	0.26	La_3In	10.4

Lattice parameters, and temperatures and entropies of fusion of III-V compounds

		Experimental		Calculated		Values used in
Compound	a_0, Å	T^F, K	ΔS^F, eu	ΔS^F, eu	T^F, K	calculation, ΔS^F, eu
AlP	5.451	2823	...	17.2	1853	17.2
AlAs	5.662	2013	...	16.7	1719	16.7
AlSb	6.136	1330	14.7	15.2	1344	14.7
GaP	5.451	1740	...	16.7	1623	16.7
GaAs	5.653	1513	16.64	16.2	1340	16.64
GaSb	6.096	985	15.8	14.7	983	15.8
InP	5.869	1333	...	15.2	1357	14.0
InAs	6.058	1210	14.52	14.7	993	14.7
InSb	6.479	797	14.3	13.2	745	13.2

Calculated values of critical temperature with and without the inclusion of coherency strain for quaternary III-V alloys

System	Highest T_c for bounding ternary, K(a)	T_c, K(b)
$Ga_xIn_{1-x}As_yP_{1-y}$	908	1081
$Ga_xIn_{1-x}As_ySb_{1-y}$	856	1428
$Al_xIn_{1-x}As_yP_{1-y}$	973	1019
$Al_xIn_{1-x}As_ySb_{1-y}$	735	1599
$Al_xIn_{1-x}P_ySb_{1-y}$	2105	2619
$Ga_xIn_{1-x}P_ySb_{1-y}$	1965	2470
$Al_xGa_{1-x}As_ySb_{1-y}$	973	973
$Al_xGa_{1-x}P_ySb_{1-y}$	1965	2045
$InP_xAs_ySb_{1-x-y}$	1319	1319
$GaP_xAs_ySb_{1-y-x}$	1996	1996
$AlP_xAs_ySb_{1-x-y}$	2105	2105
$Al_xGa_yIn_{1-x-y}P$	973	973
$Al_xGa_yIn_{1-x-y}As$	735	735
$Al_xGa_yIn_{1-x-y}Sb$	462	462

(a) Without coherency strain. (b) With coherency strain

Electronic Ceramics and Glasses

Properties of electronic ceramics

	96% alumina	Beryllia	Boron nitride	Aluminum nitride	Silicon carbide	Silicon
Knoop hardness, kg/mm^2	2000	1000	280	1200	2800	...
Flexural strength, 10^5 N/m^2	3.0	1.7 to 2.4	0.8	4.9	4.4	...
Thermal conductivity, W/m · °C	21	250	60	170 to 200	70	150
Thermal expansion, 10^{-6}/°C	7.1	8.0	0.0	4.1	3.8	3.8
Dielectric strength, kV/mm	8.3	19.7	37.4	14.0	15.4	...
Dielectric loss, 10^{-4} tan δ at 1 MHz	3 to 5	4 to 7	4	5 to 10	500	...
Dielectric constant, at 10 MHz	10	7.0	4.0	8.8	40	...

(a) Thermal properties listed are for 25 to 100 °C

Properties of selected glasses used in electronics

Application	Glass composition type	Thermal expansion coefficient, 0 to 300 °C (32 to 570 °F) 10^{-7}/K	10^{-7}/°F	Softening point °C	°F	Density g/cm^3	lb/in.^{3}	Log volume resistivity at 350 °C (660 °F)	Refractive index, n_d	Dielectric constant 1 MHz, 20 °C (70 °F)
Substrates	Fused quartz	5.5	3.1	1580	2875	2.200	0.079	10.2	1.459	3.8
	Aluminosilicates	31 to 43	17 to 24	924 to 1070	1695 to 1960	2.550 to 2.870	0.0921 to 0.104	10.1 to 11.6	...	...
	Fotoform glass (Li silicate)	84	47	450	845	2.365	0.0854	4.90	...	7.62(b)
	Fotoform opal	89	49	550	1020	2.380	0.0860	7.23	...	5.73(b)
	Fotoceram	103	57.3	750	1380	2.407	0.0870	7.07	...	5.63(b)
Solar cell cover	Alkali borosilicate	32	18	...	...	2.130	0.0770	11.2(a)	1.469	4.1
		74	41	718	1325	2.600	0.094	...	1.5281	6.69
Fiberglass	E-glass (Mg-Ca-Al borosilicate)	50	28	843	1550	2.520 to 2.620	0.0910 to 0.0947	...	1.546 to 1.560	5.8
	S-glass (Mg aluminosilicate)	29 to 30	16 to 17	860	1580	2.470 to 2.490	0.0892 to 0.0900	...	1.524 to 1.528	4.53
Microchannel plates	Tube (K-Rb lead silicate)	90.3	50.2	600	1110	3.970	0.143	...	...	...
	Carrier (K-Na lead silicate)	89.5	49.8	630	1165	3.850	0.139	...	...	...
EPROM windows	NaAl borosilicate	38	21	705	1300	2.170	0.0784	7.6	1.47	4.70
Diode enclosures	K lead silicate	91	51	578	1070	4.280	0.155	9.5	1.7	9.1
	Ba lead borosilicate	45	25	720	1330	2.880	0.104	11.3	1.55	6.0
Recording heads	Fotoceram	103	57.3	750	1380	2.407	0.0870	7.07	...	5.63(b)
	Gap glass (Ba-Ca borosilicate	80	44	770	1420	...	...	...	...	...
	Bonding glass (lead borosilicate)	80	44	490	915	...	...	...	...	...
	Na lead silicate	95	53	495	925	5.100	0.184	...	...	...
	Na-Zn borosilicate	104	57.8	580	1075	2.980	0.108	...	...	...
Resistors	Aluminosilicate	44	24	926	1700	2.640	0.0954	13.5(a)	1.547	6.6

(a) At 250 °C (480 °F). (b) At 100 kHz

Piezoelectric and pyroelectric glass-ceramics

Primary phase	Density, g/cm³	Dielectric constant	Tan δ	Specific heat, J/cm³ · K	Pyroelectric coefficient, *p*, μCurie/m² · K	M_1(a), μCurie/m²	Composition, mol%							
							SiO_2/GeO_2	B_2O_3	BaO/SrO	Li_2O	CaO	TiO_2	PbO	ZnO
$Ba_2TiSi_2O_8$(b)	3.87 to 4.0	9 to 10.5	0.001	2.04	4.5 to 6.0	0.46 to 0.60	48 to 50/…	…	25 to 33/…	…	0 to 6.9	16 to 17	1.7	…
$Ba_2TiSi_2O_8$	4.05	10	0.001	2.04	8.0	0.80 to 0.88	50/…	…	32/…	…	…	17	…	…
$Ba_2TiGe_2O_8$(c)	4.78	15	…	…	-2.0	0.13	…/33	…	33/…	…	…	33	…	…
$Sr_2TiSi_2O_8$(b)	3.53 to 3.63	10.5 to 11.5	0.001	…	7.0 to 8.0	0.66 to 0.70	47 to 50/…	…	0 to 3.3/30 to 33	…	0 to 1.7	17	…	…
$Li_2Si_2O_5$(c)	2.47	6.5	0.008	2.36	-3.0	0.46	62/…	…	…	31	…	…	…	6
$Li_2Si_2O_5$	2.32 to 2.35	7.0 to 7.5	0.05	2.39	-10 to -11	1.33 to 1.57	53 to 60/…	6 to 13	…	33	…	…	…	…

(a) M_1 equals $p \cdot K$ and is used for rapid evaluation of material. (b) Tetragonal, 4 mm (0.16 in.). (c) Orthogonal, 2 mm (0.08 in.)

Dissipation factor as a function of temperature and frequency

Frequency and temperature	Material					
	Al_2O_3, 96%	Al_2O_3, 99%	BeO, 99%	Forsterite	Steatite	Fused silica
1 Mz						
At 25 °C (75 °F)	0.0003	0.0002	0.0001	0.0004	0.002	0.0002
At 500 °C (930 °F)	0.013	0.002	0.0004	...	0.06	...
1 GHz						
At 25 °C (75 °F)	0.0003	0.0002	0.0002	...	0.0015	0.0001
At 500 °C (930 °F)	0.0015	0.0015	0.0006	...	0.015	...
10 GHz						
At 25 °C (75 °F)	0.0006	0.0001	0.0001	0.0009	0.003	0.0001
At 500 °C (930 °F)	0.002	0.0002	0.0001	0.0013	0.005	0.00009
25 GHz						
At 25 °C (75 °F)	0.0007	0.0003	0.004	...	...	...
At 500 °C (930 °F)	0.002	0.0003	0.004	...	...	...

General dielectric properties of electronic ceramics

Product type	Dielectric constant 1 MHz	Dielectric loss, tan δ, 1 MHz
9606(a)	5.6	0.017
9608(a)	6.9	0.023
LAS(b)	6.9 to 9.1	0.006 to 0.02
LAS(c)	5.5 to 7.8	0.0026 to 0.0090
	5.2 to 6.0	0.007 to 0.014
ZAS(d)	6.4 to 7.4	0.0005 to 0.0022
MAS(e)	5.6 to 7.5	0.0015 to 0.003
	5.6 to 8.5	0.0003 to 0.0005
MAS(f)	8.5	0.006
Macor type	5.9 to 6.1	0.0006 to 0.003
	4.7(g)	0.0024(g)
$SrTiO_3$(h)	~35	0.005 to 0.01
MgSiAlON-type		
E″	9.1(g)	0.0029(g)
D	8.6(g)	0.003(g)
YSiAlON-type		
YG2	10.5 to 11.3(g)	0.0031 to 0.004(g)

(a) Manufactured by Corning. (b) βeucryptite type. (c) βspodumene type. (d) Broad composition range. (e) Cordierite type. (f) Enstatite type. (g) Data are measured at 1 to 20 kHz. (h) 35% glass former

Diffusion and Electromigration

Interdiffusion in solids

Distance of diffusion	Sample	Relationship between flux and driving force	Compound formation
~1 Å	Monolayer of metallic atoms of silicon	...	...
~10 Å	Man-made superlattices	Nonlinear	Single
~1000 Å	Bilayer thin films, film/substrate	Linear	Single
~1 mm	Bulk samples	Linear	Multiple

Electromigration of selected materials

Material	Behavior
Teflon	No migration
Polystyrene	No migration
Polyethylene	No migration
Polyvinyl chloride	No migration
Hard rubber (sand-blasted)	No migration
Bakelite	No migration
Glass-filled silicone	No migration
Glass-filled epoxy	Slight surface migration
Glass microscope slide	Slight surface migration
Glazed steatite	Slight surface migration
Laminated phenolic paper	Heavy migration in fibers
Cellophane	Heavy migration
Glass-filled melamine	Heavy migration
Glass-bonded mica	Heavy surface migration
Nylon 6	Extensive surface migration

Normalized average antimony diffusivity and calculated interstitial concentration at 1100 °C

Oxidation time, min	D/D_i, measured	D/D_i^0, calculated	$(C_i)/C_i^0$	$(C_v)/C_v^0$
5	1.11	0.91	5.1	0.27
10	0.92	0.76	4.4	0.19
20	0.66	0.63	3.9	0.13
30	0.52	0.57	3.6	0.1
60	0.38	0.38	3.2	0.05

Characteristics of arsenic diffusion

Property	Result
$\Delta S_D^- >> \Delta S_D^x$	As^+V^- pair diffusion dominates at T > 1050 °C
$As^+V^=$ binding energy is ~0.25 eV less than $P^+V^=$	Fewer $As^+V^=$ pairs than $P^+V^=$ a. Less gettering of metal donors b. No emitter push above 700 °C c. No effect on $[O_i]$ precipitation
VAs_2 pair binding energy = 1.6 eV	1. Significant $[VAs_2]$ form at $n = 2 \times 10^{20}$ cm^{-3} at 1000 °C 2. Reduces *n* solubility 3. Causes retarded base diffusion in some cases
2.5 eV required to make As interstitial	Little oxidation-enhanced diffusion

Characteristics of boron diffusion

Property	Result
B^-V^+ migration energy is ~0.6 eV	Relatively fast diffuser dominated by $[V^+]$
Small tetrahedral covalent radius	1. Small diffusion entropy 2. Misfit strain a. Dislocations b. Good gettering c. Reduced diffusivity
2.26 eV required to make B interstitial	Oxidation-enhanced diffusion
Forms stable oxides B_2O_3 and HBO_2	Segregation into growing SiO_2 from Si

Characteristics of phosphorous diffusion

Property	Result
Large ΔS_D^x (11.6 k)	Diffusion via V^x is dominant at low concentrations
Large $P^+V^=$ pair binding energy (1.57 eV)	$P^+V^=$ pair dominates high concentration diffusion a. $P^+V^=$ pairs compensate monatomic P^+ b. Gettering of donor metal ions → P^+M^-
$P^+V^=$ dissociation at $E_c - E_f = 0.11$ eV	1. Emitter push effect 2. Defect shrinkage near junction 3. Reduced $[O_i]$ precipitation
Small covalent radius	1. Misfit strain a. Gettering by misfit dislocations b. Reduced diffusivity through band gap narrowing 2. $\Delta S_{P-V}^- < \Delta S_{As-V}^-$

Estimated interstitial formation energies in silicon

Element	Interstitial formation energy, eV
Si	2.2
Al^{2+}	2.21
B	2.26
P	2.4
As	2.5

Fractional interstitialcy components of diffusion via self-interstitials in silicon at 1000 to 1100 °C

Element	Fair	Antoniadis	$f_i = D_i^I / D_i^V$ Matsumoto	Gosele	Mathiot
B	0.17	0.32	0.41	0.8 to 1.0	0.18
Al	0.2	...	...	0.6 to 0.7	...
P	0.12	0.40	0.35 to 0.5	0.5 to 1.0	0.19
As	0.09	0.43	0.45 to 0.75	0.2 to 0.5	0.16
Sb	0.13	0.15	...	0.02	...

R.B. Fair, *J. Appl. Phys.*, Vol 51, 1980, p 5828; D.A. Antoniadis and I. Moskowitz, *J. Appl. Phys.*, Vol 53, 1982, p 9214; s. Matsumoto, Y. Ishikawa and T. Niimi, Extended Abstracts of 15th Conf. on Solid State Devices and Materials, in *Jap. J. Appl. Phys.*, 1983, p 19; U. Gosele and T.Y. Tan, *Defects in Semiconductors II*, edited by S. Mahajan and J.W. Corbett, North Holland Press, Amsterdam, 1983; D. Mathiot and J.C. Pfister, *J. Appl. Phys.*, Vol 55, 1984, p 3518

Devices

Lead Frames

Nominal composition of lead frame alloys

Alloy group	Designation	Nominal composition, net %
Cu-Fe	C19400	2.35Fe-0.03P-0.12Zn(a)
	C19500	1.5Fe-0.8Co-0.05P-0.6Sn(a)
	C19700	0.6Fe-0.2P-0.04Mg(a)
	C19210	0.10Fe-0.034P(a)
Cu-Cr	CCZ	0.55Cr-0.25Zr(a)
	EFTEC 64T	0.3Cr-0.25Sn-0.2Zn(a)
Cu-Ni-Si	C7025	3.0Ni-0.65Si-0.15Mg(a)
	KLF-125	3.2Ni-0.7Si-1.25Sn-0.3Zn(a)
	C19010	1.0Ni-0.2Si-0.03P(a)
Cu-Sn	C50715	2Sn-0.1Fe-0.03P(a)
	C50710	2Sn-0.2Ni-0.05P(a)
Other	C15100	0.1Zr(a)
	C15500	0.11Mg-0.06P(a)
Fe-Ni	ASTM F30 (Alloy 42)	42Ni-58Fe
Fe-Ni-Co	ASTM F15 (Kovar)	29Ni-17Co-54Fe

(a) Remainder Cu

Mechanical properties of lead frame materials

Property	Copper	Alloy 42	Kovar
Coefficient of thermal expansion, 10^{-6}/K	17.6	4.4	5.1
Tensile strength, MPa (ksi)	290 (42)	470 (68)	550 (80)
Yield strength, MPa (ksi)	250 (36)	240 (35)	345 (50)

Lead bend fatigue comparison (longitudinal orientation)

	0.2% offset yield strength	
Alloy	MPa	ksi
C15100	380	55
C19210	380	55
CCZ	430	62
C19700	450	65
C50710	450	65
C19400	475	69
K75	480	70
C50715	550	80
C19010	585	85
C7025	620	>90
ASTM F30	620	>90 (>4 to 5 cycles)

Solderabilities of lead frame alloys

Alloy group	Alloy	Solderability rating (type R flux)(a)
Cu-Fe	C19400	1 to 2
	C19500	1 to 2
	C19700	1 to 2
	C19210	1 to 2
Cu-Cr	CCZ	2
	EFTEC 64T	2
	K75	2
Cu-Ni-Si	C7025	1
	KLF-125	1
	C19010	1 to 2
Cu-Sn	C50715	2
	C50710	2
Other	C15100	1
	C15500	1
Fe-Ni	ASTM F30	3 to 4

(a) Visual ratings: class 1, uniform smooth coating; class 2, uniform smooth coating; class 2, 100% wetting, but not smooth; class 3, 2% dewetting, % pinholes; and class 4, 50% dewetting, 10% pinholes

Printed Wiring Boards

Metal usage in printed wiring board manufacturing

Metal	Application
Copper	Bulk current carrying and/or electronic signal conduction
Tin-lead	Solderability enhancement and long-term storage. Also used as an etch resist (with ammoniacal etches)
Tin	Solderability enhancement (storage life limited). Also used as an etch resist (with ammoniacal etches)
Nickel	Diffusion barrier between gold and copper. Also acts as an anvil, absorbing contact forces without deformation
Gold	Noncorrosive, low-resistance contact finish. Also used as a finish for wire bonding in chip-on-board applications
Palladium-nickel	Lower-cost alternative contact finish, also reported to be reliably solderable
Rhodium	Extremely durable precious metal contact finish, but very rarely used in applications today

Properties of some alternative printed wiring board resins

Resin	Generic type	Supplier	Relative permittivity	Loss tangent	Specific gravity	Glass transition temperature °C	°F
G-10FR(a)	FR-4 epoxy	Norplex	3.6	0.032	1.35	125	255
RSM-1212	FR-4 epoxy	Shell	3.6	0.032	...	125	255
Quatrex 5010	tetra func. epoxy	Dow	3.6	0.032	1.22	180	355
RSM-1151	tetra func. epoxy	Shell	3.6	0.032	...	180	355
G-200(a)	BT/epoxy	Norplex	3.2	0.012	1.35	185	365
BT2110F	BT	Mitsubishi	3.1	0.003	1.35	250	480
BT2060BG	modified-BT	Mitsubishi	2.9	0.001	...	220	430
Kerimid	Polyimide	Rhone Poulenc	3.2	0.02	1.4	275	525
RSM-1206(b)	Polyimide	Shell	...	...	...	...	...
RDX 64826	Cyanate ester I	Interez	3.1	0.005	1.21	260	500
XU-71787	Cyanate ester IV	Dow	2.8	0.004	1.18	260	500
	PTFE(c)	...	2.0	0.0002	2.2	...	...
UDEL	Polysulfone(c)	Union Carbide	3.0	0.004	1.24	192	375
ULTEM	Polyetherimide(c)	General Electric	3.1	0.0065	1.27	212	415
Victrex	Polyether sulfone(c)	ICI	3.5	0.0035	...	230	445
	Polyolefin(c)	...	2.3	0.002	1.18	...	...
	Polybutadiene(c)	...	2.8	0.005	1.0	...	...
	Polyphenylene oxide(c)	...	2.55	0.0007	1.1	85	185
	Polycarbonate(c)		3.0	0.001	1.2	195	385

(a) These designations refer to laminates available from Norplex Oak. The tabulated data are for the resin blends employed in these laminates. (b) This non-MDA polyimide is usually provided as a blend with an epoxy. (c) Thermoplastic polymer

Coefficient of thermal expansion measurements for Kevlar-reinforced MLBs

Printed wiring board construction	Calculated coefficient of thermal expansion, 10^{-6}/K	Composite coefficient of thermal expansion, 10^{-6}/K Per dilatometer	Per TMA
Nine-layer epoxy-Kevlar 1.7-mm (0.068-in.) thick (120 cloth)	10.6	Not measured	8.3*X*, 8.0*Y*
Nine-layer epoxy-Kevlar nonwoven, 1.62-mm (0.065-in.) thick	12.5	Not measured	17.1*X*(a), 7.0*Y*(a)
Nine-layer epoxy-Kevlar 1.65-mm (0.066-in.) thick (120 cloth)	10.6	9.8*X*, 9.0*Y*	Not measured

(a) Measurement questionable and board layup was not cross plied

Properties of fiberglass printed wiring board laminates

	Unmodified				Modified	
	BPADCy	METHYLCy	THIOCy	XU 71787	BT 2170	BPADCy
Composition						
Modification	...	...	...	...	Epoxy	Epoxy
Wt % modifier	...	...	...	...	45	60
Bromine, % of epoxy	0	0	0	0	12	16
Glass reinforcement type	7628	7628	7628	NA	NA	7628
Resin content, wt %	40±1	40±1	40±1	NA	NA	40±1

(continued)

Properties of fiberglass printed wiring board laminates (continued)

	Unmodified				Modified	
	BPADCy	METHYLCy	THIOCy	XU 71787	BT 2170	BPADCy
Cure temperature						
Platen temperature, °C (°F)	177 (350)	177 (350)	177 (350)	177 (350)	175 (347)	177 (350)
Postcure, °C (°F)	232 (450)	232 (450)	232 (450)	225 (437)	...	None
Laminate property						
T_g, °C (°F)	255 (491)	247 (477)	250 (482)	255 (491)	210 (410)	182 (360)
Thermogravimetric analysis onset, °C (°F)	405 (760)	427 (800)	400 (752)	425 (797)	305 (581)	300 (572)
Flexural strength, MPa (ksi)	448 (65)	414 (60)	443 (64)	...	483 (70)	483 (70)
Peel strength, N/cm (lbf/in.)						
At 25 °C (75 °F)	21.0 (12.0)	20.6 (11.8)	19.3 (11.0)	15.8 (9.02)	16.7 (9.54)	17.2 (9.82)
At 100 °C (212 °F)	19.1 (10.9)	18.6 (10.6)	18.7 (10.7)	12.3 (7.02)	13.7 (7.82)	16.6 (9.48)
At 200 °C (390 °F)	15.8 (9.02)	16.7 (9.54)	17.9 (10.2)	10.5 (6.00)	11.8 (6.74)	14.9 (8.51)
Steam/solder float, min.(a)	90	90	90	...	...	120
Dielectric constant at 1 MHz	4.05	3.8	4.3	4.0	4.2	4.2
Dissipation factor at 1 MHz	0.004	0.003	0.003	0.003	0.008	0.011
Flammability, UL 94	Burns	V-1	V-0/1	Burns	V-0	V-0

(a) Minutes conditioning in 121 °C (250 °F) steam passing 20-s float in 260 °C (500 °F) solder

Properties of glass-polymer printed-circuit boards

Glass-polymer composition type	Glass transition temperature, T_g °C	°F	Thermal expansion coefficient at $<T_g$ x, y 10^{-7}/K	x, y 10^{-7}/°F	z 10^{-7}/K	z 10^{-7}/°F	Dielectric constant 1 MHz	Dissipation factor, 1 MHz	Surface resistivity Ω/square	Water absorption, 24 h at RT
FR-4 epoxy-glass	120 to 130	250 to 265	160 to 200	90 to 110	500 to 700	280 to 390	4 to 5.5	0.02 to 0.03	10^{14}	0.3
Polyimide-glass	210 to 220	410 to 430	140	80	400	220	4 to 5	0.01 to 0.015	10^{14}	0.15 to 0.4

Properties of selected laminate material for SMT

Material	T_g °C	T_g °F	Coefficient of thermal expansion, 10^{-6}/K x, y	z	Thermal conductivity W/m · K	Btu · in./h · ft² · °F	Weight ratio	Dielectric constant at 1 MHz	Moisture absorption, %	Elastic modulus GPa	10^6 psi
E-glass/epoxy	125	255	12 to 16	80 to 90	0.35	2.5	1	4.8	0.1	17.4	2.5
E-glass/polyimide	200 to 260	390 to 500	11 to 14	60	0.35	2.5	1	4.5	0.4	19.6	2.8
E-glass/PTFE	75	165	24	261	0.26	1.8	1	2.3	...	1.0	0.14
Kevlar 120(W)/Quatrex	185	365	3 to 8	105	0.16	1.1	0.7	3.9	1.5	22.1 to 27.6	3.2 to 4.0
Kevlar 120(W)/polyimide	180 to 200	355 to 390	3 to 8	83	0.12	0.80	0.7	3.6	1.8	...	...
Kevlar 120(T)/Quatrex	185	365	0.75	...	...	...	0.75	...	1.5	44.8	6.5
Kevlar 108(W)/4093 resin	125 to 135	255 to 275	4 to 7	50 to 110	...	...	0.7	3.9	1.5	15.9 to 19.3	2.3 to 2.8
Kevlar 108(T)/Quatrex	185	...	0.75 to 1.25	66 to 82	...	...	0.75	3.7	1.5	37.9	5.5
Kevlar 2643(NW)/Quatrex	185	365	5 to 8	50 to 110	...	...	0.70	3.3	0.9	13.8 to 15.9	2.0 to 2.3
Quartz/polyimide	260	500	6 to 12	34	0.13	0.90	1.1	4.0	0.4	27.6	4.0
Quartz/quatrex	185	365	...	62	...	...	1.1	3.5	0.4	18.6	2.7
RO 2800	...	...	16 to 19	24 to 30	0.44	3.1	1.1	2.8	...	4.1	0.6

Packaging Materials

Coefficient of thermal expansion for some materials used in electronic packaging

Material	Coefficient of thermal expansion, 10^{-6}/K
Metals	
Aluminum and its alloys	22 to 25
Copper and its alloys	16 to 18
Gold	14.2
Gold-tin eutectic	16.0
Invar	2.0
Iron	12.0
Lead	29.0
Lead-tin eutectic	21.0
Molybdenum	5.2
Nickel	13 to 15
Silver	19
Tin-lead (5/95)	28
Titanium	10
Ceramics, glasses, and semiconductors	
Alumina	6.7
Beryllia	8.0
Gold-germanium eutectic	13.0
Gold-silicon eutectic	13
Silica glasses	0.5 to 1.0
Silicon (single crystal)	2.8
Silicon carbide	2.6
Organics (below glass transition temperature)	
Epoxies	60 to 80
Epoxy-glass (FR-4)	
In-plane	11 to 15
In through-thickness direction	60 to 80
Kevlar	−2.0
Polycarbonates	50 to 70
Polyimides	40 to 50
Polyurethanes	180 to 250
RTV	800
Sylgard	300

Thermal properties of common packaging materials at room temperature

Material	Thermal conductivity, W/m · K	Thermal conductivity, Btu · in./h · ft^2 · °F	Coefficient of thermal expansion, 10^{-6}/K
Alumina, 96%	35.1	245	6.4
Alumina, 99.5%	37.0	255	6.5
Aluminum	238	1650	23.5
Aluminum alloy 6061	178	1235	23.4
Aluminum nitride	170	1180	4.5
Beryllia	250	1735	7.5
Copper	388	2690	17.2
Epoxy	0.16	1.1	~60
Gold	295	2045	14.1
Kovar (ASTM F15 alloy)	19	130	5.5
Molybdenum	134	930	5.1
Silicon	84	580	2.5
Solder, Sn63	50	345	25

Integrated circuit packaging materials properties

Property	Kovar	Alloy 42	BeO	Cu-W	Al_2O_3	AlN	SiC	Mullite	Glass-ceramic
Modulus of elasticity, GPa (10^6 psi)	140 (20)	145 (21)	345 (50)	330 (48)	255 (37)	275 (40)	275 (40)	175 (25)	70 (10)
Dielectric constant	...	...	7 to 8	...	8 to 10	8.5	40	6.8	4.7
Thermal conductivity at room temperature, W/m · K (Btu · in./h · ft^2 · °F)	17.3 (120)	10.7 (75)	240 (1680)	210 (1470)	15 to 35 (105 to 240)	150 to 220 (1040 to 1525)	270 (1865)	5 (35)	2.5 (17)
Coefficient of thermal expansion at room temperature, 10^{-6}/K	5.9	5.4	7 to 8.5	6.0	7 to 8	3.5 to 4.5	3.7	4.2	2 to 3

Properties of monolithic metal and metal-matrix composite packaging material

Reinforcement	Matrix	Reinforcement, vol%	Density, g/cm^3	Density, lb/in.^{3}	In-plane coefficient of thermal expansion, 10^{-6}/K	In-plane coefficient of thermal expansion, 10^{-6}/°F	In-plane thermal conductivity, W/m · K	In-plane thermal conductivity, Btu/h · ft · °F
...	Aluminum	0	2.7	0.098	23	13	180	103
...	Copper	0	8.9	0.32	17	9.8	400	230
...	Kovar	0	8.3	0.30	5.9	3.2	17	10
...	Molybdenum	0	10.2	0.37	5.0	2.8	140	80
...	Tungsten	0	19.3	0.695	4.5	2.5	180	104
...	Beryllium	0	1.86	0.067	13	7.2	150	87

(continued)

Properties of monolithic metal and metal-matrix composite packaging material (continued)

Reinforcement	Matrix	Reinforcement, vol%	Density g/cm^3	Density lb/in.^{3}	In-plane coefficient of thermal expansion 10^{-6}/K	In-plane coefficient of thermal expansion 10^{-6}/°F	In-plane thermal conductivity W/m · K	In-plane thermal conductivity Btu/h · ft · °F
...	Titanium	0	4.4	0.16	9.5	5.3	16	9.5
...	Magnesium	0	1.80	0.065	25	14	54	31
Carbon fiber	Epoxy	60	1.8	0.065	-1.1	-0.6	310	180
Carbon fiber	Aluminum	28	2.5	0.092	6.5	3.6	290	168
Carbon fiber	Copper	28	7.2	0.26	6.5	3.6	400	230
Boron fiber	Aluminum	20	2.6	0.095	12.7	7.1	180	103
SiC particles	Aluminum	55	2.9	0.104	10.3	5.7	180	103

Thermal, physical, and mechanical properties of packaging materials

Property	Nylon MXD/6	Nylon 6	PET
Molded plaques			
T_m, °C (°F)	243 (470)	225 (435)	255 (490)
T_g, °C (°F)	75 (165)	62 (145)	67 (150)
Heat-deflection temperature, °C (°F)	96 (205)	57 (135)	85 (185)
Coefficient of thermal expansion, 10^{-5}/K	5.1	8	7
Tensile strength, MPa (ksi)	99 (14.4)	62 (9.0)	78 (11.3)
Flexural modulus, GPa (10^6 psi)	4.4 (0.64)	2.4 (0.35)	3.4 (0.49)
Izod impact strength, notched, J/m (ft · lbf/in.)	20 (0.37)	60 (1.1)	40 (0.75)
Rockwell hardness, M scale	108	85	106
Biaxially oriented films			
Thickness, μm (mil)	15 (0.60)	15 (0.60)	12 (0.50)
Specific gravity	1.22	1.14	1.38
Haze, %	3.1	2.1	2.6
Tensile strength, machine direction, MPa (ksi)	215 (32)	196 (28)	196 (28)
Moisture permeability, 40 °C (105 °F), 90% RH, g/m^2/day	40	260	40
Oxygen permeability, 26 °C (79 °F), 60% RH, cm^3/mm/m^2/day, atm	0.06	0.65	1.5

Films

Thick-, thin-, and polymeric-film resistor characteristics

Parameter	Thick film	Polymeric film	Ni	Tantalum nitride
Resistivity range, Ω/square	10^6 to 10^7	10^4 to 10^6	10 to 300	50 to 200
Temperature coefficient of resistance, ppm/°C (ppm/°F)	±50 to 300 (±30 to 165)	±250 to 1000 (±140 to 555)	±0 to 50 (±0 to 30)	±40 (±20)
Temperature coefficient of resistance tracking, ppm	10	50	1	1
Power handling capacity, W/cm² (W/in.²)	15 (100)	3.8 (25)	6 (40)	6 (40)
Thermal stability, at 150 °C (300 °F)/1000 h, %ΔR	≤±0.25	≤±2	≤±0.1	≤±0.2
Voltage coefficient of resistance, ppm/V	0.5 to 5	25 to 100	0	1
High-voltage stability, %ΔR	≤±0.5	≤±2	≤±1	≤±1
Noise, dB	-35 to 20	0 to 35	≤−30	<−10
Trim stability, %ΔR	≤±0.25	≤±2	≤±0.1	≤±0.01
Load life at 25 °C (77 °F)/1000 h (rated power), %ΔR	≤±0.25	≤±3	0.1	≤±0.1
Aging stability, at 25 °C (77 °F)/50% RH/1000 h, %ΔR	≤±0.25	≤±2	≤±0.1	≤±0.1
Conductor compatibility	Au, Ag, Cu, PdAg, PtAg	Au, Ag	Au, Cu	Au, Cu

Thick-, thin-, and polymeric-film dielectric comparisons

Parameter	Thick film	Thin film	Polymeric film
Insulation resistance range, Ω	$>10^{11}$	$>10^{10}$	$>10^7$
Dielectric constant range, at 1 kHz	9 to 1200	NA	4 to 6
Dissipation factor range, %	<3.5	NA	<5
Voltage breakdown range, kV/cm (V/mil)	>200 (>500)(a)	≥40 (≥100)	Low
Temperature coefficient of capacitance, ppm/°C (ppm/°F)	<140 (<250)	NA	NA
Hermeticity	Excellent	Fair	Poor
Thermal coefficient of thermal expansion range	≈96%(b)	≈99%(b)	High
Conductor compatibility	PtAu, Au	Au	Cu, Ag
Resistor compatibility	Good	Good	Good
Capacitor value range (practical), pF	≤1000	Small	Small

(a) For 0.05 mm (0.002 in.) film. (b) Aluminum oxide

Typical thick-film conductor resistivities

Composition	Resistivity at printed thickness, mΩ/square
Pt-Au	50 to 100
Ag-Pd	15 to 40
Au	3 to 5
Cu	2 to 4
Ag	1.5 to 3.0

Typical properties of thick-film dielectrics

Property	Value
Insulation resistance, Ω	$>10^{12}$
Dissipation factor	<0.5%
Breakdown voltage, kV/mm (V/mil)	>20 (>500)
Dielectric constant, κ	6 to 10

Typical thick-film conductor resistivities and temperature coefficients of resistance

Conductor compositions	Sheet resistivity(a), Ω/square	Temperature coefficient of resistance, ppm/°C	Temperature coefficient of resistance, ppm/°F	Line resistance, Ω: 25 × 0.50 mm (1 × 0.020 in.)	Line resistance, Ω: 25 × 0.15 mm (1 × 0.005 in.)
Platinum-gold	0.100	800	445	5	20
Gold	0.005	1250	695	0.15	0.6
Palladium-silver	0.035	1400	775	1.75	7
Silver	0.015	1800	1000	1	4

(a) Fired thicknesses of conductors are 10 μm (0.4 mils)

Electrical characteristics of thin-film deposited conductors(a)

Property	Line width, mm (mil)			
	125 (5)	250 (10)	380 (15)	510 (20)
Resistance, Ω				
Typical				
Per mm	0.080	0.040	0.025	0.020
Per in.	2.00	1.00	0.67	0.50
Maximum				
Per mm	0.160	0.080	0.050	0.040
Per in.	4.00	2.00	1.33	1.00
Current rating, mA				
Per mm	12	24	36	48
Per in.	300	600	900	1200
Capacitance between adjacent conductors, 125 mm (5 mils) apart, pF				
Per mm		0.040		
Per in.		1		
Resistivity, Ω/square				
Gold plating thickness of				
8 mm (70 μin.)		0.015 to 0.020		
5 mm (100 μin.)		0.010		
7 mm (144 μin.)		0.006 to 0.009		
Temperature coefficient of resistance, Ω/°C (Ω/°F)		0.004 (0.002)		

(a) Material is electroplated gold, 1 to 5 μm (40 to 200 μin.) thick; over titanium, 0.15 to 0.30 μm (6 to 12 μin.) thick; over tantalum nitride, 0.05 μm (2 μin.) thick; over alumina substrate, 0.6 mm (0.025 in.) thick

Conductor system attributes

Material	Die bondability methods: Eutectic	Die bondability methods: Solder	Die bondability methods: Organic	Wire bondability: Gold	Wire bondability: Aluminum	Solderability	Solder leach resistance	Corrosion resistance
Thick films								
Au	Good	Poor	Excellent	Excellent	Good	Poor	Poor	Excellent
PtAu	NG	Good	Excellent	Fair to poor	Fair	Excellent	Good	Excellent
PdAu	NG	Fair	Excellent	Fair to poor	Fair	Good	Good	Excellent
Ag	NG	Good	Excellent	Good	NG	Good	Poor	NG
PtAg	...	Good	Excellent	Good	NG	Good	Good	Good
PdAg	...	Good	Excellent	Good	Good	Good	Good	Good
PdPtAg	...	Good	Excellent	Good	Good	Good	Good	Good
Cu	NG	Good	Excellent	NG	Fair	Excellent	Excellent	Poor
Ni	NG	NG	Excellent	NG	NG	NG	Excellent	Excellent
Polymeric	NG	NG	Excellent	NG	NG	NG	NG	Good
W or Mo/Mn	NG	NG	Excellent	NG	NG	NG	NG	Good
Thin films								
Au	Good	NG	Excellent	Excellent	Excellent	Poor (except with Ni undercoat)	Poor	Excellent
Cu	NG	Good	Excellent	NG	NG	Excellent	Excellent	Poor
Pd	NG	NG	Excellent	NG	NG	NG	NG	Excellent
Ag	NG	NG	Excellent	NG	NG	NG	NG	Poor

NG, not good

Substrates

General substrate selection criteria

Material	Cost per in.^{2} substrate: 100	5000	100 000	Circuit technology compatibility	Applications	Remarks
Alumina	Medium	Medium	Very low	All	Multiple	General purpose
Beryllia	High	Medium	Low	All	Heat dissipating	Power products
Glazed metal	Medium	Low	Low	All	Special shapes	Becomes cost effective for large substrates
Cofired multilayer ceramic	Very high	High	Medium	All	Package	May be cost effective in large quantities
Glass	High	Medium	Low	Thick and thin film	Displays	Cost effective
Quartz	High	High	Medium	Thin film	UHF	MIC applications
Oxidized silicon	High	High	Medium	Thin film	Chip	Passive components
Sapphire	Very high	High	High	Thin film	VHF	MIC applications
Ferrite	High	High	High	Thin film	VHF/UHF	Magnetic applications
Organic film	Medium	Low	Very low	Polymeric film	Multilayer	Cost effective

UHF, ultrahigh frequency; MIC, microwave integrated circuit; VHF, very high frequency

Physical characteristics of substrate

Material	Mechanical integrity	Mechanical strength MPa	ksi	Thermal characteristics kW/m^2 · K	cal/cm^2 · s · °C	Coefficient of thermal expansion, 10^{-6}/K	Environmental compatibility	Electrical characteristics
Alumina, 96%	Hermetic	480(a) 310 to 345	70(a) 45 to 50	3.52	0.084	6.4	No limit	$\kappa = 9$
Beryllia	Hermetic	230	33	21.8	0.52	7.8	No limit	$\kappa = 6.5$
Glazed metal	Hermetic	Shock resistant		1.34	0.032	6	No limit	$\kappa = 6.4$
Cofired multilayer ceramic	Hermetic	380	55	1.47	0.035	6.0 to 6.5	No limit	$>10^{14}\ \Omega \cdot$ cm $\kappa = 7$ to 9
Glass	Hermetic	Brittle 21	 3	0.121	0.0029	11.2	No limit	$\kappa = 6$ to 7
Quartz	Hermetic	Brittle 49	 7	0.167	0.004	1	No limit	$\kappa = 4$
Oxidized silicon	Hermetic	Brittle 125 to 170	 18 to 25	Good	Good	~3	No limit	...
Sapphire	Hermetic	Brittle		37.7 to 41.9	0.9 to 1	6.7	No limit	$\kappa = 9$
Ferrite	Hermetic	Brittle		...	...	...	Dry atmosphere	...
Organic film	Nonhermetic	Flexible		Poor	Poor	12	No high temperatures	...

(a) 99.5% alumina

Thermomechanical properties of substrate materials

Substrate material	Thermal conductivity, κ W/m · K	Btu/h · ft · °F	Coefficient of thermal expansion, 10^{-6}/K x, y	z	Density g/cm^3	lb/in.^{3}	Modulus of elasticity GPa	10^6 psi	Specific thermal performance (relative to aluminum), $\kappa/\rho \div (\kappa/\rho)$Al
6061 aluminum	190	110	23.6	23.6	2.71	0.098	70	10	1.0
CDA copper	391	222	17.3	17.3	8.95	0.323	117	17	0.6
Beryllium									
High	216	125	11.5	11.5	1.86	0.067	289	42	1.66
Low	151	87	11.5	11.5	1.86	0.067	289	42	1.66
20Cu/60Invar/20Cu									
In-plane	164	95	6.02	7.7	8.45	0.305	133	19	0.28
Out-of-plane	22	12.7	6.02	7.7	8.45	0.305	133	19	0.28
16Cu/68Invar/16Cu									
In-plane	132	76	5.22	5.7	8.39	0.303	133	19	0.22
Out-of-plane	19.6	11.3	5.22	5.7	8.39	0.303	133	19	0.22
12.5Cu/75Invar/12.5Cu									
In-plane	110	64	3.69	5.3	8.31	0.30	133	19	0.19
Out-of-plane	17.8	10.3	3.69	5.3	8.31	0.30	133	19	0.19

(continued)

Thermomechanical properties of substrate materials (continued)

Substrate material	Thermal conductivity, κ W/m · K	Thermal conductivity, κ Btu/h · ft · °F	Coefficient of thermal expansion, 10^{-6}/K x, y	Coefficient of thermal expansion, 10^{-6}/K z	Density g/cm^3	Density lb/in.^{3}	Modulus of elasticity GPa	Modulus of elasticity 10^6 psi	Specific thermal performance (relative to aluminum), κ/ρ + (κ/ρ)Al
Low carbon steel	27	15	10	10	7.84	0.283	210	30	0.085
94% alumina	20.9	12.1	6.4	6.4	3.60	0.13	258	37	0.083
13Cu/74Mo/13Cu									
In-plane	208	120	6.5	8.3	9.89	0.357	272	39	0.30
Out-of-plane	159	92	6.5	8.3	9.89	0.357	272	39	0.30
99.5% BeO	208	120	6.4	6.4	2.77	0.10	272	39	1.07
FR-4, epoxy-glass	0.35	0.2	15.8	80 to 90	1.80	0.065	17.4	2.5	0.0028
Polyimide-glass	0.35	0.2	12 to 14	60	1.83	0.066	20.9	3.0	0.0027

Analysis of commercially available microelectronic substrates

Material	Thick-film substrate, %	Thin-film substrate, %
Al_2O_3	96.01	99.50
SiO_2	2.77	0.16
MgO	0.80	0.22
CaO	0.18	0.02
Fe_2O_3	0.10	0.03
Na_2O	0.08	0.04
K_2O	0.03	0.01
TiO_2	0.03	0.02

Thermal properties of various thermally conductive materials

Property	Substrate AlN	Substrate BeO, 99%	Substrate SiC	Substrate Al_2O_3, 96%
Thermal conductivity, W/m · K (Btu · in./h · ft^2 · °F)	140 to 230 (970 to 1600)	260 (1800)	270 (1870)	20 (140)
Flexural strength, MPa (ksi)	343 to 390 (50 to 57)	245 (35)	440 (64)	294 (43)
Vickers hardness, GPa (10^6 psi)	12 (1.7)	12 (1.7)	2 (0.427)	29 (4.3)

Properties of various substrate materials used in thin-film hybrids

Properties	Al_2O_3	Al_2O_3	BeO	AlN	AlN	AlN	SiC
Material							
Purity, % or model number (SC-101)	96	99.6	99.5	...	98	>99.5	SC-101
Manufacturer	Coors	Materials Research	Brush Wellman	Norton	Heraeus	Tokuyama Soda	Hitachi
Color	White	White	White	Translucent	Gray	Translucent	Gray
Major element(s) (AES, SIMS)	Al, O	Al, O	Be, O	Al, N	Al, N	Al, N	Si, C
Minor impurity (AES, SIMS)	Mg	Mg	Si	Y	Y	Ca	Be
Density, g/cm^3	3.75	3.89	2.85	3.28	3.26	3.25	3.1
Calculated density, g/cm^3	3.9869	3.9869	3.01	3.255	3.255	3.255	3.21
Surface finish, μm (μin.)	<0.6 (24)	<0.096 (3.8)	<0.36 (14)	...	0.4 to 0.6 (16 to 24)	0.4 to 0.6 (16 to 24)	...
Average surface roughness, μm (μin.)	0.49 (20)	0.14 (5.6)	0.16 to 0.33 (6.4 to 13)	0.32 (13)	0.77 (31)	0.88 (35)	0.17 (6.8)
Grain size, μm (μin.)	...	<1.5 (60)	9 to 16 (360 to 640)	...	5 to 10 (200 to 400)	...	...
Average grain diameter, μm (μin.)	9 (360)	1 (40)	16 (640)	Polished	13 (520)	17 (680)	Machined surface
Contact angle (after solvent clean), degrees	15	15 to 30	45	57	55	71	55
Thermal							
Thermal conductivity, W/m · K (Btu · in./ft^2 · h · °F)	26 (180)	3.7 (26)	250 (1700)	170 (1200)	140 to 170 (970 to 1200)	140 (970)	270 (1900)
Coefficient of thermal expansion, 10^{-6}/K	7.1	7.1	9.0	4.6	4.19	4.4	3.7
Coefficient of thermal expansion temperature range, °C (°F)	25 to 500 (80 to 930)	25 to 600 (80 to 1100)	25 to 1000 (80 to 1830)	20 to 400 (70 to 750)	20 to 400 (70 to 750)	20 to 400 (70 to 750)	20 to 400 (70 to 750)

(continued)

Properties of various substrate materials used in thin-film hybrids

Properties	Al_2O_3		BeO	AlN			SiC
Electrical							
Dielectric constant, ε_r, at 1 MHz	9.5	9.9	6.5	8.8	10	8.9	40
Dielectric loss at 1 MHz	0.0004	0.0001	0.0004	0.0005	0.002	0.001	0.05
Resistivity at RT, Ω · cm	$>10^{14}$	$>10^{14}$	10^{15}	$>10^{14}$	$>10^{11}$	10^{13}	$>10^{13}$
Dielectric strength, kV/mm (kV/in.)	23.6 (600)	25.6 (650)	9.5 (240)	...	>5 (125)	10 (255)	0.7 (18)
Dielectric strength test thickness, mm (in.)	0.635 (0.025)	0.635 (0.025)	...	...	0.5 (0.020)	...	...
Mechanical							
Flexural strength, MPa (ksi)	400 (60)	620 (90)	241 (35)	...	280 to 320 (41 to 46)	...	450 (65)
Bending strength, MPa (ksi)	...	...	...	392 (57)	...	441 (64)	490 (71)
Modulus of elasticity, GPa (10^6 psi)	...	...	345 (48.3)	304 (42.6)	300 to 310 (42 to 43)	...	400 (56)
Rockwell hardness, 15N scale	95.5	96.93	90.87	94.00	94.5	93.8	98.97

AES, Auger electron spectroscopy; SIMS, secondary ion mass spectroscopy

Relative shock resistance of substrate materials

Material	Thermal endurance factor, *F*	Coefficient of thermal expansion, from room temperature to 300 °C (570 °F), 10^{-6}/K
Fused silica	13.0	0.56
Alumina	3.7	6.8
Beryllia	3.0	7.8
Glass	0.9	9.0

Lightweight substrate materials with low coefficients of thermal expansion

Substrate material	Thermal conductivity W/m · K	Thermal conductivity Btu/h · ft · °F	In-plane coefficient of thermal expansion 10^{-6}/K	In-plane coefficient of thermal expansion 10^{-6}/°F	Density g/cm³	Density lb/in.³	Specific thermal conductivity	Modulus of elasticity GPa	Modulus of elasticity 10^6 psi
Boron-reinforced Al composites 20 vol% boron fiber	188	103	12.7	7.0	2.63	0.095	71.5	83	12
40 vol% boron fiber	120 to 130	69 to 71	7.0	3.9	2.57	0.093	50.6	145	21
50 vol% boron fiber	120 to 130	69 to 71	4.5	2.5	2.0	0.072	65.0	207	30
Boron fiber (AVCO)									
0.10 mm (4.0 mils) diameter	...	...	4.5	2.5	2.57	0.093	...	400	58
0.15 mm (5.8 mils) diameter	...	...	...	...	2.49	0.090	...	400	58
Pitch-base graphite fibers									
P55	100	58	–0.9	–0.5	1.75	0.063	57.1	379	55
P75	200	114	–1.26	–0.7	1.80	0.065	111.1	517	75
P100	550	318	–1.62	–0.9	1.86	0.067	295.7	724	105
P120	650	376							
P140									
40 vol% P100/6061	310	178	3.6	2.0	...	...	...	...	...
60 vol% P100/6061	340	195	0.3	0.17	...	...	...	...	...
60 vol% P120/6061	419	240	-0.32	-0.18	2.41	0.087	173.9	...	...
60 vol% 9P120/Cu	522	300	-0.07	0.13	6.23	0.225	83.8	...	...
40 vol% SiC_p/Al	128	74.0	12.6	7.0	2.9	0.105	44.1	...	...
50 to 55 vol% SiC_p/Al	131	75.7	7.9 to 8.6	4.4 to 4.8	2.99	0.108	43.8	...	...
Aluminum (6061)	226	391	23.6	13.1	2.71	0.098	79.7	70	10
Copper	391	226	17.6	9.8	8.94	1.323	43.7	117	17

Alternative electronic substrates

Substrate	Coefficient of thermal expansion, 10^{-6}/K	Thermal conductivity, W/m · K	Dielectric constant (usually 1 MHz)	Bending strength MPa	Bending strength ksi
Silica, amorphous	0.5 to 0.7	0.5 to 2	3 to 4	7 to 70	1 to 10
Pyrex, Corning 7740	3.3	1.2	4.7 to 5.1	7	1
Glass, Corning 7059	4.6	1.2	5	7 to 85	1 to 12
Porcelain, electrical	4 to 9	1.7	5.5	85 to 165	12 to 24
Mullite	4 to 5	4 to 7	5.4 to 6.6	125 to 275	18 to 40
Cordierite	1 to 3(a)	4	4 to 5	117	17
Alumina, >99%	6.5, 5.2(b), 6.7(b)	25 to 40, 50(b)	8 to 9	205 to 275	30 to 40
Alumina, 96%	6.3 to 9.1	12 to 26	8 to 10	205 to 345	30 to 50
AlN	3.1(c) to 4.6	60 to 230, 320(b)	8 to 10	275 to 345	40 to 50
αSi_3N_4, hot pressed	3.0 to 4.0	10 to 33, 80(b)	5 to 7	310 to 690	45 to 100
αSiC, hot pressed	2.5(c) to 4.7	33 to 270, 330(b)	40 to 100+	275 to 490	40 to 71
Boron carbide	4.5	300	...	...	...
BN, hot pressed, hexagonal	0.4 to 3.8(a)	12 to 250, 1300(b)	3.4 to 5.1	85	12
BN, cubic	4	950 to 1300	3 to 6	...	...
BeO, 99.5%	5 to 7	200 to 280, 320(b)	5.8 to 6.7	140 to 275	20 to 40
Si(b)	2.7(c) to 3.6	130 to 1240	12 to 100+	14	2

(a) Highly anisotropic. (b) Single crystal. (c) At 50 °C (120 °F)

Chemical Vapor Deposition

Electronic materials produced by chemical vapor deposition

Electronic materials	
Superconductors	Nb_3Ga
Semiconductors	Si, GaAs, ZnS
Insulators	SiO_2, Si_3N_4
Metals	W, Ta
Magnetic garnets	$Y_3GaFe_4O_{12}$
High-temperature and high-hardness materials	
Metal films	W, Mo, Ta, B, V
Carbon films	Pyrolitic carbon
Carbides	SiC, B_4C, ZrC, TaC
Oxides	Al_2O_3, Y_2O_3
Nitrides	TaN_x
Silicides	MoSi, TaSi
Alloys	Rh-W, W-Mo

Dimensionless numbers essential to chemical vapor deposition

Dimensionless number	Symbol	Physical meaning	Formula(a)	Range epi reactors
Reynolds	*Re*	$\frac{\text{Inertia force}}{\text{Viscous force}}$	$\frac{\rho VD}{v}$	25 to 200
Grashof	*Gr*	$\frac{\text{Inertia} \times \text{buoyancy}}{\text{Viscous force}}$	$\frac{\beta g \Delta T L^3}{v^2}$	100 to 1000
Prandtl	*Pr*	$\frac{\text{Momentum diffusivity}}{\text{Thermal diffusivity}}$	$\frac{v}{\alpha}$	0.4 to 0.8
Rayleigh	*Ra*	$\frac{\text{Gravity}}{\text{Thermal diffusivity}}$	$\frac{\beta g \Delta T L^3}{\alpha v}$	40 to 800

(a) ρ = density; V = velocity; D = diameter; μ = viscosity; β = temperature coefficient of expansion; ΔT = temperature difference; v = kinematic viscosity; α = thermal conductivity; g = gravitational constant; L = linear dimension

Deposition parameters for PECVD of silicon epitaxy

Deposition parameter	Value reported by: Townsend and Uddin	Suzuki and Itoh	Donahue, Burger, and Reif
Deposition temperature, °C	800	750	775
Gas ambient	SiH_4/H_2	SiH_4	SiH_4
Operating pressure, torr	0.2 to 0.6	10^{-2}	1.5×10^{-2}
SiH_4 partial pressure, torr	...	10^{-2}	1.5×10^{-2}
Discharge frequency, MHz	27	13.56	13.56
RF power, W	350	200	20 0
Deposition rate, Å/min	200	1980	450 340

W.G. Townsend and M.E. Uddin, *Solid State Electronics*, Vol 16, 1973, p 39; S. Suzuki and T. Itoh, *J. Appl. Phys.*, Vol 54, 1983, p 1466; T.J. Donahue, W.R. Burger and R. Reif, *Appl. Phys. Lett.*, Vol 44, 1984, p 346; T.J. Donahue and R. Reif, *J. Appl. Phys.*, Vol 57, 1985, p 2757

Refractory metals and silicides deposited by PECVD

Material	Gas ambient	Deposition temperature, °C	As-deposited structure(a)	Minimum resistivity (after annealing), $\mu\Omega \cdot$ cm
Tungsten	WF_6 / H_2	350	c	7
Molybdenum	$MoCl_5 / H_2$	400	a	10
W-silicide	$WF_6 / SiH_4 / He$	230	a	40
Mo-silicide	$MoCl_5 / SiH_4 / Ar/H_2$	400	a	...
Ta-silicide	$TaCl_5 / SiH_2Cl_2 / H_2$	<540	a	55
		>580	c	...
Ti-silicide	$TiCl_4 / SiH_4 / Ar$	300 to 350	a	20
Ti-silicide (nonplasma)	$TiCl_4 / SiH_4$	650 to 700	c	22(b)

(a) a, amorphous; c, crystalline. (b) As deposited

Resistivity change of LPCVD and PECVD undoped polysilicon films upon annealing(a)

Deposition process	As-deposited resistivity, $\mu\Omega \cdot$ cm	Annealed resistivity, $\mu\Omega \cdot$ cm
LPCVD	13.8	2.1
PECVD	0.78	0.96

(a) The films were deposited at 650 °C and annealed for 1 h at 1100 °C in nitrogen

Failure Analysis

Problem areas in silicon device production

Oxidation

The SiO_2/Si interface morphology and its movement as a function of processing conditions: correlation with electrical parameters
Stress distribution at the oxide edges
Integrity of the oxide after implantation doping
Availability of silicon interstitials, interaction with impurities, OSF, and OED, getting, precipitation, etc.

Ion implantation

Wafer-heating effect on amorphization
Control of amorphous/crystalline interface
Oxide-edge effect: curving of defect layer toward the junction
Oxygen recoil: effect on defect stabilization
Shallow junction degradation channeling enhance, diffused, etc.
Reliable preamorphization procedures
Dopant precipitation, reduced electrical activation

Gettering

Control of oxygen precipitation
Control of intrinsic gettering
Understanding relationship between intrinsic and extrinsic gettering
Annealing out of secondary defects in extrinsic gettering process

Silicides

Right choice of metal
Stability at processing temperatures
Degradation by oxidation
Surface preparation
Basic understanding of metal/Si → silicide/Si phase transformations

Failure mechanisms in silicon semiconductor devices

Device association	Process	Relevant factors	Accelerating factors	Acceleration (E_a = activation energy)
SiO_2 and silicon-SiO_2 interface	Surface charge accumulation	Mobile ions, V, T, Q_m	T, E	1.0 eV
	Dielectric breakdown (TDDB)	E, T	E (T)	0.35 eV
	Charge	E, T, Q_r	E, T	1.3 eV
	Hot carrier trapping	E, T, Q_{ot}	E, T	−0.06 eV
Metallization	Electromigration of Al	T, j, A, gradients of T and j, grain size	T, j	0.5 eV j^n, $n = 2 \pm 10\%$
	Electromigration of Si in Al	T, j, A	T, j	0.9 eV
	Corrosion	Contamination, H, V, T	H, V, T	Strong H effect 0.8 eV
	Contact degradation	T, metals, impurities	Varied	1.8 eV
Bonds and other mechanical interfaces	Intermetallic growth	T, impurities, bond strength	T	Al-Au: 1.0 eV
	Fatigue	T cycling, bond strength	T extremes in cycling	
Hermeticity	Seal leaks	Pressure differential, T cycling	Pressure, ΔT	

T, temperature; E, electric field; V, voltage; Q, oxide charge; j, current density; A, area; H, humidity

Common integrated circuit failure locations and mechanisms causing electrical open conditions or high resistance

Failure type or location	Failure mechanisms
Electrical damage	Fusing open of die metallization
Wires and wire bonds	Insufficient bonding pressure or energy
	Excessive bonding pressure or energy resulting in wire or bond fracturing, especially at the bond heel
	Excessive bonding pressure or energy resulting in fracturing of the semiconductor chip under the bond (cratering)
	Bond wire fracture during injection of molding compound
	Formation of brittle aluminum-gold intermetallics during component exposure to high temperatures (purple plague)
	Chip bonding pad contamination, especially oxidation
	Thermal fatigue fracture after environmental stressing from the mismatch of the thermal coefficients of expansion of the package materials
	Vibration- or shock-induced fractures of unsupported bond wires and bonds in ceramic packaged integrated circuits
	Fusing open of bond wires due to electrical overstress
Tape automated bonding	Irregular bump height or shape
	Insufficient bonding pressure or energy resulting in weak bonds
	Tape contact contamination
Chip fracture	Fracture of the semiconductor die from mechanical or thermal stress
Wafer fabrication	Lithography defects: incorrect size or shape of pattern; mechanical damage to the photoresist causing open circuits; photomask misalignment
	Film deposition defects: incomplete coverage of steps; pinholes or voids in passivation layers allowing metallization corrosion; corrosion in the presence of moisture from excess phosphorus in phosphosilica glass dielectric layers
Encapsulation	Damaged wire bonds or corrosion from package cracking due to vaporization of entrapped moisture (popcorn effect)

Common integrated circuit failure locations and mechanisms causing electrical shorted conditions or low resistance

Failure type or location	Failure mechanism
Electrical damage	Current-induced migration of metallization Latchup in complementary metal-oxide semiconductor devices Voltage punch-through of dielectric, gate oxide, or junction layers Electrostatic discharge damage
Wires and wire bonds	Lead frame movement or bond wire distortion (sweep) during injection of molding compound Misplaced or poorly shaped bond wire loops resulting in contact between wires Extra bond wires or bond wire tails of excessive length Mechanical damage during component assembly (thumbed bonds)
Chip fracture	Operation of the component or chip beyond its maximum electrical or thermal capability (outside its safe operating area) Fracturing initiated during wafer separation processes, that is, sawing or scribing, which propagates during later stressing Excessive mechanical forces or torques applied to the component package Voiding or insufficient die attach material between the die and lead frame to remove heat Mechanical damage during component assembly
Wafer fabrication	Lithography defects: incorrect size or shape of pattern; incomplete removal of photoresist causing bridging Ion implant and diffusion defects: incorrect dopant concentration or profile Etch defects: undercutting or overetching of pattern, especially metal layers Chip contamination, especially by human residue
Encapsulation material	Ionic contamination of the encapsulant Migration in the presence of moisture of halogenated fire-retardant compounds released during high-temperature exposure Moisture release or entrapment during lid sealing of ceramic packages Migration in the presence of moisture and electric fields of the die attach material, especially from silver-bearing epoxies Encapsulation contamination, especially by human residue
Lead frames	Excess material or burrs from incomplete machining or etching Mechanical damage during component assembly (for example, scratching)

Common integrated circuit failure locations and mechanisms causing parametric or functional failures other than opens or shorts

Location	Failure mechanism
Parametric drift	Resistance fluctuations due to piezoelectric effects of encapsulation materials Unstable metal or dielectric thin films Contamination or moisture Die attach material failure or voiding Incomplete step coverage Minor electrical overstress or electrostatic discharge damage to dielectric layers or junctions Electrostatic discharge induced increases in junction resistance

Manufacturing Methods

Fabrication Materials Properties

Materials for electronic fabrication

Material	Coefficient of thermal expansion, 10^{-6}/K	Thermal conductivity W/m · K	Thermal conductivity cal/cm · s · °C	Maximum temperature °C	Maximum temperature °F
Conductors					
Copper	17	390	0.94	1050	1920
Gold	14	300	0.71	1060	1940
Silver	19	420	1.0	962	1765
Aluminum	23.5	240	0.57	660	1220
Tungsten	0.44	160	0.38	3387	6130
Silicon	3.5	85 to 146	0.2 to 0.35	1414	2580
Kovar-Alloy 42	6 to 8	17	0.04	1350	2460
Insulators					
Alumina	6 to 6.5	13 to 17	0.03 to 0.04	>1000	>1830
SiC	3.5	270	0.64	>1000	>1830
AlN	5.7	60	0.14	>1000	>1830
SiO_2	0.6	1.7	0.004	>800	>1470
Epoxy(a)	10 to 30	0.4	0.001	150	300
Polyimide(a)	66	2.1	0.0005	400	750
Packaging/joining					
Au-Sn	16	250	0.6	232	450
Au-Si	10 to 13	285	0.68	370	700
Sn-Pb	24.7	50.6	0.121	183	360
Silver wax	20.4	210	0.5	...	...

(a) Also used for joining

Properties of encapsulation materials

Encapsulation materials	Dielectric strength kV/m	Dielectric strength V/mil	Dielectric constant at 1 GHz	Safe use temperature °C	Safe use temperature °F	Linear expansion 10^{-5}/K	Ultimate tensile strength MPa	Ultimate tensile strength ksi	Elongation, %	Relative adhesion (1, best; 5, poorest)
Thermoplastics										
Fluorocarbon	18	450	21	260	500	10	21	8	200	0
Polyethylene	20	500	23	115	240	17	28	4	1000	5
Polystyrene	22	550	25	85	185	7	49	7	1.5	4
Polyvinylchloride	16	400	28	100	210	5	21	3	100	3
Wax	16	400	26	55	135	20	2.1	0.3	5	4
Silicone-polyimide	60 to 112	1500 to 2800	30	400	750	54 to 85	14	2	200	1
Parylene	20 to 280	500 to 7000	28	120	250	6 to 12	70	10	200	4
Thermosets										
Alkyd	14	350	38	120	250	7	56	8	...	2
Allylester	16	400	...	100	210	7	38	5.5	...	3
Butadienestyrene	24	600	24	245	475	9	28	4	4	4
Epoxide	18	450	29	230	450	5	70	10	<1	1
Phenol-aldehyde	14	350	47	80	175	7	49	7	1.5	2
Polyester	14	350	35	165	325	11	56	8	<5	3
Silicone	24	600	28	260	500	13	17	2.5	8	4
Polyimides	136	3400	36	<425	<800	5	95 to 140	14 to 20	10 to 80	3
Silicone-epoxy	10 to 14	245 to 340	36	<200	<390	5 to 11	56	8	...	3
Elastomers										
Buna-S rubber	20	500	25	120	250	11	2.1	0.3	400	2
Chlororubber	16	400	27	...	...	16	17	2.5	500	3
Natural rubber	20	500	21	65	150	7	21	3	700	3
Silicone rubber	24	600	30	260	500	...	4.5	0.65	100	4
Thioplast	6	150	140	120	250	18	2.1	0.3	400	2
Urethane	14	350	35	90	200	18	35	5	400	1
Inorganics										
SiO_2	200	5000	35 to 40	760	1400	0.5 to 0.9	95 to 385	14 to 56	0	4
Si_3N_4	200	5000	7 to 10	370 to 760	700 to 1400	4 to 5	95 to 965	14 to 140	0	4

Soldering and Welding

Solderability test standards

Basic method	Test standard
Dip and look	ANSI/IPC-S-804 (boards) ANSI/IPC-S-805 (components) IEC 68-2-20 MIL-STD-202.750, 883 EIA RS-186-9E
Wetting balance	ANSI/IPC-S-805 (components) IEC-68-2-20 MIL-STD-883
Rotary dip	ANSI/IPC-S-804 (boards) IEC 68-2-20 BS4025 (British standard)
Globule test	ANSI/IPC-S-805 (components) IEC 68-2-20 BS 2011 (British standard) DIN 40046 (German standard)
Meniscus rise	No standards to date
Timed solder rise	ANSI/IPC-S-804 (boards)

Comparisons of typical solder systems

	Solder system					
Parameter	80Au-20Sn	88Au-12Ge	63Sn-37Pb	62Sn-36Pb-2Ag	95Sn-5Ag	50Pb-50In
Elongation, %	0	0.9	28 to 30	28 to 30	30.0	55
Electrical conductivity, IACS	4.8	6.0	11.5	14.0	12.6	5.1
Thermal coefficient of expansion, 10^{-6}/K	16	13	24.7	24.5	29.3	26.3
Solidus temperature, °C (°F)	280 (535)	356 (675)	183 (360)	180 (355)	221 (430)	180 (355)
Liquidus temperature, °C (°F)	280 (535)	356 (675)	183 (360)	185 (365)	245 (475)	210 (410)
Reflow temperature, °C (°F)	300 (570)	375 (705)	200 (390)	200 (390)	275 (525)	230 (445)
Tensile strength, MPa (ksi)	198 (28.7)	233 (33.9)	46 (6.7)	59 (8.5)	55 (8.0)	32.2 (4.67)
Shear strength, MPa (ksi)	185 (26.9)	220 (32)	40 (6)	48 (7)	45 (6.5)	18.5 (2.68)
Brinell hardness	165	139	17.0	15.4	13.7	9.6

Typical eutectic die bond alloys

	Bonding temperature		
Alloy	°C	°F	Application
Au	≤400	≤750	Si
98Au-Si	370 to 400	670 to 750	Si, GaAs, P
88Au-Sn	350	660	Si, GaAs, P
80Au-Sn	280	535	Si
95Pb-Sn	320	610	Si
63Pb-35Sn-1.8Sb	240	465	Ga
58Pb-40Sn-2Sb	230	445	Ga

Base metal solderability

	Ease of soldering	
Base metal	Ease of soldering (freshly cleaned)	Recommended fluxes
Aluminum	Poor	OA, IA, special
Brass	Good	RA, OA, IA
Copper	Excellent	R, RMA, RA, OA, IA
Copper-nickel (low nickel)	Good to excellent	RMA, RA, OA, IA
Copper-nickel (high nickel)	Fair	OA, IA
Copper-tin	Good	RA, OA, IA
Nickel	Fair to good	RA, OA, IA
Nickel-iron	Fair to poor	OA, IA
Steel	Fair	OA, IA
Stainless steel	Poor	IA
Tin-nickel	Fair	OA, IA

OA, organic acid base; IA, inorganic acid base; RA, rosin base, activated; R, rosin/alcohol; RMA, rosin base, mildly activated. Note: Synthetic activated (SA) fluxes are generally considered comparable with RA in activity

Characteristics of alloys used for microwave assembly

Composition	80Au-20Sn	88Au-12Ge	92.5Pb-5.0In-2.5Ag	90Pb-5In-5Ag	75Pb-25Sn	95Pb-5Sn	58Pb-40Sn-2Sb
Melting point, °C (°F)	280 (535) E	356 (675) E	310 (590) L 300 (570) S	310 (590) L 290 (555) S	268 (515) L 183 (360) S	314 (595) L 311 (590) S	231 (450) L 185 (365) S
Density, g/cm^3	14.51	14.673	11.02	11.00	9.96	11.06	9.17
Coefficient of thermal expansion, 10^{-6}/K	15.93	13.35	25	27	...	29.8	...
Specific heat, J/kg · K (Btu/lb · °F)	15 (0.0036)	...	...	...	...	...	...
Thermal conductivity, W/m · K (Btu · in./h · ft^2 · °F)	57 (395)	44 (305)	25 (175)	...	...	23 (160)	...

E, eutectic; L, liquidus; S, solidus

Welding techniques

Basic technique	Typical process	Application	Advantages/disadvantages
Arc Welding			
Tungsten inert gas (TIG)	Arc between a tungsten electrode and workpiece while protected by an inert gas	Thin sheets, foils, bellows, diaphragm assemblies, meshes	Equipment readily available, low cost, and simple to use; requires close-fitting joints and close distance of the torch to the workpiece
Microplasma	Same as TIG, except arc is in plasma gas	Same as TIG	Same as TIG, but distance between torch and workpiece not as critical
Percussive arc	Rapid arc generated by capacitive discharge	Wire and stud joining to components	Same as microplasma
Resistance welding			
Spot welding	Local fusion caused by electric current flow between two electrodes clamping workpieces together	Component assembly sheet fabrication	Low cost, well-proven equipment; low flexibility in joint design, high deformation (>10%), and high rate of electrode wear and contamination
Parallel seam	Roller electrodes travel in parallel across the edges, which are fused when current is pulsed intermittently across the lid to be bonded to the cavity	Microelectronic packages and sheet fabrication	Same as spot welding
Opposed electrode seam	Same as parallel seam except that there is only one roller electrode opposite a stationary electrode	Microelectronic packages and sheet fabrication	Same as spot welding
Parallel gap	Two closely spaced electrodes on the same side fusing two workpieces that are clamped down by the electrodes	Same as spot welding	Same as spot welding
Projection	Current is concentrated by projections of the workpiece interface to form weld	Microelectronic package, wire attachment component assembly, and sheet fabrication	Low-cost, well-proven process and self-fixturing; high deformation, wear of electrodes, and excess current can cause cavitation and weld splash
Butt	Ends of workpieces serve as electrodes and are fused together by electrical current and pressure	Wire and rod attachments, component fabrication	Same as spot welding
Pressure welding			
Cold	Pressure is applied so that surface oxides and contaminant are disrupted at the interface to give metal-to-metal contact, resulting in a bond	Sheets, leads, wires, and foils attachment and special packages designed for cold weld	Room-temperature bonding without flux, low-cost equipment, simple to use; only ductile materials; has high-deformation requirement (30 to 80%); cleanliness of surface is critical
Hot (thermocompression)	Same as cold pressure, except workpiece temperature is elevated prior to application of pressure	Bond wire in electronic circuits, joining of foils and sheets, and special packages	Same as cold-pressure welding
Ultrasonic	Same as cold pressure, except that ultrasonic energy is used to disrupt the surface contaminants	Same as hot-pressure welding	Same as hot-pressure welding
Thermosonic	Same as hot-pressure welding, except workpiece is heated	Same as hot-pressure welding	Same as hot-pressure welding
Friction	One workpiece is rotated over another under pressure, heating the interface by friction until a bond is made	Attachment of rods and tubes to other rods and tubes or sheets	Can weld dissimilar metals/ceramics, but has restrictive geometry and is a slow process
Electron beam welding	Electron beam is focused on the interface of the workpieces to be joined by fusion	Bellows, diaphragms, electrodes assemblies, and special electronic packages	Highly processed and controlled welds, suitable for automation; requires vacuum operation and needs safety against x-ray emission
Laser welding	Laser beam energy effects fusion in workpieces to be joined	Same as electron beam welding	High precision and controlled welds; suitable for automation; does not require vacuum or x-ray safeguards; equipment cost is high; workpiece surface reflections may be a problem

Welding processes for electronic fabrication

Process	Application
Thermocompression bonding	Wire bonds for integrated circuit interconnect (<1 mm, or 0.04 in.)
Resistance welding	Wire bonding (<25 mm, or 0.1 in.); hermetic sealing (metal packages)
Laser-beam welding	Hermetic sealing
Electron-beam welding	Component attach (has been used for large-diameter wire)
Thermosonic	Wire bonds for integrated circuit interconnect (<1mm, or 0.04 in.)

Summary of chip interconnection techniques

Wire type and diameter, mm (in.)	Attachment method	Substrate temperature °C	°F	Comments
Gold, 0.018 to 0.025 (0.0007 to 0.001)	Thermocompression	300	570	Limited repairability
Gold, 0.018 to 0.050 (0.0007 to 0.002)	Thermosonic ball or wedge	25	75	Quality of bond is improved if the substrate is raised to 125 °C (255 °F). Chips can be bonded without raising substrate temperature
Aluminum, 0.018 to 0.25 (0.0007 to 0.010)	Ultrasonic wedge	25	75	Low temperature process, restricted by tool size

Adhesives

Adhesives application methods

Type	Advantages	Disadvantages
Pin	Compact; simple; control of quality	Flat surface required; open
Screen printing	Simple; uniform; fast	Flat; uniform surface only; open
Pressure syringe	Surface profile independent uniform; closed system	Cleaning difficult; maintenance

Adhesives properties

Type	Coefficient of thermal expansion, 10^{-6}/K	Modulus of elasticity GPa	10^6 psi
Silicone, room-temperature vulcanizing	300	0.062	0.009
Modified polyimide	73	0.275	0.04
Polyimide	50	3.0	0.45

Physical properties of adhesives

Adhesives	Type	Nip roll temperature °C	°F	Maximum solder temperature °C	°F	Minimum peel strength N/mm	lbf/in.
Rogers 8145	Epoxy	204	400	316	600	0.53	3
DuPont WA	Acrylic	188	370	260	500	1.75	10
CMC 1477	Epoxy	190	375	288	550	1.58	9
CMCX-1496	Polyester	190	375	232	450	1.05	6

Plating and Etching

Surface wiring method comparison

Method	Application	Advantages	Disadvantages
Etched foil	Inner power layers Inner signal layers	Fewest process steps Excellent uniformity of line height Good uniformity of line width	Cannot use vias or PTHs
Plate and etch	Inner signal layers External signal layers	Can form vias, PTHs, and blind vias Good uniformity of line height Fair uniformity of line width	Poorest tolerance; can impact electrical performance Must tent PTHs; design must allow for registration
Pattern plating	Inner signal layers External signal layers	Can form vias, PTHs, and blind vias Excellent uniformity of line width Can easily add other electroplated metals, for example, solder	Most process steps Poor uniformity of line height

PTH, pin through-holes

Pin Through-Hole (PTH) plating selection versus aspect ratio

Aspect ratio	PTH plating selection
<5/1	Electrolytic, high speed
5/1 to 10/1	Electrolytic, low speed, or electroless
>10/1	Electroless or electrolytic, very low speed

Printed wire board (PWB) routing

S3/S4 core
- Laminate core
- Via hole formation
- Plate/circuitization
- Treat for adhesion

P1/ground core and P2/ground core
- Laminate core
- Apply and pattern resist
- Etch
- Strip resist
- Treat for adhesion

S1/S2 core and S5/S6 core
- Laminate core
- Apply and pattern resist(a)
- Etch
- Strip resist
- Treat for adhesion

Composite
- Lay up and laminate cores with prepreg
- PTH hole and blind via hole formation
- Hole clean
- Plate/circuitization(b)
- Apply solder mask
- Apply solder(c)
- Place components
- Form interconnections
- Final clean

(a) Only inner layers are circuitized; outer layers are solid copper. (b) Includes precious metal if applicable. (c) Not always required

Plate/circuitization process flow

Plate and etch	Pattern plating
Laminate copper underlay	Laminate copper underlay
Drill vias	Drill vias
Plate surface and vias	Apply photoresist
Apply photoresist	Develop photoresist
Develop photoresist	Plate vias and resist channels
Etch	Apply etch resist(a)
Strip photoresist	Strip photoresist
	Etch copper underlay
	Strip etch resist

(a) For example, immersion tin

Appendix

Abbreviations

a crack length
A ampere
A area; ratio of the alternating stress amplitude to the mean stress
Å angstrom
ABS acrylonitrile-butadiene-styrene
ac alternating current
ACM advanced cure monitor
APD avalanche photodiode
ARC accelerated rate calorimeter
at.% atomic percent
AS designation for surface-treated fiber
ATL automated tape layer
atm atmosphere (pressure)
AU designation for untreated fiber

BF_3MEA borontrifluoro-monoethylamine
BGDGE butylene glycol diglycidyl ether
BMC bulk molding compound
BMI bismaleimide (resin)
BPSG borophosphosilicate glasses
BT bismaleimide-triazine (resin)
Btu British thermal unit

c composite specific heat (C_p = constant pressure, C_v = constant volume)
C-C carbon-carbon
CAD/CAM computer-aided design/computer-aided manfacturing
CCA composite cylinder assemblage
CAT computer-aided tomography
CFRP carbon fiber reinforced plastic
cm centimeter
cpm cycles per minute
cps cycles per second
CTE coefficient of thermal expansion
CVD chemical vapor deposition
CVN Charpy V-notch (impact test or specimen)
CZ Czochralski (grown crystals)

d an operator used in mathematical expressions involving a derivative (denotes rate of change)
d depth; diameter
DADPS diaminodiphenylsufone
DAIP diallyl isophthalate
DAP diallyl phthalate
DBTT ductile-brittle transition temperature
dc direct current
DDA dynamic dielectric analysis
DGA diglycidyl aniline
DGEBA diglycidyl ether of bisphenol A
DGEBF diglycidyl ether of bisphenol F
DGT dynamic gel temperature
DH double heterostructure (lasers or LEDs)
diam diameter
DIB diiodobutane
DIP dual-in-line package
DLD dark line defects
DLP delta lattice parameter (phase diagram models)
DMA dynamic mechanical analysis
DP dissolution pit
DSC differential scanning calorimetry
DSD dark spot defect
DTA differential thermal analysis

e natural log base, 2.71828
E modulus of elasticity; Young's modulus
EL2 electrical compensation (doping)
epi epilayers, epitaxy or epitaxial
Eq equation
et al. and others
ESCA electron spectroscopy for chemical analysis

f fiber
f frequency
F force
FET field effect transistor
FMW formulated molecular weight
FP polycrystalline alumina fiber
FPP flat plastic package
FRP fiber-reinforced plastic
FRS fiber-reinforced superalloys
ft foot
FTIR Fourier transform infrared
FZ floatzone (-grown crystals)

g gram
G shear modulus
G' storage modulus
G" loss modulus
gal gallon
GPa gigapascal
GPC gel permeation chromatography
gpd grams per denier
gr grain
G_{XY} in-plane shear modulus of laminate
G_{IC} interlaminar fracture toughness (mode I, peel; mode II, shear; mode III, scissor shear)

h hour
h plate thickness of composite
H height
HERF high-energy-rate forging
HIP hot isostatic press
HM high modulus
HPLC high-performance liquid chromatography
H_r heat of reaction
HRTEM high-resolution transmission electron microscopy
HT high tensile
Hz hertz

IC integrated circuit
ID inside diameter
IM intermediate modulus
IR infrared (radiation)

J joule

k notch sensitivity factor
K Kelvin
K coefficient of thermal conductivity; modulus of elasticity
K_I stress-intensity factor
K_c plane-stress fracture toughness
K_{Ic} plane-strain fracture toughness; mode I critical stress-intensity factor
K_{Id} dynamic fracture toughness
K_{Iscc} threshold stress intensity for stress-corrosion cracking
K_t stress-concentration factor
K_t^{∞} stress-concentration factor for infinate plate
K_{th} threshold crack tip stress-intensity factor
kg kilogram
km kilometer
kPa kilopascal
ksi kips (1000 lb) per square inch
kV kilovolt

L liter; longitudinal direction
L length
lb pound
LCC leadless chip carrier
LEC liquid-encapsulated Czochralski (crystal growth)
LED light-emitting diode
LEFM linear-elastic fracture mechanics
ln natural logarithm (base e)
LPCVD low-pressure chemical vapor deposition
LPE liquid phase epitaxy

m matrix
MBE molecular beam epitaxy
MCM multichip module
MCO melt carryover
MDA methylenedianiline
MEKP methyl ethyl ketone peroxide
Mg megagram
min minute; minimum
MJ megajoule
mL milliliter
MLC multilayer ceramic (substrate or module)
mm millimeter
MMA methyl methacrylate
MMC metal matrix composite
MODFET modulation doped field effect transistor
mol % mole percent
MOS metal-oxide-semiconductor
MPa megapascal
mph miles per hour
MVT moisture vapor transmission

N Newton
N fatigue life (number of cycles)
NDE nondestructive evaluation
NDI nondestructive inspection
NDT nondestructive testing
nm nanometer
No. number

OD outside diameter
OED oxidation-enhanced diffusion
OMVPE organometallic vapor phase epitaxy
ORD oxidation-retarded diffusion
OSF oxidation-induced stacking faults
oz ounce

P applied load; pressure
Pa pascal
PAI polyamideimide
PAN polyacrylonitrile
PAS polyarylsulfone
PBI polybenzimidazole
PBT polybutylene terephthalate
PCB printed circuit board
PDCP polydicyclopentadiene
PECVD plasma-enhanced chemical vapor deposition
PEEK polyether etherketone
PEI polyetherimide
PES polyether sulfone
PET polyethylene terephthalate
PGA pin grid array
PI polyimide
PIC pressure-impregnation-carbonization
PLCC plastic-leaded chip carrier
P/M powder metallurgy

PMR *in-situ* polymerization of monomer reactants
ppb parts per billion
ppm parts per million
PPS polyphenylene sulfide
PS polysulfone
PSG phosphosilicate glass
psi pounds per square inch
psia pounds per square inch absolute
psid pounds per square inch differential
psig pounds per square inch gage
PTFE polytetrafluoroethylene
PTH pin-through-hole technology
PVA polyvinyl alcohol
PVC polyvinyl chloride

R radius; ratio of the minimum stress to the maximum stress
r rate of reaction
RA reduction of area
RDS rheometric dynamic scanning
Ref reference
RGA residual gas analysis
RH relative humidity
RIM reaction injection molding
RMS root mean square
ROM rule of mixtures; rough order of magnitude
rpm revolutions per minute
RRIM reinforced reaction injection molding
RTA rapid thermal annealing
RTD room temperature, dry
RTM resin transfer molding
RTW room temperature, wet
RTV room-temperature vulcanizing
RVE representative volume element
RDGE resorcinol diglycidyl ether

s second
SBS short beam shear
SCM single-chip module
SEM scanning electron microscope or microscopy
SF stacking fault
SMC sheet molding compound
SMT surface mount technology
S-N stress-number of cycles
SOP small outline package
SPF superplastic forming
sp gr specific gravity

t thickness; time
T transverse direction
T temperature; tenacity
TAB tape automated bonding
TCE thermal coefficient of expansion
TCL transmission cathodolumines cence (image)
TCM thermal conduction module
TEM transmission electron microscope or microscopy
TFT thin film transistor
T_g glass transition temperature
TGA thermogravimetric analysis
TGAP triglycidyl p-laminophenol
TGETPM triglycidyl ether of triphenyl methane
TLC thin-layer chromatography
T_m melting temperature
TMA thermomechanical analysis
TPI turns per inch
tan equal to ratio of the loss modulus to the storage modulus
TTU through-transmission ultrasonics

UDC unidirectional composite
UTS ultimate tensile strength
UV ultraviolet

V_f volume fraction of fiber
V_m volume fraction of matrix
V_v volume fraction of void content
VCDO vinyl cyclohexene diepoxide
vol volume
vol% volume percent
VLS vapor feed gases-liquid catalyst-solid crystalline whisker growth
VLSI very large scale integration

w whisker
W watt
W width
WPE weight per epoxide
wt% weight percent

XPS x-ray photoelectron spectroscopy

yr year

Metric Conversion Guide

This Section is intended as a guide for expressing weights and measures in the Système International d'Unités (SI). The purpose of SI units, developed and maintained by the General Conference of Weights and Measures, is to provide a basis for world-wide standardization of units and measure. For more information on metric conversions, the reader should consult the following references:

- "Standard for Metric Practice," E 380, *Annual Book of ASTM Standards*, Vol 14.02, 1987, American Society for Testing and Materials, 1916 Race Street, Philadelphia, PA 19103
- "Metric Practice," ANSI/IEEE 268–1982, American National Standards Institute, 1430 Broadway, New York, NY 10018
- *Metric Practice Guide—Units and Conversion Factors for the Steel Industry*, 1978, American Iron and Steel Institute, 1000 16th Street NW, Washington, DC 20036
- *The International System of Units*, SP 330, 1986, National Bureau of Standards. Order from Superintendent of Documents, U.S. Government Printing Office, Washington, DC 20402-9325
- *Metric Editorial Guide*, 4th ed. (revised), 1985, American National Metric Council, 1010 Vermont Avenue NW, Suite 320, Washington, DC 20005-4960
- *ASME Orientation and Guide for Use of SI (Metric) Units*, ASME Guide SI 1, 9th ed., 1982, The American Society of Mechanical Engineers, 345 East 47th Street, New York, NY 10017

Base, Supplementary, and Derived SI Units

Measure	Unit	Symbol
Base units		
Amount of substance	mole	mol
Electric current	ampere	A
Length	meter	m
Luminous intensity	candela	cd
Mass	kilogram	kg
Thermodynamic temperature	kelvin	K
Time	second	s
Supplementary units		
Plane angle	radian	rad
Solid angle	steradian	sr
Derived units		
Absorbed dose	gray	Gy
Acceleration	meter per second squared	m/s^2
Activity (of radionuclides)	becquerel	Bq
Angular acceleration	radian per second squared	rad/s^2
Angular velocity	radian per second	rad/s
Area	square meter	m^2
Capacitance	farad	F
Concentration (of amount of substance)	mole per cubic meter	mol/m^3
Conductance	siemens	S
Current density	ampere per square meter	A/m^2
Density, mass	kilogram per cubic meter	kg/m^3
Electric charge density	coulomb per cubic meter	C/m^3
Electric field strength	volt per meter	V/m
Electric flux density	coulomb per square meter	C/m^2
Electric potential, potential difference, electromotive force	volt	V
Electric resistance	ohm	Ω
Energy, work, quantity of heat	joule	J
Energy density	joule per cubic meter	J/m^3
Entropy	joule per kelvin	J/K
Force	newton	N
Frequency	hertz	Hz
Heat capacity	joule per kelvin	J/K
Heat flux density	watt per square meter	W/m^2
Illuminance	lux	lx
Inductance	henry	H
Irradiance	watt per square meter	W/m^2
Luminance	candela per square meter	cd/m^2
Luminous flux	lumen	lm
Magnetic field strength	ampere per meter	A/m
Magnetic flux	weber	Wb
Magnetic flux density	tesla	T
Molar energy	joule per mole	J/mol
Molar entropy	joule per mole kelvin	J/mol · K
Molar heat capacity	joule per mole kelvin	J/mol · K
Moment of force	newton meter	N · m
Permeability	henry per meter	H/m
Permittivity	farad per meter	F/m
Power, radiant flux	watt	W
Pressure, stress	pascal	Pa
Quantity of electricity, electric charge	coulomb	C
Radiance	watt per square meter steradian	$W/m^2 \cdot sr$
Radiant intensity	watt per steradian	W/sr
Specific heat capacity	joule per kilogram kelvin	J/kg · K
Specific energy	joule per kilogram	J/kg
Specific entropy	joule per kilogram kelvin	J/kg · K
Specific volume	cubic meter per kilogram	m^3/kg
Surface tension	newton per meter	N/m
Thermal conductivity	watt per meter kelvin	W/m · K
Velocity	meter per second	m/s
Viscosity, dynamic	pascal second	Pa · s
Viscosity, kinematic	square meter per second	m^2/s
Volume	cubic meter	m^3
Wavenumber	1 per meter	1/m

Conversion Factors

To convert from	to	multiply by
Area		
in.^{2}	mm^2	6.451 600 E + 02
in.^{2}	cm^2	6.451 600 E + 00
in.^{2}	m^2	6.451 600 E − 04
ft^2	m^2	9.290 304 E − 02
Bending moment or torque		
lbf · in.	N · m	1.129 848 E − 01
lbf · ft	N · m	1.355 818 E + 00
kgf · m	N · m	9.806 650 E + 00
ozf · in.	N · m	7.061 552 E − 03
Bending moment or torque per unit length		
lbf · in./in.	N · m/m	4.448 222 E + 00
lbf · ft/in.	N · m/m	5.337 866 E + 01
Current density		
A/in.^{2}	A/cm^2	1.550 003 E − 01
A/in.^{2}	A/mm^2	1.550 003 E − 03
A/ft^2	A/m^2	1.076 400 E + 01
Electric field strength		
V/mil	kV/m	3.937 008 E + 01
Electricity and magnetism		
gauss	T	1.000 000 E − 04
maxwell	μWb	1.000 000 E − 02
mho	S	1.000 000 E + 00
Oersted	A/m	7.957 700 E + 01
Ω · cm	Ω · m	1.000 000 E − 02
Ω circular-mil/ft	μΩ · m	1.662 426 E − 03
Energy (impact, other)		
ft · lbf	J	1.355 818 E + 00
Btu (thermochemical)	J	1.054 350 E + 03
cal (thermochemical)	J	4.184 000 E + 00
kW · h	J	3.600 000 E + 06
W · h	J	3.600 000 E + 03
Flow rate		
ft^3/h	L/min	4.719 475 E − 01
ft^3/min	L/min	2.831 000 E + 01
gal/h	L/min	6.309 020 E − 02
gal/min	L/min	3.785 412 E + 00
Force		
lbf	N	4.448 222 E + 00
kip (1000 lbf)	N	4.448 222 E + 03
tonf	kN	8.896 443 E + 00
kgf	N	9.806 650 E + 00
Force per unit length		
lbf/ft	N/m	1.459 390 E + 01
lbf/in.	N/m	1.751 268 E + 02
kip/in.	N/m	1.751 268 E + 05
Fracture toughness		
ksi $\sqrt{in.}$	MPa$\sqrt{m}$	1.098 800 E + 00
Heat content		
Btu/lb	kJ/kg	2.326 000 E + 00
cal/g	kJ/kg	4.186 800 E + 00
Heat input		
J/in.	J/m	3.937 008 E + 01
kJ/in.	kJ/m	3.937 008 E + 01
Impact energy per unit area		
ft · lbf/ft^2	J/m^2	1.459 002 E + 01
Length		
Å	nm	1.000 000 E − 01
μin.	μm	2.540 000 E − 02
mil	μm	2.540 000 E + 01
in.	mm	2.540 000 E + 01
in.	cm	2.540 000 E + 00
ft	m	3.048 000 E − 01
yd	m	9.144 000 E − 01
mile	km	1.609 300 E + 00
Length per unit mass		
in./lb	m/kg	5.599 740 E − 02
yd/lb	m/kg	2.015 907 E + 00
Mass		
oz	kg	2.834 952 E − 02
lb	kg	4.535 924 E − 01
ton (short, 2000 lb)	kg	9.071 847 E + 02
ton (short, 2000 lb)	kg × 10^3(a)	9.071 847 E − 01
ton (long, 2240 lb)	kg	1.016 047 E + 03
Mass per unit area		
oz/in.^{2}	kg/m^2	4.395 000 E + 01
oz/ft^2	kg/m^2	3.051 517 E − 01
oz/yd^2	kg/m^2	3.390 575 E − 02
lb/ft^2	kg/m^2	4.882 428 E + 00
Mass per unit length		
lb/ft	kg/m	1.488 164 E + 00
lb/in.	kg/m	1.785 797 E + 01
denier	kg/m	1.111 111 E − 07
tex	kg/m	1.000 000 E − 06
Mass per unit time		
lb/h	kg/s	1.259 979 E − 04
lb/min	kg/s	7.559 873 E − 03
lb/s	kg/s	4.535 924 E − 01
Mass per unit volume (includes density)		
g/cm^3	kg/m^3	1.000 000 E + 03
lb/ft^3	g/cm^3	1.601 846 E − 02
lb/ft^3	kg/m^3	1.601 846 E + 01
lb/in.^{3}	g/cm^3	2.767 990 E + 01
lb/in.^{3}	kg/m^3	2.767 990 E + 04
oz/in.^{3}	kg/m^3	1.729 994 E + 03
Power		
Btu/s	kW	1.055 056 E + 00
Btu/min	kW	1.758 426 E − 02
Btu/h	W	2.928 751 E − 01
erg/s	W	1.000 000 E − 07
ft · lbf/s	W	1.355 818 E + 00
ft · lbf/min	W	2.259 697 E − 02
ft · lbf/h	W	3.766 161 E − 04
hp (550 ft · lbf/s)	kW	7.456 999 E − 01
hp (electric)	kW	7.460 000 E − 01
Power density		
W/in.^{2}	W/m^2	1.550 003 E + 03
Pressure (fluid)		
atm (standard)	Pa	1.013 250 E + 05
bar	Pa	1.000 000 E + 05
in. Hg (32 °F)	Pa	3.386 380 E + 03
in. Hg (60 °F)	Pa	3.376 850 E + 03
lbf/in.^{2} (psi)	Pa	6.894 757 E + 03
torr (mm Hg, 0 °C)	Pa	1.333 220 E + 02
Specific area		
ft^2/lb	m^2/kg	2.048 161 E − 01
Specific energy		
cal/g	J/g	4.186 800 E + 00
Btu/lb	kJ/kg	2.326 000 E + 00
Specific heat capacity		
Btu/lb · °F	J/kg · K	4.186 800 E + 03
cal/g · °C	J/kg · K	4.186 800 E + 03
Stress (force per unit area)		
tonf/in.^{2} (tsi)	MPa	1.378 951 E + 01
kgf/mm^2	MPa	9.806 650 E + 00
ksi	MPa	6.894 757 E + 00
lbf/in.^{2} (psi)	MPa	6.894 757 E − 03
MN/m^2	MPa	1.000 000 E + 00
Temperature		
°F	°C	5/9 · (°F − 32)
°R	K	5/9
°F	K	5/9 · (°F + 459.67)
°C	K	°C + 273.15
Temperature interval		
°F	°C	5/9
Thermal conductivity		
Btu · in./s · ft^2 · °F	W/m · K	5.192 204 E + 02
Btu/ft · h · °F	W/m · K	1.730 735 E + 00
Btu · in./h · ft^2 · °F	W/m · K	1.442 279 E − 01
cal/cm · s · °C	W/m · K	4.184 000 E + 02
Thermal expansion		
μin./in. · °C	10^{-6}/K	1.000 000 E + 00
μin./in. · °F	10^{-6}/K	1.800 000 E + 00
Velocity		
ft/h	m/s	8.466 667 E − 05
ft/min	m/s	5.080 000 E − 03
ft/s	m/s	3.048 000 E − 01
in./s	m/s	2.540 000 E − 02
km/h	m/s	2.777 778 E − 01
mph	km/h	1.609 344 E + 00
Viscosity (dynamic and kinematic)		
poise (P)	Pa · s	1.000 000 E − 01
cP	Pa · s	1.000 000 E − 03
lbf · s/in.^{2}	Pa · s	6.894 757 E + 03
ft^2/s	m^2/s	9.290 304 E − 02
in.^{2}/s	mm^2/s	6.451 600 E + 02
Volume		
in.^{3}	m^3	1.638 706 E − 05
ft^3	m^3	2.831 685 E − 02
fluid oz	m^3	2.957 353 E − 05
gal (U.S. liquid)	m^3	3.785 412 E − 03
Volume per unit time		
ft^3/min	m^3/s	4.719 474 E − 04
ft^3/s	m^3/s	2.831 685 E − 02
in.^{3}/min	m^3/s	2.731 177 E − 07
Wavelength		
Å	nm	1.000 000 E − 01

(a) kg × 10^3 = 1 metric ton

Metric Stress or Pressure Conversions

The middle column of figures (in bold-faced type) contains the reading (in MPa or ksi) to be converted. If converting from ksi to MPa, read the MPa equivalent in the column headed "MPa". If converting from MPa to ksi, read the ksi equivalent in the column headed "ksi". 1 ksi = 6.894757 MPa. 1 psi = 6.894757 kPa.

ksi		MPa	ksi		MPa	ksi		MPa	ksi		MPa
0.14504	**1**	6.895	8.2672	**57**	393.00	33.359	**230**	1585.8	114.58	**790**	...
0.29008	**2**	13.790	8.4122	**58**	399.90	34.809	**240**	1654.7	116.03	**800**	...
0.43511	**3**	20.684	8.5572	**59**	406.79	36.259	**250**	1723.7	117.48	**810**	...
0.58015	**4**	27.579	8.7023	**60**	413.69	37.710	**260**	1792.6	118.93	**820**	...
0.72519	**5**	34.474	8.8473	**61**	420.58	39.160	**270**	1861.6	120.38	**830**	...
0.87023	**6**	41.369	8.9923	**62**	427.47	40.611	**280**	1930.5	121.83	**840**	...
1.0153	**7**	48.263	9.1374	**63**	434.37	42.061	**290**	1999.5	123.28	**850**	...
1.1603	**8**	55.158	9.2824	**64**	441.26	43.511	**300**	2068.4	124.73	**860**	...
1.3053	**9**	62.053	9.4275	**65**	448.16	44.962	**310**	2137.4	126.18	**870**	...
1.4504	**10**	68.948	9.5725	**66**	455.05	46.412	**320**	2206.3	127.63	**880**	...
1.5954	**11**	75.842	9.7175	**67**	461.95	47.862	**330**	2275.3	129.08	**890**	...
1.7405	**12**	82.737	9.8626	**68**	468.84	49.313	**340**	2344.2	130.53	**900**	...
1.8855	**13**	89.632	10.008	**69**	475.74	50.763	**350**	2413.2	131.98	**910**	...
2.0305	**14**	96.527	10.153	**70**	482.63	52.214	**360**	2482.1	133.43	**920**	...
2.1756	**15**	103.42	10.298	**71**	489.53	53.664	**370**	2551.1	134.89	**930**	...
2.3206	**16**	110.32	10.443	**72**	496.42	55.114	**380**	2620.0	136.34	**940**	...
2.4656	**17**	117.21	10.588	**73**	503.32	56.565	**390**	2689.0	137.79	**950**	...
2.6107	**18**	124.11	10.733	**74**	510.21	58.015	**400**	2757.9	139.24	**960**	...
2.7557	**19**	131.00	10.878	**75**	517.11	59.465	**410**	2826.9	140.69	**970**	...
2.9008	**20**	137.90	11.023	**76**	524.00	60.916	**420**	2895.8	142.14	**980**	...
3.0458	**21**	144.79	11.168	**77**	530.90	62.366	**430**	2964.7	143.59	**990**	...
3.1908	**22**	151.68	11.313	**78**	537.79	63.817	**440**	3033.7	145.04	**1000**	...
3.3359	**23**	158.58	11.458	**79**	544.69	65.267	**450**	3102.6	147.94	**1020**	...
3.4809	**24**	165.47	11.603	**80**	551.58	66.717	**460**	3171.6	150.84	**1040**	...
3.6259	**25**	172.37	11.748	**81**	558.48	66.168	**470**	3240.5	153.74	**1060**	...
3.7710	**26**	179.26	11.893	**82**	565.37	69.618	**480**	3309.5	156.64	**1080**	...
3.9160	**27**	186.16	12.038	**83**	572.26	71.068	**490**	3378.4	159.54	**1100**	...
4.0611	**28**	193.05	12.183	**84**	579.16	72.519	**500**	3447.4	162.44	**1120**	...
4.2061	**29**	199.95	12.328	**85**	586.05	73.969	**510**	...	165.34	**1140**	...
4.3511	**30**	206.84	12.473	**86**	592.95	75.420	**520**	...	168.24	**1160**	...
4.4962	**31**	213.74	12.618	**87**	599.84	76.870	**530**	...	171.14	**1180**	...
4.6412	**32**	220.63	12.763	**88**	606.74	78.320	**540**	...	174.05	**1200**	...
4.7862	**33**	227.53	12.909	**89**	613.63	79.771	**550**	...	176.95	**1220**	...
4.9313	**34**	234.42	13.053	**90**	620.53	81.221	**560**	...	179.85	**1240**	...
5.0763	**35**	241.32	13.198	**91**	627.42	82.672	**570**	...	182.75	**1260**	...
5.2214	**36**	248.21	13.343	**92**	634.32	84.122	**580**	...	185.65	**1280**	...
5.3664	**37**	255.11	13.489	**93**	641.21	85.572	**590**	...	188.55	**1300**	...
5.5114	**38**	262.00	13.634	**94**	648.11	87.023	**600**	...	191.45	**1320**	...
5.6565	**39**	268.90	13.779	**95**	655.00	88.473	**610**	...	194.35	**1340**	...
5.8015	**40**	275.79	13.924	**96**	661.90	89.923	**620**	...	197.25	**1360**	...
5.9465	**41**	282.69	14.069	**97**	668.79	91.374	**630**	...	200.15	**1380**	...
6.0916	**42**	289.58	14.214	**98**	675.69	92.824	**640**	...	203.05	**1400**	...
6.2366	**43**	296.47	14.359	**99**	682.58	94.275	**650**	...	205.95	**1420**	...
6.3817	**44**	303.37	14.504	**100**	689.48	95.725	**660**	...	208.85	**1440**	...
6.5267	**45**	310.26	15.954	**110**	758.42	97.175	**670**	...	211.76	**1460**	...
6.6717	**46**	317.16	17.405	**120**	827.37	98.626	**680**	...	214.66	**1480**	...
6.8168	**47**	324.05	18.855	**130**	896.32	100.08	**690**	...	217.56	**1500**	...
6.9618	**48**	330.95	20.305	**140**	965.27	101.53	**700**	...	220.46	**1520**	...
7.1068	**49**	337.84	21.756	**150**	1034.2	102.98	**710**	...	223.36	**1540**	...
7.2519	**50**	344.74	23.206	**160**	1103.2	104.43	**720**	...	226.26	**1560**	...
7.3969	**51**	351.63	24.656	**170**	1172.1	105.88	**730**	...	229.16	**1580**	...
7.5420	**52**	358.53	26.107	**180**	1241.1	107.33	**740**	...	232.06	**1600**	...
7.6870	**53**	365.42	27.557	**190**	1310.0	108.78	**750**	...	234.96	**1620**	...
7.8320	**54**	372.32	29.008	**200**	1379.0	110.23	**760**	...	237.86	**1640**	...
7.9771	**55**	379.21	30.458	**210**	1447.9	111.68	**770**	...	240.76	**1660**	...
8.1221	**56**	386.11	31.908	**220**	1516.8	113.13	**780**	...	243.66	**1680**	...

Metric Stress or Pressure Conversions (continued)

ksi		MPa	ksi		MPa	ksi		MPa	ksi		MPa
246.56	**1700**	...	278.47	**1920**	...	310.38	**2140**	...	342.29	**2360**	...
249.46	**1720**	...	281.37	**1940**	...	313.28	**2160**	...	345.19	**2380**	...
252.37	**1740**	...	284.27	**1960**	...	316.18	**2180**	...	348.09	**2400**	...
255.27	**1760**	...	287.17	**1980**	...	319.08	**2200**	...	350.99	**2420**	...
258.17	**1780**	...	290.08	**2000**	...	321.98	**2220**	...	353.89	**2440**	...
261.07	**1800**	...	292.98	**2020**	...	324.88	**2240**	...	356.79	**2460**	...
263.97	**1820**	...	295.88	**2040**	...	327.79	**2260**	...	359.69	**2480**	...
266.87	**1840**	...	298.78	**2060**	...	330.69	**2280**	...	362.59	**2500**	...
269.77	**1860**	...	301.68	**2080**	...	333.59	**2300**	...			
272.67	**1880**	...	304.58	**2100**	...	336.49	**2320**	...			
275.57	**1900**	...	307.48	**2120**	...	339.39	**2340**	...			

Metric Stress-Intensity Conversions

The middle column of figures (in bold-faced type) contains the reading (in MPa√m or ksi√in.) to be converted. If converting from ksi√in. to MPa√m, read the MPa√m equivalent in the column headed "MPa√m". If converting from MPa√m to ksi√in., read the ksi√in. equivalent in the column headed "ksi√in.". 1 ksi√in. = 1.098845 MPa√m.

ksi√in.		MPa√m	ksi√in.		MPa√m	ksi√in.		MPa√m	ksi√in.		MPa√m	ksi√in.		MPa√m
0.91005	**1**	1.0988	37.312	**41**	45.051	73.714	**81**	89.003	110.12	**121**	132.95	146.52	**161**	176.91
1.8201	**2**	2.1976	38.222	**42**	46.150	74.624	**82**	90.102	111.03	**122**	134.05	147.43	**162**	178.01
2.7301	**3**	3.2964	39.132	**43**	47.248	75.534	**83**	91.200	111.94	**123**	135.15	148.34	**163**	179.10
3.6402	**4**	4.3952	40.042	**44**	48.347	76.444	**84**	92.300	112.85	**124**	136.25	149.25	**164**	180.20
4.5502	**5**	5.4940	40.952	**45**	49.446	77.354	**85**	93.398	113.76	**125**	137.35	150.16	**165**	181.30
5.4603	**6**	6.5928	41.862	**46**	50.545	78.264	**86**	94.497	114.67	**126**	138.45	151.07	**166**	182.40
6.3703	**7**	7.6916	42.772	**47**	51.644	79.174	**87**	95.596	115.58	**127**	139.55	151.98	**167**	183.50
7.2804	**8**	8.7904	43.682	**48**	52.742	80.084	**88**	96.694	116.49	**128**	140.65	152.89	**168**	184.60
8.1904	**9**	9.8892	44.592	**49**	53.841	80.994	**89**	97.793	117.40	**129**	141.75	153.80	**169**	185.70
9.1005	**10**	10.988	45.502	**50**	54.940	81.904	**90**	98.892	118.31	**130**	142.84	154.71	**170**	186.80
10.011	**11**	12.087	46.412	**51**	56.039	82.814	**91**	99.991	119.22	**131**	143.94	155.62	**171**	187.90
10.921	**12**	13.186	47.322	**52**	57.138	83.724	**92**	101.09	120.13	**132**	145.04	156.53	**172**	189.00
11.831	**13**	14.284	48.232	**53**	58.236	84.634	**93**	102.19	121.04	**133**	146.14	157.44	**173**	190.10
12.741	**14**	15.383	49.143	**54**	59.335	85.544	**94**	103.29	121.95	**134**	147.24	158.35	**174**	191.19
13.651	**15**	16.482	50.053	**55**	60.434	86.454	**95**	104.39	122.86	**135**	148.34	159.26	**175**	192.29
14.561	**16**	17.581	50.963	**56**	61.533	87.364	**96**	105.48	123.77	**136**	149.44	160.17	**176**	193.39
15.471	**17**	18.680	51.873	**57**	62.632	88.275	**97**	106.58	124.68	**137**	150.54	161.08	**177**	194.49
16.381	**18**	19.778	52.783	**58**	63.730	89.185	**98**	107.68	125.59	**138**	151.63	161.99	**178**	195.59
17.291	**19**	20.877	53.693	**59**	64.829	90.095	**99**	108.78	126.50	**139**	152.73	162.90	**179**	196.69
18.201	**20**	21.976	54.603	**60**	65.928	91.005	**100**	109.88	127.41	**140**	153.83	163.81	**180**	197.78
19.111	**21**	23.075	55.513	**61**	67.027	91.915	**101**	110.98	128.32	**141**	154.93	164.72	**181**	198.88
20.021	**22**	24.174	56.423	**62**	68.126	92.825	**102**	112.08	129.23	**142**	156.03	165.63	**182**	199.98
20.931	**23**	25.272	57.333	**63**	69.224	93.735	**103**	113.18	130.14	**143**	157.13	166.54	**183**	201.08
21.841	**24**	26.371	58.243	**64**	70.323	94.645	**104**	114.28	131.05	**144**	158.23	167.45	**184**	202.18
22.751	**25**	27.470	59.153	**65**	71.422	95.555	**105**	115.37	131.96	**145**	159.33	168.36	**185**	203.28
23.661	**26**	28.569	60.063	**66**	72.521	96.465	**106**	116.47	132.87	**146**	160.42	169.27	**186**	204.38
24.571	**27**	29.668	60.973	**67**	73.620	97.375	**107**	117.57	133.78	**147**	161.52	170.18	**187**	205.48
25.481	**28**	30.766	61.883	**68**	74.718	98.285	**108**	118.67	134.69	**148**	162.62	171.09	**188**	206.57
26.391	**29**	31.865	62.793	**69**	75.817	99.195	**109**	119.77	135.60	**149**	163.72	172.00	**189**	207.67
27.301	**30**	32.964	63.703	**70**	76.916	100.11	**110**	120.87	136.51	**150**	164.82	172.91	**190**	208.77
28.211	**31**	34.063	64.613	**71**	78.015	101.02	**111**	121.97	137.42	**151**	165.92	173.82	**191**	209.87
29.121	**32**	35.162	65.523	**72**	79.114	101.93	**112**	123.07	138.33	**152**	167.02	174.73	**192**	210.97
30.032	**33**	36.260	66.433	**73**	80.212	102.84	**113**	124.16	139.24	**153**	168.12	175.64	**193**	212.07
30.942	**34**	37.359	67.343	**74**	81.311	103.75	**114**	125.26	140.15	**154**	169.22	176.55	**194**	213.17
31.852	**35**	38.458	68.253	**75**	82.410	104.66	**115**	126.36	141.06	**155**	170.31	177.46	**195**	214.27
32.762	**36**	39.557	69.164	**76**	83.509	105.57	**116**	127.46	141.97	**156**	171.41	178.37	**196**	215.36
33.672	**37**	40.656	70.074	**77**	84.608	106.48	**117**	128.56	142.88	**157**	172.51	179.28	**197**	216.46
34.582	**38**	41.754	70.984	**78**	85.706	107.39	**118**	129.66	143.79	**158**	173.61	180.19	**198**	217.56
35.492	**39**	42.853	71.893	**79**	86.805	108.30	**119**	130.76	144.70	**159**	174.71	181.10	**199**	218.66
36.402	**40**	43.952	72.804	**80**	87.904	109.21	**120**	131.86	145.61	**160**	175.81	182.01	**200**	219.76

Metric Energy Conversions

The middle column of figures (in bold-faced type) contains the reading (in J or ft·lb) to be converted. If converting from ft·lb to J, read the J equivalent in the column headed "J". If converting from J to ft·lb, read the equivalent in the column headed "ft·lb". 1 ft·lb = 1.355818 J.

ft·lb		J	ft·lb		J	ft·lb		J	ft·lb		J
0.7376	**1**	1.3558	28.7649	**39**	52.8769	56.7923	**77**	104.3980	129.0734	**175**	237.2681
1.4751	**2**	2.7116	29.5025	**40**	54.2327	57.5298	**78**	105.7538	132.7612	**180**	244.0472
2.2127	**3**	4.0675	30.2400	**41**	55.5885	58.2674	**79**	107.1096	136.4490	**185**	250.8263
2.9502	**4**	5.4233	30.9776	**42**	56.9444	59.0050	**80**	108.4654	140.1368	**190**	257.6054
3.6878	**5**	6.7791	31.7152	**43**	58.3002	59.7425	**81**	109.8212	143.8246	**195**	264.3845
4.4254	**6**	8.1349	32.4527	**44**	59.6560	60.4801	**82**	111.1771	147.5124	**200**	271.1636
5.1629	**7**	9.4907	33.1903	**45**	61.0118	61.2177	**83**	112.5329	154.8880	**210**	284.7218
5.9005	**8**	10.8465	33.9279	**46**	62.3676	61.9552	**84**	113.8887	162.2637	**220**	298.2799
6.6381	**9**	12.2024	34.6654	**47**	63.7234	62.6928	**85**	115.2445	169.6393	**230**	311.8381
7.3756	**10**	13.5582	35.4030	**48**	65.0793	63.4303	**86**	116.6003	177.0149	**240**	325.3963
8.1132	**11**	14.9140	36.1405	**49**	66.4351	64.1679	**87**	117.9562	184.3905	**250**	338.9545
8.8507	**12**	16.2698	36.8781	**50**	67.7909	64.9055	**88**	119.3120	191.7661	**260**	352.5126
9.5883	**13**	17.6256	37.6157	**51**	69.1467	65.6430	**89**	120.6678	199.1418	**270**	366.0708
10.3259	**14**	18.9815	38.3532	**52**	70.5025	66.3806	**90**	122.0236	206.5174	**280**	379.6290
11.0634	**15**	20.3373	39.0908	**53**	71.8583	67.1182	**91**	123.3794	213.8930	**290**	393.1872
11.8010	**16**	21.6931	39.8284	**54**	73.2142	67.8557	**92**	124.7452	221.2686	**300**	406.7454
12.5386	**17**	23.0489	40.5659	**55**	74.5700	68.5933	**93**	126.0911	228.6442	**310**	420.3036
13.2761	**18**	24.4047	41.3035	**56**	75.9258	69.3308	**94**	127.4469	236.0199	**320**	433.8617
14.0137	**19**	25.7605	42.0410	**57**	77.2816	70.0684	**95**	128.8027	243.3955	**330**	447.4199
14.7512	**20**	27.1164	42.7786	**58**	78.6374	70.8060	**96**	130.1585	250.7711	**340**	460.9781
15.4888	**21**	28.4722	43.5162	**59**	79.9933	71.5435	**97**	131.5143	258.1467	**350**	474.5363
16.2264	**22**	29.8280	44.2537	**60**	81.3491	72.2811	**98**	132.8702	265.5224	**360**	488.0944
16.9639	**23**	31.1838	44.9913	**61**	82.7049	73.0186	**99**	134.2260	272.8980	**370**	501.6526
17.7015	**24**	32.5396	45.7288	**62**	84.0607	73.7562	**100**	135.5818	280.2736	**380**	515.2108
18.4390	**25**	33.8954	46.4664	**63**	85.4165	77.4440	**105**	142.3609	287.6492	**390**	528.7690
19.1766	**26**	35.2513	47.2040	**64**	86.7723	81.1318	**110**	149.1400	295.0248	**400**	542.3272
19.9142	**27**	36.6071	47.9415	**65**	88.1282	84.8196	**115**	155.9191	302.4005	**410**	555.8854
20.6517	**28**	37.9629	48.6791	**66**	89.4840	88.5075	**120**	162.6982	309.7761	**420**	569.4435
21.3893	**29**	39.3187	49.4167	**67**	90.8398	92.1953	**125**	169.4772	317.1517	**430**	583.0017
22.1269	**30**	40.6745	50.1542	**68**	92.1956	95.8831	**130**	176.2563	324.5273	**440**	596.5599
22.8644	**31**	42.0304	50.8918	**69**	93.5514	99.5709	**135**	183.0354	331.9029	**450**	610.1181
23.6020	**32**	43.3862	51.6293	**70**	94.9073	103.2587	**140**	189.8145	339.2786	**460**	623.6762
24.3395	**33**	44.7420	52.3669	**71**	96.2631	106.9465	**145**	196.5936	346.6542	**470**	637.2344
25.0771	**34**	46.0978	53.1045	**72**	97.6189	110.6343	**150**	203.3727	354.0298	**480**	650.7926
25.8147	**35**	47.4536	53.8420	**73**	98.9747	114.3221	**155**	210.1518	361.4054	**490**	664.3508
26.5522	**36**	48.8094	54.5796	**74**	100.3305	118.0099	**160**	216.9308	368.7811	**500**	677.9090
27.2898	**37**	50.1653	55.3172	**75**	101.6863	121.6977	**165**	223.7099			
28.0274	**38**	51.5211	56.0547	**76**	103.0422	125.3856	**170**	230.4890			

Conversion of Inches to Millimeters

Inches	Millimeters	Inches	Millimeters	Inches	Millimeters
0.001	0.025	0.290	7.37	0.660	16.76
0.002	0.051	0.300	7.62	0.670	17.02
0.003	0.076	0.310	7.87	0.680	17.17
0.004	0.102	0.320	8.13	0.690	17.53
0.005	0.127	0.330	8.38	0.700	17.78
0.006	0.152	0.340	8.64	0.710	18.03
0.007	0.178	0.350	8.89	0.720	18.29
0.008	0.203	0.360	9.14	0.730	18.54
0.009	0.229	0.370	9.40	0.740	18.80
0.010	0.254	0.380	9.65	0.750	19.05
0.020	0.508	0.390	9.91	0.760	19.30
0.030	0.762	0.400	10.16	0.770	19.56
0.040	1.016	0.410	10.41	0.780	19.81
0.050	1.270	0.420	10.67	0.790	20.07
0.060	1.524	0.430	10.92	0.800	20.32
0.070	1.778	0.440	11.18	0.810	20.57
0.080	2.032	0.450	11.43	0.820	20.83
0.090	2.286	0.460	11.68	0.830	21.08
0.100	2.540	0.470	11.94	0.840	21.34
0.110	2.794	0.480	12.19	0.850	21.59
0.120	3.048	0.490	12.45	0.860	21.84
0.130	3.302	0.500	12.70	0.870	22.10
0.140	3.56	0.510	12.95	0.880	22.35
0.150	3.81	0.520	13.21	0.890	22.61
0.160	4.06	0.530	13.46	0.900	22.86
0.170	4.32	0.540	13.72	0.910	23.11
0.180	4.57	0.550	13.97	0.920	23.37
0.190	4.83	0.560	14.22	0.930	23.62
0.200	5.08	0.570	14.48	0.940	23.88
0.210	5.33	0.580	14.73	0.950	24.13
0.220	5.59	0.590	14.99	0.960	24.38
0.230	5.84	0.600	15.24	0.970	24.64
0.240	6.10	0.610	15.49	0.980	24.89
0.250	6.35	0.620	15.75	0.990	25.15
0.260	6.60	0.630	16.00	1.000	25.40
0.270	6.86	0.640	16.26	...	...
0.280	7.11	0.650	16.51	...	...

Conversion of Millimeters to Inches

Millimeters	Inches	Millimeters	Inches	Millimeters	Inches
0.01	0.0004	0.35	0.0138	0.68	0.0268
0.02	0.0008	0.36	0.0142	0.69	0.0272
0.03	0.0012	0.37	0.0146	0.70	0.0276
0.04	0.0016	0.38	0.0150	0.71	0.0280
0.05	0.0020	0.39	0.0154	0.72	0.0283
0.06	0.0024	0.40	0.0157	0.73	0.0287
0.07	0.0028	0.41	0.0161	0.74	0.0291
0.08	0.0031	0.42	0.0165	0.75	0.0295
0.09	0.0035	0.43	0.0169	0.76	0.0299
0.10	0.0039	0.44	0.0173	0.77	0.0303
0.11	0.0043	0.45	0.0177	0.78	0.0307
0.12	0.0047	0.46	0.0181	0.79	0.0311
0.13	0.0051	0.47	0.0185	0.80	0.0315
0.14	0.0055	0.48	0.0189	0.81	0.0319
0.15	0.0059	0.49	0.0193	0.82	0.0323
0.16	0.0063	0.50	0.0197	0.83	0.0327
0.17	0.0067	0.51	0.0201	0.84	0.0331
0.18	0.0071	0.52	0.0205	0.85	0.0335
0.19	0.0075	0.53	0.0209	0.86	0.0339
0.20	0.0079	0.54	0.0213	0.87	0.0343
0.21	0.0083	0.55	0.0217	0.88	0.0346
0.22	0.0087	0.56	0.0220	0.89	0.0350
0.23	0.0091	0.57	0.0224	0.90	0.0354
0.24	0.0094	0.58	0.0228	0.91	0.0358
0.25	0.0098	0.59	0.0232	0.92	0.0362
0.26	0.0102	0.60	0.0236	0.93	0.0366
0.27	0.0106	0.61	0.0240	0.94	0.0370
0.28	0.0110	0.62	0.0244	0.95	0.0374
0.29	0.0114	0.63	0.0248	0.96	0.0378
0.30	0.0118	0.64	0.0252	0.97	0.0382
0.31	0.0122	0.65	0.0256	0.98	0.0386
0.32	0.0126	0.66	0.0260	0.99	0.0390
0.33	0.0130	0.67	0.0264	1.00	0.0394
0.34	0.0134	...	...	...	...

Metric Length and Weight Conversion Factors

Unit	Inches to millimeters	Millimeters to inches	Pounds to kilograms	Kilograms to pounds
1	25.400 1	0.039 371	0.453 59	2.204 62
2	50.800 1	0.078 742	0.907 19	4.409 24
3	76.200 2	0.118 112	1.360 78	6.613 86
4	101.600 2	0.157 483	1.814 37	8.818 49
5	127.000 3	0.196 854	2.267 96	11.023 11
6	152.400 3	0.236 225	2.721 56	13.227 73
7	177.800 4	0.275 596	3.175 15	15.432 35
8	203.200 4	0.314 966	3.628 74	17.636 97
9	228.600 5	0.354 337	4.082 33	19.841 59
10	254.000 6	0.393 708	4.355 92	22.046 22

SI Prefixes-Names and Symbols

Exponential expression	Multiplication factor	Prefix	Symbol
10^{18}	1 000 000 000 000 000 000	exa	E
10^{15}	1 000 000 000 000 000	peta	P
10^{12}	1 000 000 000 000	tera	T
10^{9}	1 000 000 000	giga	G
10^{6}	1 000 000	mega	M
10^{3}	1 000	kilo	k
10^{2}	100	hecto(a)	h
10^{1}	10	deka(a)	da
10^{0}	1	BASE UNIT	
10^{-1}	0.1	deci(a)	d
10^{-2}	0.01	centi(a)	c
10^{-3}	0.001	milli	m
10^{-6}	0.000 001	micro	μ
10^{-9}	0.000 000 001	nano	n
10^{-12}	0.000 000 000 001	pico	p
10^{-15}	0.000 000 000 000 001	femto	f
10^{-18}	0.000 000 000 000 000 001	atto	a

(a) Nonpreferred. Prefixes should be selected in steps of 10^3 so that the resultant number before the prefix is between 0.1 and 1000. These prefixes should not be used for units of linear measurement, but may be used for higher order units. For example, the linear measurement, decimeter, is nonpreferred, but square decimeter is acceptable.

Symbols

- $\rightleftharpoons$ direction of reaction
- $\div$ divided by
- $=$ equals
- ˆ circumflex
- $\approx$ approximately equals
- $\neq$ not equal to
- $\equiv$ identical with
- $>$ greater than
- $\gg$ much greater than
- $\geq$ greater than or equal to
- ∞ infinity
- $\propto$ is proportional to; varies as
- $\int$ integral of
- $<$ less than
- $\ll$ much less than
- $\leq$ less than or equal to
- $\pm$ maximum deviation
- $-$ minus; negative ion charge
- $\times$ diameters (magnification); multiplied by
- $\cdot$ multiplied by
- Ω ohm
- / per
- % percent
- $+$ plus; positive ion charge
- $\sqrt{\ }$ square root of
- ~ approximately; similar to
- ° angular measure; degree
- °C degree Celsius (centigrade)
- °F degree Fahrenheit
- 0° fiber direction
- 90° perpendicular to fiber direction
- α coefficient of thermal expansion
- Δ change in quantity; an increment; a range
- η viscosity
- ϵ strain
- γ shear strain
- μin. microinch
- μm micrometer (micron)
- υ Poisson's ratio
- π pi (3.141592)
- ψ damping
- ρ density
- σ tensile stress
- τ shear stress
- θ angle

Greek Alphabet

A, α alpha	I, ι iota	P, ρ rho
B, β beta	K, κ kappa	Σ, σ sigma
Γ, γ gamma	Λ, λ lambda	T, τ tau
Δ, δ delta	M, μ mu	Y, υ upsilon
E, ϵ epsilon	N, ν nu	Φ, ϕ phi
Z, ζ zeta	Ξ, ξ xi	X, χ chi
H, η eta	O, o omicron	Ψ, ψ psi
Θ, θ theta	Π, π pi	Ω, ω omega

Temperature Conversions

The general arrangement of this table was devised by Sauveur and Boylston more than 40 years ago. The middle column of figures (in bold-faced type) contains the reading (°F or °C) to be converted. If converting from degrees Fahrenheit to degrees Centigrade, read the Centigrade equivalent in the column headed "°C". If converting from Centigrade to Fahrenheit, read the Fahrenheit equivalent in the column headed "°F". °C = 5/9 (°F − 32)

°F		°C	°F		°C	°F		°C	°F		°C	°F		°C
...	**-458**	-272.22	...	**-358**	-216.67	-432.4	**-258**	-161.11	-252.4	**-158**	-105.56	-72.4	**-58**	-50.00
...	**-456**	-271.11	...	**-356**	-215.56	-428.8	**-256**	-160.00	-248.8	**-156**	-104.44	-68.8	**-56**	-48.89
...	**-454**	-270.00	...	**-354**	-214.44	-425.2	**-254**	-158.89	-245.2	**-154**	-103.33	-65.2	**-54**	-47.78
...	**-452**	-268.89	...	**-352**	-213.33	-421.6	**-252**	-157.78	-241.6	**-152**	-102.22	-61.6	**-52**	-46.67
...	**-450**	-267.78	...	**-350**	-212.22	-418.0	**-250**	-156.67	-238.0	**-150**	-101.11	-58.0	**-50**	-45.56
...	**-448**	-266.67	...	**-348**	-211.11	-414.4	**-248**	-155.56	-234.4	**-148**	-100.00	-54.4	**-48**	-44.44
...	**-446**	-265.56	...	**-346**	-210.00	-410.8	**-246**	-154.44	-230.8	**-146**	-98.89	-50.8	**-46**	-43.33
...	**-444**	-264.44	...	**-344**	-208.89	-407.2	**-244**	-153.33	-227.2	**-144**	-97.78	-47.2	**-44**	-42.22
...	**-442**	-263.33	...	**-342**	-207.78	-403.6	**-242**	-152.22	-223.6	**-142**	-96.67	-43.6	**-42**	-41.11
...	**-440**	-262.22	...	**-340**	-206.67	-400.0	**-240**	-151.11	-220.0	**-140**	-95.56	-40.0	**-40**	-40.00
...	**-438**	-261.11	...	**-338**	-205.56	-396.4	**-238**	-150.00	-216.4	**-138**	-94.44	-36.4	**-38**	-38.89
...	**-436**	-260.00	...	**-336**	-204.44	-392.8	**-236**	-148.89	-212.8	**-136**	-93.33	-32.8	**-36**	-37.78
...	**-434**	-258.89	...	**-334**	-203.33	-389.2	**-234**	-147.78	-209.2	**-134**	-92.22	-29.2	**-34**	-36.67
...	**-432**	-257.78	...	**-332**	-202.22	-385.6	**-232**	-146.67	-205.6	**-132**	-91.11	-25.6	**-32**	-35.56
...	**-430**	-256.67	...	**-330**	-201.11	-382.0	**-230**	-145.56	-202.0	**-130**	-90.00	-22.0	**-30**	-34.44
...	**-428**	-255.56	...	**-328**	-200.00	-378.4	**-228**	-144.44	-198.4	**-128**	-88.89	-18.4	**-28**	-33.33
...	**-426**	-254.44	...	**-326**	-198.89	-374.8	**-226**	-143.33	-194.8	**-126**	-87.78	-14.8	**-26**	-32.22
...	**-424**	-253.33	...	**-324**	-197.78	-371.2	**-224**	-142.22	-191.2	**-124**	-86.67	-11.2	**-24**	-31.11
...	**-422**	-252.22	...	**-322**	-196.67	-367.6	**-222**	-141.11	-187.6	**-122**	-85.56	-7.6	**-22**	-30.00
...	**-420**	-251.11	...	**-320**	-195.56	-364.0	**-220**	-140.00	-184.0	**-120**	-84.44	-4.0	**-20**	-28.89
...	**-418**	-250.00	...	**-318**	-194.44	-360.4	**-218**	-138.89	-180.4	**-118**	-83.33	-0.4	**-18**	-27.78
...	**-416**	-248.89	...	**-316**	-193.33	-356.8	**-216**	-137.78	-176.8	**-116**	-82.22	+3.2	**-16**	-26.67
...	**-414**	-247.78	...	**-314**	-192.22	-353.2	**-214**	-136.67	-173.2	**-114**	-81.11	+6.8	**-14**	-25.56
...	**-412**	-246.67	...	**-312**	-191.11	-349.6	**-212**	-135.56	-169.6	**-112**	-80.00	+10.4	**-12**	-24.44
...	**-410**	-245.56	...	**-310**	-190.00	-346.0	**-210**	-134.44	-166.0	**-110**	-78.89	+14.0	**-10**	-23.33
...	**-408**	-244.44	...	**-308**	-188.89	-342.4	**-208**	-133.33	-162.4	**-108**	-77.78	+17.6	**-8**	-22.22
...	**-406**	-243.33	...	**-306**	-187.78	-338.8	**-206**	-132.22	-158.8	**-106**	-76.67	+21.2	**-6**	-21.11
...	**-404**	-242.22	...	**-304**	-186.67	-335.2	**-204**	-131.11	-155.2	**-104**	-75.56	+24.8	**-4**	-20.00
...	**-402**	-241.11	...	**-302**	-185.56	-331.6	**-202**	-130.00	-151.6	**-102**	-74.44	+28.4	**-2**	-18.89
...	**-400**	-240.00	...	**-300**	-184.44	-328.0	**-200**	-128.89	-148.0	**-100**	-73.33	+32.0	**±0**	-17.78
...	**-398**	-238.89	...	**-298**	-183.33	-324.4	**-198**	-127.78	-144.4	**-98**	-72.22	+35.6	**+2**	-16.67
...	**-396**	-237.78	...	**-296**	-182.22	-320.8	**-196**	-126.67	-140.8	**-96**	-71.11	+39.2	**+4**	-15.56
...	**-394**	-236.67	...	**-294**	-181.11	-317.2	**-194**	-125.56	-137.2	**-94**	-70.00	+42.8	**+6**	-14.44
...	**-392**	-235.56	...	**-292**	-180.00	-313.6	**-192**	-124.44	-133.6	**-92**	-68.89	+46.4	**+8**	-13.33
...	**-390**	-234.44	...	**-290**	-178.89	-310.0	**-190**	-123.33	-130.0	**-90**	-67.78	+50.0	**+10**	-12.22
...	**-388**	-233.33	...	**-288**	-177.78	-306.4	**-188**	-122.22	-126.4	**-88**	-66.67	+53.6	**+12**	-11.11
...	**-386**	-232.22	...	**-286**	-176.67	-302.8	**-186**	-121.11	-122.8	**-86**	-65.56	+57.2	**+14**	-10.00
...	**-384**	-231.11	...	**-284**	-175.56	-299.2	**-184**	-120.00	-119.2	**-84**	-64.44	+60.8	**+16**	-8.89
...	**-382**	-230.00	...	**-282**	-174.44	-295.6	**-182**	-118.89	-115.6	**-82**	-63.33	+64.4	**+18**	-7.78
...	**-380**	-228.89	...	**-280**	-173.33	-292.0	**-180**	-117.78	-112.0	**-80**	-62.22	+68.0	**+20**	-6.67
...	**-378**	-227.78	...	**-278**	-172.22	-288.4	**-178**	-116.67	-108.4	**-78**	-61.11	+71.6	**+22**	-5.56
...	**-376**	-226.67	...	**-276**	-171.11	-284.8	**-176**	-115.56	-104.8	**-76**	-60.00	+75.2	**+24**	-4.44
...	**-374**	-225.56	...	**-274**	-170.00	-281.2	**-174**	-114.44	-101.2	**-74**	-58.89	+78.8	**+26**	-3.33
...	**-372**	-224.44	-457.6	**-272**	-168.89	-277.6	**-172**	-113.33	-97.6	**-72**	-57.78	+82.4	**+28**	-2.22
...	**-370**	-223.33	-454.0	**-270**	-167.78	-274.0	**-170**	-112.22	-94.0	**-70**	-56.67	+86.0	**+30**	-1.11
...	**-368**	-222.22	-450.4	**-268**	-166.67	-270.4	**-168**	-111.11	-90.4	**-68**	-55.56	+89.6	**+32**	±0.00
...	**-366**	-221.11	-446.8	**-266**	-165.56	-266.8	**-166**	-110.00	-86.8	**-66**	-54.44	+93.2	**+34**	+1.11
...	**-364**	-220.00	-443.2	**-264**	-164.44	-263.2	**-164**	-108.89	-83.2	**-64**	-53.33	+96.8	**+36**	+2.22
...	**-362**	-218.89	-439.6	**-262**	-163.33	-259.6	**-162**	-107.78	-79.6	**-62**	-52.22	+100.4	**+38**	+3.33
...	**-360**	-217.78	-436.0	**-260**	-162.22	-256.0	**-160**	-106.67	-76.0	**-60**	-51.11	+104.0	**+40**	+4.44

(continued)

Temperature Conversions (continued)

°F		°C	°F		°C	°F		°C	°F		°C	°F		°C
107.6	**42**	5.56	305.6	**152**	66.67	503.6	**262**	127.78	701.6	**372**	188.89	899.6	**482**	250.00
111.2	**44**	6.67	309.2	**154**	67.78	507.2	**264**	128.89	705.2	**374**	190.00	903.2	**484**	251.11
114.8	**46**	7.78	312.8	**156**	68.89	510.8	**266**	130.00	708.8	**376**	191.11	906.8	**486**	252.22
118.4	**48**	8.89	316.4	**158**	70.00	514.4	**268**	131.11	712.4	**378**	192.22	910.4	**488**	253.33
122.0	**50**	10.00	320.0	**160**	71.11	518.0	**270**	132.22	716.0	**380**	193.33	914.0	**490**	254.44
125.6	**52**	11.11	323.6	**162**	72.22	521.6	**272**	133.33	719.6	**382**	194.44	917.6	**492**	255.56
129.2	**54**	12.12	327.2	**164**	73.33	525.2	**274**	134.44	723.2	**384**	195.56	921.2	**494**	256.67
132.8	**56**	13.33	330.8	**166**	74.44	528.8	**276**	135.56	726.8	**386**	196.67	924.8	**496**	257.78
136.4	**58**	14.44	334.4	**168**	75.56	532.4	**278**	136.67	730.4	**388**	197.78	928.4	**498**	258.89
140.0	**60**	15.56	338.0	**170**	76.67	536.0	**280**	137.78	734.0	**390**	198.89	932.0	**500**	260.00
143.6	**62**	16.67	341.6	**172**	77.78	539.6	**282**	138.89	737.6	**392**	200.00	935.6	**502**	261.11
147.2	**64**	17.78	345.2	**174**	78.89	543.2	**284**	140.00	741.2	**394**	201.11	939.2	**504**	262.22
150.8	**66**	18.89	348.8	**176**	80.00	546.8	**286**	141.11	744.8	**396**	202.22	942.8	**506**	263.33
154.4	**68**	20.00	352.4	**178**	81.11	550.4	**288**	142.22	748.4	**398**	203.33	946.4	**508**	264.44
158.0	**70**	21.11	356.0	**180**	82.22	554.0	**290**	143.33	752.0	**400**	204.44	950.0	**510**	265.56
161.6	**72**	22.22	359.6	**182**	83.33	557.6	**292**	144.44	755.6	**402**	205.56	953.6	**512**	266.67
165.2	**74**	23.33	363.2	**184**	84.44	561.2	**294**	145.56	759.2	**404**	206.67	957.2	**514**	267.78
168.8	**76**	24.44	366.8	**186**	85.56	564.8	**296**	146.67	762.8	**406**	207.78	960.8	**516**	268.89
172.4	**78**	25.56	370.4	**188**	86.67	568.4	**298**	147.78	766.4	**408**	208.89	964.4	**518**	270.00
176.0	**80**	26.67	374.0	**190**	87.78	572.0	**300**	148.89	770.0	**410**	210.00	968.0	**520**	271.11
179.6	**82**	27.78	377.6	**192**	88.89	575.6	**302**	150.00	773.6	**412**	211.11	971.6	**522**	272.22
183.2	**84**	28.89	381.2	**194**	90.00	579.2	**304**	151.11	777.2	**414**	212.22	975.2	**524**	273.33
186.8	**86**	30.00	384.8	**196**	91.11	582.8	**306**	152.22	780.8	**416**	213.33	978.8	**526**	274.44
190.4	**88**	31.11	388.4	**198**	92.22	586.4	**308**	153.33	784.4	**418**	214.44	982.4	**528**	275.56
194.0	**90**	32.22	392.0	**200**	93.33	590.0	**310**	154.44	788.0	**420**	215.56	986.0	**530**	276.67
197.6	**92**	33.33	395.6	**202**	94.44	593.6	**312**	155.56	791.6	**422**	216.67	989.6	**532**	277.78
201.2	**94**	34.44	399.2	**204**	95.56	597.2	**314**	156.67	795.2	**424**	217.78	993.2	**534**	278.89
204.8	**96**	35.56	402.8	**206**	96.67	600.8	**316**	157.78	798.8	**426**	218.89	996.8	**536**	280.00
208.4	**98**	36.67	406.4	**208**	97.78	604.4	**318**	158.89	802.4	**428**	220.00	1000.4	**538**	281.11
212.0	**100**	37.78	410.0	**210**	98.89	608.0	**320**	160.00	806.0	**430**	221.11	1004.0	**540**	282.22
215.6	**102**	38.89	413.6	**212**	100.00	611.6	**322**	161.11	809.6	**432**	222.22	1007.6	**542**	283.33
219.2	**104**	40.00	417.2	**214**	101.11	615.2	**324**	162.22	813.2	**434**	223.33	1011.2	**544**	284.44
222.8	**106**	41.11	420.8	**216**	102.22	618.8	**326**	163.33	816.8	**436**	224.44	1014.8	**546**	285.56
226.4	**108**	42.22	424.4	**218**	103.33	622.4	**328**	164.44	820.4	**438**	225.56	1018.4	**548**	286.67
230.0	**110**	43.33	428.0	**220**	104.44	626.0	**330**	165.56	824.0	**440**	226.67	1022.0	**550**	287.78
233.6	**112**	44.44	431.6	**222**	105.56	629.6	**332**	166.67	827.6	**442**	227.78	1040.0	**560**	293.33
237.2	**114**	45.56	435.2	**224**	106.67	633.2	**334**	167.78	831.2	**444**	228.89	1058.0	**570**	298.89
240.8	**116**	46.67	438.8	**226**	107.78	636.8	**336**	168.89	834.8	**446**	230.00	1076.0	**580**	304.44
244.4	**118**	47.78	442.4	**228**	108.89	640.4	**338**	170.00	838.4	**448**	231.11	1094.0	**590**	310.00
248.0	**120**	48.89	446.0	**230**	110.00	644.0	**340**	171.11	842.0	**450**	232.22	1112.0	**600**	315.56
251.6	**122**	50.00	449.6	**232**	111.11	647.6	**342**	172.22	845.6	**452**	233.33	1130.0	**610**	321.11
255.2	**124**	51.11	453.2	**234**	112.22	651.2	**344**	173.33	849.2	**454**	234.44	1148.0	**620**	326.67
258.8	**126**	52.22	456.8	**236**	113.33	654.8	**346**	174.44	852.8	**456**	235.56	1166.0	**630**	332.22
262.4	**128**	53.33	460.4	**238**	114.44	658.4	**348**	175.56	856.4	**458**	236.67	1184.0	**640**	337.78
266.0	**130**	54.44	464.0	**240**	115.56	662.0	**350**	176.67	860.0	**460**	237.78	1202.0	**650**	343.33
269.6	**132**	55.56	467.6	**242**	116.67	665.6	**352**	177.78	863.6	**462**	238.89	1220.0	**660**	348.89
273.2	**134**	56.67	471.2	**244**	117.78	669.2	**354**	178.89	867.2	**464**	240.00	1238.0	**670**	354.44
276.8	**136**	57.78	474.8	**246**	118.89	672.8	**356**	180.00	870.8	**466**	241.11	1256.0	**680**	360.00
280.4	**138**	58.89	478.4	**248**	120.00	676.4	**358**	181.11	874.4	**468**	242.22	1274.0	**690**	365.56
284.0	**140**	60.00	482.0	**250**	121.11	680.0	**360**	182.22	878.0	**470**	243.33	1292.0	**700**	371.11
287.6	**142**	61.11	485.6	**252**	122.22	683.6	**362**	183.33	881.6	**472**	244.44	1310.0	**710**	376.67
291.2	**144**	62.22	489.2	**254**	123.33	687.2	**364**	184.44	885.2	**474**	245.56	1328.0	**720**	382.22
294.8	**146**	63.33	492.8	**256**	124.44	690.8	**366**	185.56	888.8	**476**	246.67	1346.0	**730**	387.78
298.4	**148**	64.44	496.4	**258**	125.56	694.4	**368**	186.67	892.4	**478**	247.78	1364.0	**740**	393.33
302.0	**150**	65.56	500.0	**260**	126.67	698.0	**370**	187.78	896.0	**480**	248.89	1382.0	**750**	398.89

(continued)

Temperature Conversions (continued)

°F		°C	°F		°C	°F		°C	°F		°C	°F		°C
1400.0	**760**	404.44	2390.0	**1310**	710.00	3380.0	**1860**	1015.6	4370.0	**2410**	1321.1	5450.0	**3010**	1654.4
1418.0	**770**	410.00	2408.0	**1320**	715.56	3398.0	**1870**	1021.1	4388.0	**2420**	1326.7	5468.0	**3020**	1660.0
1436.0	**780**	415.56	2426.0	**1330**	721.11	3416.0	**1880**	1026.7	4406.0	**2430**	1332.2	5486.0	**3030**	1665.6
1454.0	**790**	421.11	2440.0	**1340**	726.67	3434.0	**1890**	1032.2	4424.0	**2440**	1337.8	5504.0	**3040**	1671.1
1472.0	**800**	426.67	2462.0	**1350**	732.22	3452.0	**1900**	1037.8	4442.0	**2450**	1343.3	5522.0	**3050**	1676.7
1490.0	**810**	432.22	2480.0	**1360**	737.78	3470.0	**1910**	1043.3	4460.0	**2460**	1348.9	5540.0	**3060**	1682.2
1508.0	**820**	437.76	2498.0	**1370**	743.33	3488.0	**1920**	1048.9	4478.0	**2470**	1354.4	5558.0	**3070**	1687.8
1526.0	**830**	443.33	2516.0	**1380**	748.89	3506.0	**1930**	1054.4	4496.0	**2480**	1360.0	5576.0	**3080**	1693.3
1544.0	**840**	448.89	2534.0	**1390**	754.44	3524.0	**1940**	1060.0	4514.0	**2490**	1365.6	5594.0	**3090**	1698.9
1562.0	**850**	454.44	2552.0	**1400**	760.00	3542.0	**1950**	1065.6	4532.0	**2500**	1371.1	5612.0	**3100**	1704.4
1580.0	**860**	460.00	2570.0	**1410**	765.56	3560.0	**1960**	1071.1	4550.0	**2510**	1376.7	5702.0	**3150**	1732.2
1598.0	**870**	465.56	2588.0	**1420**	771.11	3578.0	**1970**	1076.7	4568.0	**2520**	1382.2	5792.0	**3200**	1760.0
1616.0	**880**	471.11	2606.0	**1430**	776.67	3596.0	**1980**	1082.2	4586.0	**2530**	1387.8	5882.0	**3250**	1787.7
1634.0	**890**	476.67	2624.0	**1440**	782.22	3614.0	**1990**	1087.8	4604.0	**2540**	1393.3	5972.0	**3300**	1815.5
1652.0	**900**	482.22	2642.0	**1450**	787.78	3632.0	**2000**	1093.3	4622.0	**2550**	1398.9	6062.0	**3350**	1843.3
1670.0	**910**	487.78	2660.0	**1460**	793.33	3650.0	**2010**	1098.9	4640.0	**2560**	1404.4	6152.0	**3400**	1871.1
1688.0	**920**	493.33	2678.0	**1470**	798.89	3668.0	**2020**	1104.4	4658.0	**2570**	1410.0	6242.0	**3450**	1898.8
1706.0	**930**	498.89	2696.0	**1480**	804.44	3686.0	**2030**	1110.0	4676.0	**2580**	1415.6	6332.0	**3500**	1926.6
1724.0	**940**	504.44	2714.0	**1490**	810.00	3704.0	**2040**	1115.6	4694.0	**2590**	1421.1	6422.0	**3550**	1954.4
1742.0	**950**	510.00	2732.0	**1500**	815.56	3722.0	**2050**	1121.1	4712.0	**2600**	1426.7	6512.0	**3600**	1982.2
1760.0	**960**	515.56	2750.0	**1510**	821.11	3740.0	**2060**	1126.7	4730.0	**2610**	1432.2	6602.0	**3650**	2010.0
1778.0	**970**	521.11	2768.0	**1520**	826.67	3758.0	**2070**	1132.2	4748.0	**2620**	1437.8	6692.0	**3700**	2037.7
1796.0	**980**	526.67	2786.0	**1530**	832.22	3776.0	**2080**	1137.8	4766.0	**2630**	1443.3	6782.0	**3750**	2065.5
1814.0	**990**	532.22	2804.0	**1540**	837.78	3794.0	**2090**	1143.3	4784.0	**2640**	1448.9	6872.0	**3800**	2093.3
1832.0	**1000**	537.78	2822.0	**1550**	843.33	3812.0	**2100**	1148.9	4802.0	**2650**	1454.4	6962.0	**3850**	2121.1
1850.0	**1010**	543.33	2840.0	**1560**	848.89	3830.0	**2110**	1154.4	4820.0	**2660**	1460.0	7052.0	**3900**	2148.8
1868.0	**1020**	548.89	2858.0	**1570**	854.44	3848.0	**2120**	1160.0	4838.0	**2670**	1465.6	7142.0	**3950**	2176.6
1886.0	**1030**	554.44	2876.0	**1580**	860.00	3866.0	**2130**	1165.6	4856.0	**2680**	1471.1	7232.0	**4000**	2204.4
1904.0	**1040**	560.00	2894.0	**1590**	865.56	3884.0	**2140**	1171.1	4874.0	**2690**	1476.7	7322.0	**4050**	2232.2
1922.0	**1050**	565.56	2912.0	**1600**	871.11	3902.0	**2150**	1176.7	4892.0	**2700**	1482.2	7412.0	**4100**	2260.0
1940.0	**1060**	571.11	2930.0	**1610**	876.67	3920.0	**2160**	1182.2	4910.0	**2710**	1487.8	7502.0	**4150**	2287.7
1958.0	**1070**	576.67	2948.0	**1620**	882.22	3938.0	**2170**	1187.8	4928.0	**2720**	1493.3	7592.0	**4200**	2315.5
1976.0	**1080**	582.22	2966.0	**1630**	887.78	3956.0	**2180**	1193.3	4946.0	**2730**	1498.9	7682.0	**4250**	2343.3
1994.0	**1090**	587.78	2984.0	**1640**	893.33	3974.0	**2190**	1198.9	4964.0	**2740**	1504.4	7772.0	**4300**	2371.1
2012.0	**1100**	593.33	3002.0	**1650**	898.89	3992.0	**2200**	1204.4	4982.0	**2750**	1510.0	7862.0	**4350**	2398.8
2030.0	**1110**	598.89	3020.0	**1660**	904.44	4010.0	**2210**	1210.0	5000.0	**2760**	1515.6	7952.0	**4400**	2426.6
2048.0	**1120**	604.44	3038.0	**1670**	910.00	4028.0	**2220**	1215.6	5018.0	**2770**	1521.1	8042.0	**4450**	2454.4
2066.0	**1130**	610.00	3056.0	**1680**	915.56	4046.0	**2230**	1221.1	5036.0	**2780**	1526.7	8132.0	**4500**	2482.2
2084.0	**1140**	615.56	3074.0	**1690**	921.11	4064.0	**2240**	1226.7	5054.0	**2790**	1532.2	8222.0	**4550**	2510.0
2102.0	**1150**	621.11	3092.0	**1700**	926.67	4082.0	**2250**	1232.2	5072.0	**2800**	1537.8	8312.0	**4600**	2537.7
2120.0	**1160**	626.67	3110.0	**1710**	932.22	4100.0	**2260**	1237.8	5090.0	**2810**	1543.3	8402.0	**4650**	2565.5
2138.0	**1170**	632.22	3128.0	**1720**	937.78	4118.0	**2270**	1243.3	5108.0	**2820**	1548.9	8492.0	**4700**	2593.3
2156.0	**1180**	637.78	3146.0	**1730**	943.33	4136.0	**2280**	1248.9	5126.0	**2830**	1554.4	8582.0	**4750**	2621.1
2174.0	**1190**	643.33	3164.0	**1740**	948.89	4154.0	**2290**	1254.4	5144.0	**2840**	1560.0	8672.0	**4800**	2648.8
2192.0	**1200**	648.89	3182.0	**1750**	954.44	4172.0	**2300**	1260.0	5162.0	**2850**	1565.6	8762.0	**4850**	2676.6
2210.0	**1210**	654.44	3200.0	**1760**	960.00	4190.0	**2310**	1265.6	5180.0	**2860**	1571.1	8852.0	**4900**	2704.4
2228.0	**1220**	660.00	3218.0	**1770**	965.56	4208.0	**2320**	1271.1	5198.0	**2870**	1576.7	8942.0	**4950**	2732.2
2246.0	**1230**	665.56	3236.0	**1780**	971.11	4226.0	**2330**	1276.7	5216.0	**2880**	1582.2	9032.0	**5000**	2760.0
2264.0	**1240**	671.11	3254.0	**1790**	976.67	4244.0	**2340**	1282.2	5234.0	**2890**	1587.8	9122.0	**5050**	2787.7
2282.0	**1250**	676.67	3272.0	**1800**	982.22	4262.0	**2350**	1287.8	5252.0	**2900**	1593.3	9212.0	**5100**	2815.5
2300.0	**1260**	682.22	3290.0	**1810**	987.78	4280.0	**2360**	1293.3	5270.0	**2910**	1598.9	9302.0	**5150**	2843.3
2318.0	**1270**	687.78	3308.0	**1820**	993.33	4298.0	**2370**	1298.9	5288.0	**2920**	1604.4	9392.0	**5200**	2871.1
2336.0	**1280**	693.33	3326.0	**1830**	998.89	4316.0	**2380**	1304.4	5306.0	**2930**	1610.0	9482.0	**5250**	2898.8
2354.0	**1290**	698.89	3344.0	**1840**	1004.4	4334.0	**2390**	1310.0	5324.0	**2940**	1615.6	9572.0	**5300**	2926.6
2372.0	**1300**	704.44	3362.0	**1850**	1010.0	4352.0	**2400**	1315.6	5342.0	**2950**	1621.1	9662.0	**5350**	2954.4
									5360.0	**2960**	1626.7	9752.0	**5400**	2982.2
									5378.0	**2970**	1632.2	9842.0	**5450**	3010.0
									5396.0	**2980**	1637.8	9932.0	**5500**	3037.7
									5414.0	**2990**	1643.3	10 022.0	**5550**	3065.5
									5432.0	**3000**	1648.9	10 112.0	**5600**	3093.3

Periodic Table of the Elements

Metals — Nonmetals

I^a	II^a	III^b	IV^b	V^b	VI^b	VII^b	VIII	VIII	VIII	I^b	II^b	III^a	IV^a	V^a	VI^a	VII^a	0	Orbit
1 +1 H −1 1.0079 1																	2 0 He 4.00260 2	K
3 +1 Li 6.939 2 1	4 +2 Be 9.0122 2 2											5 +3 B 10.81 2-3	6 +2 +4 −4 C 12.011 2-4	7 +1 +2 +3 +4 +5 −1 −2 −3 N 14.0067 2-5	8 −2 O 15.9994 2-6	9 −1 F 18.998403 2-7	10 0 Ne 10.17$_9$ 2-8	K-L
11 +1 Na 22.9898 2 8 1	12 +2 Mg 24.312 2 8 2											13 +3 Al 26.98154 2-8-3	14 +2 +4 −4 Si 28.08 2-8-4	15 +3 +5 −3 P 30.97376 2-8-5	16 +4 +6 −2 S 32.06 2-8-6	17 +1 +5 +7 −1 Cl 35.453 2-8-7	18 0 Ar 39.948 2-8-8	K-L-M
19 +1 K 39.09 8 8 1	20 +2 Ca 40.08 8 8 2	21 +3 Sc 44.9559 8 9 2	22 +2 +3 +4 Ti 47.9 8 10 2	23 +2 +3 +4 +5 V 50.941 8 11 2	24 +2 +3 +6 Cr 51.996 8 13 1	25 +2 +3 +4 +7 Mn 54.9380 8 13 2	26 +2 +3 Fe 55.847 8 14 2	27 +2 +3 Co 58.9332 8 15 2	28 +2 +3 Ni 58.71 8 16 2	29 +1 +2 Cu 63.54 8 18 1	30 +2 Zn 65.38 8 18 2	31 +3 Ga 69.72 8 18 3	32 +2 +4 Ge 72.59 8 18 4	33 +3 +5 −3 As 74.9216 8 18 5	34 +4 +6 −2 Se 78.96 8 18 6	35 +1 +5 −1 Br 79.904 8 18 7	36 0 Kr 83.80 8 18 8	-L-M-N
37 +1 Rb 85.467 18 8 1	38 +2 Sr 87.62 18 8 2	39 +3 Y 88.9059 18 9 2	40 +4 Zr 91.22 -18-10-2	41 +3 +5 Nb 92.9064 -18-12-1	42 +6 Mo 95.94 -18-13-1	43 +4 +6 +7 Tc 98.9062 18-13-2	44 +3 Ru 101.07 -18-15-1	45 +3 Rh 102.905 18 16 1	46 +2 +4 Pd 106.4 -18-18-0	47 +1 Ag 107.868 18-18-1	48 +2 Cd 112.40 -18-18-2	49 +3 In 114.82 -18-18-3	50 +2 +4 Sn 118.69 18-18-4	51 +3 +5 −3 Sb 121.75 18-18-5	52 +4 +6 −2 Te 127.60 18-18-6	53 +1 +5 +7 −1 I 126.9045 -18-18-7	54 0 Xe 131.30 -18-18-8	-M-N-O
55 +1 Cs 132.9054 -18-8-1	56 +2 Ba 137.3 -18-8-2	57* +3 La 138.9055 -18-9-2	72 +4 Hf 178.49 -32-10-2	73 +5 Ta 180.948 -32-11-2	74 +6 W 183.85 -32-12-2	75 +4 +6 +7 Re 186.207 32 13 2	76 +3 +4 Os 190.2 -32-14-2	77 +3 +4 Ir 192.2 32 15 2	78 +2 +4 Pt 195.09 -32-16-2	79 +1 +3 Au 196.9665 -32-18-1	80 +1 +2 Hg 200.59 -32-18-2	81 +1 +3 Tl 204.37 -32-19-3	82 +2 +4 Pb 207.19 -32-18-4	83 +3 +5 Bi 208.980 -32-18-5	84 +2 +4 Po (209) -32-18-6	85 At (210) -32-18-7	86 0 Rn (222) -32-18-8	N-O-P
87 +1 Fr (223) -18-8-1	88 +2 Ra 226.0254 -18-8-2	89** +3 Ac (227) -18-9-2	104 +4 Rf (261) -32-10-2	105 Ha (262) -32-11-2	106 (263) -32-12-2													-O-P-Q

Key to chart

Atomic Number →	50	+2 +4	← Oxidation States
Symbol →	Sn		
Atomic Weight →	118.69	-18-18-4	← Electron Configuration

Transition Elements

															Orbit
*Lanthanides	58 +3 +4 Ce 140.12 -20-8-2	59 +3 Pr 140.9077 -21-8-2	60 +3 Nd 144.24 -22-8-2	61 +3 Pm 147 -23-8-2	62 +2 +3 Sm 150.4 -24-8-2	63 +2 +3 Eu 151.96 -25-8-2	64 +3 Gd 157.25 -25-9-2	65 +3 Tb 158.925 -27-8-2	66 +3 Dy 162.50 -28-8-2	67 +3 Ho 164.9304 -29-8-2	68 +3 Er 167.26 -30-8-2	69 +3 Tm 168.9342 -31-8-2	70 +2 +3 Yb 173.04 -32-8-2	71 +3 Lu 174.967 -32-9-2	-N-O-P
**Actinides	90 +4 Th 232.038 -18-10-2	91 +5 +4 Pa 231.0359 -20-9-2	92 +3 +4 +5 +6 U 238.029 -21-9-2	93 +3 +4 +5 +6 Np 237.0482 -22-9-2	94 +3 +4 +5 +6 Pu 239.052 -24-8-2	95 +3 +4 +5 +6 Am (243) -25-8-2	96 +3 Cm (247) -25-9-2	97 +3 +4 Bk (247) -27-8-2	98 +3 Cf (251) -28-8-2	99 +3 Es (254) -29-8-2	100 +3 Fm (257) -30-8-2	101 +2 +3 Md (258) -31-8-2	102 +2 +3 No (259) -32-8-2	103 +3 Lr (260) -32-9-2	-O-P-Q

Numbers in parentheses are mass numbers of most stable isotope of that element